SELECTION OF THE METHOD IN CHEMICAL ANALYSIS

SELECTION OF THE HPLC METHOD IN CHEMICAL ANALYSIS

SERBAN C. MOLDOVEANU
VICTOR DAVID

ELSEVIER

AMSTERDAM • BOSTON • HEIDELBERG • LONDON
NEW YORK • OXFORD • PARIS • SAN DIEGO
SAN FRANCISCO • SINGAPORE • SYDNEY • TOKYO

Elsevier
Radarweg 29, PO Box 211, 1000 AE Amsterdam, Netherlands
The Boulevard, Langford Lane, Kidlington, Oxford OX5 1GB, United Kingdom
50 Hampshire Street, 5th Floor, Cambridge, MA 02139, United States

Library of Congress Cataloging-in-Publication Data
A catalog record for this book is available from the Library of Congress

British Library Cataloguing-in-Publication Data
A catalogue record for this book is available from the British Library

ISBN: 978-0-12-803684-6

Publisher: John Fedor
Acquisition Editor: Kathryn Morrissey
Editorial Project Manager: Amy M Clark
Production Project Manager: Maria Bernard
Cover Designer: Christian Bilbow

Typeset by TNQ Books and Journals

Contents

7. RP-HPLC Analytical Columns

8. Polar Analytical Columns

9. Stationary Phases and Columns for Ion Exchange, Ion-Moderated, and Ligand Exchange Chromatography

10. Stationary Phases and Columns for Chiral Chromatography

11. Stationary Phases and Columns for Size Exclusion

12. Stationary Phases and Columns for Immunoaffinity Type Separations

13. Solvents, Buffers, and Additives Used in the Mobile Phase

14. Gradient Elution

15. The Practice of HPLC

Preface

High-performance liquid chromatography (HPLC) is very likely the most utilized technique in analytical chemistry. An enormous amount of information about HPLC is available in peer-reviewed journals, in books, in manufacturer catalogs, and on the Internet. The selection of the most appropriate HPLC method for a given objective from this multitude of sources is a challenging task. In addition, new possibilities for more selective and sensitive HPLC methods are continuously emerging based on the progress made in instrumentation and new chromatographic materials. Although numerous scattered publications exist, a unified and coherent presentation of the procedures for the selection and development of an HPLC method is still poorly represented in the literature. This book intends to cover this gap as well as possible. It provides criteria for the selection of an appropriate method for a particular task when such method is available in the literature, and it offers guidance for the development of a new method when the literature does not describe an adequate one for a specific objective. The book intends to provide a systematic and up-to-date material regarding the selection of an HPLC method. However, because continuous progress is made in HPLC, only the latest published literature can capture the most novel developments. The book is basically segmented into four parts. The first part presents aspects related to the collection of information necessary before the selection of a new analytical method. Comments on statistical analysis of data and validation criteria and a general discussion about alternative techniques utilized in chemical analysis are also included in this part. The first part also presents the advantages and disadvantages of selecting HPLC as a preferred method of analysis versus other analytical techniques. The second part describes various aspects of HPLC, such as instrumentation, parameters characterizing a chromatographic separation, and retention mechanism in several HPLC types. The third part of the book presents in some detail the characteristics of HPLC columns and criteria for the selection of an appropriate column for a specific analytical task. This part also includes a detailed description of possibilities to select and use a mobile phase. The last part of the book is a collection of tables with detailed information about chromatographic columns, solvents, and additives. These data provide an important source of information necessary for selecting or developing a new HPLC method. Valuable help for improving the presentation of the material from this book was provided by Owen Bussey, Crystal Byrd, Chelsea Cooke, Garry Dull, Anthony Gerardi, Christopher Junker, Carol Moldoveanu, and Wayne Scott, for which the authors are most thankful. The authors also wish to thank the editorial team from Elsevier, Amy Clark, Maria Bernard, and Kathryn Morrissey for their contribution to the publication of this book.

CHAPTER

1

Start of the Implementation of a New HPLC Method

1.1 COLLECTION OF INFORMATION AND PLANNING FOR A NEW METHOD

For good results, the collection of as much information as possible before starting the implementation of a new HPLC method of analysis is very important. It is also important to assess how reliable the various items of the collected information are. The information collected before starting a selection for an analytical method should cover a number of aspects that include the following: (1) the purpose of analysis, (2) general information about the samples, (3) the sample's constituents, (4) the required quality of the results, (5) instrumentation availability, expertise in the laboratory, and funding, (6) information about various methods of analysis, and (7) new developments in instrumentation.

The collection of information regarding the method of analysis should continue even after the method development has started. Any additional findings obtained after the development has started and the feedback from the initial trials of a specific method are valuable data for the improvement and final selection of an adequate method. For this reason, the collection of information should be considered a continuous process that may even have an iterative path, the results obtained from the first set of runs lead to the need for more "starting" information and so on until the method is well established.

The Purpose of Analysis

The purpose of analysis should be the first type of information obtained before starting the implementation of a new analytical method. Sources of information about the purpose can be very diverse, the most common being the direct request from a customer (including self-imposed requests). A number of items can be listed as important to know regarding the purpose of a chemical analysis. This information should describe why the analysis is performed, what kind of results are expected from the analysis, and possibly the utility of the analysis. A wide range of requests can be made for an analysis, and samples must be analyzed for numerous reasons. In industrial environments such purposes may include official or legal requirements, assessing the quality of raw materials, process control or troubleshooting,

Selection of the HPLC Method in Chemical Analysis
http://dx.doi.org/10.1016/B978-0-12-803684-6.00001-9

assuring the quality of finished products, research, reverse engineering, and development purposes. Samples are frequently analyzed for health-related purposes (e.g., medical analyses, analysis of pharmaceuticals, analysis of metabolites), for evaluating environmental issues, for forensic purposes, for exploratory reasons, and for fundamental research. Depending on the analysis reason, specific decisions are made about the analytical process. The purpose of analysis should describe if the analysis is related to a material, a process, or both. It is important to know if the whole sample must be analyzed (all constituents are in this case analytes) or only a specific part of the sample.

Important information must be collected indicating the required type of analysis regarding qualitative, quantitative, semiquantitative, or both qualitative and quantitative. It should be indicated if a special analysis is to be performed such as separation of enantiomers or regarding structural elucidation. The purpose should specify if the analysis has a specific target, if it is only exploratory, or the goal of performing the analysis is vague. Also, it should be known if there is a plan to perform the analysis on a repeated basis in the future or only at one time. The number of samples to be analyzed at one point or in an extended period of time should be known. The rapidity with which the results must be delivered should also be known before starting the development or implementation of a new method.

General information must specify whether a specific protocol must be followed during the analysis or that no regulations are imposed. Some analyses are required to be nondestructive, and in certain cases the analysis is done in conjunction with preparative purposes, which also should be known. There are important issues related to analyses associated with preparatory purposes since some of them modify the nature of the sample. If the biological activity of the samples must be preserved, this aspect should also be known.

The level of accuracy and precision of the analysis should be described, indicating if the analysis is geared toward major constituents of the sample, minor constituents, or traces. Also, information should be collected regarding the novelty of the planned analysis or if it was previously performed. Knowledge regarding other analyses already performed on the samples is always important, and occasionally it is useful to have information about analyses that are planned to be performed on the samples in the future.

The general information regarding the purpose of analysis allows early planning of the type of analysis to be performed on the samples. In case that very little information is available about the samples, preliminary analyses should sometimes be performed. This preliminary analysis can be qualitative or semiquantitative. If only qualitative analysis is required, high-performance liquid chromatography (HPLC) is not usually the best choice as the method of analysis. When information exists that the sample contains unique compounds or it is a mixture of very few components, nonchromatographic techniques may be considered for obtaining qualitative information. Very useful are the spectroscopic procedures such as ultraviolet (UV), infrared (IR), or nuclear magnetic resonance (NMR). When the samples have more components, chromatographic techniques are typically necessary for an initial separation of sample components. "Scan type" chromatographic techniques associated with detection by mass spectrometry such as gas chromatography−mass spectrometry (GC−MS) are probably the best tools for compound identification, provided the samples are amenable for this type of analysis (contain volatile analytes). In some cases, when the samples have a limited number of components which are known to be potentially present, HPLC or thin-layer chromatography (TLC) can be useful for identifying sample composition. Mass spectrometry hyphenated with liquid chromatography (LC) such as LC-MS or LC-MS/MS may provide qualitative information on unknown samples, although the identification of unknown compounds by LC-MS or LC-MS/MS techniques has limitations. For quantitative analysis, chromatography is typically preferred, including GC or LC (HPLC). Also, an array of techniques must be necessary for the analysis of certain samples. For large numbers of samples to be analyzed, automation must be considered.

General Information About the Samples

The information about the material(s) to be analyzed (indicated as *sample*) should cover chemical aspects of the sample, physical properties information, as well as any other contingent information. Two parts of a sample are to be considered in a chemical analysis, the *analytes* and the *matrix*. The *analytes* are the molecular or ionic species of interest in the sample and the *matrix* indicates the rest of the sample components. Similar to the case of information about the purpose of analysis, as much information as possible about the samples is desirable, although in many practical cases this information remains incomplete. A critical piece of information is related to the samples' complexity. This type of information is an early criterion related to the choice of the need for a hyphenated technique (separation + measurement) for the analysis. Information about the sampling process is also very useful, in particular indicating the samples' homogeneity, the age of the samples, and potential of contamination. If nonhomogeneous samples need to be analyzed it should be known if the analysis must be performed in a specific way (e.g., after homogenization, or selecting parts of the sample). Also, it must be known if the whole sample should be analyzed, or only a specific part (surface, soluble component, selected points, etc.). Information about the samples' physical state should be collected, indicating if the samples are gas, liquid, solid, solution, semisolid, or mixed phases. Also, knowledge should be collected whether the samples are inorganic, organic, composite, of biological origin, environmental, or of a special source. Other data about the samples should indicate the amount available (large quantity, small quantity, readily available, uniqueness, etc.) and also the value of the samples. In some cases the samples must be returned to the provider after a small amount has been used for analysis and this should also be known. Other useful information includes the samples' thermal stability and perishability, as well as any safety concerns. The information about the samples should also indicate if the samples are of a new type or if similar samples were previously analyzed.

The general information about the samples is closely related to the sampling process if the samples were not already collected. It should guide the precautions regarding sampling to avoid, for example, contamination, amount of necessary sample, the requirements of samples transportation, and the safety measures related to sampling. Nonhomogeneous samples typically need careful sampling followed by the separation of phases. Sample preparation is frequently needed. Besides the sampling process, the general information on the samples has input regarding the initial sample processing, such as physical processing (grinding, drying, homogenization), and also samples storage and preservation. Inorganic samples are typically processed differently than organic or complex samples. Sometimes, dissolution of inorganic samples is an important sample preparation step. Thermally labile samples may need special storage and handling, and HPLC type techniques are frequently used for this type of analysis. Perishability may be modified using preservatives, and the addition of such compounds must be known. The analysis of complex samples may also require sample preparation. Sample complexity also determines the use of a separation technique hyphenated with the measurement. The most common such technique is chromatography. For this reason, chromatography is the technique preferred in many organic, biological, and environmental analyses. Gases and the compounds that are thermally stable at their boiling point are preferably analyzed by GC, and sample preparation may not be necessary. Many organic compounds can be analyzed by HPLC, this technique being necessary for molecules that cannot be volatilized, but can also be used as a choice for volatile and semivolatile molecules. Appropriate measures must be taken if hazard concerns are present.

The Sample Constituents

The information about sample constituents can be placed into two categories: (1) analytical and (2) general properties. Analytical information includes data about the chemical nature of the constituents, the strength of bonding of the analytes to the matrix, range of concentrations of the analytes, etc. General properties include physicochemical data such as volatility of analytes and matrix, solubility in various solvents, acid–base character, hydrophilic–hydrophobic properties, reactivity, etc. The knowledge about the chemical nature of the analytes is essential. It is preferable to know the list of compounds (single or multiple component analysis) that must be analyzed or at least the class of the analytes (inorganic, organic, functional groups in organic compounds, ionic character, etc.). When the analytes of interest are known, it is useful to know if the analytes are volatile, nonvolatile, ionic, polymeric, weak interaction compounds, or a mixture of species of the same compound. If this information is missing, it must be known at least if the analytes are small molecules or polymeric. In the case of small analyte molecules, data regarding volatility, solubility, and reactivity are very useful. For macromolecules, a general characterization is typically necessary. Other data regarding the analytes are helpful, such as information on the estimated level of analytes in the samples (ultra-low trace, trace, minor constituent, medium levels, or major constituent). The level of analytes may be totally unknown and then preliminary analyses may be necessary. From the general information about the samples it may be possible to estimate the range of concentration for the analytes in the samples. The iterative aspect of collecting information about the samples allows for the improvement of the information about the sample composition, and as the analytical process starts to be implemented, additional information, as well as additional questions, may emerge.

Similar information to that about the analytes should be examined for the matrix. It is important to know if the matrix is homogeneous or not, if it is inorganic, organic, has an ionic character, if it is polymeric, or if it is of biological origin. Data regarding the volatility and stability of the sample matrix are also very useful. The interaction between the matrix and the analytes may play an important role in the analysis. The matrix may form stable molecular associations with the analytes (e.g., salts, clathrates including host–guest complexes, inclusion compounds), or may generate artifact analytes in specific conditions (e.g., generation of free amino acids from proteins).

This information is related to planning of different operations in the analysis and can go back to instructions about sampling (if not restricted to a sample already provided). It includes the choice of preliminary sample processing (physical processing), and the selection of a sample preparation process, such as sample cleanup and analyte concentration. The information about the chemical nature of the analytes and the matrix are the main data needed in planning a specific type of analysis. Inorganic analytes can be analyzed using chromatographic methods such as ion chromatography (IC), but techniques such as inductively coupled plasma (ICP) or atomic absorption (AA) are more frequently applied. For most organic and complex samples chromatography is the preferred technique. Samples with small volatile molecules are typically analyzed using GC (with or without sample preparation). Since GC can be easily coupled with MS as a detector, and because of the existence of large libraries of MS spectra, GC–MS is the most powerful technique for compound identification, and is also very useful for quantitation. HPLC is the most common analytical technique for the quantitation of a wide range of organic molecules. The utility of HPLC for qualitative analysis is somewhat limited even when it is coupled with MS detection. However, for quantitation purposes, HPLC is the most appropriate and commonly utilized analytical technique for organic molecules. Polymers can be analyzed by nonchromatographic methods such as IR, or they can be separated and analyzed using size exclusion chromatography (gel permeation [GPC] or gel filtration chromatography [GFC]). Sometimes the polymers are also analyzed after preliminary hydrolysis or pyrolysis that reduces the initial macromolecules to smaller molecules.

As part of the "analytical information" the knowledge about the range of the analytes contained in the samples allows for the selection of the appropriate sensitivity for the method of analysis. Since specific sensitivities are characteristic for different analytical techniques the information about the range of the analyte levels may be necessary for selecting the type of analytical technique and analytical instrumentation, and not only for adjusting method sensitivity. Data about the level of analytes may also guide the need for sample preparation such as analyte concentration procedures.

The Required Quality of the Results

The results of the chemical analysis are typically expected to answer specific questions related to the purpose of analysis. In order to have appropriate answers, the results must satisfy specific requirements regarding accuracy, precision, and the number of samples characterized. Also, it is important to know if the results are part of a larger study, if they can be compared with the results from other laboratories or with results on standard materials.

The type of necessary results has implications on various aspects of planning a new method of analysis. One starts with sampling. A specifically designed sampling protocol may be necessary when specific statistical analysis of the data is required. Also, the level of precision of the required data influences the planning for specific sensitivity of the analytical method. The use of the results in specific collaborative studies may also impose a given protocol for the method of analysis.

Instrumentation Availability, Expertise in the Laboratory, and Funding

The selection of an analytical procedure is frequently determined by various external factors besides the purpose of analysis and type of samples. Among such factors are the instrument availability, the expertise in the laboratory, the time available for developing or implementing a specific analytical procedure, and also funding. In addition to instruments, information should be collected about the materials available, indicating for example the need for preliminary purchases of chromatographic columns, reagents, solvents, and standards. Related to the safety concerns, appropriate laboratory conditions must also be assured for the analysis of certain samples. Information about all these aspects is of primary importance such that the analysis can be performed.

The information about the resources available for performing a specific analysis in a selected manner is critical for successful results. Missing the necessary instrument, for example, is an obvious barrier for planning the development of a new method analysis that would use that instrument. Potential purchase of new instrumentation must be considered in specific cases. The expertise available in the laboratory to perform the analysis is also an important factor, including information about the need for training. The performance of some analyses is a very simple task, while other analyses require considerable resources. Among the external factors that could also be considered are the effects of used chemicals from the laboratory on the environment. For this reason the utilization of environment-friendly solvents and reagents is recommended, as well as of analytical procedures that minimize the consumption of dangerous unavoidable chemicals. For example, organic solvents that could be considered "green" [1] are desirable in chromatographic analysis because their use avoids toxicity and environmental problems [2,3].

Information About Various Methods of Analysis

A large body of information is available regarding methods of analysis. This information is available in the form of articles published in peer-reviewed journals, books, company application notes and user guides, various forms of Internet information, as well as numerous courses at universities and private organizations (all further indicated as "literature"). Before starting the development or implementation of a new method of analysis a critical step is the reviewing of the available information related to the analysis of interest. For almost any analyte, there are methods of analysis that are described in the literature. The already-reported methods may not be intended for the specific type of sample on which the new method should be applied or may not satisfy certain requirements for the new method. It is also possible that the method described in the literature cannot be implemented, for example, due to the lack of instrumentation. However, this literature information always offers a valuable starting point in a new method development. When a method is available and can be utilized, it is frequently beneficial to implement the method and directly evaluate its adequacy for the new purpose. In certain cases, where no method is available in the literature, procedures for the analysis of similar compounds as those planned for new analysis are helpful.

The choice of a method of analysis requires knowledge about all the aspects previously described. When a method is already reported for the same type of sample, or for similar samples, it is typically advisable to implement that method as is, even if modifications and adjustments are to be made afterward. In most situations, multiple choices for an analytical method are available, even when the analysis has been previously practiced on the same type of analytes and samples. It is common that more than one choice can give good results, when most of the required conditions for an appropriate analysis are fulfilled. The method of analysis with all its aspects including the necessary sample processing, the type of analysis, and data processing may be more or less convenient depending on particular choices. Different analytical techniques for the same analytes and the same type of sample can offer advantages and may have disadvantages. Also, some may offer excellent qualities although not necessary needed. For example, the use of an LC-MS/MS method versus an HPLC method with UV detection may offer higher sensitivity and positive identification of the analytes, although this may not be necessary for a specific analysis. Arbitrary choices can be made for some analyses, and the quality of the result may or may not be affected. Although an optimization of the analytical method is sometimes possible, a balance must always be made between the effort of improving the method and using it as is, if the results are still satisfactory. The choice of a specific analytical method can be critical, but it is also possible that a number of alternatives lead to the same desired analytical goal and one choice or another is not critical. The selection of a method of analysis may be determined by many reasons, for example the instrument availability, personnel training, limited funding, and may be the subject of various constraints that do not allow the selection of the best possible alternative. Nevertheless, the goal of this book is to describe the best choices for selecting an HPLC method of analysis.

For HPLC analysis, a wide range of choices can be made with the end result of successfully performing a specific analysis. This book discusses various aspects of HPLC analysis that must be taken into consideration for the selection, development, and implementation of a successful method using this technique.

New Developments in Instrumentation

The improvements of analytical instrumentation, of accessories, of consumables, and the development of new technologies, and new materials is a continuous and very active process. All of these new developments may offer new possibilities for the analysis either through a variety of improvements in sensitivity of analytical instruments, media for better separations, simplifications in the necessary operations for specific analyses, cost reduction of analyses, speeding the analytical process, enhancing reliability of the analyses, etc. A variety of sources are available for providing information regarding all these new developments such as scientific meetings and conferences, literature provided by manufacturers, various Internet sites, etc.

Knowledge about all the new developments related to the analytical capabilities is very important for adopting the best path for a specific analytical task. However, the implementation in the laboratory of such new developments is not always possible. The cost of new instruments may be prohibitive, the training of the personnel may be lacking, and the new developments do not always represent better ways to solve a specific analytical problem. Nevertheless, it is always useful to be informed about novel alternatives for a chemical analysis.

1.2 OVERVIEW OF AN ANALYTICAL TECHNIQUE

Chemical analysis can be qualitative, semiquantitative, quantitative, or addressing structural problems, can be designed for the analysis of materials or for process control, analysis

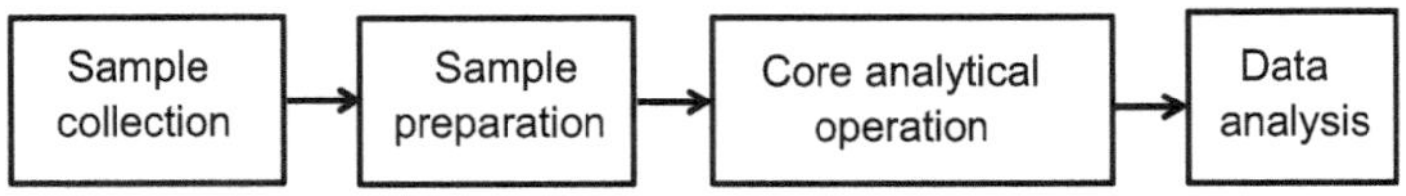

FIGURE 1.2.1 Simplified schematics for the path of a chemical analysis.

of gases, liquids, solids, or mixed phases, and can be designed for organic materials, inorganic ones, or mixed organic–inorganic, etc. It can be performed using a wide variety of procedures. Before the development of modern instruments for analysis, wet chemistry was widely used for qualitative analysis, and gravimetric and volumetric methods were used for quantitative analysis. Although these older procedures can still be utilized, this is done much less frequently, and other methods much more sensitive and selective based on various instruments are now practiced. A more systematic path for the chemical analysis is also usually implemented. An overall path for a modern chemical analysis can be schematically described by the diagram in Fig. 1.2.1. The scheme for the path of a chemical analysis can be much more complicated than the simplified scheme indicated in Fig. 1.2.1. However, the scheme describes the basic operations in a modern chemical analysis.

The decisions regarding various selections for the whole analysis are highly interrelated in the sequence: sample collection → sample preparation → core analytical procedure → data analysis. Depending on the sample, a specific sample preparation must be selected with the goal of a specific core analytical procedure. At the same time, the goal of using a specific core analytical operation is critical for the selection of the sample preparation step. An iterative selection process of sample preparation ↔ core analytical procedure is frequently necessary for the best results. Regarding sample collection, this is not always optional for the analyst, but in some instances sample collection can be adjusted to match the sample preparation and core analytical procedure.

Sample Collection

Sample collection (sampling) is the operation of obtaining the *raw sample* and delivering it in an appropriate form to the analytical process. Sampling is a very important process and a large body of information presents this subject in detail (see, e.g., [4–7]). The implications of sample collection on the selection of the method of analysis and in particular on the selection of HPLC technique are significant. For example, the collection may include a specific matrix, or may be associated with some type of concentration of the analytes which influence the further selection of the method of analysis. However, the subject of sample collection is beyond the purpose of this chapter. Starting with a given sample, various decisions must be made regarding the analysis, and such decisions will be further discussed only related to HPLC analysis.

Sample Preparation Step

The raw sample is frequently subjected to a sample preparation step. Sample preparation has the role of modifying the raw sample to make it more amenable for the core analytical operation. The operation of sample preparation can range from a minor step to very complex processing of the *raw sample* (sample as obtained from sampling). Among the potential operations in sample preparation are the routine manipulations of the sample such as homogenization, reduction of the sample size, drying, weighing, volume measuring, and sample dissolution. Also these operations may include cleanup procedures when the matrix of the sample is simplified, fractionation, concentration of the analytes, and chemical modifications. Various separation operations can be applied to the raw sample such as mechanical (e.g., filtration), phase transformation of the mixture of components (e.g., distillation, dissolution, crystallization), extraction with solvents, separations using solid phase extraction, migration in a field for separation (gravitational, magnetic, electric), etc. The initial raw sample can also be subjected to various chemical modifications in order to make it more amenable for the core analysis. Chemical modification may include reactions with one or more reagents (derivatizations), chemical degradations (applied for example to polymers), and pyrolysis (thermal degradation). All these operations will deliver a *processed sample* that is more amenable for further operations of separation and detection that compose the core analytical operation. Sample preparation is always performed considering the planned core analytical processes that follow in the analysis (further component separation and detection). Since sample preparation is performed with the goal of rendering the raw sample more amenable for the core analytical process, the characteristics of the core analytical operation are always taken into consideration during sample preparation.

Besides the preparation of the raw sample in order to deliver a processed sample, other operations must be considered in this stage of the analysis. These include, for example, the preparation of standards for the generation of calibrations or for obtaining preliminary information on the levels of materials that are further used in the analysis. Most analytical methods include a sample preparation step such that sample preparation is an important part of the whole analysis. For this reason, the subject of sample preparation is extensively presented in the literature and dedicated books, and other materials present as a standalone subject the methodologies for sample preparation (see, e.g., [8–11]). Sample preparation cannot be disconnected from the selection of the core analytical procedure, and depending on the extent and nature of sample preparation, an adequate core analytical method is selected. However, this book presents the specifics for the selection of an HPLC analysis when a given sample must be analyzed after preliminary processing. For this reason, the subject of sample preparation is not further discussed, although within a complete strategy for a chemical analysis sample preparation is an integral part of the analytical process.

Sample preparation is frequently a time- and manpower-consuming step, although it renders to the core analytical method a cleaner and possibly more concentrated processed sample. For this reason, the tendency to use simpler and shorter sample preparation techniques is common. However, these simpler techniques may result in a processed sample having a less simplified matrix and with a modest or no increase in the analyte concentration. As a result, the core analytical technique (e.g., HPLC) has a more difficult task in providing accurate results, and the development of a new and better core analytical technique is a common task.

The Core Analytical Operation

The core analytical operation receiving the processed sample has the role of delivering the analytical results that are further evaluated in the data analysis step for providing the

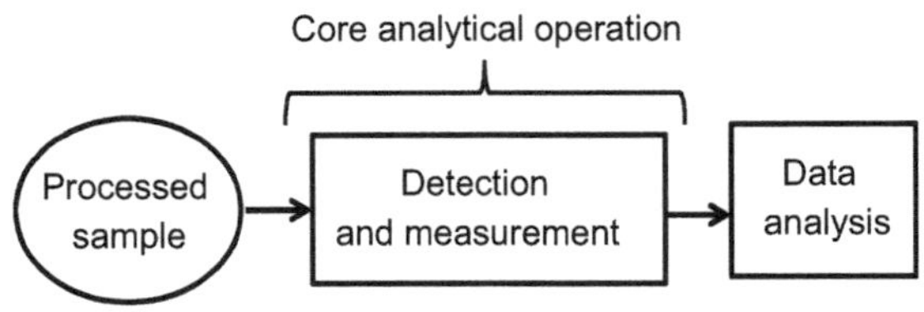

FIGURE 1.2.2 Schematic descriptions of the flow of operations in a *nonhyphenated* core analytical operation.

necessary information about the sample. The core analytical operation is usually performed using instrumental techniques that may involve spectroscopic, electrochemical, radiochemical, thermal, or other instrumental procedures. When applied on pure compounds, or when a compound in a mixture has unique physicochemical properties, the analysis can be performed without having a separation step added to the core analytical operation (sample preparation should be considered different from a core analytical procedure). Such techniques can be indicated as *nonhyphenated* (separation for different ions could be imbedded in the technique itself). Fig. 1.2.2 shows the schematic descriptions of the flow of operations in a nonhyphenated analytical operation.

Because of the necessity to analyze more and more complex samples, many modern analytical techniques that are able to address such analyses have a separation step imbedded in the *core analytical operation*. These techniques can be indicated as *hyphenated*. Sample preparation may use separation techniques for sample processing (see, e.g., [8,9]), but they are not online and are not part of the core analytical step. The processed sample may have a less complex matrix than the raw sample or may have a higher concentration of analytes, but it is common that further separations are needed for analytical purposes. When such a separatory step is necessary and it is imbedded (online) in the core analytical procedure, the technique can be indicated as *hyphenated*. Fig. 1.2.3 shows the schematic descriptions of the flow of operations in an analytical technique that contains the core analytical operations in a separation step.

The separation can be performed by several procedures, but the most common is chromatography. *Chromatography* designates several similar techniques that allow the separation of different molecular species from a mixture. For the chromatographic separation, the sample is placed in a fluid called *the mobile phase*, which carries the sample through a material called the *stationary phase*. The separation in chromatography is achieved because the stationary phase retains stronger or weaker different molecular species from the flowing mobile phase and releases them separately in time back into the mobile phase (different molecular species will have different *retention times*). The process of moving the analytes through the stationary phase is known as *elution*. The detection and measurement in analytical chromatography is usually based on a certain physicochemical property of the separated molecule (UV-absorption, refractive index, fluorescence, molecular mass and fragmentation in a

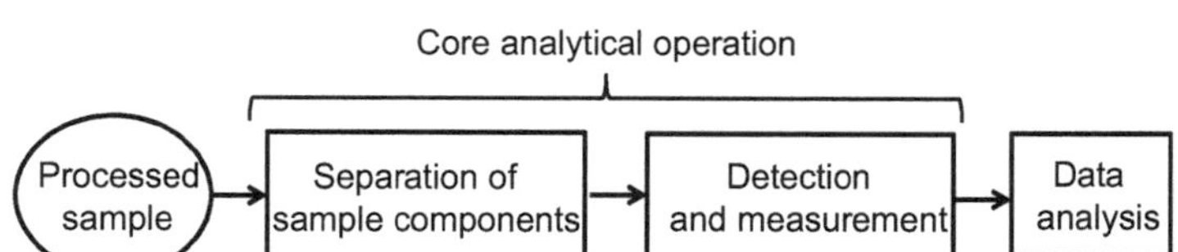

FIGURE 1.2.3 Schematic descriptions of the flow of operations in an analytical technique that contains in the core analytical operations a separation step (*hyphenated* technique).

mass spectrometer), a property which is different from that of the mobile phase. For analyte detection, many instrumental techniques developed independent of a chromatographic separation step were adapted to be used in analytical chromatography. As a result, numerous chromatographic detectors are based on UV-absorption, refractive index, fluorescence, molecular mass, and fragmentation in a mass spectrometer, etc.

Besides chromatography, other separation techniques can be connected online with an instrumental technique for detection. Among these electro-separations and membrane separations can be mentioned. Also, more complicated patterns of operations can be encountered, such as combinations of hyphenated techniques (e.g., in bidimensional separations).

Important decisions are made regarding the selection of the type of core analytical procedure. Depending on several factors, such as properties of the analytes, properties of matrix, requested turnaround time of the analysis, available equipment, etc., a technique including a separation step (a hyphenated technique) is or is not selected. The subject is further discussed in this book.

Data Evaluation

Data evaluation, also a part of a chemical analysis, is discussed in many published materials. Data evaluation (or analysis) typically involves statistical evaluations (with numerous chemometrics tools), but includes more general operations such as inspecting, cleaning, transforming, and modeling of the data with the goal of discovering useful information. Data analysis is a much wider field than applied to chemical analysis, but even restricted to this subject, data analysis will be only briefly presented in this book (see Section 1.3). Detailed information regarding statistical data processing can be found in many publications (see, e.g., [12–15]. For details regarding data analysis, the dedicated literature for this subject is recommended.

Qualitative and Quantitative Analysis

The information typically generated in a chemical analysis can be related to the finding of the nature of the analytes (qualitative analysis), level of the analyte (quantitative analysis), or both. Qualitative analysis can refer to various degrees of knowledge beginning with a brute formula, indications of the compound class, presence of specific functional groups, up to a true identification of the compound (possibly including stereo-isomeric structure, detailed structure of the molecule, etc.). Quantitative analysis describes how much analyte is present in a specific sample, and this information may also range from a semiquantitative analysis to a very precise one. The capability of different analytical techniques to provide qualitative and/or quantitative information about the samples is further discussed in Chapters 2 and 3.

For the quantitative analysis, the ideal case is when the signal of the analytical instrument depends linearly on the analyte concentration in a given volume (or amount) of sample. The concentrations C of interest can be determined from the signal y using the relation:

$$C = by \tag{1.2.1}$$

The value b is the slope for the dependence and can be obtained using calibrations with standards of known concentration. The standards are typically used as pure compounds and the calibrations are done independently of the sample. The calibration standards can be considered external standards. (An *external standard* is analyzed in a different run from the sample, while an *internal standard* is added and analyzed together with the sample.) In many practical applications it is preferable to make the calibrations by adding different levels of the calibration compound to a blank sample that does not contain the analyte. This procedure makes the analysis of the samples containing the calibration standards as close as possible to the analysis of a real sample and allows the subtraction of the overall influence of the matrix in the analysis. For compounds that have similar structures, the calibration curve for only one of the compounds is utilized sometimes, and different compounds are quantitated based on the same calibration. However, this is not a recommended practice and calibration curves are usually necessary for each analyte that must be quantitated.

In order to better mimic the influence of the matrix on the analyte response, it is common to use in the analyses an internal standard at constant concentration that will generate a signal $y_{i.s.}$ for any standard or sample. The calibrations are then generated using the formula:

$$C = b\frac{y}{y_{i.s.}} \tag{1.2.2}$$

The use of expression 1.2.2 for calibration instead of expression 1.2.1 is expecting that any influence of the matrix on the value of y will also affect $y_{i.s.}$ such that the matrix effects will be compensated.

Some linear calibrations do not pass through the origin, and the calculation of the concentration from the response is done using a relation of the form:

$$C = a + by \tag{1.2.3}$$

This type of dependence may indicate some problems with the particular analytical method, such as sample decomposition or loss of sample in the measurement process when a is negative, or a background interference when a is positive.

Besides linear dependencies between the concentration of the analyte and the signal y, other types of dependencies are also encountered. For example, quadratic dependence is rather common for signal vs. concentration dependence, mainly for wide ranges of concentrations of the analyte.

A different quantitation technique besides the external calibration is that of standard addition. The standard addition method can be used to analyze an unknown sample without the use of a calibration curve obtained in separate runs. The sample to be analyzed has an unknown concentration $C_0 = q_0/V_0$, where q_0 is the unknown amount in the sample and V_0 is the known volume of the sample to be analyzed. It must be assumed, however, that the dependence of the signal on the concentration follows expression 1.2.1 (and not expression 1.2.3). A set of known amounts of analyte $\{q_i\}$ $i = 1, 2\ldots n$ are added to the unknown sample, leading to the concentrations $C_i = (q_0 + q_i)/(V_0 + V_i)$, where V_i is the volume of the added solution with the standard and $C_0 = q_0/V_0$. The relation between the concentration C_i and the signal y_i is in this case given by the relation:

$$C_i = by_i \tag{1.2.4}$$

The values for C_0 and b (as unknown parameters) can be obtained from the added amounts $\{q_i\}i = 1\ldots n$ and the signal measurements $\{y_i\}i = 0, 1\ldots n$ using, for example, least-squares fitting. The standard addition method can be used even with a single added amount to the unknown sample. If one single addition q_1 is made to the sample, two signals y_0 and y_1 are generated corresponding to C_0 and C_1. The two equations of the form 1.2.4 are as follows:

$$\begin{cases} q_0/V_0 = b\, y_0 \\ (q_0 + q_1)/(V_0 + V_1) = b y_1 \end{cases} \tag{1.2.5}$$

The value for q_x can be obtained from expression 1.2.5 and the value for $C_x = C_0$ is given by the formula:

$$C_0 = \frac{q_1 y_1}{(V_0 + V_1) y_1 - V_0 y_0} \tag{1.2.6}$$

When the addition of the standard does not dilute the sample ($V_1 = 0$), expression 1.2.6 can be written in the form:

$$C_0 = \frac{c_1 y_1}{y_1 - y_0} \tag{1.2.7}$$

The accuracy of standard addition is significantly better when more than one standard addition is performed. The calculation of concentration using standard addition using only one addition is in some respects equivalent to using a calibration curve with only two points.

Other procedures can also be used for quantitation (see, e.g., [8]). One of these procedures uses a response factor for quantitation. This response factor is calculated as the ratio of the responses y_A and y_B of two compounds A and B at the same concentration, A being the analyte and B the internal standard. The ratio of the two signals is usually obtained as an average of several measurements giving the response factor:

$$f_A = \frac{y_A}{y_B} \tag{1.2.8}$$

Ideally, the value for f_A remains constant for an interval of values for the pairs of concentrations of the standard and the analyte. The unknown concentration $C_{A,x}$ of compound A is then obtained by measuring in the same run the signal of the compound A to be analyzed (at unknown concentration) and the signal of the standard B (at known concentration C_B) using the formula:

$$C_{A,x} = f_A \frac{y_{A,x}}{y_B} C_B \tag{1.2.9}$$

Typically, it is recommended that the analyte A and the internal standard B are chemically similar or even identical but isotopically labeled.

The selection of a specific quantitation procedure depends on numerous factors such as properties of the matrix, concentration of the analyte in the sample, and availability of isotopically labeled standards. Although ideally all procedures should generate the same result, some differences exist between different procedures. The calibration with pure standards has the disadvantage that the matrix may influence the instrument response, and for real samples the response to the same concentration of the analyte may be different than the response for pure solutions. The use of an internal standard and use of expression 1.2.2 for calibration requires that the internal standard response must be constant in every sample.

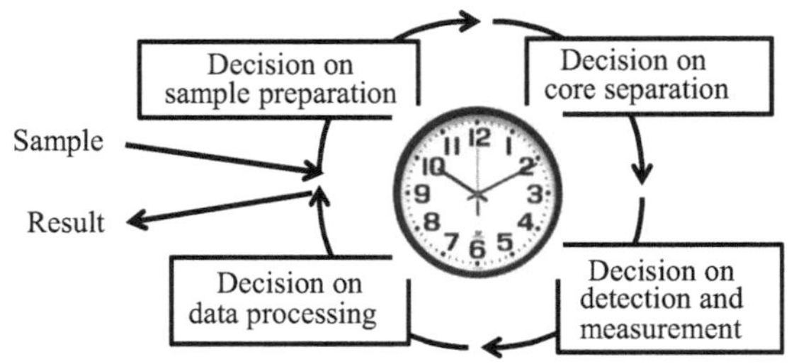

FIGURE 1.2.4 Diagram suggesting the iterative process of establishing a method for analysis.

For the case of analyzing multiple analytes and one internal standard, any error in the internal standard measurement will affect all the other measurements, which can be a disadvantage. The use of standard addition takes into account the influence of sample matrix, and when used with multiple standard additions provides sufficient reliability of the procedure. However, conditions such as the linearity of the dependence and zero value for the response in the absence of the analyte (dependence of the form 1.2.1) must be satisfied for its applicability. The use of a response factor for quantitation requires identical behavior of the compound used as internal standard *B* and of the analyte *A*.

Decision Process in Selecting a Method of Analysis

The path for selection of a method of analysis is an iterative process. Each part of the analysis has its own characteristics and the sample characteristics, decisions on sample preparation and each component of the core analytical method must be taken into consideration with implications from one phase to another. This process is schematically illustrated in Fig. 1.2.4.

The cycle indicated in Fig. 1.2.4 can be followed only once, but it is possible to be repeated as decisions at one step are influenced by those from previous steps. In the decision process all the preliminary information should be included, but also any additional information gathered along the decision process and possibly results from preliminary experiments. The time factor is also to be taken into consideration in the decision process, the implementation of an analytical method usually has a deadline.

1.3 STATISTICAL EVALUATION OF DATA AND CRITERIA FOR METHOD VALIDATION

The results from both qualitative and quantitative analysis can be processed using statistical concepts (see, e.g., [8]), although statistics is more frequently applied for the evaluation of the quantitative results. Accuracy (nearness to the true value) and precision (repeatability of the results) are the most common parameters generated using statistical evaluations. However, other aspects related to analytical data characterization have a statistical interpretation.

Precision and Accuracy in Quantitative Chemical Analysis

In most quantitative analytical determinations the measurement of the amount or of the concentration of an analyte is performed using the dependence of a measured signal y on the concentration (or quantity of the analyte) x:

$$y = \mathfrak{F}(x) \tag{1.3.1}$$

The amount or the concentration x can be calculated based on the dependence described by relation 1.3.1 using a number of procedures that generate a calibration function:

$$x = \mathfrak{F}^{-1}(y) \tag{1.3.2}$$

When more than one measurement is performed, the results for x are typically scattered around a specific value, the measurements being always affected by errors. The errors of measurements are classified as systematic (determinate) or random (see, e.g., [16]). Systematic errors are generated by a specific cause, and they affect the accuracy of the results. Random errors do not have an assigned cause, and they affect the precision of the measurement. *Precision* refers to the reproducibility of measurement within a set, indicating the scatter or dispersion of a set of measurements about their central value [12]. For example, repeated determination of the amount or concentration of an analyte generates for each measurement j a value x_j. If the number of measurements is n, they will generate a set of measurements also indicated in statistics as a *sample* $\{x_1, x_2 \ldots x_n\}$ of measurements (not to be confused with sample = specimen in a chemical analysis). For this set, an *average* (or the *mean*) m, and a *standard deviation* (SD) s can be obtained using the following formulas:

$$m = \frac{\sum_{j=1}^{n} x_j}{n} \tag{1.3.3}$$

and:

$$\mathrm{SD} = s = \sqrt{\frac{\sum_{j=1}^{n} (x_j - m)^2}{n - 1}} \tag{1.3.4}$$

The standard deviation s shows the distribution of measurements about the mean and characterizes *precision*. A small standard deviation indicates good precision for the set of measurements. Both m and s are expressed in the same units. A relative standard deviation % (*RSD*%) is frequently used instead of standard deviation s for the characterization of precision, and it is expressed by the formula:

$$\mathrm{RSD\%} = 100 \frac{s}{m} \% \tag{1.3.5}$$

Instead of standard deviation s, it is also common to use for precision characterization the value s^2 called *variance*.

In statistics, any obtained set of measurements $\{x_1, x_2 \ldots x_n\}$ represents a *sample* of values taken by a random variable x. This *sample* is a part of an infinite set of values that variable x may have. This infinite set of values is known as a *population*. Similar to the average m

and standard deviation s for one *sample*, the *population* is characterized by the mean $\underline{\mu}$ of the *population* and the standard deviation σ of the *population* which are given by the formulas:

$$\underline{\mu} = \lim_{n\to\infty} m_n \tag{1.3.6}$$

$$\sigma = \lim_{n\to\infty} s_n \tag{1.3.7}$$

Since an infinite number of measurements cannot be performed, the true values for $\underline{\mu}$ and σ are not known, and they are approximated with m and s for large number of measurements n. The true (ideal) value for $\underline{\mu}$ represents the ideal result for all the measurements, but does not necessarily represent the true value of the measured quantity. The difference between $\underline{\mu}$ and the true value (if known) of a quantity is called the *bias*. This true value can be, for example, the level of an analyte in a standard (regardless of the fact that making of the standard can also be affected by errors). Another convention for a true value is the assumption that it is the one "generally accepted." Accepting m_0 as a true value of the measurement and taking the population mean $\underline{\mu} \approx m$, the *bias* is approximated by the following formula:

$$bias = m - m_0 \tag{1.3.8}$$

A method is considered accurate when the bias is very small (or even zero).

For the same quantity, more than one set of measurements can be performed (e.g., sets of measurements of the same sample performed in different laboratories), generating a number of *samples* of data, each one with its own mean $m_1, m_2, \ldots m_k$. Although expected to be close to each other, these means are not necessarily identical. A total mean $\boldsymbol{m}$ can be obtained from these partial means, $\boldsymbol{m}$ being equal with the average of all measurements (addition is associative). The standard deviation of the mean $\boldsymbol{m}$ is known a standard error of the mean s_m and it is expressed by the formula:

$$s_m = \frac{s}{n^{1/2}} \tag{1.3.9}$$

One question regarding the distribution of measurements about their mean is the expected frequency of occurrence of an error as a function of the error magnitude. It is expected that larger random errors will be less frequent than small errors. The most commonly utilized function, which describes well the relative frequency of occurrence of random errors in large sets of measurements, is given by the Gauss formula:

$$f(x) = \frac{1}{(2\pi\sigma^2)^{1/2}} \exp\left[-\left(x - \underline{\mu}\right)^2 / 2\sigma^2\right] \tag{1.3.10}$$

This frequency function $f(x)$ (known as Gaussian density function) shows that the point of maximum frequency is obtained for the mean (when $x = \underline{\mu}$), the distribution of positive and negative errors is symmetrical, and as the magnitude of the deviation from the mean increases, an exponential decrease in the frequency takes place. The errors with the relative frequency of occurrence given by relation 1.3.10 have a so-called normal distribution $N(\underline{\mu},\sigma)$. Besides Gaussian density functions, other frequency functions are known (e.g., Student frequency function).

For x with a normal distribution $N(\underline{\mu},\sigma)$, the following substitution:

$$y = \left(x - \underline{\mu}\right) / \sigma \tag{1.3.11}$$

leads to a variable y with a normal distribution $N(0,1)$.

The value $(x - \underline{\mu})$ is so-called mean-centered value, and by division with σ, y is expressed in σ units (or it is standardized). Mean-centered standardized variables y (standardized variates) are commonly used in statistical data processing.

The area under the curve $f(x)$ for $x < \alpha$, with $f(x)$ given by formula 1.3.10, will give a cumulative frequency distribution expressed by the formula:

$$F(\alpha) = \int_{-\infty}^{\alpha} f(x)dx \tag{1.3.12}$$

The cumulative frequency distribution is equal to the probability P for x to have a value below α in any measurement. The integral of $f(x)$ over the whole space gives $P = 1$. The values of function $F(\alpha)$ (for $f(x)$ a Gaussian function) are known and tabulated (e.g., [14]) or given in computer statistical packages (e.g., [15]).

The mean m_k of a sample of n values $\{x_1, x_2 \ldots x_n\}$, is itself a value of a random variable $\mathcal{m}$. Assuming that x has a normal distribution $N(\underline{\mu},\sigma)$, the random variable $\mathcal{m}$ (m_k is one of the possible values of $\mathcal{m}$) takes a continuous range of values with a normal distribution $N(\underline{\mu},\sigma/n^{1/2})$.

A mean-centered standardized variable defined by the formula:

$$z = \left(\mathcal{m} - \underline{\mu}\right) \Big/ \left(\sigma/n^{1/2}\right) \tag{1.3.13}$$

will have an N(0,1) distribution. With the help of variable z it is possible to evaluate how close the values of $\underline{\mu}$ and m_k are for a certain population. For the variable z given by relation 1.3.13, two values $-z_{\alpha/2}$ and $z_{1-\alpha/2}$ can be found such that the probability for z of being outside the interval ($-z_{\alpha/2}$, $z_{1-\alpha/2}$) is equal to α (areas $\alpha/2$ indicated in Fig. 1.3.1). The interval ($-z_{\alpha/2}$, $z_{1-\alpha/2}$) is indicated as confidence interval. For z being inside the interval ($-z_{\alpha/2}$, $z_{1-\alpha/2}$), or:

$$-z_{\alpha/2} < \left(\mathcal{m} - \underline{\mu}\right) \Big/ \left(\sigma/n^{1/2}\right) < z_{1-\alpha/2} \tag{1.3.14}$$

the value for probability will be $P = 1 - \alpha$.

Expression 1.3.14 can indicate the limits for m_k with the probability $P = 1 - \alpha$. For this purpose, it should be noted that $z_{\alpha/2} = z_{1-\alpha/2}$, and rearranging expression 1.3.14 the result is as follows:

$$\underline{\mu} - z_{1-\alpha/2}\left(\sigma/n^{1/2}\right) < m_k < \underline{\mu} + z_{1-\alpha/2}\left(\sigma/n^{1/2}\right) \tag{1.3.15}$$

The values for $z_{1-\alpha/2}$ for a selected probability P and Gaussian distribution are available in the literature (see, e.g., [14,17]). Such values are given for $P = 1 - \alpha$ (two-sided normal distribution) or for $P = (1 - \alpha)/2$ (one-sided normal distribution). Fig. 1.3.1 shows the curve N(0,1) on which are indicated two values $z_{\alpha/2}$ and $z_{1-\alpha/2}$ such that the probability for z of being outside the confidence interval ($-z_{\alpha/2}$, $z_{1-\alpha/2}$) is equal to α (area under the curve). Larger values for α have correspondingly smaller values for P and smaller values for $z_{1-\alpha/2}$.

Sensitivity and Limit of Detection

The sensitivity of a quantitative analytical method can be defined as the slope of the curve that is obtained when the result of a series of measurements is plotted against the amount (or

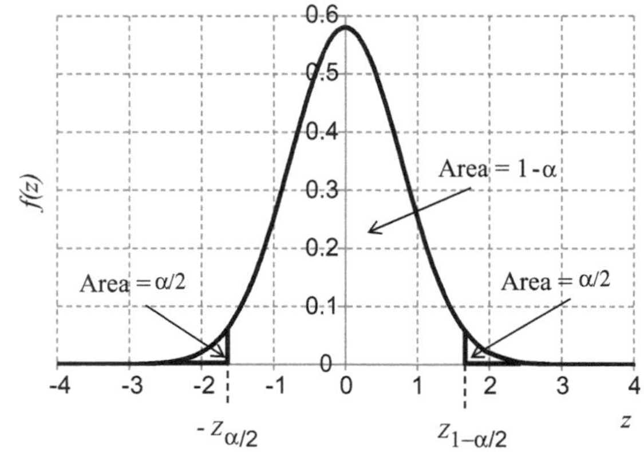

FIGURE 1.3.1 Gaussian curve N(0,1), showing two values, $-z_{\alpha/2}$ and $z_{1-\alpha/2}$, such that the probability for z of being outside the interval $(-z_{\alpha/2}, z_{1-\alpha/2})$ is equal to α (and area under the curve is $P = 1 - \alpha$ showing probability inside the same interval).

concentration) that is to be determined. For the dependence described by expression 1.3.1, the sensitivity is defined as the first derivative of the function $\mathfrak{F}(x)$ or:

$$\mathcal{S} = d\mathfrak{F}(x)/dx \tag{1.3.16}$$

It is common that the dependence described by expression 1.3.16 is a linear function (see expression 1.2.3) and has the expression:

$$\mathfrak{F}(x) = a + bx \tag{1.3.17}$$

In this case, the sensitivity $\mathcal{S}$ is equal to the constant b. Sensitivity can therefore be determined from the calibration curve for the method. For nonlinear dependencies the definition still can be applied, but the sensitivity is not constant for all concentrations. Many methods have only a range of linear dependence, with an upper region and a lower region that are not linear. In the nonlinear regions, the sensitivity varies, and for this reason sensitivity is not necessarily a convenient way to characterize an analytical method. It is common that concentrations below a certain value show a sensitivity decrease. The deviation from linearity at lower concentrations in chromatographic analysis is likely due to a combined effect of the loss of a small amount of sample that is adsorbed irreversibly in the chromatographic system.

The concentration of an analyte below which the detection is not possible is typically indicated as limit of detection. For a quantitative characterization, the detection limit can be discussed in terms of signal and transformed in terms of amount or concentration using the calibration function. The signal y for an analytical measurement (see relation 1.3.1) is in fact the difference between the signal for a sample y_s and that of a blank y_b, both affected by errors. The signals for the blank (or the noise) are assumed to have a mean μ_b and the signals from the sample to have a mean μ_s. The standard deviations for both the blank and the sample can be considered equal to the same value $\sigma_y = \sigma$. A specific value of the signal can be selected (arbitrarily) and considered as not generated by the noise. This value is indicated as decision limit L [18] where the signal is noticeable and:

$$L = \underline{\mu}_s \tag{1.3.18}$$

Assuming that the errors in the signal of the blank and sample have a normal distribution, the probability to obtain signals from the blank higher than the decision limit L is given by the expression:

$$\alpha = \int_{L}^{\infty} f(y_b)dy_b \quad (1.3.19)$$

In expression 1.3.19, the function f is given by formula 1.3.10 (Gaussian density function) that has $\underline{\mu} = \underline{\mu}_b$ (σ, the standard deviation is the same for both the blank and the sample). The decision limit L in the integral can be expressed in terms of signals for the blank using the expression:

$$L = \underline{\mu}_b + k\sigma \quad (1.3.20)$$

The probability to consider a noise as being signal from the analyte for a value higher than L is given by $P = \alpha$. For $k = 2.33$ the resulting value for the one-sided normal distribution gives $\alpha = 0.01$ (or 1% if expressed in percent). Therefore, if the signal is higher than $\underline{\mu}_b + 2.33\ \sigma$, the probability of false positives (signal from the background to be considered analyte) is 1%.

A signal y with $\underline{\mu}_s = L$ has, however, the problem of possibly generating false negatives. The probability of false negatives is given by the expression:

$$\beta = \int_{-\infty}^{L} f(y_s)dy_s \quad (1.3.21)$$

In expression 1.3.21, $f(y_s)$ is a Gaussian with $\underline{\mu} = \underline{\mu}_s$. This probability is $P = 0.5$ (or 50%) because the normal curve is symmetrical around the mean. Therefore, the possibility of false negatives is very high, and 50% of the true positives will be considered noise. This situation is depicted in Fig. 1.3.2.

A higher signal (higher values for $\underline{\mu}_s$) and maintaining $L = \underline{\mu}_s$ will continue to diminish the chances for false positives since the area α will become smaller and smaller as the second Gaussian moves to higher values of y. However, maintaining $L = \underline{\mu}_s$ does not modify the probability of false negatives. For a chosen probability $P = 0.01$ of obtaining a false negative, the corresponding value of the signal can be calculated. This value is noted by D, should be smaller than $\underline{\mu}_s$, and has the expression:

$$D = \underline{\mu}_s - k'\sigma \quad (1.3.22)$$

For the probability $\beta = 0.01$, the resulting value for the one-sided normal distribution is $k' = 2.33$.

For L $= D$, the subtraction of relation 1.3.22 from relation 1.3.20 leads to the expression:

$$\underline{\mu}_s = \underline{\mu}_b + (k + k')\sigma \quad (1.3.23)$$

The overlapping of frequency functions for the two probabilities is shown in Fig. 1.3.3.

The result of these considerations is that in order to have a probability of 1% for a false-negative signal, and a probability of 1% for a false-positive signal requires $k + k' = 4.66$, and in this case it can be written in the following formula:

$$\underline{\mu}_s = \underline{\mu}_b + 4.66\sigma \quad (1.3.24)$$

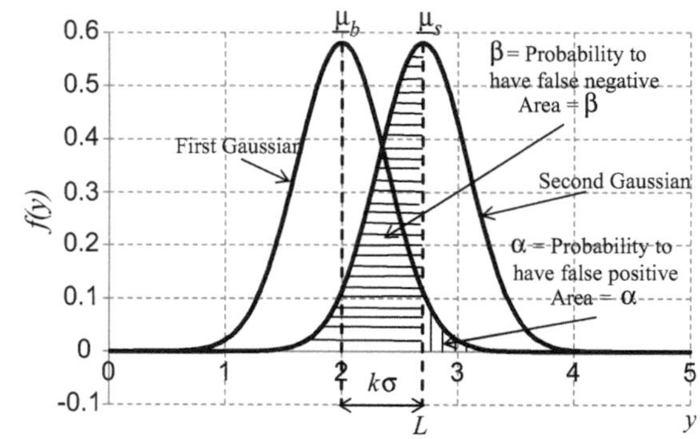

FIGURE 1.3.2 Graph showing on the horizontal axis the signal y, the values $\underline{\mu}_b$ and $\underline{\mu}_s$ as well as the decision limit $L = \underline{\mu}_s$. The vertical axis gives the frequencies $f(y)$ of occurrence for the value of a measurement, where f is given by relation 1.3.10 and describes a normal distribution.

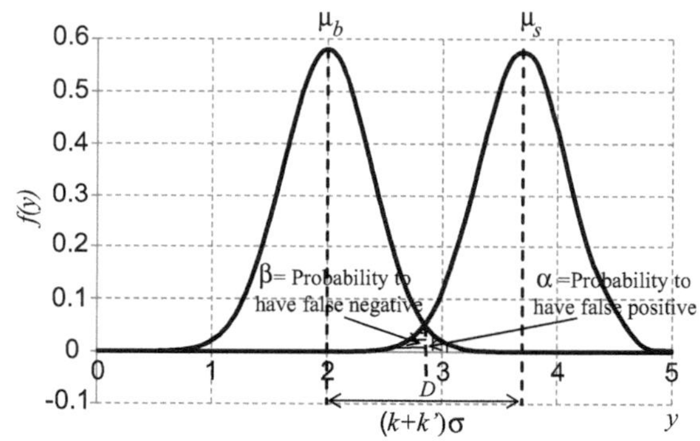

FIGURE 1.3.3 Graph showing on the horizontal axis the signal y, the values $\underline{\mu}_b$ and $\underline{\mu}_s$ as well as the decision limit $D = \underline{\mu}_s - k'\sigma$ The vertical axis gives the frequencies $f(y)$ of occurrence for the value of a measurement where f is given by relation 1.3.10 and describes a normal distribution.

If the signal is considered $\mathcal{S} = \underline{\mu}_s - \underline{\mu}_b$ and the noise is taken as $\mathcal{N} = \sigma$, formula 1.3.24 is equivalent with the formula:

$$\mathcal{S}/\mathcal{N} \approx 4.66 \tag{1.3.25}$$

For the detection to be possible in analytical practice, a value of $\mathcal{S}/\mathcal{N} \approx 3$ that is slightly smaller than 4.66 was considered to be sufficient (with lower probability than 1% for false negatives, and higher than 1% to give false positives). The value $\mathcal{S}/\mathcal{N} = 3$ has been adopted in analytical practice for the definition of *detection limit or limit of detection* (LOD) [19–21]. The estimation of $\mathcal{S}/\mathcal{N} = 3$ in a set of chromatographic measurements can be done directly on the plot of chromatograms (typically using the capabilities of the data-processing software of the chromatographic instrument). The concentration (or amount) generating $\mathcal{S}/\mathcal{N} = 3$ is taken as LOD.

Although the theory supports the selection of the value $\mathcal{S}/\mathcal{N} \approx 3$ for LOD, in chromatographic practice this definition is not very informative. The noise is typically measured as the variation in detector response at the baseline (difference between the minimum and the maximum of the baseline signal), and the signal $\mathcal{S}$ for a peak is measured as the difference from the maximum of the peak and the average value of the baseline for a range around the peak. The simplest way to check the baseline noise $\mathcal{N}$ is to measure the variation of the detector response for a chromatogram without making an injection [22]. It is also common to measure the variation of the detector response in any part of the chromatogram where no extraneous peaks are present. The place where the noise is measured still remains arbitrary, in spite of some recommendations that the noise should cover a specific range of time, wider than the chromatographic peak of the analyte. In addition to that, the noise can be different from one chromatogram to another. The signal and noise measurements are exemplified in Fig. 1.3.4 that shows the measurement of two signals and two selections for the noise leading to two different values $\mathcal{N}_1$ and $\mathcal{N}_2$.

A better procedure for LOD evaluation is based on repeated measurements and the use of the approximations $\mu_s - \mu_b \approx m$ and $\sigma = s$ (s is also indicated as SD). The choice $\mathcal{S}/\mathcal{N} = 3$ is equivalent with the expression:

$$m = 3s \tag{1.3.26}$$

Expression 1.3.26 indicates that for detection to be possible, a concentration (or level) at least equal to 3s for a set of measurements at low concentration is necessary. Based on expression 1.3.26 it results that for a set of measurements at low analyte concentration the definition of LOD should be based on the following expression [23,24]:

$$\text{LOD} = 3s \tag{1.3.27}$$

The convention of taking LOD equal with 3 increases the probability P of obtaining a false positive and decreases the probability P of obtaining a false negative (in Fig. 1.3.3 a smaller coefficient for σ moves the point D toward lower values). To address this problem, an additional parameter indicated as *limit of quantitation* (LOQ) has been introduced. The definition of LOQ is the following:

$$\text{LOQ} = 10s \tag{1.3.28}$$

LOQ indicates the minimum amount (or concentration) that can be quantitated by a specific method. The choice of 10s for LOQ decreases the probability P for obtaining a false positive but increases the probability P of obtaining a false negative. To prevent the problem

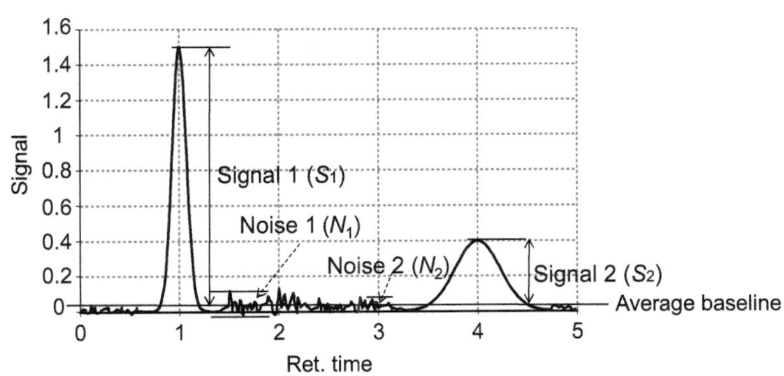

FIGURE 1.3.4 Example of measurement of two signals and two selections for the noise leading to two different values, $\mathcal{N}_1$ and $\mathcal{N}_2$.

of generating false negatives, the measurements with results below 10*s* are not labeled as "analyte absent" but "analyte below LOQ."

Practical LOD and LOQ (PLOD and PLOQ)

The establishing of LOD and LOQ for an analytical method is very important. Some variation in the procedure for the establishing of LOD and LOQ is described with each analytical method. Although it is common to measure *s* (or SD) for a set of low standards, there is a known fact that it is possible for the analysis of standards to generate cleaner chromatograms than the analysis of the same analytes in a matrix. Even when sample preparation offers procedures for matrix cleanup, in many instances only a partial cleanup is performed, and besides the analytes some additional matrix compounds may remain in the processed sample (residual matrix). The residual matrix can be tolerated when it does not interfere significantly with the analyte measurement, but this does not necessarily imply that at very low level of the analyte the residual matrix has no effect on sensitivity. For this reason, in certain analyses, although the measurement is accurate for the analytes above a certain level, the determinations close to the LOQ (LOQ determined with standards) is difficult or even not possible. For this reason, for some procedures instead of reporting LOD and LOQ obtained with standards, (practical) PLOD and PLOQ are reported. In these cases, the measurement of standard deviation *s* or the signal to noise $\mathcal{S}/\mathcal{N}$ are measured for a low level of analyte in the sample with residual matrix components (possibly after partial cleanup). This is not always possible since samples with low levels of analyte may not be available. The concentration of the analyte in the sample can be measured as usual, using standards or standard addition procedures. The values of PLOD and PLOQ are more useful for practical application of a method when they are different from LOD and LOQ.

Linearity of the Instrumental Response and Least Square Regression

Linear dependence of an analytical signal and the concentration (or the amount) of an analyte is the common procedure for quantitative analysis. The dependence typically has the form of a regression line with the equation given by the following formula (see expression 1.2.3):

$$y = a + bx \tag{1.3.29}$$

The same form is valid for y, an analytical signal and x, the analyte concentration, or for y, the analyte concentration, depending linearly on a signal x (e.g., peak area in a chromatogram). In practice, for a number of values for x, $\{x_1, x_2 \ldots x_n\}$ a number of responses (measurements) $\{y_1, y_2 \ldots y_n\}$ are obtained. The values of a and b must be obtained such that to minimize (in absolute value) for all i the differences between experimental data y_i and calculated data $a + bx_i$:

$$r_i = y_i - (a + bx_i) \tag{1.3.30}$$

The values r_i are called residuals, and for the minimization of all r_i ($i = 1, 2, \ldots n$), the following expression must be minimal:

$$E(a,b) = \sum_{i=1}^{n} (y_i - a - bx_i)^2 = \sum_{i=1}^{n} r_i^2 = min \tag{1.3.31}$$

Since the procedure involves the minimization of the errors in the calculated values for y, it is known as linear minimum mean-square (least square) error fitting. For expressing further results related to the minimization of expression $E(a,b)$, several notations must be introduced. The average of x_i values will be noted with m_x, and the average of y_i values will be noted with m_y. Also several partial sums are noted as follows:

$$S_{xx} = \sum_{i=1}^{n} (x_i - m_x)^2 \tag{1.3.32a}$$

$$S_{yy} = \sum_{i=1}^{n} (y_i - m_y)^2 \tag{1.3.32b}$$

$$S_{xy} = \sum_{i=1}^{n} (x_i - m_x)(y_i - m_y) \tag{1.3.32c}$$

The minimum is achieved for $E(a,b)$ when its partial derivative with respect to a and b is set to zero. This is expressed by the formulas:

$$\frac{\partial E(a,b)}{\partial a} = 0 \quad \frac{\partial E(a,b)}{\partial b} = 0 \tag{1.3.33}$$

Expression 1.3.33 leads to a pair of equations in a and b. The solution of these two equations leads to the following results for the minimization of $E(a,b)$:

$$b = S_{xy}/S_{xx} \quad a = m_y - bm_x \tag{1.3.34}$$

The same minimization generates results for the standard deviations for the residuals s_r, for the slope s_b, and for the intercept s_a.

$$s_r^2 = \frac{S_{yy} - b^2 S_{xx}}{n-2} \tag{1.3.35}$$

$$s_b^2 = \frac{s_r^2}{S_{xx}} \tag{1.3.36}$$

$$s_a^2 = \frac{s_r^2 \sum_{i=1}^{n} x_i^2}{n S_{xx}} \tag{1.3.37}$$

An additional parameter called *coefficient of determination*, R^2, is utilized for the characterization of the regression line. The formula for R^2 is the following:

$$R^2 = 1 - \frac{\sum_{i=1}^{n} r_i^2}{S_{yy}} \tag{1.3.38}$$

Perfect fit of experimental data $\{y_1, y_2 \ldots y_n\}$ and calculated data $\{a + bx_1, a + bx_2, \ldots a + bx_n\}$ is indicated by $R^2 = 1$ when $\sum_{i=1}^{n} r_i^2 = 0$. For larger residuals, the value of R^2 decreases and can have values between 1 and 0. The calculations for a and b for the regression line as well as of s_r, s_b, s_a, and R^2 are available in Excel computer packages (see, e.g., [25]).

The minimum mean-square error fitting can be applied not only to linear dependences, but also to other types of dependence such as the quadratic one. The value for R^2 is identical for the dependence $y = \mathfrak{F}(x)$ and the dependence $x = \mathfrak{F}^{-1}(y)$ when $\mathfrak{F}$ expresses a linear dependence, but the R^2 values are not identical for other types of dependence.

Also, the minimum mean-square error fitting can be applied to a set of data using a weighted least square fitting. In a data set obtained, for example, for the calibration content in analyte vs. instrumental response, some $\{x,y\}$ values are small, while others are large. Assuming, for example, that the errors in y_i are about 5%, a much larger error will occur for the larger y_i values as compared to the smaller y_i values. Since the least square errors apply to the differences r_i (see expression 1.3.30) the impact of larger $\{x_i,y_i\}$ values will be larger than those for small $\{x,y\}$ values in generation of the regression line (or curve). This problem can be corrected by generating a weighed least square curve. For this reason a set of "weights" $\{w_i\}$ is utilized and the minimization from expression 1.3.30 takes the following from:

$$\sum_{i=1}^{n} w_i r_i^2 = min \tag{1.3.39}$$

Common selections for $\{w_i\}$ are $\{1/x_i)$ or $\{1/x_i^2\}$, although other selections can be made, if the importance of some points is higher than for other. The Excel computer package provides functions and formulas capable of calculating weighed least square fitting, as well as nonlinear least square fitting. Also, the data-processing capabilities of the analytical instruments provide various procedures for linear, nonlinear, and weighed least square calibrations.

Statistical Comparison of the Results from Two Methods

Various statistical tests have been developed for the comparison of accuracy and precision (see, e.g., [14]) of analytical methods. A common procedure is the "*t*-test" that determines (with a selected value for probability) if the difference between a known true value and the experimental mean obtained by a new method is caused by random errors or by a bias. If caused by random errors, the new method can be considered accurate. Also, statistical tests may determine if the accuracy of the methods is the same or different. The comparison of precisions of two different methods can also be done using statistical evaluations (see, e.g., [9]). Statistical evaluation of the results is a very important part of a chemical analysis, but the detailed description of the subject is outside the main purpose this book. Dedicated monographs are recommended for a proper description of the subject precision (see, e.g., [14,26,27]).

Validation of an Analytical Method

The validation of an analytical method describes the measures taken to ensure that it provides accurate and reproducible results that are precise and suitable for the application intended. The formal validation also includes documented evidence indicating that the analytical process is performed consistently. The validation involves internal confirmation or external confirmation by other laboratories, the use of other methods, and the use of reference materials in order to evaluate the suitability of the chosen methodology. Validation issues have been addressed by several public and private organizations such as the International Organization for Standardization (ISO), the US Food and Drug Administration (FDA), the US Environmental Protection Agency (EPA), and the Association of Official Analytical Chemists (AOAC), etc. Specific international laboratory accreditation standards, such as ISO/IEC 17025, assure the quality of analyses through external accreditation processes accepted by national accreditation organizations operating within the ISO umbrella. Only some basic aspects regarding the validation are discussed in this material, while detailed information can be found in a number of publications dedicated to this subject [28–32]. Method validation is typically performed based on the verification of a number of parameters regarding the method [33]. Among these parameters are the following.

1) Specificity, which refers to the quality of the method to produce a response for a single analyte in the presence of other components in the matrix. In chromatographic methods, specificity is a combined result of the separation of sample components and selectivity of the detector. Some detectors used in chromatography are "universal" or not selective, and the specificity must be achieved from the chromatographic separation. Other detectors, such as mass spectrometric ones may provide excellent specificity. With such detectors, a good chromatographic separation is sometimes no longer required since even compounds in merging peaks can be individually detected. However, even when mass spectrometry is used for detection it may be necessary in some cases to use a separation. In mass spectrometry ion suppression effects may occur when a sample component present in larger concentration elutes from the chromatographic column together or close to the analyte. The mass spectrometric signal of the analyte can be significantly decreased in such cases. For this reason, even if the retention times of the two compounds are different, calibration curves for quantitation are recommended to be obtained in the presence of the potentially interfering compound. The lack of interferences in an analytical method is very important for generating correct (accurate) analytical results. Specificity is typically verified for a specific type of sample on which the method is recommended. Sample preparation is an important tool for enhancing specificity. The progress made in chromatographic separations and the use of specific detectors (such as mass spectroscopic) are able to produce methods with very good specificity.

2) Selectivity, which refers to the quality of a method to respond to a limited number of chemical compounds. Selectivity, similar to specificity, is affected by the sample preparation procedure and specificity of the detector.

3) Precision, which refers to the repeatability of measurements within a set, indicating the scatter or dispersion of the set about its central value (mean). The scatter is characterized by the standard deviation SD (see expression 1.3.4) and relative standard deviation RSD% (see expression 1.3.5). Precision is sometimes indicated as repeatability. It is common for some published methods of analysis to describe precision as SD for a set of average standards. Such values can in some cases be much better than those obtained on "real life" samples or even on low standards, where the matrix may influence the results. For this reason, a much better parameter for assessing the quality of an analytical method is reproducibility. In case that reproducibility is not reported (comparison of results from different laboratories), the reported precision must be obtained by running a set of samples low in the analyte concentration at different days [32].

4) Reproducibility, which is typically considered the precision of results from the same method, obtained in different laboratories. Other associated parameters are the intermediate

precision, which refers to long-term variability within a single laboratory, and repeatability that refers to the precision obtained over a short period of time with the same equipment (in the same laboratory). Also, the repeatability of the analysis on different matrices and different concentrations of the analyte may be indicated as reproducibility.

5) Accuracy, which can be considered as an experimental value that approximates the *bias* (see expression 1.3.8). *Bias* is the difference between an accepted (or true) value for an amount or a concentration analyzed and the average result of the analysis. A small or zero bias indicates a good accuracy. Accuracy is an important characteristic of an analytical method, and agreement with results obtained on values published for standard materials or the same type of material by other laboratories is an important tool of comparison. In cases when such comparisons are not possible, the analysis of samples with the same matrix (preferred with no analyte) and spiked with known levels of analyte are sometimes used for comparison. Analysis by two different methods of the same material and the generation of the same results can also be used to prove accuracy. For materials that are available as standards with known content in a specific analyte, their analysis can be utilized for assessing accuracy [32].

6) Linearity, which indicates the linear dependence between the signal and the concentration or amount and is characterized by coefficient of determination R^2 (see expression 1.3.38). Quantitation should be done within the calibration range, but in the case of linear dependence, the data slightly outside the calibration range still can be considered as reliable. In some instances, the dependence between the signal and the concentration or amount of the analyte is not linear. In such cases, the calibration can be obtained using a quadratic, higher polynomial, or even other type of dependence (e.g., in case of evaporative scattering detection). Such calibrations are acceptable when proven to generate accurate results [34]. Also, no quantitation outside the calibration range should be done using nonlinear calibrations.

7) Linear range, which is the interval between the upper and lower levels that have been demonstrated to be determined with precision and accuracy. The range with linear response is the linear range. As indicated in the comment on linearity, calibration using quadratic, higher polynomial, or even other types of dependence are sometimes used for quantitation [35].

8) Limit of detection (LOD), which is the concentration (or amount) corresponding to the average signal that is within three standard deviations of a low sample (see expression 1.3.27), or the signal three times higher than the average of the blank signal ($\mathcal{S}/\mathcal{N} \approx 3$). Limit of detection (and limit of quantitation) are very important parameters for the characterization of an analytical method. However, some ambiguity is still present regarding the procedures for measuring this parameter. The value of the blank signal may vary considerably from region to region of the chromatogram and depends on the time interval where the average of the signal is taken. Some recommendations indicate that the average of the noise must be taken next to the peak of the analyte in a region five times wider than the peak width. However, such a width may not be present when the peaks of other analytes are close to the analyte. For these reasons, the reported values for LOD may be significantly different from those obtained when the method is implemented in a different laboratory.

A more informative LOD can be obtained using the value for 3 *SD* (3*s*). For this calculation, the lowest calibration standard should be used, and the value of *s* should be obtained from at least three separate analyses of the lowest calibration standard. This procedure provides a better characterization of the analytical method, although differences between LOD and PLOD can be significant for some samples.

9) Limit of quantitation (LOQ), which is the minimum concentration (or amount) that produces quantitative measurements with acceptable precision corresponding to the average signal that is within 10 standard deviations of a low sample (see expression 1.3.28), or with a signal about 10 times higher than the blank ($\mathcal{S}/\mathcal{N} \approx 10$). The same discussion applied for LOD can be made for LOQ, and a preferred procedure for indicating an LOQ for an analytical method is to use the value of 10*s* for the lowest calibration standard.

10) Recovery, which is the ratio (in percent) between a known amount of an analyte added to a sample and the measured amount following sample preparation and analysis. Good recoveries (95–105%) are typically associated with good accuracy of an analytical method. However, extraction of the analytes from a sample may not be equivalent with spiking the sample with a known amount of analyte. Recovery is useful for verifying no losses during core analysis, but it is not necessarily an absolute proof of accuracy. In situations where, for example, the analyte is "trapped" in the matrix and not well extracted, while the added amount of the analyte is placed in a soluble form on the sample, it is possible to obtain good recoveries but low accuracy [36].

11) Robustness, which refers to the quality of an analysis to not be influenced by small experimental modifications during the performance of the process. The evaluation of robustness is rather complex and consists of several tasks such as: (1) identification of the factors to be tested, (2) evaluation of the importance for these factors, (3) selection of an experimental design to test robustness, (4) design of an experimental protocol, (5) execution of the experiments, (6) calculation of the effects of different changes, (7) statistical and/or graphical analysis of the changes, and (8) drawing the relevant conclusions from the statistical analysis and, if necessary, taking measures to improve the performances of the method [37]. One specific aspect of robustness can be considered the lack of variation in the quality of the analysis results when the sample matrix is not identical from sample to sample. Small variations in the matrix are common for different samples, and the results of the analytical method must not be influenced by these variations. However, some analytical methods may generate correct results with a wider range of variation in the matrix components, while other methods are easily influenced by changes in the matrix. Robustness to changes in the matrix components is an important characteristic of an analytical method, which should always be evaluated for the range of matrices expected in the analyzed samples.

12) Ruggedness, which is the degree of reproducibility under a variety of conditions such as different laboratories, analysts, or instruments [38,39].

13) Stability, which indicates that the same results are obtained in time and under different conditions. At the same time, measurement uncertainty should be included, which is a statistical parameter that describes the possible fluctuations of experimental parameters [40].

Various degrees of validation can be applied to an analytical method, and it is common for the description of a method to have a full description of the validation procedure that may be considered acceptable. Analytical method development has the goal of delivering reliable measurements within a given application, and method validation is integrated in the development process because it ascertains the method capabilities and demonstrates its fitness for the projected purpose [41]. Validation is necessary not only for new analytical methods, but also for methods newly implemented from the literature, and even for methods transferred from one laboratory to another [42].

The analytical data reported by a laboratory should be achieved under good laboratory practices (GLP). Quality assurance (QA) should include all of the activities associated with ensuring that the chromatographic measurements are made properly, interpreted correctly, and reported with appropriate estimates of error and confidence levels. QA activities also include the maintenance of appropriate records of specimen/sample origins and history (sample-tracking), procedures, raw data, and results associated with each specimen/sample. Generally, the main conditions for quality assurance should take into consideration the following: (1) the practice of standard operating procedures, (2) the use of statistical procedures for data evaluation, (3) instrumentation validation, (4) reagent/materials certification, (5) analyst certification, (6) laboratory facilities certification, and (7) specimen/sample tracking [43,44].

References

[1] C. Capello, U. Fischer, K. Hungerbühler, What is a green solvent? A comprehensive framework for the environmental assessment of solvents, Green Chem. 9 (2007) 927–934.
[2] J. Namiesnik, Green analytical chemistry—some remarks, J. Sep. Sci. 24 (2001) 151–153.
[3] M. Tobiszewski, J. Namiesnik, Developments in green chromatography, LC GC Eur. 27 (8) (2014) 405–408.
[4] G.E. Baiulescu, P. Dumitrescu, P.G. Zugravescu, E. Sampling, Horwood, Chichester, 1991.
[5] M. Thomson, M.H. Ramsey, Quality concepts and practices applied to sampling-an exploratory study, Analyst (London) 120 (1995) 261–270.
[6] L.H. Keith (Ed.), Principles of Environmental Sampling, ACS Professional Ref. Book, ACS, 1988.
[7] J.M. Bayona (Ed.), Sampling theory and methodology, in: J. Pawliszyn (Ed.), Comprehensive Sampling and Sample Preparation, vol. 1, Elsevier, Amsterdam, 2012.
[8] S.C. Moldoveanu, V. David, Sample Preparation in Chromatography, Elsevier, Amsterdam, 2002.
[9] S.C. Moldoveanu, V. David, Modern Sample Preparation for Chromatography, Elsevier, Amsterdam, 2015.
[10] J. Pawliszyn (Ed.), Comprehensive Sampling and Sample Preparation, Elsevier, Amsterdam, 2012.
[11] R.E. Majors, Sample Preparation Fundamentals for Chromatography, Agilent Technology, Wilmington, 2014.
[12] J.C. Miller, J.N. Miller, Statistics for Analytical Chemistry, third ed., Ellis Horwood, PTR Prentice Hall, New York, 1993.
[13] A. Papoulis, Probability, Random Variables, and Stochastic Processes, WCB/McGraw-Hill, Boston, 1991.
[14] C. Liteanu, I. Rica, Teoria si Metodologia Statistica a Analizei Urmelor, in: Scrisul Romanesc (Ed.), Craiova, 1979, or its English Translation, Statistical Theory and Methodology of Trace Analysis, E. Horwood, Chichester, 1980.
[15] Statistica, ninety eight ed., Stat Soft Inc., Tulsa, 1998.
[16] F.M. Garfield, E. Klesta, J. Hirsch, Quality Assurance Principles for Analytical Laboratories, third ed., AOAC International, Gaithersburg, 2000.
[17] D.J. Koosis, Statistics. A Self-Teaching Guide, John Wiley, New York, 1997.
[18] L.A. Currie, Detection and quantification limits: origins and historical overview, Anal. Chem. Acta 931 (1999) 127–134.
[19] C. Liteanu, I. Rica, Mikrochim. Acta, Über die Definition der Nachweisgrenze 61 (1973) 745–757.
[20] C. Liteanu, I. Rica, On the frequentometric estimation of the detection limit, Mikrochim. Acta 64 (1975) 311–323.
[21] C. Liteanu, E. Hopirtean, I.C. Popescu, Detection limit on ion-sensitive membrane-electrodes: the electrodic function in the nonlinear domain, Anal. Chem. 48 (1976) 2013–2019.
[22] M.D. Nelson, J.W. Dolan, UV detector noise, LC GC 77 (1999) 12–15.
[23] U.S. EPA-Title 40: Protection of environment: Par 136-Gidlines establishing test procedures for the analysis of pollutants: appendix B to part 136-Definition and procedures for the determination of the method detection limit-revision 1.11.
[24] http://www.agilent.com/cs/library/technicaloverviews/public/5990-8341EN.pdf.
[25] E.J. Billo, Excel for Chemists, Wiley, Hoboken, 2011.
[26] J.N. Miller, J.C. Miller, Statistics and Chemometrics for Analytical Chemistry, Sixth ed., Pearson Ed. Lim., Harlow, 2010.

[27] D.B. Hibbert, J.J. Gooding, Data Analysis for Chemistry: An Introductory Guide for Students and Laboratory Scientists, Oxford University Press, Oxford, 2006.
[28] L. Huber, Validation and Qualification in Analytical Laboratories, Interpharm Press, Inc., Buffalo Grove, 1999.
[29] U.S. FDA, Technical Review Guide: Validation of Chromatographic Methods, Center for Drug Evaluation and Research (CDER), Rockville, 1993.
[30] U.S. EPA, Guidance for Methods Development and Methods Validation for the Resource Conservation and Recovery Act (RCRA) Program, Washington, 1995.
[31] AOAC Peer-Verified Methods Program, Manual on Policies and Procedures, Arlington, 1993.
[32] M.E. Swartz, I.S. Krull, Handbook of Analytical Validation, CRC Press, Boca Raton, 2012.
[33] M. Rambla-Alegre, J. Esteve-Romero, S. Carda-Broch, Is it really necessary to validate an analytical method or not? that is the question, J. Chromatogr. A 1232 (2012) 101–109.
[34] K. Araujo, Key aspects of analytical method validation and linearity evaluation, J. Chromatogr. B 877 (2009) 2224–2234.
[35] L. Cuadros-Rodríguez, M. Gracia Bagur-González, M. Sánchez-Viñas, A. González-Casado, A.M. Gómez-Sáez, Principles of analytical calibration/quantification for the separation sciences, J. Chromatogr. A 1158 (2007) 33–46.
[36] L.E. Vanatta, D.E. Coleman, Calibration, uncertainty, and recovery in the chromatographic sciences, J. Chromatogr. A 1158 (2007) 47–60.
[37] Y. Vander Heyden, A. Nijhuis, J. Smeyers-Verbeke, B.G.M. Vandeginste, D.L. Massart, Guidance for robustness/ruggedness test in method validation, J. Pharm. Biomed. Anal. 24 (2001) 723–753.
[38] B. Dejaegher, Y. Vander Heyden, Ruggedness and robustness testing, J. Chromatogr. A 1158 (2007) 138–157.
[39] Y. Vander Heyden, J. Smeyers-Verbeke, Set-up and evaluation of interlaboratory studies, J. Chromatogr. A 1158 (2007) 158–167.
[40] V.R. Meyer, Measurement uncertainty, J. Chromatogr. A 1158 (2007) 15–24.
[41] D. Stöckl, H. D'Hondt, L.M. Thienpont, Method validation across the disciplines - critical investigation of major validation criteria and associated experimental protocols, J. Chromatogr. B 877 (2009) 2180–2190.
[42] E. Rozet, W. Dewé, E. Ziemona, A. Bouklouze, B. Boulanger, P. Hubert, Methodologies for the transfer of analytical methods: a review,, J. Chromatogr. B 877 (2009) 2214–2223.
[43] http://osha.europa.eu/en/good_practice.
[44] J.P. Seiler, Good Laboratory Practice: The Why and the How, Second ed., Springer Science + Business Media, 2005.

CHAPTER

2

Short Overviews of Analytical Techniques Not Containing an Independent Separation Step

2.1 SUMMARY CLASSIFICATION OF ANALYTICAL TECHNIQUES NOT CONTAINING AN INDEPENDENT SEPARATION

A large number of instrumental analytical techniques do not require an independent separation step before the measurement of the analytes (and are indicated as *nonhyphenated*). This is possible in a number of instances. One common case is that of instrumental techniques that have an intrinsic capability of differentiating molecules of different nature, for example by the radiation they emit or by the mass of the ions they form. Other cases include those utilized for the analysis of pure compounds for the determination of structure or nature, and those analyses where a physicochemical property of the analytes differentiates them enough from the matrix or from other analytes that a separation is not necessary. Some such nonhyphenated techniques can also be used for obtaining a "fingerprint" of a mixture without the separation of its components. Many instrumental analytical techniques that can be used nonhyphenated can be adapted as detectors in hyphenated techniques where one part of the instrument is a component performing separation [such as a gas or liquid chromatograph (LC)] and the other part is the measuring component (such as a mass spectrometer). A short discussion of several common nonhyphenated techniques used in chemical analysis is further presented.

Types of Instrumental Techniques Not including an Independent Separation (Nonhyphenated Techniques)

The techniques applied in chemical analysis without being online with a separation step have a large number of applications and are covered by numerous scientific publications. Their applicability depends on numerous factors, the complexity of the matrix and the number of analytes in the sample being important elements. Sample preparation steps may clean up the sample and include fractionations with the result of a processed sample amenable for analysis. Among the most common instrumental nonhyphenated techniques

Selection of the HPLC Method in Chemical Analysis
http://dx.doi.org/10.1016/B978-0-12-803684-6.00002-0

are the following: (1) optical spectroscopic techniques, (2) optical nonspectroscopic techniques, (3) mass spectrometry, (4) X-ray methods, (5) nuclear magnetic resonance, (6) radiochemical techniques, (7) thermal methods, and (8) electrochemical methods. A variety of techniques have been developed for each category of the methods, and a schematic view of several common techniques is given in Fig. 2.1.1. A short discussion on more common instrumental analytical techniques that do not use a separation step online with detection is given in this chapter. Some of these techniques are used for detection in chromatography or in other techniques that use separation in their core analytical procedure, but are not based on chromatography.

2.2 OPTICAL TECHNIQUES

Optical techniques are the basis for a large number of very useful analytical methods. They are based on the measurement of the modifications of light path or intensity or on the measurement of light absorption and emission at a specific wavelength. This difference allows the classification of these techniques in nonspectroscopic and spectroscopic. Spectroscopic techniques can be based on light absorption or emission. Methods based on light absorption include techniques such as: ultraviolet and visible spectroscopy (UV–Vis), infrared spectroscopy (IR), and atomic absorption. The methods based on light emission include techniques such as: flame photometry, inductively coupled plasma (ICP), arc and spark emission spectroscopy, luminescence, etc. Other techniques involve both light absorption and emission, such as molecular fluorescence. Also, light scattering can be utilized in a useful analytical method.

Light Absorption Spectroscopy

1) UV–Vis spectra are generated by electronic transitions of the molecules that absorb energy in the form of ultraviolet or visible light going from the ground electronic state into excited states, from where the energy is further dissipated by nonradiative processes such as collisions with other molecules. The light absorption in UV–Vis for a molecular species in solution takes place as a relatively wide band since vibrational and rotational energy levels are superimposed on the electronic energy levels and the systems have a continuous density of states distribution. Because of these wide absorption bands, UV spectra are, in general, not diagnostic for the identification of a specific compound. As an example, the UV spectra of paracetamol, propyphenazone, and caffeine are shown in Fig. 2.2.1.

Two related quantities, transmittance T and absorbance A, are measurable for the light passing through the solution at a specific wavelength λ. Transmittance is defined as follows:

$$T = \frac{I_1}{I_0} \tag{2.2.1}$$

In relation (2.2.1) I_0 is the intensity of the radiant energy incoming to the sample and I_1 is the intensity of the emerging light (T can also be expressed as a percent). As expected, T is a function of the wavelength λ of the radiation that is absorbed (or of frequency $\nu = c_{light}/\lambda$,

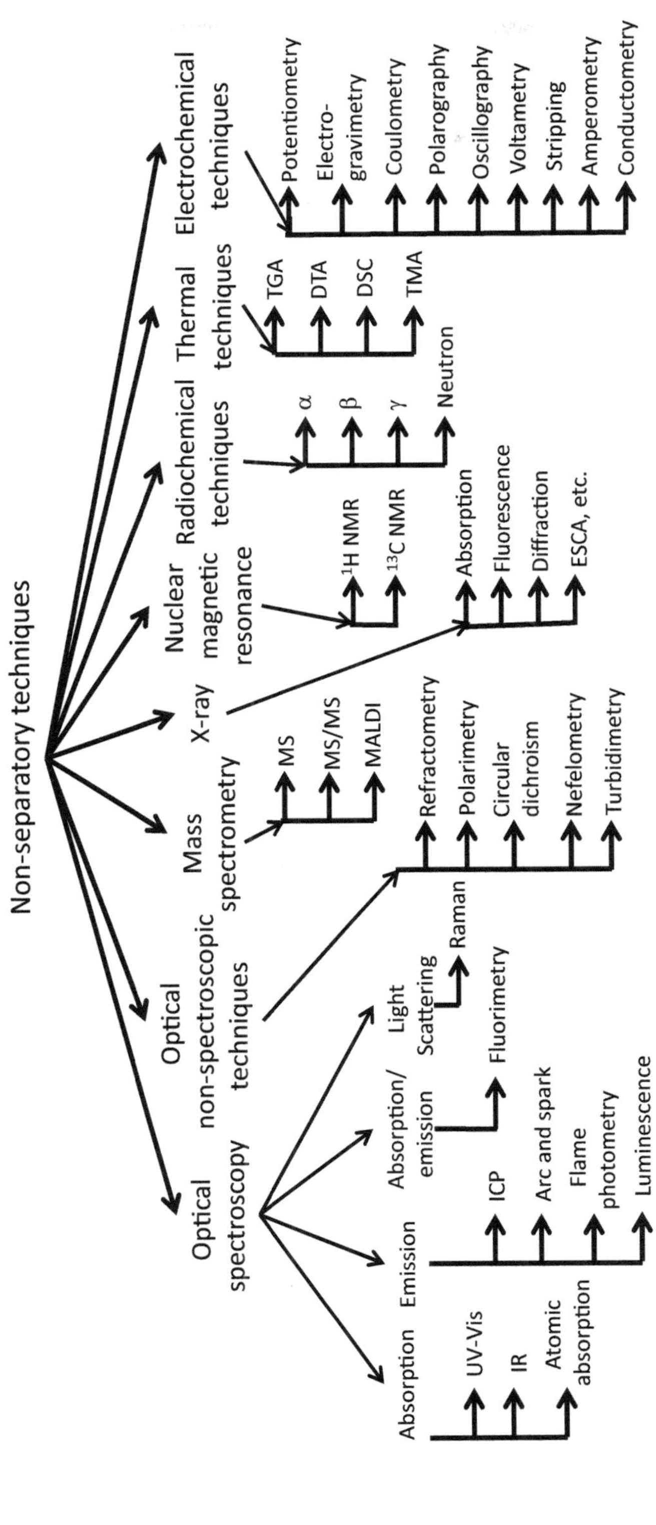

FIGURE 2.1.1 Common nonhyphenated analytical techniques.

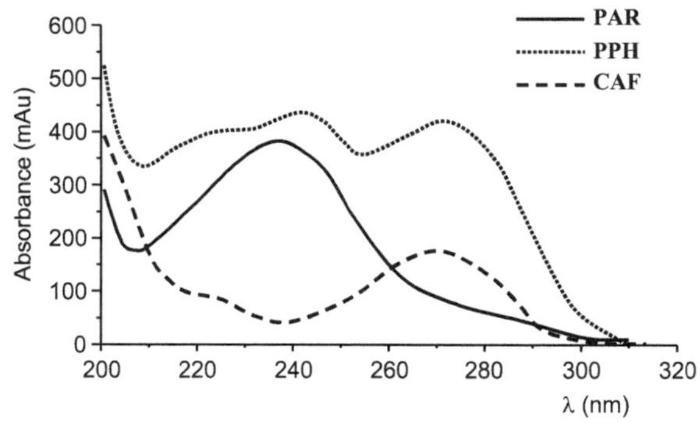

FIGURE 2.2.1 UV spectra of paracetamol (PAR), propyphenazone (PPH), and caffeine (CAF), in acetonitrile (10%) and aqueous phosphate buffer with pH = 3 (90%) [1].

where c_{light} is the speed of light). Absorbance A_λ is defined by the logarithm in base 10 of the inverse of transmittance as follows:

$$A_\lambda = \log\frac{1}{T} = \log\frac{I_0}{I_1} \tag{2.2.2}$$

Absorbance is related to the molar concentration C_X of the absorbing molecular species X by Lambert–Beer law:

$$A_\lambda = \varepsilon_\lambda C_X L \tag{2.2.3}$$

where ε_λ is the molar absorption (absorbance) coefficient at the specific wavelength λ, and L is the path length of the light through the sample. For quantitation purposes, the absorbance is commonly used, because it is proportional with the concentration of the analyte X. UV–Vis spectroscopy can be used in analytical chemistry for the quantitative determination of different analytes usually present in a solution, although solids and gases can also be studied by this technique. The technique can be used for quantitative purposes when the compound of interest has significantly higher absorbance compared to the other matrix components. For mixtures, qualitative characterizations have been developed using chemometrics techniques, for example, for quality control (see, e.g., [2]). However, if more absorbing species with relatively similar molar absorption are present in a solution, a separation step is necessary. The UV–Vis absorption of light (in the range 190–800 nm) is frequently used as a detection/measurement system for HPLC. The absorbance of the liquid leaving the chromatographic column is typically measured at specific (small) time intervals. The absorption of light in gases in the range between 120 and 200 nm (vacuum UV) can be used as a detection technique in gas chromatography [3].

2) Another important light absorption technique is infrared spectroscopy, this technique being used for many analyses of organic compounds such as various polymers (thermoplastics, fibers, paints, adhesives, elastomers, etc.). However, the use of IR hyphenated with high-performance liquid chromatography (HPLC) is not common. Water that is frequently present in the mobile phase in HPLC has very strong absorption in IR, and IR detection is possible only after the elimination of this solvent. However, gas chromatographic detection by IR has useful applications [4–6].

3) Atomic absorption is another spectroscopic technique based on light absorption. However, this technique is used for metallic ion analysis, and it is not used online with chromatographic or other type of separations.

Light Emission Spectroscopy

Several optical spectroscopic techniques are based on light emission. Some of these techniques are further discussed here.

1) Fluorescence (FL) is the process of emission of light by a molecule after absorbing an initial radiation (excitation light). A molecule M goes from a lower energetic state (commonly ground state) to an excited state M* by absorbing energy. The emission process may take place by the molecule bouncing back to the initial state without the change in the wavelength of the absorbed light. In this case the process is difficult to use for analytical applications. However, it is possible that part of the energy of the excited molecule M* is lost by nonradiative processes. In this case, the electron may go to another excited electronic state with lower energy and then, emitting a photon, reach the ground state. It is also possible that no intermediate electronic state is present, but the molecule acquires a lower vibrational energetic level and jumps to the ground state by emitting a photon of lower energy than the absorbed one. In both these cases, fluorescence is observed with the fluorescence radiation having a lower frequency than the excitation radiation. Fluorescence by emission of radiation at higher frequency than the absorbed one is also possible (anti-Stokes radiation) but is uncommon. The average lifetime of an excited state of a molecule M* undisturbed by collisions is about 10^{-8} s, and fluorescence can take place within this length of time. For some special compounds, the molecules can remain for a longer time in a metastable excited state. In this case fluorescence can be observed long after the initial radiation is interrupted. This type of fluorescence is commonly called phosphorescence. Fluorescence is less frequently observed than expected because a nonradiative loss of energy usually takes place. The theory of fluorescence emission shows that the intensity of fluorescence F_{int} at the emission wavelength λ_2 can be expressed as a function of the intensity I_0 of the excitation radiant energy with wavelength λ_1 incoming into the sample, by the expression:

$$F_{\text{int},\lambda_2} = I_{0,\lambda_1}[1 - \exp(\varepsilon_{\lambda_1} C_X L)]\Phi \tag{2.2.4}$$

In relation (2.2.4), Φ is the (quantum) fluorescence yield (*quantum yield*) of the process, the other parameters being the same as defined for UV–Vis. For low concentrations the following approximation is valid:

$$[1 - \exp(\varepsilon_{\lambda_1} C_X L)] \approx \varepsilon_{\lambda_1} C_X L \tag{2.2.5}$$

and the intensity of fluorescence F_{int,λ_2} is related to the concentration C_X by the approximation relation:

$$F_{\text{int},\lambda_2} = I_{0,\lambda_1} \varepsilon_{\lambda_1} C_X L \Phi \tag{2.2.6}$$

In reality only a part of emitted fluorescence is measured in the analytical instrument and the intensity of this measured fluorescence $F'_{\text{int},\lambda_2}$ is given by the expression:

$$F'_{\text{int},\lambda_2} = a I_{0,\lambda_1} C_X \tag{2.2.7}$$

where *a* is a constant coefficient that incorporates all the other constants including the losses due to the measurement of only a part of total fluorescent radiation. Measurement of fluorescence intensity (usually at the maximum of the emission band) is the base of quantitation of the fluorescent species. In practice, the fluorescence intensity is measured using sensitive light detectors that generate an electrical signal of intensity depending on a calibration constant for the instrument. The output is given in luminescence (or light) units (LU) that are arbitrary units proportional with the fluorescence intensity, but specific for the measuring instrument [7]. The measurement by fluorescence encounters several difficulties because of nonlinearity of fluorescence due to self-absorption effects, difficulty in discriminating between overlapping broad spectra of interfering molecules, quenching produced by oxygen dissolved with the solvent, etc.

2) Luminescence (chemiluminescence) is a method based on the measurement of emitted light, usually from the analyte in solution upon reaction with a specific reagent [8,9]. Certain compounds achieve excited energy states in specific chemical reactions and emit light following a transition to ground state. The wavelength of the light emitted by a molecule in chemiluminescence is the same as in its fluorescence, the energy levels involved in fluorescence or in luminescence being the same. The difference comes from a different excitation process. If the energy of the chemical reaction is lower than required for attaining an excited state, the chemiluminescence does not occur. Also, the deactivation of the excited molecule by nonradiative processes such as collisions with other molecules takes place for chemiluminescence similarly to fluorescence. Because no excitation light is needed in chemiluminescence, the interference from an excitation light scattering by trace particles is nonexistent. In addition, the development of detectors virtually able to detect single photons makes the technique highly sensitive. Levels as low as a few hundred attomoles were detected using chemiluminescence for certain analytes. However, the luminescent molecules are not very common.

3) Light scattering at molecular level with the specific wavelength selection includes techniques such as Raman spectroscopy used as a standalone analytical procedure. Attempts were also made to use this technique as a detection procedure for liquid chromatography [10,11].

4) Among light emission techniques are several that are used mainly for metallic ion analysis such as flame photometry, ICP, and arc and spark emission spectroscopy. These techniques can be practiced without a hyphenated separation based on the uniqueness of atomic emission spectra (sample preparation may include such separations). Detection in gas chromatography can take advantage of this uniqueness of atomic spectra, and a specific detector for gas chromatography (GC) has been developed based on emission of light from plasma (AED detector). In this type of detector, a microwave-induced plasma produces atomic spectra from the eluting molecules and allows the measurement of compounds containing carbon, sulfur, nitrogen, hydrogen, chlorine, phosphorus, oxygen, etc.

Although optical spectroscopy includes techniques of choice for solving many analytical problems, the analysis of samples consisting of complex mixtures of organic molecules (or organic and inorganic molecules) may encounter difficulties using these techniques without a preliminary separation of sample components. This is caused by the superposition of complex spectra of sample components. Although in some cases the analysis is still possible, for example when a spectral characteristic is unique to a compound of interest, in most cases of the analysis of mixtures, a separation of sample components is necessary. Separation techniques such as chromatographic methods are frequently used in such situation. The most successful and commonly used analytical procedures are those hyphenating (connecting online) a chromatographic separation with a spectroscopic detection technique as are, for example, analytical HPLC and gas chromatography. A large body of information is available regarding optical spectroscopic methods, including numerous books, dedicated journals (*Spectroscopy*, *Journal of Molecular Spectroscopy*, *Journal of Applied Spectroscopy*, and many others), and web information.

Optical Nonspectroscopic Techniques

Optical nonspectroscopic methods of analysis are also commonly used and they include refractometry, polarimetry, circular dichroism, nephelometry, turbidimetry, and many other less common optical methods based on light absorption and emission, that have a narrower field of applications. In some instances, these techniques can be accompanied with a spectroscopic component such as optical rotatory dispersion that measures rotation of light depending on its wavelength, circular dichroism spectroscopy, etc. The measurement in the variation of refractive index, for example, can be used for the measurement of the concentration in a solution. However, the dependence of the refractive index on the solute concentration is not necessarily linear depending on the nature of the solute and the solvent. Refractive index is also influenced by the temperature (and the wavelength of the light used for the measurement).

Turbidimetry is based on the measurement of the loss of intensity of transmitted light in an emulsion (or solution containing fine particles) due to the scattering effect of particles suspended in it. Nephelometry is based on the measurement of scattered light by a solution containing fine particles. Light scattering can also be used for the measurement of concentration in polymer solutions.

2.3 MASS SPECTROMETRY

Mass spectrometry (or spectroscopy) is based on converting the compounds of the sample into gas-phase ions and on measuring the mass-to-charge ratio (m/z) and abundance of these ions, their various fragments, or their adducts with other molecules. The plot of the ion abundance as a function of the mass-to-charge ratio generates the mass spectrum that can be used for compound identification and evaluation of their amounts (related to the concentration in the sample). Mass spectrometry can be used as a nonhyphenated technique, but its combination with either gas chromatography for the analysis of gases and molecules that can be transferred into vapors without decomposition, or with liquid chromatography (usually HPLC), lead to two of the most powerful analytical techniques, GC-MS and LC-MS (and LC-MS/MS).

The most common procedures for the generation of ions in mass spectrometry for gases and vapors are electron impact (EI) and chemical ionization (CI). Those techniques are used when the mass spectrometer receives the sample, for example, from a gas chromatograph as in the GC-MS technique (other sources of transferring the sample to the mass spectrometer are also known, such as a pyrolyzer). The ions and their fragments are further separated in the mass spectrometer, by several techniques.

Electron Impact Ionization

During electron impact ionization (EI+), the electrons interact with the analyte molecules M brought from the gas chromatograph (GC) by the carrier gas. The molecules M eject an additional electron leaving a positively charged species that is a molecular ion of the type $M^{\bullet+}$ (odd-electron $OE^{\bullet+}$):

$$M + e^- \rightarrow M^{\bullet+} + 2e^- \tag{2.3.1}$$

The molecular ion obtained during the electron impact can decompose with formation of fragment ions A_i^+ and radicals B_i (the radicals are not detected in the mass spectrometer). The fragment ions A_i^+ are commonly even electron ions EE^+ (odd-electron fragments $OE^{\bullet+}$ are also possible), the fragmentation reaction taking place as follows:

$$M + e^- \rightarrow M^{\bullet+} + 2e^- \rightarrow A_i{}^+ + B_i{}^\bullet \tag{2.3.2}$$

More than one fragmentation path is possible for a given molecule, but the nature and abundance of fragments is characteristic for a given compound. The fragmentation is a complex process depending on the nature of the initial molecule, the stability of the ion products, and the reaction pathway for the ion formation. The energy of ionization of 70 eV typically used in EI+ mass spectrometry (in fact a range of energies between 30 eV and 100 eV with a maximum intensity at 70 eV) produces molecular ions $M^{\bullet+}$ in an excited state. The formation of these ions takes place with a range of internal energies. Those that have lower energies than necessary for decomposition will not decompose before detection and will appear as $M^{\bullet+}$ (molecular ion) in the spectrum. The $M^{\bullet+}$ ions frequently decompose by a variety of energy-dependent reactions, each of which results in the further formation of an ion and a neutral species. The abundance of a fragment ion in the spectrum is dependent on the abundance of its precursor, its stability, and on the energy acquired during ionization. If a fragment has higher internal energies than necessary for its decomposition, it will suffer further fragmentations. The ions generated from this fragmentation are separated in the mass spectrometer and recorded to produce a plot representing the relative abundance of each ion as a function of its mass/charge (m/z) value. This plot is known as a *mass spectrum*. More than one reaction path may lead to the same fragment ion, and the abundance of each fragment ion depends on the rates of decomposition reactions leading to that ion. The fragmentation (when done in standard EI+ conditions with electron impact having 70 eV energy) generates typical patterns for each molecular species. As an example, the mass spectrum of silylated (3β)-3-hydroxy-lupa-18,20(29)-dien-28-oic acid (betulenic acid) obtained in EI+ mode at 70 eV is given in Fig. 2.3.1. The figure also shows the molecular structure of several fragments seen in the spectrum.

The specificity of the numerous fragments generated by electron impact (e.g., at 70 eV) can be used as a fingerprint of the molecular species that generated it and allow the identification of compounds by matching it with standard spectra found in mass spectral libraries. The development of large libraries with standard spectra (over 675,000) and of algorithms for automatic library searches made the use of this technique an invaluable tool for compound identification, such as Wiley Registry 9th [13] and NIST 14 [14].

Some compounds may also form negative radical ions under electron bombardment:

$$M + e^- \rightarrow M^{\bullet-} \tag{2.3.3}$$

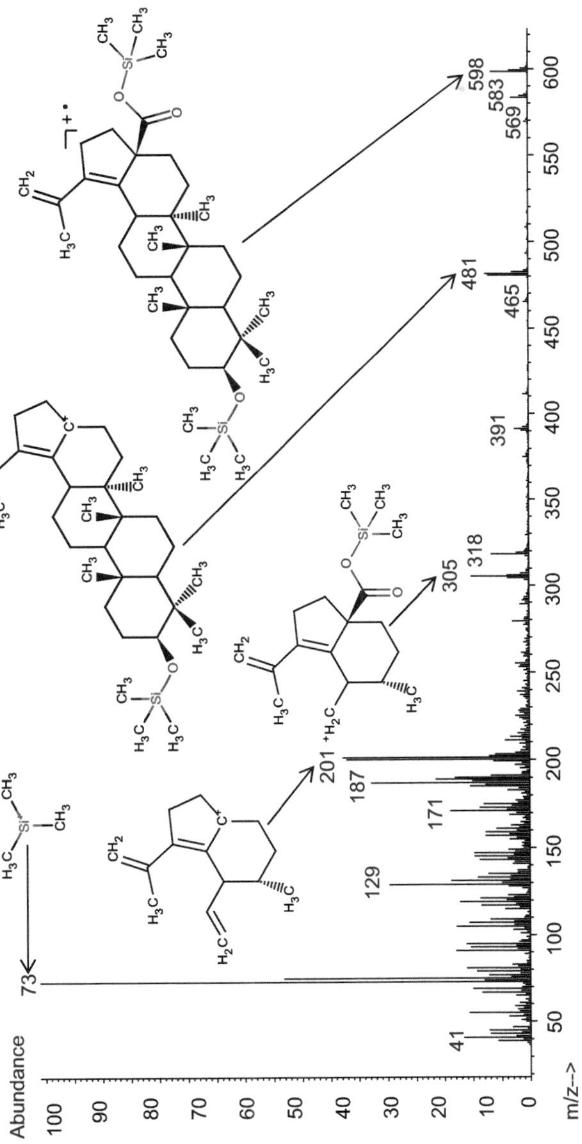

FIGURE 2.3.1 EI+ mass spectrum of silylated betulenic acid and the molecular structure of several fragments [12].

Negative ion operation (with detection and use for analytical purposes) involves altering the source, the focus potential, and some other parameters in the mass spectrometer, in order to allow negative ions to enter the ion separation section of the mass spectrometer and then to be detected. For this reason, the mass spectrometer instruments are set to work either in EI+ or EI− mode.

Chemical Ionization

In chemical ionization, in addition to the carrier gas and the analytes coming from the GC, a reagent gas (methane, butane, ammonia, etc.) is injected into the ion source of the mass spectrometer at a typical pressure of 10^{-2} to 10^{-3} mbar. The initial step in the ionization takes place by the electron impact on the molecules of the reagent gas (which is in large excess compared to the analyte molecules) forming $R^{\bullet+}$ ions, assuming positive CI (CI^+). Following this step, the $R^{\bullet+}$ ions react with other R molecules generating $R\text{-}H^+$ ions (and radicals R_1). These $R\text{-}H^+$ ions further react with the analyte forming $M\text{-}H^+$ ions by the following type of reactions:

$$R + e^- \rightarrow R^{\bullet+} + 2e^- \quad (2.3.4a)$$

$$R^{\bullet+} + R \rightarrow R\text{-}H^+ + R_1^{\bullet} \quad (2.3.4b)$$

$$R\text{-}H^+ + M \rightarrow M\text{-}H^+ + R^{\bullet} \quad (2.3.4c)$$

The ions $M\text{-}H^+$ formed in reaction (2.3.4c) tend to remain unfragmented because they have a much lower energy than the ions $M^{\bullet+}$ generated by direct electron impact. Besides $M\text{-}H^+$ ions, ion clusters containing the analyte molecule and one or more molecules of the reagent gas may be formed during chemical ionization. The ions formed in the ion source are then accelerated and shaped by electric lenses into an ion beam. Some particular compounds are able to generate enough negative ions in chemical ionization that make the CI^- operation a convenient mode of operation.

Electrospray Ionization (ESI) and Atmospheric Pressure Chemical Ionization (APCI) in LC-MS

For liquid samples delivered, for example, from an HPLC instrument, electrospray ionization (ESI) and atmospheric pressure chemical ionization (APCI) are the most common procedures for ion formation. ESI and APCI are typically used in connection with an HPLC system, but other sources of liquid samples can be utilized. In the ESI source, the effluent is introduced through a capillary at 3–5 kV potential toward an opposite plate. The spray is changed into small droplets and some of the solvent is vaporized using a current of a heated gas. The charged heated droplets lose most (but not all) of the solvent. Due to ionic repulsion the droplets generate individual ions, most of them still solvated. In APCI, the effluent from the HPLC column is sent through a capillary heated and having added a flow of gas, but not under an electrical potential. The jet of molecules of solvent and analyte flow further near a needle charged at a high voltage (3–5 kV) in order to generate ions. The generation of ions in ESI and APCI is schematically shown in Figs. 4.1.5 and 4.1.6 (see Chapter 4). Since water is typically present in the effluent and H^+ ions are abundant for both ESI and APCI, the production of positive ions is the result of formation of molecular ions of the type $M\text{-}H^+$, where M is the molecules of the analyte. Similar to the formation of positive ions, molecules of the type M-H can be ionized to generate negative ions M^-. In gas phase, the proton affinity is characterized by the gas-phase basicity (GPB). This is the energy (enthalpy) released in the reaction:

$$M + H^+ \rightarrow [M\text{-}H]^+ \tag{2.3.5}$$

The ionization reaction (2.3.5) is characterized by the free enthalpy (Gibbs free energy) ΔG^0_{GPB} given by the expression:

$$\Delta G^0_{GPB} = \Delta G^0([M\text{-}H]^+) - \Delta G^0(M) - \Delta G^0(H^+) \tag{2.3.6}$$

These reactions are always exergonic in the gas phase. However, proton affinities are conventionally quoted with the opposite sign from other thermodynamic properties (exothermic reactions having assigned a negative free enthalpy), and a positive value of ΔG^0_{GPB} indicates a release of energy by the system. The higher the proton affinity, the stronger is the base and the weaker is the conjugate acid in the gas phase. Usually, ΔG^0_{GPB} for most organic compounds lies between 500 and 1000 kJ/mol.

Similar to the formation of positive ions, molecules of the type M-H can be ionized to generate negative ions in a reaction as follows:

$$M\text{-}H \rightarrow M^- + H^+ \tag{2.3.7}$$

Reaction (2.3.7) is related to the tendency of losing a proton, a property also known as gas phase acidity (GPA), described by the variation of free enthalpy ΔG^0_{GPA}:

$$\Delta G^0_{GPA} = \Delta G^0(M^-) + \Delta G^0(H^+) - \Delta G^0(M\text{-}H) \tag{2.3.8}$$

The values for ΔG^0_{GPA} for most organic compounds lie between 1300 and 1650 kJ/mol. The larger values for ΔG^0_{GPA} than for ΔG^0_{GPB} are probably responsible for the lower sensitivity of negative ionization than of positive ionization in LC-MS. The gas phase basicity or acidity is energetically significantly different from the same ionization process in solutions, due to the energy contribution of hydration of ions in aqueous phase.

The difference in the polarity between the solvent and the analyte molecules favors the formation of positive ions (or negative ions when working in negative mode) from the analyte and much less from the solvent.

Adducts between the analyte molecule and different ions reaching the MS interface are often observed in LC-MS. They are formed by ion-dipole, ion-induced dipole, hydrogen bonds, and even by van der Waals interactions. In the adduct formation, the molecules of the solvent *Solv* are frequently involved. In some instances the adducts are stable enough to be seen in the mass spectrum instead of the analyte molecular ion. Besides adducts with solvent molecules, other adducts can be formed between the analyte M and other species in the eluted material. For example, in positive ionization mode, ions such as $[M\text{-}Na]^+$, $[M\text{-}K]^+$, $[M\text{-}NH_4]^+$, $[M\text{-}H_2O\text{-}Na]^+$, $[M\text{-}Solv\text{-}Na]^+$, and $[M\text{-}2Solv\text{-}Na]^+$ can be seen. Negative ionization is less favorable to adduct formation, but still possible with ions of the following types: $[2M]^-$, $[3M]^-$, $[2M\text{-}Na]^-$, etc.

Due to the high concentration of the solvent (although it is less ionized than the analyte), it is common that some of its molecules are also ionized to form ions $[Solv\text{-}H]^+$ (or in the case of molecules of the type *Solv*-H to form by negative ionization [*Solv*]$^-$). Further ionization of the analyte can be described as a proton transfer from the ionized solvent to the molecule of the analyte.

$$\text{M} + [\textit{Solv}\text{-H}]^+ \rightarrow \text{M-H}^+ + \textit{Solv} \tag{2.3.9}$$

or in case of negative ionization:

$$\text{M-H} + [\textit{Solv}]^- \rightarrow \text{M}^- + \textit{Solv}\text{-H} \tag{2.3.10}$$

The proton transfer reactions show that the solvent has an important role in the ionization process of the analytes in LC-MS. The gas-phase basicities (or acidities) of the solvent molecules $\Delta G^0_{\text{GPB}}(\textit{Solv})$ (or $\Delta G^0_{\text{GPA}}(\textit{Solv}\text{-H})$) are important values regarding the formation of ionized solvent and the proton transfer reactions. The process of ion formation is complex, and more than one process participates in the formation of analyte ions. The proton transfer process as a source of analyte ions can have an important contribution to the ionization efficiency of the analyte and therefore is related to the sensitivity of mass spectral analysis. For this reason, the sensitivity of the MS detectors depends on the nature of the mobile phase.

The ions are further separated and analyzed in the mass spectrometer. Depending on the specific properties of each molecule and the exact conditions of ionization (ionization voltage, temperature, etc.), in both ESI and APCI sources mainly the molecular ions are formed. Further fragmentations of the analytes for MS/MS applications are achieved in a collision cell.

Other Ionization Techniques

Other procedures can be used in mass spectroscopy to form ions from the analyte molecule such as photoionization, glow discharge, field desorption (FD), fast atom bombardment (FAB), thermospray, desorption/ionization on silicon (DIOS), secondary ion mass spectrometric ionization (SIMS), spark ionization, and thermal ionization (TIMS). An important ionization technique that can be applied for the ionization of large molecules is matrix-assisted laser desorption and ionization (MALDI). In MALDI, the sample is mixed with a suitable matrix material and applied to a metal plate. A pulsed laser irradiates the sample, producing vaporization of the sample and of the matrix material. The vaporized molecules are ionized by being protonated or deprotonated in the hot plume produced by the laser. Various matrix materials can be used for MALDI, such as a solution in water and an organic solvent of 3,5-dimethoxy-4-hydroxycinnamic acid, α-cyano-4-hydroxycinnamic acid, or 2,5-dihydroxybenzoic acid.

Ion Suppression in LC-MS

The proton transfer reactions described between the molecules of the analyte and those of the solvent may also take place between the molecules of the analyte and those of a component of the sample matrix, in case that this component is not well separated from the analyte. For an analyte M (or M-H) and a matrix component *Matrix*-H (or *Matrix*) the transfer can be described by reactions of the type:

$$\text{M} + [\textit{Matrix}\text{-H}]^+ \leftrightarrow \text{M-H}^+ + \textit{Matrix} \tag{2.3.11}$$

or in case of negative ionization:

$$\text{M-H} + [\mathit{Matrix}]^- \leftrightarrow \text{M}^- + \mathit{Matrix}\text{-H} \tag{2.3.12}$$

These types of reactions may occur in one direction or the other, and can contribute significantly to the change in ion concentration of the analyte in the LC-MS source. For this reason, even if the matrix compound has a different m/z from the analyte, the response of the detector can be influenced when the chromatographic separation is not very good. In case of very large concentration of a matrix constituent, if the apex of the chromatographic peak of the analyte is different but not very far from that of the matrix constituent, even the tail of the matrix constituent may interfere in sensitivity.

Ion Separation in Mass Spectrometry

The ions generated in the source are further separated by the mass analyzer and measured using a detector. The separation power of the mass analyzer is characterized by its resolution expressed by the formula:

$$R = \frac{M}{\Delta M} \tag{2.3.13}$$

where ΔM is the closest spacing of two peaks [in mass units (m.u.)] of equal intensity with the valley between them less than a specified fraction of the peak height (e.g., 50%), and M is the mass of the second peak. Sometimes ΔM is replaced with FWHM (full width at half maximum) characterizing a width of a single peak for a specific mass M of an ion. Based on their resolution, mass spectrometers are usually classified as "low resolution" and "high resolution" (sometimes "medium resolution" and "ultrahigh resolution" are recognized). The low-resolution instruments typically discriminate ions with a difference in mass of 0.2 m.u. ($R \approx 1000$ for $M = 200$), while a high resolution instrument can discriminate ions with a difference in mass of 0.0001 m.u. ($R \approx 2{,}000{,}000$ for $M = 200$).

Several types of mass analyzers are utilized, such as (1) quadrupole, (2) ion trap, (3) time-of-flight, and (4) other ion separation techniques such as magnetic sector, Wein filter, ion cyclotron resonance, etc. The mass analyzers can be used hyphenated with separation instruments such as gas chromatographs (GC), LCs, supercritical fluid chromatographs (SFC), etc.

1) Quadrupole type mass spectrometers are common mass analyzers which separate the ions by passing them along the central axis of four parallel equidistant rods that have a fixed voltage (DC) and an alternating (RF) voltage applied to them. The field strengths (voltage) can be set such that only ions of one selected mass can pass through the quadrupole, while all other ions are deflected to strike the rods. By varying (with a precise rate) the strength and frequencies of the electric fields, different masses can be filtered through the quadrupole. With the quadrupole instruments a low-resolution type spectrum is obtained. For an m/z = 200 for example, the minimum mass difference that can be separated could be around 0.2 m.u. (resolution 1000). The mass range for the quadrupoles can go as high as 2000 Dalton, but common commercial quadrupole instruments have a mass range between 2 and 1100.

2) The ion trap mass spectrometer (ITMS) consists of a hyperbolic cross-section center ring electrode (a doughnut) and two hyperbolic cross-section endcap electrodes. The ion trap works also in cycles. A cycle starts with the application of a low RF amplitude and fixed

frequency to the ring electrode (no DC), while the endcaps are grounded. A pulse of electrons is injected into the ion trap to ionize and fragment the gaseous sample coming from the transfer line. The ring electrode at low RF amplitude traps all the ions formed during the ionization pulse. The RF amplitude is then increased, and ions of increasing mass (in fact m/z) are sequentially ejected from the trap and detected. A key parameter to the operation is the gas pressure inside the trap (usually He), which must be maintained at 10^{-3} mbar by appropriate gas flow from the GC. This gas forces the ions toward the middle of the trap and provides a better sensitivity. A small AC voltage of fixed frequency and amplitude is also applied to the endcaps during the analysis part of the cycle. Then the process is repeated. The ion traps generate low-resolution mass spectra but can have very good sensitivity. Some problems have been reported regarding the similarity between the spectra generated by ion traps and other types of mass spectrometers when the concentration of a certain compound is high. For high concentration, some ion traps do not generate typical EI+ spectra. A special type of ion trap used for generating high-resolution separations is the Orbitrap (resolution 2,000,000 for $M = 200$). This is an ion trap mass analyzer consisting of two electrodes of special shape that traps ions in an orbital motion around the inner electrode having a frequency of rotation that depends on the m/z of the ion. The current from the trapped ions is detected and converted to a mass spectrum using the Fourier transformation of the frequency signal.

3) The time-of-flight (TOF) instruments separate the ions of different m/z ratio based on the fact that the ions accelerated with a fixed potential have different velocities. Therefore, they travel a fixed distance in different amounts of time. TOF spectrometers work on scans using a pulsed ion source. The ions are sent through an electrostatic sector, which acts as a velocity filter going into a long flight tube. At the end of the flight tube the ions are detected based on their "arrival" time. Some TOF instruments utilize a variable accelerating potential, which allows the light ions, which are faster, to encounter a weaker field, and the heavier ions to encounter a stronger field, increasing in this way their separation. The frequency of pulses in a TOF instrument can be high, and the time of a scan can be much shorter than the time required for a cycle with a magnetic sector, quadrupole, or ion trap system. The number of acquired spectra per second for a TOF instrument can be as high as 500. For GC–MS systems, it is useful to have more scans per chromatographic peak, and this can be better achieved using a TOF spectrometer. Another advantage of TOF MS instruments is that their mass range can be very large, similar to those of magnetic sector instruments but at a lower cost. TOF instruments working at low resolution and others working at high resolution ($R \approx 60{,}000$) are commercially available.

4) A number of other techniques for the separation of ions are known, some leading to excellent resolution. Their use is, however, less common in hyphenated techniques.

After the separation by m/z, the ions are detected using different procedures. Two classes of detectors are known: point ion detectors, which detect the arrival of all ions sequentially at one point, and array detectors, which detect all ions simultaneously along a surface. These detectors record the number (abundance) of individual ions at each m/z. Several point ion detectors are available, such as Faraday cup, electron multiplier, and scintillator. The array detectors are commonly made of separate point detectors (of miniature dimensions) clustered together in the area exposed to the incoming ions. The array detectors have the advantage of collecting simultaneously the signal for a series of ions, but their dynamic range and sensitivity are usually lower. The signal from the detector is further processed (amplified and analyzed) by an electronic data system. Problems such as mass/time calibration, linearity of the detector response, etc., are usually transparent when using modern mass spectrometers, but they are very important in their construction. More details about the ion sources, ion optics, detectors or data processing systems of different mass spectrometers can be found in the literature dedicated to mass spectrometry (see, e.g., [15]).

In addition to single mass spectrometers, tandem mass spectrometers are powerful analytical instruments. In tandem mass spectrometry, also known as MS/MS, multiple steps of mass spectrometry separations are used, with some form of fragmentation occurring in between the MS stages. Multiple stages of mass analysis separation can be accomplished with individual mass spectrometer elements separated in space or using a single mass spectrometer with the MS steps separated in time. In tandem mass spectrometry in space, the separation elements are physically separated. These elements can be sectors, transmission quadrupole, or time-of-flight. When using multiple quadrupoles, they can act as both mass analyzers and collision chambers. For tandem mass spectrometry in time, the separation is accomplished with ions trapped in the same place, with multiple separation steps taking place over time.

A common tandem mass spectrometer type is the triple quadrupole. In a triple quadrupole the first quadrupole (Q1) has the role of separating the (precursor or parent) ions generated in the ion source. These molecular ions are selected for further interactions in the collision cell (Q2) where they can be fragmented. For this purpose, a gas (such as N_2 or Ar) is introduced in the cell, and a specific voltage is applied to the Q2 quadrupole (or hexapole in some instruments). Depending on the collision gas pressure and on the voltage applied to the collision cell, the parent ions undergo different degrees of fragmentation (fragmentation by collisionally activated dissociation [CAD]). The fragmentation strongly depends on the structural characteristics of the analyte. The third quadrupole Q3 is used for the separation of the resulting (product or daughter) ions following fragmentation in the collision cell. Following Q3 the ions are detected by procedures similar to those used in LC-MS.

Based on the exceptional sensitivity of mass spectrometry the technique has been used for solving numerous analytical problems. Although in mass spectrometry (spectroscopy) the measured ions are separated, the technique is less successful in identifying the components of a mixture, unless an ion is unique to a compound of interest. This is the case, for example, in the technique known as ICP-MS that is used for metal ion identification. The introduction of mixtures of compounds directly into the mass spectrometer generates by fragmentation a complex combined spectrum which is sometimes used for identifications. This is the case, for example, when coupling the MS directly with a pyrolysis instrument (see, e.g., [16]). However, for simplifying the spectrum complexity, instead of EI+ ionization that generates many fragments, milder ionization techniques must be used when no separation is done for the compounds sent to the mass spectrometer. This can be achieved either using chemical ionization or photoionization, and fewer fragments or only the molecular ion are generated in this way. Of significantly more utility for the analysis of complex mixtures of organic molecules is the use of a chromatographic separation technique before the mass spectral detection.

2.4 ELECTROCHEMICAL METHODS

Electrochemical methods include several techniques that use electrical properties of a solution for the measurement of analytes. Several procedures are indicated as electrochemical.

These consist of methods where the electric potential of a specific electrode is measured as a function of the concentration of the analyte (potentiometric methods), methods where an electrolysis takes place with its product measured (electrogravimetric, coulometric), methods based on current–voltage dependence (polarography, voltammetry, oscillographic polarography, stripping methods), methods based on the measurement of current intensity during an oxidation or reduction (amperometric methods at fixed or pulsed potential), and methods based on measurement of electrical conductivity of a solution. Electrochemical techniques have a number of applications for analyte detection and measurement after a chromatographic separation. Electrochemical analytical techniques are commonly used for analysis in bulk solution (see, e.g., [17]), but they can be easily adapted for HPLC detection.

The device capable of either generating electrical energy from chemical reactions or facilitating chemical reactions through the introduction of electrical energy is indicated as an electrochemical cell. A cell can be composed of a working electrode coupled with a nonpolarizable electrode (one that does not modify its potential upon passing of a current). This electrode is known as the reference electrode, and examples are the saturated calomel electrode (SCE) and Ag/AgCl electrode. More frequently, a three-electrode cell arrangement is used. In this arrangement, the current is passed between a working electrode (made, for example, from glassy carbon) and an auxiliary electrode, while the potential of the working electrode is measured relative to a separate reference electrode.

Potentiometric Methods

In oxidation–reduction reactions a transfer of electrons takes place, the loss of electrons or an increase in oxidation state being indicated as oxidation, and the gain of electrons or a decrease in oxidation state by a molecule indicated as reduction. Any overall reaction comprises two independent half-reactions, and their cell potential can be broken into two individual half-cell potentials. The half-reaction of interest that takes place at the working electrode surface can be either an oxidation or a reduction. A simple reduction reaction is written as follows:

$$\mathrm{Ox} + \mathrm{ne}^- \rightarrow \mathrm{Red} \tag{2.4.1}$$

The electrode potential E for this half-reaction is reported to the potential of a reference standard hydrogen electrode (NHE), which is taken as zero. Experimental measurements are commonly done with an SCE or Ag/AgCl reference electrode. The potential of an SCE electrode vs. NHE is +0.242 V, and the potential of an Ag/AgCl electrode is +0.197 V vs. NHE. If a compound accepts electrons from a standard hydrogen electrode, it is defined as having a positive redox potential, and a compound donating electrons to the hydrogen electrode is defined as having a negative redox potential. A high positive E^0 indicates an oxidant and a low negative E^0 indicates a reducing compound. The reaction takes place spontaneously in a cell with a positive resulting potential.

Considering a reversible reduction that has a very rapid electron transfer, and assuming that both Ox and Red are soluble species, the molar concentrations C_{Ox} and C_{Red} at the electrode surface ($x = 0$) are governed by Nernst equation:

$$E = E^0 + \frac{RT}{nF} \ln \frac{C_{\mathrm{Ox}}(x = 0)}{C_{\mathrm{Red}}(x = 0)} \tag{2.4.2}$$

where E^0 is the standard electrochemical potential of the half-cell, R is the gas constant (in SI units $R = 8.31451$ J/deg mol), T is the temperature (in K deg.), n is the number of electrons involved in the electrochemical reaction, and F is the Faraday constant (9.6485309×104 C/mol). Tables of standard electrochemical potentials in specific media are available in the literature [18] and the half-reactions are expressed as reductions.

Expression (2.4.2) is applied, for example, for potentiometric measurements such as those performed with an ion-selective electrode. The best-known ion-selective is the glass electrode used for pH measurements (pH $= -\log C_{H+}$, where C_{H+} is the molar concentration of hydronium ions). A glass electrode consists of a glass bulb membrane (ion-exchange type of glass), which separates an internal solution and an Ag/AgCl electrode from the studied solution. A second Ag/AgCl electrode is immersed directly in the studied solution. The two electrodes are connected to a special voltmeter, the pH meter. The pH meter measures the electrical potential difference across the glass membrane. The dependence of this electrical potential on the pH is given by an expression of the form (based on Nernst equation):

$$E = E' - \frac{RT}{nF}\mathrm{pH} \tag{2.4.3}$$

In relation (2.4.3), E' is a constant potential that is obtained experimentally using the calibration of the pH meter with solutions of known pH. Other ion-selective electrodes are made on a similar principle as the glass electrode, having an ion-specific membrane (e.g., an ion-exchange resin membrane) separating the internal solution of the electrode from the studied solution. The electrical potential difference across the membrane is measured with a special voltmeter, and its potential is given by an expression similar to relation (2.4.3):

$$E = E' + \frac{RT}{nF}\log C_X \tag{2.4.4}$$

Electrochemical ion-selective detectors can also be combined with a biological element (e.g., enzymes, antibodies, etc.) that interacts (binds or recognizes) with the analyte. The product resulting from the interaction of the analyte with the biological element is further measured and quantified electrochemically. These types of combinations are known as "biosensors."

Amperometric and Coulometric Methods

In amperometric techniques, the current intensity is measured in an electrochemical cell when a specific potential is applied between two electrodes. For the very rapid electron transfer at the electrode surface, the rate v of the electrochemical reaction (expressed in $\mathrm{mol^{-1}\,s^{-1}\,cm^{-2}}$) is proportional with the current intensity I and inversely proportional with the electrode area A:

$$v = \frac{I}{nFA} \tag{2.4.5}$$

This rate is determined by the mass transfer in solution and is given by the rate at which the electroactive species are brought to the surface of the electrode. In the absence of convection (for bulk solutions) and of migration under the influence of the electric field, diffusion is the only mechanism for mass transfer, and the rate $v_{\text{mass transfer}}$ (flux) can be approximated by the following expression (equivalent with Fick's first law):

$$v_{\text{mass transfer}} = m_{\text{Ox}}\left[C^*_{\text{Ox}} - C_{\text{Ox}}(x = 0)\right] \tag{2.4.6}$$

where m_{Ox} is a proportionality coefficient called mass transfer coefficient, and C^*_{Ox} is the concentration of oxidized species Ox in the bulk solution. The largest rate of mass transfer for Ox occurs when $C_{\text{Ox}}(x = 0) = 0$. The value of the current in these conditions is called the limiting current I_l, and its value is given by the following expression:

$$I_l = nFAm_{\text{Ox}}C^*_{\text{Ox}} \tag{2.4.7}$$

The expression for $C_{\text{Ox}}(x = 0)$ can be written now from relations (2.4.5)–(2.4.7) as follows:

$$C_{\text{Ox}}(x = 0) = \frac{I_l - I}{nm_{\text{Ox}}FA} \tag{2.4.8}$$

When the reducing species Red is absent in the bulk solution, $C^*_{\text{Red}} = 0$, and using a relation similar to relation (2.4.8) for the reduced species, the expression for C_{Red} can be written as follows:

$$C_{\text{Red}}(x = 0) = \frac{I}{nm_{\text{Red}}FA} \tag{2.4.9}$$

With relations (2.4.8) and (2.4.9), Nernst equation (2.4.2) can be written as follows:

$$E = E^0 + \frac{RT}{nF}\ln\frac{m_{\text{Ox}}}{m_{\text{Red}}} + \frac{RT}{nF}\ln\frac{I_l - I}{I} \tag{2.4.10}$$

Expression (2.4.10) represents the relation between the potential E and the current intensity I for a reversible redox reaction with very rapid electron transfer at the electrode surface.

For the current intensity equal to half of the limiting current, $I = 1/2\, I_l$, the last term in relation (2.4.10) is null, and the corresponding potential $E_{1/2}$ is independent of the concentrations of the oxidant or reduced species. This $E_{1/2}$ potential is known as "half wave potential" and it is specific for a molecular (or ionic) species. The cell potential is then given by the expression:

$$E = E_{1/2} + \frac{RT}{nF}\ln\frac{I_l - I}{I} \tag{2.4.11}$$

In amperometric detection, the current passing through the cell is measured at a fixed potential E, commonly chosen higher in absolute value than $E_{1/2}$ specific for the analyte. In these conditions, the desired electrochemical process takes place, but also all other species present in solution and having $E_{1/2}$ lower (in absolute value) than the chosen E value can become electrochemically active species. This may include even the solvent if the working potential E is very high. For eliminating this type of interference, compounds with low redox electrochemical potentials (in absolute value) are preferred for electrochemical detection.

For the case of the electrochemically active species flowing over the surface of an electrode (which is the case of electrochemical detection in HPLC), the current-potential dependence is the function of a convective diffusion process since the flow contributes to the concentration changes. This makes the limiting current intensity for a Nernstian process dependent on the mobile phase flow rate and on channel and electrode geometry. For a rectangular channel flow electrode in steady-state laminar flow with the working electrode with area A at one wall, the limiting current intensity is given by the relation:

$$I_l = 1.467\, nFC^* \left(\frac{AD}{b}\right)^{2/3} U^{1/3} \qquad (2.4.12)$$

In relation (2.4.12), C^* is the bulk concentration of the analyte, D is the diffusion coefficient, b is the channel height, and U is the volumetric flow rate (e.g., of the mobile phase in HPLC). For different channel and electrode shapes, the expression for the current intensity is different. In HPLC, amperometric detection is frequently used for oxidation reactions. The quantitation can be done by calibration of the measured current I versus different concentrations of analyte, while maintaining strictly controlled flow conditions. Also, instead of a constant oxidation potential, a pulse amperometric detection (PAD) can be used, alternating the oxidation analytical potential with a reducing pulse used for cleaning the electrode. The application of different working potentials is done at specific time intervals, and the measurement is made only when the active species are oxidized.

In coulometric detection, the amount of electricity (in coulombs) is measured during the electrochemical process. Coulometric measurements are typically used for large samples in static solutions.

Conductometric Methods

The conductivity (conductometry) detectors are also used for the measurement of concentrations of electrolytes in aqueous solutions. The molar concentration of an analyte that produces solution conductivity can be obtained from the measured electrical resistance of the solution. The dependence of concentration of the resistance is given by the formula:

$$C = \mathrm{Const}_{\mathrm{cell}} \frac{1}{\Lambda_m} \frac{1}{\mathrm{Res}} \qquad (2.4.13)$$

where $\mathrm{Const}_{\mathrm{cell}}$ is a constant dependent on the measuring cell, Res is the electrical resistance measured with the instrument (from Ohm's law $\mathrm{Res} = I/V$ and at constant voltage V the intensity I measurement allows the calculation of Res), and Λ_m is the equivalent conductivity for the ionic species. Although Λ_m can be taken for practical purposes as constant, it varies with the concentration following Kohlrausch's law:

$$\Lambda_m = \Lambda_m^0 - \Theta \sqrt{C} \qquad (2.4.14)$$

where Θ is a constant and Λ_m^0 is the limiting molar conductivity specific for each ion. Molar conductivity is temperature-dependent. Tabulated values for limiting molar conductivities can be found in reference [19].

2.5 OTHER ANALYTICAL TECHNIQUES NOT INCLUDING THE SEPARATION OF SAMPLE COMPONENTS

A multitude of other analytical techniques using instrumentation has been developed for the study and analysis of materials and processes. Only a few of these techniques, that are more common in chemical analysis, are briefly presented in this section.

X-Ray Spectroscopy

X-ray methods include a group of spectroscopic techniques of considerable importance. These methods can be used similar to optical methods based on absorption, emission, and diffraction of X-rays. Also, similarly to optical spectroscopy, the X-ray methods are more successful when used on pure compounds, but they can be used on mixtures when a unique characteristic exists for the mixture components such as the X-ray fluorescence wavelength for elements with the atomic number higher than 8 (oxygen). Besides X-ray diffraction and X-ray fluorescence, other X-ray techniques are utilized in practice for analytical purposes. Among these are several electron spectroscopy techniques (X-ray photoelectron spectroscopy or ESCA, electron impact spectroscopy, Auger spectroscopy). The chromatographic separations are very seldom connected online with techniques based on X-rays. Among several unique cases is the use of X-ray study of compounds containing phosphorus, bromine, or sulfur separated by thin-layer chromatography followed by X-ray fluorescence [20].

Nuclear Magnetic Resonance

Nuclear magnetic resonance (NMR) is another spectroscopic technique with many practical applications mainly related to structural elucidations. NMR can be used for the study of any molecule containing atoms with an odd number of protons and/or neutrons such as ^{1}H, ^{13}C, ^{15}N, ^{19}F, etc., but it is most commonly applied for the study of molecules containing ^{1}H or ^{13}C. Similar to other spectroscopic techniques, structural information should be generated from pure compounds and not from mixtures. A separation technique is therefore necessary when mixtures of compounds are of interest to be analyzed by NMR. For this reason, a number of attempts were made to couple HPLC with NMR [21]. However, the coupling of HPLC with NMR encounters several problems. One of these is the sensitivity of NMR, which is not very high (and about 5000 lower for studying ^{13}C compared to ^{1}H). Advances in the modern high-field NMR instruments have improved NMR sensitivity, but also significantly increased the instrument cost. Another problem is the difficulty of using solvents with molecules containing ^{1}H as a mobile phase for the HPLC, when the signal of the mobile phase would generate a very intense signal compared to the analytes. This would make the analyte detection virtually impossible with common NMR instruments. The problem can be overcome using, for example, fully deuterated solvents as mobile phase in the HPLC, or special techniques for the suppression of the solvent signal. The coupling of the HPLC and an NMR instrument has been achieved using various techniques including "stop flow" mode that allows a longer time for the NMR signal acquisition and an increase in sensitivity, collection of fractions of HPLC eluate that contain the analyte of interest in special loops that are subsequently analyzed by NMR, use of solid-phase extraction traps for specific eluates that are later eluted with deuterated solvents and analyzed, etc. HPLC-NMR hyphenated instruments are not common but are available from some vendors.

Radiochemical Methods

Radiochemical methods are based on the measurement of various types of radiation such as alpha (α, helium nuclei carrying a 2+ charge), beta (β, consisting of electrons), gamma (γ, electromagnetic radiation having the wavelength less than 10^{-3} nm), and neutrons. Radioactivity can be detected and measured at extremely low intensities. Radiochemical methods have a variety of applications, for example, in tracing the presence of a specific analyte when the sample is spiked with a small quantity of the analyte in radioactive form. This tracing procedure can be used in connection with chromatographic separations. β-Ray absorption is used in the electron capture detector (ECD). This is a device used in gas chromatography, which detects molecules in a gas by their property of attaching electrons via electron capture ionization.

Thermal Methods of Analysis

Thermal methods are analytical procedures used for the study of various properties of samples as a function of temperature. Different thermal methods study the relation with temperature of various sample properties such as mass variation with the increase in temperature (thermogravimetric analysis, TGA), temperature difference between the sample and a standard during heating (differential thermal analysis, DTA), difference in the heat absorption or emission of the sample compared to a standard (differential scanning calorimetry, DSC), mechanical properties variation (thermomechanical analysis, TMA), etc. The variation in thermal conductivity of the gas effluent from a chromatographic column in comparison with a reference flow of a carrier gas is used in the thermal conductivity detector in GC. The change in thermal conductivity is generated by the different components of the sample leaving the chromatographic column. This change produces changes in the temperature of a filament leading to modifications of its electrical resistance, which can be easily measured.

2.6 SELECTION OF A NONHYPHENATED METHOD OF ANALYSIS VERSUS ONE CONTAINING A SEPARATION STEP

The choice between a nonhyphenated method of analysis and a method that includes a separation step in the core analytical procedure such as chromatography depends on many factors related mainly to the purpose of analysis and the sample properties. However, all the information collected about the sample (see Section 1.1) is usually necessary for the decision regarding the selection of an analytical method. Depending on whether the analysis is qualitative or quantitative, one or another technique is used for analysis. The information available in the literature regarding the analysis frequently guides the choice between nonhyphenated methods and methods with a separation in the core analysis. Also, the available instrumentation in the laboratory influences this choice.

Nonhyphenated methods of analysis can be selected in situations where a certain physicochemical property of the analyte is specific enough compared to the same property exhibited by the other components of the sample, and by its measurement the analyte can be detected and/or measured. The measurement can be done using solutions of the sample (generated by sample preparation), or the sample as is. For sample solutions, autosamplers were specifically developed to allow the injection of a large number of samples for measurement. One specific procedure used for multiple sample analysis is, for example, flow injection analysis (FIA). In this technique the analysis is accomplished by injecting a plug of sample solution into a flowing liquid carrier stream followed by the continuous measurement by a technique such as spectrophotometry, fluorescence, atomic absorption, electrochemical, mass spectrometric, or other instrumental methods. Specific reagents can also be injected in the carrier stream (or present in it) such that a desired property is developed (e.g., color for spectrophotometric measurement). Similar to FIA is segmented flow analysis (SFA), where air is injected into the carrier stream separating the sample plugs.

Analysis of metals and metallic ions is frequently performed using nonhyphenated techniques such as ICP and X-ray fluorescence. The identification of the atomic and molecular structure of crystals is typically done using X-ray crystallography, and unknown molecular structure of pure compounds is determined using NMR and IR. Polymeric materials that may have a uniform repetitive structure are frequently analyzed using IR. Also, pyrolysis instrumentation directly connected to a mass spectrometer has been used for polymer analysis. Biomolecule analysis covers a wide range of possibilities including both nonhyphenated methods and methods that involve an online separation. Biomolecules even in their cellular matrix can be analyzed without separation using, for example, MALDI or MALDI imaging. In MALDI imaging, the sample, often a thin tissue section, is moved in two dimensions while the mass spectrum is recorded and the results analyzed. Many other analyses are performed using nonhyphenated techniques even when the analytes are in a complex matrix. This is possible, however, with the condition that the specific signal generated by the analyte is well distinguished from that of the matrix or of the other analytes in the sample. Samples that have a complex matrix, and/or more than one analyte that is of interest, frequently require separations before measurement. These separations, including matrix clean up and fractionations, are in some cases achieved during the sample preparation process, and the processed sample may be amenable to a measurement without the need for including a separation in the core analysis.

A separation step in the core analytical procedure is very frequently utilized either because the analysis cannot be otherwise performed, or even for convenience reasons. Modern analytical instrumentation having a separation technique imbedded in the core analytical procedure is frequently capable of analyzing complex samples containing a "dirty" matrix and multiple analytes, requires small amounts of sample that can be analyzed repeatedly, provides very high sensitivity being able to detect and measure traces of components, and frequently includes automation capability that allows the analysis of a large number of samples without involving too much manpower. Among the main methods containing a separation step in the core analytical procedure are three chromatographic methods, GC, HPLC, and supercritical fluid chromatography, although some other chromatographic methods have useful applications. Among the chromatographic techniques, HPLC is the most widely used. It combines separation of compound mixtures with detection by instrumentation also applied in nonhyphenated techniques, extending in this way the capability of instrumental techniques to the analysis of complex samples.

References

[1] F. Soponar, D. Staniloae, G. Moise, B. Szaniszlo, V. David, Simultaneous determination of paracetamol, propyphenazone and caffeine from pharmaceutical preparations in the presence of related substances using a validated HPLC-DAD method, Rev. Roum. Chim. 58 (2013) 433–440.

[2] M.J. Culzoni, M.M. De Zan, J.C. Robles, V.E. Mantovani, H.C. Goicoechea, Chemometrics-assisted UV-spectroscopic strategies for the determination of theophylline in syrups, J. Pharm. Biomed. Anal. 39 (2005) 1068–1074.

[3] K.A. Schug, I. Sawicki, D.D. Carlton Jr., H. Fan, H.M. McNair, J.P. Nimmo, P. Kroll, J. Smuts, P. Walsh, D. Harrison, Vacuum ultraviolet detector for gas chromatography, Anal. Chem. 86 (2014) 8329–8335.

[4] T.A. Sasaki, C.L. Wilkins, Gas chromatography with Fourier transform infrared and mass spectral detection, J. Chromatogr. A 842 (1999) 341–349.

[5] G.T. Reedy, S. Bourne, P.T. Cunningham, Gas chromatography/infrared matrix isolation spectrometry, Anal. Chem. 51 (1979) 1535–1540.

[6] S. Bourne, G.T. Reedy, P.T. Cunningham, Gas chromatography/matrix isolation/infrared spectroscopy: an evaluation of the performance potential, J. Chromatogr. Sci. 17 (1979) 460–463.

[7] A.K. Gaigalas, L. Li, O. Henderson, R. Vogt, J. Barr, G. Marti, J. Weaver, A. Schwartz, The development of fluorescence intensity standards, J. Res. Nat. Inst. Stand. Technol. 106 (2001) 381–389.

[8] S.J. Woltman, W.R. Even, E. Sahlin, S.G. Weber, Chromatographic detection of nitroaromatic and nitramine compounds by electrochemical reduction combined with photoluminescence following electron transfer, Anal. Chem. 72 (2000) 4928–4933.

[9] S.C. Moldoveanu, V. David, Sample Preparation in Chromatography, Elsevier, Amsterdam, 2002.

[10] B.J. Marquardt, P.Q. Vahey, R.E. Synovec, L.W. Burgess, A Raman waveguide detector for liquid chromatography, Anal. Chem. 71 (1999) 4808–4814.

[11] R.J. Dijkstra, C.J. Slooten, A. Stortelder, J.B. Buijs, F. Ariese, U.A.Th Brinkman, C. Gooijer, Liquid-core waveguide technology for coupling column liquid chromatography and Raman spectroscopy, J. Chromatogr. A 918 (2001) 25–36.

[12] S.C. Moldoveanu, W.A. Scott, Analysis of four pentacyclic triterpenoid acids in several bioactive botanicals with gas and liquid chromatography and mass spectrometry detection, J. Sep. Sci. 39 (2016) 324–332.

[13] F.W. McLafferty (Ed.), Wiley Registry of Mass Spectral Data, Wiley, Hoboken, 2014.

[14] NIST Standard Reference Database 1A, NIST, Gaithersburg, 2014.

[15] E. de Hoffmann, V. Stroobant, Mass Spectrometry: Principles and Applications, John Wiley, Hoboken, 2007.

[16] S.C. Moldoveanu, Analytical Pyrolysis of Natural Organic Polymers, Elsevier, Amsterdam, 1998.

[17] A.J. Bard, L.R. Faulkner, Electrochemical Methods, Fundamentals and Applications, J. Wiley, New York, 1980.

[18] B.E. Conway, Electrochemical Data, Elsevier, Amsterdam, 1952.

[19] https://www.grc.com/dev/ces/tns/Conductivity_v_Concentration.pdf.

[20] R.A. Libby, Quantitative thin layer chromatography with X-ray emission spectrometry, Anal. Chem. 40 (1968) 1507–1512.

[21] M. Sandvoss, L.H. Pham, K. Levsen, A. Preiss, C. Mügge, G. Wünsch, Isolation and structural elucidation of steroid oligoglycosides from the starfish *Asterias rubens* by means of direct online LC-NMR-MS hyphenation and one- and two-dimensional NMR investigations, Eur. J. Org. Chem. 2000 (2000) 1253–1262.

CHAPTER

3

Short Overviews of the Main Analytical Techniques Containing a Separation Step

3.1 SEPARATION TYPES USED IN THE CORE ANALYTICAL TECHNIQUES

The separations imbedded in the core analytical technique have the goal of delivering compounds without interferences (e.g., pure compounds in a mobile phase) online to a detection device. Several separation techniques can be used in the core analytical operation in online instruments used for separation and detection (measurement) of the components of a processed sample. These include (1) chromatographic techniques, (2) solid phase extraction, (3) electro separations, (4) membrane separations, and (5) other separations. A schematic view of common separation techniques that can be used online with measurement is given in Fig. 3.1.1.

Besides the separations shown in Fig. 3.1.1, many other separation techniques are used in practice, such as mechanical, based on physical properties of sample components such as volatility (distillation, sublimation) and capacity to crystallize (crystallization), solubility (precipitation), or even retention on solid phases (solid-phase extraction) and ion exchange. Such techniques are used for various practical purposes including industrial (preparative) as well as analytical. However, with few exceptions, they are not amenable for direct coupling with a measuring device and are not used in hyphenated analytical instruments. A short discussion about the more common separation analytical techniques commonly used online with detection is given below.

Chromatographic Separations

Chromatographic separations include a number of separation procedures that have in common the selective retention of analytes on stationary phase and their sequential release in the mobile phase. These techniques have numerous laboratory and industrial applications. The terminology used in chromatography is well described in the literature [1].

Selection of the HPLC Method in Chemical Analysis
http://dx.doi.org/10.1016/B978-0-12-803684-6.00003-2

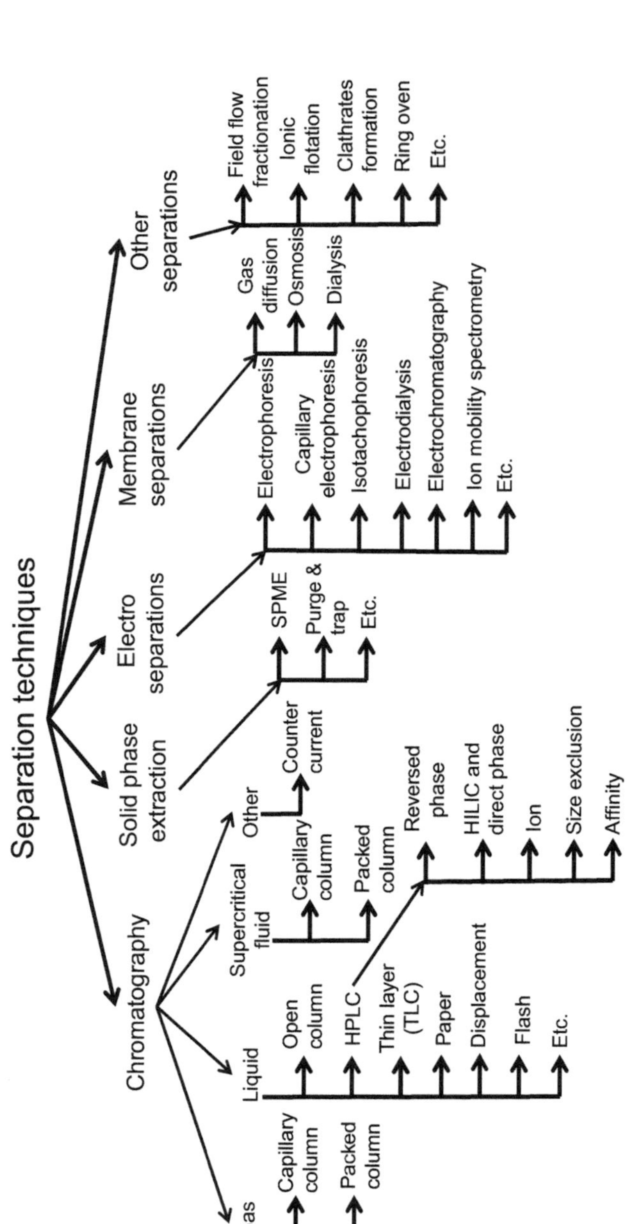

FIGURE 3.1.1 Common separation analytical techniques used online with measurement of separated analytes.

Chromatographic separations can be differentiated based on several criteria such as: (1) the purpose of the chromatographic separation, (2) the format in which the stationary phase is placed, (3) the physical state of the mobile phase, (4) the scale of the operation, and (5) the separation mechanism. Other criteria of differentiating chromatographic techniques are also known, and some types of chromatography have unique characteristics that make them distinct from the other types [e.g., chiral chromatography, countercurrent chromatography (CCC), etc.].

1) The purpose of chromatographic separation allows the differentiation of: (1a) analytical chromatography and (1b) preparative chromatography.

1a) Analytical chromatography is performed for the identification and measurement of various analytes in a sample, and it is hyphenated with a detection and measurement capability (see Fig. 1.2.1). The detection performed with specially designed detectors is typically translated into an electrical signal, and the graphic XY output in of this signal (X being time and Y signal intensity) is known as a chromatogram. The components of a mixture separated in time are displayed as peaks in the *chromatogram*. Many analytical chromatographic techniques, although they may have differences from each other, have a common set of components (parts of a chromatographic instrument). These components include a continuous source of a mobile phase (gas, liquid, or supercritical fluid), an injector that allows the sample to be placed into the mobile phase, a stationary phase on which the separation takes place, and a detector for the components that are leaving the chromatographic column (the *eluates*). The whole analytical set-up is typically connected to a control and data processing unit (e.g., a computer). This unit can control the composition or pressure of the mobile phase, the conditions of sample injection, the temperature of the stationary phase, and the parameters for the detector, and receives the data from the detector for processing and visualization. This set-up is schematically indicated in Fig. 3.1.2.

Each component in the set-up from Fig. 3.1.2 may be different for different chromatographic techniques. For example, for GC the source of mobile phase is a gas tank followed by a series of flow and pressure controllers, while in HPLC it is a source of pressurized liquid. The injectors are typically dedicated to the injection of samples present in a solvent but "solventless" injectors are also known. A variety of detectors can also be used, most of them being based on measuring a physicochemical property of the analytes as those described regarding nonhyphenated methods of analysis. Also, more complex arrangements are possible for a chromatographic instrument, for example, adding collection capability for the separated sample components, addition of more separation units, etc. However, the basic scheme from Fig. 1.2.4 is common for many analytical chromatographic instruments.

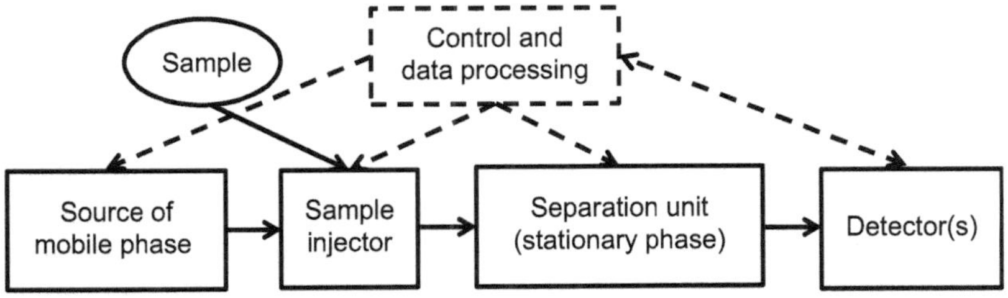

FIGURE 3.1.2 Schematic set-up of an analytical chromatographic instrument.

1b) Preparative chromatography has a wide range of applications and covers a range of quantities of materials subject to separation. That starts with small quantities at laboratory scale to large quantities in industrial applications (see, e.g., Ref. [2]).

2) Chromatography types based on the format in which the stationary phase is placed include the following: (2a) *column chromatography*, (2b) *planar chromatography*, and (2c) CCC, which is a special type of chromatography where both the mobile phase and the stationary phase are "mobile" (see, e.g., Refs. [3,4]).

2a) Column chromatography contains the stationary phase placed in a bed within a tube. The stationary phase can be in the form of small particles (e.g., between 1.8 and 5 μm diameter) or a porous rod indicated as a *monolith*. The particles are typically porous with the active part on the surface. Small solid particles coated with a liquid can also be used as stationary phase. It is possible, as well, that the stationary phase is a coating of the inside of a capillary column, similar to the columns used in gas chromatography (GC) (see Section 3.2), or as small particles placed in a capillary tube. The dimension of the column can vary significantly depending on many factors.

2b) In planar chromatography the stationary phase is present as a plane (a layer of material having a small thickness). This layer can be made of small solid particles and the technique is indicated as *thin-layer chromatography* (TLC), or can be made of a special porous sheet of paper (*paper chromatography*). The separation in column chromatography can be performed in only one direction, while separation in planar chromatography can be performed either in one direction or in two directions.

2c) CCC is a form of liquid–liquid chromatography that uses a liquid stationary phase that is held in place by a centrifugal force. The technique includes several related techniques such as support free CCC, droplet CCC, hydrodynamic CCC, etc.

3) Chromatography types based on the physical state of the mobile phase include the following: (3a) *gas chromatography* (GC), (3b) *liquid chromatography* (LC), and (3c) *supercritical fluid chromatography* (SFC).

3a) In gas chromatography, the mobile phase is a gas, typically hydrogen or helium. Not commonly because it has some disadvantages, nitrogen can also be used as a carrier gas. Gas chromatography using capillary columns inside coated with the stationary phase is the most common in practice. The column in GC is heated at a specific temperature. When this temperature is kept constant during the GC separation, the technique is labeled as *isothermal*. The GC column can also be heated at different temperatures during the separation (chromatographic run) and the technique is labeled as *gradient*. GC is used for the separation of compounds that are in gas phase or can be vaporized without decomposition at the temperature range of the GC column.

3b) Liquid chromatography includes several types of techniques including both column and planar types. Column chromatography can use a low pressure for the mobile phase (e.g., in open column LC), but for analytical purposes a relatively high pressure of the mobile phase is typically used and the technique is referred to as *high-pressure liquid chromatography* (HPLC). The term *high-performance liquid chromatography* is also used for HPLC because of the technique's high separation capability. The chemical composition of the liquid mobile phase can be kept constant and the procedure is indicated as *isocratic*, or can be modified (e.g., by increasing the content of one solvent in a mixture) when the separation is indicated as with *gradient*.

3c) Supercritical fluid chromatography (SFC) is a technique where the mobile phase is a fluid in supercritical state. The most common such fluid is carbon dioxide, in pure form or with addition of other compounds indicated as modifiers (e.g., a certain proportion of methanol, isopropanol, etc.). However, other compounds can be used as supercritical fluids.

4) The scale of operation differentiates the following: (4a) small-scale chromatography (microscale HPLC, or typical analytical HPLC), (4b) semipreparative chromatography, and (4c) preparative or large-scale chromatography. The scale of operation typically refers to the amount of sample that is used for separation and it is closely related to the purpose of separation. Each type covers, in fact, a range of quantities of sample components that are separated. The amount of sample in small-scale chromatography serves usually only for analysis purposes and does not exceed 1−2 mg, with the lower limit controlled only by the detector sensitivity. In semipreparative and preparative chromatography, larger amounts of sample components are separated, and the purpose is to generate enough of a specific material to be used for a variety of purposes.

5) The classification based on separation mechanism can generate a considerable number of chromatographic types. In gas chromatography, the compounds can be separated based on the following criteria: (5a) the differences in the boiling point and (5b) the difference in the strength of the interaction of different molecular species with the stationary phase (e.g., solubility in the stationary phase). Compounds that boil at higher temperatures will stay much longer in the stationary phase and move to the mobile gas phase when the column temperature becomes high enough. Also, stronger interactions between the sample components and stationary phase determine longer retention. The interaction of sample components with the stationary phase weakens as temperature increases, and temperature increase also accelerates the leaving of the chromatographic column (the *elution*).

Among the most important mechanisms of retention and separation in chromatography are those based on the balance between the polarity of the analytes, the stationary phase, and the mobile phase (in the case of HPLC and SFC). Polarity is caused by the separation of positive and negative charge densities in the molecule. This separation can be permanent as indicated by the permanent dipole moment μ of the molecule, and can be induced under the influence of an external field created by other molecules and is indicated as polarizability α. For two point opposite charges, the dipole moment is defined as $\mu = qd$, where q is the charge and d is the distance separating the positive and negative charges; the dipole moment is in fact a vector oriented from the negative to the positive charge. Besides a quantitative characterization of molecules based on μ and α values, it is common to qualitatively assess a polar character by the presence in the molecules of polar functional groups such as $-OH$, $-COOH$, $-NH_2$, $>NH$, $-SO_3H$, etc., since these groups bring both permanent dipole moments and polarizability to molecules. The retention mechanism is frequently evaluated in terms of "polarity." In liquid chromatography, in particular for HPLC, a variety of types of retention mechanisms can be indicated. These types include the following: (5a′) partition and (5b′) adsorption both based on polarity, (5c′) equilibria involving ions, (5d′) equilibria based on size exclusion, (5e′) affinity interactions, etc. Further discussion regarding mechanisms of separation in HPLC will be discussed in detail in this book.

Solid-Phase Extraction Online With Analytical Chromatography

Solid-phase extraction is a technique commonly used for sample preparation (matrix cleanup, concentration). However, some of these techniques are connected online with a chromatographic instrument that also has measurement capability. For this reason, techniques such as solid-phase microextraction (SPME) or purge and trap (P&T) can be viewed either as sample preparation procedures or part of the core separation step (see, e.g., Ref. [5]).

Electro Separations

Electro separations include electrophoretic techniques, electrochromatography, electrodialysis, electrofiltration, and a special technique indicated as ion mobility. Electrophoretic separations include a number of techniques based on the differences in the migration rate of different molecules in an electric field through a specific medium. The separation time in these techniques as well as in electrochromatography ranges from 30–60 s to 5–10 min. Typical drift times for ion mobility are 1–1.5 s. The electrophoretic techniques in a free liquid include moving boundary electrophoresis, isotachophoresis, microscopic electrophoresis, etc. Of special interest is *capillary electrophoresis*, in which the analytes move in a capillary filled with a specific buffer in an electric field applied across the capillary. The technique is a very efficient separation procedure when narrow capillaries with less than 0.1 mm inner dimension (i.d.) and high electric field intensity are used. Separation by capillary electrophoresis can be detected by several detection devices based, for example, on UV or UV–Vis absorbance or even mass spectrometry. The migration of charged particles under the influence of the electrical field in a support medium (in the shape of a column or a plate) that minimizes convection also includes zone electrophoresis, isoelectric focusing, electrophoresis in gels with high density, etc. Other classifications are based on electric field intensity (low and high voltage) and type of support (paper, polyacrylamide, etc.). Electrophoresis is frequently used for protein separation and one variant of this technique, known as immunoelectrophoresis, has numerous applications in the medical field (see, e.g., Ref. [6]).

Electrochromatography uses the migration in an electric field associated with separation based on differential partition of the analytes between a stationary and the mobile phase similar to chromatography. It is common that electrochromatographic separations use size exclusion type media. In such separations, the molecules are separated by size due to the gel filtration mechanism and by electrophoretic mobility due to the gel electrophoresis mechanism. Detection can be performed similarly to that used in analytical chromatography. Capillary electrochromatography uses a capillary column filled with the stationary phase, with the mobile phase driven through the chromatographic bed by electroosmosis. Micellar electrokinetic chromatography is a modification of capillary electrophoresis (CE), where the samples are separated by differential partitioning between micelles (pseudostationary phase) and a surrounding aqueous buffer solution (mobile phase) (see, e.g., Ref. [7]).

A special type of electro separation is based on ion mobility [IM or ion mobility spectroscopy (IMS)]. In this technique, the ions of the analyte move in an electrical field against a carrier buffer gas [8]. This type of separation is typically connected to a mass spectrometric detection. The separation in IM occurs at a time scale of milliseconds (in the range 10^{-1} to 10^{-3} s), while in time-of-flight mass spectrometers (MS-TOFs) the detection occurs on a microseconds scale making the MS-TOF a suitable detection procedure. In this technique, the ions are generated, for example, using electrospray ionization or a corona discharge. Positive or negative working mode for the separation can be selected, depending on the charge of the ions of interest. At specified intervals of time, a sample of ions is introduced into a drift chamber. The ions are continuously generated, but a gating mechanism is used such that an ion pulse of precise width is admitted into the drift chamber. The ions are moved

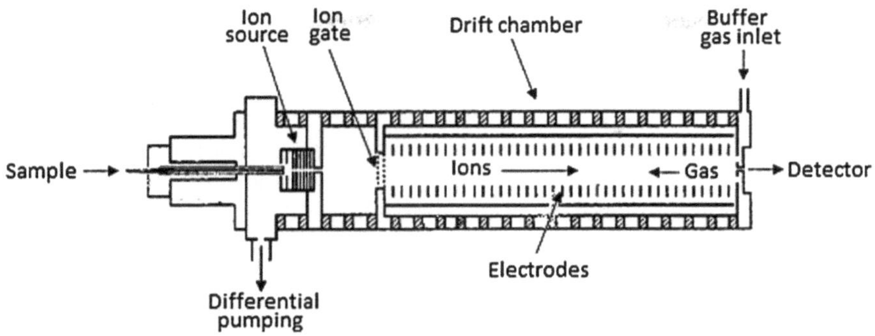

FIGURE 3.1.3 Schematic diagram of an ion mobility spectrometer.

through the drift chamber against a flow of a buffer gas. The buffer gas can be He, N_2, or other common gases and can have ambient pressure or reduced pressure (a few torr). The schematic diagram of an ion mobility spectrometer is shown in Fig. 3.1.3.

There are three common procedures to move the ions along the drift chamber that differentiates the types of ion mobility: drift time (DT-IMS), traveling wave (TW-IMS) and differential mobility spectroscopy (DMS). In DT-ion mobility the ions are moved in one direction in a linear electric field gradient through the buffer gas. The electric field gradient is generated using a set of electrodes. In TW-ion mobility, the ions are moved in one direction by a transport potential wave (also applied to the electrodes). This transport potential is in the shape of a well that travels along the drift chamber as illustrated in Fig. 3.1.4. The ions generated by the ion gate are trapped in the potential well and move along the drift tube in the opposite direction of the flow of the stream of neutral molecules (buffer gas). By properly adjusting the "potential depth" of the well, some molecules are transported less efficiently than others by "skipping" from one well into the previous one following collision with the buffer gas molecules. In this way, molecular species can be separated based on their mass, charge, size, and shape, depending on the interactions with the buffer gas (e.g., collisions that have higher

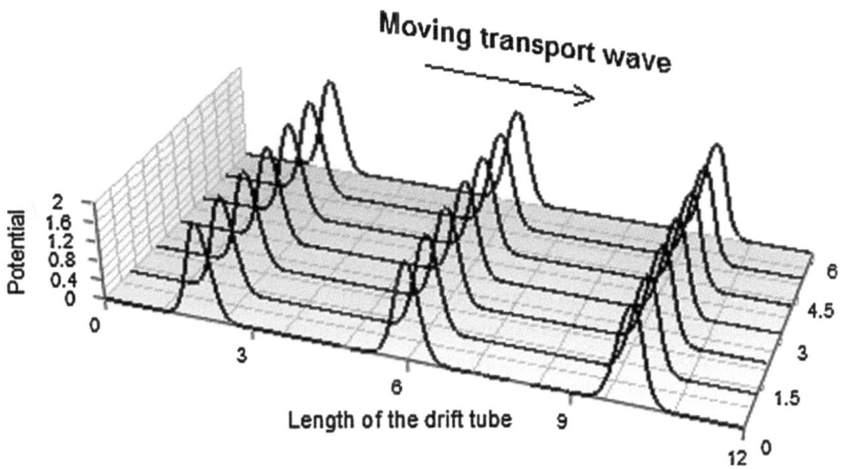

FIGURE 3.1.4 Illustration of the moving potential well of the transport wave.

probability for larger molecules). For this reason, higher pressures of the buffer gas lead to a higher separation selectivity due to a higher rate of ion–molecule interactions (collisions). A more complicated separation system is used in DMS [9]. Ion mobility separation can be coupled for sample input with an HPLC or a GC unit, followed by the ionization process. In some systems, a quadrupole mass spectrometer generates the ions that will have a specific mass (precursor/parent ions for the analytes). After the ion mobility drift tube, another mass spectrometer can be used as a detector. Before the detector it is also possible to include a collision cell that produces fragmentation (product/daughter ions) useful for further analyte differentiation. The separation in IMS takes place at a small time scale. For this reason, to take advantage of the separation, the detection using a mass spectrometer must be synchronized with the ion generation from the ion mobility device. The ion products may be used for identification and quantitation. Such hyphenated instruments are very powerful analytical tools.

Membrane Separations

Several separation processes are based on the differences in the rate of penetrating a specific barrier when the analytes to be separated are driven through the barrier by forces such as mechanical pressure, chemical potential, electrical field, etc. Among these separation techniques can be listed gas diffusion, osmosis and reverse osmosis, dialysis, electrodialysis, and electrofiltration (the last two techniques being already indicated as electro separations). A number of analytical methods of membrane separation hyphenated with detection have been developed and utilized for specific analyses. Among these are membrane introduction mass spectrometry (MIMS) [10], membrane extraction with a sorbent interface (MESI) [11], etc.

Selection of a Separation for the Core Analysis

The selection of a separation technique for the core analytical procedure is an important step in the chemical analysis. A number of factors must be considered for this selection. As indicated in Section 1.1, these include the following: (1) the purpose of analysis, (2) the information about the sample in general, (3) the sample constituents, (4) the required quality of the results, (5) the instrumentation available, and (6) the available information about methods for the analysis of identical or similar samples.

1) Based on the information regarding the purpose of analysis it should be known if the analysis should be qualitative, quantitative, semiquantitative, both qualitative and quantitative, and if special structural information or enantiomer separation is required. This information is not sufficient alone for making a decision regarding the selection of a specific analytical procedure, since it must be combined with that related to the sample and sample constituents. For example, for qualitative analysis of small organic molecules the best information is obtained using mass spectrometry, while IR spectroscopy is best for synthetic polymers. Quantitative analysis of organic molecules is frequently performed using HPLC, while quantitative analysis of metal ions in a solution is best performed using ICP.

2) The information about the sample, in particular related to sample complexity, is important in determining the need for a separation. Samples with a complex matrix and samples with multiple analytes frequently require the addition of a separation technique in the method of analysis. The use of sample preparation may simplify the sample matrix and generate some

fractionations, reducing to a certain extent the need for a separation hyphenated with the measuring capability. However, sample preparation is frequently not sufficient for simplifying the sample enough to make it amenable for direct measurement. In specific cases, the use of a separation can be eliminated from the core analytical procedure, and in other cases, even if it is not absolutely necessary, a separation is included for convenience, instrument availability, or other reasons.

3) As more information about the analysis and the sample is accumulated, the clearer it becomes if a separation is needed in the core analytical process. The information about sample constituents and their level (concentration) frequently adds sufficient data regarding the need for a separation in the core analytical procedure. Complex samples with many organic compounds present in the matrix or as analytes are frequently analyzed using a separation in the core analytical procedure. Also, the information acquired up to this point allows in general a decision to be made regarding the type of separation necessary. Many compounds that are volatile can be analyzed using a gas chromatographic separation. Because the combination of gas chromatography with mass spectrometry (e.g., in GC–MS instruments) provides excellent capability for both compound identification and quantitation, this technique is very common. For compounds that are not volatile but through derivatization can be changed into volatile compounds (e.g., by replacing polar hydrogens with trimethylsilyl groups), it is possible to still select GC–MS for analysis. Many organic compounds that are not volatile can be analyzed using HPLC. This is extended for many biological compounds, small or large molecules, and even to the analysis of inorganic ions for which other procedures can also be selected.

4) The information regarding the quality of results further refines the decision regarding the addition of a separation technique to the core analytical procedure. For the analysis of compounds at very low levels, for example, the separation is very useful for enhancing signal-to-noise ratio of the detection.

5) The information regarding other factors about the chemical analysis, such as availability of instrumentation, may be important for further decisions, but this is an aspect depending on individual laboratories.

6) Availability in the literature of methods using separation for samples identical or similar with the one of interest is an important guide for choosing a separation in the analytical procedure. However, the particularities of the samples of interest may require different approaches.

3.2 GAS CHROMATOGRAPHY

Gas chromatography (GC) is one of the most common analytical techniques having a separation chromatographic step associated with a measuring capability. Gas chromatography uses a gas as mobile phase, and as stationary phase a coating inside a long capillary column or, less commonly, small particles of a solid material packed in a column. In GC the sample should have the capability to be evaporated such that in gas phase it flows with the gaseous mobile phase. When the sample is retained in the stationary phase it does not move toward the end of the column. By this process, the more volatile compounds will be for a longer time in the mobile gas phase and will elute faster, while the compounds that are less volatile and retained more strongly by the stationary phase will have longer retention times. The temperature gradient to which the chromatographic column is exposed (the increase from the initial relatively low temperature to higher temperatures) is frequently used for accelerating the elution of less volatile compounds, which otherwise would elute at very long times. The signals generated by the detectors for the eluting components of the sample are used for qualitative and quantitative analysis.

Typical Gas Chromatography Instrumentation

In principle, the GC instrument consists of (1) a source of gas (carrier gas), (2) an injector, (3) a chromatographic column placed in an oven, and (4) a detector. The source of a gas (most commonly helium or hydrogen) must deliver it at a constant desired pressure. The purity of the carrier gas is important for assuring a low noise background of the detectors. The use of hydrogen is becoming more common, and its use improves the chromatographic separation without affecting the sensitivity of the detection. The carrier gas flows through an injection port, which has the role of introducing the sample into the mobile phase. A variety of injection systems have been developed for GC. One group of such injectors is dedicated to the injection of samples present in a solvent (usually volatile). Other injection systems are "solventless." The solventless injection systems are designed to introduce into the GC flow the sample in gas form.

Injectors for Gas Chromatography

For the injection of liquid samples, a measured volume of the sample is injected in the injection port. The injection port is set at a specific temperature and has the role to vaporize the solvent and the dissolved sample and to introduce them in the gas flow. The injected sample is vaporized due to the elevated temperature of the injection port and transferred as gas/vapors into the chromatographic column. A number of injector types are available for solution injection in the GC, and some of the more common ones are listed in Table 3.2.1.

The temperature of the injection port is a very important parameter in gas chromatography and for split or splitless injection can be selected in a wide range from around 100–400°C. This temperature must be selected such that the solvent and the analytes are vaporized and at the same time that no decomposition of the analytes takes place. The injection is not done without specific modifications of the sample composition. Some components of the sample are not volatized and remain in the injection port (analytes must not be among the compounds remaining in the injection port). Also, some discrimination between less volatile and more volatile compounds may occur, depending on the injection type and conditions. Some solvents and some specific compounds must be avoided for injection. Water is typically not recommended as solvent in gas chromatography. The stationary phases used in the chromatographic column are affected adversely by water.

Solventless injection systems make the injection from the sample already present in gas (vapor) form. In some applications, the initial sample is already gaseous. However, in addition to a gaseous sample, a variety of raw samples (liquid or solid) can be analyzed, for example by sampling the headspace of the material emitting volatile components. Headspace analysis techniques can be classified into two general groups, (1) static headspace and (2) dynamic headspace. Static headspace techniques are those that inject the analytes from a

TABLE 3.2.1 Types of Injection Techniques for Liquid Samples in Gas Chromatography

Injection Technique	Description	Injection Volume (μL)	Percentage in Column
Direct	The sample passes directly from the syringe into a hot inlet, where it vaporizes. Entire sample enters the GC column	0.1–2	~100
Split	The sample passes from the syringe into a hot inlet, where it is "flash" vaporized. Only a fraction of sample enters the GC column	0.1–2	0.01–10
Splitless	The sample passes from the syringe into a hot inlet, where it is flash vaporized, and the bulk of sample enters the column for a duration of 0.3–2 min (Pulse splitless is a variant of splitless injection)	0.1–2	80–95
Cold on-column	The sample passes from the syringe into the column or its extension as a liquid. The portion of column or its extension accepting the sample is kept relatively cool during injection	0.1–1	100
Programmed-temperature split	The sample passes as a liquid from the syringe into a cooled inlet that is subsequently heated to vaporize the sample. Only a fraction of sample enters the column	0.1–2	0.01–10
Programmed-temperature splitless	The sample passes as a liquid from the syringe into a cooled inlet that is subsequently heated to vaporize the sample. The bulk of sample enters the column for a duration of 0.5–1 min	0.1–2	80–95
Solvent elimination without splitting	The sample passes as a liquid from the syringe into a cooled inlet. The solvent is allowed to evaporate through a vent leaving a nonvolatile residue behind. The inlet is subsequently heated to vaporize the residue, which enters the column as in splitless injection	1–100	80–95

closed vessel where the sample is assumed to be in equilibrium with its vapors at a specific temperature and pressure. An aliquot of the headspace can be directly transferred to the mobile phase (gas) stream, or in several variations of static headspace a step can be included where the headspace vapors are first accumulated, for example in a sorbent. The material accumulated in the sorbent is further transferred to the injection port of a chromatographic instrument where it is desorbed. This desorption can be achieved, for example, by thermally releasing the analytes directly into the GC port. The thermal release of the analytes from the sorbent has the advantage of not using a solvent that produces further dilution of the sample but typically requires dedicated instrument attachments (see, e.g., Ref. [12]).

Dynamic headspace is practiced using various procedures. These procedures can involve a pump that aspirates the gas containing the volatile analytes or can use a gas that flows through a liquid sample or over the sample at a specific pressure and flow rate to generate the volatile analytes. In both techniques, the measurement of gas flow and collection time are important for assessing the analyte concentration. For solid samples, various types of glassware are available for containing the sample and allowing the gas to pass through, as well as for the sample heating. The analytes can be concentrated by using trapping similar to the case of static headspace. The dynamic headspace technique where the volatiles are first released from the sample and then retained on a sorbent followed by their subsequent release into the gas flow of the GC is known as purge and trap.

Another common solventless technique that allows the introduction of samples in a GC system is SPME. In SPME a small amount of a stationary phase is exposed to the compounds in the headspace of a sample or to a sample solution with the purpose of accumulating the analytes. After the analyte collection, the stationary phase of the fiber can be directly desorbed in the injection port of the GC instrument using heat. This technique has a large number of practical applications (see, e.g., Ref. [13]).

Among other techniques used to introduce a sample into a GC system is pyrolysis. In pyrolysis, the sample is heated at a high temperature such that it suffers chemical transformations. The pyrolysis products being typically small molecules that are volatile, can be directly introduced into a GC system in a hyphenated technique indicated as pyrolysis-GC (Py-GC) (see, e.g., Ref. [14]). Pyrolysis as well as SPME, P&T, and other similar techniques can be viewed as sample preparation procedures and not as simple solventless injection techniques for GC (see, e.g., Ref. [12]).

Gas Chromatography Oven and Columns

The oven of the GC provides a controlled temperature for the chromatographic column. In most GC systems, the set temperature can be kept within 0.1°C, and a range between −100 to 400°C can be achieved using a cryogenic agent (liquid N_2 or CO_2). Also, the GC ovens are commonly able to provide temperature gradients such that a sequence of isotherm and gradient portions (usually three or four ramps) are available. The temperature program is designed to increase the oven temperature, which increases the rate of migration of compounds in the chromatographic column such that by achieving different migration rates the compounds separate and elute within an acceptable chromatographic run time.

The mixture of compounds injected in the chromatographic instrument is further separated in the chromatographic column. The column is the main component of the system where the separation of the sample components is achieved. The columns are classified in terms of tubing dimensions and type of material that makes the stationary phase. Packed columns containing the stationary phase as particles are typically 1.5−10 m in length and 2−4 mm i.d. They are generally made of stainless steel or glass. More common are the capillary columns (open tubular columns), which are usually made from capillary tubes of fused silica strengthened by an outside polyimide coating. Coating of the silica with a metal (e.g., aluminum) or making the body of the column from stainless steel lined inside with silica is also possible. The capillary usually contains the stationary phase as a coating film on the inner wall of the tube, the commonly used film thickness in GC columns ranges from 0.1 to 5.0 μm. It is also possible to have the tube wall of the capillary lined with a thin layer of a solid adsorbent. The capillaries typically have a 0.1−0.5 mm i.d. and can be 10−100 m long. The stationary phase used for coating the inside of the capillary column can be a bonded phase, which is immobilized and/or chemically bonded within the tubing (e.g., a crosslinked polymer), or a nonbonded phase, which is simply coated on the capillary wall. Generally, the bonded phases are preferred because they exhibit less bleed during use and can stand higher temperatures. The capability of the stationary phase to separate the sample compounds depends on the type of interactions that can be established between the phase and the sample molecules. These interactions are determined by the nature of the stationary phase and of the sample components. The stationary phase of the capillary column is typically characterized by its "polarity." Polarity of a stationary phase is usually characterized using McReynolds constant [15]. This constant was further modified by L. Rohrschneider [16,17] and is sometimes known as McReynolds/Rohrschneider constant. The constant for a specific stationary phase is obtained with the help of Kováts retention indexes [18,19]. Kováts retention index I of a compound is a measure of its retention in a GC column relative to normal alkanes. Using Kováts indexes, an average McReynolds/Rohrschneider constant P is defined, and a classification of stationary phases can be done, from the least polar phase that is squalane, to the most polar, which is 1,2,3-tris(2-cyanoethoxy)propane (TCEP). The average McReynolds index P has values 0 for squalane and 830.1 for TCEP. The polarity P of the column can be normalized by the value for TCEP (and multiplied by 100) and described as "polarity" with values between 0 and 100 [20].

The selection of the chromatographic column depends on various factors, such as the sample nature and the compounds that must be separated. Detailed description of various stationary phases and their selection for particular separation are extensively described in the literature (see, e.g., Ref. [21]). The ability of the GC chromatographic column to separate the analytes depends on many factors, including the nature of the analytes, the nature of the stationary phase of the column, the physical characteristics of the column (for capillary columns the length, the internal diameter, the thickness of the coating with stationary phase), and the GC oven temperature program. By adjusting these parameters correctly, the separation capability in GC is extremely high (see Section 4.2 for the characterization of a chromatographic separation).

Detectors for Gas Chromatography

The detector of a GC is an important part of the instrument. The detectors respond to the presence of the components in the carrier gas as they elute from the chromatographic column. The detectors typically generate an electrical signal, preferably proportional with the amount of the analyte. This electrical signal is amplified by the electronics of the instrument and its variation with time generates the chromatogram. Some detectors are nonselective and do not have the capability of qualitative identification of the eluting compounds. Some detectors are element-specific and can determine if the eluting compounds contain, for example, nitrogen or sulfur. Detectors such as a mass spectrometer or an infrared spectrophotometer offer the capability of qualitative identification of the eluting compounds. Elaborate descriptions of different detectors can be found in the literature (e.g., Refs. [22–24]). Some of the detectors known in gas chromatography and their characteristics are indicated in Table 3.2.2.

Among the GC detectors the most frequently utilized are the FID and the MS. The MS detector offers the capability of compound identification. Extensive literature is available regarding gas chromatography/mass spectrometry (GC–MS) analysis of organic molecules. The MS detector typically generates a total ion chromatogram (TIC), which is a plot of the

TABLE 3.2.2 Main Types of Gas Chromatography Detectors, Their Sensitivity and Limits of Detection

Detector Type	Abbre-viation	Sensitivity[a]	LOD/mg Analyte Injected[b]	Linear Range	Noise	Type of Selectivity
Thermal conductivity	TCD	10 mV mL/mg	2×10^{-5} to 10^{-8} (50 mL gas/min)	10^6	0.01 mV	Nonselective
Flame ionization	FID	0.01 C/g	2×10^{-8} to 10^{-11}	10^7	10^{-14} A	Nonselective
Nitrogen phosphorus	NPD		10^{-8} to 10^{-12}			Nitrogen, phosphorus
Thermoionic ionization	TID		10^{-9} or lower			Specific
Electron capture	ECD	40 A mL/g	10^{-10} to 10^{-13}	10^4	2×10^{-12} a	Halogen, some carbonyl
Flame photometry	FPD	4×10^{-10} A	10^{-6} to 10^{-10}	10^3	2×10^{-12} g/sec	Some specificity
Photoionization	PID		10^{-5} to 10^{-12}	10^7		Nonselective (with some exceptions)
Electrolytic conductivity	Hall		10^{-6} to 10^{-11}			Halogen, sulfur, nitrogen
Sulfur chemiluminescence			10^{-6} to 10^{-14}			Sulfur
Nitrogen chemiluminescence			10^{-5} to 10^{-13}			Nitrogen, NO
Aroyl-luminescence	ALD		Very low			Some specificity
Atomic emission	AED					Specific elements
Helium ionization	HID	100 C/g	4×10^{-10}	5×10^3	3×10^{-14} A	Some specificity
Vacuum UV	VUV		15–246 pg	10^3–10^4		Qualitative and quantitative
Infrared	IRD		Instrument-dependent 10^{-6} to 10^{-5}	10^3		Qualitative and quantitative
Mass selective	MSD		Instrument-dependent 10^{-9} to -10^{-11}	10^5		Qualitative and quantitative

[a]*The sensitivity units are dependent on the nature of detector. TCD detector responds to changes in concentration, while FID, NPD, etc., respond to mass of material entering the detector per unit of time. Also, the flows and the parameters in the detector may strongly influence sensitivity.*

[b]*A common injection volume is 1 μL, but a range between 0.2 μL and 2–3 μL can be performed in split or splitless modes. Other injectors can accommodate larger injection volumes.*

total ion count (detected and processed by the data system) as a function of time. The single ion chromatogram (extracted from TIC or obtained using single ion monitoring or SIM) representing the intensity of one ion (m/z value) as a function of time is used to increase the sensitivity of MS detection.

3.3 SUPERCRITICAL FLUID CHROMATOGRAPHY

Supercritical fluid chromatography (SFC) is a technique that uses as a mobile phase a supercritical fluid (see, e.g., Ref. [25]). A fluid becomes supercritical when its temperature and pressure are above the critical point (critical temperature T_c and critical pressure p_c). In these conditions, the forces from the kinetic energy of the molecules exceeds the intermolecular forces that produce condensation such that no distinct liquid phase exists. The values for T_c and p_c are characteristic for every compound. Some critical values including density for compounds that can be used as mobile phase in SFC are given in Table 3.3.1.

Among the compounds that can be used as supercritical fluids chromatography, CO_2 is by far the most common. In a supercritical fluid, intermolecular forces are lower than the molecular kinetic energy and therefore are insufficient for the condensation of the compound into a liquid. However, the number of molecules per unit volume (consequently the density) can be that of a liquid, while the tendency to expand as the volume increases is gas-like. The density, viscosity, and diffusion coefficient of supercritical fluids are intermediate between liquids and gases, as can be seen in Table 3.3.2.

In an SFC instrument, the mobile phase is a compound that can be easily changed from liquid phase into supercritical fluid. This is the form in which the mobile phase exists in the chromatographic column, but not necessarily in the whole chromatographic system. For example, at the delivering point (pumps) the mobile phase can be kept as liquid and at the detector it can be a gas. In principle, a block diagram for an SFC system is not very different from that of an HPLC system (see Section 2.3) and it is shown in Fig. 3.3.1.

TABLE 3.3.1 Critical Properties of Some Extraction Fluids (Temperature, Pressure, and Density at Critical Point)

Solvent	T_c (°C)	p_c (atm)	ρ_c (g/mL)
CO_2	31.1	72.9	0.47
N_2O	36.5	71.7	0.45
NH_3	132.5	112.5	0.24
$CHClF_2$ (Freon-22)	96	48.5	–
CHF_3 (Freon-23)	25.9	46.9	–
CH_3OH	240	78.5	0.27
i-C_3H_7OH	235.2	47	0.27
H_2O	374.1	218.3	0.32
SF_6	45.6	37	0.74
C_3H_8	96.7	41.9	–
n-C_4H_{10}	152.0	37.5	0.23
n-C_6H_{14}	234.7	29.9	–

TABLE 3.3.2 Approximate Ranges for Density, Viscosity, and Diffusion Coefficient of Liquids, Supercritical Fluids, and Gases

Fluid	Density (g/cm^3)	Viscosity (g/cm s)	Diffusion coefficient (cm^2/s)
Gas	0.6×10^{-3}–2×10^{-3}	1×10^{-4}–3×10^{-4}	0.1–1
Supercritical fluid	0.2–0.9	1×10^{-3}–3×10^{-3}	0.1–5×10^{-4}
Liquid	0.6–1.6 (except metals)	2×10^{-3}–3×10^{-2}	0.2×10^{-5}–3×10^{-5}

In a typical SFC system with CO_2 the mobile phase is delivered by pumps maintaining the CO_2 in a liquid state where a specified flow rate can be measured. To this liquid, various modifiers can be added, such as methanol, ethanol, isopropanol, etc. Also, in the injector, the dissolving power of a liquid is necessary. For this reason, the CO_2 is initially used as liquid and it becomes supercritical only post the injector. The CO_2 is supercritical in the column that is kept in an oven. The oven has temperatures exceeding 40°C, which is maintained such that the supercritical state is achieved. Also, a specific pressure is maintained in the column using a backpressure regulator to achieve supercritical state and not to change the CO_2 in gas. For the detector, the mobile phase can be kept amenable for detectors typical for HPLC or allowed to become a gas, which can be analyzed by detectors typical for gas chromatography.

SFC as a chromatographic process has been likened to a process having the combined properties of the power of a liquid to dissolve a matrix, with the chromatographic interactions and kinetics of a gas. For supercritical fluids the solvation capability is highly related to the density of the fluid, showing a significant increase as the density increases toward the critical density. At critical temperature T_c and critical pressure p_c, the critical density for CO_2 is $\rho_c = 0.468$ g/mL and critical volume $V_c = 94$ mL/mol. The density of supercritical fluids can be varied in a relatively wide range by the variation of pressure, and higher pressures lead to an increase in density. (Note: It is common to express the properties of a supercritical fluid using reduced parameters such as reduced temperature $T_r = T/T_c$, reduced pressure $p_r = p/p_c$, reduced volume $V_r = V/V_c$, and reduced density $\rho_r = \rho/\rho_c$.) At higher density ρ_r of CO_2 many analytes are much better dissolved. As a result, using the higher pressures p_r of CO_2, in SFC it is

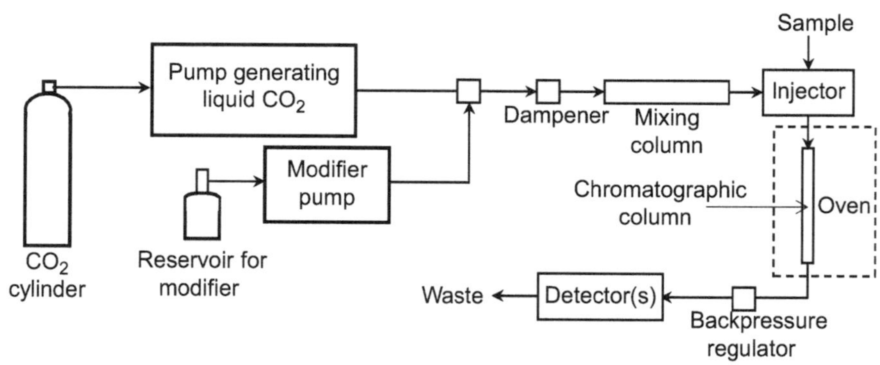

FIGURE 3.3.1 Schematic diagram of an supercritical fluid chromatography instrument.

possible to place a larger mass of analyte on column per injection as compared to GC and still maintain a high chromatographic efficiency. Typically, gradient elution is employed in analytical SFC using a polar cosolvent such as methanol (or other modifiers), possibly with a weak acid or base at low concentrations ($\sim 1\%$). Computer software is used to set mobile phase flow rate, cosolvent composition, system back pressure, and column oven temperature. If the outlet CO_2 is captured, it can be recompressed and recycled, allowing for >90% reuse of CO_2.

Two types of columns can be used in SFC: (1) open tubular (derived from GC columns) and (2) packed columns (derived from HPLC). The open tubular columns are usually with very small diameter (50 μm i.d.) and lengths of 10–20 m. The packed columns are filled with porous particles (3–5 μm diameter) and are 10 m × 4.6 mm i.d. Typical silica-based chemically bonded phases can be used as stationary phase. Improvements in column manufacturing and use of a range of solvents and mobile phase pressure led to improvements of older SFC systems in a so-called "convergence chromatography (CC)" [26].

Similar to HPLC, SFC uses a variety of detection methods including UV/Vis, evaporative light scattering, mass spectrometry, and similar to GC, SFC can use FID. The mass spectrometry detection in SFC is similar to that in HPLC, and electron impact (EI+) detection used in GC is not applicable directly to SFC. The volume of mobile phase in SFC is too large to generate by electron impact typical fragmentation for EI+ that produces spectra searchable in common mass spectral libraries.

3.4 HIGH-PERFORMANCE LIQUID CHROMATOGRAPHY

HPLC is the most common analytical technique having a separation chromatographic step associated with a measuring capability. In this technique the mobile phase is a liquid, which typically is pressurized to flow through a separation column filled with the stationary phase. The elevated pressure of the mobile phase differentiates HPLC from other liquid chromatographic techniques (such as low-pressure liquid chromatography). The variety of stationary and mobile phases that can be selected for HPLC allows the separation of a wide variety of compounds starting with small molecules and extending to polymers such as proteins and nucleic acids. Depending on the detector selected for the measurement of the analytes (and on the analyte properties), the range of sensitivity for HPLC analysis may go as low as picograms/mL of analyte in the solution subject to analysis. HPLC can also be applied to concentrated samples. The capability of identification of unknown analytes in HPLC (even with MS detection) is not as good as that of GC with MS detection, but the technique is unmatched by any other analytical technique regarding quantitation capabilities. For this reason, HPLC is the method of choice in such a wide range of analyses. The multitude of possibilities of selecting a mobile and a stationary phase led to the development of various types of HPLC. Several such types are further presented.

Classification of High-Performance Liquid Chromatography Types Based on Separation Mechanism

A variety of HPLC types were differentiated, some of these types being significantly different from each other, but some having important similarities. The differentiation can be based on various criteria such as the nature of the stationary phase, the nature of the mobile phase, the type of interactions assumed to lead to the separation, but also the range of concentration of specific solvents in the mobile phase (e.g., of water), etc. Because different HPLC types have different characteristics and different applications, it is important to understand their differences and select the appropriate HPLC type for solving a specific separation/analysis problem.

1) Reversed-phase HPLC (or RP-HPLC) is the most common HPLC technique. This type of chromatography is performed on a nonpolar stationary phase with a polar mobile phase. A variety of nonpolar stationary phases is available. The stationary phase for RP-HPLC can be obtained, for example, by chemically bonding long hydrocarbon chains on a solid surface such as silica. One common type of chain bound to silica is C18 (it contains 18 carbon atoms), the long hydrocarbon chain having a high hydrophobic character. However, the bonded phase hydrophobicity may be varied by changing the nature of the substituent, for example making the bonding with C8 groups, with phenyl groups, etc. The C18 bonded phase has a higher hydrophobicity than C8 bonded phase, which has higher hydrophobicity than phenyl. Polymeric materials are also used as RP-HPLC stationary phase. The mobile phase in RP-HPLC is typically a mixture of an organic solvent (CH_3CN, CH_3OH, isopropanol, etc.) and water, with a range of content in the organic solvent. Small amounts of buffers can also be added to the mobile phase in RP-HPLC. The elution in RP-HPLC is more rapid when the mobile phase contains more organic solvent and less water. The interactions in RP-HPLC are considered to be caused by the energies resulting from the disturbance of the dipolar structure of the mobile phase containing water (hydrophobic interactions).

The types of molecules that can be analyzed by RP-HPLC cover a very large range. This includes small molecules with a variety of polarities, but also large molecules, such as proteins and nucleic acids. RP-HPLC has wide applications in the analysis of environmental samples (pollutants of many types such as pesticide residuals, pollutants resulting from incomplete combustion of organic materials, industrial waste, landfill leachates, etc.), analysis of pharmaceuticals, numerous biological analyses including that of metabolites, proteins, nucleic acids, analysis of compounds from food, beverages, agricultural products, as well as analysis of organic compounds from many other sources (detergents, cosmetics, polymer additives, tobacco and smoke, etc.).

2) Ion-pair chromatography (IP or IPC) is applied in particular to ionic or strongly polar compounds. This type of chromatography is very similar to RP-HPLC, with the difference of having a special mobile phase (ion pair RP). In the mobile phase of ion-pair chromatography, a reagent is added, which interacts with the ions of the analytes and forms less polar compounds that can be separated based on hydrophobic interactions with the stationary phase. For example, acids that are ionized (or very polar) can be coupled with a reagent (ion pair agent or IPA) that produces "ion pairs" amenable to be separated by RP-HPLC. The reagent must have a basic group to bind to the acid and a hydrophobic group to allow RP-HPLC separation. For amines with strong basic character, IPA should have an acidic moiety and a hydrophobic one, such as an alkylsulfonic acid. Elution in IPC follows the same rule as in RP-HPLC, with solvents having a higher content in the organic component being capable to elute faster the analytes.

3) Hydrophobic interaction chromatography (HIC) is a type of RP-HPLC, sometimes indicated as a milder RP-HPLC, applied to the separations of proteins and other biopolymers.

The technique is based on interactions between nonpolar moieties of a protein with solvent-accessible nonpolar groups (hydrophobic patches) on the surface of a hydrophilic stationary phase (e.g., hydrophobic ligands coupled on crosslinked agarose that contains many OH polar groups). Both the protein and the stationary phase have specific hydrophobic regions. The promotion of the hydrophobic effect by addition of salts (such as ammonium sulfate) in the mobile phase forces the adsorption of hydrophobic region of the protein on the hydrophobic region of the stationary phase. The desorption of the protein from the solid support is achieved by decreasing the concentration of salts in solution (the reduction of the salting out effect).

4) Nonaqueous reversed-phase chromatography (NARP) is an RP-HPLC type utilized for the separation of very hydrophobic molecules such as triglycerides. In this type of chromatography, the stationary phase is nonpolar (similar to RP), while the mobile phase, although more polar than the stationary phase, is nonaqueous (usually a mixture of less polar and polar organic solvents are used) and capable of dissolving the hydrophobic molecules.

5) Hydrophilic interaction liquid chromatography (HILIC) is a type of HPLC applied for polar, weakly acidic, or basic samples. In this type of HPLC the stationary phase is polar and the mobile phase is less polar than the stationary phase, but it contains water, and water-soluble solvent such as CH_3OH or CH_3CN. HILIC is the "reverse" of RP-HPLC. For HILIC, the polar stationary phase is typically made by chemically bonding on a solid support molecular fragments with a polar end group (diol, amino, special zwitterionic, etc.). The chromatography performed on bare silica support with free silanol (Si-OH) groups can also be considered as HILIC, depending on the mobile phase. The separation is based on the difference in polarity between the molecules. Ion–polar interactions may also play a role in separation. Viewed as having the separation equilibrium based on the interaction of a solid surface with the molecules from a liquid, HILIC is a type of adsorption chromatography. However, a (polar) bonded phase may be seen as a stationary liquid phase, and in this case HILIC is a type of partition chromatography. When the separation is done on zwitterionic phases, HILIC chromatography is sometimes indicated as ZIC. Elution of the compounds retained on an HILIC column is stronger when the mobile phase contains more water and weaker when it contains more organic solvent.

HILIC separations can also be performed on an ion exchange stationary phase with the mobile phase containing a high proportion of an organic solvent. This type of separation is sometimes indicated as eHILIC or ERLIC (from electrostatic repulsion hydrophilic interaction chromatography). This technique can be cationic eHILIC or anionic eHILIC, depending on the nature of the ion exchange stationary phase. In this type of chromatography, the ionic stationary phase repels the similar ionic groups of the analyte and allows HILIC type interactions with the neutral polar molecules of the analyte.

6) Normal-phase chromatography (NP-HPLC or NPC) is a chromatographic type that uses a polar stationary phase and a nonpolar mobile phase for the separation of the compounds. The nonpolar mobile phases used in this type of chromatography are solvents such as hexane, CH_2Cl_2, tetrahydrofuran, etc. The difference between NPC and HILIC is that NPC uses hydrophobic mobile phases, while in HILIC it is sufficient for the mobile phase to be less polar than the stationary phase. The analytes separated by NPC typically include lipids, fat-soluble vitamins, steroids, and nonionic surfactants. These compounds are typically too strongly adsorbed on RP-HPLC type stationary phases and do not elute with polar mobile

phases. In addition to that, these compounds are not soluble in polar solvents which impose restriction to the nature of solvent that is injected in the mobile phase. The solvent used for dissolving the sample must dissolve in the mobile phase and should be nonpolar. Normal-phase chromatography does not have a major difference from HILIC. The difference consists in the use in HILIC of a mobile phase that contains some proportion of water. A polar organic normal phase is sometimes mentioned as a type of chromatography when the nonaqueous solvent contains polar additives such as trifluoroacetic acid.

7) Aqueous-normal-phase chromatography (ANPC or ANP) is a technique performed on a special stationary phase (silica hydride), and the mobile phase covers the range including the types used in reversed-phase chromatography and those used in normal-phase chromatography. The mobile phases for ANP are based on an organic solvent (such as methanol or acetonitrile) with a certain amount of water such that the mobile phase can be both "aqueous" (water is present) and "normal" (less polar than the stationary phase). Polar solutes are most strongly retained in ANP, with retention decreasing as the amount of water in the mobile phase increases. ANP has rather limited utilization but can be applied to a variety of analytes.

8) Cation-exchange chromatography is a type of HPLC used for the separation of cations (inorganic or organic). In this HPLC type the retention is based on the attraction between ions in a solution and the opposite charged sites bound to the stationary phase. In ion exchange chromatography (IEC or IC) the ionic species are retained on the column based on coulombic interactions. In cation exchange chromatography the ionic compound consisting of the cationic species M^+ in solution is retained by ionic groups covalently bonded to a stationary support of the type $R\text{-}X^-$. The ion exchange material (e.g., an organic polymer with ionic groups) is not electrically charged, and therefore the initial form of the cation exchange already has an ionically retained cation in the form $R\text{-}X^-\ C^+$. The separation is achieved when different molecules in solution have different acidic or basic strength.

9) Anion-exchange chromatography is a type of HPLC used for the separation of anions (inorganic or organic). This HPLC is similar in principle to cation exchange type, but the anionic species B^- from solution are retained by covalently bonded ionic groups of the type $R\text{-}Y^+$. Similarly to cation exchange stationary phases, an anion exchange is initially in the form $R\text{-}Y^+\ A^-$. For an anion exchange material the anion A^- previously bound is replaced on the resin by the anion B^-, and two different anions $B1^-$ and $B2^-$ are separated based on their different retention strengths. The mobile phase in ion-exchange chromatography frequently consists of buffer solutions.

10) Ion exchange on amphoteric or zwitterionic phases is a type of IEC very similar in principle to the cation-exchange or anion-exchange IEC. The stationary phase for this type of IEC contains groups that have an amphoteric character or, in the case of zwitterionic phases, both anionic and cationic groups. The mobile phase in this type of chromatography also consists of buffer solutions. The technique is used for the separation of analytes that are weak ionic species.

11) Ion exclusion (or ion moderated) chromatography is an HPLC technique in which an ion exchange resin is used for the separation of neutral species between them and from ionic species of molecules. In this technique, ionic compounds from the solution are rejected by the selected resin (through the Donnan effect), and they are eluted as nonretained compounds. Nonionic or weakly ionic compounds penetrate the pores of the resin and are retained selectively as they partition between the liquid inside the resin and the mobile phase. (Donnan effect or Gibbs–Donnan effect describes the distribution of ions in solution in two

compartments separated by a semipermeable membrane). In ion exclusion chromatography, weakly charged anions are separated on cation exchange resins and weakly charged cations are separated on anion exchange resins.

12) Ligand exchange chromatography is a type of chromatography in which the stationary phase is a cation exchange resin loaded with a metal ion (e.g., of a transitional metal) that is able to form coordinative bonds with the molecules from the mobile phase. These types of analyte molecules are therefore retained on the stationary phase. The elution is done with a mobile phase able to displace the analyte from the bond with the metal, and the separation is based on the differences in the strength of the interaction (of coordinative type) of these solutes with the bonded metal ion.

13) Immobilized metal affinity chromatography is closely related to ligand exchange chromatography and uses a resin containing chelating groups that can form complexes with metals such as Cu^{2+}, Ni^{2+}, Zn^{2+}, etc. The metal ions loaded on the resin still have coordinative capability for other electron donor molecules such as proteins. The retained analytes can be eluted by destabilizing the complex with the metal, e.g., by pH changes or addition of a displacing agent such as ammonia in the mobile phase.

14) Ion-moderated chromatography is an HPLC technique similar to ligand exchange chromatography with the difference that the stationary phase loaded with the metal ion (e.g., Ca^{2+}, Na^{+}, K^{+}, Ag^{+}, or even H^{+}) does not form coordinative bonds with the analyte, the interactions being based mainly on polarity.

15) Gel filtration chromatography (GFC) is a type of size-exclusion chromatography (SEC) in which the molecules are separated based on their size (more correctly, their hydrodynamic volume). In gel-filtration an aqueous (mostly aqueous) solution is used to transport the sample through the column and is applied to molecules that are soluble in water and polar solvents. Size-exclusion chromatography uses porous particles with a variety of pore sizes to separate molecules. Molecules that are smaller than the pore size of the stationary phase enter the porous particles during the separation and flow through the intricate channels of the stationary phase. Small molecules have a long path through the column and therefore a long transit time. Some very large molecules cannot enter the pores at all and elute without retention (total exclusion). Molecules of medium size enter only some larger pores and not the small ones and are only partly retained, eluting faster than small molecules and slower than the very large ones. The separation of small molecules between themselves is not typically achieved, and the technique is utilized mainly for the separation of macromolecules and of macromolecules from small molecules. GFC is sometimes indicated as aqueous SEC. One important use of size-exclusion chromatography (GFC and GPC) is the evaluation of molecular weight (*Mw*) of polymers by calibrating the retention time in SEC as a function of *Mw* (see Section 4.3).

16) Gel permeation chromatography (GPC) is another type of SEC, the only difference from gel filtration being the mobile phase, which in this case is an organic solvent. The technique is used mainly for the separation of hydrophobic macromolecules (such as solutions of certain synthetic polymers). GPC is sometimes indicated as nonaqueous SEC. GPC has numerous applications for studying polymers regarding their *Mw* and molecular weight distribution (distribution of chain length). The distribution of chain length can be evaluated, for example, using a temperature-rising elution and fractionation (TREF) technique [27].

17) Displacement chromatography is a chromatographic technique where all the molecules of a sample are initially retained on a chromatographic column (loading phase). After the

sample is loaded, a "displacement" reagent dissolved in the mobile phase is passed through the column and elutes the specific retained molecule. The method is more frequently applied as a preparative chromatographic technique than as an HPLC analytical method.

18) Affinity chromatography is a liquid chromatographic technique typically used for protein and other biomolecules separation and commonly indicated as bioaffinity chromatography. It can be practiced on a variety of specifically made stationary phases that allow selective retention of the analytes based on affinity interactions.

19) Chiral chromatography on chiral stationary phases is a type of HPLC used to separate chiral compounds. Only specific applications require the separation of chiral compounds, and regular chromatography is much more common than chiral chromatography. Chiral chromatography still has numerous applications, particularly in the analysis of pharmaceutical compounds. The technique typically requires chiral stationary phases containing chiral selector groups.

20) Chiral chromatography on achiral stationary phases is also possible for some chiral solutes, by using chiral modifiers in the mobile phase, although the stationary phase is not chiral.

21) Multimode HPLC is a type of chromatography in which the column contains by purpose more than one type of stationary phase, for example, some with bonded nonpolar groups (e.g., C18) and some with ionic groups (e.g., SO_3^-). This type of character can be encountered unintentionally on columns made using as a stationary phase a silica support covered with silanol groups, and also with hydrophobic groups (such as C18). In most cases, the presence of two types of interactions (e.g., polar and hydrophobic) is not desirable, but in some instances dual properties of a stationary phase can be used to the advantage of the separation. For example, specific proteins can be separated using multimode HPLC.

Other Classifications of Analytical High-Performance Liquid Chromatography Techniques

Besides the mechanisms that differentiate the types of HPLC techniques, several other criteria are used to indicate specific characteristics of analytical HPLC. One such criterion is related to the differentiation between HPLC and ultra-performance liquid chromatography or UPLC. HPLC uses typical injection volumes between 1 and 25 μL, columns with particles having 3–5 μm diameter, i.d of the column of 3–10 mm and typical length in the range 50–250 mm. The flow rates for HPLC are typical between 0.3 and 2 mL/min, and the pumping system delivers pressures up to 400 bar (6000 psi). UPLC indicates an HPLC technique performed on a column with stationary phase consisting of smaller particles (1.7–1.8 μm in diameter), narrower columns (e.g., 2.1 mm i.d.) and a mobile phase with a precise composition at pressures up to 1030 bar (15,000 psi). Also, the injection volume of sample in UPLC can be smaller than 1 μL. These modifications of the operating conditions in HPLC result in better chromatographic performance (narrower peaks, shorter chromatographic runs).

With the continuous increase in the sensitivity of detectors, in particular based on very high sensitivity of MS and MS/MS detection, it became possible to decrease the amount of analyte reaching the detector and still achieve limit of detection (LOD) values as good or even better than in standard HPLC. As a result, it was possible to decrease the volume of the injected sample, the dimensions of the LC column, and the flow rate of the mobile phase in techniques indicated as micro-LC and for even smaller parameters nano-LC. In micro-LC the injection volume range is 15 nL to 10 μL, the flow rate is in the range of 5–200 μL/min, and the columns have diameters in the range 300 μm to 1 mm, with length from 30 to 150 mm. In nano-LC the injection volume ranges from 1 to 50 nL, the flow rate is in the range from 1 to 50 μL/min, and the columns have diameters around 75 μm and are placed in special cartridges for convenient connection. Commercial micro-LC and nano-LC systems are available, as well as experimental ones, having excellent sensitivity and reproducibility. In connection to nano-LC systems are also available chip-based microfluidic systems. Micro- and nano-LC systems are more demanding regarding the cleanliness of the matrix of the sample and provide less reliability when a large number of samples must be analyzed in series.

3.5 ELECTROPHORESIS AND ELECTROCHROMATOGRAPHY

The electrophoretic separation of the components of a sample is based on the migration of charged molecules of different compounds at different rates in an electric field through a specific medium. The electrophoretic techniques in a free liquid include moving boundary electrophoresis, isotachophoresis, microscopic electrophoresis, etc. Of special interest is capillary electrophoresis, in which the analytes move in a capillary filled with a specific buffer in an electric field applied across the capillary. Electrochromatography is based on the migration of analytes in an electric field associated at the same time with differential partition of the analytes between the stationary and the mobile phases.

Electrophoresis

In electrophoresis, the charged analyte particles (or molecules) move in the solution under the influence of an applied electric field $\mathcal{E}$ (expressed as a vector). Because in the solution various ions are present (e.g., from a buffer), these ions can be adsorbed on the surface of the analyte. The adsorption can be considered as consisting of two layers. One layer (Stern layer) of solution ions is immobilized on the analyte surface (and charged opposite to the analyte charge). This layer is continued with a second diffuse layer (Gouy–Chappman layer) of solution ions (more diluted). The charged analyte j and surrounding ions create an electric field in solution. The electric potential characterizing this field decreases with the distance from the analyte j. At the surface of the Stern layer, the electric potential has a specific value ζ_j known as zeta potential. The value of ζ_j can be evaluated from the charge z_j of the analyte, dielectric constant ε of the medium, temperature, etc. (see, e.g., Ref. [12]).

Various forces affect the movement of the analyte in the applied electric field $\mathcal{E}$. The movement takes place with a constant velocity v (expressed as a vector) when the vectorial sum of the forces applied to the analyte is zero. One such force is the electrophoretic attraction $z_j \mathcal{E}$ of the particle with the charge z_j in the electric field. Another force is the Stokes friction force $-6\pi\eta r_j v$ opposing the particle movement (where η is the dynamic viscosity of the medium and r_j is the diameter of the moving particle/analyte). One additional force is created by the interaction of the field generated in the double layer surrounding the analyte with the applied electric field $\mathcal{E}$. This force is known as electrophoretic retardation and its value is $-(z_j - \varepsilon\zeta_j r_j)\mathcal{E}$ (where ε is the dielectric constant of the medium). When the sum of all the interaction forces is equal to zero, the velocity of the movement of the analyte can be expressed by the formula:

$$v = \frac{\varepsilon \zeta_j}{6\pi\eta} \mathcal{E} \tag{3.5.1}$$

This expression shows that the movement of a particle or molecule in an electric field is proportional with the field intensity $\mathcal{E}$(V/cm), the proportionality constant ($\varepsilon\zeta_j/6\pi\eta$) being known as *electrophoretic mobility* (of the analyte *j*). The notation for electrophoretic mobility is μ.

A number of electrophoretic techniques have been developed, their classification being typically done considering the nature of the migrating medium. The electrophoretic techniques in a free liquid include moving boundary electrophoresis, isotachophoresis, electrophoresis in gels with high density, microscopic electrophoresis, etc. Other classifications of electrophoretic techniques are based on electric field intensity (low and high voltage) and type of support (paper, polyacrylamide, etc.). The migration of charged particles under the influence of the electrical field in a support medium minimizes convection.

Of special interest for analytical purposes is capillary electrophoresis, in which the analytes move in a capillary filled with a specific buffer in an electric field applied across the capillary. The technique provides a very efficient separation procedure when narrow capillaries with less than 0.1 mm i.d. (25–75 μm i.d.), 10–50 cm long, and a high intensity electric field are used. Capillary techniques include zone electrophoresis (CZE), isoelectric focusing, isotachophoresis, and micellar electrokinetic capillary chromatography.

In capillary electrophoresis (CE), besides the electrophoretic migration of the analyte molecules, an additional process known as electroosmotic flow takes place. The fused silica capillaries have silanol groups that become ionized in the buffer, in particular when the $pH > 3$. The negatively charged SiO^- ions attract positively charged cations, which form two layers: a stationary and a diffuse cation layer (on the capillary internal surface). In the presence of an applied electric field, the diffuse layer migrates toward the negatively charged cathode creating an electroosmotic flow that drags bulk solvent along with it. Anions in solution are attracted to the positively charged anode, but get swept to the cathode as well. This provides a flow of materials past the detector, as if driven by a pump. The rate of the electroosmotic flow is governed by an equation similar to Eq. (3.5.1), with the corresponding zeta potential ζ_c corresponding to the cations from the solution. The combined migration due to electrophoretic mobility and the electroosmotic flow has the result that the cations with the largest charge-to-mass ratios separate out first, followed by cations with reduced ratios, neutral species, anions with smaller charge-to-mass ratios, and finally anions with greater ratios. The electroosmotic velocity can be adjusted by altering pH, the viscosity of the solvent, ionic strength, voltage, and the dielectric constant of the buffer. For weak acids or bases, the apparent electrophoretic mobility μ is also a function of the ionization degree of the molecule (since ζ depends on the molecular charge) and therefore of the pH of the solution of the analyte (see Section 4.4).

The instrumentation used to perform capillary electrophoresis consists of two reservoirs containing an electrolyte buffer, a capillary, electrodes, a high-voltage power supply (5–20 kV), a detector, and a data output and handling device. The source reservoir, the destination reservoir, and the capillary are filled with an electrolyte such as an aqueous buffer solution. The electrolyte buffer may contain a gelling agent such as polydimethylacrylamide. One problem with CZE is electrostatic binding of cationic substances to the walls of the tubing. This effect is observed with proteins when operating in a buffer that has a pH below the pK_a of the analyte. To avoid this problem, it is common to perform the separation using a buffer at least two pH units above the pK_a of the protein. The use of specially treated capillaries can also reduce the wall binding.

For sample introduction, the capillary inlet is placed into a vial containing the sample. The sample is introduced into the capillary via capillary action, pressure, siphoning, or electrokinetically, and the capillary is then returned to the source reservoir. The migration of the analytes is initiated by an electric field that is applied between the source and destination reservoir and is supplied to the electrodes by the high-voltage power supply. The analytes separate as they migrate due to their differences in electrophoretic mobility and are detected near the outlet end of the capillary. Most common commercial systems use UV or UV—Vis absorbance as their primary mode of detection. Fluorescence detection can be used if the analyte is fluorescent. In these systems, a small transparent section of the capillary itself is used as the detection cell. Capillary electrophoresis can also be coupled with a mass spectrometer detector. For MS detection, the destination reservoir is eliminated and the contact is made, for example, using a metal electrode. Several procedures are used to transfer the liquid from the capillary into the MS source, at a constant flow rate. The most common procedure uses the addition at the end of the capillary of a sheath of liquid flowing coaxially with the silica capillary through a metal capillary tubing, such that both the electroosmotic flow and the flow from the additional sheath of liquid are going into the MS source. The subject of electrophoresis and related techniques is presented in a large body of literature including original peer-reviewed publications (some published in the dedicated journal *Electrophoresis*) and books [28,29].

Electrochromatography

Most commonly, electrochromatography is applied as a combination of size-exclusion chromatography and gel electrophoresis. The two separation mechanisms, one based on size exclusion and the other on migration in an electric field, allow the differentiation of molecules by size and by electrophoretic mobility. In capillary electrochromatography, the mobile phase is driven through the chromatographic bed by electroosmosis. The stationary can be a typical HPLC stationary phase. A considerable amount of information is dedicated to the subject of electrochromatography and capillary electrochromatography [7,30].

3.6 SELECTION OF GC, SFC, CZE, OR HPLC

Selection of GC as the Method for Analysis Versus HPLC

Gas chromatography coupled with various injection and detection procedures is a technique (or better described as a group of techniques) widely utilized for both qualitative and quantitative analysis. This is explained by its excellent versatility regarding the possibility to use the technique on samples of various natures (origin, composition, physical state) and the solving power regarding both qualitative and quantitative problems. The main limitation of GC is caused by the requirement of being applied for the analysis of compounds that can be volatilized without decomposition. Due to the excellent capabilities of GC, a considerable number of sample preparation techniques involving derivatization are dedicated to the transformation of nonvolatile compounds into volatile derivatives (see, e.g., Refs. [12,31]). This procedure is applied, for example, for compounds that contain active hydrogens, such as those from groups such as $-COOH$, $-OH$, $-SH$, or $-NH_2$. These molecules are typically indicated as "polar" molecules. The asymmetrical charge distribution in a molecule (separation of the center of positive charges from that of negative charges) makes the molecule act as an electric dipole and it is defined as "polarity." To this intrinsic polarity another "polar" effect is added caused by the dipole moment induced by surrounding molecules. The polar compounds tend to have stronger interactions with surrounding molecules and they display higher boiling points and higher interactions with the stationary phase in chromatographic columns. Compounds containing active hydrogens are in particular known to be highly "polar." The "active" hydrogens also have the tendency to form hydrogen bonds that add to lack of volatility and lead to stronger retentions. For this reason, for an analyte Y-H derivatization reactions with a reagent R-X are sometimes necessary. This type of reaction can be written in a simplified form as follows:

$$Y - H + R - X \rightarrow Y - R + HX \tag{3.6.1}$$

For GC analysis the group R in the reagent is typically a low-molecular-mass fragment with low polarity. The groups typically used are CH_3 or C_2H_5, a short-chain fluorinated alkyl in alkylation reactions, $Si(CH_3)_3$ or other silyl groups in silylations, $COCH_3$ or short-chain fluorinated acyl groups in acylations, C(O)-OR groups following reactions with chloroformates, etc. The replacement of active hydrogens has multiple effects on molecular characteristics. One effect is the elimination of the possibility of forming hydrogen bonds that are important in the retention process in the chromatographic column. Another effect is the potential decrease in polarity expressed by the lowering of the dipole moment of the molecule. This is caused by the redistribution of the partial atomic charges in the molecule. The end effect of the elimination of hydrogen bond formation and of the decrease of polarity is typically translated into a significant increase in the volatility and amenability to GC analysis.

Of particular usefulness for chemical analysis capabilities is the GC–MS technique. GC–MS is a very common tool in many laboratories and can be applied to the analysis of all samples amenable to be analyzed by GC. In this technique the high separation power of GC combined with the sensitivity and compound identification capability of MS detection led to an exceptionally useful tool for both qualitative and quantitative analysis. The GC is able to deliver the separated components of the sample in gas form, which is a good match for the vacuum conditions necessary in a mass spectrometer. The mass spectra obtained using 70 eV for the ion formation are unique for many compounds, providing a powerful tool for compound identification. The identification is possible even for traces of compounds at levels as low as 0.2–0.3 μg/mL or even lower concentration in the injected solution in the GC. This identification is facilitated by computer-performed searches in mass spectral libraries, available for about 675,000 compounds (see Section 1.2). Also dedicated programs are available for assisting in evaluating peak purity for the GC separation such that the MS spectrum is correctly identified following the library search, such as MassLib from Max-Plank Institute,

MassWorks from Cerno Bioscience LLC, and AMDIS (Automated Mass Spectral Deconvolution and Identification System) from NIST.

The high sensitivity of GC–MS also makes this technique an excellent quantitation procedure. The GC–MS/MS technique, where the detection for the GC is performed using tandem mass spectrometry, lowers the detection limit for the analytes at levels below ng/mL in the injecting solution. The use of concentration techniques such as P&T and SPME further increases the sensitivity of GC–MS techniques. Due to its high capability of compound identification, GC–MS is unsurpassed by any other technique in compound identification. Combining this quality with its high sensitivity and reproducibility, GC–MS is the analytical technique of choice for many analyses.

Based on the capabilities on GC in particular with MS detection, GC and GC–MS methods of analysis can be preferred to HPLC when the compounds are amenable to be analyzed by these techniques. Volatile compounds with small molecules in particular are more frequently analyzed by GC and not by HPLC. The lack of volatility of larger and more polar molecules makes them impossible to be analyzed by GC, and for such molecules HPLC is the preferred technique of analysis.

Selection of SFC as the Method for Analysis Versus HPLC

As a separation technique, SFC is a hybrid of gas and liquid chromatography. Similar to HPLC, the variation of the mobile phase composition affects separation. SFC has a number of advantages versus GC and versus HPLC, as well as certain disadvantages. Among the advantages can be listed the use of a solvent that has less environmental impact as compared to HPLC, and a typical shorter time for the chromatographic run. The high fluidity of the mobile phase in SFC may lead to narrow peaks that can be integrated more precisely. In SFC, the mobile phase affinity for the analyte is also a function of mobile phase density, which is controlled by modifying the system pressure, and of temperature. SFC has numerous applications, in particular for the analysis of nonpolar analytes. Pure supercritical CO_2 has a polarity close to hexane, although depending on pressure and temperature this polarity may vary. This limits to a certain extent the type of compounds that can be dissolved in the supercritical CO_2 and eluted from a chromatographic column. Even varying the mobile phase composition with polar modifiers (e.g., MeOH), or by changing the density and the temperature, the "polarity" of the mobile phase in SFC remains relatively low. For this reason, SFC is not adequate for the analysis of polar samples, although successful analyses on a variety of compounds were obtained using SFC [32]. SFC is a good replacement for normal phase HPLC (see Section 8.4) and has applications, in particular with the improvements brought by convergence chromatography to the quality of the separation media (new improved chromatographic columns). Ultra-performance convergence chromatography (UPC^2) uses supercritical CO_2 as the predominant mobile phase solvent but the polarity of this mobile phase can be modified not only through pressure and temperature, but also by adding a variety of modifiers (such as methanol, isopropanol, etc.). The use of small-particle stationary phases offers a wide range of selectivity and the technique has been successfully utilized in the analysis of lipids, fat-soluble vitamins, steroids, nonionic surfactants, and in some chiral separations [33]. The introduction of mass spectrometric detection (e.g., QDa from Waters [34]) improved the capabilities of SFC type analysis although the ions are obtained using an ESI type source and the generated spectra do not show the fragmentation similar to EI+ spectra from GC–MS. For most polar molecules, HPLC is still the preferred analytical method.

Selection of CZE as the Method for Analysis Versus HPLC

Capillary zone electrophoresis (CZE) is a technique successfully used for the separation of proteins, peptides, and nucleic acids. Other applications where CZE may be useful include analysis of inorganic anions and cations, such as those typically separated by ion chromatography. Small charged molecules can also be separated using CZE. The quantitation using CZE is less reliable compared to HPLC. A number of issues such as irreproducibility of the injection, adsorption of the samples on the capillary walls, heating of the capillary during separation, difficulties in assuring a uniform flow of sample when using MS detection, the need of extreme cleanliness of the system, alignment in the detection window when using UV or fluorescence detection, and other similar problems reduce the reproducibility and robustness of this technique [35].

Advantages and Disadvantages in Selecting High-Performance Liquid Chromatography as a Method of Analysis

HPLC is the most common analytical technique, and because it consists of a collection of different similar techniques, it offers excellent flexibility for the analysis of a variety of samples and for numerous different purposes. The main utilization of HPLC is quantitation, which can be applied to a wide range of types of molecules, including small molecules of different polarities, as well as larger molecules such as proteins, nucleic acids, polymeric carbohydrates, synthetic polymers, etc. HPLC can also be used for molecular weight evaluation, and to a certain extent, for the identification of unknown species. Other applications of HPLC are related to the estimation of specific molecular properties such as hydrophobicity and dissociation constants [36,37]. HPLC capabilities for the separation of the analytes are very good, and with the combination of stationary-phase characteristics, mobile phase composition, and conditions of separation (e.g., gradient of the mobile phase, temperature) it is possible to obtain very detailed separations. Using chiral stationary phases, even the separation of stereoisomers can be done with very good results. The present material is dedicated to analytical HPLC, but scaling up of an HPLC separation can be successfully done such that large amounts of materials can be separated.

Quantitation in HPLC can be performed using several procedures such as calibration curves, standard addition, and use of a response factor (see Section 1.2). Precision in HPLC can be very good, depending on the utilized detector, and reproducibility/repeatability is typically excellent. The limit of detection (and quantitation) in HPLC depends on numerous factors such as the nature of the compound, the sample preparation procedure, and the detector sensitivity, which is a critical factor. Among the most sensitive are the fluorescence detector and the mass spectrometric detector. For example, for certain compounds, the mass spectrometric detection (MS/MS working in MRM acquisition mode) can be as low as several fg/mL. In addition to that, most HPLC methods report excellent linearity, robustness, ruggedness, and stability. Combining the wide range of compounds that can be quantitated using HPLC with the excellent conditions for potential validation of HPLC methods, the technique is widely applied to the quantitation of analytes in samples such as pharmaceuticals, environmental samples, pollutants, biological samples, food and agricultural products, and many other materials and/or processes.

The use of HPLC for molecular weight evaluation is related to size-exclusion chromatography. A discussion regarding this type of application can be found in Sections 4.3 and 11.3 and in various publications dedicated to this subject (see, e.g., Ref. [38]).

Qualitative analysis with the identification of unknown compounds encounters several problems in HPLC. Depending on the detector utilized in the HPLC set-up, some are universal detectors and do not provide any qualitative information (e.g., refractive index and light-scattering detectors, see Section 4.1). Other detectors such as a fluorescence detector or UV—Vis detector can be selective and the use of a DAD (diode array detector for UV measurements) can provide UV spectra of the analytes, which in certain cases can be diagnostic in the sense that once the UV spectrum of a compound is known, it can be useful for positive identification. However (with a few exceptions), these detectors are not useful for the identification of unknown compounds.

The most useful for compound identification are the mass spectral detectors. However, a series of problems are encountered for unknown identification using LC-MS or LC-MS/MS. The ionization techniques typically used in LC-MS and LC-MS/MS are ESI and APCI. In both these ionization modes, mainly the molecular ions of the analytes are formed. LC-MS detection indicates the molecular ion mass, which is valuable information but not a diagnostic parameter for the identification of analyte structure. Further fragmentation of the molecular ions takes place in the collision cell when MS/MS detection is employed (see Section 2.3). This fragmentation depends on the nature of the analyte, but also on the conditions applied in the collision cell (nature of the collision gas, gas pressure, and collision voltage). In addition to that, fragmentation may differ for MS/MS instruments from different manufacturers due to differences in collision cell construction. Because of the variability in fragment formation, standard spectra for each analyte are difficult to have. The measurement of the intensity of one of the most intense fragments generated in MS/MS provides an excellent procedure for quantitation, and other fragments typically provide proof for positive identification of the analyte. This made LC-MS/MS one of the best analytical techniques for quantitation. However, the lack of universally accepted collision conditions, and the existence of only limited LC-MS/MS mass spectral libraries (e.g., Ref. [39]) do not make the LC-MS/MS a universal method for the identification of unknown compounds as GC/MS is. Although the fragments may provide indication for the analyte structure, this is applicable for some molecules but not for all. Previous results regarding the mass spectral fragmentation of a specific molecule can always be used for molecular identification, and local mass spectral libraries obtained in specific conditions are useful for unknown identification when appearing in other samples.

Other routes are frequently utilized for unknown identification in LC-MS/MS. The two main procedures utilized for this purpose are based on obtaining high-resolution masses for the molecular ion and fragments, and on verifying the isotopic intensity distribution of spectral peaks. Dedicated programs are available to generate from the precise mass M of the molecular ion (or fragment) a molecular formula (numbers of each type of atom in the molecule, with no information on structure). Based on the molecular formula and known isotopic abundance of each element, the expected intensity of the ions $M+1$, $M+2$, etc. can be predicted. This intensity can be verified in the mass spectrum for the unknown (see, e.g., MassWorks program package from Cerno Bioscience LLC [40]) to validate (or not) the molecular formula. However, the next step from the molecular formula to a structural formula is much more difficult, not having the same information about the fragmentation of the molecule as obtained in GC–MS. For this reason, unknown identification using LC-MS/MS is usually done using additional information besides mass spectral. Such information should significantly reduce the possible alternatives for a structural formula and use the mass spectral information only for validation.

References

[1] R.E. Majors, J. Hinshaw, The chromatography and sample preparation terminology guide, LC/GC N. Am. 34 (S2) (February 2016) 6–90.
[2] G. Guiochon, D.G. Shirazi, A. Felinger, A.M. Katti, Fundamentals of Preparative and Nonlinear Chromatography, Elsevier B.V., Amsterdam, 2006.
[3] J.-M. Menet, D. Thiebaut (Eds.), Countercurrent Chromatography, Marcel Dekker, New York, 1999.
[4] A. Berthod, T. Maryutina, B. Spikav, O. Shpigun, A. Sutherland, Countercurrent chromatography in analytical chemistry, Pure Appl. Chem. 81 (2009) 355–387.
[5] J. Pawliszyn (Ed.), Handbook of Solid Phase Microextraction, Elsevier, Amsterdam, 2012.
[6] D. Sheehan, Physical Biochemistry: Principles and Applications, second ed., John Wiley and Sons, Chichester, 2009.
[7] F. Svec, Z. Deyl (Eds.), Capillary Electrochromatography, Elsevier Science B.V., Amsterdam, 2001.
[8] A.B. Kanu, P. Dwivedi, M. Tam, L. Matz, H.H. Hill, Ion mobility-mass spectrometry, J. Mass Spectrom. 43 (2008) 1–22.
[9] J.C. May, J.A. McLean, Ion mobility-mass spectrometry: time-dispersive instrumentation, Anal. Chem. 87 (3) (2015) 1422–1436.
[10] R.C. Johnson, R.G. Cooks, T.M. Allen, M.E. Cisper, P.H. Hemberger, Membrane introduction mass spectrometry: trends and applications, Mass Spectrom. Rev. 19 (2000) 1–37.
[11] M.J. Yang, S. Harms, Y.Z. Luo, J. Pawliszyn, Membrane extraction with a sorbent interface for capillary gas chromatography, Anal. Chem. 66 (1994) 1339–1346.
[12] S.C. Moldoveanu, V. David, Sample Preparation in Chromatography, Elsevier, Amsterdam, 2002.
[13] J. Pawliszyn, Solid-Phase Microextraction, Theory and Practice, Wiley-VCH, New York, 1997.
[14] S.C. Moldoveanu, Pyrolysis of Organic Molecules with Applications to Health and Environmental Issues, Elsevier, Amsterdam, 2010.
[15] W.O. McReynolds, Characterization of some liquid phases, J. Chromatogr. Sci. 8 (1970) 685–691.
[16] L. Rohrschneider, Eine methode zur chrakterisierung von gaschromatographischen trennflüssigkeiten, J. Chromatogr. A 22 (1966) 6–22.
[17] L. Rohrschneider, Eine methode zur charakterisierung von gas-chromatographischen trennflüssigkeiten: II. Die berechnung von retentionsverhältnissen, J. Chromatogr. A 39 (1969) 383–397.
[18] E. Kováts, Gas-chromatographische Charakterisierung organischer Verbindungen. Teil 1: Retentionsindices aliphatischer Halogenide, Alkohole, Aldehyde und Ketone, Helv. Chim. Acta 41 (1958) 1915–1932.
[19] L.S. Ettre, The Kováts retention index system,, Anal. Chem. 36 (8) (1964) 31A–41A.
[20] Chromatography Product Guide 14/15. Phenomenex. www.phenomenex.com.
[21] R.L. Grob, E.F. Barry (Eds.), Modern Practice of Gas Chromatography, fourth ed., Wiley, Hoboken, 2004.
[22] R.P.W. Scott, Introduction to Analytical Gas Chromatography, M. Dekker, New York, 1997.
[23] C.H. Hartmann, Gas chromatography detectors, Anal. Chem. 43 (2) (1971) 113A–125A.
[24] R. Buffington, M.K. Wilson, Detectors for Gas Chromatography, Hewlett-Packard, 1987, pp. 16–18.
[25] G.K. Webster (Ed.), Supercritical Fluid Chromatography. Advances and Applications in Pharmaceutical Analysis, CRC Press, Boca Raton, 2014.
[26] http://www.njcg.org/yahoo_site_admin/assets/docs/UPC2_NJCDG__6_13.16993835.pdf.
[27] A.G. Boborodea, D. Daoust, A.M. Jonas, C. Bailly, An improved analytical temperature-rising elution fractionation system for automated analysis of polyethylenes, LCGC Asia Pac. 7 (2004) 40–43.

[28] C.D. Garcia, K.Y. Chumbimuni-Torres, E. Carrilho, Capillary Electrophoresis and Microchip Capillary Electrophoresis: Principles, Applications, and Limitations, Wiley, Hoboken, 2013.
[29] M.L. Marina, A. Ríos, M. Valcárcel (Eds.), Analysis and Detection by Capillary Electrophoresis, Elsevier, Amsterdam, 2005.
[30] R.D. Strickland, Electrochromatography, Anal. Chem. 34 (1962) 31R–34R.
[31] S.C. Moldoveanu, V. David, Modern Sample Preparation for Chromatography, Elsevier, Amsterdam, 2015.
[32] A.G.-G. Perrenoud, J.-L. Veuthey, D. Guillarme, Comparison of ultra-high performance supercritical fluid chromatography and ultra-high performance liquid chromatography for the analysis of pharmaceutical compounds, J. Chromatogr. A 1266 (2012) 158–167.
[33] http://www.waters.com/waters/en_US/Convergence-Chromatography/nav.htm?cid=134717180&locale=en_US.
[34] http://www.waters.com/webassets/cms/library/docs/720004799en.pdf.
[35] http://afdaa.org/2013/wp-content/uploads/2013/01/AFDAA-CE-Updated.pdf.
[36] R. Kaliszan, Quantitative Structure-Retention Relationships QSRR in Chromatography, Academic Press, San Diego, 2000.
[37] V. David, A. Medvedovici, Structure-retention correlation in liquid chromatography for pharmaceutical applications, J. Liq. Chromatogr. Rel. Technol. 30 (2007) 761–789.
[38] A. Striegel, W.W. Yau, J.J. Kirkland, D.D. Bly, Modern Size-Exclusion Liquid Chromatography: Practice of Gel Permeation and Gel Filtration Chromatography, second ed., Wiley, Hoboken, 2009.
[39] Smile MS: Small Molecule Identification and Library Exploration Using LC-MS/MS, Wiley, Hoboken, 2011.
[40] www.cernobioscience.com.

CHAPTER

4

Basic Information Regarding the HPLC Techniques

4.1 BASIC INFORMATION ABOUT INSTRUMENTATION IN HPLC

The basic construction of a high-performance liquid chromatography (HPLC) instrument includes the following parts: (1) a solvent supply system (solvent containers and degasser), (2) a high-pressure pumping system, (3) an injector, frequently included in an autosampler, (4) a thermostatted column holder, (5) a chromatographic column (possibly with a guard column or precolumn), (6) one or more detectors, and (7) the instrument control and data processing unit (typically a computer with dedicated programs). The whole system is controlled in modern HPLC instruments by a computer system also capable of data collection and processing. Optionally, other instrument components can be included, such as a cooling system for the autosampler/injector, a fraction collector, and column switching for back-flow. A diagram of a simple type of construction for an HPLC is shown in Fig. 4.1.1.

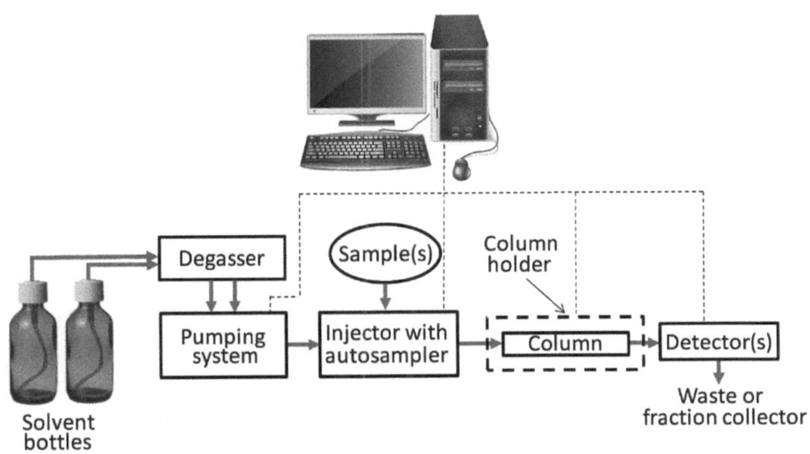

FIGURE 4.1.1 Schematic diagram of a simple configuration for an HPLC system.

http://dx.doi.org/10.1016/B978-0-12-803684-6.00004-4

The construction of an HPLC instrument may vary from one manufacturer to another and depends on the intended function and size of the HPLC process. Also, modern HPLC instrumentation can be rather sophisticated, and the instruments are in continuous development (see, e.g., Ref. [1]). Some details on each of the components of an HPLC system are further discussed in this section. The description of HPLC instrumentation is not the main goal of this book, and more information on the subject can be obtained from various other sources such as instrument manuals (e.g., Refs. [2,3]) or from other dedicated publications (e.g., Refs. [4–6]). The basic diagram of an UPLC system is the same as for the HPLC, and only some characteristics of the instrument components are different. For this reason, all further discussions about HPLC are applicable to UPLC, with specification which characteristics are different.

Solvent Supply System

The solvent supply plays the role of providing the solvent(s) necessary as a mobile phase for the HPLC (or UPLC). The solvent supply system should also have the capability to remove the gases dissolved in solvents (degassing capability). Degassing can be done in two stages. The first stage, which is optional, consists of sparging at a low flow rate an inert gas (N_2, He) through the solvents. This procedure is useful for the reduction of the oxygen content, however it may pose a problem in the case of premixed solvents. Premixed solvents are frequently used as one of the mobile phase components in HPLC. If the premixed solvent contains a volatile component, the solvent composition may be changed in time by the preferential evaporation of the volatile component due to sparging. In particular, when ammonia or trifluoroacetic acid is used in a mobile phase to adjust the pH of the solution, sparging is not recommended since drastic changes in the pH occur in time by the preferential elimination of these volatile species from the solution.

From the solvent containers (bottles), the solvent(s) are transferred through low-pressure tubing to the pumping system. The tubes used for passing the mobile phase through the system need to fulfill mainly the requirement of being inert to the utilized solvents and to stand pressures up to about 50 psi (1 psi = 6.89,476 kPa = 6.89,476 10^{-2} bar = 6.80,460 10^{-2} atm; 1 bar = 14.5,037,738 psi). Fluorocarbon polymers such as Teflon are common materials used for this type of tubing, but polypropylene is also used. The solvent supply system of an HPLC has one or more reservoirs for the solvents used as mobile phase. For HPLC performed in isocratic conditions and using a pure or a premixed solvent, only one reservoir is necessary. However, it is common in HPLC to use gradient separations, or to use an isocratic separation but to generate the mixture of solvents using the pumps. In this case, two (or more) solvents that are mixed with the pumping system in variable proportions are required. For this reason, two or more solvent reservoirs are common in modern HPLC instruments. The reservoirs must be clean and inert to the solvents they contain. The solvents from the reservoirs must be free of particles, and they are either purchased as HPLC grade or/and filtered through 0.45-μm filters before use. The filter selected for the filtration must be inert to the solvent. In the case of solvent mixtures containing a buffer, the general rule indicates that the buffer solution is made in water, then preferably filtered, and only after that mixed with the organic solvent (assuming correct concentrations and no precipitation after the organic solvent addition, due to lower solubility of the buffer in the mixture of solvents). The tubing transferring the liquid to the pump(s) typically has a frit at its mouth.

The second degassing in an HPLC system is performed in a degasser. A degasser apparatus is a device in which the solvent passes through a piece of special polymeric tubing placed in a vacuum chamber. The tubing material (membrane) has selective permeability to gases, and the (mild) vacuum created by a small pump reduces the content of the gases from the solvent (see, e.g., Ref. [7]). The degassers, although popular in HPLC equipment, may pose problems in specific applications. The polymeric tubing may absorb selectively specific components from the solvents and may be a source of contamination when changing from one solvent to another. Also, it should be noted that large bubbles coming from the solvent reservoir cannot be eliminated by the degasser apparatus, and these bubbles make their way into the pumps affecting their function.

The degassing is necessary because even very pure solvents may have dissolved in them small quantities of oxygen. This oxygen may be released in the form of very small bubbles in the HPLC system when a drop in pressure occurs (e.g., between the chromatographic column and the detector), or when a solvent with high solubility for oxygen (e.g., water) is mixed with another solvent with low solubility for this gas. When this mixing is done before the high-pressure pump, the gas bubbles may lead to variations in the pressure of the liquid delivered by the pump (pressure fluctuations). The pressure fluctuation should be low and in modern HPLC systems it is typically below 0.1% of the nominal pressure. When the gases are not removed from the solvents, even if the pumps are working properly, pressure fluctuations of 4−6% of the nominal pressure can be noticed. Dissolved gases in the mobile phase may also influence the injection volume when small sample volumes (e.g., 1−2 μL) are injected. Also, the reading of the detectors can be perturbed by dissolved gases. For example, oxygen may affect the reading of electrochemical detectors, the fluorescence intensity of certain compounds, and the UV absorption at very low wavelength range.

A special type of solvent delivery system can be used in ion chromatography. This system is known as an eluent generator (Thermo Scientific/Dionex) (see Section 13.5).

Pumping System and the Mobile Phase

The main pumping system consists of pump(s) able to deliver a constant flow of mobile phase through the injector, chromatographic column, and through the detector(s). A full discussion on the selection of mobile phase composition is given in Chapters 13 and 14. The pumps must be able to generate a high pressure of the mobile phase, which is needed mainly to overcome the resistance to flow of the chromatographic column. This flow is characterized by the volumetric flow rate U (in old systems set manually, but usually set using a computer that controls the HPLC system). In conventional HPLC systems, the pumps are usually capable of delivering U between 0.1 mL/min to 10 mL/min fluid and generate up to 6000 psi (about 400 bar; 1 bar = 14.5,037,738 psi). A shut-off system is typically present to stop the flow if the pressure attains the maximum allowed. New developments in using very fine particles in the chromatographic column require higher pressure and sometimes capability to produce flows at less than 0.1 mL/min. The new instruments can generate up to 8500 psi (about 600 bar) or higher (e.g., 1200 bar for up to 5 mL/min) and are indicated as UPLC or U-HPLC. For the HPLC systems used for other purposes than analytical, pumping parameters can vary significantly. Also, special HPLC systems can generate higher pressure but they are not common. The flow from the pumps (volumetric flow rate U) must be constant, without fluctuations (in fact with very small ones). This requirement is necessary mainly for the detectors, where the signal may fluctuate when the flow rate varies.

Most high-pressure pumps used in analytical HPLC are reciprocating pumps. A single piston reciprocating pump consists of a cylinder with a reciprocating plunger in it and two valves mounted in the head of the cylinder. The liquid enters the cylinder through an inlet (suction) valve and is pushed through a discharge valve. During the suction the plunger retracts and the inlet valve opens causing the admission of fluid into the cylinder, while the discharge valve is closed. In the forward stroke the plunger pushes the liquid out through the discharge valve while the inlet valve is closed. However, the fluid flow from a single piston reciprocating pump (and therefore the pressure in the system) has a pulsating profile. When the piston is moved by a circular motion of a driving cam, the flow rate has a half sinusoid shape. This type of flow is not suitable for HPLC. Dual-piston pumps consisting of two reciprocating pumps that alternate the forward stoke are able to generate flow with only one zero flow point per cycle. However, with this setup the flow is still fluctuating. The use of specially shaped driving cams or of stepper-driven motors allows the generation of an almost continuous flow of liquid. The pumps in the dual-pump system may be connected in parallel or in series. The series-type system requires two pistons but only three valves in order to achieve the task of generating a continuous flow. Modern systems are able to deliver flow with a precision of about 0.07% relative standard deviation (RSD%) and a flow accuracy of less than 1% from the nominal value of U. Because the pumps must deliver flow at high or very high pressure, their construction requires special materials such as inert steel body, sapphire or ceramic pistons, high-precision valves that do not have any leaks, and special polymeric seals. For ion chromatography, the whole pumping system is typically made of strong polymeric materials such as polyetheretherketone (PEEK).

In addition to the pumps and valves specially designed, the flow without fluctuations from the pumping system can be achieved using a pressure pulsation damper. Pressure dampening is done, for example, by passing the fluid through a cell with a diaphragm wall that compensates the pressure variation. The pulsation can be reduced to less than 0.1% with dampening combining the stepper-driven motors electronically controlled. Various models of dampers are available, and most of them have a volume of around 500 μL in order to assure a delay volume as small as possible in the delivered fluid. A typical delay volume for an analytical HPLC system is around 700 μL. In some instances, this delay volume is undesirable and the pulsation damper may be removed (on the account of larger pulsations).

A dual-piston pump can handle only one solvent and can be applied for isocratic separations that use a pure or a premixed solvent. However, since in HPLC it is frequently necessary to use gradient separation, instruments that handle more than one solvent were developed. This type of instrument is also frequently used to generate a solvent mixture to make a mobile phase of a desired composition, even when this composition is not changed during the separation (isocratic conditions). Two common procedures are used to physically achieve the mixing of solvents: (1) low-pressure mixing, where the solvents necessary for the gradient are premixed with a low-pressure pump connected to the high-pressure pump and (2) high-pressure mixing that uses two high-pressure pumps with each one dedicated to one solvent and with the mixing of the flows in a low-volume mixer (see Fig. 4.1.2). A third type of

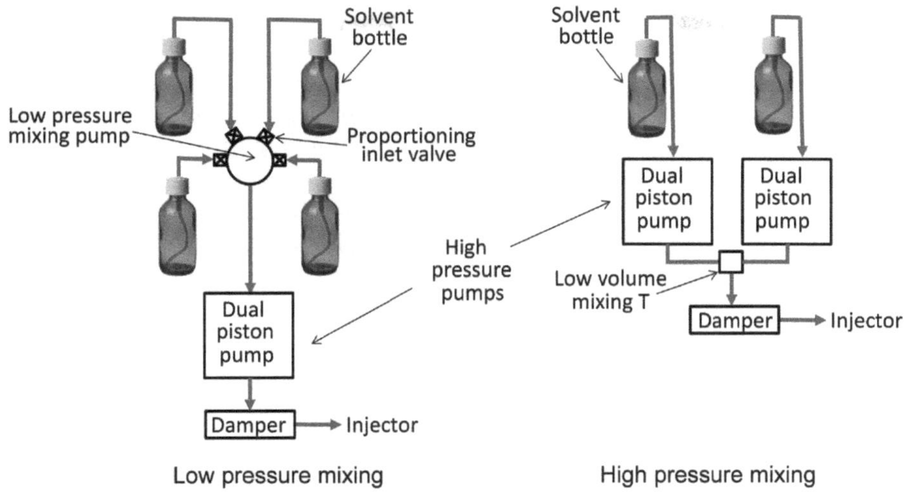

FIGURE 4.1.2 Schematic description of low-pressure mixing and high-pressure mixing of solvents in HPLC.

system known as hybrid mixing uses a high-pressure pump with two or three proportioning inlet valves.

In low-pressure mixing, two or more (usually four) solvents can be blended at the desired composition by using a low-pressure pump with several *proportioning inlet valves* controlled by a computer. The valves open repeatedly for a short period of time (typically less than 1 s) the duration of time the valve is opened being proportional with the desired mobile phase composition. The mixed solvents are delivered further to one (dual-piston) high-pressure pump. Low-pressure mixing has the advantage of using a single high-pressure pump (that is typically expensive) and has more flexibility in choosing a variety of solvents (in systems with four proportioning valves). However, the changes in the mobile phase composition when a low-pressure mixing system is used, are taking place more gradually than for the other systems (the change in composition is not instantaneous). Also, low-pressure mixing may be prone to the formation of small bubbles of gas in the mixed solvent, when the solubility of oxygen, for example, is higher in one solvent than in the mixture. These bubbles may enter the high-pressure pump and generate flow fluctuations as high as 4–6% from the nominal pressure.

In high-pressure mixing, two high-pressure (dual-piston) pumps are used, and the ratio of solvents in the mobile phase is controlled by the flow rate of the high-pressure pumps. The mixing of the two solvents A and B can be achieved either in a specially designed mixer (typically with volume of less than 500 μL) or in a mixing T (with virtually zero volume). High-pressure mixing is considered to provide a more precise control of the composition of the mobile phase (with a typical composition precision of less than 0.15% RSD% at 1 mL/min flow), and does not have the problem of formation of bubbles caused by the difference in solubility of oxygen in the mixed solvent compared to that in one of the components. However, the cost of high-pressure pumps is a disadvantage for this type of mixing.

Most modern HPLC systems with high-pressure mixing have two pumps and the capability of switching between two solvent pairs (A1, B1, and A2, B2).

Since instruments with low-pressure mixing and instruments with high-pressure mixing are common in laboratories, when an established analytical technique is transferred from one instrument to another, attention should be paid to the type of instrument. In both types of chromatographic systems, there is a specific volume passing through the system from the point at which the mobile phase solvents are mixed until they reach the head of the chromatographic column. This volume is known as the dwell volume, V_D. The dwell volume is typically different in low-pressure mixing systems (2–4 mL) and in high-pressure mixing systems (1–3 mL). Special instruments, such as those used in microscale HPLC may have smaller dwell volumes (less than 300 μL). For certain applications using gradient separations on common HPLC systems, differences can be seen when working with one type of instrument or with the other, although the gradient program is the same. This is in particular caused by the differences in the dwell volume from one system to another.

Hybrid mixing uses a single dual-piston high-pressure pump with inlet from two or three proportioning inlet valves that determine the solvent composition. Hybrid systems are not very common [8].

In order to be able to modify the composition of the mobile phase in gradient HPLC, the solvents that are blended in specified proportions should be perfectly miscible. Particular care must be paid to the solubility differences of certain additives present in one solvent when another solvent is added. For example, it is common in HPLC to use buffer solutions. These solutions can be easily made in water by adding mixtures of salts and acids or salts and bases. When a water solution containing these types of additives is mixed with an organic solvent (such as CH_3OH or CH_3CN), the solubility of the additives in the mixed organic/aqueous solution is drastically diminished. The formation of precipitates following the mixing must be carefully avoided and only buffers at low concentration of salts (typically less than 100 μmol) should be used when organic solvents are to be added to the aqueous buffer. This is especially important when the proportion of the organic solvent is high (or increases during the gradient).

During the separation, the composition of the mobile phase can be kept constant for some intervals of time and modified for other periods of time. The modern instruments are commonly controlled using a computer with a dedicated program that assists in generating a specific gradient using a programmable timetable. Based on the gradient timetable, the computer controls the pumping system that physically generates the desired mobile phase composition by mixing in the correct proportion of the solvents from the solvent supply system. The gradient starts when the sample is injected. After the gradient ends, the HPLC chromatograph is made ready for the next injection. The total run time of the chromatogram, starting with the moment of injection until the end of the chromatographic run, is sometimes referred to as the total cycle time. In a gradient separation, the dwell volume V_D of the system creates a dwell time t_D. Because of the dwell time, there is a delay between the change of composition at the point of solvent mixing (set in the timetable) and the change in composition at the head of chromatographic column. Therefore, attempts to modify the retention time of a peak by using a "stronger" solvent should be done in the gradient timetable ahead of the peak retention time.

The modification in concentration between two changing points of the gradient can be linear. For a gradient starting at time t_1 with the concentration C_1 of component A and ending at time t_2 with the concentration C_2, at an intermediate point at time t, the concentration of component A can be obtained using the formula:

$$C(t) = C_1 + \frac{(t - t_1)}{(t_2 - t_1)}(C_2 - C_1) \tag{4.1.1}$$

In most HPLC separations the mobile phase during the chromatographic run is changed from one content in an organic modifier to a higher or lower value. The change in the concentration of the organic modifier in a period of time is indicated as gradient slope. For a linear change in concentration, the gradient slope can be defined by the expression:

$$\Delta = \frac{C_2 - C_1}{t_2 - t_1} \tag{4.1.2}$$

Some HPLC pumping systems allow both a linear change in the gradient and a nonlinear modification of the concentration (see, e.g., Ref. [5]). However, nonlinear changes are less common.

Much less frequently than reciprocating pumps, syringe pumps are sometimes used in HPLC. Only the recent developments in UPLC made the syringe pumps more attractive. UPLC requires low flow rates (of the order of 0.1 mL/min) and less solvent compared to conventional HPLC. In syringe pumps, a cylinder is loaded with the mobile phase, which is delivered at a specified flow rate by the movement of a piston. The flow from a syringe high-pressure pump can be virtually fluctuation free as compared to that from a reciprocating pump. This is an advantage compared to piston pumps at low flow rates since it produces a stable baseline for the detectors. The price of a syringe pump can also be lower. Even lower flow rates (of the order of 10 μL/min) are required when using capillary HPLC columns. Such flows are typically possible only when using dedicated HPLC equipment and the syringe pumps are adequate for such flows [9].

The selection of a pumping system, when possible, is related mainly to the pressure that can be generated by the system. Some other aspects such as the stability of the pressure (lack of pulsations), the minimum and maximum flow rate, and the type of mixing (low-pressure mixing or high-pressure mixing) are also elements to consider. The selection of an HPLC (8500 psi/600 bar maximum pressure) or a UPLC (15,000 psi/1300 bar maximum pressure) is made depending on the type of chromatographic column. The columns produce backpressures which increase as the flow rate increases. The modern chromatographic columns made with smaller particles have higher backpressure compared to older-type columns that were made with larger particles. For a method requesting a specific flow rate U through the column, it should be assured that the pumping system does not reach its maximum pressure during the run, the pressure varying during the gradient. It is important to consider the limitations of the pumping system when selecting a specific analytical method. Also, the differences between pumping systems with low-pressure mixing and high-pressure mixing should be considered, mainly when changing from one instrument type to another for the same analytical method. Few analytical methods require a very sharp gradient change, and in these cases a high-pressure mixing system should be selected.

Injectors and Autosamplers

The role of the injector is to place in the mobile phase (in an accurate and reproducible manner) a small, precisely measured volume of a solution containing the sample. Conventional HPLC systems have injectors capable to inject between 1 μL up to 100 μL sample solution, typical volumes for injection being between 1 μL and 25 μL. A larger injection volume places in the chromatographic column more analyte, allowing an increase in sensitivity, but in some cases this affects the shape of the chromatographic peak. Some special injection systems may allow large injection volumes which are used, in particular, in semipreparative or preparative HPLC. For UPLC, the injection volumes can be between 0.05 and 2.0 μL, and when capillary HPLC columns are used, the injection volume can be even smaller. For volume lower than 0.1 μL, specially designed injectors may be necessary. Reproducibility of injection is of particular importance, and modern injectors typically show less than 0.5% RSD% in the injected volume. The accuracy errors in injection volume are important mainly when comparing different instruments since for the same instrument the use of standards for quantitation may compensate for small variations from a nominal volume. However, for a specific method it is recommended to keep the same injection volume when injecting different samples in order to avoid accuracy problems.

From different vendors, a number of different injector models are available [10]. One such model consists of a loop of a precise volume that is filled first with the sample and then connected to the flow circuit using a switching valve. This system allows only the injection of a fixed volume of sample, equal with the loop volume, and it is typical for manual injectors. The switching valve can be actuated manually or electronically. However, the volume of sample solution to be injected for different analyses may need to be changed and the replacement of a fixed loop can be inconvenient. Injection of different sample volumes can be achieved using a larger loop (e.g., of 100 μL) that is only partially filled with sample (partial loop injection). The injection is performed by loading the sample from a vial in an injector needle that is part of the sample loop. The loop is filled initially with the mobile phase, and then the sample is introduced in the loop, occupying only a small portion of its volume, with the sample volume typically controlled using a computer. The placing of the loop in the mobile phase main circuit is then done in a similar manner as for fixed-volume loops.

Injector systems with automation capability for injecting consecutively a number of samples placed in a tray are common (the samples are typically present in vials with a small septum that can be punctured by the sampling needle). From a large number of samples in different vials (or well plates), these automatic systems (computer-controlled autosamplers/autoinjectors) have the capability to select the desired sample vial and to repeat the injection at a specified time or upon receiving an electrical signal from the computer. In autosamplers, since several samples are injected one after the other, carryover effects are possible. Carryover effects represent the contamination of a sample with small amounts of components from the previous sample that remained in the autosampler after an injection. This problem is typically solved using a needle wash.

Some autoinjectors have the capability of mixing the sample with specific reagents from different vials, in case derivatization is necessary before the separation and detection of analytes. Such an autoinjector can load in the injector loop different volumes of solution of the sample and reagents, and assure that they are mixed properly. Specific intervals of time between mixing and injection can be selected for assuring the completion of the reaction between the analytes and the reagent used for derivatization.

Also, cooling and heating capability is frequently present in modern autoinjectors. Keeping the samples at a low temperature (e.g., 3–7°C) may be important to avoid decomposition for thermally sensitive samples, or to diminish evaporation from a vial that was already sampled and has its septum already punctured.

Usually the injection operation is unselective. However, special online sample preparation methods require more special injection techniques. In the case of a multidimensional HPLC, for example, a fraction from eluted sample from a first column can be transferred by an interface to become an injected sample to a second column. Injection devices normally do not affect retention parameters unless operation problems occur. Also, in most HPLC applications, one injection is done for each chromatographic run. However, multiple injections in a single run (MISER) are possible for specific analyses [11]. In this technique the peak of the analyte(s) of interest can be detected without the interference from the matrix peaks of the previous sample (for example that elute later), and a larger number of samples can be analyzed within a specific interval of time and with a lower solvent usage.

Two important parameters must be selected by the operator regarding injection: (1) the nature of the solvent for the sample, and (2) injection volume. Besides the obvious requirement that the solvent for the sample should dissolve it completely, this solvent must also be soluble in the mobile phase. When the sample solvent and the mobile phase are different, they also may have different capabilities to elute the analytes. For very small injection volumes this difference may not be very important. However, when the sample solvent can easily elute the analytes and the injection volume is larger, this affects the peak shape. Some details on the selection of the sample solvent (diluent) will be discussed in Section 13.6.

The injection volume is selected depending on a number of factors including the type of instrumentation (HPLC, UPLC, detector type, etc.), the sensitivity of the detector, the loading capacity of the column (maximum amount of sample that still allows separation), and the effect of sample solvent on peak shape. There are problems with both too small volumes and with too large volumes of injection. Too small volumes may lead to problems with injection reproducibility, sensitivity of the detector, or sample loss in the chromatographic column, but offer better peak shape sometimes resulting in better separation. Too large volumes may affect the peak shape (broadened, with flat top, asymmetrical), which affects separation. A large volume of sample does not necessarily mean a large amount of analyte (e.g., when the sample is very diluted), but when the large volume is also associated with too much analyte, this can be associated with an overload of the column (problems with the separation) and of the detector (leading to a nonlinear response). The injection volume depends on various factors and is different in micro-HPLC application, current analytical HPLC, and in preparative HPLC.

Besides the common procedure of placing the sample in the mobile phase using liquid injections, other procedures have been described. One such procedure is the use of an SPME (solid phase microextraction) fiber that is placed in the HPLC mobile phase using a special interface (see, e.g., Refs. [12,13]).

Tubing and Connectors

The sample is ideally introduced in the mobile phase flow as a zone (plug) with the concentration of the sample following a perfectly rectangular profile. However, even in a laminar flow the shape of the sample plug is changing and generates a parabolic profile. Tubing with smooth walls are specifically designed to avoid turbulent flow useful in particular in UPLC. This process, as well as the diffusion and other convection effects, contribute to deviations of the chromatographic peak from the ideal Gaussian shape, introducing tailing and asymmetry. The tubing used after the high-pressure pumps must be inert and also must withstand the high-pressure generated by the pumps. Typical materials for the tubing are stainless steel (316 stainless steel) and PEEK. Stainless steel is inert in most solvents, while PEEK may become very stiff after using solvents such as tetrahydrofuran or dimethylsulfoxide. Also, stainless steel tubing can be used even at very high pressure, while some restrictions to pressure are applied to the PEEK tubing (as function of internal diameter). However, for IC chromatography, PEEK is the material of choice for tubes and connectors. Tubes of several internal diameters (i.d.) are available, such as 0.065 mm i.d. (0.0025 in), 0.12 mm i.d. (0.005 in), 0.17 mm i.d. (0.007 in), 0.25 mm i.d. (0.010 in), and 0.50 mm i.d (0.020 in). For both stainless steel tubing and PEEK tubing a color code is available to designate the i.d. The choice of the tubing after the injector starts to play a role in the shape of the sample plug. Tubing with 0.12 mm i.d. (0.005 in) is in general recommended to connect the injector with the chromatographic column. This tubing has a volume of about 0.13 μL/cm such that a sample of 5 μL will spread over about 38 cm, diminishing the effects of sample plug shape modifications. Another factor contributing to dilution and modifications in sample plug shape is the "void spaces" in fittings that connect the injector and the chromatographic column. Loss of resolution by peak broadening due to large void spaces and also due to turbulent flow either along the tubing or in fittings must be avoided. The fittings typically use a nut that connects to a port and a ferule used to secure the end of the tubing in the fitting port. Void spaces may appear at the fitting point when the connection is not done properly between the end of the tube and the entrance in the fitting port. The whole HPLC system should have no leaks. The tightness of the HPLC system must withstand the backpressure, which may become more critical in UPLC systems that can reach up to 1300 bar.

Chromatographic Columns

The chromatographic column is designed for performing the separation of the sample components. Only a simplified overview regarding the chromatographic column is given in this section, the subject being further discussed in detail in other sections of this book. The body of the column typically consists of a tube made from metal (stainless steel) or plastic (e.g., PEEK) that is filled with a stationary phase. Common internal diameters for the analytical columns are between 2 and 10 mm, and length between 50 and 250 mm. Other dimensions are also possible. In UPLC the utilized columns are typically narrow (e.g., 2 mm i.d.) and short (e.g., 50 mm). Capillary HPLC columns (e.g., used in ion chromatography) have a 0.4 mm i.d., and similar lengths as the common analytical columns.

The stationary phase usually consists of small solid particles coated with the active material involved in the separation. This active material can be a bonded phase such as long hydrocarbon chains or other organic groups. Besides small particles, also monolithic materials can be used as support for stationary phase. The stationary phase made from small particles or from a porous monolithic material has specific physical and chemical characteristics to allow separations. The particles can be of three main types: porous, superficially porous (core–shell), and pellicular. Because the solid particles are covered with the active part of the stationary phase, the surface area of the particles (or of the monolithic rods) is an important physical characteristic, being directly related to the retention in the column of the compounds to be separated. The particle size is also important for the separation. It influences the eddy diffusion, which appears when local small streams of liquid follow different channels of different length in the column. This generates a broadening of the chromatographic peaks which is not a desired feature. The small porous particles are frequently obtained from an inert substrate material (such as silica) that is covered with the active phase.

Porous particles are still the most common type of stationary phase used in HPLC. They are made from particles usually of 1.7–5 μm diameter with a specific porous structure. Core–shell particles are made from particles with similar diameters having a solid core and 0.3–0.5 μm depth of a porous shell. The most common porous structure is obtained using porous silica. This is due to the capability of silica materials to be made porous with a large surface covered with reactive groups (silanol Si-OH) and at the same time to have a high rigidity and resilience to crushing. Chemical reactions are possible with the silanol groups present on silica surface, such that desired organic group attachments through covalent bonds are possible. These attachments (ligands) form a layer that further acts as a stationary phase necessary for the HPLC separation. The high chemical stability and the rapid mass transfer effects obtained with these bonded phases played a major role in the development of HPLC techniques. The columns having a hydrophobic active phase, for example with octadecyl groups (C18 or ODS) or with octyl groups (C8) bonded on silica, indicated as reversed-phase (RP) columns, are among the most commonly used. For other types of chromatography, the stationary phase may be made in various forms and have a variety of interactions with the analytes and mobile phase.

The nature of the stationary phase is selected based on the HPLC type of chromatography that is utilized for the separation. A large assortment of types of stationary phases is available, the column selection depending usually on the chromatography type. Table 4.1.1 shows a summary of HPLC types and the corresponding type of stationary phase utilized for separation (see Section 3.4 for a discussion on HPLC classification).

The nature of the surface of the stationary phase may have active moieties in addition to those intended for the separation. For example, when C18 groups are attached on a porous silica surface, a number of silanol groups still remain unaffected. The effort to reduce the number of residual silanols on the silica matrix is a common procedure in the preparation of chemically bonded stationary phases. This procedure is known as "end-capping" and it is performed by a separate procedure of blocking residual silanols by derivatization with small groups, such as trimethylsilyl. The process is applied after the initial silica derivatization with the desired bonded phase.

The subject of the nature of stationary phases used in various analyses is fully discussed further in this book (see Chapters 7–12). In some systems, more than one chromatographic column is necessary for achieving the desired separation. In size-exclusion chromatography, for example, two to four columns may be connected in series. In other types of separation, the use of more than one column is less common. The nature of the columns used in series may be the same or may be different. More than one column is also used in multidimensional HPLC, where a portion of the initial separation is further submitted for a second separation in a different column.

TABLE 4.1.1 Types of Stationary Phases Used in Various HPLC (or UPLC) Separations

Type of HPLC Mechanism	Type of Stationary Phase in the Column
1. Reversed phase (RP)-HPLC **2.** Ion pair (chromatography) (IPC) **3.** Nonaqueous reversed phase (NARP)	Solid supports covered with nonpolar moieties (e.g., long-chain hydrocarbon groups bound on a silica surface)
4. Hydrophilic interaction (liquid chromatography) (HILIC) **5.** Aqueous normal phase (ANP)	Solid supports covered with organic moieties containing polar groups (e.g., polar groups bound on a silica surface)
6. Hydrophobic interactions (chromatography) (HIC)	Organic moieties containing nonpolar groups (hydrophobic patches) on the surface of a hydrophilic stationary phase
7. Normal phase (NPC) **8.** Aqueous normal phase (ANP)	Polar materials (e.g., bare silica)
9. Cation exchange	Cation exchange materials
10. Anion exchange	Anion exchange materials
11. Ion exchange on amphoteric and zwitterionic phases	Materials with amphoteric character
12. Ion exclusion	Special materials with ion exchange capability
13. Ligand exchange **14.** Immobilized metal affinity **15.** Ion-moderated	Cation exchange materials loaded with a metal ion
16. Gel filtration **17.** Gel permeation	Porous particles with special pore sizes
18. Displacement	Various materials
19. Bioaffinity	Stationary phases specifically made to allow selective retention of the analytes based on affinity interactions
20. Chiral chromatography	Chiral phases
21. Multimode	Mixed phases

In order to protect the stationary phase from the analytical HPLC column against small heterogeneous particles from injected samples or from the delivered solvents, it is common to use small pore frits (e.g., with 0.45-μm pores) as well as guard columns (cartridges) in the path of the mobile phase before the column. The frits have the capability of mechanical filtration of the mobile phase. For column protection, more important than frits are the guard columns. The guard (cartridge) columns are selected to match the stationary phase of the analytical column (same active material), but their length is much shorter (e.g., a few mm), and in some cases their stationary phase has larger particle size. In the analyzed samples, there are sometimes components that are very strongly retained by a specific stationary phase. These components do not elute and tend to accumulate at the head of the column, deteriorating its performance in time. The use of a guard column allows selective retention of these components without affecting the efficiency of the column, retention time, backpressure, or the level of analytes. Guard columns are changed from time to time, while the analytical columns have longer service life (see Section 6.3).

Some chromatographic columns require a specific temperature for performing a good separation, and for this purpose special column ovens are used. Common column ovens have the capability to control the column temperature in a range from about 10°C below ambient to 80–100°C. Higher temperatures can be achieved with special ovens used in high-temperature HPLC. In addition to heating the column, the ovens typically are able to heat the solvent entering the column, since peak shape distortions may be noticed when the column and the entering solvent have different temperatures [14].

The selection of a specific method of analysis, including that of the mobile phase and its delivery as isocratic or gradient, of the solvent for the sample, of the injection volume, and especially the selection of the chromatographic column are subjects fully discussed further in this book, but not in this section.

The Main Characteristics of the Detectors Used in HPLC and UPLC

After having its components separated in the chromatographic column, the mobile phase passes through a detector (or several detectors) capable of performing measurements. The measurements are based on the fact that the molecules of the sample have physicochemical properties different from those of the mobile phase. The choice of a specific property for detection depends on factors such as the extent of difference from that of the mobile phase, sensitivity of the detector to the specific property, and availability of the detector. The selection of a specific detector is also correlated with the separation conditions used for the analysis. Some detectors have specific requirements for the nature of mobile phase, selection of isocratic or gradient separation, selection of a specific temperature for the column and mobile phase, etc.

The detector response (output) is typically dependent either on the instantaneous concentration of the detected species or on the instantaneous mass (amount) of analyte present in the detector. For this reason the quantitation is based on the area of the chromatographic peak. The output appears as an electrical signal recorded in the form of a chromatogram. In addition to the signal, all detectors are affected by the "noise," which is the random oscillation of the detector (electric) output. The measurement of the noise is done on a flat area of the chromatogram (baseline) close to the place of a peak, and the measurement of the signal is done from the middle of baseline noise to the top of the peak. A signal is typically considered as usable for detection when the value of the signal S is at least three times larger than the value for the noise $\mathcal{N}$ ($\mathcal{S}/\mathcal{N} > 3$). Other disturbances in the detector signal may include long-term noise and the drift. The long-term noise appears as a fluctuation of the signal with a wider frequency. The drift is a continuous variation (increase or decrease) of the signal for a period of time comparable with the length of the chromatogram. Drift may be caused not by the detector itself, but by changes, for example, in the composition of the mobile phase during gradient elution.

The qualities of the detectors should include the following: (1) capability to provide qualitative information, (2) good selectivity, (3) good sensitivity, (4) reproducible response, (5) large dynamic range, (6) linearity in a wide range of concentrations of sample, (7) low baseline noise and no baseline drift, (8) capability to make detection in a small volume of sample, (9) capability to not contribute to peak broadening, (10) stability to changes in flow and environmental parameters, and (11) high frequency of data collection.

1) Some detectors have the capability to generate only quantitative information, while others offer both qualitative and quantitative information, such as the MS detectors. Although the response of the detectors used in HPLC is typically not diagnostic, depending on the analyte and sample characteristics, the identification of compounds is possible using LC-MS, LC-MS/MS, and in some cases even other detectors.

2) The selectivity of the detector is related to the response to only a specific type of analyte. However, some detectors are designed to respond to all analytes (such as the RI detector, or ELSD) and are indicated as "universal" detectors. When the HPLC separation is good, the universal detectors are very useful since they account for all the components. Selective detectors respond to only a class of compounds or they can be set to detect only specific compounds. The detector selectivity adds one more capability for detecting separately certain analytes, and this can be a significant advantage when the HPLC separation is not possible (or is incomplete). However, this characteristic may also have a disadvantage. For example, the MS detection which is highly selective does not provide information about a coeluting compound with the analyte, unless the detector is also set for this compound. The coeluting compound can produce interferences such as ion suppression or detection of incorrect high signal, affecting the accuracy of quantitation.

3) Detector sensitivity is a very important factor in HPLC analysis. Since in UPLC the injection volumes (and the amount of analyte) are typically small, the detector sensitivity becomes even more important. This sensitivity depends on several factors such as analyte properties, sample matrix, mobile phase properties, and also on detector settings (e.g., electronic amplification of the signal), and detector manufacturer. Therefore, a specific discussion on detector sensitivity is difficult to be made. For these reasons, in analytical methods using HPLC, parameters such as limit of detection (LOD) and limit of quantitation (LOQ) are reported. These characterize globally the sensitivity of the method and consider a number of factors, including the detector sensitivity. The ranges of sensitivities of various detectors are indicated in Table 4.1.2.

TABLE 4.1.2 Ranges of Sensitivity for Various Types of HPLC Detectors

Type of Detector	Minimum Mass Injected (Typical)	Minimum Mass Injected (Extreme)
UV–Vis	0.1–1 ng	0.01 ng
Fluorescence	1–10 pg	10 fg
Refractive index	100 ng–1 μg	10 ng
Electrochemical amperometric	0.1–1 ng	100 fg
Electrochemical conductometric	0.5–1 ng	0.5 ng
Mass spectrometry	1 pg–1 ng	10 fg
Evaporative light scattering	5 μg	0.5 μg
FT-IR	1 μg	0.5 μg

Depending on the injected volume of the sample, the concentration of the analyte may vary in the injected solution. A typical injection volume of 10 μL would require 100 times higher concentrations per mL compared to the values indicated in Table 4.1.2.

4) Reproducibility of the response to the same injected sample is an important characteristic of any detector. Reproducibility of a detector is mainly related to the quality of detector construction (e.g., its electronics).

5) The dynamic range of a detector is related to the range where the detector has a response dependent on the analyte amount or concentration. The detector may not have a linear response for the entire dynamic range, but even in the range where the response is not linear, a positive dependence should exist between the signal versus the amount of analyte. The dynamic range of a detector can be independent of the nature of the analyte (for universal detectors), but also can be highly dependent on the analyte nature. The dynamic ranges of several types of detectors are given in Table 4.1.3.

6) Detector linearity indicates the range in which the dependence of the detector signal depends linearly on the amount or concentration of the analyte. Similar to the detector dynamic range, the linear range of a detector can be either independent of the nature of the analyte or highly dependent on this nature. The typical linear ranges for various detector types are given in Table 4.1.3.

7) The low baseline noise and no baseline drift are properties typically related to the quality of detector construction (e.g., its electronics).

8) The capability to make detection in a small volume of sample is usually related to the volume of a flow-through cell or other parts in the instrument construction. For UV–Vis detectors, for example, instrument manufacturers may offer cells of different dimensions, with longer path of the beam of light in the flow-through cell for achieving higher sensitivity of the instrument, but with slightly larger cell volume, or cells with shorter path for the light beam, but also with smaller volumes. Microcells are also available for certain instruments that are

TABLE 4.1.3 Typical Dynamic Ranges and Linear Ranges for Various Types of HPLC Detectors

Type of Detector	Dynamic Range of Concentrations	Linear Range Concentrations
UV–Vis	10^{+5}	10^{+4}
Fluorescence	10^{+4}	10^{+3}
Refractive index	10^{+4}	10^{+3}
Electrochemical amperometric	10^{+4}	10^{+3}
Electrochemical conductometric	10^{+4}	10^{+3}
Mass spectrometry	10^{+4}	10^{+3}
Evaporative light scattering	10^{+3}	10^{+2}
FT-IR	10^{+3}	10^{+3}

designated for micro-HPLC systems. Special cells are also necessary when using capillary HPLC columns and low flow rates.

9) The capability to not contribute to peak broadening is also related to the HPLC construction. The absence of dead volumes, small flow-through cells, laminar flow within the detector, and adjustment of the total volume of mobile phase in the detector in accordance with the range of injection volume for the sample are some of the steps toward achieving little or no contribution to peak broadening.

(10) The stability of the detector response to changes in flow of mobile phase may be an intrinsic characteristic. For some detectors such as UV−Vis the response is more stable to changes in flow rate since this detector responds to instantaneous analyte concentration. For other detectors, the stability cannot be achieved without a constant flow since they are responsive to the amount of analyte reaching the detector and not to the instantaneous analyte concentration. For example, for MS detection, the response is dependent on the amount of analyte in the source and therefore is dependent on the flow rate. In addition, the dependence of response on flow rate in MS is not linear, and higher flows beyond an optimal point may be in some cases detrimental to sensitivity. Other detection techniques, such as RI are also flow-dependent. In RI the dependence is related to baseline stability. The stability of a detector to changes in flow must be known before modifying the flow rate.

(11) Most modern detectors used in HPLC do not generate a continuous signal based on measured physicochemical property. The signal generated by the detector is typically obtained at short time intervals, and a measurement frequency (sometimes indicated as sampling frequency) is characteristic for each detector. It is possible in some instruments that such frequency can be set by the operator, but in other detectors this frequency is fixed. It is important for the measuring frequency of the signal to be high enough for obtaining an accurate representation of chromatographic peak. A sampling frequency of about five points per second could be sufficient for peaks with 4−5 s width (or larger). However, for narrow peaks, the sampling frequency must be higher, such as 20 or 50 Hz. The measurement frequency must be high in particular for narrow chromatographic peaks. The curve representing the peak shape is generated by connecting each measurement point, and sparse points do not account properly for this shape and introduce errors, in particular regarding peak areas.

Types of Detectors Used in HPLC and UPLC

The detection of the molecular species eluted from the chromatographic column can be done using a variety of principles and techniques, many of these techniques being also used in analytical instruments not hyphenated with chromatography (see Section 2.2). Due to the importance of detection in HPLC, some of the more common detection techniques are further discussed. Among these are the following: (1) spectrophotometric UV−Vis absorption, (2) fluorescence, (3) chemiluminescence, (4) refractive index (RI), (5) mass spectrometry, (6) various types of electrochemical detection, (7) evaporative light-scattering (ELS), and (8) other detection techniques.

1) Spectrophotometric detectors are basically UV−Vis spectrophotometers equipped with a flow-through cell. They respond to the instantaneous concentration of the solute passing through the detector. In these instruments, a beam of monochromatic light (more correctly

a beam of light with a narrow wavelength range) is sent through the eluent flowing through a cell of small volume (e.g., 1 μL for a microcell, 5 μL for a semimicro, and 14 μL for a standard cell, with path length of 5 mm–10 mm depending on the cell, some cells having a conical shape). The UV–Vis detectors may have a fixed wavelength, a variable wavelength capability with unique wavelength detection, or variable wavelength capability with multiple wavelength detection (such as the diode array detector [DAD]). The variable wavelength range is typically between 195 nm and 600 nm. Below 195 nm common solvents used as mobile phase have significant absorption, rendering the range unfit for utilization. The UV–Vis detector can be used for the measurement of a large number of compounds that have absorption bands in the range 200–600 nm. Because the absorption bands in solution are typically quite broad, the specificity of UV–Vis detectors is low. The monochromatic beam generated in the detector may pass only through the flow-cell or may have a beamsplitter, such that one part of the monochromatic beam goes through the cell with the sample and to the detector and the other to a reference detector. Optionally, in some instruments the beam directed to the reference detector may also pass through a reference cell. The baseline is obtained from the reading for the eluate when no solute is emerging from the column. Various models of spectrophotometric detectors are commercially available. A schematic of a detector with variable wavelength capability and optional reference cell for pure mobile phase is shown in Fig. 4.1.3.

Similar to a common spectrophotometric measurement, the instrument is capable of measuring the absorbance A_λ of the eluent, which is related to instantaneous concentration of the analyte (see Eq. 2.2.3). The quantitation can be done using calibration curves or a standard addition method (see Section 1.2). The absorbance of the liquid eluting from the separation system is typically measured at specific (small) time intervals (not continuously),

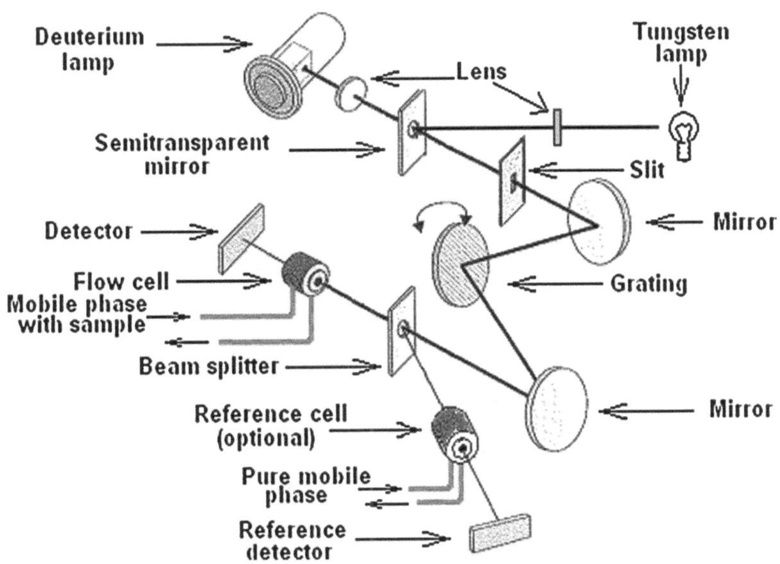

FIGURE 4.1.3 Schematic of a spectrophotometric HPLC detector with variable wavelength and reference cell for pure mobile phase.

generating points that form the chromatogram. The peaks in the chromatogram indicate an increase in the absorbance A_λ when an absorbing species elutes from the chromatographic column. For variable-wavelength detectors, the wavelength can be selected (by rotating the grating) and is usually set where a strong absorption of light by the analyte occurs (possibly at the maximum). This type of detector is one of the most commonly used in HPLC. Older detectors were made to measure the absorbance at a fixed wavelength such as 254 nm. This wavelength corresponds to the maximum emission (253.7 nm) of a mercury lamp that has been used as a UV source in simpler spectrophotometers. The area under the chromatographic peak of the analyte being proportional with the analyte amount injected into the column, it is used for quantitation with the help of a (linear) calibration curve, or a response factor between peak area and the analyte concentration. The peak height can also be used for quantitation assuming equal peak widths for all concentrations. If the entire UV–Vis absorption spectrum is measured in different points across a chromatographic peak (as it is done with the UV–Vis DADs), the maximum absorption can be selected for quantitation. Also, the absorption spectrum can be used for the evaluation of peak purity and can be a guide for qualitative identification of the analyte (although UV–Vis spectra are very seldom sufficient for compound identification). The UV region of the spectrum starts at about 190 nm, and modern UV–Vis instruments have a working range between 190 nm and 600 nm. However, the common range of practical utility in UV spectrophotometric measurements starts at about 205 nm or higher. At lower values than this wavelength, a strong light absorption usually takes place because the solvents used as mobile phase start absorbing. The wavelength cut-off of various solvents can be found in tables, e.g., Ref. [15]. Depending on the nature of the analyzed material, the detection limit of the UV–Vis detection in HPLC can be 0.1–1.0 ng, with a linear range of five orders of magnitude. With an appropriate solvent that does not absorb in the range of UV–Vis measurement, the use of elution gradient can be applied for separation.

2) Fluorescence detectors (FLDs) are another common type of detector in HPLC. Their use is however limited to the measurement of analytes that display fluorescence. Similar to UV–Vis detection in HPLC, the fluorescence is measured in a flow-through cell that is connected to the flow of the eluent incoming from the chromatographic column. Modern fluorescence detectors may have the capability of recording a tridimensional emission spectrum of the analyte by stopping the mobile phase flow at a chosen retention time (selected for the analyte) and performing a scan of the entire UV range used for excitation and for the entire emission band. In this way, the tridimensional fluorescence spectrum is displayed as a dependence of fluorescence intensity on emission wavelength and excitation wavelength. It is common in modern fluorescence HPLC detectors that the excitation beam is generated by a high-power lamp that flashes at a specific number of times per second (e.g., 296 times in an Agilent 1200 Ser. detector), such that the signal is modulated in time. Also, the systems typically have a reference detector that measures the excitation light and corrects the flash lamp fluctuations. The detection in fluorescence methods encounters several difficulties because of nonlinearity of fluorescence due to self-absorption effects, difficulty in discriminating between overlapping broad spectra of interfering molecules, quenching produced by oxygen dissolved with the solvent, etc. Because the intensity of fluorescence increases linearly with the intensity of the initial radiation, laser-induced fluorescence (LIF) detection is a successful technique applied in HPLC. For HPLC, lasers are a convenient excitation source because they have

intense light focused into a small volume, they are highly monochromatic, and the associated Raman light has a well-defined wavelength that can be avoided with the monochromator used for observing fluorescence. However, LIF is still affected by background interference commonly arising from the Raman effect in the blank (molecular scattering) or from low level of solid impurities in the solvent producing Rayleigh light scattering. Detectors with constant excitation wavelength and variable absorption or with variable wavelength excitation and absorption are commercially available. Depending on the nature of the analyzed material, the detection limit using FLD can be as sensitive as 10^{-2}–10^{-3} ng/mL, with a linear range of four orders of magnitude. When appropriately selected, the use of elution gradient can be applied for separation without interfering with the fluorescence. Different factors related to the mobile phase influence fluorescence such as pH, solvent nature, temperature (as much as 2%), presence of impurities, as well as the flow rate. Fluorescence detectors are not universal detectors, and their specificity is higher than that for UV–Vis detectors.

3) Chemiluminescence (CL) detectors are also used in HPLC when the analyte may be involved in emission of light as a result of a chemical reaction. The chemiluminescence intensity follows the same law as fluorescence with the difference that quantum yield Φ from fluorescence (see Eq. 2.2.4) must be replaced with a different quantum yield Φ_{CL}, which is defined as the proportion of analyte molecules that emit a photon during chemiluminescence. The value of Φ_{CL} increases with the efficiency of the chemical reaction producing the excitation (such as the oxidation process). Higher energies required by molecules to achieve the excited state diminish Φ_{CL}. In analytical uses of chemiluminescence, one more factor that must be taken into account is the time frame of the light emission. Certain chemiluminescent systems, although with very good Φ_{CL}, may emit the light for a period of 40–50 min. Much shorter times can be achieved using a catalyst. Because no excitation light is needed in chemiluminescence, the interfering light from Raman effect or light scattering by trace particles is nonexistent. In addition, the development of detectors virtually able to detect single photons makes the technique highly sensitive. Concentrations as low as a few hundred amol/mL of material were detected using chemiluminescence for certain analytes [16]. However, the luminescent molecules are not very common and usually the chemiluminescence is generated by postcolumn derivatization with proper reagents [17].

4) Refractive index detectors (RI or RID) are also frequently used in HPLC. Due to the modification of the refractive index of a solution as a function of the concentration of the solute, RI can be used for the quantitation of a variety of analytes. RI is a "universal" detector and it is not selective to specific compounds. The response in RI is generated from various concentrations of any compound with a refractive index different from that of the mobile phase and it is sensitive to temperature changes in the detection cell. For this reason, this type of detection can be applied without the need for chromophore groups, fluorescence-bearing groups, or other specific properties in the molecule of the analyte. A schematic diagram of an RI detector is shown in Fig. 4.1.4. The RI detector is based on the deviation of the direction of a light beam when passing under an angle from one medium to a medium with a different refractive index. This deviation depends on the difference in the refractive index between the two media. The change in the location of the beam on the (photoelectric) detector is made to generate a difference in the detector output. This output is electronically modified to provide a signal proportional to the concentration of the solute in the sample cell. The refractive index depends on the wavelength of the incident beam, and the most accurate RI

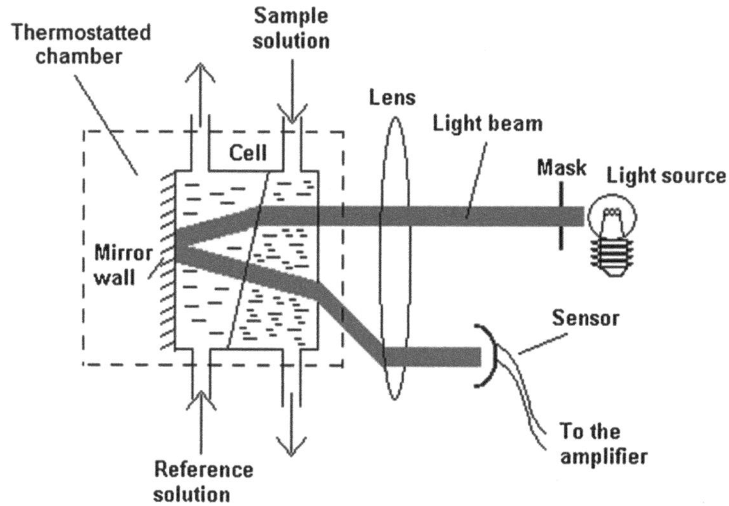

FIGURE 4.1.4 Schematic diagram of a refractive index detector (see, e.g., Ref. [5]).

measurements are done with monochromatic light (usually 589 nm, the sodium D line). With optical corrections, white light can still be used for the measurements. Refractive index is sensitive to temperature changes, and a constant temperature must be maintained during measurements. The response of the detector is given in arbitrary RI units, depending on the detector settings, but proportional with the concentration of the analyte.

In many cases the sensitivity of RI detection is not as good as that of other types of detection. Also, it is not possible to use elution with gradient, since this is associated with large variations in the refractive index of the mobile phase. Even the generation of an eluent with a given composition by using two solvents and achieving the desired composition using the pumps of the HPLC system is not recommended. A "waving" of the baseline is typically seen when the mobile phase is not obtained from a unique solvent. When a unique solvent is used as mobile phase, the HPLC pumping system may generate a slight pressure fluctuation, but not a change in composition, and therefore the baseline of the detector is more stable.

5) Mass spectrometric detectors combined with HPLC separations offer one of the best tools for chemical analysis [18,19]. This type of detector responds to the instantaneous amount of solute passing through the detector. LC-MS and LC-MS/MS can provide exceptional sensitivity and selectivity compared to other detection techniques. The capability to easily measure ng/mL levels of compounds in the sample, to differentiate between molecules with different mass and fragmentation patterns, as well as the potential identification capability make LC-MS and LC-MS/MS invaluable techniques. The good resolving power required for the separation in other detection techniques can become less necessary with MS detection when coeluting compounds can be differentiated by their MS signal (the ions selected for measurement can be different for different analytes). This is of particular interest when using deuterated internal standards for quantitation. The HPLC separation between the nonlabeled molecule and the labeled molecule is typically very small and the two peaks coelute. Without a detection that measures the mass of specific ions, the two peaks cannot be

distinguished. However, since the *m/z* of the two molecules are different, the separate detection is possible.

The progress in using MS detection in HPLC was made possible by the development of interfaces able to convert efficiently the dissolved analyte from the mobile phase of the LC into gas-phase ions. This is usually done by electrospray ionization (ESI) or atmospheric pressure chemical ionization (APCI) (see Section 2.3). In these techniques, the goal is to generate charged molecules of the analyte, while the molecules of the solvent remain neutral. The charged molecules are further attracted toward the ion mass analyzer entrance, while the solvent is as much as possible eliminated. The ionization process in LC-MS is mild compared to electron impact (EI) ionization typically used in GC–MS and is similar to CI ionization. In the ESI source, the effluent from the HPLC flows through a capillary and it is sprayed into the ion source (which is kept at atmospheric pressure). The capillary has a large voltage applied to it, such that the aerosol particles become charged. The spray is heated and diluted with an inert gas such as N_2. The charged droplets suffer a desolvation process, and only the ionized molecules are attracted toward the entrance of the mass analyzer while most of the inert molecules including the solvent are eliminated. A schematic drawing of an electrospray (ESI) source (based on an AB-Sciex source) is shown in Fig. 4.1.5 [5]. In the design indicated in Fig. 4.1.5 the direction of the spraying of the effluent from the HPLC is perpendicular to the direction of the ion path into the mass spectrometer in order to avoid neutral molecules entering the mass spectral source. Other source designs are commercially available for different instruments (e.g., a Z-type shape of ion path from the capillary to the mass spectrometers in instruments manufactured by Waters).

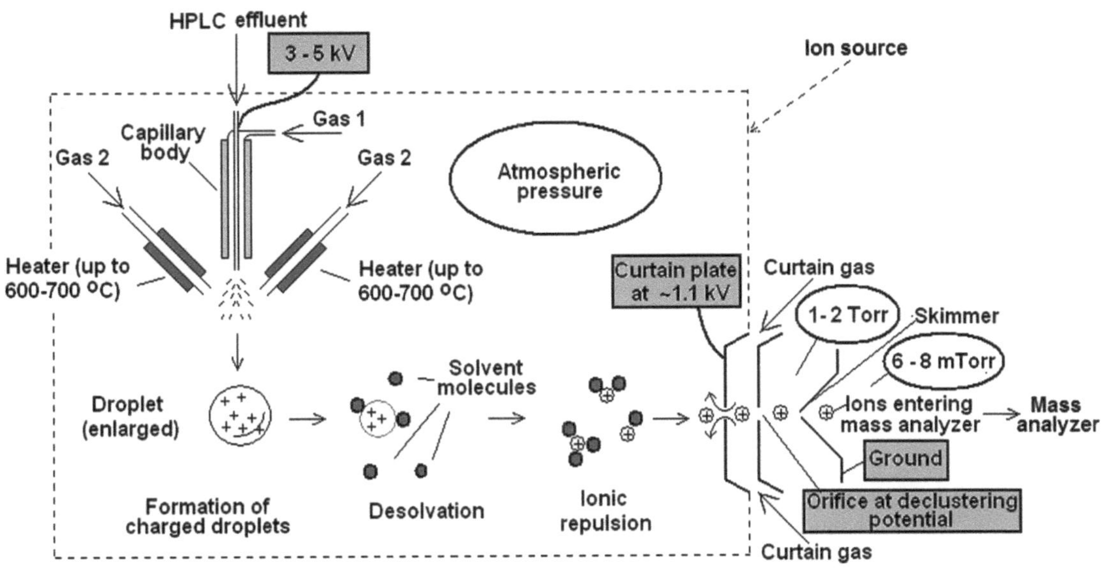

FIGURE 4.1.5 Schematic diagram of an electrospray ionization (ESI) inlet for an LC/MS instrument (positive ion mode). The decrease in electrical potential and in pressure are also indicated on the diagram (based on an AB-Sciex source).

In the ESI source shown in Fig. 4.1.5, the curtain gas diminishes the amount of remaining neutral molecules entering the mass analyzer, and the declustering potential allows the elimination of solvent molecules that form clusters with the analyte ions. A decrease in stages of the pressure from atmospheric to the low pressure in the mass spectrometer is assured by vacuum pumps. ESI ion sources with different constructions are used by different instrument manufacturers, but the basic construction of these sources remains the same.

In APCI, the effluent from the HPLC column is sent through a capillary heated and in a flow of gas, but not under an electrical potential. The jet of molecules of solvent and analyte in gas form and those of added gas (N_2, O_2) flow by a needle charged at a high voltage (3–5 kV) that generates a corona discharge and forms ions. Some of the molecules are loaded with positive charges (when working in positive mode). Due to the difference in polarity between the molecules of the analyte on one hand and those of solvent and gases on the other, the charges have the tendency to migrate to the analyte molecules. These charged molecules are attracted toward the curtain plate of the ion source and further into the ion mass analyzer, as previously described for an ESI source. A schematic drawing of an APCI source (based on an AB-Sciex source) is shown in Fig. 4.1.6.

The ionization in LC-MS can be conducted to form either positive ions or negative ions from the analyte neutral molecules. The choice between positive or negative ionization mode depends on the nature of the compounds to be analyzed. Molecules with a basic character (e.g., amines, heterocycles containing nitrogen, etc.) have a tendency to form positive

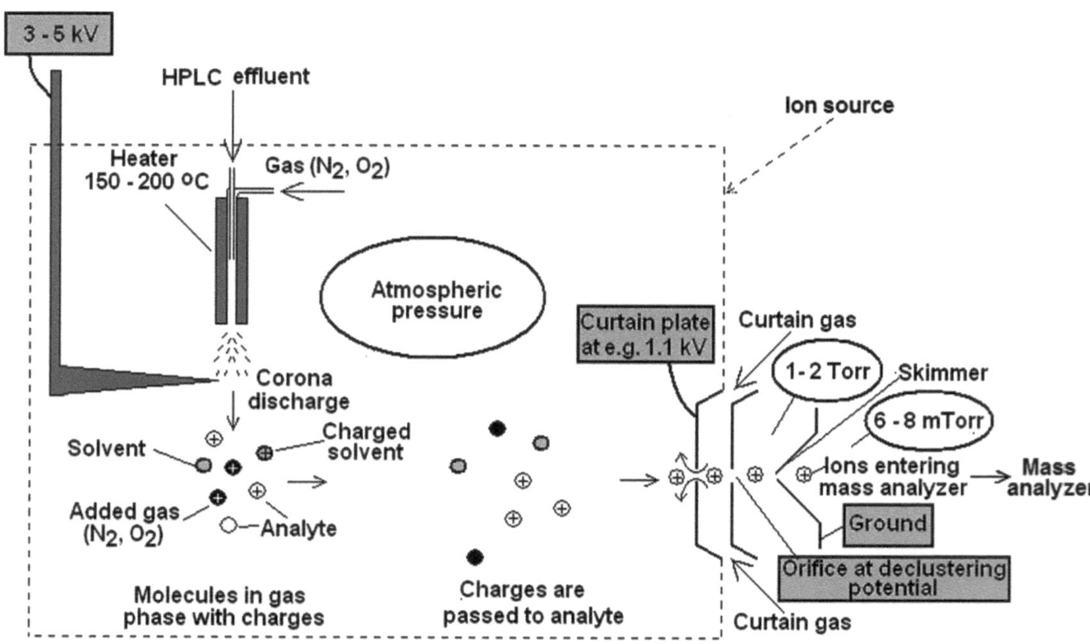

FIGURE 4.1.6 Schematic drawing of an atmospheric pressure chemical ionization (APCI) source inlet for an LC/MS instrument (positive ion mode). The decrease in electrical potential and in pressure are also indicated on the diagram (based on an AB-Sciex source).

ions and are typically analyzed in positive mode. Other molecules such as acids and some oxygenated or halogenated compounds are typically analyzed in negative mode. Both ESI and APCI ionization techniques offer very reproducible generation of ions, can be used with a wide range of solvents as HPLC eluent, can work in a range of flow rates (e.g., 0.05–1.0 mL/min), and do not involve problems with capillary plugging. However, ionization depends on the nature of mobile phase, and a change in the mobile phase composition may affect significantly the ionization yield (and therefore the detector response). In addition to composition, the sensitivity can also be affected by flow rate. Both ESI and APCI being evaporative-type detectors, the mobile phase flow and the heater temperatures (see Fig. 4.1.5) should be adjusted properly to assure maximum yield of ions.

The typical response $\mathcal{S}$ for the mass spectrometer detector is related to the instantaneous mass q of the analyte entering the detector by the expression (a and b are parameters depending on detector and detector settings):

$$\mathcal{S} = a + b\,q \tag{4.1.3}$$

In certain applications the dependence between the response $\mathcal{S}$ and quantity q becomes nonlinear, and several derived response functions, such as log–log are used in the calibration process [20]. Other procedures are used to form ions from the analyte molecule from the HPLC effluent (see Section 2.3). One of these is the use of an intense beam of UV light for the analyte ionization instead of a corona discharge. This technique is known as atmospheric pressure photoionization (APPI). Also, in an effort to generate for the analytes in the LC eluent mass spectra that are similar to EI+ mass spectra (searchable in standard mass spectral libraries such as NIST8), special instrumentation is being developed such as LC-MS with supersonic molecular beam and electron ionization capability [21–23].

Older techniques to interface an LC with a mass spectrometer include the following: (1) particle beam (PB), which consists of an aerosol generator from the LC flow (at 0.1–1 mL/min) followed by a desolvation chamber and a separator that directs the aerosols through a series of apertures separating the volatile compounds including the solvent from the solid aerosols, (2) continuous flow fast atomic bombardment (FAB), where the effluent is introduced directly into a vacuum region (with a flow rate of 5–10 μL/min) mixed with a matrix material such as glycerol, and the ionization is achieved using a beam of ions at 5–8 keV, and (3) thermospray.

The ions generated in the source of the LC-MS or LC-MS/MS instrument are further separated by mass and measured using a mass analyzer. Mass separation is usually achieved using either a quadrupole or an ion trap. The abundance of ions of a measured analyte is proportional with the analyte concentration, and the MS response can be calibrated for quantitative measurements. However, the absence of molecular fragment formation from the molecular ion in LC/MS limits the identification capability of this technique. The schematic drawing of an LC-MS system (based on a QDa analyzer from Waters) is shown in Fig. 4.1.7.

Very useful in the analysis of eluates in LC are the MS/MS analyzers. Several utilization alternatives are common for MS/MS analyzers, such as (1) product ion scan, when the whole range of ions generated by fragmentation of the precursor ions (parent ions) is analyzed, (2) precursor ion scan, when only one ion is selected for the detector by Q3, while Q1 is scanning

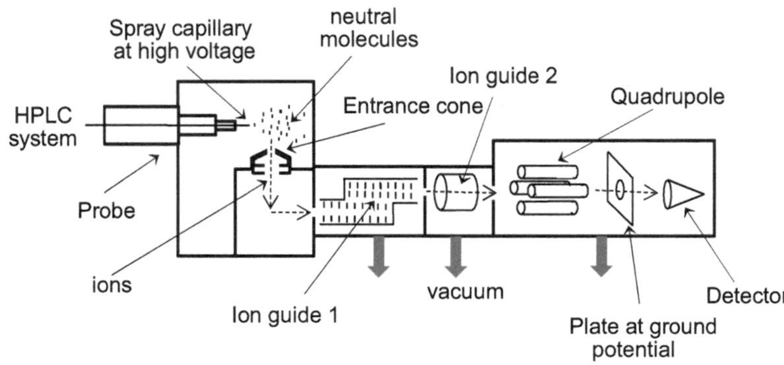

FIGURE 4.1.7 Schematic diagram of an LC-MS system (based on a QDa analyzer from Waters) with a Z-type shape of ion path from the capillary to the entrance of the mass analyzer.

the whole range of ions produced by the source, (3) neutral loss scan, when the instrument scans for a specific mass difference between the ions from Q1 and Q3, and (4) multiple reaction monitoring (or MRM), where a specific precursor ion is selected by Q1 and a specific fragment (product ion) is detected by Q3 (more than one pair of ions can be analyzed by MRM at the same time). Qualitative information from LC-MS/MS can be generated based on fragmentation of the parent ions. The fragmentation in LC-MS/MS is characteristic for each molecule. However, the nature of fragment ions in LC-MS/MS is typically used for the confirmation of a specific analyte, and less for compound identification as it is done in GC−MS. There are several reasons for this. The pattern of fragments obtained in LC-MS/MS is typically less complex that that obtained in GC−MS with EI+ ionization at 70 eV. Also, the fragmentation strongly depends on the conditions in the collision cell (collision gas nature and pressure, collision energy, collision cell construction). Since a universally standardized set of conditions for obtaining fragments in LC-MS/MS is not available, the fragmentation is not standard. For these reasons, large libraries with LC-MS/MS spectra to be used for any compound identification are not available. New successful developments regarding the use of LC-MS/MS for qualitative analysis are in progress for specific classes of compounds (e.g., peptides).

The use of LC-MS (and LC-MS/MS) as a highly selective and sensitive detector, has benefits but also it has some disadvantages, in particular related to unknown compound identification [24]. Also, when monitoring only specific analytes (e.g., in MRM mode), these can be either separated or not separated by the HPLC column, and some coeluting compounds may not be detected. While this can be a big advantage when the HPLC separation is difficult, some matrix interference may occur without being detected. For this reason, recoveries must be evaluated in LC-MS and LC-MS/MS in the presence of all possible matrix constituents. Other limitations are also known for LC-MS and LC-MS/MS detectors. One such limitation is related to the requirement that all the components in the mobile phase must be volatile. This imposes that the buffers that can be used in the mobile phase must be volatile, and typically formic acid or ammonium formate are used for both enhancing ionization ($HCOOH$ for positive, but also for negative ionization, and $HCOONH_4$ or CH_3COONH_4 mainly for negative ionization) as well as for obtaining a desired pH. The formation of fine solid particles in the ionization chamber of the MS leads to a decrease in ionization efficiency and also generates a very unstable signal in the MS instrument (high background noise). For this reason, salts such as KH_2PO_4 or K_2HPO_4 cannot be used to obtain a buffer solution to be used in LC-MS. Also, the presence of an organic phase in the mobile phase is almost always necessary. For this reason, totally aqueous mobile phase is less frequently used in LC-MS. Another problem with LC-MS using ESI or APCI sources is related to the problem of ion suppression when the amount of ions generated in the source is high. These high levels of ions can be caused by a (relatively) high level of analyte, which leads to a rather narrow dynamic range of the detector. Another source of high level of ions may result from unseparated components of the matrix that can be ionized, resulting in the reduction of the signal of the analyte.

The LC-MS/MS is more frequently used for highly selective and sensitive quantitative analysis on known analytes. Quantitative information using the MRM mode is characterized by exceptional sensitivity (as low as fmol/mL concentration in the analyzed solution). Other instrument developments are continuously made in the field of LC-MS/MS. Examples are coupling an ion trap with a collision cell and a quadrupole, coupling a triple quadrupole with a fourth ion analyzer such as an Orbitrap (see, e.g., Ref. [25]) that achieves high sensitivity and also high resolution (e.g., $M/\Delta M \approx 60{,}000$ for $m/z = 400$ or even up to 100,000), or using special techniques such as Fourier transform ion cyclotron resonance (FT-ICR-MS) capable of 1,000,000 resolution.

6) Electrochemical detectors of different types are also used in HPLC. Among these are amperometric, coulometric, potentiometric and conductometric detectors. The ones more commonly applied in HPLC are the amperometric and conductometric. These types of detectors are useful in particular for ion chromatography. Electrochemical detectors typically consist of a flow cell with the mobile phase passing by the detectors. A schematic diagram of a flow cell for an amperometric detector is shown in Fig. 4.1.8.

In the amperometric detector (see Section 2.4) the electrical current measured can be obtained, for example, by the oxidation of the analyte at the surface of an inert electrode (such as gold used for the analysis of carbohydrates). In a flow-through cell, the measured current intensity is related to the analyte concentration by expression (2.4.7). The products of the oxidation reaction may also "poison" the surface of the electrode, and this surface

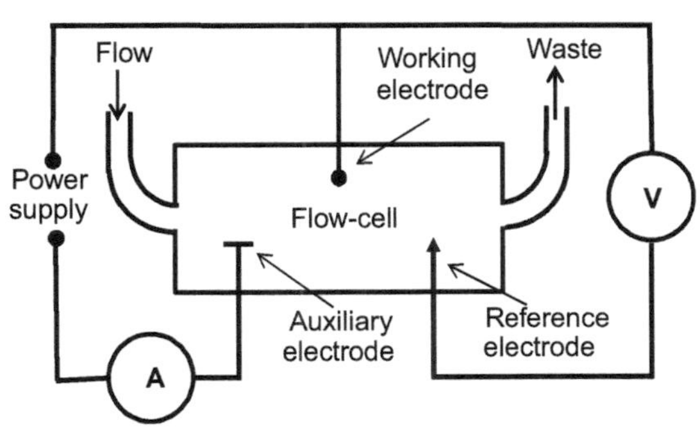

FIGURE 4.1.8 Schematic diagram of an electrochemical flow cell for amperometric measurements. The voltage (V) and the current intensity (A) are measured.

has to be cleaned between the measurements. This is accomplished by first raising the potential of the electrode to a level sufficient to completely oxidize the electrode surface, which causes desorption of the analyte oxidation products. The electrode potential is then lowered to reduce the electrode surface back to its initial metal form (pure gold). A sequence of three potentials is therefore used for *pulsed amperometric detection*. The three potentials are applied for fixed durations with the analyte oxidation current measured at a specific time in the cycle. The detection can achieve excellent signal-to-noise ratios and for carbohydrates, for example, it is possible to analyze as low as 10-picomole analytes.

Conductometric detectors are used for the measurement of ionic species that increase the conductivity of the mobile phase when they elute. The conductivity measurement is performed in a flow-through cell with two electrodes separated by a specific distance. Special cells are necessary when using capillary IC columns and low flow rates. The measurement of the analyte concentration can be based on expression (2.4.13), which connects the value of conductivity with the concentration of conducting species in solution. However, besides the analytes, the mobile phase may contain acids, bases, or salts (such as carbonates) that may interfere with the conductometric detection. For this reason, before reaching the detector it is common to use a suppressor, which can reduce the conductivity caused by the eluent components by virtually eliminating the ions belonging to the mobile phase and increasing the conductivity only due to the analytes. Two general types of chemical suppressors are commonly used for IC for conductometry measurements (electronic suppression is also used with some instruments). One type of chemical suppressor is based on the addition in the flow of an ion exchange cartridge (e.g., commercially available from Metrohm or Altech Co.) that eliminates the conductive ions of the mobile phase. The other type is based on semipermeable membranes (commercially available from Thermo Scientific/Dionex). For ion-exchange suppressors, special devices are commercially available that use a pair of ion-exchange cartridges placed alternately in the eluent flow. One cartridge is used for suppressing the eluent ions while the other is regenerated, such that no interruption in the operation is incurred. Various models of suppressors based on semipermeable membrane technology are also available (e.g., self-regenerating suppressor [SRS], micromembrane suppressor [MMS], capillary electrolytic suppressor [CES], or Atlas electrolytic suppressor [AES], which are commercially available from Thermo Scientific/Dionex [26]). In a separation where the eluent is, for example, a solution of NaOH or KOH, the suppressor, which can be a resin or a semipermeable membrane, provides immobilized (or not passing through the membrane) $R\text{-}SO_3H$ groups. In this case, the reaction of the NaOH from the eluent through the suppressor is described by the following scheme:

$$NaOH + R\text{-}SO_3H \rightarrow R\text{-}SO_3Na + H_2O \tag{4.1.4}$$

At the same time, an analyte in the form of a Na salt (e.g., a chloride) passing through the suppressor undergoes the following change:

$$NaX + R\text{-}SO_3H \rightarrow R\text{-}SO_3Na + HX \tag{4.1.5}$$

As a result of suppression, the conductivity caused by the NaOH in the mobile phase is eliminated with the compound being changed into H_2O, while the conductivity caused by the analyte NaX is not modified since NaX is changed into HX which is also dissociated and increases conductivity. Aqueous buffers containing $NaHCO_3$ and Na_2CO_3 are also frequently used as eluents in IC. The conductivity caused by such buffers is eliminated using a chemical suppressor by transforming the alkaline carbonates into H_2CO_3 that decomposes into H_2O and CO_2, which do not contribute to conductivity. Other anions (e.g. F^-, Cl^-, SO_4^{2-}) generate strong acids that are easily detected based on conductivity. Various other suppression techniques are used in practice [27], including electrolytic suppressors used in anion exchange LC.

In the case of a cation analysis, suppression of conductivity from the mobile phase can be achieved with a resin in OH^- form. For example, H_2SO_4 or methanesulfonic acid (MSA) can be used as an additive to the mobile phase and is retained by the anion exchange resin (in OH^- form) with the formation of H_2O in the solution. At the same time, a salt (e.g., of an alkaline ion such as Na^+ or K^+) passing through the resin can generate a strong base (NaOH or KOH) with high electric conductivity. Sulfuric acid or MSA are also used in the mobile phase for semipermeable membrane ion suppression. Semipermeable membrane for cation exchange separations may use a flow of tetrabutyl-ammonium hydroxide as a reagent on the other side of the membrane. The reagent interacts with the mobile phase acids. The OH^- ions of the reagent in the semipermeable membrane replace the SO_4^{2-} or $CH_3SO_3^-$ ions from the mobile phase. In this way, water is formed with the H^+ of H_2SO_4 or CH_3SO_3H. Other ions (e.g. Na^+, K^+, etc.) generate hydroxides that have high conductivity (NaOH or KOH are highly dissociated). Conductometric detection does not have selectivity, any ion present in the measuring flow-cell of the detector generating a signal. For this reason, the separation by the HPLC should assure the elimination of interferences [28].

7) An evaporative light-scattering detector (ELS or ELSD) is another type of detector used in HPLC, in particular for compounds that do not have good light absorbance in UV, are not fluorescent, and may be difficult to ionize (see, e.g., Refs. [29,30]). ELSD uses the formation of particles that do not evaporate and can scatter light, while the mobile phase forms a gas by evaporation. In this technique, the eluent is injected in the form of a spray from a nebulizer into a drift tube where a nebulizer gas is also introduced. The drift tube is heated and the solvent is evaporated while nonvolatile molecules form a fine mist. This mist passes through a cell that is illuminated with a beam of light and the scattered light from the mist is recorded. The gas generated from the solvent does not influence light scattering. A schematic diagram of an ELSD is shown in Fig. 4.1.9.

The intensity of the scattered light is dependent on the analyte concentration (within a certain range of concentrations since the linearity is not followed for a wide range). This detector has the advantage over the RI of being usable with gradient elution. However, the presence of any salts or nonvolatile materials in the mobile phase disturbs the measurements. Also, changes in the temperature of the eluent do not affect ELSD, while with RI detectors careful temperature control must be applied since the refractive index varies with temperature. ELSD can be more sensitive than RI in specific applications. Modifications of ELSD were attempted to further improve its sensitivity, such as by adding a saturated stream of solvent to the mist of the analyte in order to grow the particles by nucleation and detect them better (the technique is known as CNLSD) or to use lasers as light source (LLSD). Light scattering detectors do not have a linear response to analyte concentration for all analytes, in particular when the concentration is relatively high when changes can be seen in the particle size generated by the detector. For this reason, in some analyses the calibration is performed

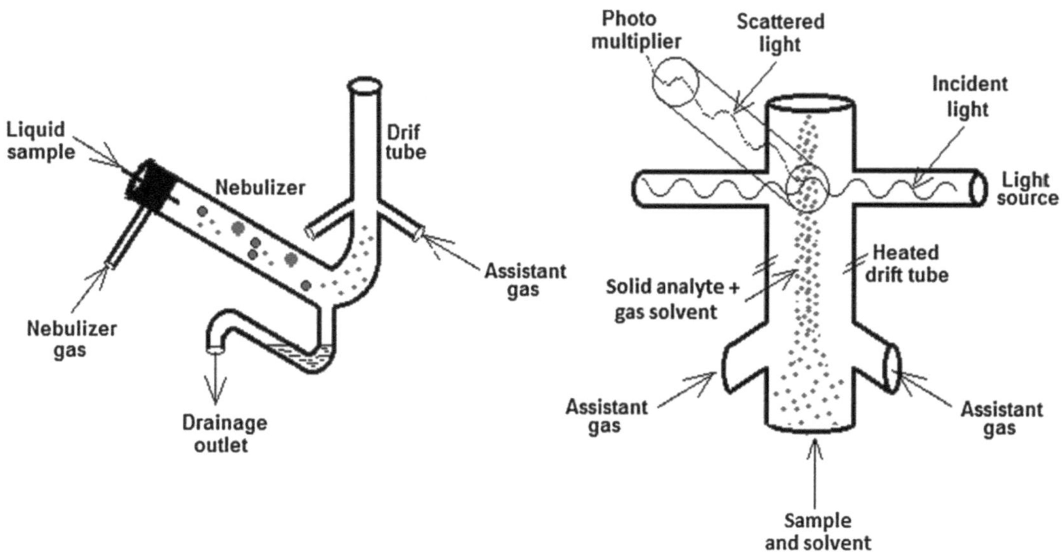

FIGURE 4.1.9 Schematic diagram of an ELSD setup.

using a log/log dependence. Using the notations C for the analyte concentration and y the peak area in the chromatogram, this dependence can be written as follows [31]:

$$\log C = a + b \log y \tag{4.1.6}$$

An alternative to light-scattering detection is a corona charged aerosol detector (CAD or cCAD) [32]. This detector is based on nebulization of column effluent (e.g., with N_2) and drying of resulting droplets to remove the mobile phase components, producing analyte particles. A secondary stream of N_2 is made positively charged by passing it by a high-voltage platinum corona wire, which is sent in the stream of analyte particles to generate charged aerosol particles. The charged particles obtained by the interaction of the aerosols with the charged nitrogen gas are detected, generating an electric signal with the intensity dependent on the amount of analyte eluted from the column. The typical response of a CAD detector follows the formula:

$$\mathcal{S} = a\, q^b \tag{4.1.7}$$

where $\mathcal{S}$ is the detector signal, a and b are related to the response of the detector and q is the instantaneous mass of the analyte entering the detector. For a constant flow of the mobile phase, the value for q is dependent on concentration C such that formula (4.1.7) is in fact equivalent with formula (4.1.6) (with other values for a and b). The CAD is more sensitive than ELSD and has a wider dynamic range [33]. Similarly to ELSD, one of the drawbacks of CAD is its inability to accurately quantify volatile compounds. Because CAD can measure only analytes that form particles, it cannot detect compounds that are volatile or do not readily form particles. In this situation, detection can be achieved by other specific detectors, such as chemiluminescent nitrogen detector [34]. CAD detection is reported to be possible in a wide range of concentrations.

8) Other types of detectors can be used in HPLC, but they are less common than those previously discussed. For example, in the case that the target compounds contain at least one nitrogen atom, the chemiluminescent nitrogen detector (CLND) can be employed. The principle of this detector relies on the combustion of the column effluent in a high-temperature furnace that converts the N-containing compounds into NO. The dried gas stream is passed into a chamber where it reacts with O_3, a reaction that is associated with chemiluminescence (measured by a photomultiplier). This detector has a high sensitivity but is not compatible with acetonitrile in the mobile phase [35]. Other known detection techniques include Fourier-transform infrared spectrometry (FTIR) [36,37], nuclear magnetic resonance (NMR) [38], inductively coupled plasma-mass spectrometry (ICP-MS) [39], circular dichroism (CD), optical rotatory dispersion, polarimetry, radioactivity, plasmon resonance [40], and multiangle light scattering applied directly to solution of polymers used, e.g., in connection with size-exclusion chromatography [41,42].

A promising detection system is the combination of an ion-mobility spectrometer (see Section 3.1) with a mass spectrometer, known as IMS-MS detection (see, e.g., Ref. [43]). Commercial instruments combining HPLC with IMS-MS are available from various vendors (see, e.g., Ref. [44]). This type of detection for HPLC allows a tridimensional type of differentiation of analytes (chromatographic, by ion mobility, and by mass spectrometry) and allows for each compound separated by HPLC discrimination based on molecular size, shape, charge, as well as mass.

It is common in modern HPLC systems that they have more than one detector available. When more than one detector is used, the detectors are typically connected in series. For example, UV−Vis and fluorescence detectors are frequently coupled in series, although not necessarily used simultaneously. Since the flow through a detector and through the connecting tubing may pose some backpressure, when using detectors coupled in series, it should be verified that the flow-cell of the detector upstream can handle the backpressure generated by the second detector. It is common that UV−Vis detectors stand higher pressure than fluorescence detectors, which stand slightly more pressure than RI detectors. Typical series of detectors, when more than one detector is used, is UV−Vis, fluorescence, RI. The connecting tubing between the detectors also should be verified for generating some backpressure. For avoiding any problems with additional backpressure from the tubing, it is typically recommended to use short tubing and possibly with 0.25 mm i.d. (0.010 in). Attention should also be given to potential blockage of the tubing at the connections, which may create overpressure in the cell of the upstream detector. For the detector downstream it should be verified that undesirable peak broadening does not occur because of the upstream detector and the connecting tubing. The coupling of detectors in parallel is also possible, but care must be taken to assure that appropriate flow goes through all the detectors. Since different detectors may pose different backpressures to the flow, the risk exists that most (or even all) of the flow goes to a single detector.

Selection of a Detector in HPLC

The selection of a specific detector is a very important step for an HPLC analysis, and for this reason several factors must be considered. These include the following: (1) availability of instrumentation, (2) the purpose of analysis, (3) the chemical nature of the analytes and of the matrix, (4) detector performance including its sensitivity, (5) the required quality of the results, (6) type of elution in the selected method, (7) the properties of the mobile phase in the selected method, (8) detector reliability, and (9) special characteristics of the detector. The choice of a specific detector is not always possible because of availability limitations. For this reason, it is necessary in some instances to change the analytical method in order to use the detector that is available. There are also cases when a recommended method can be modified to use a better detector in order to increase sensitivity or selectivity. Also, since more than one detector can be present in an HPLC system, the detection can be done in more than one way, which may offer some advantages. Several criteria for detector selection are further presented.

1) The availability of instrumentation is a straightforward requirement. When several instrumental setups are available, besides other factors, the instrumental cost, cost per analysis, cost of instrument maintenance, and expertise available in the laboratory in using the specific detector should be considered.

2) The purpose of analysis is essential in the instrument selection in general. This influences the selection of pumps, injectors, etc., but one important part affected by the purpose of analysis is the selection of the detector. All detectors are designed to allow quantitative measurements, but some are not designed for providing any qualitative information, such as the RI detectors. Other detectors provide some information related to the qualitative nature of the analytes, but this information is not diagnostic or sufficient for providing positive identification of the analyte. In this group can be included the UV detectors (in particular those that can generate an absorption spectrum [e.g., DAD detectors]). Mass spectrometric detectors can provide information that can be used for compound identification, but even these detectors have limitations. When only a quantitative analysis is necessary and the nature of the analyte is known, the peak identification for the known compound can be done based on the retention time alone (obtained from running standards), the only concern being an efficient chromatographic separation with no interference. Interferences may be detected using the comparison of UV spectra of the peak of pure compound with that of the corresponding peak from the sample. However, mass spectrometric detection (MS or even more reliable MS/MS) is capable of selecting only the measurement for the analyte. Setting the MS detection for a specific mass (usually molecular ion) is an important procedure for selecting only the desired compound for measurement. For MS/MS detection, in particular when this is performed in MRM mode, the precursor ion and the product ion can be selected such that the possibilities for interferences are much reduced.

For qualitative analysis of a sample with unknown composition, even HPLC in tandem with MS is not always able to provide decisive information in structural identification. The mass spectra in LC, when using a single mass spectrometer as a detector, do not offer sufficient information about the fragments of an analyzed molecule, typically indicating only the molecular ion. Better information is obtained with the use of MS/MS instruments, but the fragmentation provided can vary considerably depending on the instrument and acquisition method. Although some library searches are available (e.g., SmileMS) as well as some mass spectral libraries for LC, the qualitative information on unknown compounds is not always conclusive. As an alternative to searching based on molecular fragmentation, the nature of an unknown compound can also be obtained from its precise molecular mass. Instruments generating high-resolution mass spectra (see Section 2.3) are often equipped with search programs (e.g., MassFrontier) that help in identifying the molecular composition and potential structures. For qualitative analysis of samples containing a mixture of compounds with a suspected structure that requires only confirmation, techniques such as mass spectrometry (MS or MS/MS) must be used, working in specific modes (e.g., MRM).

3) The chemical nature of the analytes and of the matrix of the sample are important factors in the selection of the method of analysis. All these are included in method selection, which can be previously reported in the literature or planned to be utilized for the analysis. Various aspects of this selection are further discussed in this book. The method selection includes the sample preparation before analysis, the separation, and also the detection technique. The physical and chemical characteristics of the analytes, as well as their difference from other sample components and from the mobile phase, must be carefully evaluated in order to select the property best used for detection. For example, the presence or absence of chromophore groups typically determines the use of UV/visible absorption detectors (fixed wavelength, variable wavelength, or DAD), which are among the most common detectors used in HPLC. For compounds that can fluoresce, it is common to select this property for detection. In some cases, e.g., in the analysis of carbohydrates, the absorption in UV is very low (except for very low wavelengths) and the compounds are not fluorescent. In such cases other properties such as change in refractive index of eluting solution, electrochemical properties, or formation of ions capable to be analyzed by mass spectroscopy can be selected. Also, techniques that do not require a specific molecular structure such as evaporative light scattering can be used. The derivatization is frequently applied during sample preparation for modifying the initial analyte properties such that it can be amenable for a specific detection. The detection technique is always described for an analytical method when this is reported in the literature.

4) Based on the properties of the analytes, a detector with specific capabilities must be selected, such that it is sensitive to the specific analyte property that allows measurement. Some detectors respond to most analytes and are indicated as universal detectors [45]. Other detectors are specific for a class of compounds, and some have relatively limited applicability (e.g., radioactivity detectors, or circular dichroism [CD] detectors). Refractive index detectors are typically universal detectors. Other detectors used for compounds that do not contain chromophores or fluorophores are electrochemical (e.g., amperometric), ELSD, CAD or cCAD, etc. MS detectors are also universal, but their sensitivity is also dependent on the nature of the analyte.

The selection of the settings of the detectors is also important in attaining a required sensitivity, and the settings depend on the nature of the analyte and its concentration in the sample. This may include the choice of wavelength of absorption for UV, the choice of excitation and emission wavelength for FLD, or the choice of several parameter settings and masses to be monitored for MS.

Detector sensitivity is another factor important for selection. This sensitivity selection should be related to the concentration of the analyte in the analyzed sample. The detectors used in LC can have very different sensitivities, which depend on the type of detector and can be different toward different analytes depending on the analyte nature. For example, the sensitivity of RI detectors is typically much lower than that of fluorescence detectors for a fluorescent compound, but in the absence of fluorescence, the RI detector can be utilized while the FLD is useless. Detector settings, model, and manufacturer play an important role regarding the detector sensitivity.

5) The required quality of the results is a factor determining the needed precision and accuracy a detector is capable of attaining. This needed precision and accuracy is also dependent on the nature of the analyte, the level of the analyte in the sample, and on the methods used for sample preparation. Depending on the purpose of analysis, the selection of a detector or of a detector setting must be done in such a way to cover the analysis needs and be capable of achieving a required LOQ. The sensitivity of various types of detectors can vary significantly. When a method has been already selected, e.g., from the literature, the choice of the detector is indicated in the method. However, it is common that better sensitivity is necessary when adopting a method from the literature. In such a case, the selection of a different detector can be attempted. Also, when a new method of analysis is developed, the detector should be selected based on detector capability.

6) The type of elution also plays a key role in the selection of a detector, and in many cases the choice is made the other way around, the available detector determining the choice of isocratic or gradient elution. Several techniques are not applicable for gradient elution or do not respond very well to the change of the composition of mobile phase. Refractive index detector, for example, must be used only with isocratic elution. Other detectors, such as ELSD or CAD, can be used with gradient elution, but the solvents must be volatile (should not contain any nonvolatile additives) and the change in solvent composition may also generate some drift in the baseline. Even MS and MS/MS detectors may show differences in sensitivity at one solvent composition or the other, and the choice of gradient vs. isocratic elution is sometimes influenced by this difference. Detectors such as UV can be used with gradient separation without any problem (as long as the mobile phase does not have absorption by itself).

7) The properties of the mobile phase also contribute to the decision regarding the choice of a specific detector. Mobile phase composition plays a crucial role in most separations, and the subject of mobile phase selection is presented in detail in Chapters 13 and 14. A specific mobile phase may determine the type of detector that can be used, and in some instances, the method is developed particularly to be used with a specific detector.

Physical and chemical properties of the mobile phase must be considered in relation to the requirements of a specific detector. For example, when detection is done in UV, the cutoff wavelength of the solvent must be considered, such that the solvent is "transparent" at the measuring wavelength. For the detection using ELSD or CAD, the mobile phase should be totally volatile, and in the case of the need for salt buffers in the mobile phase, these detectors are not usable. In LC/MS (and LC-MS/MS) the mobile phase composition influences the detection, and the presence of nonvolatile salts in the mobile phase is not recommended. Since LC/MS and LC-MS/MS detection may offer particular benefits to the analysis (very good sensitivity, qualitative information) it is common that the selection of the mobile phase is done such that it is suitable for the MS detector, and not the other way around (choose the detector to accommodate the mobile phase). Also, the sensitivity of the detection may be drastically influenced by the mobile phase composition in MS and MS/MS. When a method has been developed for another type of detection (e.g., using nonvolatile buffers) and must be applied with MS detection, for example for enhancing sensitivity or for obtaining some qualitative information on the analytes, it is common to modify the mobile phase to be suitable for the new detection technique.

8) Detector reliability is another factor to be considered in its selection. Some detectors such as those based on UV absorption are typically extremely stable and reliable. Other detectors may be more prone to problems, which also may depend on instrument age, manufacturer, environmental conditions, etc. For example, electrochemical detectors may show some drift or unreproducible results depending on the cleanliness status of the measuring electrode. In some instances, a choice between sensitivity and stability of the detector must be made. For this reason, other characteristics also must be considered in detector selection, such as stability of the signal, frequency of the measurement per unit time (that assures an accurate evaluation of the peak shape), resistance to acids and bases in the mobile phase, capability to be used in series with other detectors, and acceptable backpressure in the case of connection in series with another detector.

9) Some detectors require special maintenance and more effort for establishing the proper operating conditions, such as the MS or MS/MS detectors. Other detectors are very simple to operate and require virtually no adjustments. The advantages of such detectors must be weighed vs. disadvantages, such as loss of sensitivity or lack of qualitative information.

Because of the importance of the detector selection in an HPLC method, the subject will continue to be discussed in this book.

Other Devices That Can Be Part of the HPLC System

Various analysis needs may require an increase in the complexity of the HPLC instrumentation. For example, when a more complex separation is needed, this may be achieved using a bidimensional separation. In such cases, switching devices are used to divert, for a specific time interval, the eluent from the first column into the second column. The volume of diverted eluent from the first column containing the unseparated analytes of interest is indicated as a "heart-cut." After the collection of the heart-cut, this is submitted into the second column for further separation. The choice of the columns for multidimensional separation is typically done such that they have very different selectivity. The HPLC separations performed on different types of column and using different solvents that lead to a different separation are indicated as *orthogonal*.

Other auxiliary devices may include systems for column switching. Such a system can be used, for example, for analyte concentration at the head of a first column in a given mobile phase. This is followed by the change of the flow direction and of mobile phase composition, which elutes the analytes and sends them to a second column where the separation takes place.

Another device sometimes used in HPLC is a flow splitter. This device is necessary when the flow from a conventional HPLC pump is too high to be used either with a capillary LC column or with a specific detector (such as a mass spectrometer). A flow setting below a specific limit at a conventional HPLC pump may lead to undesirable fluctuations or to a difficult to control gradient program. The flow splitter allows the selection of a desired fraction from the total flow from the pump by using a waste outlet with controllable backpressure.

Some HPLC techniques require the addition to the eluent flow from the chromatographic column a reagent or an additional solvent. This can be done by including a "mixing T" to the tubing connecting the column and the detector and connecting the mixing T, besides the detector, to an additional pump. The flow from this additional pump may contain, for example, a derivatization reagent that requires time for interacting with the analytes. In such cases, a reactor-type device that allows heating of the effluent may be necessary. Simply, a longer tubing may be sufficient, which allows a longer time of contact between the analytes and reagent before reaching the detector. These types of experimental setups are described in various special methods of analysis [46].

After some HPLC separations, some fractions of interest may need to be collected as the effluent exits the detectors. This collection can be done using a fraction collector. Automated systems for fraction collection are available. They direct the flow emerging from the last detector to specified vials either at a given time or upon receiving a signal from the detector (indicating an eluting peak).

More Complex HPLC Setups

More complex HPLC setups can be designed for various purposes that are not limited to the basic system made from: (1) solvent supply, (2) pump, (3) injector, (4) column, and (5) detector. Among such setups that require two different chromatographic columns, each one with its own solvent supply and pumping system, include those that allow the separation of a heart-cut (segment of time during the chromatographic run) from the eluate of the first column to be sent for further separation to a second column. This type of setup is usually indicated as *bidimensional* (or two-dimensional), and the columns used in such systems are selected to be very different and perform different types of separation being indicated as *orthogonal* [47,48].

A similar setup is used for comprehensive two-dimensional separations where heart-cuts are taken continuously for specific interval of times. The schematics of the flow for a comprehensive HPLC with cuts of 100 μL is shown in Fig. 4.1.10. With a flow in the first column of 0.1 mL/min, the loop is filled in 1 min. Every 1 min, the switch valve changes the flow to the other loop. When one loop is being filled, the content of the other loop is being analyzed by

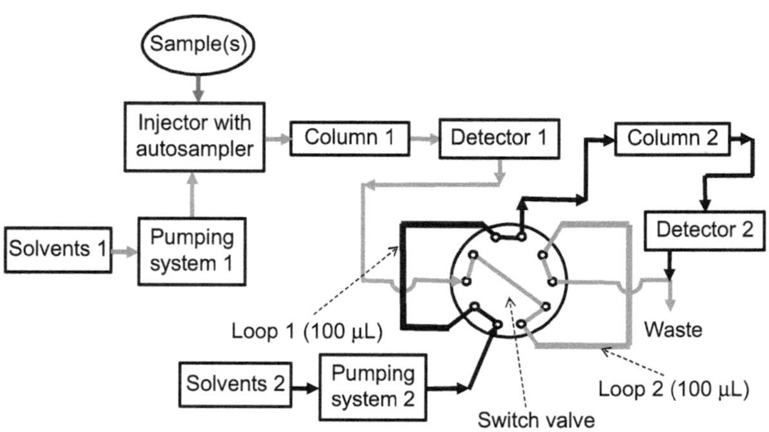

FIGURE 4.1.10 Diagram of a comprehensive two-dimensional HPLC system (the valve is switched at regular time intervals).

sending it through the second column. The separation in the second column should have a short run time (of 1 min) obtained, for example, using a faster flow (e.g. 0.5 mL/min) and short retention times. Such a system can be used, for example, for the analysis of protein hydrolysates. A computer program may be used for "putting together" the whole chromatogram when the loop switch occurs in the middle of one peak.

Other systems are designed, for example, for sample enrichment in analytes. In this use, the flow with the sample is sent to a column where the analytes are strongly retained for the specific solvent composition and, possibly, the matrix components are less retained. The flow to this column may carry a large volume of a diluted sample. The analytes are accumulated usually at the head of the column and the flow is sent to waste. After enough analytes are accumulated, the flow in the column is reversed (with a switching valve), and the solvent is changed such that the analytes are eluted from the head of the first column and sent to a different column where the separation takes place. The flow from this second column is sent to the detectors.

Instrument Control and Data Processing Unit

Part of modern HPLC (UPLC) instrumentation is a computer with dedicated programs for instrument control and data processing. Instrument manufacturers developed sophisticated such programs that are necessary for the control of HPLC pumps, the autosamplers, the temperatures of column compartment, and the parameters for the detectors. Also, the programs include capabilities for data processing, peak detection, peak integration, and other data manipulation such as data storing and recovery, peak purity verification, data smoothing, calibrations using linear or nonlinear curves, etc. Some programs also incorporate libraries of spectra that allow qualitative identification of the compound from each peak. Dedicated literature describes in detail the capabilities of such programs (e.g., instrument manuals).

Selection of the HPLC System and Transition From HPLC to UPLC

The selection of an HPLC system is frequently limited by the instrument availability. The main differences between instruments can be considered the type of pumping (low-pressure mixing or high-pressure mixing), and the type of detector. When possible, the selection of the instrument should be done based on the requirements of the analysis (as previously discussed for pumping system selection and detector selection). The instruments can be identified as HPLC and UPLC, although the UPLC instruments can be used in UPLC mode as well as HPLC mode. The UPLC instruments are characterized by the capability of pumps to deliver higher pressures, a higher precision of the injection system that can inject correctly as low as 0.1 μL sample, and detectors with small dead volumes (e.g., small volume cells for UV absorption measurements). Most UPLC systems can be used as simple HPLC without taking advantage of certain UPLC features. For this reason, the transition from HPLC to UPLC on an instrument capable to be used as UPLC can be a gradual process. The key element that decides the separation of UPLC use from HPLC use is the chromatographic column. The columns with small and very small particles (e.g., 1.6–1.8 μm diameter) require higher pressures from the pumps. For this reason the use of columns with small particles requires a UPLC instrument. Such columns provide a better resolution but sometimes do not have large loading capability. This implies that small injection volumes are required and they should be very precise. For small injection volumes the detectors must be sensitive, should involve small dead volumes, and should have high sampling frequency. The increased use of mass spectrometry detectors usually involves lower flow rates (0.1–0.6 mL/min) and the precision of gradient composition of the mobile phase must be precise. In addition to the previously listed features, attention must be given in UPLC to the use of small i.d. tubing (between 0.065 mm i.d. [0.0025 in] and 0.12 mm i.d. [0.005 in]) and correct fittings that do not produce void volumes.

The transition from HPLC use to UPLC use generally provides better sensitivity and shorter run times for the separations. Depending on the detector and the nature of the analytes, the use of UPLC can routinely allow the measurement of analyte concentrations as low as a few pg/mL, and produce chromatograms with run times of 2–3 min.

4.2 PARAMETERS DESCRIBING THE CHROMATOGRAPHIC PROCESS

The chromatogram is the visual output of the electrical signal from the chromatographic detector. This signal $y(t)$ is dependent on the instantaneous concentration $C(t)$ (for some detector amount $m(t)$) of the analyte that is flowing through the detector. The elution of the analytes from the chromatographic column is represented as chromatographic peaks. Fig. 4.2.1 illustrates a model of a chromatogram.

A number of parameters are used in HPLC for characterizing the separation, and the peak elution time and shape are important for this characterization. Some of the parameters of interest in a chromatographic separation are discussed in this section.

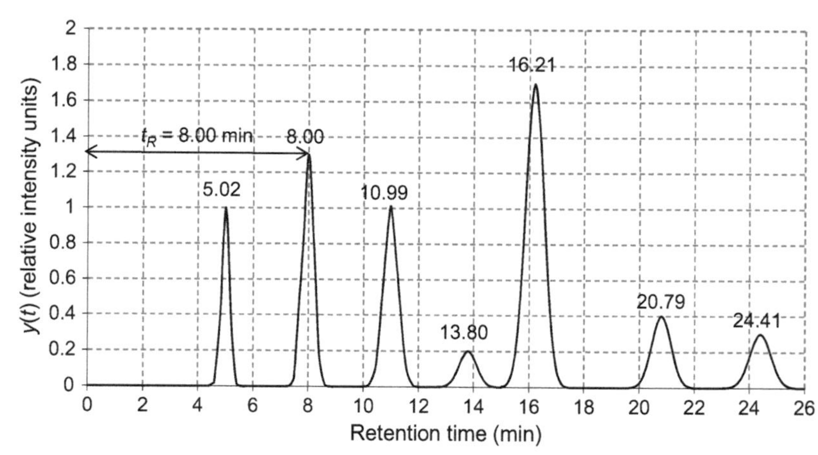

FIGURE 4.2.1 Model of a chromatogram showing several peaks and their retention times.

Flow Rate of the Mobile Phase

The flow of the mobile phase in the chromatographic column can be described by the *volumetric flow rate* (U) and by the *linear flow rate* (u). The volumetric flow rate, U, is the volume of fluid that flows per unit time (expressed, e.g., in mL/min) through the chromatographic column. This parameter is set in the HPLC instrument by the user (see Section 3.1).

The linear flow rate, u, is the velocity of a point in the fluid passing through the column (expressed as length per time), and can be considered the velocity of an unretained molecule in the chromatographic column. The linear flow rate is related to the volumetric flow rate by an expression of the form:

$$u = \frac{U}{A_c} \tag{4.2.1}$$

where A_c is the area of the channel in which the flow takes place. It should be noticed that A_c is not the surface area of the circular cross-section of the empty column since the column is filled with the stationary phas.e For a column with the inner diameter d, surface area of the empty column is $(\pi/4)d^2$ and $A_c = \varepsilon^*(\pi/4)d^2$, where ε^* is a constant depending on column porosity.

Retention Time

The peak *retention time* t_R (X) is the time (usually measured in min) from the injection of the sample into the chromatographic system to the time of elution of the compound X. In Fig. 4.2.1 the peak retention times are indicated above each peak. The time is taken at the maximum (the apex) of the chromatographic peak. The retention time $t'_R(X)$ is an important chromatographic characteristic for a molecular species X, since it has a constant value as long as the other parameters of the chromatographic separation are kept constant. In this way after establishing the retention time for a specific compound, for example using a standard, the retention time can be used for the compound identification (assuming no changes in the chromatographic conditions and no interference from other compounds).

The retention time $t'_R(X)$ can be separated into two components: the time analyte X spends in the mobile phase moving through the column known as *dead time* or *void time* t_0, and the time the analyte is retained on the stationary phase $t'_R(X)$ known as *reduced retention time*. In this way, the retention time is given by the expression:

$$t_R(X) = t'_R(X) + t_0 \tag{4.2.2}$$

Retention time in a chromatogram is determined by the properties of the compound X, the retention capability of the column, the nature of the mobile phase, but also by the flow rate of the mobile phase. The dead time t_0 is not dependent on the compound X. It depends on column construction and on the flow rate of the mobile phase. This is expressed by the formula:

$$t_0 = \frac{L}{u} \tag{4.2.3}$$

where L is the length of the chromatographic column (a very small fraction from the value of t_0 represents the time taken for the sample to flow through the tubing from the injector to the column and to the detector).

The value for t_0 can be obtained experimentally. For example, an approximation of t_0 can be obtained by measuring the time for the elution of a compound that is virtually not retained (very slightly retained, since it can be difficult to find a compound that is not retained at all on a chromatographic column). For example, the solvent used for injecting the sample (when it is different from the mobile phase) can be such a compound, and the retention time of this solvent peak can be taken as dead time. Similarly the retention time of a deuterated solvent analogous to a mobile phase component can be used for measuring t_0. Several other procedures for the estimation of t_0 are known [49,50]. One procedure uses the minor disturbances in the background signal created by the sample injection. For RP-HPLC, the use of uracil or of inorganic salts that are assumed not to be retained on a hydrophobic column is also a common procedure for t_0 estimation. Another more elaborate procedure involves the use of a plot of retention times for a homologous series of compounds that are retained less and less as the number of carbon atoms is decreasing and extrapolated to zero.

Run Time

The time for the whole chromatographic separation is indicated as the *run time*. The total time necessary for completing a chromatographic separation is slightly longer than the retention time of the last peak in the chromatogram. This time is sometimes referred to as the total run time, or length of the chromatogram. In practice, when multiple samples are analyzed, the total run time is an important parameter since its value is related to the number of samples analyzed within a certain length of time.

Retention Volume

For a specific molecular species X the *retention volume* $V_R(X)$ is defined as the volume of mobile phase flowing from the time of injection until the corresponding retention time $t_R(X)$ of a molecular species. The values for V_R and t_R are related by the simple formula:

$$V_R(X) = U\, t_R(X) \tag{4.2.4}$$

The retention volume corresponding to the dead time t_0 is known as dead volume V_0 or void volume. This volume corresponds to the volume of liquid in the column (and in the transfer lines from the injector to the column and from the column to the detector). The chromatographic column has a "volume not occupied by the stationary phase," which is the space between the stationary-phase particles and inside their pores. The not-retained molecule has to travel through the tubing from the injector to the column (which is very small), through the volume not occupied by the stationary phase, and through the tubing from the column to the detector (also very small), which accounts for the t_0 and for the dead volume V_0. Corresponding to the reduced retention time t'_R, a *reduced retention volume* V'_R can be defined by the formula:

$$V'_R = V_R - V_0 \tag{4.2.5}$$

TABLE 4.2.1 Typical Values for the Void Volume of Packed HPLC Columns

Dimensions (i.d. × Length in mm)	Empty Volume (mL)	Void Volume V_0 (mL)	Dimensions (i.d. × Length in mm)	Empty Volume (mL)	Void Volume V_0 (mL)
2.1 × 100	0.35	0.24	4.6 × 250	4.15	2.90
2.1 × 150	0.52	0.37	4.6 × 300	4.99	3.49
2.1 × 250	0.87	0.61	10.0 × 100	7.85	5.50
2.1 × 300	1.04	0.73	10.0 × 150	11.78	8.25
4.6 × 100	1.66	1.16	10.0 × 250	19.63	13.75
4.6 × 150	2.49	1.75	10.0 × 300	23.56	16.49

The dead volume V_0 of a chromatographic column can be found by multiplying the dead time with the volumetric flow ($V_0 = t_0U$) but also by direct measurements. For this purpose, a column is sequentially filled with solvents of different densities and weighed (pycnometric method). From the difference in the weight and the difference in the density of the solvents, it is possible to calculate the dead volume of the column.

The value of V_0 can be considered proportional with the volume of the empty column, the proportionality constant depending on the dimension (and shape) of the stationary phase particles, and also on the way they are packed. For a column of length L and inner diameter d, the empty column volume is $(\pi/4)d^2L$. However, the column is filled with the stationary phase, and only a fraction of this volume will give the dead volume. This can be expressed with the use of column packing porosity ε^*, a constant with the approximate value $\varepsilon^* \approx 0.7$ (for 5-μm particles column, u in mm/min, L and d in mm, and U in mm^3/min). Since the column i.d. and the column length are typically expressed in mm while the volume is usually expressed in mL (cm^3), a factor of 10^{-3} must also be included in the calculation when these units are used. The value for ε^* may vary depending on the stationary-phase particle size and structure such that values that are somewhat different from $\varepsilon^* \approx 0.7$ are possible. Including the packing porosity, the column dead (void) volume is $V_0 = \varepsilon^*(\pi/4)d^2L$. Table 4.2.1 gives some typical values for the empty volume and dead volume for an HPLC column, depending on its dimensions. At $U = 1$ mL/min the void time t_0 is numerically equal to the void volume V_0 as given in Table 4.2.1. Precise void volume of a column must be experimentally measured with an unretained compound.

The estimation of the void volume of the column is important for the calculation of several parameters in chromatography (see further retention factor). Also, during gradient operation, the void volume must be considered for evaluating when a certain mobile phase concentration is reaching the end of the column (dwell time is typically much smaller than void volume).

Migration Rate

While the linear flow rate u in the chromatographic column is the migration rate of an unretained molecule, the velocity at which the species X moves through the column is indicated by a parameter $u_R(X)$ known as the *migration rate* of species X. The value of $u_R(X)$ is a constant only when the mobile phase composition does not change during the separation (isocratic conditions).

The migration rate multiplied with retention time equals the length of the column for both retained and not retained compounds, and the following expression can be written:

$$u_R(X)t_R(X) = u\,t_0 = L \tag{4.2.6}$$

If during the separation all the molecules of compound X would all the time be in the mobile phase, then $u_R(X)$ is equal to u. However, some of the molecules are intermittently retained and when retained they do not move, such that only a fraction of molecules of compound X that are present in mobile phase are moving. The value of $u_R(X)$ is determined by this fraction. Assuming that during the separation process the number of molecules of compound X that are in the mobile phase is $\nu_{mo}(X)$ and in the stationary phase is $\nu_{st}(X)$, then $u_R(X)$ will be given by the expression:

$$u_R(X) = \frac{\nu_{mo}(X)}{\nu_{mo}(X) + \nu_{st}(X)}u \tag{4.2.7}$$

With the notation:

$$\nu(X) = \frac{\nu_{mo}(X)}{\nu_{mo}(X) + \nu_{st}(X)} \tag{4.2.8}$$

Expression (4.2.7) can be written in the form:

$$u_R(X) = \nu(X)u \tag{4.2.9}$$

Expressions (4.2.6 and 4.2.9) indicate that for the retention time this formula is valid:

$$t_R(X) = \frac{1}{\nu(X)}t_0 \tag{4.2.10}$$

Relation (4.2.10) shows that for a solute present only in the mobile phase, $\nu(X) = 1$ and $t_R(X) = t_0$, and for a solute completely retained on the stationary phase $\nu(X) = 0$ and $t_R(X) = \infty$. The typical situation is in between these two limits and, for example, if in a separation 25% of the molecules of an analyte are present in the mobile phase, $t_R(X) = 4\,t_0$. A similar relation with Eq. (4.2.10) is valid between the retention volume $V_R(X)$ and the dead volume V_0.

Equilibrium Constant and Phase Ratio in HPLC Separations

During the chromatographic process, the molecules that are separated can be considered as being in a continuous equilibrium between the mobile phase and the stationary phase. For a molecular species X, this equilibrium can be written as follows:

$$X_{mo} \leftrightarrow X_{st} \tag{4.2.11}$$

and can be considered governed by an equilibrium constant $K(X)$, defined as:

$$K(X) = \frac{C(X)_{st}}{C(X)_{mo}} \tag{4.2.12}$$

where $C(X)_{mo}$ is the molar concentration of species X in the mobile phase and represents the amount (in moles) of X in the volume V_0 of the mobile phase in the chromatographic column. This amount is proportional with the fraction ν of molecules in the volume V_0 of the mobile phase. Similarly, the concentration in the stationary phase $C(X)_{st}$ can be considered as representing the amount in moles of X from the stationary phase (proportional with $1 - \nu$) in the volume V_{st} of the stationary phase. As a result, the equilibrium constant $K(X)$ can be written in the form:

$$K(X) = \frac{(1 - \nu(X))/V_{st}}{\nu(X)/V_0} = \frac{1 - \nu(X)}{\nu(X)} \frac{V_0}{V_{st}} \tag{4.2.13}$$

As shown in formula (4.2.13), $K(X)$ depends on the ratio between the fraction of molecules of compound X that are present in the stationary phase (equal with $1 - \nu(X)$) and the fraction of molecules that is present in the mobile phase (equal with $\nu(X)$), as well as the ratio of the volumes V_0 and V_{st}. Since the ratio $(1 - \nu(X))/\nu(X))$ is kept constant only in isocratic conditions in a chromatographic separation, it becomes obvious that $K(X)$ is a constant only for isocratic chromatographic separations (unchanged composition of the mobile phase). The dependence of $K(X)$ on the value for $\nu(X)$ in formula (4.2.13) can be easily replaced with a dependence on retention time. This can be achieved considering formula (4.2.10), such that the value for $K(X)$ can be written in the form:

$$K(X) = \frac{t_R(X) - t_0}{t_0} \frac{V_0}{V_{st}} \tag{4.2.14}$$

The ratio V_0/V_{st} in formula (4.2.14) depends on the chromatographic column construction. Its inverse is indicated as phase ratio Ψ:

$$\Psi = \frac{V_{st}}{V_0} \tag{4.2.15}$$

Phase ratio is an important chromatographic parameter. For its evaluation, the necessary value V_0 has been previously described, and for the estimation of stationary phase volume V_{st} a number of procedures have been reported in the literature [51,52]. However, the correct evaluation of V_{st} is not a simple task, since there is no sharp boundary between the mobile phase and the stationary phase such that the volume is difficult to assess. In addition the separation boundary depends on the mobile phase composition. Other procedures to evaluate Ψ not using the values for V_0 and V_{st} have been reported [53]. Values for Ψ for C8 and C18 columns vary between 0.15 and 0.55 depending on column construction.

Retention Factor

The ratio $(t_R(X) - t_0)/t_0$ from formula (4.2.14) is an important parameter used for the description of chromatographic process and it is indicated as *retention factor* (or capacity factor). The retention factor is noted $k'(X)$ (or in some texts $k(X)$) and is given by the formula:

$$k'(X) = \frac{t_R(X) - t_0}{t_0} = \frac{t'_R(X)}{t_0} \tag{4.2.16}$$

Based on formula (4.2.14) for $K(X)$, and formula (4.2.15) for Ψ, the expression for the retention factor is the following:

$$k'(X) = K(X)\Psi \tag{4.2.17}$$

Expression (4.2.17) indicates that the retention factor depends on two parameters governing the separation in HPLC. One is the equilibrium constant $K(X)$, which is determined by the nature of analyte X and the nature of stationary and mobile phases in the chromatographic column. The other parameter is the phase ratio Ψ that depends on the characteristics of the stationary phase but also on other factors.

The retention factor $k'(X)$ has the advantage of being dimensionless and independent of the flow rate of the mobile phase or the dimensions of the column, and for this reason it is a very common and useful parameter for peak characterization. From expression (4.2.16), for example, it can be seen that reduced retention time $t'_R(X)$ is related to the retention factor by the formula:

$$t'_R(X) = k'(X)t_0 \tag{4.2.18}$$

Expression (4.2.18) can be written in the form:

$$t_R(X) = t_0 + k'(X)t_0 \tag{4.2.19}$$

The retention factor k' can be used to establish a relation between t_R and t'_R. The ratio of expressions (4.2.18) and (4.2.19) leads to the following formula (compound X not specified):

$$\frac{t'_R}{t_R} = \frac{k'}{k' + 1} \tag{4.2.20}$$

Expression (4.2.19) shows that retention time in HPLC depends on $k'(X)$ and therefore on the equilibrium constant for the analyte $K(X)$, the chromatographic column construction (phase ratio Ψ), and t_0. The retention factor k' is frequently expressed in logarithmic form in base 10 or in base e (either $\log_{10} k' = \log k'$, or $\log_e k' = \ln k'$). Similar expressions to (4.2.18), (4.2.19), and (4.2.20) relating t_R, t'_R, t_0, and k' can be established between V_R, V'_R, V_0, and k'. For example, $V'_R(X) = k'(X)V_0$ and $V_R(X) = V_0 + k'(X)V_0$.

The retention factors k' should have values between 1.1 and 10, obtained by selecting the column and the mobile phase for a set of analytes. These values for k' are necessary for performing the separation at acceptable retention times. From the void volume of a column given by expression $V_0 = \varepsilon^*(\pi/4)d^2L$ where $\varepsilon^* \approx 0.7$, for a flow rate $U = 1$ mL/min it can be evaluated that for a column with dimensions (i.d. • length) of 4.6 • 150 mm, the dead time $t_0 \approx 1.75$ min, and for a compound with $k' = 10$ the result is $t_R \approx 19.25$ min. Once the column is selected, the k' value in the desired range can be adjusted by modifying the mobile phase composition. Retention factors $k'(X)$ below 1 indicate poor retention of component X. Low values for $k'(X)$ generate short retention times. Since many compounds from the sample matrix may also be poorly retained, a low $k'(X)$ for the analyte poses the risk of interference from matrix constituents. If the separation of the analytes is still good, lower k' values typically indicate lower t_R for the peaks of interest, which is desirable. However, it is preferable to achieve low t_R values by using columns with smaller dimensions, small particles in the chromatographic column, special particles (such as core–shell), and elevated column temperature to speed up the separation that shorten t_0 while affecting k' as little as possible. Generally, retention factors exceeding a value of 10 indicate strong retention. The corresponding peaks elute after a longer time and are typically wide. Retention factors up to 20 are sometimes necessary, mainly when very complex samples are studied, but such values for k' indicate excessively long run times for the chromatogram.

Since the retention factor k' depends on equilibrium constant $K(X)$ (see expression 4.2.17) and the equilibrium constants depend on temperature, k' also depends on temperature. This dependence is further discussed in Section 4.3 (see expression 4.3.9).

Characteristics of an Ideal Peak Shape in Chromatography

The sample is injected in the HPLC system as an extremely narrow band (plug), but during the chromatographic run this band is broadened. Further description about band broadening does not account for the width of the sample "plug" injected in the chromatographic column. The peak broadening is an important (negative) effect taking place during the chromatographic process, and its study is necessary. Broadening is generated, for example, by the ordinary diffusion in time of the analyte molecules, such that the ideal shape of the chromatographic peaks can be described by a Gaussian bell curve. The transport of the band of the analyte across the HPLC system is supposed to not affect directly the peak broadening, which in reality is not the case. A number of other effects contribute to peak broadening but, at first, the present discussion will be limited to diffusion. Being a diffusion process, peak broadening can be studied based on Fick's laws. Fick's second law (see, e.g., Ref. [54]) for diffusion in one direction (longitudinal diffusion in the direction of x) has the expression:

$$\frac{dC}{dt} = D\frac{\partial^2 C}{\partial x^2} \tag{4.2.21}$$

In expression (4.2.21), t is time, C is concentration expressed in units of mass per units of length, and D is the diffusion coefficient (of the diffusing species in a specific solvent and at a specific temperature, expressed typically in $cm^2\ s^{-1}$). Fick's law can be directly applicable, for example, to the diffusion in a tube or a rectangular channel where the concentration C varies only along the channel length and is the same across the channel. On the basis of the assumption that for $t = 0$ (initial condition) the concentration is described by a given function $C\ (x,0)$, the solution of Eq. (4.2.21) can be written as follows (see, e.g., Ref. [55]):

$$C(x,t) = \frac{1}{\sqrt{\pi D t}} \int_{-\infty}^{+\infty} C(\eta, 0) \exp\left[\frac{-(\eta - x)^2}{4Dt}\right] d\eta \tag{4.2.22}$$

With the assumption that D is constant and the whole amount m of material was initially (at $t = 0$) contained in one point at $x = 0$, upon integration expression (4.2.22) leads to the following expression (see, e.g., Ref. [56]):

$$C(x,t) = \frac{m}{2\sqrt{\pi D t}} \exp\left(\frac{-x^2}{4Dt}\right) \tag{4.2.23}$$

In expression (4.2.23), t is the time of diffusion, x is the distance from the initial point of application of the amount m, and C is the concentration (as mass per length). By introducing in expression (4.2.23) the notation:

$$\sigma_L^2 = 2Dt \tag{4.2.24}$$

the formula for $C(x, t)$ (where $t = \text{const.}$ and C depends on t through σ_L) can be written as follows:

$$C(x,t) = \frac{m}{\sqrt{2\pi\sigma_L^2}}\exp\left(\frac{-x^2}{2\sigma_L^2}\right) \tag{4.2.25}$$

Expression (4.2.25) characterizes a typical Gaussian bell curve (as a function of x while t is fixed but part of the value of σ), is called a normal probability density function, and is used to describe a random process. For example, the observational random errors in an experiment are assumed to follow such a normal distribution. The parameter σ_L^2, usually noted simply σ^2, is called the variance, and σ (or σ_L) is called standard deviation. The parameter σ describes the width of the Gaussian curve, larger σ leading to wider bell shapes as seen in Fig. 4.2.2, where the ratio $C(x)/m$ as a function of x is plotted. The graphs show the Gaussian bell shape of the concentration distribution. The whole amount m of material was initially contained at $x = 0$. The apex of the Gaussian curve is obtained for $x = 0$ and $C_{max}/m = (2\pi\ \sigma^2)^{-1/2}$. The value for migration distance x corresponding to a concentration C, with $0 < C < C_{max}$, can be obtained based on expression (4.2.25), using the formula:

$$x = \pm\sqrt{-2\sigma^2 \ln\left[(C/m)\sigma\sqrt{2\pi}\right]} \tag{4.2.26}$$

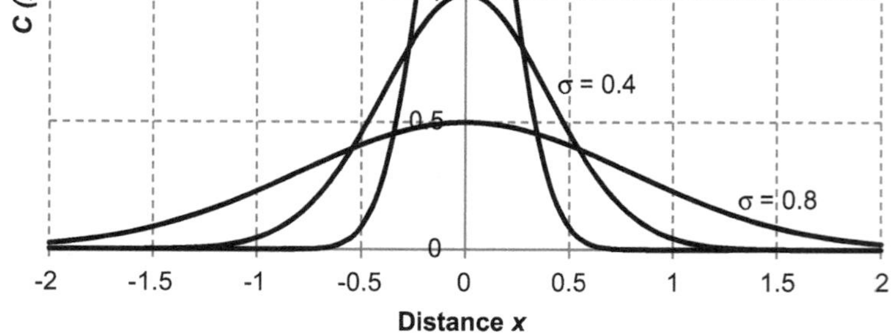

FIGURE 4.2.2 The variation of $C(x)/m$ as a function of distance x for three values of σ, namely $\sigma = 0.2$ (corresponding for example to $D = 10^{-2}$ and $t = 2$), $\sigma = 0.4$, and $\sigma = 0.8$.

From Eq. (4.2.26), the bell width $W = 2|x|$ is obtained as an increasing function of σ. Of particular importance in chromatography are the width at half height W_h and the width at the inflection point W_i of the Gaussian curve. Taking $C = 0.5$, $C_{\max} = 0.5\, m\, (2\pi\, \sigma^2)^{-1/2}$ in Eq. (4.2.26), the following result is obtained for the Gaussian width:

$$W_h = 2(2 \ln 2)^{1/2}\sigma \tag{4.2.27}$$

From the second derivative (as a function of x) of the expression (4.2.25) (that gives the inflection of Gaussian curve), the value for W_i is obtained as:

$$W_i = 2\sigma \tag{4.2.28}$$

The value for W_i for a given compound in a specific mobile phase can be obtained from expression (4.2.24) using the value for D and the diffusion time. As an example, for the diffusion of aniline in water at 25°C, $D = 1.05\ 10^{-5}\ \mathrm{cm^2/s}$. For a diffusion time of 5 min $W_i \approx$ 1.6 mm.

Up to this point, the diffusion process was considered for a "static material" that diffuses in one dimension, and formula (4.2.25) gives the concentration C as a function of distance x from the point of application. However, in a chromatographic process the compound of amount m not only diffuses (along the length x), but also advances (is eluted) along the chromatographic column due to the movement of the mobile phase. At the same time as the analyte is diffusing and also moves with the distance ξ, The expression for the concentration C, as a function of the distance x from the point of application, after the diffusion zone was eluted with the distance ξ, is given by:

$$C(x) = \frac{m}{\sqrt{2\pi\sigma^2}} \exp\left(\frac{-(x-\xi)^2}{2\sigma^2}\right) \tag{4.2.29}$$

This expression can be used for the understanding of peak broadening in a chromatographic process. For a separation where the mobile phase has a linear flow rate u, the distance from the origin to the center of the moving zone is $\xi = u\ t_R$. Therefore, from expression (4.2.24) for a given ξ, the resulting σ^2 is given by the expression:

$$\sigma^2 = 2D\frac{\xi}{u} = 2Dt_R \tag{4.2.30}$$

Formula (4.2.30) shows that at larger values for the distance ξ (equivalent to larger t_R), the value of σ^2 is larger (σ is expressed in length since D is expressed in $\mathrm{cm^2/s}$ and can be indicated as σ_L). Expression (4.2.29) for C as a function of x (from the application point) is in this case the following:

$$C(x) = \frac{m}{2\sqrt{\pi D t_R}} \exp\left(\frac{-(x - u\, t_R)^2}{4Dt_R}\right) \tag{4.2.31}$$

The same Gaussian curves shown in Fig. 4.2.2 but including the movement with the distance ξ (due to the moving of the diffusing band as the mobile phase flows) are shown in Fig. 4.2.3.

The broadening in units of length of the chromatographic peak previously discussed should now be converted into a broadening in units of time. In the chromatographic processes the measured parameter is the retention time and not the length of the path in the column, and for this reason, σ (function of distance σ_L) should be replaced with a σ_t a function of time (time broadening). This time broadening is given by the formula:

$$\sigma_t = \frac{\sigma_L}{u}\frac{t_R}{t_0} = \frac{\sigma_L}{L}t_R \tag{4.2.32}$$

With this replacement, expression (4.2.31) can be written in the form:

$$C(t) = \frac{m}{\sqrt{2\pi\sigma_t^2}}\exp\left(\frac{-(t-t_R)^2}{2\sigma_t^2}\right) \tag{4.2.33}$$

The maximum instantaneous concentration along the chromatographic peak is obtained for $t = t_R$. At this point, the concentration (measured by the detector) is given by the formula:

$$C_{max} = \frac{m}{\sigma_t\sqrt{2\pi}} \tag{4.2.34}$$

The maximum (instantaneous) concentration C_{max} in the effluent passing the detector in chromatography is translated into a maximum height h_{max} of the chromatographic peak, the height being also dependent on instrument sensitivity.

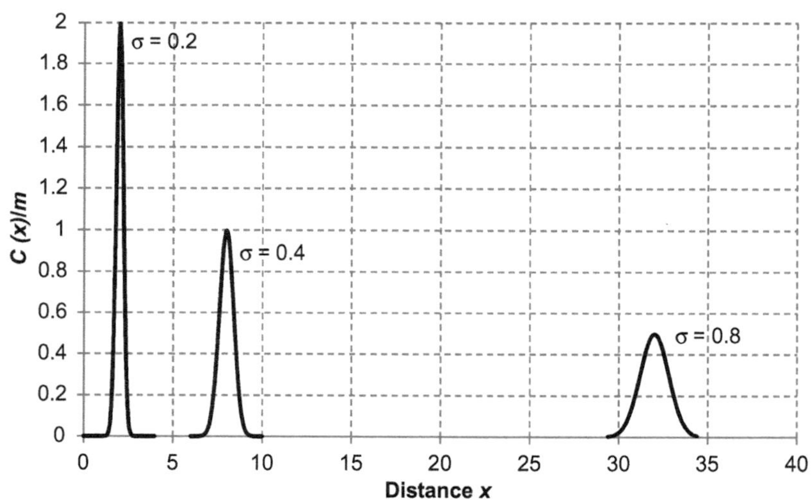

FIGURE 4.2.3 Peak broadening (in units of length) of an eluting analyte along a chromatographic column with the same σ values as those shown in Fig. 4.2.2: $\sigma = 0.2$, $\sigma = 0.4$, and $\sigma = 0.8$. These values for σ can be obtained, for example, for $D = 0.01$ and $\xi/u = 2$, 4 and 32, respectively (arbitrary units).

Similar to peak broadening as a function of distance x (expressed in units of length), a peak broadening in units of time can be measured. Time is typically the X-axis of a chromatogram, and broadening must be described in time units. For example, half height width W_h and the width at the inflection point W_i of the Gaussian curve are needed in units of time. Also, a width at the baseline $W_b = 2W_h$ is commonly measured in time units. The peak width W_b is measured between the points of intersection of baseline with the tangents to the curve at the inflexion points to the Gaussian curve representing the chromatographic peak. Fig. 4.2.4 shows the measurements of t_R, W_i, W_h, and W_b on a (model) chromatographic peak having the X-axis expressed as time.

The calculation of σ_t from expressions (4.2.27) and (4.2.28) generates the following formulas (where σ_t, W_h, W_i, and W_b are in time units):

$$\sigma_t = (8 \ln 2)^{-1/2} W_h = 0.5\, W_i = 0.25\, W_b \tag{4.2.35}$$

Broadening of the chromatographic peaks is not caused only by diffusion. Other processes with random character also contribute to peak broadening. These processes also produce a Gaussian bell curve distribution of retention times of individual molecules around the apex of the chromatographic peak, and the theory previously developed for diffusion remains valid with the difference that formula (4.2.30) for σ^2 is no longer valid. At the same time, expressions (4.2.33) and (4.2.35) remain valid with the condition that the correct σ values are used. The variance σ^2 combining all the random processes contributing to peak broadening can in this case be written as follows:

$$\sigma^2 = \sum_n \sigma_n^2 \tag{4.2.36}$$

where σ_n^2 are the variances generated by each of these random independent processes. All the broadening effects should also be included in σ_t from Eq. (4.2.35) that relates σ_t to the peak broadenings W_h, W_i, and W_b (expressed in time units).

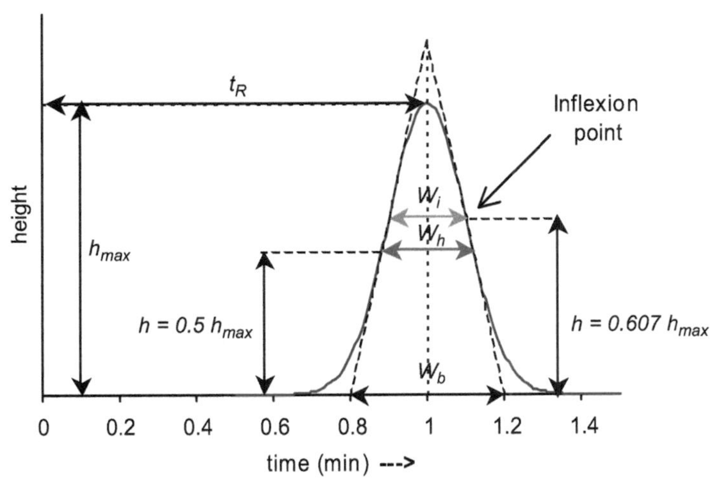

FIGURE 4.2.4 Measurements of retention time t_R and peak broadening W_i, W_h, and W_b on a (model) chromatographic peak with time as X-axis.

Efficiency of a Chromatographic Column

The value of σ_L (space broadening) is related to another parameter used to characterize zone spreading, namely the height equivalent to a theoretical plate H (HETP), which is defined as:

$$H = \frac{\sigma_L^2}{L} \tag{4.2.37}$$

This parameter is very useful in chromatography for the characterization of peak broadening per unit length of the column. Expressed as a function of time, formula (4.2.37) can be written in the form:

$$H = \frac{\sigma_t^2 L}{t_R^2} \tag{4.2.38}$$

In addition to H, the peak broadening characterization in a column can be obtained using the *theoretical plate number* N. For a column of length L, N is defined as:

$$N = \frac{L}{H} \tag{4.2.39}$$

The theoretical plate number N can be expressed as a function of length and σ_L^2 by a simple substitution of Eqs. (4.2.37) into (4.2.39) to give the expression:

$$N = \frac{L^2}{\sigma_L^2} \tag{4.2.40}$$

The value of σ_L^2 in expression (4.2.40) is expressed in length and should be further related to peak broadening in time W_b. For this purpose, σ_L will be replaced with σ_t using expression (4.2.32), and L will be expressed as a function of time as $L = t_0\, u$. With these two substitutions in expression (4.2.40) the following formula for N is obtained:

$$N = \frac{t_R^2}{\sigma_t^2} = 16\frac{t_R^2}{W_b^2} \tag{4.2.41}$$

Similar relations can be established between N and W_i or W_h (time units) with the exchange of coefficient 16 with 4 or 5.5452, respectively.

In addition to the theoretical plate number N, an effective plate number $\boldsymbol{n}$ is defined by using t_R' in Eq. (4.2.41) instead of t_R. The formula for $\boldsymbol{n}$ will be the following:

$$\boldsymbol{n} = 16\frac{t_R'^2}{W_b^2} \tag{4.2.42}$$

The value for $\boldsymbol{n}$ is smaller than that for N since $t_R' < t_R$. Since t_R (and t_R') as well as W_b are chromatographic parameters dependent on the eluting compound (index X was omitted in previous formulas), the values for N and $\boldsymbol{n}$ are also compound-dependent. Either expressions (4.2.41) or (4.2.42) can be used to measure the theoretical plate number or effective plate number based on experimental data obtained with a given column. This measurement is useful in practice to select columns (higher N gives lower peak broadening) and also to assess the loss in performance of a column after a certain period of usage when the N values start to decrease. Because N is related to the important characteristic of peak broadening, it is common to indicate it as a parameter to characterize the efficiency of a column. The values for N for HPLC columns can be given either for a specific column or reported as efficiency per meter (N/m). Both N and $\mathbf{n}$ are used for the characterization of *column efficiency*, and $\mathbf{n}$ is sometimes named simply *efficiency*.

For modern HPLC analytical columns the efficiency per meter N/m can be between 20,000 and 150,000. In HPLC, depending on the nature of the column, this can vary (per m) between 40,000 and 120,000 (or even higher), and for core–shell columns N/m can be as high as 300,000 (common column length L is between 50 mm and 250 mm) (see Section 5.1). The values for N (per unit length of the column) are influenced by physical properties of the stationary phase such as dimension of stationary-phase particles, homogeneity of the particle dimensions, and structure of particles.

Factors Contributing to Peak Broadening and van Deemter Equation

As indicated by expression (4.2.36), besides the longitudinal diffusion, a number of effects contribute to peak broadening. These effects include the following: (1) longitudinal diffusion (already discussed, producing the variance σ_L), (2) eddy diffusion, (3) lateral movement of material due to convection, (4) the differences between individual molecules in the mass transfer rate in and out of the stationary phase, and (5) the presence of random spots of stagnant mobile phase in the porous material of the column (mass transfer in and out of the mobile phase). The inclusion of the contribution of all these factors to the value of plate height H requires the replacement of expression (4.2.37) with a new formula of the form:

$$H = \frac{\sigma_L^2 + \sigma_E^2 + \sigma_C^2 + \sigma_T^2 + \sigma_S^2}{L} = H_L + H_E + H_C + H_T + H_S \tag{4.2.43}$$

The value of H_L can be obtained in theory from expression (4.2.24), but since the diffusion medium is not homogeneous and the packing material is "obstructing" to a certain extent the diffusion, the value for H_L should include a correction factor γ such that the corrected formula can be written in the form:

$$H_L = \gamma \frac{2Dt}{L} = \gamma \frac{2D}{u} \tag{4.2.44}$$

The value for γ depends on the column packing and is typically around 0.625. The values for the diffusion coefficient D (in $cm^2\ s^{-1}$) are reported in the literature for various solutes and solvents [57].

Eddy diffusion is caused by the fact that in a packed material the flow occurs through a tortuous channel system with various path lengths. The molecules of the same solute may randomly take different paths. This path will depend on the average diameter of a particle, and using the notation d_p for an average value for the particle diameter in the stationary phase, the contribution to the plate height of the eddy diffusion can be written in the form:

$$H_E = \Lambda d_p \tag{4.2.45}$$

The coefficient Λ depends on how irregular is the particle shape and also on other packing characteristics.

Another random process contributing to peak broadening is the lateral movement of material due to convection, which is stronger at higher linear flow rates u of the mobile phase. Convection depends on column packing (through a parameter Γ) and increases when the particle diameter d_p increases, decreasing when the diffusion coefficient D increases. The contribution to the plate height for convection is given by the expression:

$$H_C = \frac{\Gamma d_p^2}{D} u \tag{4.2.46}$$

The rate of transfer of solute into and out of the stationary phase is controlled by the rate of diffusion in the liquid stationary phase or by the adsorption–desorption kinetics in the case of adsorption processes. It can be shown using a random walk model that, for a distribution process, the contribution to the plate height H_T due to this effect can be expressed by the formula:

$$H_T = \Theta \frac{d_f^2}{D_s} u \tag{4.2.47}$$

In formula (4.2.47), Θ is a proportionality constant, d_f is the depth of the stationary phase on the solid support, and D_s is the diffusion coefficient of the analyte in the stationary phase.

The contribution to the plate height from the mass transfer in the stagnant mobile phase in the porous material is given by an expression similar to Eq. (4.2.47) where the parameter Θ is about the same, but d_f should be replaced by d_p (diameter of the particle which is related to the depth of the pores), and D_s should be replaced by D the diffusion coefficient in the mobile phase. This effect, known as mobile phase mass transfer contribution, gives the following increased H_S to the theoretical plate height:

$$H_S = \Theta \frac{d_p^2}{D} u \tag{4.2.48}$$

The combination of all the above contributions leads to the following formula for the plate height:

$$H = \Lambda d_p + \gamma \frac{2D}{u} + \left(\Gamma \frac{d_p^2}{D} + \Theta \frac{d_f^2}{D_s} + \Theta \frac{d_p^2}{D} \right) u \tag{4.2.49}$$

Expression (4.2.49) gives a general formula describing the dependence of H on the linear flow rate u and on various parameters related to the stationary phase and mobile phase. This equation is known as van Deemter equation [58] and can be written in the form:

$$H = A + \frac{B}{u} + Cu \tag{4.2.50}$$

The expressions for the coefficients A, B and C in Eq. (4.2.50) can be obtained by comparison with the equivalent expression (4.2.49). An example of a plot for the van Deemter equation for $A = 4$ μm, $B = 500$ μm^2/s, and $C = 0.0005$ s is given in Fig. 4.2.5.

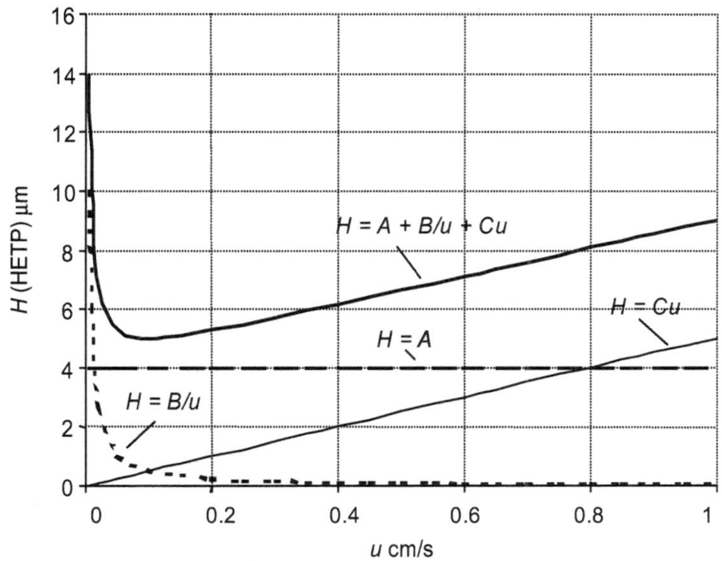

FIGURE 4.2.5 The plot of van Deemter equation and of its components for $A = 4$ μm, $B = 500$ μm^2/s, and $C = 0.0005$ s.

The shape of dependence of H on u given by van Deemter equation indicates that a minimum value for H is obtained at a specific u value that can be obtained from the condition $dH/du = 0$. This condition gives the optimum linear flow rate at $u_{opt} = \sqrt{B/C}$. This value included in formula (4.2.50) gives the optimum H_{min} that can be achieved with a specific column for u_{opt}. From the expressions for B and C given by expression (4.2.49) and neglecting the H_T term, the formula for the H_{min} becomes:

$$H_{min} = A + 2\sqrt{BC} = d_p\left(\Lambda + 2\sqrt{B'A'}\right) \tag{4.2.51}$$

where A' and B' are parameters independent of D. It can be roughly estimated that $\Lambda + 2\sqrt{B'A'} \approx 2$ *to* 3, and $H_{min} \approx (2\text{–}3)\ d_p$. A too slow flow or an increase in flow rate beyond u_{opt} leads to a decrease of column performance due to the increase in H. Faster flows than u_{opt} (and U_{opt} with values obtained from u_{opt} based on expression 4.2.1) may be sometimes needed for generating shorter run times.

Experimental attempts to verify Eq. (4.2.50) showed that some deviations from the theoretical model occur. The main explanation for this effect is that eddy diffusion and mobile phase mass transfer are not totally independent effects. The differences in the intraparticle velocity for the local streams affect the mobile phase mass transfer and in return the eddy diffusion. The coupling of the two processes is captured in a different expression for H, given by Knox equation [59,60].

Besides the effects previously discussed to produce peak broadening, in practice, the chromatographic peaks can also be broadened by other causes. Such causes may include the presence of void volumes in the chromatographic system before the sample reaches the stationary phase. Also, larger injected sample volumes lead to peak broadening such that the ideal theoretical plates number decreases compared to its optimum value obtained for a very narrow injection. Besides peak broadening, an experimental peak compression is also possible, but this compression does not affect the five effects (longitudinal diffusion, eddy diffusion, etc.) previously discussed. The peak compression may compensate for effects such as large injection volumes and void volumes in the HPLC system. This is achieved by the accumulation of the analytes at the front end of the column before the elution process starts, by using a sample solvent and an initial mobile phase that does not elute the sample from the column head (see Section 13.5).

Peak Asymmetry

Theoretically, chromatographic peaks should have a Gaussian shape. However, some analytes are present in the mobile phase in more than one form, such as molecular and ionic (e.g., for acids or for amines). These different species may interact differently with the stationary phase, in particular when the retention does not occur based on a unique mechanism. This different retention generates asymmetrical peaks. To the peak asymmetry may also contribute various minor interactions that cannot affect at the same time all the molecules of the analyte, column contaminations, column overload with sample, etc. Two parameters are used for the characterization of peak asymmetry. One is the *asymmetry parameter* $As(X)$, and the other is the *tailing factor* $TF(X)$, both dependent on the nature of the analyte X. The asymmetry $As(X)$ is defined as the ratio of the rear r vs. front f segments cut on the chromatographic peak by a parallel to the baseline at 10% peak height (some other choices are possible), and separated by the perpendicular from the apex, and it is given by the formula:

$$As(X) = \frac{r}{f} \quad (4.2.52)$$

The rear r and front f segment cuts at 10% of the height are shown on an asymmetrical peak in Fig. 4.2.6.

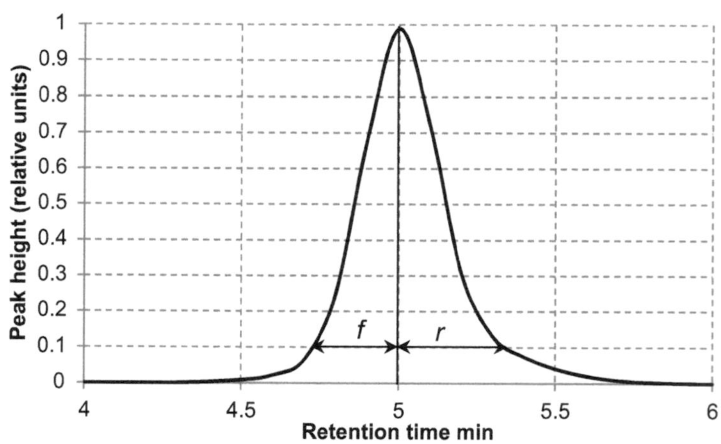

FIGURE 4.2.6 Front f and rear r in an asymmetrical chromatographic peak.

The peak tailing is defined by the formula:

$$TF(X) = \frac{f' + r'}{2r'} \quad (4.2.53)$$

where f' and r' are measured in the same way as f and r, but at 5% of the peak height. Peak asymmetry may be very different from compound to compound, since the interaction with the stationary phase may involve different mechanisms depending on the compound nature.

Another parameter used for the description of peak asymmetry is the skew. This parameter takes into consideration that the Gaussian shape of a chromatographic peak (with variance σ^2) is due to the random spreading of the injection plug in the chromatographic column, but a modification is necessary for accounting for other broadening effects. This modification is based on an exponential decay of the form $h(t) = h(t_0)\exp(-t/\tau)$ and depends on a parameter τ which is the mean lifetime (time at which the initial value is reduced by the factor $1/e$). Skew is defined by an expression depending on the ratio σ/τ (see, e.g., Ref. [5]). Other functions besides Gaussian are also used for modeling the shape of chromatographic peaks [61], and peak characteristics can be calculated from these shapes. Deviation from Gaussian shape of the front of the chromatographic peak is typically indicated as *fronting*.

Selectivity (Separation Factor)

Selectivity (or *separation factor*) is another empirical parameter used for column characterization and is related to the separation of two compounds X and Y. The notation for selectivity is usually α, and this parameter is defined by the ratio:

$$\alpha(X, Y) = \frac{t'_R(X)}{t'_R(Y)} \quad (4.2.54)$$

where $t'_R(X) > t'_R(Y)$. Parameter α indicates the ratio of the distances in time between the apexes of two chromatographic peaks (for compounds X and Y). The selectivity (separation factor) is usually of interest for compounds that give adjacent peaks, since peaks that are well distanced do not pose separation problems. Using expressions (4.2.16) or (4.2.17) it can be easily noticed that α can also be expressed by the formula:

$$\alpha(X, Y) = \frac{k'(X)}{k'(Y)} = \frac{K(X)}{K(Y)} \quad (4.2.55)$$

In any chromatographic separation, larger α values are desirable for a better separation. However, the value of α alone cannot describe how good the separation of two compounds is. Even when α is large, peak broadening can be so large that the separation can be poor. The values for α are solute-dependent, but also depend on the nature of the stationary phase and of the mobile phase. The chemical nature of the stationary phase is one of the two main factors influencing the separation for a given set of analytes, the second main factor being the choice of mobile phase. For this reason, the choice of a chromatographic column is frequently based on its selectivity α toward the analytes being separated. A value for $\alpha > 1.2$ is typically necessary for an acceptable separation (provided the peaks are not very broad).

Resolution

Regardless of how far apart are the apexes of two chromatographic peaks (as described by α), if the peaks are broad their separation can be compromised. A parameter that truly characterizes peak separation is the resolution R. This parameter is defined by the formula:

$$R(X, Y) = \frac{2[t_R(X) - t_R(Y)]}{W_b(X) + W_b(Y)} \quad (4.2.56)$$

The difference in the retention times $t_R(X) - t_R(Y) = \Delta t_R$ can be replaced with $t'_R(X) - t'_R(Y)$. The values for formula (4.2.56) can be obtained from the chromatogram as illustrated in Fig. 4.2.7. A good peak separation is typically considered acceptable when $R > 1.0$ and good when $R > 1.5$ when the two peaks are separated at the baseline ($2\ \Delta t_R > W_b(X) + W_b(Y)$).

The widths at the baseline of the two peaks can be different (as also shown in Fig. 4.2.7), but as an approximation it is possible to take $W_b(X) = W_b(Y) = W_b$. With these assumptions, the formula for R can be written in the form:

$$R = \frac{\Delta t_R}{W_b} \quad (4.2.57)$$

The difference in retention time between two analytes Δt_R can be written as a function of selectivity α in the following form:

$$\Delta t_R = (\alpha - 1)t'_R(Y) \quad (4.2.58)$$

The expression for R with this substitution is the following:

$$R = (\alpha - 1)t'_R(Y)/W_b \quad (4.2.59)$$

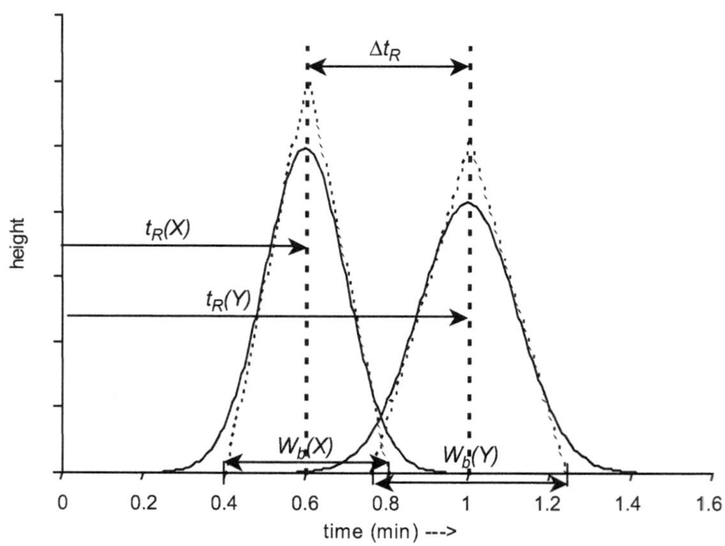

FIGURE 4.2.7 An idealized chromatogram showing the measurable parameters used for the calculation of resolution R.

The value for $t_R'(Y)$ can be replaced with $t_R(Y)$ using expression (4.2.20) to obtain the following formula:

$$R = (\alpha - 1)\frac{k'(Y)}{1 + k'(Y)}\frac{t_R(Y)}{W_b} \tag{4.2.60}$$

From expression (4.2.41), the value for $t_R(Y)/W_b$ can be expressed as a function of N, and the value for R will become the following:

$$R = \frac{1}{4}(\alpha - 1)\frac{k'(Y)}{1 + k'(Y)}N^{1/2} \tag{4.2.61a}$$

Formula (4.2.60) indicates the resolution relative to the "previous peak." A similar formula for R can be obtained relative to the "next peak" by using the following substitution:

$$\Delta t_R = \frac{(\alpha - 1)}{\alpha}t_R'(X) \tag{4.2.62}$$

In this case, the resulting formula for R is the following:

$$R = \frac{1}{4}\frac{(\alpha - 1)}{\alpha}\frac{k'(X)}{1 + k'(X)}N^{1/2} \tag{4.2.61b}$$

Both Eqs. (4.2.61a) and (4.2.61b) are approximations and the two values should be close to each other. Both expressions are obtained by taking the peak width at the baseline as equal for the two peaks, the theoretical plate number N being measured for one compound (Y in Eq. 4.2.61a) or for the other (X in Eq. 4.2.61b). The value for R is most sensitive to the parameter α, which is critical for obtaining a good separation. Larger k' values are also useful, but as k' increases, its importance for the increase in R is diminished. This is exemplified in Fig. 4.2.8, for a chromatographic column with $N = 18{,}000$.

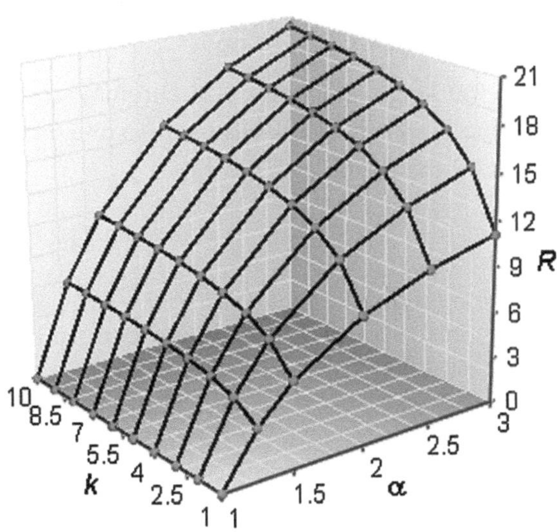

FIGURE 4.2.8 Graph showing the variation of R as a function of α and k' assuming a chromatographic column with $N = 18{,}000$ [5].

From the dependence of R on α and k', it is simple to obtain the formula for the dependence of R on $K(X)$, $K(Y)$ and Ψ. Using in formu (4.2.58) simple substitutions for α (expression 4.2.54) and for k' (expression 4.2.17), the following formula for R is obtained:

$$R(X,Y) = \frac{1}{4}\left[\frac{K(X)}{K(Y)} - 1\right]\frac{K(Y)}{1/\Psi + K(Y)}N(Y)^{1/2} \tag{4.2.63}$$

Expression (4.2.63) shows that a large α (large $K(X)/K(Y)$), a large Ψ, and a large N are desirable for obtaining a large R. The requirement for the value of resolution R to be higher than 1.0 in order to have a good separation is translated for the theoretical plate number N in the requirement to satisfy the relation:

$$N > 16\frac{(1+k')^2}{k'^2}\frac{\alpha^2}{(\alpha-1)^2} \tag{4.2.64}$$

In many practical applications, the separation factor α between an analyte and other components of a specific matrix may be too close to unity. The increase in the number of theoretical plates N of the column can be helpful for enhancing separation in these cases.

Peak Capacity

The efficiency of an HPLC separation can be characterized by a parameter known as peak capacity P. This parameter gives the number of peaks in a chromatogram that can be separated from one another with a resolution $R = 1$ within a specified window of time defined by the longest retention time $t_{R,max}$. The theory of a maximum peak capacity is based on the fact that an ideal peak (Gaussian shape) has the peak width $W_b = 4\sigma_t$ (see expression 4.2.35). As the peak width changes with the retention time, the peak capacity P can be defined by the formula:

$$P = 1 + \int_{t_0}^{t_{R,max}} \frac{dt}{4\sigma_t} \tag{4.2.65}$$

Expression (4.2.65) can easily be interpreted as the length (in time) of a chromatogram ($\int_{t_0}^{t_{R,max}} dt$) divided by one peak width (W_b). The value of σ_t can be related using expressions (4.2.41) and (4.2.16) to the theoretical plate number by the formula:

$$\sigma_t = \frac{t_0}{\sqrt{N}}\left(k'_e + 1\right) \tag{4.2.66}$$

where k'_e is the retention factor k' at the point of elution (and can be taken as $k'_e = t/t_{0\_}1$). For isocratic separations, introducing expression (4.2.66) in the integral (4.2.65) and performing the integration, the resulting value for the peak capacity P is given by the formula:

$$P = 1 + \frac{\sqrt{N}}{4}\ln\left(\frac{t_{R,max}}{t_0}\right) \tag{4.2.67}$$

It should be specified that expression (4.2.67) is valid only for isocratic separations.

Peak Characteristics for Gradient Separations

Some of the formulas for the empirical peak characteristics were developed assuming that the HPLC separation takes place in isocratic conditions (constant composition of the mobile phase during the chromatographic run). In isocratic conditions, the retention factor $k'(X)$ and the equilibrium constant $K(X)$ are constant during the chromatographic run. In gradient separations, the mobile phase composition changes, and the migration rate of species X changes during the chromatographic run. For this reason, it is no longer possible to define a constant $u_R(X)$. In gradient, the migration rate of compound X becomes a function $u(X,t)$ where t varies in the interval $t_0 < t < t_R(X)$. As previously discussed, the migration rate $u_R(X)$ is related to the value of $K(X)$ (see expressions 4.2.7 and 4.2.13) and therefore to the values of $k'(X)$. Therefore, both $K(X)$ and $k'(X)$ are changing during the chromatographic run when the separation is performed using gradient. In reversed-phase HPLC, for example, the change in mobile phase composition consists usually in changing the proportion of an organic component (such as acetonitrile or methanol) and that of water. The volumetric fraction of organic component ϕ in the mobile phase in a binary mixture with water is given by the formula:

$$\phi = \frac{V_{org}}{V_{org} + V_w} \tag{4.2.68}$$

where V_{org} indicates the volume of the organic component and V_w the volume of water (ϕ can be related to molar concentration C_{org} of organic phase, and using the notations Mw_{org} for the molecular weight and ρ_{org} for the density of the organic phase, $\phi = C_{org}Mw_{org}/1000\rho_{org}$). An empirical formula has been established between $k'(X)$ and a retention factor $k'_w(X)$, where $k'(X)$ refers to the retention in the mobile phase consisting of a mixture of an organic solvent with water, and $k'_w(X)$ refers to retention in pure water. This formula can be written as follows:

$$\log k'(X) = \log k'_w(X) - S(X)\phi \tag{4.2.69}$$

The value for $k'_w(X)$ is not usually known and is obtained from extrapolation to 100% from the values of $k'(X)$ as the content of water increases. Formula (4.2.69) is only an approximation (see, e.g., Ref. [5]), but this approach is useful for the understanding of variation of $k'(X)$ in gradient conditions. The parameter $S(X)$ is specific for a solute X, a solvent mixture, and a column, but does not depend on ϕ.

For a linear gradient performed for a period of time equal to t_{grad}, the composition of the mobile phase changes following the expression:

$$\phi = \phi_0 + \frac{\Delta\phi}{t_{grad}} t \tag{4.2.70}$$

where $\Delta\phi/t_{grad}$ is the gradient slope (see Eq. 3.1.2). With ϕ given by expression (4.2.70), the expression for the retention factor $k'(X)$ as given by expression (4.2.69) becomes the following:

$$\log k'(X) = \log k'_w(X) - S(X)\left(\phi_0 + \frac{\Delta\phi}{t_{grad}} t\right) \tag{4.2.71}$$

Expression (4.2.71) shows that in linear gradient conditions, the capacity factor $k'(X)$ depends across the chromatogram not only on the nature of the compound X but also on the instantaneous mobile phase composition. The variation of $k'(X)$ during a gradient run time for two compounds having different $k'_w(X)$ values and the same S is illustrated in Fig. 4.2.9. As the values for k' approach zero ($k'(X) \approx 0$), the compound is no longer retained and moves along the chromatographic column at the linear flow rate of the mobile phase u and is fast eluted. The variation of k' for the two compounds shown in Fig. 4.2.9 indicates, for example, a significant difference in the migration rate of the two compounds X and Y at lower ϕ values, while as the ϕ increases above 0.7, the two compounds migrate at rates close to that of the mobile phase, but the compounds are already separated.

The gradient slope $\Delta\phi/t_{grad}$ is typically included in another parameter indicated as *gradient steepness* defined by the formula:

$$b = \frac{S\Delta\phi}{t_{grad}} t_0 \tag{4.2.72}$$

With the use of gradient steepness, expression (4.2.72) is written in the form:

$$\log k'(X) = \log k'_w(X) - S(X)\phi_0 - b\frac{t}{t_0} \tag{4.2.73}$$

Because a compound has different k' values during the chromatographic run, it is not possible to use k' in gradient separations for the calculation of a number of parameters such as α or R, as previously described for the characterization of the chromatographic process. Since k' changes as the compound migrates across the chromatographic column, it was useful to define an effective value for k'. This *effective value* is the gradient retention factor k^*, and is obtained using various approximations such as averages. Further discussion on

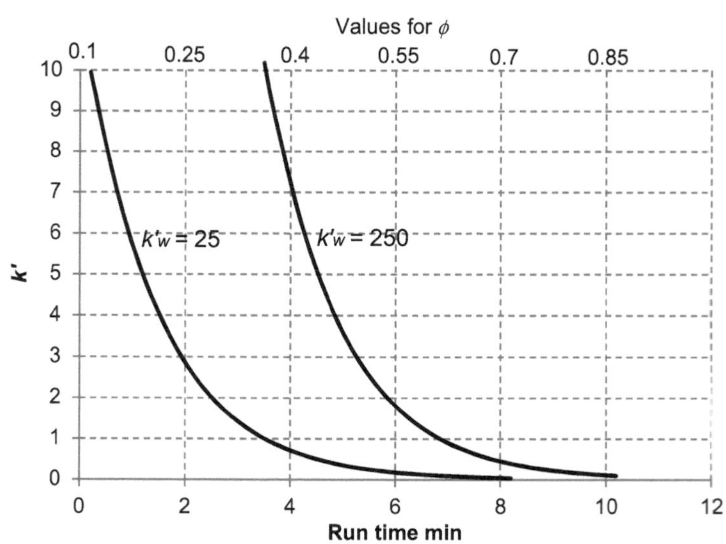

FIGURE 4.2.9 Variation of $k'(X)$ during a linear gradient for two compounds with $k'_w(X) = 25$ and $k'_w(Y) = 250$, $S = 4$, $\phi_0 = 0.1$, $\Delta\phi = 0.6$, $t_0 = 0.2$ min, and $t_{grad} = 10$ min.

gradient elution, including the calculation of k^* and of other parameters for gradient elution can be found in Section 14.2.

Summary of Chromatographic Peak Characteristics

A summary of peak characteristics can be obtained by processing the peaks in a chromatogram either manually or more commonly by using the capability of data-processing programs from the computer that controls the HPLC. A summary of such characteristics is given in Table 4.2.2.

Quantitation in HPLC

The integration of $C(t)$ given by expression (4.2.33) for t between $-\infty$ and $+\infty$ leads to the following result:

$$\int_{-\infty}^{+\infty} C(t)dt = m \tag{4.2.74}$$

Expression (4.2.33) is based on the fact that the following relation is true:

$$\int_{-\infty}^{+\infty} \exp\left(\frac{-(t-t_R)^2}{2\sigma_t^2}\right)dt = \sqrt{2\pi\sigma_t^2} \tag{4.2.75}$$

At the same time, the following expression is true:

$$\int_{-\infty}^{+\infty} C(t)dt = A \tag{4.2.76}$$

where A *is* the total peak area (A_{peak} for the compound X) under the curve representing the function $C(t)$. In chromatographic instruments, this peak area is obtained using instrumental detection/amplification procedures and the value for A in a chromatogram will become only proportional with the amount of material injected in the HPLC system. Area measurement is typically performed using dedicated programs for data processing with numerical procedures used for area integration expression [62].

The injections are performed using a given volume of the solution of the sample, and the mass of injected material is $m = C_{init}\, V_{inj}$ where C_{init} is the concentration of X in the injected sample and V_{inj} is the injection volume. As a result, the following expression can be used in HPLC for quantitation (measurement of an unknown C_{init}):

$$C_{init} = ct.\frac{A_{peak}}{V_{inj}} \tag{4.2.77}$$

The proportionality constant $ct.$ depends on experimental conditions, and a common quantitation procedure is the use of calibration curves between peak area A_{peak} and the concentration of an analyte to determine $ct.$ These calibration curves are obtained using standards. In practice, deviations from Eq. (4.2.77) may sometimes be encountered. For example, the relation between sample concentration and peak area can be of the form:

$$C_{init} = ct._1 + ct._2\,\frac{A_{peak}}{V_{inj}} \tag{4.2.78}$$

TABLE 4.2.2 Typical Peak Characteristics Obtained Using the Capability of the Data-processing Program From the Computer That Controls the HPLC

No.	Characteristic	Notation	Common Range
1	Retention time	t_R	1–30 min (longer times for complex chromatograms)
2	Void (dead) time	t_0	1–3 min
3	Retention factor	k'	2–10
4	Peak height	h	Depends on detector settings
5	Peak area	A	Depends on detector settings
6	Peak width at half height	W_h	0.05–1 min
7	Peak width at baseline	W_b	0.1–2 min
8	Peak start	t_{start}	Depends on t_R and W_b
9	Peak end	t_{end}	Depends on t_R and W_b
10	Symmetry (at 10% height)	*As*	1–1.3
11	Tailing	*TF*	1–1.3
12	Skew	*Skew*	1–1.3
13	Noise at peak baseline		Depends on detector settings
14	Signal to noise ratio	S/N	Depends on detector settings, compound nature and concentration, etc.
15	Integration type		Base to base, base to shoulder, etc.
16	Data points per peak		Depends on the rate of detector measurements of signal
17	Theoretical plate number (plates/column)	N	4000–40,000
18	Theoretical plate number (plates/meter)	N	30,000–300,000
19	Efficiency (plates/column)	n	4000–40,000
20	Efficiency (plates/meter)	n	30,000–300,000
21	Foley–Dorsey plates/column[a]		N corrected for asymmetry (lower than N)
22	Foley–Dorsey plates/meter		N corrected for asymmetry (lower than N)
23	Selectivity to previous peak	α	1.5–10 (depending on separation)
24	Selectivity to next peak	α	1.5–10 (depending on separation)
25	Resolution to previous peak	R	1.5–15 (depending on separation)
26	Resolution to next peak	R	1.5–15 (depending on separation)

[a]*Note: The plate number can be corrected for peak asymmetry by the Foley–Dorsey formula:* $N = 41.7(t_R/W_{0.1})^2(f/r + 1.25)$ *where* $W_{0.1}$ *is the peak width at 10% of its height, f is front and r is rear of the peak, as shown in Fig. 4.2.6.*

In some cases, a quadratic equation fits better the relation between the concentration and peak area. These deviations from formu (4.2.77) are caused by the background response of the detector or by its nonlinear response. The quantitation should be performed always within the calibration range, and for the quadratic dependence between the concentration and peak area, the values outside the calibration range may generate large errors in the quantitation.

Quantitation based on standard addition or peak area ratios with isotopic labeled compounds (used, e.g., with MS detection) are also practiced and are based on proportionality given by expression (4.2.77) (standard addition quantitation is not applicable if the dependence between the analyte concentration and peak area follow expression (4.2.78) and $ct._1$ is not known).

Besides the peak area, peak height can also be used for quantitation. Expression (4.2.34) shows that $C(t)$ in a chromatogram is maximum when $t = t_R$. Because $m = C_{init}\,V_{inj}$, the following expression is valid:

$$C_{max} = \frac{C_{init}V_{inj}}{\sigma_t\sqrt{2\pi}} \tag{4.2.79}$$

The value C_{max} is the concentration of the analyte at the maximum of the peak (apex) in the chromatogram, and $C_{max} = ct.\ h_{max}$, where the proportionality constant $ct.$ depends on the detector response factor. This observation shows that the peak height is also proportional with the initial concentration C_{init} of the solute in the injected sample. The quantitation using peak height also requires the use of calibration curves. Although the measurement of the concentration of the injected sample seems to be equally possible using the peak area or the peak height, there are some differences between the two procedures. The formula (4.2.79) has been developed based on the assumption that the peak has an ideal Gaussian shape. This is not the case in all practical situations. Even if the peak shape deviates from Gaussian, expression (4.2.77) remains valid, and the peak area in the chromatograms remains proportional with the amount of sample injected in the HPLC system. For peaks with a shape different from Gaussian, more variability regarding the proportionality between the peak height and the amount of the injected sample is typically seen.

The maximum concentration C_{max} for the apex of a chromatographic peak is an important parameter since it determines the maximum signal generated by the detector (h_{max}) and therefore is related to the detection limit of an analytical method. The ratio C_{init} to C_{max} is indicated as *dilution* D and C_{max} depends on C_{init} based on the following formula:

$$C_{max} = \frac{C_{init}}{\mathfrak{D}} \tag{4.2.80}$$

Expression (4.2.80) shows that the maximum (instantaneous) concentration of the analyte passing the detector is directly proportional with the initial concentration injected in the HPLC instrument and inversely proportional with dilution D. It is desirable to obtain in a separation as high C_{max} as possible and therefore to have a small D. The value for D can be obtained by combining expressions (4.2.79) and (4.2.32). By taking $\sigma_L = L/\sqrt{N}$, the following formula can be written as follows:

$$\mathfrak{D} = \frac{\sqrt{2\pi}}{V_{inj}\sqrt{N}} t_R \tag{4.2.81}$$

Expression (4.2.81) indicates that the larger instantaneous concentration of analyte C_{max} at the apex of the chromatographic peak can be obtained when using large injection volumes (that puts more material in the chromatographic system for a given concentration C_{init}), columns with high number N of theoretical plates, and fast elution of the compound (short t_R). Such choices lead to larger detector signals.

Sample Volume and Amount Injected in the Chromatographic Column

The injected sample in an HPLC system is characterized by its volume, usually in the range of 1−25 μL for standard analytical HPLC. However other injection volumes are used. For certain UPLC applications or for micro-HPLC, volumes in the range of 20−500 nL are common. Large volume injections up to 1 mL can also be used in special applications [63,64]. The injection volume must be precise and the injection must be reproducible. The injection volume V_{inj} is directly related to the amount of sample $m_i = C_i\, V_{inj}$ delivered to the HPLC system. Since the area A_i of the chromatographic peak (detector response) is proportional with the amount of sample m_i (see expression 4.2.77), a larger sample volume will generate a larger signal (peak area). This indicates that for samples of lower concentration, a larger injection volume may be desirable. However, a large injection volume may affect the chromatographic peak width [65]. In developing the theory about peak broadening, the value of W_b did not consider any broadening due to injection volume. In such ideal conditions, when an extremely small (narrow) injection is made, the ideal peak broadening is given by expression (4.2.41), which can be written in the form:

$$W_b = \frac{4t_R}{N_0^{1/2}} \tag{4.2.82}$$

where the notation N_0 indicates the theoretical plate number in ideal conditions, without any peak enlargement because of the injection volume of the sample. For the time length W_b in the chromatogram, the volume of mobile phase passing through the column is given by the expression:

$$V_{0,peak} = W_b U \tag{4.2.83}$$

The volume $V_{0,peak}$ is further indicated as the "ideal peak volume." For a sample volume V_{inj} the peak volume increases and the new volume can be approximated using the expression [4]:

$$V_{peak} = \left(1.333\ V_{inj}^2 + V_{0,peak}^2\right)^{1/2} \tag{4.2.84}$$

In expression (4.2.84), the coefficient 1.333 is an empirical value to account for additional broadening during the injection. From expression (4.2.84), the new value for N obtained due to the larger sample volume can be calculated using formula (4.2.82):

$$N = \frac{16\,(t_R U)^2}{1.333\,V_{inj}^2 + V_{0,peak}^2} \tag{4.2.85}$$

Expression (4.2.85) allows the evaluation of the decrease in the column efficiency (as measured by the theoretical plate number) as a function of the volume of the injected sample. The value of N is changed from N_0 by the formula:

$$N = N_0 \frac{V_{0,peak}^2}{1.333\,V_{inj}^2 + V_{0,peak}^2} \tag{4.2.86}$$

Formula (4.2.86) allows to estimate that the loss of efficiency less than 10% corresponds to a value for the injected volume $V_{inj} < 0.3\,V_0$, and for a 1% loss of efficiency a value $V_{inj} < 0.1\,V_0$.

An estimation of $V_{0,peak}$ is now necessary for an estimation of an optimum injection volume that would affect only to a small extent the peak width. For this purpose expression (4.2.83) will be utilized and W_b must be estimated. This can be done using estimations for t_R in expression (4.2.82). The value for t_R is obtained from $t_R = t_0\,(1 + k')$ with an estimation for t_0. The value for t_0 can be obtained from expression (4.2.3) as $t_0 = (\varepsilon^* \pi\, d^2 L)/4U$. Putting together all these estimations, the result is expressed by the formula:

$$V_{0,peak} \approx \varepsilon^* \pi d^2 L(1 + k')/N^{1/2} \tag{4.2.87}$$

For a sample injection that produces a loss of efficiency less than 10% ($V_{inj} < 0.3\,V_0$) with $\varepsilon^* \approx 0.64\;10^{-3}$, the injection volume should be selected such that:

$$V_{inj} < 6 \cdot 10^{-4} d^2 L(1 + k')/N^{1/2} \tag{4.2.88}$$

A few examples of column dimensions and properties with the maximum allowed injection volume required to achieve less than 10% change in the optimum column efficiency (as measured by N) and assuming $k' = 1$ are given in Table 4.2.3.

TABLE 4.2.3 Examples of Maximum Injection Volume of Sample in a Column With the Condition of Not Decreasing Efficiency by More Than 10% (Assuming $k' = 1$).

Length L (mm)	Diameter i.d. d (mm)	Efficiency N	Max. V_{inj} μL
150	4.6	9000	40.2
150	3.0	9000	17.1
100	4.6	10,000	25.5
100	2.1	10,000	5.5
50	2.1	8000	3.0

Although the increased injection volume may produce a reduction of column efficiency, larger volumes than recommended are sometimes used when the resolution is still satisfactory and an increase in analysis sensitivity is necessary for the measurement of compounds present in traces that do not provide a good detector signal. Columns with higher value for N are more sensitive to the volume of the injected sample since the peaks from these columns are narrower. In practice, when the sensitivity of the detector is not high enough, larger injection volumes than described in Table 4.2.2 are utilized.

The amount of material injected in the chromatographic column is an additional parameter that affects the separation. An excessive amount of sample in the chromatographic column leads to stationary phase overload. In such a case, the stationary phase becomes saturated with a specific analyte, and the retention cannot take place within the narrow region occupied by the zone in the column containing the sample. The mobile phase carries the unretained solute on a "fresh" portion of the column where the analyte is retained, but this effect is associated with an apparent lowering of k' values (shorter retention times), increase in peak width, and tailing. The amount (in μg) that can be loaded in an analytical chromatographic column can be roughly estimated using the expression:

$$m = x\,V_0(1+k')N^{-1/2} = \frac{x\,(\varepsilon^*\pi\,d^2L)}{4\,N^{1/2}}(1+k') \tag{4.2.89}$$

In expression (4.2.89), x is a constant depending on the nature of the stationary phase, V_0 is the dead volume of the column, k' is the retention factor, N is the number of theoretical plates of the column, d is column diameter, and L (in mm) is the column length. For conventional analytical columns x varies between 0.05 and 0.2 and depends on the nature of the stationary phase.

4.3 RETENTION AND ELUTION MECHANISMS IN DIFFERENT TYPES OF HPLC

HPLC includes a number of similar techniques that also have specific differences allowing various classifications. The subject of classification of HPLC techniques is covered in many publications (see, e.g., Ref. [5]), and a summary of the subject can be found in Section 2.3. One important criterion that differentiates HPLC techniques is their retention/separation mechanism. For the appropriate selection of a specific HPLC method, it is important to understand these mechanisms. The subject of retention/separation mechanisms is further discussed in the present section for several types of HPLC. A basic discussion on the thermodynamic aspects of retention and separation is also included in this section.

Partition Equilibrium and Its Thermodynamic Aspects

Regardless of the intrinsic mechanism, the retention and separation in HPLC are the result of the equilibrium established between the molecules from the mobile phase and those present in the stationary phase. For several types of chromatography (RP-HPLC, IP, HIC, NARP, HILIC, NP-HPLC, etc.), this equilibrium can be described as a partition process of the analytes between the liquid mobile phase and the stationary phase. Other types of equilibria can be assumed to exist in the chromatographic column, such as an adsorption/desorption on a solid surface. However, the basic conclusions when considering such equilibria remain the same as those obtained from the partition process. Also, special types of equilibria exist in ion chromatography and size-exclusion chromatography and these will be discussed separately. A few basic aspects of the thermodynamics of a partition process are described below.

The equilibrium of molecular species X between mobile phase and stationary phase is indicated by an equilibrium of the form $X_{mo} \leftrightarrow X_{st}$ as shown by formula (4.2.11). This equilibrium is governed by an equilibrium constant $K(X)$ (see expression 4.2.12). The value of $K(X)$ is related to the retention factor (see expression 4.2.17), and is determined by the energies involved in the retention and elution process. This can be shown considering that when the equilibrium is attained for the distribution of the compound X between phases *st* and *mo*, the chemical potentials $\mu_{X,st}$ and $\mu_{X,mo}$ for transferring of the analyte X from the phase *mo* into phase *st* must be equal. Based on the expression of chemical potential (see, e.g., Ref. [66]) given by the expression:

$$\mu_X = \mu_X^0 + RT \ln a_X \tag{4.3.1}$$

where μ_X^0 is the standard chemical potential of compound X and a_X is the activity of analyte X, the following equality can be written as follows:

$$\mu_{X,st}^0 + RT \ln a_{X,st} = \mu_{X,mo}^0 + RT \ln a_{X,mo} \tag{4.3.2}$$

In expression (4.3.2), $\mu_{X,st}^0$ and $\mu_{X,mo}^0$ are the standard chemical potentials of compound X in the phases *st* and *mo*, and $a_{X,st}$ and $a_{X,mo}$ are the activities of analyte X in the two phases *st* and *mo*, respectively, T is absolute temperature, and R is the gas constant ($R = 8.31{,}451$ J deg^{-1} mol^{-1} = 1.987 cal deg^{-1} mol^{-1}). The rearrangement of formuls (4.3.2) leads to the expression:

$$\ln \frac{a_{X,st}}{a_{X,mo}} = -\left(\mu_{X,st}^0 - \mu_{X,mo}^0\right)/(RT) = -\Delta\mu_X^0/(RT) \tag{4.3.3}$$

In expression (4.3.3), $-\Delta\mu_X^0$ represents the change in standard chemical potential when analyte X is transferred from the mobile phase *mo* into the stationary phase *st*. The expression shows that for a constant temperature T the ratio of the activities of the analyte X in the two phases is always a constant $K^{therm}(X)$ (named the thermodynamic distribution constant for the partition process). The activities a_X ($a_{X,st}$ and $a_{X,mo}$) are proportional with the molar concentrations C_X through the relation:

$$a_X = \gamma_X C_X \tag{4.3.4}$$

where γ_X is the activity coefficient for species X (and C_X is the molar concentration). With the replacement of formula (4.3.4) for activity in expression (4.3.3), the ratio of concentrations of species X in the two phases is given by the formula:

$$\frac{C_{X,st}}{C_{X,mo}} = \frac{\gamma_{X,mo}}{\gamma_{X,st}} \exp\left(-\frac{\Delta\mu_X^0}{RT}\right) \tag{4.3.5}$$

The ratio $C_{X,st}/C_{X,mo}$ is the constant $K(X)$ (see expression (4.2.12) for the equilibrium $X_{mo} \leftrightarrow X_{st}$), and taking $\gamma_{X,mo} = \gamma_{X,st}$, expression (4.3.5) becomes the following:

$$K(X) = \exp\left(-\frac{\Delta\mu_X^0}{RT}\right) \tag{4.3.6}$$

For a constant pressure and temperature $\Delta\mu^0 = \Delta G^0$, where ΔG^0 is the variation in the standard free enthalpy (Gibbs free energy) for the process of transferring of the analyte X from the phase *mo* into phase *st*. Therefore, expression (4.3.6) can be written in the form:

$$\ln K(X) = -\frac{\Delta G^0}{RT} \tag{4.3.7}$$

Formula (4.3.7) describes the dependence of the equilibrium of the analyte X between the mobile and the stationary phase as a function of the free energy change of the system. The retention factor k' for the analyte X can be written using formula (4.3.7) as follows:

$$\ln k'(X) = -\frac{\Delta G^0}{RT} + \ln \Psi \tag{4.3.8}$$

The free enthalpy ΔG^0 of the system is related to $\Delta \boldsymbol{H}^0$, which is the change in standard enthalpy, and ΔS^0, which is the change in the entropy by the formula $\Delta G^0 = \Delta \boldsymbol{H}^0 - T\Delta S^0$ (see, e.g., Ref. [66]). As a result, expression (4.3.8) can be written in the form:

$$\log k'(X) = \frac{-\Delta H^0 + T\Delta S^0}{2.303\,RT} + \log \Psi \tag{4.3.9}$$

Formula (4.3.9) is known as the van't Hoff equation and shows that the retention factor k' for a chromatographic separation depends on the change in enthalpy $\Delta \boldsymbol{H}^0$ and entropy ΔS^0 of the system in equilibrium when X is transferred from the phase *mo* into phase *st* (as well as on the column phase ratio Ψ). A larger $\Delta \boldsymbol{H}^0$ (in absolute value since $\Delta \boldsymbol{H}^0$ is negative for exothermic processes) indicates a stronger retention. For a system at constant pressure and volume, as the chromatographic conditions are supposed to be, the enthalpy and energy are equal ($\Delta \boldsymbol{H}^0 = \Delta E^0$), and for this reason it is common when discussing the equilibria in HPLC to indicate changes in energy instead of changes in enthalpy. The contribution of changes in entropy during the separation process is typically lower than the energy component (except for size exclusion separations). Formula (4.3.9) also shows the variation of $\log k'$ with the temperature.

The direct calculations for the values of ΔE^0 are difficult to make. Different types of interactions exist between molecules (van der Waals, hydrogen bonding, ionic, covalent). The ranges of values for various types of interacting energies are given in Table 4.3.1 (1 kJ/mol = 0.239 kcal/mol). Additional types of interaction are known, such as stacking and inclusion interactions, their energies depending on the interacting components.

The interaction energies are directly related to the values of retention times $t_R(X)$ on specific chromatographic systems, but while retention times are easily measured, this is not the case for the energies. For this reason, inversely, the values for the energies can be obtained from the chromatographic parameters that can be measured for a specific separation (retention times and phase ratios) (see also [5]). This is done using the plots showing the linear

TABLE 4.3.1 Typical Values for Different Interaction Energies

Interaction Type	Energy (kJ/mol)	Energy (kcal/mol)
Dispersion	8–30	2–7
Dipole–induced dipole	4–8	1–2
Dipole–dipole	4–13	1–3
Hydrogen bonding	5–40	1–10
Donor–acceptor	10–130	2–30
Ion–dipole	60–130	15–30
Ionic	200–800	50–200
Covalent	200–400	50–100

variation of log k' (or ln k') with inverse absolute temperature, $1/T$, known as van't Hoff plots [5,67–69]. Deviations from the linearity in van't Hoff plots also have been studied and reported in the literature (see, e.g., Ref. [70]).

The contribution to the free enthalpy change for the transfer of X from the phase *mo* into phase *st* comes from the changes suffered by the entire system, and not only from the interactions generated by the analyte with the mobile or stationary phase. These changes must also include any energetic modification in the mobile phase upon the transfer of X from the mobile phase into the stationary phase. As a result, the following formula can be written:

$$\Delta G^0 = \Delta G^0_{analyte-eluent} + \Delta G^0_{eluent-eluent} \tag{4.3.10}$$

With the help of formula (4.3.10), the partition model of the chromatographic equilibrium provides a means to estimate the dependence of its equilibrium constant on the organic content in the mobile phase when this phase is made, for example, from a mixture of water and an organic solvent. For this purpose, it should be assumed that the free energy ΔG^0_{eluent} can be separated into the contributions of the two solvents ($\Delta G^0_{H_2O}$ and ΔG^0_{org}) as follows:

$$\Delta G^0_{eluent-eluent} = (1 - x)\Delta G^0_{H_2O} + x\Delta G^0_{org} \tag{4.3.11}$$

In expression (4.3.11), x is the molar fraction of the organic component in the mobile phase. (Molar fraction for a component X in a solution is the number of moles n_X of X divided by the sum of the total number of moles for the solution components, and it is given by the formula $x_X = n_X / \sum_i^{tot} n_i$. For a diluted solution x_X is related to molar concentration C_X and the average molecular weight of the solvent Mw_{mo} by the formula: $x_X \approx \gamma_X C_X Mw_{mo}/(1000\rho_{mo})$, where γ_X is the activity coefficient for X and ρ_{mo} is the density of the solvent.) Including expressions (4.3.11) and (4.3.10) in expression (4.3.8), the following expression can be written:

$$\ln k'(X) = -\frac{\Delta G^0_{analyte-eluent} + (1 - x)\Delta G^0_{H_2O} + x\Delta G^0_{org}}{RT} + \ln \Psi \tag{4.3.12}$$

The terms in formula (4.3.12) can be grouped as the free term $-\left(\Delta G^0_{analyte-eluent} + \Delta G^0_{H_2O}\right)/RT$ and the coefficient for x which is $\left(\Delta G^0_{H_2O} - \Delta G^0_{org}\right)/RT$, both terms being constant at a constant temperature. Since ln Ψ is also a constant (for a specific separation system) and because x and the volume fraction of the organic solvent in mobile phase ϕ are proportional (see expression 4.2.68), the expression for ln $k'(X)$ (or log $k'(X)$) from formula (4.3.12) can be written in the form:

$$\log k'(X) = a - b\phi \tag{4.3.13}$$

Expression (4.3.13) is equivalent with expression (4.2.69) previously discussed for gradient separations. The linear dependence is verified experimentally for many compounds and solvents [71], although deviations from expression (4.3.13) were also encountered [72]. These deviations were found in particular for compounds with more complex interactions with mobile phase (e.g., acetonitrile) and the stationary phase, but the partition mechanism is still considered as predominant in the separation process.

Adsorption Equilibrium

Another model describing the equilibrium in a chromatographic column is based on the hypothesis that an adsorption/desorption process is the basis of retention-elution. A simplification assumption for the adsorption/desorption process is that the stationary phase is covered by a monolayer of molecules of either solute X or of mobile phase A. It can be assumed in this case, that in order for a molecule X of the analyte to be retained, it should displace one or more molecules of the solvent A that were already adsorbed on stationary phase. Therefore, the distribution of X between mobile phase and stationary phase can be expressed by the displacement equilibrium:

$$X_{mo} + nA_{st} \leftrightarrow X_{st} + nA_{mo} \tag{4.3.14}$$

A constant describing such equilibrium will be given by the formula:

$$K_{ad} = \frac{C_{X,st}}{C_{X,mo}}\left(\frac{C_{A,mo}}{C_{A,st}}\right)^n \tag{4.3.15}$$

In formula (4.3.15) it is possible to replace the concentration ratios with the ratios of molar fractions x, and for simplification it can be assumed that $n = 1$ (one analyte molecule replaces one solvent molecule). In this case, formula (4.3.15) will be written as follows:

$$K_{ad} = \frac{x_{X,st}(1 - x_{X,mo})}{x_{X,mo}(1 - x_{X,st})} \tag{4.3.16}$$

Typically $x_{X,mo}$ is very low and $(1 - x_{X,mo}) \approx 1$, and formula (4.3.16) can be rearranged as follows:

$$x_{X,st} = \frac{K_{ad}x_{X,mo}}{1 + K_{ad}x_{X,mo}} \tag{4.3.17}$$

Formula (4.3.17) is known as Langmuir isotherm. For low concentrations of X in the mobile phase and for relatively low values for K_{ad}, $1 + K_{ad}\, x_{X,mo} \approx 1$ and expression (4.3.17) becomes basically identical with expression (4.2.12) ($K_{ad} = x_{X,st}/x_{X,mo}$) developed for the partition process.

The adsorption model of the chromatographic equilibrium also provides a means to estimate the dependence of equilibrium constant on the organic content in the mobile phase. For this purpose, it is necessary first to replace in formula (4.3.15) the concentrations of the solvent A with molar fractions. In this case, the expression (4.3.15) can be written as follows:

$$K_{ad} = \frac{C_{X,st}}{C_{X,mo}} \left(\frac{x_{A,mo}}{x_{A,st}} \right)^n \qquad (4.3.18)$$

The assumption that the surface of the stationary phase is basically covered mainly with solvent molecules gives $x_{A,st} = 1$. For $C_{X,st}/C_{X,mo} = K_A(X)$ where $K_A(X)$ represents the equilibrium constant for species X in pure solvent A, after multiplication with phase ratio Ψ to change equilibrium constants into retention factors ($K_{ad}\ \Psi = k'_{ad}$ and $K_A\ \Psi = k'_A$), and after taking the logarithm, the following formula is obtained:

$$\log k'_{ad}(X) = \log k'_A(X) - n \log x_{A,mo} \qquad (4.3.19)$$

Formula (4.3.19) is known as the Soczewinski (or Snyder–Soczewinski) equation and indicates that if the chromatographic equilibrium is based on adsorption/desorption, the logarithm of retention factor should depend linearly on the logarithm of the molar fraction of solvent A in the mobile phase [73]. The molar fraction x being proportional with volume fraction $\phi_{A,mo}$, expression (4.3.19) shows that the logarithm of retention factor $\log k'$ for the adsorption/desorption mechanism depends linearly on $\log \phi_{A,mo}$ (the logarithm of volume fraction of solvent A), and not on $\phi_{A,mo}$ as indicated by expression (4.3.13) (which is valid for the partition model).

The Role of Polarity in Separation Mechanisms

The concept of polarity is frequently utilized in chromatographic separations, and the term "polarity" is typically used with a wide (sometimes ill-defined) meaning. In strict terms polarity refers to the values of permanent dipole moment μ and of polarizability α of a molecule. However, very frequently the term is used with a qualitative meaning. For small molecules, polarity can be assessed from the presence of polar groups in their structure. (Polarity of the groups can be evaluated based on the separation of the center of partial positive charges from that of partial negative charges. The capability of formation of hydrogen bonds is also usually considered.) The term is also extended to larger molecules to moieties in solid phase structures, or even to mixtures of molecules in solvents, although in many cases the dipole moment μ and polarizability α cannot be defined for such systems. For example, a stationary phase containing on the surface acidic -OH groups is considered polar, and when the stationary phase contains on its surface long hydrocarbon groups, it is considered not polar (hydrophobic). Also, a mixture of solvents containing a large proportion of water is considered polar, and when it contains a large proportion of an organic solvent, it is considered hydrophobic. This classification is not based strictly on the values for dipole moment or polarizability (for water $\mu = 3.12$ D and $\alpha = 1.51\ 4\pi\varepsilon_0(\text{Å})^3$ and for acetonitrile $\mu = 3.39$ D and $\alpha = 4.27\ 4\pi\varepsilon_0(\text{Å})^3$, but water is considered polar and acetonitrile less polar).

The polar character of small molecules can be characterized well using the octanol/water partition constant (coefficient or parameter). The octanol/water partition constant K_{ow} (in some texts denoted as P) represents the ratio of molar concentrations of a (not ionized) compound distributed between two phases, one being octanol and the other water, and it is expressed by the formula (see also Section 5.1):

$$K_{ow}(X) = \frac{C(X)_{octanol}}{C(X)_{water}} \tag{4.3.20}$$

A higher value for K_{ow} indicates a low polarity (a high hydrophobic character of the molecule), while a low value for K_{ow} indicates a polar molecule. Hydrocarbons are typical examples of hydrophobic molecules, while compounds with groups such as $-COOH$, $-OH$, and $-NH_2$ are polar. The octanol/water partition constant has a widespread utilization in separation science and also in other important fields of science, such as drug design and environmental studies. For this reason, values for K_{ow} are available in the literature for a large body of compounds [74,75], and also can be calculated using dedicated computer programs (e.g., MarvinSketch 5.4.0.1, ChemAxon Ltd. [76], EPI Suite [75]). The log K_{ow} values can be successfully utilized for the selection of HPLC type and even for obtaining information about the separation process in RP-HPLC. For example, the octanol/water partition coefficient for a compound X was proven experimentally to correlate well with the retention factor log k' in RP-HPLC (see Section 5.3).

The characterization of polarity with octanol/water partition coefficient is frequently extended from single small molecules to solvent mixtures and even to stationary phases. However, such K_{ow} values for solvent mixtures are only hypothetical (K_{ow}^{hyp}) and are generated with the goal of providing estimations or models (see Section 5.1). They serve the purpose of providing only some guidance without the capability to give an accurate value for K_{ow}. For stationary phases the concept of "polarity" is used to describe the polarity of the moieties of the stationary phase interacting with the analytes. Models for a stationary phase were produced to simulate the phase with a relatively small molecule such that a value K_{ow}^{model} can be generated and used to compare phases (see Section 6.2). In spite of being vaguely defined, the use of the concept of "polarity" for both mobile and stationary phases is very common.

Mechanism in Reversed-Phase HPLC

Reversed-phase HPLC (RP-HPLC) is performed on a nonpolar stationary phase with a polar mobile phase that contains water. The technique is applied to molecules that have in their structure hydrophobic moieties, but can also have some polar parts (such as groups $-OH$, $-NH_2$, $>C{=}O$, $-COOH$, $-CONH_2$, and $-Cl$). This is the case for numerous types of organic molecules, which explains the widespread use of RP-HPLC as an analytical technique. A common type of stationary phase in RP-HPLC consists of a nonpolar material bonded on a porous silica support (silica offers a large contact surface with the solvent). Examples of bonded materials on silica are hydrocarbon chains containing 8 carbon aliphatic chains (C8), 18 carbon aliphatic chains (C18 or ODS), or phenyl groups. Less common RP-HPLC stationary phases are polymeric organic materials such as polystyrene crosslinked with divinylbenzene (PS-DVB), and other inorganic supports (e.g., zirconia) coated with organic groups.

The mobile phase in RP-HPLC is typically partially aqueous partially organic solvent and it is more polar than the stationary phase (e.g., based on a K_{ow}^{hyp} value). The solute molecules are in equilibrium between the more polar mobile phase and the hydrophobic stationary phase, the direction of the equilibrium determining a stronger or a weaker retention of the analytes. This retention depends mainly on three factors: structure of the analyte molecule, how polar is the solvent, and what (nonpolar) stationary phase is used.

The retention in RP-HPLC is, with very good approximation, assumed to have a partition mechanism [77,78]. Only the separation on RP phases of polymeric molecules, such as proteins, is more dominated by an adsorption process. The energetic details of the process are best explained with the help of solvophobic theory [79,80]. This theory shows that, for retention, a hydrophobic molecule is "expelled" from a polar solvent and is "accepted" in a hydrophobic stationary phase. This process is caused by the fact that in order to dissolve a molecule having nonpolar moieties in a polar solvent, it is necessary to eliminate polar↔polar interactions between the solvent molecules. Dissolution produces a disruption of polar bonds in the mobile phase. This disruption takes place when the space is generated for placing the solute molecules in the polar solvent. (The polar↔polar interactions include the following: dispersion, dipole–induced dipole, dipole–dipole, as well as hydrogen bonding, with energies indicated in Table 4.3.1). Because the interactions between polar molecules of the solvent are much stronger than the interactions between the solvent and the solute, it is energetically favorable to have the solute eliminated from the mobile phase and placed in the stationary phase (which is hydrophobic). In other words, for RP-HPLC the following expression can be written for the retention process (vertical bars indicate absolute value):

$$\left|\Delta G^0_{eluent-eluent}\right| \gg \left|\Delta G^0_{analyte-eluent}\right| \qquad (4.3.21)$$

vertical bars indicate absolute value and are necessary because free enthalpy has negative values for the displacement of the equilibrium $X_{mo} \leftrightarrow X_{st}$ to the right. Expression (4.3.21) is caused by the formation of more polar↔polar (i.e., solvent↔solvent) interactions in the mobile phase when the solute is placed in the stationary phase.

The interactions nonpolar↔nonpolar between the analyte and the hydrophobic stationary phase are in fact weaker than nonpolar↔polar ones (of the analyte with the solvent), and therefore it would appear counterintuitive that nonpolar molecules would prefer a nonpolar medium. However, the nonpolar↔polar interactions (of the analyte with the solvent) are much weaker than polar↔polar interactions among the solvent molecules, and such polar↔polar interactions are generated when the solute leaves the polar solvent. The replacement of nonpolar↔polar interactions with polar↔polar interactions is energetically favorable.

For solvents containing water or other solvents that form hydrogen bonds, polar↔polar interactions may have energies in the range of several kcal/mol. Many analyte molecules have both polar and nonpolar moieties, and although soluble in a polar solvent (e.g., partially aqueous partially organic), they offer weaker interactions than those occurring between the strongly polar solvent molecules. For this reason, the analyte is easily displaced from the mobile phase and placed in the hydrophobic stationary phase. This process is schematically pictured in Fig. 4.3.1 (polar interactions may include different types as described in Table 4.3.1). The process pictured in Fig. 4.3.1 is supported by direct calculations of the

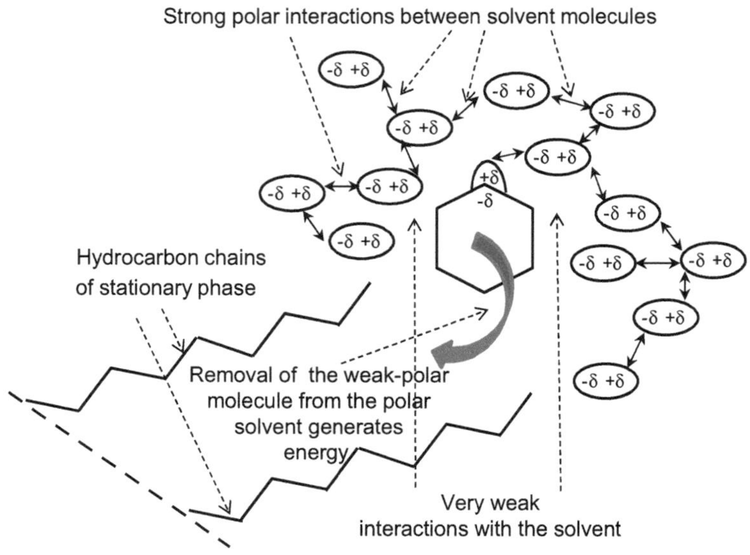

FIGURE 4.3.1 Schematic description of solvophobic effect where the removal of a weak-polar molecule from the polar solvent and placing it in the stationary phase is energetically favorable.

interactions in the reversed-phase retention [81]. In RP-HPLC, the more hydrophobic is a molecule, the higher is its tendency to remain in the stationary phase since the gain in energy by its removal from the mobile phase is higher. Because the "retention" of analytes with nonpolar moieties in RP-HPLC is in fact a "rejection" from the mobile phase, the role of the mobile phase is very important in this technique [82–85].

By decreasing the polarity of the mobile phase (with the addition of an organic component), the intensity of the interactions between the solvent molecules decreases, the free enthalpies in expression (4.3.21) become equal, and then the first term becomes smaller (in absolute value) than the second one, such that the compound elutes. A mobile phase in HPLC capable of eluting the analytes at shorter retention times compared to another one is typically indicated as "stronger." The "strength" of common solvents (related to their hydrophobic character) in RP-HPLC decreases in the order: acetonitrile > isopropanol > ethanol > methanol > water. The fine equilibrium based on polarity/hydrophobicity between the analyte, the mobile phase, and the stationary phase allows the separation of numerous types of molecules. Depending on hydrophobicity, the analytes are rejected more or less efficiently from the mobile phase, and they can be separated generating different free enthalpies ΔG^0 and, as a result, different k' values and different retention times.

Although the chromatographic retention and elution in RP-HPLC is dominated by the process of hydrophobic molecules being "expelled" from a polar solvent and "accepted" in the hydrophobic stationary phase, as explained by the solvophobic theory, other interactions also have contributions to the separation, and they are accounted for in solvophobic theory (see, e.g., Ref. [5]). Among these are the interactions of the analyte with the stationary phase. A stationary phase with very low polarity will still be more accessible to hydrophobic molecules than a less hydrophobic stationary phase. This explains why not all hydrophobic stationary phases respond equally to analyte molecules with no polar groups such as hydrocarbons (aliphatic, aromatic, etc.).

Because the retention/elution process in RP-HPLC is basically dominated by one mechanism, and the mass transfer process between the liquid mobile phase and the solid phase is rapid, the column efficiency in RP-HPLC is better than in any other HPLC technique, and the peak shape is frequently close to the ideal Gaussian [86,87].

When the mobile phase in RP has a higher content in an organic solvent, the elution is faster since the solutes are more easily transferred back into the mobile phase. The variation of log k' with the concentration of the organic phase is predicted by expression (4.3.13). The decrease of k' with the increase in the content of organic solvent is illustrated in Fig. 4.3.2 using three compounds (benzene, toluene, and ethylbenzene) separated on a Chromolith Performance RP-18 100 × 4.6 mm, pore size 130 Å column, with the mobile phase water/methanol, at different methanol proportions. As shown in Fig. 4.3.2, log k' depends very close to linear on ϕ (% CH_3OH), indicating a partition process. The graphs shown in Fig. 4.3.2 also indicate the decrease of log k' as the organic component increases in the mobile phase, as predicted by expression (4.3.13). Partition is the dominant process in most RP-HPLC separations.

The increase in the proportion of organic phase in the mobile phase is performed during the chromatographic run in gradient elution (see Section 4.2). Using gradient, the compounds less retained will be separated at a lower content of organic phase, while the higher proportion of organic phase, which produces faster elution, is seen by the compounds with higher retention, such that their elution is accelerated. Based on this procedure, the run time of the chromatogram is shortened and the gradient is selected such that the separation is achieved for both less retained and higher retained compounds.

Since most molecules separated by RP-HPLC frequently have polar (or even ionic) groups in addition to the hydrophobic moiety, other interactions also contribute to the retention/elution process. Molecules with polar moieties may develop several types of interactions (acceptor H-bonding or donor H-bonding) with the silanol −OH groups still present in C18 or C8 silica-based columns, ion−ion or ion−dipole interactions with the ionized −OH groups (in the form $-O^-$) also from the silanols, or interactions with the traces of metallic impurities from the silica. Besides these interactions, the retention/elution process can be influenced by π−π interactions [88] as well as by steric hindrances. Also, more metal impurities in

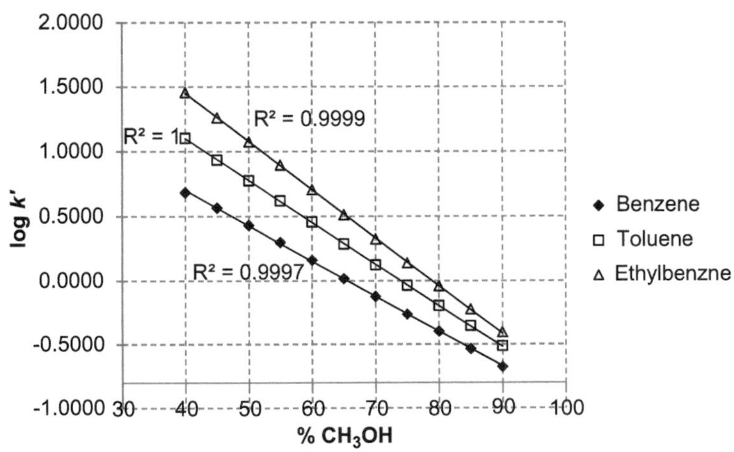

FIGURE 4.3.2 Dependence of log k' on % organic component in the mobile phase for benzene, toluene, and ethylbenzene, separated on a C18 monolithic column, with water/methanol mobile phase.

the silica-base of the column may lead to more unexpected interactions. Polar interactions are stronger when the RP-HPLC columns are not end-capped, when they are end-capped with polar groups (e.g., Synergi Hydro type column from Phenomenex, Accucore aQ C18, Hypersil Gold aQ C18, Syncronis aQ C18 from Thermo Scientific.), or when they contain embedded polar groups (see Chapter 7).

Retention Factor in RP-HPLC as Predicted by Solvophobic Theory

Solvophobic theory [79] has been successfully utilized for the explanation of the process of retention and elution in RP-HPLC [80]. The theory also allows a description of several parameters related to the analyte, stationary phase, and mobile phase that affect retention in RP-HPLC. However, the direct calculation of a retention factor based on the theory is impractical, because of the use of several molecular parameters with values not commonly available.

The separation in RP-HPLC can be considered with good approximation to be a partition process between an immobilized liquid S acting as stationary phase, and the liquid mobile phase L. With this assumption, expression (4.3.7) for $\ln K(X)$ shows that equilibrium constant for a partition process depends on ΔG^0, the variation in the standard free enthalpy (Gibbs free energy) for the process of transferring of the analyte X from one phase to the other. Solvophobic theory allows the calculation of ΔG^0. Assuming that no volume changes occur during the process, the free enthalpy ΔG^0 is taken as equal to the free energy of the process ΔA^0 and expression (4.3.8) can be written in the form:

$$K(X) = \exp\left(-\frac{\Delta A^0_{X,S} - \Delta A^0_{X,L}}{RT}\right) \tag{4.3.22}$$

In Eq. (4.3.22), the free energy $\Delta A^0_{X,S}$ result from the change in energy necessary for placing the molecular species X into the solvent formed by molecules S. Likewise, the process of placing the molecular species X in solvent L generates the free energy $\Delta A^0_{X,L}$. (Symbol Δ and index 0 for standard expressions of energy will be further omitted for the simplification of notation.) The change in energy is caused by the free energy required for the creation of the cavity in the solvent to accommodate the species X indicated as A^{cav} plus the free energy of van der Waals interactions between the molecule X and the surrounding molecules of the solvent, indicated as A^{vdW} (the molecule X is assumed to not be ionic). In conclusion, each free energy ($A_{X,S}$ and $A_{X,L}$) should have an expression of the form:

$$A^{sol} = A^{cav} + A^{vdW} \tag{4.3.23}$$

The free energy of van der Waals interactions can be further separated into two different terms, one caused by electrostatic forces A^{es} and the other due to dispersion forces A^{disp} such that A^{sol} can be written for each solvent as follows:

$$A^{sol} = A^{cav} + A^{es} + A^{disp} \tag{4.3.24}$$

Formulas for each term in expression (4.3.24) are described in the literature [79–81]. For the solvent S, the formula for A^{cav} is the following:

$$A^{cav}_{X,S} = \mathcal{N}\left[1 + (\kappa^e_S - 1)(V_S/V_X)^{2/3}\right]\mathcal{A}_X\gamma'_s(1 - W_S) \tag{4.3.25}$$

In expression (4.3.25), $\mathcal{N}$ is Avogadro number ($\mathcal{N} = 6.022{,}141 \cdot 10^{+23}$), V_S and V_X are the molar volumes for the solvent S and for the molecule X, respectively, $\mathcal{A}_X$ is the surface area of the cavity in the solvent necessary to accommodate the molecule X, γ_S' is the surface tension of solvent S, W_S is a correction factor with value close to 0 that will be further neglected, and κ_S^e is a coefficient with the following formula:

$$\kappa_S^e = \frac{\mathcal{N}^{1/3}\boldsymbol{\Delta H}_{vap,S}}{V_S^{2/3}\gamma_S'\left(1 - \dfrac{d\ln\gamma_S'}{d\ln T} - \dfrac{2}{3}C_{exp\,a,S}T\right)} \tag{4.3.26}$$

In formula (4.3.26), $\boldsymbol{\Delta H}_{vap,S}$ is the heat of evaporation of the solvent and is available in the literature, and $C_{expa,S}$ is the coefficient of thermal expansion for the solvent S.

The expression for $A_{X,S}^{es}$ is obtained from the Onsager reaction field (see, e.g., Ref. [5]), and is given by the formula:

$$A_{X,S}^{es} = -\frac{\mathcal{N}^2\mu_X^2}{2V_X}\mathcal{D}_S\mathcal{P}_{X,S} \tag{4.3.27}$$

In expression (4.3.27), μ_X is the dipole moment of molecule X, and $\mathcal{D}_S$ and $\mathcal{P}_{X,S}$ have the following formulas:

$$\mathcal{D}_S = \frac{2(\varepsilon_S - 1)}{2\varepsilon_S + 1} \quad \mathcal{P}_{X,S} = \frac{V_X}{4\pi\varepsilon_0(V_X - \mathcal{N}\mathcal{D}_S\alpha_X)} \tag{4.3.28}$$

where ε_S is the dielectric constant for solvent S and ε_0 is vacuum permittivity ($\varepsilon_0 = 8.854 \cdot 10^{-12}\ \mathrm{CV^{-1}m^{-1}}$).

The expression for $A_{X,S}^{disp}$ is obtained by using an effective pair potential that contains a correction from the gas phase potential of a Lennard–Jones-type interaction [5]. Following a relatively elaborate calculation [79] the following formula is obtained:

$$A_{X,S}^{disp} = -\frac{27(1-x)}{8\pi}\left(Q_{X,S}' + Q_{X,S}''\right)\mathcal{Y}_{X,S}D_X D_S \tag{4.3.29}$$

In expression (4.3.29), x is a proportionality constant with the typical value of $x \approx 0.436$, $Q_{X,S}'$ is a function of molecules X and S diameters and reduced vapor pressures, $Q_{X,S}'' \approx 0.1\ Q_{X,S}'$, $\mathcal{Y}_{X,S}$ is a function depending on ionization potentials of X and S, and D_X and D_S are the Clausius–Mosotti functions for the molecular species X and S, dependent on molecular volume and polarizability α of each molecule, and given (for X) by the formula:

$$D_X = \frac{4\pi}{3V_X}\mathcal{N}^2\alpha_X \tag{4.3.30}$$

Similar expressions as those indicated for solvent S are valid for solvent L. Placing together in expression (4.3 22) all the formulas for A^{cav}, A^{es}, and A^{disp} for the two solvents S and L the final result can be written as follows:

$$\ln K(X) = \boldsymbol{a}\mathcal{A}_X + \boldsymbol{b}(V_X)^{-2/3}\mathcal{A}_X + \boldsymbol{c}_X\mu_X^2 + \boldsymbol{d}_X\boldsymbol{\alpha}_X \tag{4.3.31}$$

Parameters $\boldsymbol{a}$, $\boldsymbol{b}$, $\boldsymbol{c}_X$, and $\boldsymbol{d}_X$ have rather complicated expressions which are indicated below:

$$\boldsymbol{a} = \frac{\mathcal{N}}{RT}\left(\gamma'_L - \gamma'_S\right) \tag{4.3.32}$$

$$\boldsymbol{b} = \frac{\mathcal{N}}{RT}\left[\left(\kappa^e_L - 1\right)\gamma'_L (V_L)^{2/3} - \left(\kappa^e_S - 1\right)\gamma'_S (V_S)^{2/3}\right] \tag{4.3.33}$$

$$\boldsymbol{c}_X = -\frac{\mathcal{N}^2 V_X}{8\pi\varepsilon_0 RT}\frac{\mathcal{D}_L - \mathcal{D}_S}{(V_X - \mathcal{N}\mathcal{D}_L\alpha_X)(V_X - \mathcal{N}\mathcal{D}_S\alpha_X)} \tag{4.3.34}$$

$$\boldsymbol{d}_X = -\frac{2.791\mathcal{N}^2}{RTV_X}\left(\mathcal{Q}'_{X,L}\mathcal{Y}_{X,L}D_L - \mathcal{Q}'_{X,S}\mathcal{Y}_{X,S}D_S\right) \tag{4.3.35}$$

In expression (4.3.31), the surface area $\mathcal{A}_X$ of the cavity in the solvent necessary to accommodate the molecule X can be estimated as equal with van der Waals area of molecule, when the whole molecule is hydrophobic. The notation $\mathcal{A}_X$ will further indicate the van der Waals surface area of molecule X.

Some simplifications can be made to expression (4.3.31) based on the fact that the polarizabilities α_X of many compounds are proportional with van der Waals surface area of the molecule [81]. With the use of a proportionality constant C_t and with the notation $\boldsymbol{d}^*_X = C_t \cdot \boldsymbol{d}_X$, formula (4.3.31) can be further simplified as follows:

$$\ln K(X) = \left(\boldsymbol{a} + \boldsymbol{d}^*_X\right)\mathcal{A}_X + \boldsymbol{b}(V_X)^{-2/3}\mathcal{A}_X + \boldsymbol{c}_X\mu^2_X \tag{4.3.36}$$

The value for $\boldsymbol{d}^*_X$ was found to vary only slightly from compound to compound (although it is dependent on the solvents S and L). Further simplification can be made by neglecting the term $\boldsymbol{c}_X\mu^2{}_X$, which is very small [81]. As a result, the expression for ln K(X) can be written as follows:

$$\ln K(X) = \left(\boldsymbol{a} + \boldsymbol{d}^*_X\right)\mathcal{A}_X + \boldsymbol{b}(V_X)^{-2/3}\mathcal{A}_X \tag{4.3.37}$$

Formula (4.3.37) is valid only assuming the molecule of interest is hydrophobic. In case the molecule has polar groups these will by solvated, and from the value of van der Waals area $\mathcal{A}_X$ a part corresponding to each of these polar groups must be subtracted. Including this subtraction (for several groups "j"), and changing the natural logarithm into decimal logarithm, expression (4.3.37) can be written in the following form:

$$\log K(X) = a'\mathcal{A}_X - \sum_j b''_j \tag{4.3.38}$$

In expression (4.3.38), a' is a coefficient depending only on the nature of phases S and L, and b''_j are constants with specific values for each type of polar functionality present in the molecule X (e.g., −OH, −COOH, =CO, $-NH_2$), and with $b'' = 0$ for hydrocarbons (since their whole molecule is hydrophobic).

Several conclusions can be obtained based on solvophobic theory: (1) retention in RP-HPLC is larger for molecules with a larger van der Waals surface area, (2) the presence of polar groups in the molecule is decreasing the retention (as expected), (3) a larger surface tension γ'_L of the mobile phase leads to a stronger retention (see formula 4.3.32), and (4) the dipole moment μ_X of the analyte plays a minor role in the retention process. In theory, expression (4.3.38) should also allow the calculation of log $k'(X)$ for a system where a', b''_j, and Ψ are known ($k'(X) = K(X)\Psi$), but the correct values for these parameters are difficult to have.

Mechanisms in Ion-Pair Chromatography (IP)

This technique is applied for the separation on hydrophobic columns of compounds with ionic or partially ionic character that cannot be separated by conventional RP-HPLC since they do not have enough hydrophobicity to be retained by the column. However, a modification can be performed such that RP-type chromatographic columns and polar solvents can be utilized for the analysis of these types of compounds. The modification consists of the addition of a reagent in the mobile phase, this reagent being indicated as an ion-pairing agent (*IPA* or hetaeron). The ion pairing agent, *IPA*, is an ionic species selected such that it has opposite charge to the analyte and is able to form molecular association with it, but at the same time has a hydrophobic moiety. Compounds used as *IPA* can be, for example, quaternary amines (with various length of the hydrocarbon chain) in case of analyzing acids, or strong organic acids (e.g., sulfonic acid with various hydrocarbon chain substituents) in the case of analyzing amines. The *IPA* must have both a hydrophobic moiety, in order to participate in the equilibrium mobile phase/stationary phase in RP-HPLC, and an ionic (or strongly polar) group to interact with the analyte. As shown in Table 4.3.1, the ionic interactions (ion–dipole or ion–ion) may generate considerable energy that bonds the analyte and *IPA*.

There are two limiting model theories explaining the mechanism of retention in IPC: (1) a partition model where the mechanism assumes the formation of ion pairs in solution, with the ion pair formed between the ionic or partially charged molecules of the analyte and the *IPA* with an opposite charge, followed by the retention of the preformed ion pair by an RP-HPLC mechanism, and (2) an electrostatic model, which assumes that *IPA* is first bound to the hydrophobic stationary phase by its hydrophobic moiety, and once adsorbed it interacts with the components of the mobile phase by ionic/polar forces. Both mechanisms indicate similar dependencies of the retention factor k' on the hydrophobicity of the analyte, concentration of *IPA*, polarity of *IPA*, and pH (as well as phase ratio Ψ) [89,90].

In partition model, once the ion pair is formed, the separation can be considered as based on the interaction of the ion pair with the hydrophobic stationary phase in a polar solvent. The association mechanism of the analyte and the *IPA* is rather complicated. According to the partition model, three equilibria characterize the partition process of an ion pair between an aqueous phase (*mo*) and an immiscible organic (bonded) phase (*st*) that is immobilized as stationary phase. For a basic compound B^+OH^-, the following equilibria must be considered:

$$\begin{aligned} (B^+OH^-)_{mo} &\leftrightarrow (B^+OH^-)_{st} \\ B^+ + IPA^- &\leftrightarrow B^+IPA^- \\ (B^+IPA^-)_{mo} &\leftrightarrow (B^+IPA^-)_{st} \end{aligned} \tag{4.3.39}$$

Taking into account the equilibrium constant K_{BOH} for free *BOH* between stationary and mobile phase, the basicity constant K_b for *BOH*, and the stability constant $K_{B^+IPA^-}$ of the complex $B^+ IPA^-$, the result for the constant D_B describing the total concentration of species B between the stationary and mobile phase is given by the formula [5] (K_w is the ion product of water):

$$D_B = K_{BOH} + K_{B^+IPA^-}\frac{K_b 10^{-pH}}{K_w + K_b 10^{-pH}} \tag{4.3.40}$$

Formula (4.3.40) shows that for a basic compound and a strong acid *IPA*, the formation of the complex in the solution is favored by higher retention of the analyte on the stationary phase (higher K_{BOH}), higher stability constant $K_{B^+IPA^-}$ of the complex, higher basicity of the analyte K_b, and a lower pH. For an acidic analyte, the same parameters will influence the formation of the ion pair, but the *IPA* should be an ionizable base, and the equilibrium is favored by a stronger acidity constant of the analyte and a higher pH. The equilibrium constant governing the retention in IPC is still given by an expression similar to Eq. (4.3.7), and depending on the model used to explain the retention, the source of interaction enthalpy ΔH^0 is either similar to that from RP-HPLC, or is based on polar interactions between the solute and the *IPA* retained on the stationary phase.

The formula for the capacity factor ($\ln k'$) in ion pair chromatography as obtained in the electrostatic model is the following [91]:

$$\ln k' = \ln k'(0) - \frac{z z_{IPA}}{z_{IPA}^2 + 1}\left[\ln C_{IPA} + \ln\frac{n_{IPA,max}^{s.p.} K_{IPA}}{\kappa} + \ln\frac{F^2}{\varepsilon_0 \varepsilon_{mo} RT} + 1\right] \tag{4.3.41}$$

In formula (4.3.41), $k'(0)$ is the capacity factor of the analyte as an ionic species in the absence of *IPA*. If the analyte even in its ionic form has some retention on the hydrophobic column, this contributes to the retention. The charge z of the analyte and z_{IPA} the charge of the *IPA* also have importance in analyte−IPA complex retention. Since these charges must be opposite (e.g., 1 and −1), the second term in expression (4.3.41) is positive. This second term depends on several parameters of the separation. One parameter is the concentration of *IPA* that increases the retention when it is higher (through $\ln C_{IPA}$). Another parameter is K_{IPA} which is the constant governing the adsorption of *IPA* on the stationary phase. A higher affinity of *IPA* for the stationary phase (higher K_{IPA}) increases the retention. As shown in expression (4.3.41) higher K_{IPA} leads to higher k', although the value of K_{IPA} is multiplied in by $n_{IPA,max}^{s.p.}$, which is the maximum value for the *IPA* molar fraction on the surface of the stationary phase, and it is divided by the parameter κ (Debye length) given by the formula:

$$\kappa = F\left(\frac{\sum_j z_j^2 C_j}{\varepsilon_0 \varepsilon_{mo} RT}\right)^{1/2} \tag{4.3.42}$$

In expression (4.3.42), F is Faraday constant, z_j the charges of all ionic species in the mobile phase, C_j their concentrations, ε_0 is the electrical permittivity of vacuum, and ε_{mo} is the dielectric constant of the mobile phase.

The values of $\ln k'$ may be difficult to directly calculate based on formula (4.3.41), but this formula provides valuable guidance regarding which parameters affect the retention in ion pair chromatography. Similar predictions as from formula (4.3.43) are made by the partition model which indicates a stronger retention when *IPA* has higher concentration and has a larger hydrophobic moiety [92]. A faster elution occurs when the mobile phase contains more organic component and is less polar.

Other more complex treatments of the retention process in IP were developed that attempt to include in the equilibria both the formation of ion pairs in solution as well as retention of the analyte after the *IPA* is adsorbed on the stationary phase [93,94].

Mechanisms in Normal-Phase HPLC (NP-HPLC or NPC) and in HILIC

Normal-phase HPLC and HILIC are two types of chromatography in which the stationary phase is polar while the mobile phase is less polar than the stationary phase. NPC uses only nonpolar or weak polar solvents as mobile phases, while in HILIC the mobile phase, although less polar than the stationary phase, contains a certain proportion of water. The two techniques are used for very different types of analytes. Because NPC uses nonaqueous solvents, it is useful for the analysis of strongly hydrophobic molecules such as fats, and other highly lipophilic compounds that are not water-soluble. On the other hand, HILIC is typically used for the analysis of very polar (or partially ionic) compounds that have too small hydrophobic moieties to be analyzed by RP-HPLC (typically with log K_{ow} <0). The polar interactions (including hydrogen bonding) between the analyte molecule and the stationary phase can be considered the main source of energy (interaction enthalpy, ΔH^0) responsible for the analyte retention, although the process is not limited to these interactions. Electrostatic/ionic interactions are also likely to contribute to the retention process (several polar types of interactions are listed in Table 4.3.1.) In other words, for NPC and HILIC, the following expression can be written for the free enthalpy of the system during retention:

$$\left|\Delta G^0_{analyte-eluent}\right| \ll \left|\Delta G^0_{analyte-stationary\ phase}\right| \tag{4.3.43}$$

The source of ΔG^0 (or ΔH^0) in HILIC is therefore different from the source of ΔG^0 for RP-HPLC. The driving interaction for the retention is therefore that between the analyte and the stationary phase. The stationary phase in NP-HPLC and HILIC is typically made from materials such as bare silica, or silica covered with polar groups (such as NH_2, diol, amide, or zwitterionic groups.). On specific columns such as those containing a nonpolar handle (or spacer) for attaching the polar group to the silica-base, hydrophobic interactions also may play some role when the analyte has larger nonpolar moieties in addition to the polar ones. In addition, some polar groups (e.g., amine) can develop ion exchange-type interactions. As a result, in HILIC, the interaction mechanism with the stationary phase includes the following: (1) polar (and hydrogen bonding), (2) ionic, and (3) hydrophobic interactions. Besides the main interaction types, other interactions such as steric hindrance may also play a role in HILIC separations. Because these complex interactions between the analyte and the stationary phase may take place with somewhat different rates, and also because the main driving force of the separation is solid-phase↔analyte interaction (and not interaction between free molecules in a liquid as it is in RP-HPLC) the column efficiency in HILIC and in NPC is usually lower than in RP-HPLC (the peaks are usually broader) [86]. The elution of the analytes is performed with more polar solvents eluting the analytes faster than more hydrophobic solvents (inversely than in RP-HPLC). When the solvent is polar enough, it offers interactions equal to or stronger than the stationary phase can offer and the elution is possible. The "strength" of solvents in HILIC (related to their polar character) is water > methanol > ethanol > isopropanol > acetonitrile. The pH of the mobile phase in HILIC is also very important. Many polar molecules may be present in molecular or ionic form in solution, and the polarity of the ionic form is much higher than that of neutral molecule. As a result polar molecule retention is different depending on mobile phase pH.

The equilibrium process in both normal-phase HPLC and in HILIC is also a combination including both partition and adsorption. This can be seen from the type of dependence of log k' on the solvent composition of the mobile phase. In some cases, the two processes, partition and adsorption, can be noticed separately. One such variation is exemplified in Fig. 4.3.3 showing the variation of log k' with the amount of water added to the mobile phase in an HILIC separation.

Fig. 4.3.3 shows that the increase in water proportion in the mobile phase leads up to a point to the decrease in retention factor k', as expected for these types of HPLC. However, beyond a specific water concentration, an increase in retention factor can be noticed, in particular for less polar compounds. This effect indicates that HILIC separation is a more complex process than simply based on polar analyte–stationary phase interactions. The increase in log k' for an excess of water indicates that some hydrophobic interactions may also have a contribution to the retention/elution process. The HILIC columns do not contain only polar moieties, and it is likely that at some point the interactions between solvent molecules provide more energy than the interaction between the analyte and the solvent, similar to the case of RP-HPLC. It is also likely that the layer of water adsorbed on the polar surface of a polar stationary phase plays an important role in the separation, and may be the medium where the retention takes place. The linear interval of dependence of log k' on log ϕ_w is typically interpreted as a region of water concentration in the mobile phase where the retention equilibrium

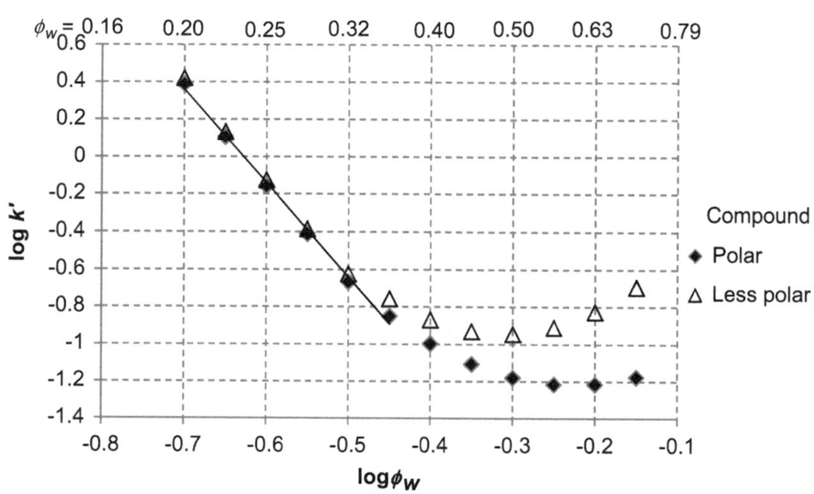

FIGURE 4.3.3 Variation of log k' with the amount of water (ϕ_w is volume fraction of water in the mobile phase) in the mobile phase in an HILIC separation for a polar compound and a less polar one, where the partition process and the adsorption process can be seen separately.

is based on adsorption and not partition because it follows expression (4.3.19) and not expression (4.3.13) [95].

In other cases of HILIC separations, the two types of process (partition and adsorption) take place simultaneously, and the variation of log k' with the content of volumetric fraction of water content $\phi_w = V_w/(V_w + V_{org})$ follows an equation of the type [96]:

$$\log k' = a - b\phi_w - c \log \phi_w \tag{4.3.44}$$

Parameters a, b, and c depend on the column, analyte, and mobile phase. The increase in water content leads to a decrease in the value of log k'. Other equations, polynomial or including both linear and logarithmic dependence, were suggested for describing the variation of log k' with the content of water ϕ_w for HILIC separations [97].

The combined partition and adsorption mechanisms for HILIC separations are also reflected in the equation showing the dependence of log k' on temperature. For a partition process, this dependence is given by van't Hoff Eq. (4.3.9), which can be written in the form:

$$\log k' = a + \frac{b}{T} \tag{4.3.45}$$

For HILIC separations, the temperature dependence of log k' showed deviations from Eq. (4.3.45), and better fits with experimental data were obtained, for example, by the addition of a quadratic term (c/T^2) to Eq. (4.3.45) [97].

Equilibria in Ion-Exchange Chromatography

Ion-exchange chromatography is dedicated to the separation of ionic species of the sample present in the mobile phase (water or water with a given proportion of a polar solvent). The ion species can be inorganic ions, but also organic ions such as organic acids, amines, amino acids, proteins, or oligonucleotides. The stationary phase is an ion-exchange resin, i.e., a polymer that contains acidic groups (in the case of a cation exchange) or basic groups (in the case of an anion exchange). For example, a cation exchange material may contain covalently bonded sulfonic groups $(-SO_3^-)$,or carboxylic groups $(-COO^-)$, and an anion exchange may contain covalently bonded quaternary amine groups $(-N(CH_3)_3^+)$, tertiary amines, etc. The counter-ion of the acidic groups in the cationic resin can be H^+, Na^+, K^+, etc., and the counter-ion of the basic groups in the anionic resin can be OH^-, Cl^-, CO_3^{2-}, etc. These ionic species are reversibly retained by ionic groups covalently bonded to the stationary phase. The mechanism of retention and separation in ion chromatography consists of ionic equilibria and is based on the difference in the affinity for the column of the ionic species that are separated. For the retention, the ions of the analyte should have higher affinity for the column than the counter-ions preexistent in the resin. The elution uses a mobile phase that contains competing ions (driving ions) that replace the analyte ions and "push" them out from the stationary phase. For an ion M^{z+} in the mobile phase and a cation exchange in the form H^+, for example, the following equilibrium takes place:

$$z\, resin^-H^+ + M^{z+} \leftrightarrow (resin^-)_z M^{z+} + z\, H^+ \tag{4.3.46}$$

This equilibrium is governed by a thermodynamic equilibrium constant $K_{M,H}^{therm}$ given by the expression:

$$K_{M,H}^{therm} = \frac{a_{resin}(M^{z+})}{a(M^{z+})}\left(\frac{a(H^+)}{a_{resin}(H^+)}\right)^z \tag{4.3.47}$$

In formula (4.3.47), $a_{resin}(M^{z+})$ represents the activity of M^{z+} in the resin, and $a(M^{z+})$ the activity of M^{z+} in the mobile phase (similar notations for H^+). Activities can be replaced with molar concentrations (as shown in expression (4.3.4) in dilute solutions $\gamma \approx 1$). When the activity is replaced with concentration, the constant $K_{M,H}^{therm}$ can be replaced with an ion exchange equilibrium constant $K_{M,H}$ for the pair M^{z+},H^+. A higher value for $K_{M,H}^{therm}$ (or for $K_{M,H}$) indicates a higher affinity of M^{z+} for the column, and therefore a stronger retention.

For a monovalent ion M^+, and with concentrations in place of activities, expression (4.3.47) can be written in the form:

$$K_{M,H} = \frac{C_{resin}(M^+)}{C(M^+)}\frac{C(H^+)}{C_{resin}(H^+)} \tag{4.3.48}$$

Considering now the partition of species M^+ between the resin and the mobile phase, this is characterized by the distribution constant $K_d(M^+)$ given by the expression:

$$K_d(M^+) = \frac{C_{resin}(M^+)}{C(M^+)} \tag{4.3.49}$$

From expressions (4.3.48) and (4.3.49), the following expression can be written:

$$K_d(M^+) = K_{M,H}\frac{C_{resin}(H^+)}{C(H^+)} \tag{4.3.50}$$

For a chromatographic IC separation, with the retention factor $k'(M^+) = K_d(M^+)\Psi$, where $K_d(M^+)$ is given by the expression (4.3.50), the following formula can be written:

$$k'(M^+) = K_{M,H}\frac{C_{resin}(H^+)}{C(H^+)}\Psi \tag{4.3.51}$$

Expression (4.3.51) shows that the retention of species M^+ depends on the constant $K_{M,H}$ but also on the concentration of the species H^+. When the species H^+ is used as a driving ion, expression (4.3.51) shows that a higher concentration of H^+ (a lower pH) leads to a smaller $k'(M^+)$, which indicates that even an ion with lower affinity for the stationary phase (such as H^+) can replace a strongly retained one if it is present at high enough concentration in the mobile phase. This process of modifying the concentration of the driving ions is used in IC for achieving retention and elution. Initially, the driving ion, which has a lower affinity for the resin than the analyte, is present at low concentration in the mobile phase, and the analyte ions with stronger affinity are retained. By increasing the concentration of the driving ions, the retained ions are eluted, the order of elution depending on their affinity for the resin.

The main interactions taking place in ion exchange are ionic interactions (although other types may be present). The ions that are exchanged are assumed to be distributed freely over the two phases, resin and solution. A boundary between the two phases can be visualized as a semipermeable membrane for all free ionic species (not for the acidic or basic functional groups that are ionic but are covalently connected to the skeleton of the resin). The ion exchange equilibrium can be in this case evaluated based on the Donnan membrane equilibrium theory. This theory refers to the uneven distribution of ions on the two sides of a semipermeable membrane separating solutions. When an electrolyte $Large^{-}M^{+}$ is dissolved on one side of the membrane, and one ion is small enough to pass through the membrane while the other is too large or immobilized to penetrate the membrane, a difference in the electrical potential between the two sides of the membrane is generated. The expression for this potential is given by the formula (see, e.g., Ref. [5]):

$$E_{Donnan} = \frac{\left(RT \ln \frac{a_{resin}}{a} - \Pi V\right)}{zF} \tag{4.3.52}$$

In expression (4.3.52), a_{resin} is the activity of the electrolyte in the resin and a is the activity in the solution around the resin. Other parameters include Π, which is swelling (osmotic) pressure for the resin, z is the charge of exchanging ions, V is the partial molar volume of the ions, and F is Faraday's constant.

In an ion exchange process following Eq. (4.3.46), there are two types of ions that are exchanged by the resin, H^+ and M^{z+}, each creating a Donnan potential. At equilibrium, the two potentials must be equal. As a result, for Eq. (4.3.46), the following expression is obtained:

$$RT \ln\left[\frac{a_{resin}(M^{z+})}{a(M^{z+})}\left(\frac{a(H^+)}{a_{resin}(H^+)}\right)^{z}\right] = \Pi(zV_H - V_M) \tag{4.3.53}$$

By substituting expressions (4.3.47) into (4.3.53), the result is the following:

$$\ln K_{M,H}^{therm} = \Pi(zV_H - V_M)/RT \tag{4.3.54}$$

A higher value of $K_{M,H}^{therm}$ is obtained for higher charge z of the ions M^{z+}, smaller V_M, and larger Π. The initial ions on the stationary phase (e.g., H^+) are typically the same as those used in the mobile phase as driving ions, and the column is conditioned with the mobile phase before starting the separation. By increasing the concentration of these ions in the mobile phase, the elution is accelerated as is also indicated by expression (4.3.47) (since $K_{M,H}^{therm}$ is constant, higher $a(H^+)$ leads to modifications such as higher $a_{resin}(H^+)$ and higher $a(M^{z+})$ in solution).

The retention and elution in ion exchange chromatography is in fact a more complex process than described by Eq. (4.3.54). Polar interactions, for example, may occur in addition to ion–ion interactions. The ion exchange stationary phases are frequently obtained from a polymeric material that contains covalently bonded ionic groups such as $-SO_3H$, or $-NR_3{}^+$. Although the ionic groups will cause the main types of interaction with the ionic molecules from solution, the organic polymeric material may provide a hydrophobic phase that can influence the separation process for molecules containing, in addition to the ionic groups, a hydrophobic moiety. Including various other influencing factors in the retention/elution theory it can be shown (see, e.g., Ref. [5]) that the retention factor k'_M for an ion M^{z+} depends on the concentration of the driving ions by a relation of the following form:

$$\log k'_M = \alpha - \beta \log C(H^+) \tag{4.3.55}$$

In expression (4.3.55), α and β are constants that can be determined experimentally for a given column, and $C(H^+)$ is the molar concentration of the driving ions (H^+ in this case). A similar theory as previously described for a cation exchange can be developed for the anion exchange process.

Mechanism in Size-Exclusion Chromatography (SEC)

Size-exclusion chromatography is differentiated in two main techniques, gel filtration and gel permeation. In both techniques the separation is based on molecular size (more correctly, molecular hydrodynamic volume), and it is applied mainly for the separation of macromolecules and of macromolecules from small molecules. The difference between gel filtration (GFC) and gel permeation (GPC) is in the nature of mobile phase, which is water (or mostly aqueous) in GFC, and an organic (or partially organic) solvent in GPC. In size exclusion, the small molecules from the mobile phase penetrate the pores of the stationary phase and have a long path through the column and therefore long retention times. Very large molecules cannot enter the pores at all and elute without retention (total exclusion). The macromolecules of medium size enter only some larger pores and are only partly retained, eluting faster than small molecules and slower than the very large ones, thus achieving separation. This process is schematically shown in Fig. 4.3.4.

In SEC, it is common to use retention volumes V_R (which are equivalent with retention times) for the characterization of separations. From formula (4.2.19), the following expression can be written:

$$V_R(X) = V_0 + K(X)V_{st} \tag{4.3.56}$$

In SEC the partition process takes place as a migration of the analytes between the interstitial volume of the column filled with the packing and the pores of the packing. In expression (4.3.56) the dead volume V_0 can be indicated for this process as the interstitial volume V_{inter}, and the volume V_{st} can be considered as equal to the volume of the pores of the packing V_{pores}. Therefore, the general Eq. (4.3.56) can be written for SEC separation in the form:

$$V_R(X) = V_{inter} + K_{SEC}(X)V_{pores} \tag{4.3.57}$$

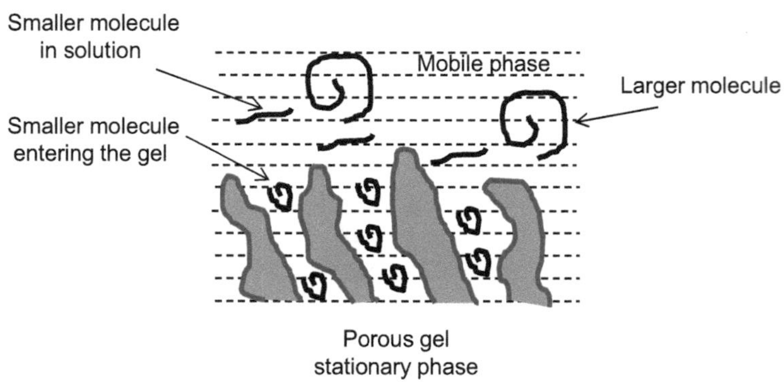

FIGURE 4.3.4 Schematic description of the process in size separation showing smaller molecules entering the porous gel while larger molecules flow with the mobile phase.

The expression for K_{SEC} should still be given by a formula of the type shown in Eq. (4.3.7) describing an equilibrium, with $\Delta G^0 = \Delta H^0 - T\Delta S^0$, where ΔH^0 is the standard enthalpy and ΔS^0 the entropy change for the transfer of the analyte from the mobile to the stationary phase. However, in the ideal case of SEC the analyte has no interaction with column packing (no forces exist between the analyte and the column). For this ideal case, it should be assumed that the value for ΔH^0 is zero for SEC, and the separation mechanism is controlled exclusively by the entropy of the process. Therefore the expression for a "pure" size-exclusion process described by the constant K_{SEC}^{pure} is given by the formula:

$$K_{SEC}^{pure} = \exp\frac{\Delta S^0}{R} \quad (4.3.58)$$

The change in the entropy during the size-exclusion process can be explained by the fact that when a molecule enters the pore due to the flow or in order to equalize concentrations outside and inside the pore, the macromolecules change their shape, and their conformational entropy decreases. The macromolecules inside the pore are contracted and lose part of their conformational entropy. A schematic representation of this process is illustrated in Fig. 4.3.5.

Size exclusion of macromolecular analytes in the absence of any energetic interactions with the stationary phase is therefore an entropy-controlled process. Such a process is usually indicated as an entropic partition [98]. In many cases, the separation mechanism in SEC also has an enthalpic contribution besides the entropic change (interactions with enthalpic contribution are present). This happens in practice when some interactions between the packing and the macromolecular analytes are taking place, and for this reason the SEC partition constant can be defined by two terms:

$$K_{SEC} = K_{SEC}^{pure} K_{SEC}^{interaction} \quad (4.3.59)$$

where $K_{SEC}^{interaction}$ is given by an expression of the form in Eq. (4.3.7) (with both $\Delta H^0 \neq 0$ and $\Delta S^0 \neq 0$). For the SEC separations where $K_{SEC} > 1$, there is always an enthalpic contribution to the separation.

The loss of entropy when the molecules are trapped inside the stationary phase causes ΔS^0 to have negative values [99]. Expression (4.3.55) indicates that temperature should not influence significantly the exclusion processes. This was proven experimentally in several eluents that are good solvents for polymers [100]. The change in entropy is more significant for larger

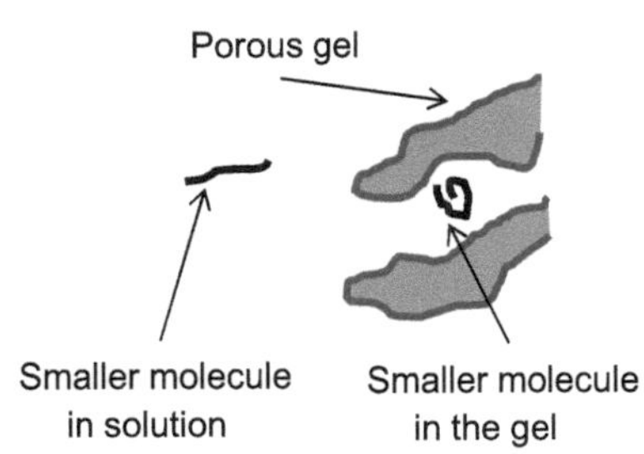

FIGURE 4.3.5 Schematic description of entering of a molecule inside the pore of stationary phase, leading to a loss of entropy.

molecules (of polymers) and less important for smaller ones. This can be understood by starting with the following expression for entropy:

$$S = k_B \ln \Omega \tag{4.3.60}$$

where k_B is Boltzmann constant and Ω is the number of possible (equally probable) micromolecular states. The number of ways in which the individual molecules can occupy the space within the pore of a stationary phase is significantly larger for a small molecule than for a large one. This indicates that the number of micromolecular states for a small molecule in the pore is considerably larger than for a large molecule (Ω is larger for small molecules and small for larger molecules). Starting with a similar number of states for large and small molecules in solution, the large molecules will have a smaller S value in the stationary phase, and therefore a considerable loss of entropy. At the same time, the small molecules will have only a minor loss of entropy. The result is that during the adsorption process, the large molecules will have a larger (in absolute value) negative ΔS^0. From expression (4.3.58), it can be seen that a ΔS^0 larger in absolute value (and negative) leads to a smaller K_{SEC}^{pure} and consequently to a smaller V_R in expression (4.3.57) for large molecules.

Depending on the molecule shape, its dimension that influences K_{SEC}^{pure} can be characterized in different ways. For example, for random coiled molecules the shape can be described by a mean radius $\bar{r}$. The relation between K_{SEC}^{pure} and the radius $\bar{r}$ can be described by two formulas, depending on whether the radius $\bar{r}$ is smaller or larger than the mean pore radius $\bar{d}$ of the stationary phase. These formulas are indicated below:

$$\begin{aligned} K_{SEC}^{pure} &= 1 - \frac{2\bar{r}}{\sqrt{\pi}\,\bar{d}} && \text{for } \bar{r} \ll \bar{d} \\ K_{SEC}^{pure} &= \frac{8}{\pi^2}\exp\left[-\left(\frac{\pi\bar{r}}{2\bar{d}}\right)^2\right] && \text{for } \bar{r} > \bar{d} \end{aligned} \tag{4.3.61}$$

The dimension of a molecule is also related to molecular weight, and as a result it is expected that V_R will depend in some way on the molecular weight (Mw) of the polymeric material. This dependence is verified in practice, and the following formula has been proven as valid for a certain range of Mw for various polymers:

$$V_R(X) = A - B \log(Mw(X)) \tag{4.3.62}$$

In expression (4.3.62), A and B are constants for polymers with different molecular weights but of the same type, and an ideal variation of V_R with the molecular weight is shown in Fig. 4.3.6.

In practice, expression (4.3.62) is frequently used for the evaluation of Mw for polymers [101]. However, expression (4.3.62) is valid only for a certain range of Mw values, and different molecular structures lead to different slopes for the dependence of V_R on log (Mw). The deviations from the ideal dependence usually include linearity for only a narrow range of Mw, although a decrease in the retention time remains valid for larger molecules. For being used in Mw determination, the dependence between Mw and the elution volume must be calibrated in the molecular range of interest using polymer standards with known molecular weight and having similar structures (shape) to those that are analyzed. As indicated by expression (4.3.61), the K_{SEC}^{pure} depends in fact not on *Mw* but on the mean radius $\bar{r}$ of the molecule (or of the molecule hydrodynamic volume). Since molecules with different shape may have the same values for *Mw* but different molecular hydrodynamic volumes, only a calibration of Eq. (4.3.62) with polymer standards of the same shape as the analyte can provide correct results for the *Mw*.

In many cases, enthalpic interactions may also be encountered such that the more interacting polymers with the stationary phase appear with lower *Mw* than their actual value. The presence of enthalpic contributions to the retention process can be detected based on chromatographic peak shape (sharp leading edge followed by tailing), unexpected retardation of the peak of polymers with high *Mw*, poor reproducibility of retention time, or large changes in the retention time upon solvent changes.

Mechanism in Chiral Chromatography

Molecular species having the same molecular formula but differences in their structure are called isomers. The isomers can be structural with the atoms bonded in different ways, and steric with the same bond structure but having differences in the arrangement in space of their atoms. The stereoisomers have different interatomic distances between certain atoms that are not directly bound. Stereoisomers that are mirror images to each other and are not superimposable although the atomic distances are the same in the molecules are called enantiomers. These compounds have the property called chirality. Chirality, which is needed for the existence of enantiomers, is commonly caused by the existence in the molecule of one or more tetrahedral carbon atoms substituted with groups that are different and are bonded spatially different. The chirality in an enantiomer is characterized using the symbols R and S. For the assignment of a symbol R or S to a chiral carbon that has four different substituents, its substituents are at first ranked in a sequence a, b, c, d. The ranking of the substituents is based on specific rules. For example, for the four atoms directly attached to the asymmetric carbon, a higher atomic number outranks the lower one (e.g., O > N > C > H), and a higher atomic mass outranks the lower. For the same atoms directly attached to the asymmetric

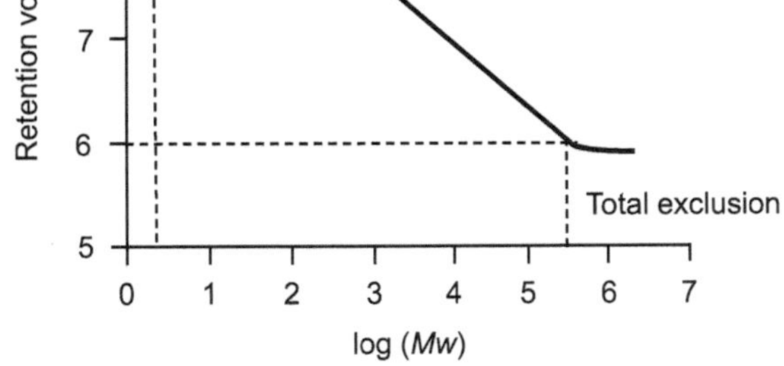

FIGURE 4.3.6 Ideal variation of elution volume V_R as a function of log (*Mw*) in SEC.

carbon, the priorities are assigned at the first point of difference. Double and triple bonds are treated by assuming that each such bonded atom is duplicated or triplicated (e.g., C*=O is treated as O–C*–O–C). More detailed rules are reported in the literature (see, e.g., Ref. [102]). Viewed in space and keeping the substituent d in the back of the page plane, when substituents a, b, c are seen counterclockwise the carbon is labeled S, and when they are seen clockwise the carbon is labeled R, as indicated in Fig. 4.3.7.

Besides having a chiral carbon, chiral molecules may be generated from other elements such as phosphorus or sulfur that are bonded to different substituents. Also, not only a chiral center (e.g., an asymmetric carbon) generates enantiomers. A chiral axis or a chiral plane also can lead to enantiomers (helicoidal chirality is also known). More than one asymmetric carbon can be present in a molecule. The stereoisomers generated by more than one asymmetric carbon can be a mirror image of one another (enantiomers) or may have different steric arrangements being diastereoisomers. For example, for a molecule with two chiral centers, these can be SS, SR, RS, and RR. The first couple of enantiomers is SS and RR and the second couple is SR and RS. The molecules SS and RS (or SR) are diastereoisomers, as well as RR and SR (or RS), which are also diastereoisomers. Diastereoisomers can be separated on common stationary phases. The separation of enantiomers cannot be performed on nonchiral phases. For their separation, a chiral stationary phase must be used, or a chiral modifier (capable of interacting with the analytes) must be present in the mobile phase. Several interaction mechanisms were suggested as responsible for the separation of chiral molecules. Several such mechanisms can take place in the interaction with chiral stationary phases and are further described.

For chiral stationary phases, the separation can be based on differences in the intermolecular interactions between the enantiomers and the stationary phase. Most chiral phases have been developed to be used in a mobile phase less polar than the stationary one, similar to NPC and HILIC phases. The interactions causing the separation are of polar nature, including polar–polar, hydrogen bonding, π-donor–π-acceptor, and π–π stacking. Since the enantiomer molecules have very similar physical properties, such as polarizability and dipole moments, the reason for differences in the interactions with the stationary phase are of geometric (steric) nature, such as the spatial access to the chiral polar phase, access which is different for the two analyte enantiomers. Different moieties with specific space orientation are present in both chiral analyte and chiral stationary phase. Each moiety is supposed to offer specific types of interactions. A single point or two points of interaction (e.g., two hydrogen bonds formation) between the chiral stationary phase and the chiral analyte are not sufficient for providing a difference in the enantiomer separation. For one-point interaction (a ↔ A) as

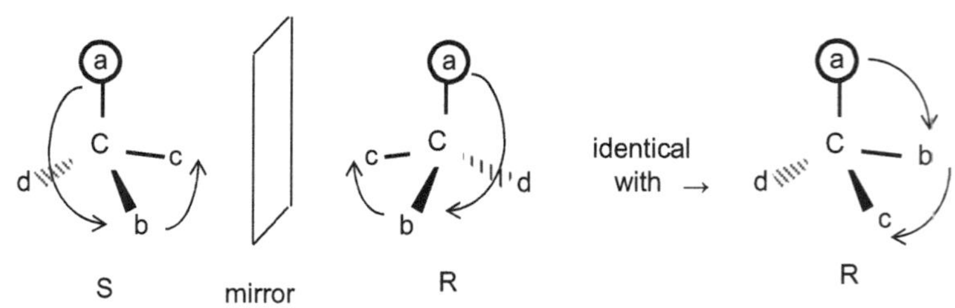

FIGURE 4.3.7 S and R molecules as mirror images, with one chiral carbon and four different substituents.

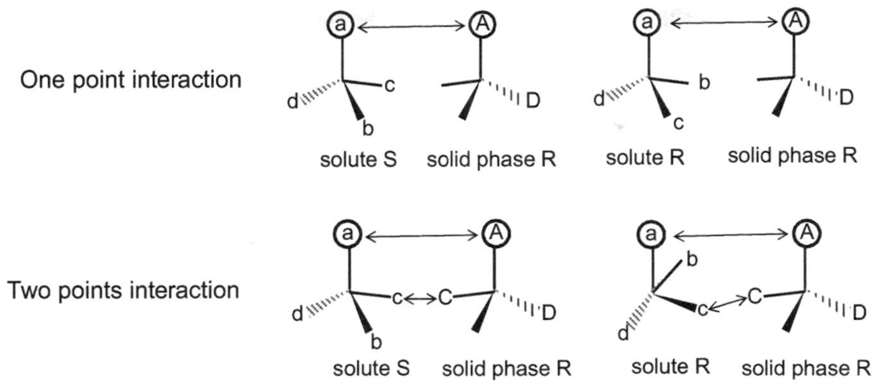

FIGURE 4.3.8 Illustration showing that one-point interaction (a ↔ A) and two-point interaction (a ↔ A and c ↔ C) are equal for the R and S enantiomers and do not lead to separation.

well as for two-point interaction (a ↔ A and c ↔ C), both R and S enantiomers behave similarly, as illustrated in Fig. 4.3.8.

In the case of a three possible points interaction, solute S can establish all three points (a ↔ A, b ↔ B, and c ↔ C) of interactions, while solute R establishes only two points (a ↔ A and c ↔ C), such that the retention of the enantiomers is different (the third point of interaction for R may still be present but it is very weak) [103]. This is illustrated in Fig. 4.3.9. Three points of interaction that are different in nature (e.g., through one donor-type hydrogen bond, an acceptor-type hydrogen bond, and π-donor−π-acceptor interactions) are necessary for obtaining better differentiation. The type of structure offering such possible interactions from the stationary phase is built, for example, in Pirkle-type stationary phases (see, e.g., Refs. [104,105]).

Another type of chiral stationary phase is offered by cellulose and cellulose derivatives. Cellulose-based polymers have a linear structure, but the cellulosic polymeric chains form strong hydrogen bonds between them, which leads to an interstitial structure that allows molecular inclusions. The main types of interactions remain hydrogen bonding between stationary phase and the molecules of the analyte, but the inclusion leads to restraints in the movement of molecules. The restrained movement combined with the specific orientation of OH groups in cellulose lead to multipoint interactions that are different between the enantiomers.

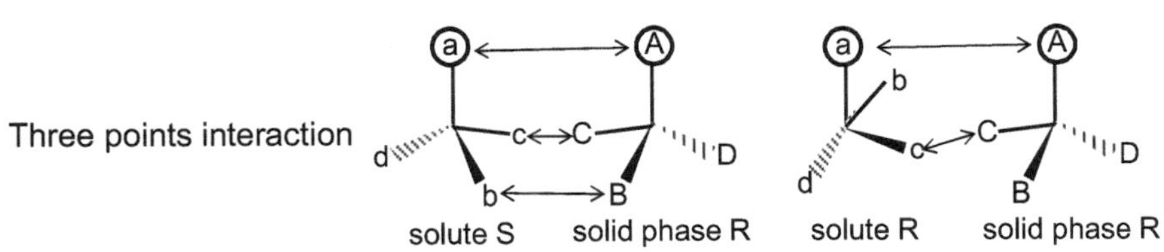

FIGURE 4.3.9 Illustration showing three points of interaction for solute S and two points of interaction for solute R, leading to the separation of enantiomers on a chiral stationary phase.

A somewhat different type of mechanism in separation also involves differences in the inclusion properties in the stationary phase, but these inclusion properties are more important while the hydrogen bonds are weaker compared to cellulose. The phases displaying this type of interaction are cyclodextrins, crown ethers, and amylose-type phases. These phases have the ability to produce inclusion complexes with numerous molecules, depending on the guest molecule geometry. Cyclodextrins, for example, have a truncated conical cavity that forms an inclusion similar to a pseudorotaxane with one part of the chiral molecule. The inside of the cavity of a cyclodextrin is relatively hydrophobic, giving the cyclodextrins the ability to accept a wide variety of molecular guests and limit the rotation of the included molecule. The still-existing hydroxyl groups offer enough chiral centers that have different interactions with the remaining substituents of the chiral molecule. These interactions are of the types polar−polar and hydrogen bonding. Derivatized cyclodextrins (e.g., with acetyl groups) provide further differentiation in the types of interactions with the chiral molecule. Due to the chirality of the guest, different groups from the two enantiomers and from the cyclodextrin come in close proximity. The result is that the overall molecular interactions of the enantiomers R and S with the host molecule are different. The type interactions are not specifically localized as in the case of Pirkle-type phases but possibly still based on three-point interactions. As an example, Fig. 4.3.10 shows a β-cyclodextrin, partially derivatized with acetyl groups and anchored, e.g., to a silica surface, in interaction with two enantiomers. The figure shows one enantiomer having more interactions than the other, and therefore being stronger retained.

Other mechanisms can be used for chiral separations, such as the one based on ligand exchange chromatography using a chiral resin loaded with a transitional metal (e.g., Cu^{2+}) capable of forming at the same time complexes with the solutes (enantiomeric analytes). The separation is based on the differences in the strength of the interactions (of coordinative and ionic type) of the solutes with the bonded metals ion in the asymmetric resin. Multiple mechanisms are involved in other materials that can be used as chiral stationary phases such as proteins or macrocyclic antibiotics that allow in certain cases the separation of enantiomers [106].

One special type of chiral separation involves the addition of a chiral additive in the mobile phase with separation on a nonchiral stationary phase. Two different mechanisms are suggested for this type of separation: (1) the chiral additive forms diastereoisomeric complexes with the chiral analytes and these can be separated on achiral stationary phases, and (2) the chiral additive is strongly retained on the stationary phase creating a chiral environment on which the diastereoisomers can be separated.

For diastereoisomer formation in solution, the differences in the relative stability of these compounds and the differences between the partition between the mobile phase and the stationary phase allow the separation [107]. The formation of compounds in solution between the analyte and the chiral additive can be based on complexation, formation of various types of adducts, or ion pairing.

In the case of the retention of the chiral additive on the stationary phase, the retention and elution can be considered as caused by the same types of interactions as those previously discussed for chiral stationary phase. The chiral additive retained by the stationary phase should have differences in the interactions with different enantiomer analytes. Theoretical models for describing the separation of enantiomers based on the use of chiral additives in the mobile phase can be found in the literature (see, e.g., Ref. [5]).

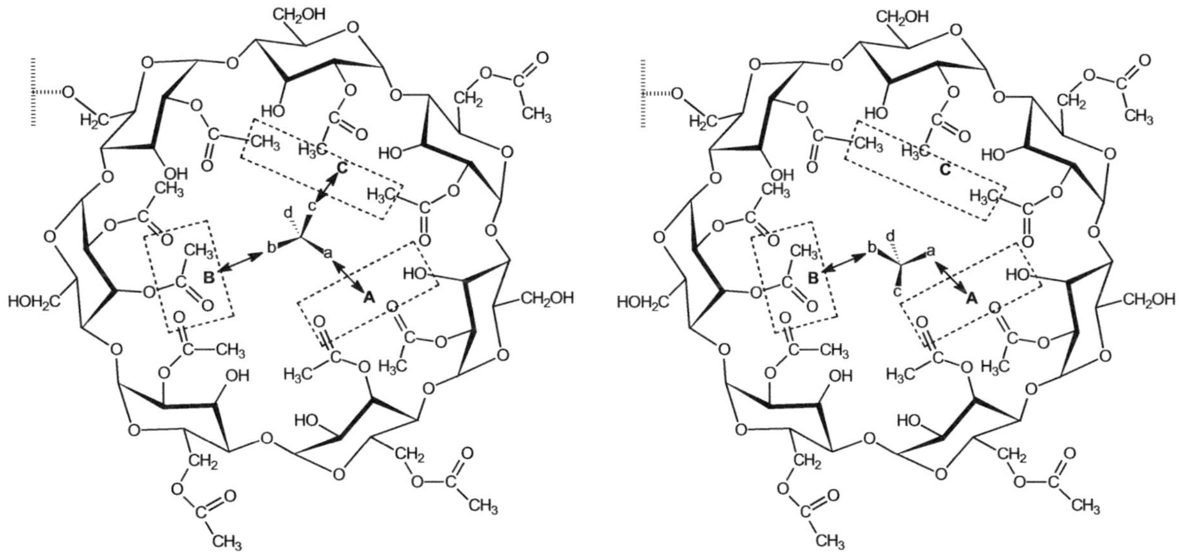

FIGURE 4.3.10 Types of interactions of a partially acetylated β-cyclodextrin and an S vs. an R guest molecule (d substituent is along the cyclodextrin channel). Stationary phases are obtained by attaching the β-cyclodextrin on a silica gel surface.

Retention Process in Other Chromatography Types

Besides the main chromatographic techniques previously discussed, a number of other HPLC types are used in practice (see Section 2.3). Among these are bioaffinity chromatography, hydrophobic interaction chromatography (HIC), electrostatic repulsion hydrophilic interaction chromatography (eHILIC or ERLIC), aqueous-normal-phase chromatography (ANPC or ANP), ion exclusion (or ion moderated), ligand exchange, immobilized metal affinity chromatography (IMAC), displacement chromatography, and multimode HPLC. These techniques frequently involve complex types of interactions, including hydrophobic, polar, or ionic types. Bioaffinity chromatography, for example, which is used mainly for peptides and protein separation, is based on reversible interactions between a protein and a ligand covalently attached to a solid support with complex and difficult-to-capture types of mechanisms. For each HPLC type, the adjustment of retention and elution depends on the specific type of interactions existent between the analyte, the stationary phase, and the mobile phase. For this reason, before applying any of these HPLC types it is important to understand the type of mechanisms involved in its use. Description of different mechanisms for each of these HPLC types can be found in the dedicated literature (see, e.g., Ref. [5]).

4.4 THE INFLUENCE OF PH, TEMPERATURE, AND ADDITIVES ON RETENTION EQUILIBRIA

For many separations, the pH is an important parameter that must be considered. The structure of many compounds is influenced by the pH of their solution (e.g., as being in molecular or ionic form). Also, the pH of the mobile phase influences the equilibrium of the analyte between the mobile and stationary phases. Many types of stationary phases are stable only within a specific pH range of mobile phase. For this reason, the mobile phase pH and its influence on separation are further discussed in Chapters 13 and 14. In this section, only the basic concept of the influence of pH on retention equilibria will be presented.

Temperature is another parameter that influences equilibria and for this reason is important in HPLC. The way the equilibrium analyte/stationary phase/mobile phase is influenced by temperature is described for partition equilibria by van't Hoff Eq. (4.3.9). In this section, the importance of pH, temperature, and of additives not involved directly in the retention equilibrium are discussed.

The Influence of pH on Retention Equilibria

The pH was initially defined for water as the solvating medium. This parameter characterizes the molar concentration C of H^+ ions or more precisely of H_3O^+ in an aqueous solution. Several notations can be used for the molar concentrations. For a molecule HX it is possible to use the notation C_{HX}, but also the "bracket" notation $[HX]$ is used. The same notations are used for the molar concentration of ions and, for example, the molar concentration of the ion H^+ can be expressed as C_{H+} or as $[H^+]$.

The pH is a parameter defined as the logarithmic (in base 10) of the activity of hydrogen ions. The pH expression is the following:

$$pH = -\log a_{H_3O^+} = -\log C_{H_3O^+} - \log \gamma_{H_3O^+} \quad (4.4.1)$$

where $(\gamma_{H_3O^+})$ is the activity coefficient of hydronium ions and $C_{H_3O^+}$ is their molar concentration. The reason for the replacement of hydrogen ions H^+ with hydronium $H3O^+$ or even with higher molecular species (e.g., $H_9O_4{}^+$) is to account for solvation of protons. All these ions describe in fact the same entity. In an infinitely diluted water solution $\gamma_{H_3O^+} = 1$, and the second term from expression (4.4.5) is zero. In such diluted solutions, or when the solution is approximated as infinitely diluted, expression (4.4.1) takes the form:

$$pH = -\log C_{H_3O^+} = -\log[H_3O^+] \quad (4.4.2)$$

In water, the pH scale is between 0 and 14 at 25°C. The pH for water is 7 based on the ionic product of water given by $K_w = C_{H+}\, C_{OH^-} = 10^{-14}$ at 25°C (like any equilibrium constant K_w is temperature-dependent). Since $C_{H^+} = C_{OH^-}$ in neutral solution, the concentration for the protons is $C_{H^+} = 10^{-7}$ and pH = 7. The pH of solutions in water of a specific compound depends on the concentration of the C_{H^+} ions generated by the compound.

For electrolyte solutions, the activity factor γ in expression (4.4.1) can be estimated with Debye–Hückel equation, written in the form:

$$\log \gamma_{H_3O^+} = -\frac{A\sqrt{I}}{1 + a_s B\sqrt{I}} \quad (4.4.3)$$

In expression (4.4.3), I represents the ionic strength of the solution, a_s is an ion size parameter, and A and B are parameters dependent on the solvent and temperature. The values of A and B are tabulated from empirical data. The ionic strength in a solution depends on the molar concentration C_i of various molecular or ionic species i, and their net charges, z_i, by the formula:

$$I = \frac{1}{2}\sum_i C_i z_i \quad (4.4.4)$$

The definition of pH and the comments about activity coefficient indicate that correct pH values depend on several factors. These factors are even more important when an organic solvent is added to the water solution, or when the pH is considered for a nonaqueous solution. For example, in methanol the ionic product $C_{H^+}\, C_{CH_3O^-} = 5.18 \times 10^{-14}$ and therefore the neutral pH = 6.64.

The dissociation in water of simple acids undergoing the equilibrium $HX \leftrightarrow H^+ + X^-$ is described by the dissociation constant K_a given by the formula:

$$K_a = \frac{C_{H^+} C_{X-}}{C_{HX}} \quad (4.4.5a)$$

With "bracket" notation for molar concentration, expression (4.4.5a) is written in the form:

$$K_a = \frac{[H^+][X^-]}{[HX]} \quad (4.4.5b)$$

The values for K_a are frequently expressed in "log" form, with the notation:

$$\log K_a = -\mathrm{p}K_a \tag{4.4.6}$$

Assuming $C_{H^+} = C_{X^-}$, expression (4.4.5) gives the following expression for C_{H^+} in a solution of a simple acid:

$$[H^+] = C_{H^+} = \sqrt{K_a C_{HX}} \tag{4.4.7}$$

Expression (4.4.7) shows that the concentration C_{H^+} (and the pH) in a solution of an acidic compound is a function of the nature of the compound (characterized by the acidity constant K_a), as well as of the concentration of the compound in solution (the theory can be easily extended to polyprotic acids). Expression (4.4.5) also shows that the dissociation of an acid in a solution is influenced by the solution pH (C_{H^+} value). Fig. 4.4.1 shows the content of undissociated benzoic acid and of benzoate ion in a solution as function of pH (pK_a = 4.202 for benzoic acid). At very low pH (high C_{H^+}, e.g. produced by the addition of a strong acid), the dissociation of an organic acid is low (the concentration C_{X^-} is small).

A number of acids, indicated as "strong" are totally dissociated in aqueous solutions (e.g., some inorganic acids such as HCl, HBr, HNO_3). For these compounds, the concentration C_{H^+} is given by $C_{H^+} = C_{HX}$.

Similar to acids, the theory is applied to bases. A simple base YOH undergoing the equilibrium $Y^+ + H_2O \leftrightarrow YOH + H^+$ has the dissociation constant K_a given by the expression:

$$K_a = \frac{C_{H^+} C_{YOH}}{C_{Y^+}} \tag{4.4.8}$$

Since in expression (4.4.8) $C_{H+} = C_{YOH}$, formula (4.4.7) for the calculation of C_{H+} remains valid for the case of bases where C_{HX} is replaced with C_{YOH}.

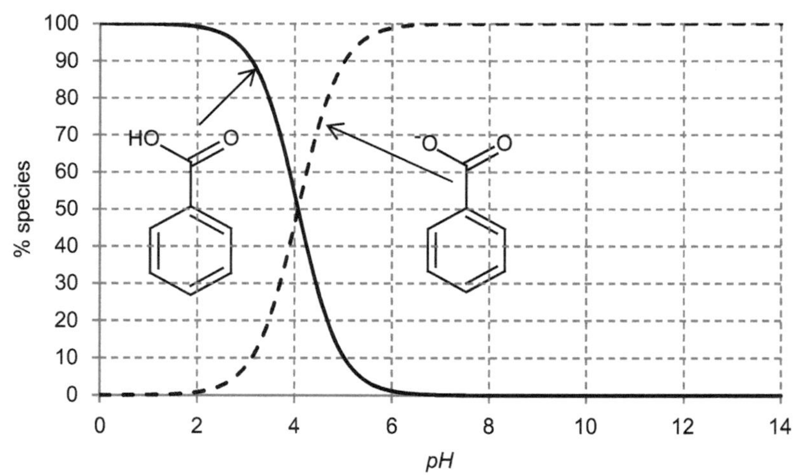

FIGURE 4.4.1 Variation in the % content of undissociated benzoic acid and benzoate ion as a function of solution pH.

For a simple partition equilibrium in HPLC for a compound X that is present in only one form in solution, the equilibrium constant is given by expression (4.2.12). However, many compounds may be present in solution in more than one form. For example, acidic compounds can be present as free or ionized molecules (see, e.g., Fig. 4.4.1 for benzoic acid at pH $\approx$ 4.0). In this case, more than one species of X is present in solution (e.g., $X_1 \equiv XH$ and $X_2 \equiv X^-$), each one with its own equilibrium constant. Since it is very common that $K(X_1) \neq K(X_2)$, the chromatographic trace may generate two peaks for compound X (or more peaks in the case of more species). It is also possible that one species is different enough from the other that it is not at all retained or it is not at all eluted. If the two species of the same compound are similar (but not identical) regarding their retention properties on an HPLC separation, they may generate distorted chromatographic peaks formed from the two compounds' coelution. For this reason, the pH of the mobile phase, and also of the sample solution, must be always properly adjusted, with the goal of having as much as possible one species of the analyte at a predominant level.

Depending on the type of chromatographic separation, however, it is possible that all species of the compound are retained and eluted together, if the pH of the mobile phase is adjusted within a specific range. For example, the separation of several organic acids (acetic, malic, citric, quinic, pyruvic, lactic, fumaric) can be done with a mobile phase at pH = 2.9 on an RP-HPLC column (Synergi 4u Hydro-RP, 4.6 × 250 mm from Phenomenex, CA, USA) with unique peaks (see Fig. 7.5.1), although the pK_a of these acids are as follows: acetic pK_a = 4.76, malic pK_{a1} = 3.40, citric pK_{a1} = 3.13, quinic pK_a = 4.33, pyruvic pK_a = 2.50, lactic pK_a = 3.86, fumaric pK_{a1} = 3.03. The separation in unique peaks is possible, although at pH 2.9, pyruvic acid is present at about equal levels as free acid and pyruvate ion, and citric and fumaric acids are present as free acids at about the 80% level.

The Influence of Temperature on Retention Equilibria

In the discussion on thermodynamic aspects of partition process (see Section 4.3), it was shown that an expression for the partition constant k' is given by van't Hoff equation (see expression 4.3.9). This equation can be written in the form:

$$\ln k' = a + b\frac{1}{T} \tag{4.4.9}$$

where

$$a = \Delta S^0/R + \ln \Psi \quad \text{and} \quad b = \Delta H^0/R \tag{4.4.10}$$

Expression (4.4.9) indicates that for typical separations the value of k' decreases as the temperature increases. Some separations can be improved by decreasing the column temperature because k' increases. In the case that the separation is good, the increase in the column temperature may be used to accelerate the separation and even to improve the peak shape.

When the stationary phase, the analyte, and the mobile phase properties do not change with a temperature change, ΔH^0 and ΔS^0 can be considered temperature invariants. However, this assumption is only an approximation. Corrections for the change with temperature T of thermodynamic functions ΔH^0 and ΔS^0 are reported in the literature (see, e.g., Ref. [108]).

Also, additional effects may influence ΔH^0 and ΔS^0 for a separation when temperature changes (see, e.g., Section 13.2). The plots of ln k' (or log k') as a function of $1/T$ are known as van't Hoff plots. When these plots are linear, as expected, they allow the estimation of ΔH^0 and ΔS^0 from the values of a and b, obtained from the plots (with ln Ψ assumed to be known). Deviation from linearity of van't Hoff plots are encountered for some compounds, and the deviation can be caused by various factors such as the presence of the compound in several forms in the mobile phase (see, e.g., Ref. [5]).

The changing of the temperature in an HPLC separation also has applications in specific analytical applications. Column temperature can play a role in reducing analysis time and modifying retention. Also, due to the decrease in viscosity of the mobile phase as temperature increases, this effect can be used for decreasing the column backpressure when such an effect is needed [109]. The modification of solvent viscosity also affects the diffusion coefficient of the mobile phase, affecting column efficiency (see Section 13.2).

Influence of Additives Not Involved in the Equilibrium

The addition of additives in the mobile phase is a subject further discussed in Chapter 13 regarding mobile phase composition. Additives may play a variety of roles regarding the separation, and the use of IPAs and of chiral modifiers in the mobile phase were already mentioned in Section 4.3. One additional influence of additives in the mobile phase is that of affecting the effective concentration (thermodynamic activity) of the analytes. For a compound X, activity a_X can be defined based on expression (4.1.3). Activity is related to the concentration based on expression (4.3.4). The activity coefficient γ_X can be related to the change in the enthalpy of mixing when the compound X is placed in a solvent and does not form an ideal solution but a regular solution (solutions with mixing enthalpy different from zero) [110]. For such a regular solution, the expression for the chemical potential of compound X is given by the expression:

$$\mu_X = \mu_X^0 + \frac{1}{n_X}\left(\boldsymbol{H}_X - \boldsymbol{H}_X^0\right) + RT \ln C_X \tag{4.4.11}$$

where n_X is the number of moles of compound X and $\boldsymbol{H}_X - \boldsymbol{H}_X^0$ is the change in enthalpy caused by the interactions during mixing. The combination of expression (4.4.11) with expression (4.3.1) where $a_X = \gamma_X\, C_X$ (see expression 4.3.4) leads to the following formula for the activity coefficient:

$$\ln \gamma_X = \frac{1}{n_X RT}\Delta \boldsymbol{H}_{mix,X} \tag{4.4.12}$$

The change in the enthalpy of mixing depends on the interactions of the solute with the solvent and the solvent can be a pure compound, a mixture of solvents, and also may contain additives. The nature of the solvent (solvents) as well as additives influence the interactions and these affect the activity coefficient. One type of additive that is known to strongly affect the interactions in solution is known as chaotropes.

Chaotropic Salts Influence on Equilibria

A chaotropic agent is a molecule in water or water/organic solvent solutions that can disrupt the hydrogen bonding and solvation of ions. Chaotropic solutes interfere with intramolecular interactions, such as hydrogen bonding, van der Waals forces, and hydrophobic effects, and as shown by formula (4.4.12), they affect the activity coefficient of the analytes. Chaotropic salts that dissociate in solution can produce a shielding of ionic charges. Also, since hydrogen bonding is stronger in nonpolar media, and salts increase the chemical polarity of the solvent, hydrogen bonding is affected. Chaotropic salts can be added in some mobile phases to influence the activity of the analytes and therefore the separation [111]. Some inorganic ions can act as chaotropes with the increase in their disruptive character in the following order: $H_2PO_4^- < HCOO^- < CH_3SO_4^- < CF_3COO^- < BF_4^- < ClO_4^- < PF_6^-$.

References

[1] http://www.hplcweb.com/HPLC_Equipment/HPLC_Manufacturers/.

[2] Agilent 1200 Series, User Manuals, Agilent Technologies, 2006.

[3] http://www.waters.com.

[4] L.R. Snyder, J.J. Kirkland, J.W. Dolan, Introduction to Modern Liquid Chromatography, 3rd ed., Wiley, Hoboken, 2010.

[5] S.C. Moldoveanu, V. David, Essentials in Modern HPLC Separations, Elsevier, Amsterdam, 2013.

[6] J. De Vos, K. Broeckhoven, S. Eeltink, Advances in ultrahigh-pressure liquid chromatography technology and system design, Anal. Chem. 88 (2016) 262–278.

[7] A. Medvedovici, V. David, Pump/solvent delivery system design for HPLC, in: J. Cazes (Ed.), Encyclopedia of Chromatography, third ed., CRC Press, New-York, 2009, pp. 1946–1962.

[8] S. Ahuja, High-pressure liquid chromatography (Chapter 15), in: S. Ahuja, N. Jespersen (Eds.), Modern Instrumental Analysis, vol. 47, Elsevier, Amsterdam, 2006, p. 502.

[9] L.R. Snyder, J.J. Kirkland, J.L. Glajch, Practical HPLC Method Development, second ed., John Wiley & Sons, New York, 2012, pp. 41–43.

[10] V. David, A. Medvedovici, Sample introduction techniques for HPLC, in: J. Cazes (Ed.), Encyclopedia of Chromatography, third ed., CRC Press, New York, 2009, pp. 2067–2076.

[11] C.J. Welch, X. Gong, W. Schafer, E.C. Pratt, T. Birkovic, Z. Pirzada, J.F. Cuff, B. Kosjek, MISER chromatography (multiple injections in a single experimental run): the chromatogram is the graph, Tetrahedron Asym. 21 (2010) 1674–1681.

[12] J. Chen, J. Pawliszyn, Solid phase coupled to high-performance liquid chromatography, Anal. Chem. 67 (1995) 2530–2533.

[13] V.K. Gaurav, A. Kumar, A.K. Malik, P.K. Rai, SPME-HPLC: a new approach to the analysis of explosives, J. Hazard. Mater. 147 (2007) 691–697.

[14] C. Zhu, D.M. Goodall, S.A.C. Wren, Elevated temperature HPLC: principles and applications to small molecules and biomolecules, LCGC Asia Pacific 8 (March 2005) 48–59.

[15] P.G. Sadek, The HPLC Solvent Guide, J. Wiley, New York, 1996.

[16] T.J. Novak, M.L. Grayeski, Acridinium-based chemiluminescence for high-performance liquid chromatography detection of chlorophenols, Microchemical J. 50 (1994) 151–160.

[17] A. Medvedovici, A. Farca, V. David, Derivatization reactions in liquid chromatography for drug assaying in biological fluids, in: Advances in Chromatography, vol. 47, 2009, pp. 283–322.

[18] A.L. Yergey, C.G. Edmonds, I.A.S. Lewis, M.L. Vestal, Liquid Chromatography/Mass Spectrometry: Techniques and Applications, Springer, New York, 2014.

[19] O. Nunez, H. Gallart-Ayala, C.P.B. Martins, P. Lucci, Fast Liquid Chromatography-mass Spectrometry Methods in Food and Environmental Analysis, World Scientific, Singapore, 2015.

[20] L.D. Asnin, Peak measurement and calibration in chromatographic analysis, Trends Anal. Chem. 81 (2016) 51–62.

[21] B. Seemann, T. Alon, S. Tsizin, A.B. Fialkov, A. Amirav, Electron ionization LC-MS with supersonic molecular beams-the new concept, benefits and applications, J. Mass Spectrom. 50 (2015) 1252–1263.
[22] A. Cappiello, G. Famiglini, P. Palma, A. Siviero, Liquid chromatography–electron ionization mass spectrometry: fields of application and evaluation of the performance of a direct-EI interface, Mass Spectrom. Rev. 24 (2005) 978–989.
[23] P. Palma, G. Famiglini, H. Trufelli, A. Cappiello, Towards a universal detector for small molecule applications: direct-EI LC-MS, LC-GC North Am. 29 (2011) 68–75.
[24] M.S. Smile, Small Molecule Identification and Library Exploration Using LC-MS/MS, Wiley, Hoboken, 2011.
[25] LTQ XL, Orbitrap Operation Manual, Thermo-Fisher Scientific, 2009.
[26] http://www.dionex.com/en-us/webdocs/4270-DS-Eluent_Suppressors-23Nov10-LPN1290-10.pdf.
[27] J.S. Fritz, D.T. Gjerde, Ion Chromatography, Wiley-VCH, Weinheim, 2009.
[28] C.A. Pohl, J.R. Stillian, P.E. Jackson, Factors controlling ion-exchange selectivity in suppressed ion chromatography, J. Chromatogr. A 789 (1997) 29–41.
[29] T.H. Mourey, L.E. Oppenheimer, Principles of operation of an evaporative light-scattering detector for liquid chromatography, Anal. Chem. 56 (1984) 2427–2434.
[30] http://www.waters.com/webassets/cms/support/docs/71500121802rb.pdf.
[31] V.L. Cebolla, L. Membrado, J. Vela, A.C. Ferrando, Evaporative light-scattering detection in the quantitative analysis of semivolatile polycyclic aromatic compounds by high-performance liquid chromatography, J. Chromatogr. Sci. 35 (1997) 141–150.
[32] L.-E. Magnusson, D.S. Risley, J.A. Koropchak, Aerosol-based detectors for liquid chromatography, J. Chromatogr. A 1421 (2015) 68–81.
[33] T. Górecki, F. Lynen, R. Szucs, P. Sandra, Universal response in liquid chromatography using charged aerosol detection, Anal. Chem. 78 (2006) 3186–3192.
[34] Y. Jiang, D. Hascall, D. Li, J.H. Pease, A charged aerosol detector/chemiluminescent nitrogen detector/liquid chromatography/mass spectrometry system for regular and fragment compound analysis in drug discovery, J. Chromatogr. A 1411 (2015) 63–68.
[35] E.M. Fujinari, L.O. Courthaudon, Nitrogen-specific liquid chromatography detector based on chemiluminescence: application to the analysis of ammonium nitrogen in waste water, J. Chromatogr. A 592 (1992) 209–214.
[36] K.L. Norton, A.J. Lange, P.R. Griffiths, A unified approach to the chromatography-FTIR interface: GC-FTIR, SFC-FTIR, and HPLC-FTIR with subnanogram detection limits, J. High Res. Chromatogr. 14 (1991) 225–229.
[37] G.W. Somsen, C. Gooijer, U. A. Th. Brinkman, Liquid chromatography-Fourier-transform infrared spectrometry, J. Chromatogr. A 856 (1999) 213–242.
[38] K. Albert, Liquid chromatography-nuclear magnetic resonance spectroscopy, J. Chromatogr. A 856 (1999) 199–211.
[39] L.K. Sutton, J.A. Caruso, Liquid chromatography-inductively coupled plasma mass spectrometry, J. Chromatogr. 856 (1999) 243–258.
[40] G. Cepria, J.R. Castillo, Surface plasmon resonance-based detection. An alternative to refractive index detection in high-performance liquid chromatography, J. Chromatogr. A 759 (1997) 27–35.
[41] G.E. Fredheim, S.M. Braaten, B.E. Christensen, Molecular weight determination of lignosulfonates by size-exclusion chromatography and multi-angle laser light scattering, J. Chromatogr. A 942 (2002) 191–199.
[42] S.-H. Yoo, J. Jane, Molecular weights and gyration radii of amylopectins determined by high-performance size-exclusion chromatography equipped with multi-angle laser-light scattering and refractive index detectors, Carbohydrate Polym. 49 (2002) 307–314.
[43] A.B. Kanu, P. Dwivedi, M. Tam, L. Matz, H.H. Hill Jr., Ion mobility-mass spectrometry, J. Mass Spectrom. 43 (2008) 1–22.
[44] http://www.waters.com/webassets/cms/library/docs/720004681en.pdf.
[45] B. Zhang, X. Li, B. Yan, Advances in HPLC detection – towards universal detection, Anal. Bioanal. Chem. 390 (2008) 299–301.
[46] W.-Y. Lee, T.A. Nieman, Determination of dansyl amino acids using tris(2,2′-bipyridyl)ruthenium (II) chemiluminescence for post-column reaction detection in high-performance liquid chromatography, J. Chromatogr. A 659 (1994) 111–118.
[47] D.R. Stoll, X. Li, X. Wang, P.W. Carr, S.E.G. Porter, S.C. Rutan, Fast, comprehensive two-dimensional liquid chromatography, J Chromatogr. A 1168 (2007) 3–43.

[48] J. Pól, M. Kivilompolo, T. Hyötyläinen, Comprehensive two-dimensional liquid chromatography (LC×LC): a review, LCGC Europe 24 (2011) 232–237.
[49] C.A. Rimmer, C.R. Simmons, J.G. Dorsey, The measurement and meaning of void volumes in reversed-phase liquid chromatography, J. Chromatogr. A 965 (2002) 219–232.
[50] I. Rustamov, T. Farcas, F. Ahmed, F. Chan, R. LoBrutto, H.M. McNair, Y.V. Kazakevich, Geometry of chemically modified silica, J. Chromatogr. A 913 (2001) 49–63.
[51] K.B. Sentell, J.G. Dorsey, On the calculation of the stationary phase volume in reversed phase chromatography, J. Liq. Chromatogr. 11 (1988) 1875–1885.
[52] S. Bocian, P. Vajda, A. Felinger, B. Buszewski, Excess adsorption of commonly used organic solvents from water on nonend-capped C18-bonded phases in reversed-phase liquid chromatography, Anal. Chem. 81 (2009) 6334–6346.
[53] S. Moldoveanu, V. David, Estimation of the phase ratio in reversed-phase high-performance liquid chromatography, J. Chromatogr. A 1381 (20150) (2015) 194–201.
[54] R.S. Berry, S.A. Rice, J. Ross, Physical Chemistry, Oxford Univ. Press, Oxford, 2000, pp. 526–527.
[55] V.I. Smirnov, Kurs Vischei Matematiki, vol. 2, Gosudarstvennoe Izd., Moscow, 1952.
[56] B.L. Karger, L.R. Snyder, C. Horvath, An Introduction to Separation Science, J. Wiley, New York, 1973.
[57] W. Hayduk, H. Laudie, Prediction of diffusion coefficients for non-electrolytes in dilute aqueous solutions, AIChE J. 20 (1974) 611–615.
[58] J.J. van Deemter, F.J. Zuiderweg, A. Kinkenberg, Longitudinal diffusion and resistance to mass transfer as causes of nonideality in chromatography, Chem. Eng. Sci. 5 (1956) 271–289.
[59] P.A. Bristow, J.H. Knox, Standardization of test conditions for high performance liquid chromatography columns, Chromatographia 10 (1977) 279–289.
[60] J.H. Knox, in: P.R. Brown, E. Grushka (Eds.), Advances in Chromatography, vol. 38, M. Dekker, New York, 1998.
[61] V.B. Di Marco, G.G. Bombi, Mathematical functions for the representation of chromatographic peaks, J. Chromatogr. A 931 (2001) 1–30.
[62] M. Otto, Chemometrics. Statistic and Computer Application in Analytical Chemistry, Wiley-VCH, Weinheim, 1999, p. 61.
[63] A. Holm, P. Molander, E. Lundanes, T. Greibrokk, Determination of the pesticide rotenone in river water utilizing large-volume injection packed capillary column switching liquid chromatography with UV and time-of-flight mass spectrometric detection, J. Chromatogr. A 983 (2003) 43–50.
[64] A.C. Hogenboom, M.P. Hofman, S.J. Kok, W.M. Niessen, U.T. Brinkman, Determination of pesticides in vegetables using large-volume injection column liquid chromatography-electrospray tandem mass spectrometry, J. Chromatogr. A 892 (2000) 379–390.
[65] B.L. Karger, M. Martin, G. Guiochon, Role of column parameters and injection volume on detection limits in liquid chromatography, Anal. Chem. 46 (1974) 1640–1647.
[66] W.J. Moore, Physical Chemistry, second ed., Prentice-Hall, Inc., Englewood Cliffs, 1955.
[67] T.L. Chester, J.W. Coym, Effect of phase ratio on van't Hoff analysis in reversed-phase liquid chromatography, and phase ratio independent estimation of transfer enthalpy, J. Chromatogr. A 1003 (2003) 101–111.
[68] J. Petre, V. Iancu, V. David, Thermodynamic study of the liquid chromatography retention process for some pesticides on C8 and C18 silica-based stationary phases, Rev. Roum. Chim. 58 (2013) 425–432.
[69] V. David, C. Bala, L. Rotariu, Thermodynamic parameters of the reversed-phase liquid chromatography retention for some lipid-soluble vitamins, Chem. Anal. (Warsaw) 49 (2004) 191–199.
[70] T. Galaon, V. David, Deviation from van't Hoff dependence in RP-LC induced by tautomeric inter-conversion observed for four compounds, J. Sep. Sci. 34 (2011) 1423–1428.
[71] T. Galaon, A. Medvedovici, V. David, Hydrophobicity parameter (log K_{ow}) estimation for some phenolic compounds of pharmaceutical interest from retention studies with mobile phase composition in RP-LC, Sep. Sci. Technol. 43 (2008) 147–163.
[72] A.P. Schellinger, P.W. Carr, Isocratic and gradient elution chromatography: a comparison in terms of speed, retention reproducibility and quantitation, J. Chromatogr. A 1109 (2006) 253–266.
[73] W. Markowski, K.L. Czapinska, M. Blaszczak, Determination of the constants of the Snyder-Soczewinski equation by means of gradient multiple development, J. Liq. Chromatogr. 17 (1994) 999–1009.

[74] C. Hansch, A. Leo, D. Hoekman, Exploring QSAR, Hydrophobic, Electronic and Steric Constants, ACS, Washington, 1995.
[75] http://www.epa.gov/oppt/exposure/pubs/episuite.htm.
[76] http://www.chemaxon.com.
[77] A. Vailaya, C. Horváth, Retention in reversed-phase chromatography: partition or adsorption? J. Chromatogr. A 829 (1998) 1–27.
[78] K.A. Dill, The mechanism of solute retention in reversed-phase liquid chromatography, J. Phys. Chem. 91 (1987) 1980–1988.
[79] O. Sinanoğlu, in: B. Pullman (Ed.), Molecular Associations in Biology, Academic Press, New York, 1968, pp. 427–445.
[80] C. Horvath, W. Melander, I. Molnar, Solvophobic interactions in liquid chromatography with nonpolar stationary phases, J. Ch.romatogr. 125 (1976) 129–156.
[81] S.C. Moldoveanu, V. David, Dependence of the distribution constant in liquid–liquid partition equilibria on the van der Waals molecular surface area, J. Sep. Sci. 36 (2013) 2963–2978.
[82] P.W. Carr, J. Li, A.J. Dallas, D.I. Eikens, L.C. Tan, Revisionist look at solvophobic driving forces in reversed-phase liquid chromatography, J. Chromatogr. A 656 (1993) 113–133.
[83] L.C. Tan, P.W. Carr, Revisionist look at solvophobic driving forces in reversed-phase liquid chromatography - II. Partitioning vs. adsorption mechanism in monomeric alkyl bonded phase supports, J. Chromatogr. A 775 (1997) 1–12.
[84] P.W. Carr, L.C. Tan, J.H. Park, Revisionist look at solvophobic driving forces in reversed-phase liquid chromatography - III. Comparison of the behavior of nonpolar and polar solutes, J. Chromatogr. A 724 (1996) 1–12.
[85] J.H. Park, Y.K. Lee, Y.C. Weon, L.C. Tan, J. Li, L. Li, J.F. Evans, P.W. Carr, Revisionist look at solvophobic driving forces in reversed-phase liquid chromatography – IV. Partitioning vs. adsorption mechanism on various types of polymeric bonded phases, J. Chromatogr. A 767 (1997) 1–10.
[86] F. Gritti, Quantification of individual mass transfer phenomena in liquid chromatography for further improvement of column efficiency, LC/GC North Am. 32 (12) (2014) 928–940.
[87] F. Gritti, G. Guiochon, Comparison between the intra-particle diffusivity in the hydrophilic interaction chromatography and reversed phase liquid chromatography modes. Impact on the column efficiency, J. Chromatogr. A 1297 (2013) 85–95.
[88] K. Croes, A. Steffens, D. H Marchand, L.R. Snyder, Relevance of π–π and dipole–dipole interactions for retention on cyano and phenyl columns in reversed-phase liquid chromatography, J. Chromatogr. A 1098 (2005) 123–130.
[89] J.H. Knox, R.A. Hartwick, Mechanism in ion-pair liquid chromatography of amines, neutrals, zwitterions, and acids using anionic hetaerons, J. Chromatogr. A 204 (1981) 2–21.
[90] C. Horvath, W. Melander, I. Molnar, P. Molnar, Enhancement of retention by ion-pair formation in liquid chromatography with nonpolar stationary phases, Anal. Chem. 49 (1977) 2295–2305.
[91] A. Bartha, J. Ståhlberg, Electrostatic retention model of the reversed-phase ion-pair chromatography, J. Chromatogr. A 668 (1994) 255–284.
[92] M. Radulescu, V. David, Partition versus electrostatic model applied to the ion-pairing retention process of some guanidine based compounds, J. Liq. Chromatogr. Rel. Technol. 35 (2012) 2042–2053.
[93] T. Cecchi, F. Pucciarelli, P. Passamonti, Extended thermodynamic approach to ion interaction chromatography, Anal. Chem. 73 (2001) 2632–2639.
[94] T. Cecchi, Extended thermodynamic approach to ion interaction chromatography. Influence of the chain length of the solute ion: a chromatographic method for the determination of ion-pairing constants, J. Sep. Sci. 29 (2005) 549–554.
[95] D.V. McCalley, U.D. Neue, Estimation of the extent of the water-rich layer associated with the silica surface in hydrophilic interaction chromatography, J. Chromatogr. A 1192 (2008) 225–229.
[96] J. Urban, V. Škeříková, P. Jandera, R. Kubíčková, M. Pospíšilová, Preparation and characterization of polymethacrylate monolithic capillary columns with dual hydrophilic interaction reversed-phase retention mechanism for polar compounds, J. Sep. Sci. 32 (2009) 2530–2543.
[97] M.R. Euerby, J. Hulse, P. Petersson, A. Vazhentsev, K. Kassam, Retention modelling in hydrophilic interaction chromatography, Anal. Bioanal. Chem. 407 (2015) 9135–9152.

[98] D. Berek, Size exclusion chromatography – a blessing and a curse of science and technology of synthetic polymers, J. Sep. Sci. 33 (2010) 315–330.
[99] A.L. Myers, Characterization of nanopores by standard enthalpy and entropy of adsorption of probe molecules, Colloid. Surface A Phys. Eng. Asp. 241 (2004) 9–14.
[100] A.A. Gorbunov, A.M. Skvortsov, Statistical properties of confined macromolecules, Adv. Colloid Interface Sci. 62 (1995) 31–108.
[101] A. Striegel, W.W. Yau, J.J. Kirkland, D.D. Bly, Modern Size-Exclusion Liquid Chromatography: Practice of Gel Permeation and Gel Filtration Chromatography, second ed., Wiley, Hoboken, 2009.
[102] A. Berthod, Chiral Recognition in Separation Methods: Mechanisms and Applications, Springer, Heidelberg, 2010.
[103] M. Lämmerhofer, Chiral recognition by enantioselective liquid chromatography: mechanisms and modern chiral stationary phases, J. Chromatogr. A 1217 (2010) 814–856.
[104] W.H. Pirkle, Y. Liu, On the relevance of face-to-edge π-π interactions to chiral recognition, J. Chromatogr. A 749 (1996) 19–24.
[105] M.H. Hyun, J.-J. Ryoo, W.H. Pirkle, Experimental support differentiating two proposed chiral recognition models for the resolution of N-(3,5-Dinitrobenzoyl)-α-arylalkylamines on high-performance liquid chromatography chiral stationary phases, J. Chromatogr. A 886 (1–2) (2000) 47–53.
[106] A. Fortuna, G. Alves, A. Falcão, Chiral chromatographic resolution of antiepileptic drugs and their metabolites: a challenge from the optimization to the application, Biomed. Chromatogr. 28 (2014) 27–58.
[107] L. Yu, S. Wang, S. Zeng, Chiral mobile phase additives in HPLC enantioseparations, Method. Mol. Biol. 970 (2012) 221–231.
[108] S.C. Moldoveanu, V. David, Sample Preparation in Chromatography, Elsevier, Amsterdam, 2002.
[109] S. Heinisch, J.-L. Rocca, Sense and nonsense of high-temperature liquid chromatography, J. Chromatogr. A 1216 (2009) 642–658.
[110] H. Hildebrand, R.I. Scott, The Solubility of Non-Electrolytes, Dover Pub., New York, 1964.
[111] R. LoBrutto, Y.V. Kazakevich, Chaotropic effects in RP-HPLC, in: E. Grushka, N. Grinberg (Eds.), Advances in Chromatography, vol. 44, CRC Press, Hoboken, 2005, pp. 291–315.

CHAPTER

5

Properties of Analytes and Matrices Determining HPLC Selection

5.1 PHYSICOCHEMICAL PROPERTIES RELATED TO SEPARATION

The physicochemical properties of the analytes and of the matrix are critical factors in selecting a method of analysis. Although other factors (see Section 1.1) must also be considered for a method selection, the physicochemical properties of a compound are important and at the same time simple to obtain since they can be found in the literature. The physicochemical properties affect not only the core analytical process but also the sample preparation step preceding the core analysis. The subject of sample preparation is frequently discussed in the literature (see, e.g., Refs. [1–4]); however the subject is beyond the purpose of this book.

For the core analysis, some properties of the analytes and of the matrix are of particular importance for the selection of the appropriate separation conditions. The properties discussed in this section are related to the analytes in the processed sample. The sample preparation step may be designed to change some of the initial properties of the analyte and make it more amenable for HPLC analysis. These changes can refer to physicochemical properties related to separation and also related to detection. A discussion regarding several physicochemical properties pertinent to the separation of any compound present in the (processed) sample is given in this section. However, this section discusses only more general properties of analyte and matrix components. Additional physicochemical parameters, specifically developed for the characterization of solvents or of stationary phases, and important for the separation, will be discussed in other sections (see Chapters 6–12 in relation to stationary phases and Chapters 13 and 14 in relation to mobile phases). Some of the physicochemical properties are molecular properties (e.g., molecular weight, dipole moment, polarizability, van der Waals volume, and surface area) and other are bulk properties (e.g., acidic or basic character in solution, octanol/water partition coefficient, and solubility.) However, all these properties are determined by the chemical structure of the molecule.

Chemical Composition and Structure

Chemical compounds are typically classified as organic or inorganic, depending on their elemental composition. Organic compounds are usually considered those that contain in their

Selection of the HPLC Method in Chemical Analysis
http://dx.doi.org/10.1016/B978-0-12-803684-6.00005-6

molecule carbon and, in the majority of organic compounds, also hydrogen. A small number of compounds that are considered organic do not contain hydrogen, and also several compounds that contain carbon are not included in the large class of organic compounds (such as metal carbonates, carbides, and the oxides of carbon). The samples in chemical analysis can contain both organic and inorganic compounds, or may be only organic or only inorganic.

The simplest class of organic compounds is that of saturated hydrocarbons (linear or branched). Following this class are saturated cyclic hydrocarbons, unsaturated hydrocarbons with one or more double bonds, unsaturated hydrocarbons with triple bonds, aromatic hydrocarbons, etc. Combinations of all these structures are possible. On the hydrocarbon backbone various functional groups can be attached. They can be classified based on the nature of atoms in the functional group or other criteria (such as monofunctional, bifunctional, trifunctional). This procedure will differentiate halogenated compounds, alcohols, enols, phenols, ethers, peroxy compounds, thiols, sulfides, amines, imines, a variety of other nitrogenous compounds (nitro, oximes, etc.), aldehydes, ketones, carboxylic acids, various derivatives of organic acids (such as esters, lactones, acyl chlorides, etc.), derivatives of carbonic acid (such as ureas, cyanates), sulfonic acids, etc. More than one functional group and more than one type of group can be attached on the hydrocarbon backbone. A special class of compounds is that of heterocycles (aromatic or not). Heterocycles can be classified based on the heteroatoms in the ring such as oxygen (furans, pyrans), nitrogen (pyrole, pyrazole, imidazole, triazole, pyridine, pyrazines, etc.), sulfur (thiophene), or different heteroatoms (oxazoles, thiazole, oxadiazoles, etc.). Organic molecules may have very complex structures and a simple classification is typically hard to make and at the same time less relevant. Specific classes of organic compounds, with similar properties, are also known, such as carbohydrates, amino acids, lipids, steroids, nucleotides, triterpenes, flavonoids, and others. In addition to all the classes of compounds containing C, H, O, N, or S, other types of organic compounds include those containing boron, silicon, arsenic, antimony, and metallic elements (organometallic compounds). A large number of natural polymers such as polymeric carbohydrates, proteins, nucleic acids, and a wide variety of synthetic polymers are also organic compounds. Polymers that contain inorganic and organic components such as silicones are also known.

Inorganic compounds also cover a wide range of possibilities of combinations. These possibilities start with binary compounds such as oxides, sulfides, halides, to very complex structures. Acids, bases, salts, and coordination compounds, are among common inorganic materials. Inorganic polymers are also common in nature. The chemical nature of the analytes and of the matrix components is an important factor in the selection of the method of analysis. This is related to other properties determined by the chemical structure, such as polarity, hydrophobicity, and acid–base character, that are further discussed.

Molecular Weight

The molecular mass of one molecule is its mass expressed in unified atomic mass units [1/12 of the mass of one atom of the isotope carbon-12, sometimes named a dalton (Da)]. The molecular weight (*Mw*) of a molecule is the ratio of the mass of the molecule to 1/12 of the mass of isotope carbon-12 (*Mw* is dimensionless). *Mw* and molecular mass are numerically equal, but they are not the same parameter, although the terms are frequently used interchangeably.

The molecular weight represents an important parameter in molecular characterization, and it is related to many other molecular properties. A common purpose of *Mw* is the differentiation between small molecules and macromolecules. As a common definition, macromolecules are chemical compounds formed from at least 1000 atoms linked by covalent bonds. However, instead of number of atoms, a *Mw* higher than about 5000 is commonly used to indicate a macromolecule, and a *Mw* lower than about 2000 to indicate a small molecule. Between these two limits is a gray area where, among others, the molecules known as oligomers are placed. Molecular weight plays a major role in the chromatographic size-exclusion separations (gel filtration chromatography and gel permeation chromatography).

Acidic or Basic Character of Solutes

Many types of molecules contain functional groups likely to lose or gain protons under specific circumstances. The molecules capable of donating a proton are indicated as "Brønsted" acids, and those capable of accepting a proton are indicated as "Brønsted" bases. For most organic compounds, the dissociation with the formation or acceptance of protons is an equilibrium process (see Section 4.4). This dissociation is characterized by the acidity constant (or constants in the case of polyprotic molecules) commonly expressed as $pK_a = -\log K_a$. Tables with acidity constants are readily available in the literature for common acids and bases (see Appendix 7i). Similar to any equilibrium constant K_a is temperature-dependent and its values are typically listed for 25°C. Various techniques are available for pK_a calculation. Computer programs are also available for the calculation of pK_a (e.g., MarvinSketch [5]). For macromolecules with multiple groups ionizable in solution, the individual pK_a values for each functionality are not relevant parameters, the basic or acidic character of the molecule being a global property. The acidity or basicity constant for a compound is a very important parameter in the decision of the type of chromatography that should be used for its separation. Also, the pH of the mobile phase is selected in many separations as dependent of the pK_a of the analytes, with the purpose of keeping the analyte in mainly one form or another.

Isoelectric Point

The isoelectric point (pI) is the pH value at which the molecule carries no electrical charge. The concept is particularly important for zwitterionic molecules such as amino acids, peptides, and proteins. For an amino acid, the isoelectric point is the average of pK_a values for the amine and the carboxyl group. In the case of amino acids with multiple groups ionizable in solution (e.g., lysine with two amino groups or aspartic acid with two acid groups), the isoelectric point is given by the average of the two pK_a of the acid and base that lose/gain a proton from the neutral form of the amino acid. This can be extended to the definition of pI of peptides and proteins. The pI value can be used to indicate the global basic or acidic character of a zwitterionic molecule, and compounds with $pI > 7$ can be considered basic, and those with $pI < 7$ can be considered acidic. For complex molecules such as proteins, the isoelectric point is useful in the description of acidic or basic character, where individual pK_a values are not relevant.

Dipole Moment as Part of Molecular Polarity

A few comments on polarity were previously made related to ionization of molecules in mass spectrometry and also related to chromatographic retention. Polarity is an important characteristic of molecules, and it is frequently used with a wide meaning. The strict definition of polarity refers to the separation of the center of partial positive charges from that of partial negative charges, leading to an electric dipole of the molecule. The electric dipole of a polar molecule is characterized by the dipole moment μ (the notation for chemical potential is also μ and the differentiation must be based on context). The definition of dipole moment is:

$$\vec{\mu} = q\,\vec{d} \tag{5.1.1}$$

where $\vec{d}$ is a vector with the length equal to the distance (in nm) between the two separated point charges $+q$ and $-q$ directed from the negative charge to the positive charge. The vector character of the dipole moment is frequently neglected and only its value is indicated (μ). The unit of dipole moment is the debye (D), and $1\,D = 3.336 \times 10^{-30}$ C m (coulomb meter). Besides molecular dipole moments, the value for μ can be calculated for chemical bonds and also for molecular fragments. Extensive tables with dipole moments are available in the literature [6].

Polarizability as Part of Molecular Polarity

Under the concept of polarity is frequently included the polarizability. Polarizability indicates the tendency of molecules to develop a dipole moment under the influence of an electric field, usually created by other molecules from the environment. Polarizability α is defined by the ratio:

$$\alpha = \frac{\vec{\mu}_{ind}}{\vec{E}} \tag{5.1.2}$$

where $\vec{E}$ is the electric field generating the induced dipole moment. As defined by Eq. (5.1.2), polarizability is a scalar quantity indicating that the electric field produces polarization only in the direction of the field. However, polarizability can also be defined as a tensor when the electrical field generates moments of dipole in different directions from that of field $\vec{E}$.

Besides a quantitative characterization of molecules based on μ and α values, it is common to qualitatively assess a polar character by the presence in the molecules of polar functional groups such as $-OH$, $-COOH$, $-NH_2$, $>NH$, $-SO_3H$, since these groups bring both permanent dipole moments and polarizability to molecules.

Hydrogen Bonding Capability as Part of Molecular Polarity

Hydrogen bonding is the attractive interaction that occurs between a hydrogen atom bound in a molecule to an electronegative atom such as oxygen or nitrogen and another electronegative atom from a different molecule (intermolecular) or even from the same molecule (intramolecular). The energy of hydrogen bonds can be between 10 and 40 kJ/mol (see Table 4.3.1) and can be stronger than van der Waals type interactions. The hydrogen bonding has features of both electrostatic interaction and covalent bonding. It is directional and strong, leading to the formation of interatomic distances shorter than the sum of van der Waals radii of the participating molecules. The capability of formation of hydrogen bonds contributes to the polar character of a molecule to an extent that can exceed the contribution of dipole moment and polarizability (responsible for van der Waals interactions). For example, the polar character is considered to be in the order $H_2O > CH_3OH > CH_3CN$, although water has the lowest μ and α and acetonitrile the highest.

Octanol/Water Partition Constant and Its Use for Polarity Estimation

A common characterization of polarity of molecules is based on a "bulk" property and not on molecular properties. This characterization is obtained using the octanol/water partition constant $K_{ow}(X)$ (P is also a common notation for K_{ow}). Octanol/water partition constant is used for the characterization of hydrophobicity of a compound X, which is the opposite of polar character. The expression for $K_{ow}(X)$ has been given by Eq. (4.3.20). The polar and/or hydrophobic character of a molecule is related to its K_{ow} value, with positive values for log K_{ow} indicating hydrophobic character, and with larger values for log K_{ow} showing more hydrophobicity. Molecules with low log K_{ow}, even if the values are positive, show the presence of some polar character, and negative values clearly indicate polar properties. However, there is not a direct relation between K_{ow} and the charge distribution in the molecule. Octanol/water coefficient is a "bulk" property referring to the whole material, while charge distribution (dipole moment, polarizability) is properties at molecular level. The application of the findings at the molecular level to the bulk level is not straightforward. Several properties of the continuum, such as dielectric constant of the bulk that affects interactions, should also be considered for interpreting the findings at the molecular level.

Octanol/water parameters have a widespread utilization in separation science and also in other important fields of science, such as drug design and environmental studies. Values for K_{ow} are available for many compounds [7,8], can be calculated using computer programs (e.g., MarvinSketch 5.4.0.1, ChemAxon Ltd. [5], EPI Suite [8], ClogP [9], ACD/logPdb, KowWin [10], and SciLogP/Ultra), and can be obtained using additive fragment methodology [1,11]. Other methods for the estimation of K_{ow} are based on physicochemical molecular properties such as van der Waals molecular surface area [12], solvatochromic parameters [13], etc.

Octanol/water partition coefficient can also be used for the characterization of compounds that can be partially present in the form of ions or can exist in more than one form. These compounds are characterized by a related parameter to K_{ow}, which is the distribution coefficient D_{ow}. For example, the compounds having groups ionizable in solution exist as a mixture of different forms (some ionic and some neutral). In this case, a distribution coefficient $D_{ow}(X)$ for a given compound X (existent in more than one form) is defined by the formula:

$$D_{ow}(X) = \frac{C_{1,o} + C_{2,o} + \ldots + C_{n,o}}{C_{1,w} + C_{2,w} + \ldots + C_{m,w}} \tag{5.1.3}$$

where $\{C_{i,o}\}$ $i = 1, 2, \ldots n$ represent all the forms of the compound X present in octanol, and $\{C_{i,w}\}$ $i = 1, 2, \ldots m$ represent all the forms of compound X present in water. (Note: the terms partition and distribution are used interchangeably, and the difference must result from the meaning of the parameter). For compounds having no ionizable groups or that exist in only one form, $K_{ow} = D_{ow}$. The distribution coefficient D_{ow} for species with groups ionizable in solution depends on pH. For example, for an acid HX expression (5.1.3) is reduced to the formula:

$$D_{ow}(HX) = \frac{C_{HX,o}}{C_{HX,w} + C_{X^-,w}} \tag{5.1.4}$$

The value for $C_{X^-,w}$ based on expression (4.4.5) (acidity constant for HX being K_a) can be written as follows:

$$C_{X^-,w} = K_a \frac{C_{HX,w}}{[H^+]} \tag{5.1.5}$$

The replacement of expression (5.1.5) in (5.1.6) leads to the formula:

$$D_{ow}(HX) = \frac{C_{HX,o}}{C_{HX,w}(1 + K_a/[H^+])} = K_{ow}(HX)\frac{10^{-pH}}{K_a + 10^{-pH}} \tag{5.1.6}$$

In Eq. (5.1.6), the standard expression for $K_{ow}(HX) = C_{HX,o}/C_{HX,w}$ has been utilized. Expression (5.1.6) indicates that at high pH, the value for 10^{-pH} is extremely small and $D_{ow}(HX)$ is expected to be small. In acidic conditions, $K_a < 10^{-pH}$ and the value for $D_{ow}(HX)$ become closer to that for $K_{ow}(HX)$. Similar calculations can be performed for bases. Dedicated computing programs can be used for the evaluation of D_{ow} for molecules with ionic groups (e.g., MarvinSketch 5.4.0.1 [5]). As an example, the variation of log D_{ow} for nicotine as a function of pH, obtained with such a program, is given in Fig. 5.1.1. The figure also shows the variation of % content of different forms (ionized and not ionized) of nicotine as a function of pH.

The values of log K_{ow} (and log D_{ow}) are very useful for the characterization of polar or hydrophobic character of small molecules. However, the same technique cannot be applied for quantitative characterization of polarity of polymers, where the log K_{ow} value loses its meaning. Proteins, for example, have the capability of folding, and the polar groups have the tendency to congregate in such a manner to maximize electrostatic interactions with the polar solvating medium. In a polar solvent like water or aqueous solutions of acids, the protein may change its tertiary and quaternary structure and expose polar side chains toward the solvent, the hydrophobic moieties being congregated toward a more hydrophobic core. The opposite effect may take place in the presence of organic solvents. This behavior would lead to a variable octanol/water partition.

The extension of polarity characterization with octanol/water partition coefficient can be done to solvent mixtures and even to stationary phases, although such values do not have a true experimental meaning. For example, for a mixture of solvents a hypothetical K_{ow}^{hyp} can be utilized for the description of the apparent hydrophobicity of the mixture. Such a value can be taken as being the weighted (by content) average of the K_{ow} values of the participating solvents (see Section 13.1). An experimental value for a solvent mixture cannot be obtained since the components of a solvent mixture (e.g., acetonitrile + water) would distribute independently between octanol and water and will not remain as an initial mixture. However, hypothetical values are still useful for hydrophobicity estimation.

For the extension of K_{ow} value to the characterization of a stationary phase, one possibility is to simulate the stationary phase with a model small molecule that contains the main characteristics of the phase (e.g., taking the bonded moiety and a small portion of a silica surface). The stationary phase can be characterized by a K_{ow}^{model} although the value of K_{ow}^{model} does not

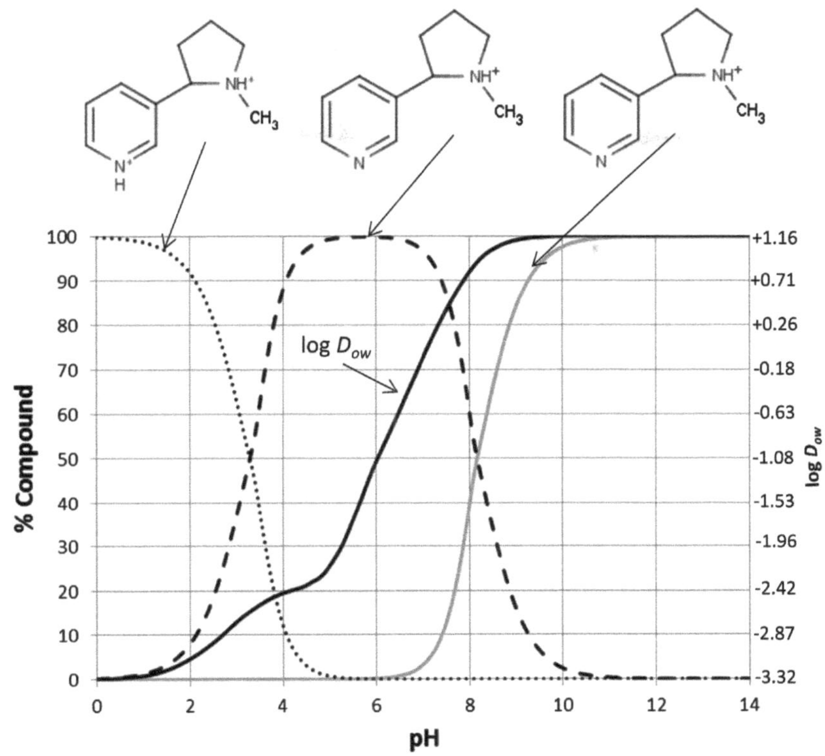

FIGURE 5.1.1 The variation of % content of different forms (ionized and not ionized in solution) of nicotine as a function of pH and the variation of log D_{ow}.

correspond to an experimental property. The use of K_{ow}^{model} for stationary-phase characterization is further discussed in Section 6.2.

Molar Volume

Molar volume V_{mol} (sometimes noted only by V) is the volume occupied by 1 mol of a compound (or element) at a given temperature and pressure. The expression for V_{mol} is given by the formula:

$$V_{\text{mol}} = \frac{Mw}{\rho} \tag{5.1.7}$$

In expression (5.1.7), Mw is the molecular weight and ρ is the density of the compound. Molar volume is not a parameter used for a specific selection in HPLC, but its value is related to the calculation of various other parameters used in separations.

Solubility of Nonelectrolytic Compounds

Solubility is a property related to separation, since the analyte must be soluble in the mobile phase for achieving retention and elution. In addition to that, the sample should be soluble in the solvent used for injection. The solubility is defined as the maximum amount of solute that dissolves in a fixed volume of a solvent at a given temperature and can be expressed as mol/L of solute, mole fraction x, and some other ways. This maximum amount of dissolved compound is attained when equilibrium is established between the solid substance and its dissolved form in a saturated solution. The equilibrium can be written as follows:

$$X_{solid} \leftrightarrow X_{solution} \tag{5.1.8}$$

The molar concentration in a solid $C(X)_{solid} = 1$, the constant governing the equilibrium (5.1.8) can be written in the form:

$$K_{solub} = C(X)_{solution} \tag{5.1.9}$$

Expression (5.1.9) shows that the molar concentration of a saturated solution $C(X)_{solution}$ is constant (for a constant temperature). In a saturated solution in the presence of solid, by addition of solvent more solid dissolves (until the solution is not saturated anymore). The elimination of solvent leads to the formation of more solid.

The dissolution process of nonelectrolytes can be viewed hypothetically as formed from two steps. The first step is melting (fusion) of the compound X to form a supercooled liquid. The second step is the mixing of this liquid with the solvent S. This indicates that the free enthalpy of dissolution ΔG_{dis} can be written in the form:

$$\Delta G_{dis} = \Delta G_{fus} + \Delta G_{mix} \tag{5.1.10}$$

From the expression $\Delta G = \Delta \boldsymbol{H} - T\Delta S$, and considering that at melting temperature $\Delta G_{fus}(T_{fus}) = 0$, it is obtained: $\Delta \boldsymbol{H}_{fus} = T_{fus}\,\Delta S_{fus}$ (the value for ΔS_{fus} is typically taken using the approximation $\Delta S_{fus} \approx 13$ cal/mol deg). From the standard expression $\Delta G = \Delta \boldsymbol{H} - T\Delta S$ and using the value $\Delta S_{fus} = \Delta \boldsymbol{H}_{fus}/T_{fus}$, the expression for ΔG_{fus} at a given temperature T becomes:

$$\Delta G_{fus}(T) = \Delta \boldsymbol{H}_{fus}\left(1 - \frac{T}{T_{fus}}\right) \tag{5.1.11}$$

The formula for the free energy of mixing is given by the expression:

$$\Delta G_{mix} = \Delta \boldsymbol{H}_{mix} - T\Delta S_{mix} \tag{5.1.12}$$

For regular solutions (see, e.g., Ref. [14], the entropy of mixing is given by the expression:

$$\Delta S_{mix} = -R\sum_{j} n_j \ln x_j \tag{5.1.13}$$

In Eq. (5.1.13), j is one component in the solution, n_j is the number of moles for j, and x_j is the mole fraction of component $j\left(x_j = n_j \Big/ \sum_i^{tot} n_i\right)$. Considering only two components (solute X and solvent S), estimating $x_S \approx 1$ (and $\ln x_S \approx 0$), and taking 1 mol of compound $n_X = 1$, the expression for ΔS_{mix} is the following:

$$\Delta S_{mix} = -R \ln x_X \tag{5.1.14}$$

At equilibrium between solid and solution, for constant pressure and temperature $\Delta G_{dis} = 0$. As a result, $\Delta G_{fus} + \Delta G_{mix} = 0$. Formulas for both ΔG_{fus} and ΔG_{mix} are previously indicated. The replacement of ΔG_{fus} with its expression given by Eq. (5.1.11) and of ΔG_{mix} with its expression given by Eq. (5.1.12) (where ΔS_{mix} is given by Eq. (5.1.14)) leads to the following result: $\Delta \boldsymbol{H}_{fus}(1 - T/T_{fus}) + \Delta \boldsymbol{H}_{mix} + TR \ \ln x_X = 0$. From this expression the following formula can be generated for the maximum amount of solute that dissolves in a fixed volume of solvent:

$$\ln x_X = -\frac{\Delta \boldsymbol{H}_{mix}}{RT} - \frac{\Delta \boldsymbol{H}_{fus}}{R}\left(\frac{1}{T} - \frac{1}{T_{fus}}\right) \tag{5.1.15}$$

In Eq. (5.1.15), $\Delta \boldsymbol{H}_{mix}$ and $\Delta \boldsymbol{H}_{fus}$ must be estimated. As previously shown, the value for $\Delta \boldsymbol{H}_{fus}$ can be approximated from the formula $\Delta \boldsymbol{H}_{fus} \approx 13T_{fus}$. Therefore, a low value for the temperature of melting T_{fus} is favorable to solubility (for nonpolar compounds). The increased temperature T also favors solubility ($T < T_{fus}$).

The evaluation of ΔH_{mix} can be done for nonpolar compounds based on the use of *Hildebrand solubility parameter*. This approximation considers first the dissolution in a solvent S of a vaporized molecule X. This dissolution can be viewed as the reverse of vaporization. The energy of vaporization can be separated into two terms, one accounting for the removal of a molecule from the liquid, and the other for the formation of new interactions between X molecules remaining in solution. Assuming that a molecule in the liquid is surrounded by n other molecules and the energy for each interaction is E_{XX}, the energy term of removal will be nE_{XX}. The formation of new interactions $X - X$ in solution will have the same energy E_{XX}, but the number of such interactions will be $n/2$. The vaporization energy $\Delta E_{vap,X}$ is therefore given by the expression:

$$\Delta E_{vap,X} = \mathcal{N}\frac{nE_{XX}}{2} \tag{5.1.16}$$

where $\mathcal{N}$ is the Avogadro constant, $\mathcal{N} = 6.022 \times 10^{+23}$ mol^{-1}. When the process of dissolution of a molecule X in the solvent S takes place, $n/2$ interactions $S - S$ will be broken, and n' interactions $X - S$ will be created (and $n' \approx n$). Therefore, indicating by E_{XS} the interaction energy between a molecule of solute and one of solvent, the dissolution energy is given by the expression:

$$\Delta E_{sol,XS} = \mathcal{N}\left(\frac{nE_{SS}}{2} - nE_{XS}\right) \tag{5.1.17}$$

In the solution, an equal number of intermolecular interactions per unit volume can be assumed, such that $\mathcal{N}n = C_t V_X$, where V_X is the molar volume of species X ($V_X = Mw_X/\rho_X$ where ρ_X is the compound density) and C_t is a constant. With this assumption, Eq. (5.1.16) takes the form:

$$\Delta E_{vap,X} = \frac{C_t V_X E_{XX}}{2} \tag{5.1.18}$$

Eq. (5.1.18) can be written in the form:

$$E_{XX} = \frac{2\Delta E_{vap,X}}{C_t V_X} \tag{5.1.19}$$

Eq. (5.1.19) indicates that the molecular interaction E_{XX} in the liquid is proportional with the ratio of vaporization energy per unit molar volume. This ratio can be used to give a specific solubility parameter with the notation δ_X that has the expression [15]:

$$\delta_X^2 = \frac{\Delta E_{\text{vap},X}}{V_X} \tag{5.1.20}$$

Parameter δ_X is known as *Hildebrand solubility parameter* and the units for δ are $(\text{cal/cm}^3)^{1/2}$. The value of δ_X can be estimated using a number of procedures [15]. These include calculation from heats of vaporization $\Delta \boldsymbol{H}_{\text{vap},X}$, which is related to $\Delta E_{\text{vap},X}$ by the formula:

$$\Delta E_{\text{vap},X} \approx \Delta \boldsymbol{H}_{\text{vap},X} - RT \tag{5.1.21}$$

where $\Delta \boldsymbol{H}_{\text{vap},X}$ is either measured or estimated. Hildebrand solubility parameter δ_X can also be estimated from superficial tension γ' with the formula:

$$\delta_X \approx 4.1\left(\frac{\gamma'_X}{V_X^{1/3}}\right)^{0.43} \tag{5.1.22}$$

Values for Hildebrand solubility parameter are available in the literature [16], and some such values for a number of solvents are given in Appendix 7a.

This solubility parameter can be used as a measure of the intermolecular interactions per unit volume of a pure liquid based on the following relation:

$$E_{XX} = \frac{2\delta_X^2}{C_t} \tag{5.1.23}$$

For the solvent S the parameter δ_S is defined similarly to δ_X and E_{SS} is expressed by a formula similar to Eq. (5.1.23). The energy for the interaction $X - S$ can be approximated as the geometric mean of E_{XX} and E_{SS} such that:

$$E_{XS} = \sqrt{E_{XX}E_{SS}} = \frac{2\delta_X\delta_S}{C_t} \tag{5.1.24}$$

The energy of mixing of a mol of X with a large quantity of pure S to form a dilute solution should be equal with the sum of $\Delta E_{\text{vap},X}$ and $\Delta E_{\text{sol},XS}$ given by Eqs. (5.1.16) and (5.1.17). The use of the values for E_{XX}, E_{SS}, and E_{XS} as function of Hildebrand solubility parameters in this sum leads to the expression for the energy of mixing in the form:

$$\Delta E_{\text{mix},XS} = \mathcal{N}\left(\frac{n}{C_t}\right)\left(\delta_X^2 + \delta_S^2 - 2\delta_X\delta_S\right) \tag{5.1.25}$$

Assuming no volume variation during mixing at constant pressure, the energies $\Delta E_{\text{mix},XS}$ can be taken as equal with the enthalpy (heat) of mixing $\Delta \boldsymbol{H}_{\text{mix}}$. In conclusion, Eq. (5.1.25) can be written (after including $\mathcal{N}n = C_t V_X$) for 1 mol of solute X in the form:

$$\Delta \boldsymbol{H}_{\text{mix}} = V_X(\delta_X - \delta_S)^2 \tag{5.1.26}$$

For a solution where the concentrations of X and S are comparable, Eq. (5.1.26) must be replaced by the similar relation [17]:

$$\Delta \boldsymbol{H}_{\text{mix}} = (x_X V_X + x_S V_S)(\delta_X - \delta_S)^2 \phi_X \phi_S \tag{5.1.27}$$

where x_X and x_S are the mole fractions and ϕ_X and ϕ_S are the volume fractions of X and S, respectively (see Eq. (4.2.68)). When x_X very small, Eq. (5.1.27) is reduced to Eq. (5.1.26), although the transformation is not straightforward due to the approximations that are involved. Eqs. (5.1.26) and (5.1.27) indicate that the variation in the heat of mixing is always positive, which shows that the enthalpy of mixing is always unfavorable to the process. The enthalpy term is higher when the difference between the solubility parameters ($\delta_X - \delta_S$) is larger, which is in agreement with the experimental observation that similar compounds dissolve one in the other, while different ones are less likely to mix. The use of Eq. (5.1.15) with the values for ΔH_{mix} given by Eq. (5.1.26) allows an estimation of solubility of a nonelectrolyte. The simple principle of "like-to-like" describes the conditions when the dissolution is favored. The presence of other compounds in the solution may influence the solubility through other interactions. Solubility of electrolytes is a separate subject (see, e.g., Ref. [18]).

Van der Waals Molecular Volume and Surface Area

Other parameters not directly used for a specific selection in HPLC, but important for the evaluation of certain separation characteristics are van der Waals molecular volume $\mathcal{V}$ and van der Waals surface area $\mathcal{A}$. The calculation of both van der Waals molecular volume and area starts with the concept of van der Waals radius r of an atom. This is the radius of an imaginary sphere used to model the atoms describing its finite size. The radius is obtained based on results from gas kinetic collision cross-sections, gas critical volumes, crystal densities (extrapolated at 0 K), liquid state properties, and X-ray diffraction data. Standard values for van der Waals radii of elements are reported in the literature [19,20]. From the value of the radius, the surface of an atom is immediately obtained as $\mathcal{A} = 4\pi r^2$ and the volume is obtained as $\mathcal{V} = 4/3\pi r^3$. For the calculation of $\mathcal{V}$ or $\mathcal{A}$ for molecules, the atoms are placed at the bond distance, which is typically shorter than the sum of the van der Waals radii of the connected atoms. For this reason, the van der Waals molecular volume is smaller than the sum of the volume of each component atom. This is also true for the molecular surface. The calculation of $\mathcal{V}$ or $\mathcal{A}$ for diatomic molecules is simple, by subtracting the volume (or the area) of overlapping. However, for more complex molecules, the calculation becomes more elaborate, but computer programs are available for such calculations (see, e.g., Refs. [5,21]).

The correlation between van der Waals surface area $\mathcal{A}$ and retention factor $k'(X)$ for a compound X, in reversed phase chromatography, has been discussed in Section 4.3 (see Eq. (4.3.38) for $K(X)$ in a partition process involving hydrophobic interactions). The use of this correlation has been reported in the literature [22–24], and also has been used for the estimation of phase ratio of RP-HPLC type columns [25].

5.2 PHYSICOCHEMICAL PROPERTIES RELATED TO DETECTION

The detection in HPLC is based on certain physicochemical properties that are different between the eluted molecules and the mobile phase such that the analytes can be detected and measured. For this reason, the choice of the type of detection in HPLC is determined by such particular physicochemical properties. Among these properties are UV-absorption, refractive index, fluorescence, molecular mass, and fragmentation in a mass spectrometer, etc. and a large number of instrumental procedures have been developed for measuring such properties, as described in Sections 2.2 and 4.1. Some of these properties are further discussed in this section related to the analytes and matrix.

UV–Vis Absorption

Light absorption in UV and visible region (with wavelength λ between 200 nm and 800 nm) is caused by the electronic transitions in the absorbing molecule that goes from one energetic state into a higher energetic state. Molecules commonly present in the lowest electronic energy state (fundamental state) when irradiated with UV or visible light may absorb the energy $\Delta E = h\nu$ and change into an excited electronic state (h is Planck's constant, ν is light frequency with $\nu = c_{\text{light}}/\lambda$). The electronic states of molecules are discrete and known as electronic levels. The electronic levels are classified as bonding (with lower energy), nonbonding, and antibonding (with higher energy). Depending on the molecule, the number of electronic energy levels can be quite large. For a closed-shell molecule, for example, the number of bonding levels, each occupied by two electrons and described by a molecular orbital, is usually equal to half of the number of electrons in the molecule. The electronic transitions responsible for UV and visible light absorption take place preferentially between the energy levels corresponding to the highest occupied molecular orbital (HOMO) to the lowest unoccupied molecular orbital (LUMO) (see, e.g., Ref. [26]). The electronic energy levels of the molecules are commonly classified based on the symmetry of the molecule [27]. For linear and planar molecules the molecular orbitals corresponding to a given energy level can be classified as σ, π, δ,..., and the corresponding bonding energetic levels are noted σ, π,... while the corresponding antibonding energetic levels are noted σ^*, π^*,... For each electronic level in a molecule there are several vibrational/rotational levels. For this reason, electronic transitions have a fine structure covering a range of wavelengths. This fine structure is not detectable for molecules in solution or in solid state due to other molecular interactions, such that the absorption spectra typically appear as broad bands. The schematic diagram of electronic transitions during UV or visible light absorption is shown in Fig. 5.2.1. The molecules in the excited state return in fundamental state by nonradiative processes when the energy is dissipated typically as heat. The energies ΔE involved in $\sigma \rightarrow \sigma^*$ or in $n \rightarrow \sigma^*$ transitions are usually quite large and correspond to short wavelengths λ in the far UV region (below 220 nm). The energy involved in $n \rightarrow \pi^*$ or $\pi \rightarrow \pi^*$ transitions may correspond to either UV absorptions at higher wavelength than 200–220 nm or fall in the visible region (above 360 nm). The transitions $\pi \rightarrow \pi^*$ are characteristic to molecules that contain in the molecule π bonds and therefore for molecules with groups such as >C=C<, >C=O, –C≡, >C=S, –N=N–, etc. Such groups are typically indicated as chromophores. Conjugated π bonds have an even stronger chromophore character.

The absorption of light by a solution of a compound is described by Eq. (2.2.3), $A_\lambda = \varepsilon_\lambda C_X L$ where ε_λ is the molar absorption (absorptivity) coefficient at the specific wavelength λ, L is the path length of the light through the sample, and C_X is the molar concentration of the absorbing molecular species. The higher is the value for ε_λ, the higher is the absorption A_λ (for the same C_X and L), and therefore the better is the sensitivity of the analysis. The coefficient ε_λ is related to the Einstein coefficients for photo absorption and induced emission which are intrinsic properties of the two energy levels that undergo the electronic transition (see, e.g., Ref. [28]). The Einstein coefficients may be zero for certain transitions

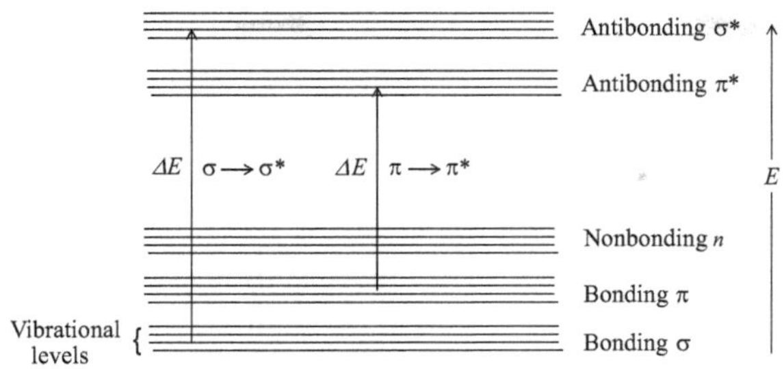

FIGURE 5.2.1 Schematic diagram of electronic transitions during UV or visible light absorption.

due to restrictions imposed by the symmetry of the molecules, an effect that simplifies the absorption spectra of molecules. Although there are theoretical paths for the calculation of Einstein coefficients, the values for ε_λ are in most cases obtained experimentally. Various lists (more or less comprehensive) of values for the maximums of absorption wavelength λ_{max} for organic compounds and the corresponding molar absorptivity ε_λ can be found in the literature (see, e.g., Refs. [29,30]) or on the web (see, e.g., Ref. [31]).

While a high UV absorption is desirable for the analytes, the opposite is the case for the mobile phase components. The solvents utilized as mobile phase should have very low absorption in the wavelength range (UV or visible) where the measurement of the analytes is performed. The solvent must be "transparent" (absorb very little light) in the region where the absorption of the analyte is measured. This "transparency" is characterized by the UV cut-off value, defined as the wavelength at which the absorbance A of the solvent versus air, in a 1 cm cell, is equal to unity (see Eq. (2.2.3)). The UV absorption increases significantly when the wavelength decreases, and this cut-off value indicates that at lower wavelength the absorption of light is too strong to allow utilization. The UV cut-off values for several common solvents are specifically discussed in Section 13.3.

Fluorescence

As indicated in Section 2.2, the measured fluorescence intensity is expressed by Eq. (2.2.7), $F'_{int,\lambda_2} = a\, I_{0,\lambda_1} C_X$, where parameter a depends on molecular characteristics of the compound of interest, and also on other constants including the losses due to the measurement of only a part of the total fluorescent radiation (C_X is the analyte concentration and $I_{0,\lambda}$ is the intensity of the excitation radiant energy). Due to the various factors influencing the measured intensity of the fluorescence of a specific compound, the correlation between the structure of the analyte and its fluorescence properties is usually established empirically (see, e.g., Ref. [32]). Empirically it was noticed that compounds such as coumarins (substituted at C6 or C7), oxazoles, benzofurans, acridines, quinoxalines, xanthenes, polyaromatics, and other specific compounds have moieties that carry fluorophore properties. However, most compounds do not fluoresce. Because fluorescence measurement can lead to very high sensitivity in an analytical method, derivatization procedures are utilized when possible to attach fluorophore groups to the analyte and use fluorometric detection (see, e.g., Ref. [1]). Besides the dependence on the compound properties, fluorescence can be affected (usually in the negative way) by the composition of the mobile phase when the compound of interest elutes.

Chemiluminescence

Chemiluminescence is the emission of light as the result of a chemical reaction, and not a property of a specific compound. One such typical reaction is that between luminol (5-amino-2,3-dihydro-1,4-phthalazinedione) and hydrogen peroxide in basic medium, with the formation of 3-aminophthalate and emission of light. Luminol moiety can be attached by derivatization to different analytes and chemiluminescence can be used for detection. Another type of chemiluminescence reaction is that involving peroxyoxalates (POCL). Several molecules containing the moiety of a peroxylate have been synthesized [33]. The peroxyoxalate reacts with hydrogen peroxide, and the activated derivative of oxalic acid, in the presence of a fluorescent substance, produces chemiluminescence. This reaction is classified as a sensitized or indirect chemiluminescent reaction since the light is emitted from the fluorescent molecule (fluorophore). The fluorophore gains its excitation energy from intermediates appearing along the reaction path of peroxyoxalates and hydrogen peroxide. Only specific fluorescent molecules are used in these POCL type reactions, but various analytes can be measured by this procedure after appropriate derivatizations (see, e.g., Ref. [1]).

Refractive Index

For visible light most transparent media have refractive indices between 1 and 2. The exact value of the refractive index depends on the wavelength of the refracted light. The difference between the refractive index of a mobile phase in HPLC and that of an analyte can be used for detection (see Section 4.1). The refractive index of many compounds can be found in the literature (see, e.g., Ref. [34]), and also can be estimated with relatively good precision using the expression:

$$n = \left(\frac{Mw + 2R_m\rho}{Mw - R_m\rho}\right)^{1/2} \tag{5.2.1}$$

where R_m is the molar refraction of the compound (Mw is molecular weight and ρ density). Molecular refraction is an additive parameter that can be obtained from the contribution of atoms and bonds in a molecule, following the expression:

$$R_m = \sum(\text{atomic contributions}) + \sum(\text{bond contributions}) \tag{5.2.2}$$

The contributions of various atoms and types of bonds are tabulated in various publications (see, e.g., Refs. [1,35,36]). The refractive index of various solvents used as mobile phase in HPLC are further discussed in Section 13.3.

Mass Spectra

Mass spectra obtained by EI+ ionization of the analytes with electron impact at 70 eV, practiced in GC–MS, generates "standard spectra" (see Section 2.3). Large libraries of such spectra are available (see Section 2.3). For LC-MS, ionization of analyte molecule is generated differently, with ESI type ionization introducing the flow of the LC into the MS source through a capillary at 3–5 kV potential or by APCI type ionization. In positive type ionization the formation of ions can be written as follows (see Eq. (2.3.5)):

$$M + H^+ \rightarrow M - H^+ \qquad (5.2.3)$$

The equilibrium constant K governing the reaction (5.2.3) depends on gas-phase basicity (GPB) of molecule M and is given by the expression:

$$K = \exp\frac{-\Delta G^0_{GPB}}{RT} \qquad (5.2.4)$$

where the free enthalpy of formation ΔG^0_{GPB} is given by Eq. (2.3.6). In case of the formation of negative ions, molecules of the type M-H are ionized as indicated by reaction (2.3.7), and the gas-phase acidity (GPA) is characterized by the variation of free enthalpy ΔG^0_{GPA} given by Eq. (2.3.8) (see Section 2.3). The values for ΔG^0_{GPB} and for ΔG^0_{GPA} can be useful in the prediction of ionization and fragmentation under CI-like conditions, and the generally larger values of ΔG^0_{GPA} can explain why the formation of positive ions takes place more easily than that of negative ions in LC-MS [37].

An important property of the analyte molecules related to the formation of ions is their polarity. The difference in the polarity between the solvent and the analyte molecules favors the formation of positive ions (or negative ions when working in negative mode) particularly from the analyte and much less from the solvent. A volatile acid such as HCOOH is typically added in the mobile phase to favor the process of positive ion formation. The positive ions are attracted toward the curtain plate in the ion source while the solvent molecules that are not charged are not attracted. Further desolvation and elimination of solvent molecules occur as the positive ions are directed toward the skimmer and further into the ion mass analyzer. For negative ion formation, salts such as $HCOONH_4$ or CH_3COONH_4 are added to favor the ionization of the analyte.

The formation of adducts between analyte molecules and different ions reaching the MS interface in ESI and APCI was previously discussed (see Section 2.3). The adducts with the solvent have an important role in the ionization process, and in some cases they can be seen in the mass spectrum instead of the analyte molecular ion. These equilibria are influenced by ΔG^0_{GPB} or ΔG^0_{GPA} values for the analytes and the solvents and by the concentration of *Solv*-H^+, or $Solv^-$, the electric fields applied into the source, etc. Other adducts besides those with the solvent can be formed between the analyte M and certain species in the eluted material such as ions of Na^+, K^+, and $NH_4{}^+$. Negative ionization is less favorable to adduct formation, but the formation is still possible.

Electrochemical Properties

Two types of electrochemical properties of analytes can be used for detection, including oxidation–reduction properties and conductivity properties (see Sections 2.4 and 4.1). Oxidation reduction reactions involve the equilibrium between two species, each characterized by the standard electrochemical potential of its half reaction ($Ox + ne- \leftrightarrow Red$, or $Red - ne- \leftrightarrow Ox$). This potential is measured for aqueous solution of the analyte of 1 M concentration for each ion participating to the reaction, at 25°C, a partial pressure of 1 atm for each gas that is part of the reaction, and with metals in their pure state. From Eq. (2.4.2), in such conditions, $E = E^0$. Tables of standard electrochemical potentials in specific media are available in the literature [38–40]. These values can be used for deciding if an electrochemical detector can be used for the analyte measurement by a redox process, and if the detection should be done by an oxidation or a reduction. Similar to other procedures, the mobile phase components should have the E^0 value significantly different from that of the analyte. For example, for the reduction of water by the reaction:

$$H_2O(l) + 2e^- \rightarrow H_2(g) + 2OH^-(aq) \tag{5.2.5}$$

the value for standard potential is $E^0 = -0.83$ V. Any attempt to reduce an analyte in water solution with a potential lower than −0.83 V would result in the reduction of water and not of the analyte. The same is valid for oxidation reactions, where the oxidation of the solvent must be more difficult to achieve compared to the analytes.

Conductivity measurements are used mainly for the analysis of ionic species that are completely dissociated in aqueous solution. The equivalent conductivity for the ionic species Λ_m is the parameter depending on the analyte properties. Tables with conductivity of electrolyte solutions at different concentrations are available in the literature [41].

5.3 SELECTION OF THE HPLC SEPARATION BASED ON SAMPLE PROPERTIES

During the selection process for an HPLC method, several decisions regarding the analysis characteristics must be made, such as the need to focus on quantitative measurements or on both quantitative and qualitative ones, to decide about the analysis precision and select a high or low detection sensitivity. Related only to the separation, another set of decisions must be made, such as: (1) what type of HPLC must be used, (2) what type of column must be used, and (3) what mobile phase must be used. These decisions are interconnected to the selection of sample preparation and with the type of detection to be used in the method. More detailed aspects of the separation should also be considered such as: (1) to use the core HPLC as a main separation of sample components without preliminary cleanup, or to use for the core HPLC a preprocessed and relatively clean sample (considering a more elaborate sample preparation step), (2) to use a separation with a high or a low resolution between target analytes, (3) to choose a short or a long chromatographic run, (4) to select a reduced solvent consumption for mobile phase, (5) to use the same chromatographic column to a large or a reduced number of analyzed samples, (6) to include semipreparative purpose in the separation, and (7) to include other goals.

The evaluations of analyte and of matrix properties are key steps for the selection of the HPLC separation method. Among the analyte and matrix main characteristics to consider are the following: (1) function of the analyte in everyday life, (2) chemical nature of the analyte, (3) polarity of the analyte, (4) other physicochemical properties of the analyte, and (5) estimated content of the analyte in the sample.

Classification of Samples Based on Their Role in Everyday Life

The role of the sample in everyday life is one of several criteria very useful for HPLC selection. This role is related to the chemical class of compounds from the sample, the required precision of measurements, the typical levels for the analyte in the assigned group of compounds, the number of samples to be analyzed, the length of acceptable turnaround time for the results, etc. The role of the sample in everyday life is related to both the analyte and the type of matrix of the sample. For example, an analyte that is an active ingredient in a pharmaceutical drug can be found in large quantity in the pharmaceutical product having an excipient as the matrix, in traces and in a different matrix in the biological fluids after the drug was administered to a human subject, and also as a pollutant in the environment with a different matrix than in the pharmaceutical product or the biological fluid. Knowing the role of the sample in everyday life also provides indications related to the expected level of the analyte. For example, the level of a pesticide in a material used for pest control is expected to be considerably larger as compared to the same pesticide compound as a pollutant in the environment. The selection of a rapid analysis versus a relatively longer turnaround time for analysis is also related to the role of the sample. The use of a rapid analysis may require less sample preparation but more robust or selective core analytical process, such that samples with a less clean matrix and lower analyte concentration can be analyzed. The choice of a rapid method may require the selection of a more sensitive (and possibly more expensive) detection and the need of more frequent chromatographic column replacements, the columns having a "shorter life." The role of the sample in everyday life also provides important information for the selection of sample preparation to make it more amenable for the core analysis (e.g., for HPLC). Sample preparation is an integral part of the analysis and the selection of sample preparation must be done in close relation to the selection of the core analytical method (see Section 1.2).

In addition, important information about analytical methods can be obtained based on the analyte classification on the role in everyday life since the materials with a specific role in everyday life are the subject of dedicated journals and books. Also, such literature describes sources of standards, recommended protocols for analysis, and descriptions regarding typical matrix simplification (see, e.g., Refs. [42–45]).

The role in everyday life allows several classifications of samples and of their containing analytes. One such classification includes the following: (1) environmental sample, (2) pharmaceutical and nutraceutical samples, (3) biological samples, (4) food and beverages samples, (5) agricultural products, (6) synthetic polymers, (7) other synthetic chemicals, and (8) other types of samples. Each group can be further divided into a number of subgroups, the variety of such subgroups can be separated based on different criteria. Significant overlapping exists between different classifications. Also, some of the listed groups are considerably larger than others.

1) Environmental samples can be classified based on their origin as being obtained from air, water, sediments, or soil. The type of analytes of interest can be the main components of the sample (e.g., for air the analysis can address the level of oxygen, nitrogen, ozone, carbon monoxide, carbon dioxide, nitrogen oxides, for soil the analysis can address the level of minerals and of natural organic compounds, for water it can address the level of natural anions and cations). The aim of analysis can also be the content of specific foreign compounds that are not part of the natural background composition and may include pollutants such as pesticides, herbicides, fungicides, persistent pollutants, and pollutants having different specific origin. The selection of an HPLC technique for environmental samples needs to consider the type of analytes, the nature of matrix, but at the same time the number of samples needed to be analyzed, turnaround time for the analysis, the required accuracy for the results, etc.

2) Pharmaceuticals include many compounds starting with synthetic drugs with relatively small molecules and ending with biopharmaceuticals that include recombinant proteins, vaccines, blood products, gene therapy, and cell therapy (e.g., stem cell therapies). Pharmaceuticals are also drugs of natural origin such as plant or minerals, drugs obtained from chemical modifications of natural products, and drugs of microbial origin (antibiotics). Besides the pharmaceuticals as an end product, many compounds including intermediate compounds in their synthesis/preparation as well as byproducts of these syntheses that are undesirable in drugs should be associated with the analysis of pharmaceuticals. The importance of knowing that an analyte is a pharmaceutical drug is related to its matrix, but also to the fact that analytical methods have been frequently developed for the analysis of pharmaceuticals and are likely to be available in the literature. Also, analysis of pharmaceuticals in the evaluation of absorption, distribution, metabolism, and excretion (ADME), with the goal of identification of active compounds and prediction of in vivo properties of the compound typically require a large number of analyses and therefore high-throughput analytical procedures. The number of samples and the need for a short turnaround time are important factors to be considered in developing an analytical method. The determination of impurities in pharmaceutical products can be a quite difficult task. Usually the impurities are present at low concentrations and some of them are difficult to be separated by chromatography. Analysis of trace active ingredients is equally a challenging task. Comprehensive information on recent trends in analytical perspectives on active pharmaceutical ingredients and on their degradation and impurities can be found in various reviews and dedicated books [46–50].

Nutraceuticals include specific nutrients (some natural and some synthetic), antioxidants, dietary supplements, herbal products, specific foods for diets (e.g., medicinal food), and other materials. They can also be differentiated by chemical families, such as lipids, vitamins, proteins, glycosides, phenolic compounds, and others. The analysis of nutraceuticals is also complex, some of these materials having a biological origin (e.g., fruits, vegetables, plants, microalgae, cereals, milk) and a very complex matrix. Similar to the case of the analysis of pharmaceuticals the information about the matrix is also related to the sample preparation procedure for the sample [51].

3) Biological samples include a large number of analytes of very different structures such as complex proteins, peptides, nucleic acids, metabolites, and biodegradation products, but also of smaller molecules such as amino acids, carbohydrates, or even inorganic ions (e.g., Na^+, K^+, Ca^{2+}, and Cl^-). The matrix of these analytes can be human tissues, body fluids (urine, bile, breast milk, gastric acids, pleural fluid, peritoneal fluid, lymph, mucus, tears, semen, synovial fluid, vaginal secretion, various exudates, etc.), and other materials related to various life forms. One of the most studied biomatrices is blood and its derivatives, such as plasma, serum, and dried blood. In the same class of biological samples are included breath condensate, volatiles from skin, etc. In intermediate place between samples of biological origin and food are the animal products, their classification in one type or another depending usually on the purpose of the analysis. A number of special aspects of the analysis are related to biological samples such as handling, preservation, and storage. Also, some analytical techniques must be associated with a preparative step and the preservation of a specific biological activity.

Frequently, the analysis of biological samples is performed using specific methods, involving for example affinity chromatography. The separation in affinity chromatography is based on the interaction of a specific (biological) compound present in the mobile phase with its natural biological complement immobilized on the surface of a stationary phase [52]. Examples of such interactions are between an antigen and its antibody, between lectins and glycoproteins, between an immobilized metal ion and proteins containing amino acid residues that have affinity for the metal ion (e.g., histidine), and between biotin and avidin. Common chromatography (e.g., RP-HPLC) is also frequently used in the analysis of biological samples.

4) Food is considered any substance ingested (by humans or animals) that provides nutritional support for the body, producing energy, maintaining life, and stimulating growth. Food can be of plant or animal origin, and in many cases there is an overlapping in classification between food and biological samples. Beverages are liquids specifically prepared for human consumption (and typically do not refer to water), some natural (natural juices, milk), some made from natural ingredients such as tea, coffee, wine, and beer, and others man-made such as carbonated drinks, energy drinks, antiaging waters, and herbal nutritional supplements. A large volume of dedicated literature is related to food analysis (see journals such as *Journal of Agricultural and Food Chemistry*, *Journal of Food Composition and Analysis*, or various books [53–55]). Analysis of food typically covers various aspects including the following: (1) analysis of nutrients such as carbohydrates, fats, proteins, (2) analysis of micronutrients such as vitamins, minerals, (3) analysis of biologically active ingredients, such as enzymes, antioxidants, (4) analysis of additives such as preservatives, (5) analysis of toxins and toxic components, (6) analysis of allergens, (7) analysis of food pathogens, (8) analysis of flavors, and (9) other analyses. The types of analysis performed on food depend on purpose of analysis, food composition, regulatory requirements of using specific analysis protocols, etc. For this reason, besides the classification as "analysis of food" specific information must be added for having a useful criterion that helps in the selection of an analytical method in food analysis. The role of a sample as food also provides important information regarding the matrix (related to the food type). Specific methods are reported in the literature related to the sample preparation for different types of food (e.g., fruits and vegetables, fats and oils, bread products, meat, milk, wine, honey) and the appropriate core analytical method.

5) Agricultural products include those related to animals, plants, fungi, and other life forms that are used for food, fibers, biofuel, medicinals, and other products used to sustain and enhance human life. Due to their importance, agricultural products are analyzed for a variety of purposes, in various forms, and in a wide range of detail. The analyses are performed on raw materials, finished products, or associated agricultural materials. Similar to food analysis, various aspects are covered by agricultural products analysis, and it may refer to basic components, active components (vitamins, antioxidants, enzymes), contaminants, and other food components. Particular analytical procedures are developed for the analysis of specific products, and more details about the sample are important for deciding the path for the analysis. For example, when applying chromatographic methods, specific procedures are dedicated to the analysis of small carbohydrates in plant material (see, e.g., Ref. [56]), and different types of procedures are used for the analysis of lipids (see, e.g., Ref. [57]). The differences in the selected methods are caused by the difference in the analyte structure, in spite of the similar plant origin. The subject is covered in numerous journals, dedicated monographs (see, e.g., Ref. [58]), and web publications.

6) Synthetic polymers represent a large group of materials. The analysis of synthetic polymeric materials is not limited to the analysis of the chemical polymeric structure (further discussed from the point of view of chemical nature of the analyte is available in Ref. [59]). In everyday utilization, polymers frequently contain fillers, plasticizers, antioxidants, coloring agents, slip agents for reducing the friction with processing equipment, and unreacted monomers. All these components may require special evaluation, and HPLC is frequently utilized in these analyses.

7) Numerous other synthetic chemicals are manufactured and their analysis involves various aspects (see, e.g., Ref. [60]). Among the synthetic chemicals are many pharmaceutical products, but these form a separate group and they were previously discussed. Other synthetic chemicals include the pesticides, herbicides, and fungicides as they are manufactured (and not as pollutants). Numerous other chemicals are produced to be used as raw materials for other syntheses, for use in agriculture, or for specific industrial applications such as industrial cleaning, and water treatment. Similar to the case of chemical analyses related to pharmaceuticals, these compounds require analysis of bulk, analysis of impurities, and analysis of the compounds from which they are synthesized. The same chemicals are classified differently when they appear as pollutants or are present in food, contaminants in pharmaceuticals, or other matrices.

8) Many other smaller groups of materials can be classified based on their role in everyday life. Examples can include the following: (1) flavors and fragrances, (2) cosmetics, (3) petroleum products, (4) solvents, (5) dyes, pigments, and inks, (6) surfactants, (7) preservatives, (8) tobacco products and cigarette smoke, and (9) chemical warfare agents.

Flavors, fragrances, and odors include many volatile compounds, and they may have relatively high stability to heating. For this reason, GC and GC–MS methods are frequently utilized for their analysis. However, in many cases, HPLC can be used as an analytical procedure. Flavor and fragrance analysis can be related to a variety of other problems such as impurities that may exist and produce allergic reactions, or decomposition products that are formed in time. Dedicated literature related to this group of compounds includes journals (e.g., *Flavor and Fragrances Journal*) and books (e.g., Ref. [61]).

Cosmetics include various types of materials such as creams, lotions, powders, lipsticks, fingernail and toenail polish, eye and facial makeup, hair colors, hair sprays and gels, deodorants, hand sanitizers, baby products, bath oils, bubble baths, bath salts, and many other types of products. This variety of products requires diverse types of analyses. Some analyses address bulk components, others address active ingredients, fragrances, impurities, colorants. For each type of component it is likely that different approaches for analysis must be taken.

Petroleum products form another groups of compounds with specific characteristics and analysis requirements, including that of composition, additives, and other material potential components. Liquid and gaseous petroleum products are frequently analyzed by GC (GC–MS), although HPLC can also be used for specific analyses such as that of additives. The application of GC (GC–MS) is also common for the analysis of solvents, either in bulk or as traces (residual solvents) in other materials.

Dyes, pigments, and inks represent a group of materials utilized for changes of color of another material on which they are deposited or in which they are incorporated. Analysis of materials used as dyes and pigments is also complex, referring to the colored material only, but frequently to other components associated with the materials used as dyes, pigments, or inks such as solvents, polymeric carriers, antioxidants, and surfactants. Depending on the purpose of analysis, a variety of techniques are utilized for the characterization of this group of materials.

Many commercial surfactants are nonvolatile mixtures of members of a homologous series, and the utilization of HPLC for their analysis is common. The analysis of surfactants can be geared toward the evaluation of their main components, or toward contaminants resulting from their synthesis. Their emulsifying properties make their analysis rather difficult [62]. Surfactants may appear as pollutants in the environment and multiresidue methods for their analysis in the environment and for their degradation products are of considerable interest [63].

The analysis of preservatives is usually performed in a variety of matrices including food, cosmetics, pharmaceuticals, and plastic materials. The separation from the matrix is the main problem in preservative analysis, and a variety of procedures are utilized for this purpose. Chromatographic techniques such as GC (GC–MS), HPLC, and SFC are used for preservatives analysis, depending on their nature, volatility, or even the type of sample preparation adopted for their recovery.

Among other consumer products, tobacco products and cigarette smoke were subject to numerous types of analyses. A large body of publications covers this subject including dedicated journals (e.g., *Beitraege Tabakforschung International*, *Tobacco Science*) and books (see, e.g., Ref. [64]). Various types of analyses are performed on tobacco and cigarette smoke, including basic analysis of tobacco composition for assuring product integrity, analysis of additives, and more importantly analyses of compounds that are harmful to human health [65].

Analysis Selection Based on the Role of the Sample in Everyday Life

The role of samples in everyday life is an important criterion for selecting a specific type of analysis. This is related to several aspects of the analysis. The role of samples in everyday life is usually correlated with a specific set of analytes and a specific set of matrices. In samples related to the environment, it is common that the "role" of analytes is that of pollutants. For pharmaceuticals, specific analyses for active ingredients are frequently required. For food, specific methods are developed for the analysis of carbohydrates, fats, organic acids, specific pesticides, etc., depending on the type of food.

The matrix in a group of samples classified in the same group based on their role in everyday life is frequently similar (if not the same). For example, for environmental samples, typical types of matrices are air, water, or soil. For biological samples, the matrix is a biological medium (e.g., a tissue), and for agricultural products the common type of matrix is a plant material. It is common that methods for sample preparation are related to the type of matrix, and such methods are described in the literature for specific types of samples.

It is also common that depending on the field of everyday life utilization a large set of methods and protocols for the analysis are available. Dedicated literature related to the utilization of the sample in everyday life is widely available. For example, for food analysis a number of dedicated journals are published such as *Journal of Food Composition and Analysis* (Elsevier), *Food Chemistry* (Elsevier), *Food and Chemical Toxicology* (Elsevier), *Journal of Agricultural and Food Chemistry* (ACS Publications), and *Journal of Food Science* (Wiley). For food analysis, besides this body of journals, numerous books and web publications are also available. Similar publications are available for other fields such as environmental, pharmaceutical, and biological. The literature related to the role of the sample in everyday life is a primary source of information regarding the selection of an analytical method for the sample.

5.4 SELECTION OF THE HPLC SEPARATION BASED ON ANALYTE PROPERTIES

The type and quantity of the analyte is one of the main factors dictating the selection of an appropriate analysis method. In this selection the main factors to be considered are the following: (1) the chemical nature of the analyte, (2) the pK_a of the analyte in the case of acids and bases, (3) the octanol/water partition coefficient of the analyte, and (4) the analyte concentration in the sample. Each of these characteristics will be further evaluated in more detail.

Role of Chemical Nature of the Analyte

The chemical nature of the analyte is related to various chemical and physical properties previously discussed in Sections 5.1 and 5.2. For this reason, the classification of molecules based on their chemical nature is overlapping with the classification based on specific physical and chemical properties that are determined by the chemical nature. Chemical nature is important for the choice of core analysis, details regarding the type of HPLC selection (when HPLC is used for analysis), but also for other aspects of the analysis such as sample preparation, conditions of storage, sample handling, and precautions regarding safety. Using the chemical nature, the following groups of analytes can be differentiated, related to the HPLC procedures adequate for analysis: (1) inorganic compounds, (2) polymeric organic compounds, (3) organic molecules with no or very low polarity, (4) organic molecules with low to medium polarity, (5) highly polar compounds, (6) ionic compounds, (7) compounds with amphiphilic properties, and (8) chiral compounds. A special group of analytes is that of chiral molecules (when these must be separated as enantiomers).

It is possible that the analytes in a sample belong to one of the previously listed groups, but it is also very possible that different analytes belong to different chemical groups. For example, the analytes may consist of a mixture of organic and inorganic compounds, or can be organic compounds with a wide range of polarities. In such cases, a unique analysis may be attempted, but also it is possible that two or even more analyses must be performed on the same sample, targeting separately each specific group of analytes.

1) Inorganic analytes represent a large group of molecules and ions that are subject to various types of analyses such as ICP, X-ray fluorescence, spectrophotometry, and atomic absorption. HPLC is less frequently used for the analysis of heavy metallic ions, but was proven successful for the analysis of ions such as Li^+, Na^+, K^+, NH_4^+, Ca^{2+}, etc. Also, HPLC is frequently used for the analysis of common anions such as Cl^-, Br^-, $CO_3{}^{2-}$, $SO_4{}^{2-}$, $NO_3{}^-$, $NO_2{}^-$, $ClO_4{}^-$, $PO_4{}^{3-}$, etc. A specific branch of HPLC, ion chromatography, is dedicated to such analyses. Ion chromatography utilizes specific columns (see Section 9.3), specific mobile phases (see Section 13.5), and very frequently a specific detection technique involving electrochemical detectors.

The analysis of inorganic compounds may involve extensive sample preparation procedures for rendering the samples amenable for analysis (either nonchromatographic or chromatographic). Among these procedures are aggressive chemical treatments for solubilization of silicates, oxides, or other insoluble inorganic materials, and derivatization for improving analyte detection. Such procedures typically modify the sample matrix, and the nature of compounds formed by the ions of interest.

2) Polymeric organic analytes represent another large group of compounds that may require analysis. Polymeric organic molecules can be natural, synthetic, or chemically modified natural molecules. Natural organic polymers include various classes of biopolymers such as polyterpenes (e.g., rubber), polymeric carbohydrates, proteins and peptides, nucleic acids (that are of significant biological importance), polymeric lipids, lignins, tanins, and browning polymers (resulting from Maillard reaction). Synthetic organic polymers include thermoplastics, fibers, paints, adhesives, elastomers, and other synthetic materials. Various aspects of polymer analysis are encountered in practice. For example, one subject of interest is the molecular mass distribution where techniques such as light scattering, viscometry, and size-exclusion chromatography are utilized. For water-soluble type polymers the chromatographic technique used for molecular mass distribution evaluation is gel filtration (GFC), while for the organic solvent soluble type polymers the molecular mass distribution is measured by gel permeation (GPC). Size-exclusion chromatography also allows quantitation of polymers in a solution by using calibration with appropriate standards.

Polymer structure is frequently determined using techniques such as IR and Raman spectroscopy, NMR, and X-ray diffraction, thermal methods, MALDI, electrophoresis, as well as chromatography. Chromatographic techniques applied for polymer analysis include pyrolysis-GC-MS (see, e.g., Refs. [66,67]), LC-MS, and LC-MS/MS. Entire fields of science, such as proteomics and genomics, make full utilization of HPLC separation techniques and of MS detection capability. Polymer quantitation also utilizes HPLC, as well as other techniques such as capillary electrophoresis. For quantitation, the detection can be UV absorption, mass spectrometry, as well as other techniques depending on polymer structure. For qualitative and quantitative purposes a variety of HPLC types are used, these include in particular ion exchange chromatography (IC), HILIC, as well as reversed phase chromatography (RP).

A very large field regarding the analysis of polymeric organic compounds is occupied by the analysis of proteins. There are numerous books dedicated to this subject (see, e.g., Refs. [68–71]), as well as several peer-reviewed journals [e.g. *Journal of Proteome Research* (ACS Publication), *Journal of Proteomics* (Elsevier), *Proteomics* (Wiley)]. Protein analysis utilizes a variety of HPLC separations including IC, HILIC, and RP, as well as combinations of these techniques.

3) Organic molecules with no or very low polarity represent another large group of compounds including hydrocarbons, lipids (glycerolipids, sterol and stanol lipids, prenol lipids, phosphoglycerides, sphingolipids), carotenoids, tocopherols, and other classes of compounds that are not soluble in water or polar solvents such as methanol or acetonitrile. Such analytes require a nonpolar solvent for sample solubilization and for elution. These types of analytes are typically analyzed using normal-phase chromatography (NPC) or nonaqueous reversed-phase chromatography (NARP) (see Section 2.3). Other procedures for this type of analysis use SFC (under the name "convergence chromatography").

4) Organic molecules with low to medium polarity represent a large group of molecules including numerous types such as pharmaceuticals, pollutants, biological small molecules, components of food, flavorants and fragrances, preservatives, and many other categories of compounds. This group of molecules is frequently analyzed using reversed-phase HPLC (RP-HPLC). A wide variety of RP-HPLC types of columns are commercially available, as further

discussed in Chapter 7. The nature and composition of the mobile phase in RP-HPLC plays an important role in the separation, and the selection of the mobile phase is further discussed in Section 13.5. The mechanism of the separation in RP-HPLC (see Section 3.3) is best explained by solvophobic theory, which shows that a hydrophobic molecule is "expelled" from a polar solvent and is "accepted" in a hydrophobic stationary phase. This explains how the balance between the polar and hydrophobic character of the molecules allows their separation. Since the mass transfer process between the liquid mobile phase and the solid phase is rapid, and a unique mechanism dominates the separation, RP-HPLC frequently offers the best type of separation as compared to other HPLC techniques. The wide range of compounds and the qualities of RP-HPLC make it the most common analytical technique currently utilized. Molecules with medium toward high polarity can still be analyzed on special RP-HPLC columns, such as those with polar imbedded groups or end-capped with polar moieties (see Section 7.4).

5) Highly polar compounds include many molecules with polar groups. Examples of compounds with polar groups are organic acids, amines, carbohydrates, nucleobases (adenine, guanine, thymine, uracil, cytosine) and the corresponding nucleosides (cytidine, uridine, adenosine, guanosine, thymidine, inosine), and deoxynucleosides. The polar character can be present in molecular form but can become ionic (with even higher polarity) depending on the pH of the medium. For example, an organic acid R-COOH may be considered a polar compound but not ionized in a strongly acidic medium. At the same time, in a basic medium the compound can be found in $RCOO^-$ form. At very high pH even carbohydrates may be present as anions. For amines and some heterocyclic compounds containing nitrogen, basic condition may assure that the compound is not ionized, while in acidic conditions such compounds are ionized. As an example, the R-NH_2 groups can be changed into $R - NH_3^+$ groups (see, e.g., ionization of nicotine in Fig. 5.1.1). The polar molecules can be analyzed by a variety of HPLC techniques such as HILIC, NP, and ion pairing (IPC). Depending on the presence of larger hydrophobic moieties, even RP-HPLC can be used, and for these separations special RP-columns are recommended, such as those with polar imbedded groups or end-capped with polar moieties. The medium used to dissolve the analyte as well as the pH of the mobile phase are important parameters in case the analyte can be present in molecular form (polar) or in ionic form. When the compound is in molecular form (not ionized), its separation can be performed differently than in the case the compound is in ionic form.

6) Ionic compounds include both "permanent ions" such as metallic ions but also many organic compounds that can be present in ionic form, depending on the pH of the medium. For compounds in ionic form, the HPLC types may include ion exchange chromatography (IEC or IC), HILIC, or ion pairing. Maintaining appropriate pH conditions for the mobile phase, IC can be performed on columns that are available commercially in a wide variety (see Chapter 9). The ion pairing chromatographic separations are performed on RP-HPLC columns.

7) Molecules with amphiphilic properties possess both hydrophilic and lipophilic properties. Such molecules include both natural compounds such as phospholipids as well as synthetic molecules such as various detergents. These types of molecules are in general difficult to analyze, and the more common procedures are based on NP-HPLC and HILIC (see, e.g., Refs. [72,73]).

8) Many molecules have chiral properties, but the separation between their enantiomers is not always necessary. However, when the enantiomers of chiral molecules must be separated, special chromatographic conditions are useful for the task. When the molecule volatility allows, GC separation can be applied, but more frequently HPLC is the technique of choice for the separation of enantiomers. The separation of chiral molecules in HPLC can be done in some cases using a chiral

agent in the mobile phase and an achiral stationary phase. However, in most applications, chiral stationary phases must be utilized, details about such columns being given in Chapter 10.

The case of analytes belonging to different chemical groups is common to many analyses. When separate analyses are selected for each group, the guidance previously discussed will be applied. Some separation procedures may be necessary in the sample preparation step of the analysis. For example, when a sample contains proteins and small molecules, it is common to eliminate the proteins (deproteinization of the samples) and analyze the remaining solution for small organic molecules. The proteins may be analyzed separately. The analysis of both groups may involve a separation using size-exclusion chromatography. This technique allows, for example, the evaluation of molecular weight of proteins and separates the small molecules that elute late. The collection of the group of small molecules can be followed, for example, by RP-HPLC for their individual analysis.

Role of pK_a in the Case of Acids and Bases

The strength of ionic character of compounds with ionizable groups such as $-COOH$, $-SO_3H$, and $-NH_2$. is characterized by the pK_a value (or pK_b value) (see Section 5.1). The value of pK_a offers a criterion for the selection of the type of chromatography, and further it is important in column and mobile phase selection. For example, when the separation for a set of acids is intended to be performed a nonpolar column (taking advantage of the hydrophobic moiety of the organic acid), it is recommended to have the acids in molecular form (not dissociated). For such cases, the mobile phase must have a pH lower with 1.5–2 units than the lowest pK_a of the acids. In such conditions, the acids do not dissociate more than 1–2%. For the use of a polar or ion-exchange type column for the separation of acids, the mobile phase must have a high pH (within the acceptance pH range for the chromatographic column, which may suffer degradation when extreme pH values are used for the mobile phase). In case of separation of bases, the high pH maintains the analytes nonionized, while low pH favors ionization. The pK_a of the analytes is also important in the selection of an ion pairing reagent when IP type separations are intended.

Role of Octanol/Water Partition Coefficient

Polarity of the analytes is determined by their chemical nature, and for this reason part of the role of polarity in the selection of HPLC has been previously discussed related to the analyte chemical nature. However, polarity as described by octanol/water partition coefficient K_{ow} is a very useful parameter in the selection of an HPLC technique for analysis. Octanol/water partition coefficient K_{ow} can be used as a qualitative measure of polarity with positive values for log K_{ow} indicating hydrophobic character, with larger values for log K_{ow} for more hydrophobicity, and lower log K_{ow} for polar character (see Section 5.1). The polarity as described by log K_{ow} has the following roles in HPLC: (1) allows selection of the type of HPLC that can be used in a specific separation, (2) in RP-HPLC the value of log K_{ow} is closely related to the retention factor log k', (3) in other types of HPLC separations based on polarity, log K_{ow} still may be used as a guidance for the expected separation results.

1) Selection of the HPLC type can be guided from log K_{ow} values which are readily available in the literature [7] and can be calculated using computer programs (e.g., MarvinSketch 5.4.0.1, ChemAxon Ltd. [5], EPI Suite [8]). Based on log K_{ow} values (obtained from a neutral water solution), the selection of HPLC type can be made using the scheme shown in Fig. 5.4.1.

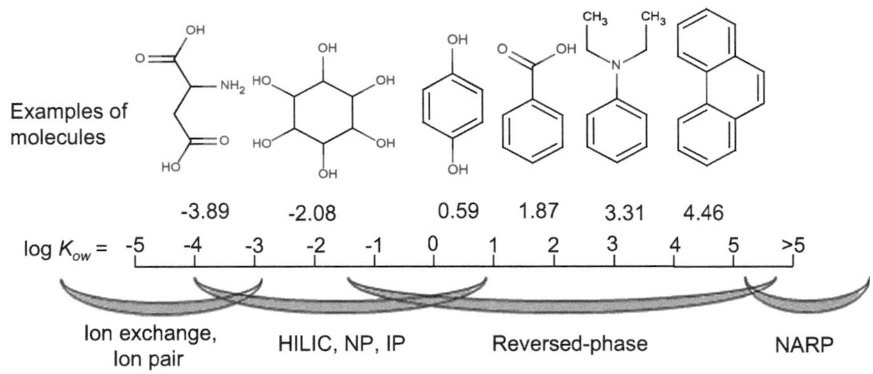

FIGURE 5.4.1 Different HPLC types preferentially utilized depending on log K_{ow} of the analyte molecule.

Since the compounds with ionizable groups may have in solution both neutral molecules and the molecule in ionized state, instead of log K_{ow}, the distribution coefficient D_{ow} (or log D_{ow}) is used for such compounds for describing their equilibrium between octanol and water (see Section 5.1). The pH of the mobile phase in an HPLC separation can have different pH values, and D_{ow} (or log D_{ow}) is pH-dependent. In such cases, for example for acidic or basic compounds, the selection must be guided by the log D_{ow} value at a specific pH. For example, benzoic acid in a buffer at pH = 8 will have log $D_{ow} \approx -1.55$, and therefore will have strong hydrophilic properties. In such a case, to ensure that the benzoic acid is kept in molecular form (and not ionic) and an RP type column can be used, the pH of the mobile phase must be kept lower (pH = 3.0–3.5). Alternatively, for a compound that is intended to be analyzed in ionic form and having a very low log D_{ow}, the pH of the mobile phase must be adjusted appropriately.

2) For reversed-phase chromatography, the value for log K_{ow} (or log D_{ow}) provides even more detailed information regarding the separation since it can be easily related to retention factor log k' for the analyte. In Section 4.3 it was shown that for a partition process, the equilibrium constant can be expressed by Eq. (4.3.38). Therefore this formula should be valid for both the equilibrium of a molecule X distributed between octanol and water and characterized by log K_{ow}(X), as well as for molecule X distributed between the stationary and mobile phase in a reversed-phase HPLC separation and characterized by log k'(X). The use of Eq. (4.3.38) leads for log K_{ow}(X) and log k'(X) to the following two expressions:

$$\begin{cases} \log K_{ow}(\mathrm{X}) = a_1 \mathcal{A}(\mathrm{X}) - b_1 \\ \log k'(\mathrm{X}) = a_2 \mathcal{A}(\mathrm{X}) - b_2 \end{cases} \tag{5.4.1}$$

b_1 being the notation for the corresponding $\sum_j b_j''$ for the first partition system and b_2 being the notation for $\sum_i b_i'' - \log \Psi$ for the second partition system. The elimination of parameter

$\mathcal{A}(X)$ (van der Waals surface area) between the two Eq. (5.4.1) leads to the following dependence:

$$\log k'(X) = \frac{a_2}{a_1}\log K_{ow}(X) + \left(\frac{a_2 b_1}{a_1} - b_2\right) \tag{5.4.2}$$

As indicated in Section 4.3, parameters a_1 and b_1 are dependent on functionalities present in the molecule X (e.g., $-OH$, $-COOH$, $=CO$, $-NH_2$, etc., or no functionality in hydrocarbons) and for a_2 and b_2 also on the nature of mobile and stationary phase in the RP-HPLC separation. With the notations $c_1 = a_2/a_1$ and $c_2 = a_2 b_1/a_1 - b_2$, expression (5.4.2) can be written in the simpler form as follows:

$$\log k'(X) = c_1 \log K_{ow}(X) + c_2 \tag{5.4.3}$$

Parameters c_1 and c_2 depend on the column and mobile phase, but they are with a good approximation constant for series of compounds with similar structure (see, e.g., Ref. [74]). A theoretical proof for expression (5.4.3) has been obtained based on solvophobic theory [25]. The linear correlation between $\log k'(X)$ and $\log K_{ow}(X)$ is shown in Fig. 5.4.2 for several types of compounds (indicated in Table 5.4.1) separated on a LiChrospher 100 RP-18e column, 125 × 4 mm, 5-μm particles, and a mobile phase acetonitrile/water at three different concentrations (20%, 40%, and 60%).

The linear dependence of $\log k'$ on $\log K_{ow}$ is a good guidance for the estimation of the elution order of a group of compounds in RP-HPLC. Small differences between $\log K_{ow}$ values indicate that the compounds will elute close to each other. Although some compounds may have equal $\log K_{ow}$ values, this does not necessarily indicate that they cannot be separated by RP-HPLC. Expression (5.4.2) is only an approximation for $\log k'$, and although it is very well fulfilled for some compounds, some

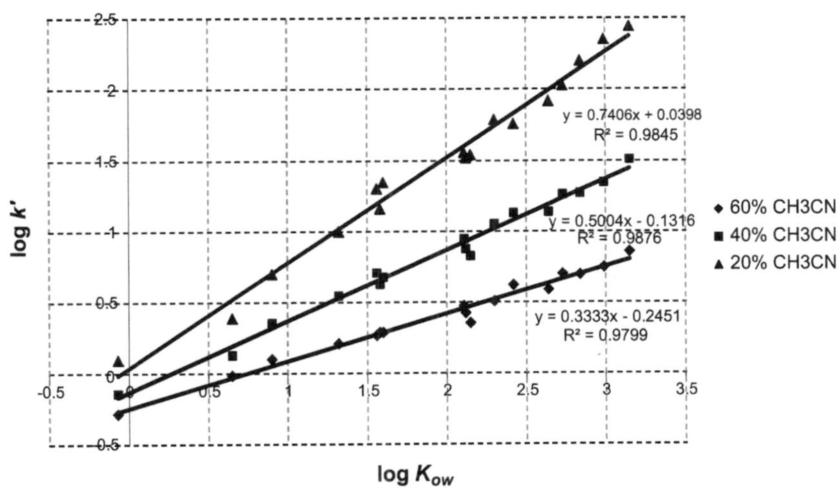

FIGURE 5.4.2 Variation of $\log k'$ as a function of $\log K_{ow}$ for several types of compounds (see Table 5.4.1) separated on a LiChrospher 100 RP-18e column, 125 × 4 mm, 5 μm particles, and a mobile phase acetonitrile/water at three different concentrations (20%, 40%, and 60%).

TABLE 5.4.1 The Values for log K_{ow} for Several Types of Compounds Separated on a LiChrospher C18 Column

Compound	log K_{ow}	Compound	log K_{ow}
Caffeine	−0.07	α-Naphtylamine	2.15
Pyridine	0.65	o-Nitrotoluene	2.3
Aniline	0.9	N,N-Dimethylaniline	2.42
o-Toluidine	1.32	Ethyl benzoate	2.64
Benzyl cyanide	1.56	Toluene	2.73
Acetophenone	1.58	Chlorobenzene	2.84
Dimethyl phthalate	1.6	Bromobenzene	2.99
Anisole	2.11	Ethylbenzene	3.15
Methyl benzoate	2.12		

deviations are encountered for other compounds. The theory developed for proving expression (5.4.2) takes into account only solvophobic theory, while in the separation by RP-HPLC, other interactions may also be present, even if they are not the major contributors to the separation. Such interactions occur with the free silanol groups still present on an RP-HPLC column made using silica as a support for the active stationary phase, with the metal impurities possibly still present in the silica, or with polar groups intentionally present in the RP column such as polar end-capping or polar imbedded groups. For such columns, the retention process is more complex, and simple use of log K_{ow} values of the analytes is less relevant for the separation.

For compounds that can be in molecular form or in ionic form, the medium used to dissolve the analyte and the pH of the mobile phase can change their structure as previously indicated. This change is also associated with the change in log D_{ow}, as exemplified for nicotine in Fig. 5.1.1. Since the choice of HPLC separation and the order of elution (through log k') depend on log D_{ow}, and log D_{ow} values are low for ionic forms and higher for the molecular form of the molecule, depending on the selected separation, a basic or an acidic pH for the HPLC separation must be selected to keep the molecules of the analyte to be ionized. For example, for the analysis of nicotine (100 μg/mL standard) on an RP-HPLC column (XTerra RP 18, 5 μm, 4.6 × 150 mm from Waters, MA, USA) the mobile phase uses a gradient of acetonitrile and an aqueous buffer. At pH = 4.40, log $D_{ow} = -2.34$ for nicotine, and the separation on an RP type column shows little retention as exemplified in Fig. 5.4.3A. Also the peak shows undesirable broadening. At pH = 6.80, log $D_{ow} = -0.24$ and the retention time is longer and with a better peak shape as exemplified in Fig. 5.4.3B. At pH = 9.90, log $D_{ow} = 1.13$ for nicotine and the separation retention time is longer, with excellent peak shape as shown in Fig. 5.4.3C.

3) For other types of chromatography based on polarity, the values for log K_{ow} (or log D_{ow}) of the analyte still may be a useful parameter. In ion pair (IP) chromatography, where the separation is performed on an RP column, the retention of the analyte in the absence of the IPA

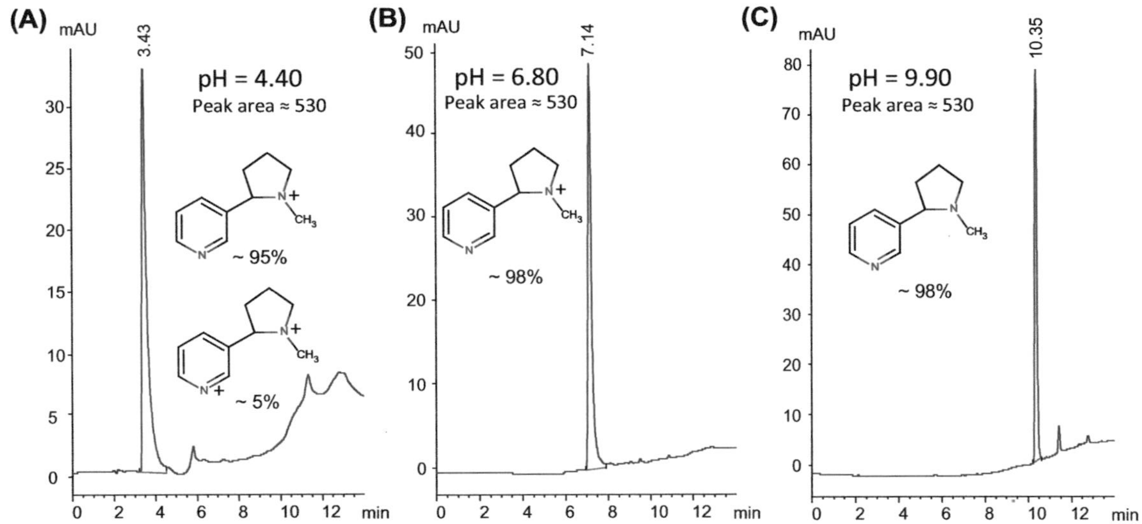

FIGURE 5.4.3 Nicotine (at 100 μg/mL) separated on a C18 column with acetonitrile/aqueous buffer mobile phase at different pH values. (Note: the pH refers to the value measured for the aqueous buffer.)

(IP agent) is related to log K_{ow}. Also, the hydrophobicity of the IPA is related to its log K_{ow} value, the IPA agents with a large hydrophobic moiety having higher log K_{ow} values and leading to complexes that are retained stronger on an RP column.

For HILIC, the value of log K_{ow} can give an indication regarding how polar the analyte is. Reports are available in the literature indicating even a linear response (with a negative slope) between log K_{ow} and log k'. It has been shown, for example, that for the separation of about 30 compounds on an Atlantis HILIC silica column, with a mobile phase containing acetonitrile and aqueous solution of ammonium formate at $pH = 3.0$, a correlation (with a negative slope) can be obtained between the values of log D_{ow} and log k', with R^2 values between 0.689 (for 95% acetonitrile) to 0.751 (for 85% acetonitrile) [75]. Although the R^2 values for the correlation are not very high, the results indicate that polarity as described by log D_{ow} is a good indicator for describing polar ↔ polar interactions (larger log D_{ow} indicating less polar compounds). The linearity is not as good as in RP-HPLC since the separation in HILIC is a more complex process than based solely on polar ↔ polar interactions, and some hydrophobic interaction may play a role in the separation. The value for log D_{ow} can still be used as a guidance criterion to judge how strongly or weakly a compound can be retained on an HILIC column.

For ion exchange chromatography, when the separation is based exclusively on ion ↔ ion interactions, the retention should not be related at all to the log K_{ow} value of the analytes. However, as indicated in Section 3.3, for organic molecules with both ionic groups and hydrophobic moieties, additional interactions may exist during ion exchange separations, besides ion ↔ ion. The manufacturers of columns to be used in IEC typically indicate the minimization of column hydrophobic character. It is useful to also evaluate the hydrophobicity of the ionic analyte before attempting to perform an ion exchange separation. This may help in the understanding of unexpected behavior of the analyte, when unexpected retention patterns occur in the separation process.

In size-exclusion separations, the polarity or the hydrophobic character of the analyte should not affect the retention, as long as no enthalpic contribution exists in the separation. However, most size-exclusion separations have an enthalpic component due to various types of interactions between the analyte and the stationary phase. Such interactions are usually of the polar ↔ polar type, and molecules that are small and expected to elute very fast from the size-exclusion chromatographic column may be retained longer and interfere with some macromolecules. For this reason, the polarity of the small molecule type analytes must be inspected when they are present together with polymers separated by size exclusion. Also, the nature of the stationary phase used in SEC must be characterized regarding other potential interactions with the analytes.

The Role of Analyte Concentration in the Selection of HPLC Separation

The concentration of the analyte in the sample is usually unknown (exceptions include, for example, analyses of standards for verifying the accuracy of a method). However, for some samples the range of analyte concentration can be anticipated. The same incertitude is common regarding the matrix, which can be totally unknown or expected to be complex or simple. Based on information such as the role of the analyte in everyday life, or in the case of preprocessed sample on the information from the cleanup step, it can be usually indicated if the matrix will pose a significant challenge for the analysis or not. According to this information, the selection of the HPLC separation must be made with the goal of separating large amounts of sample or small amounts. HPLC chromatographic columns are available in a variety of dimensions, containing a more or less active stationary phase. When larger amounts of analyte and matrix must be separated, it is usually recommended to use HPLC columns containing a larger amount of stationary phase. More difficult problems appear when the analytes are in trace amounts, while the matrix is complex and represents most of the sample. In such cases, a sample preparation step may be recommended, for making the initial sample more amenable for the core separation. The decision regarding the selection of an HPLC column with a larger mass of stationary phase must also consider how different is the matrix from the analytes, and if the matrix is not retained on the column, a smaller amount of stationary phase may be acceptable since it is used only for the analyte retention and elution.

5.5 SELECTION OF THE HPLC SEPARATION DEPENDING ON THE MATRIX

The nature of the sample matrix is essential to the selection of sample preparation and sample analysis. Sample preparation based on the nature of sample matrix is presented in numerous publications (see, e.g., Refs. [1–4]), and the subject is beyond the purpose of this book. The selection of the core analytical process is strongly related to the sample preparation. After sample preparation, a processed sample is further subject to the core analytical process. The processed sample usually has a matrix less complex than the raw sample, and may contain the analytes at a higher concentration. However, sample preparation does not necessarily generate a perfectly "clean" sample, and the separation of the analytes and of the residual matrix are frequently part of the role of core analytical process. It is important to notice that, for most samples, variations are encountered in the matrix composition. Even if the matrix is of the same type, the differences from the matrix of one sample to another may influence the analysis results differently. For this reason, particular attention must be given to the range of possible components of the matrix, for developing robust analytical methods. Some properties of the residual matrix and their connection to the selection of an HPLC separation are discussed further in this section.

Role of the Chemical Nature of the Matrix

The chemical nature of the matrix components can be similar to that of the analyte (e.g., both analytes and the matrix components are small molecules), but also can be very different (e.g., the analytes are small molecules while the matrix is polymeric). The classification of the chemical nature of the matrix is similar to that of the analytes: (1) inorganic matrix, (2) polymeric organic matrix, (3) matrix consisting of organic molecules with no or very low polarity, (4) matrix of organic molecules with low to medium polarity, (5) highly polar compounds in the matrix, (6) ionic compounds in the matrix, and (7) compounds with amphiphilic properties. While the interest in a chemical analysis appears to be focused on analyte separation, a separation must also be performed between the matrix and analytes, although the matrix components do not need separation among themselves. The attention to the matrix components is in particular important since these components may vary significantly from sample to sample, even for the same type of samples. The separation of the analytes from the matrix components is also addressed in the sample preparation step, but it is common that some matrix components remain in a processed sample and must be separated in the core chromatographic analysis. The type of chromatography that separates the matrix components from the analytes is very frequently the same as the one used for analyte separation. When the analytes and the matrix have a similar nature, this is a straightforward task, although the separation must be good enough to avoid interferences. In some separations, the matrix components are somewhat different from the analytes, and the matrix components are not well retained by the chromatographic column and are eluted at the beginning of the chromatogram. One typical example of residual matrix components is those remaining from biological sample deproteinization. The protein matrix can be removed from animal or human biological samples using various procedures, involving different physical techniques such as heating or using chemical reagents. Besides the compounds of interest, after deproteinization other molecules, which may interfere in the chromatographic separation and detection, remain in the supernatant resulting after protein removal. These molecules may be oligoproteins, amino acids resulting from partial protein hydrolysis, or various metabolites. The interference of these molecules can be seen, for example, in the HPLC chromatograms mainly when detection is achieved using UV with wavelength below 240 nm. An example of such interferences of residual plasma matrix after deproteinization carried out with acetic anhydride can be seen in the chromatogram illustrated in Fig. 5.5.1 for an HPLC analysis of metformin with separation on an Intersil ODS column using ion pairing with $C_8H_{17}SO_3Na$ and detection at 232 nm [76]. The chromatogram was obtained from a deproteinized plasma sample spiked with 1 μg/mL metformin and 1 μg/mL methylbiguanidine.

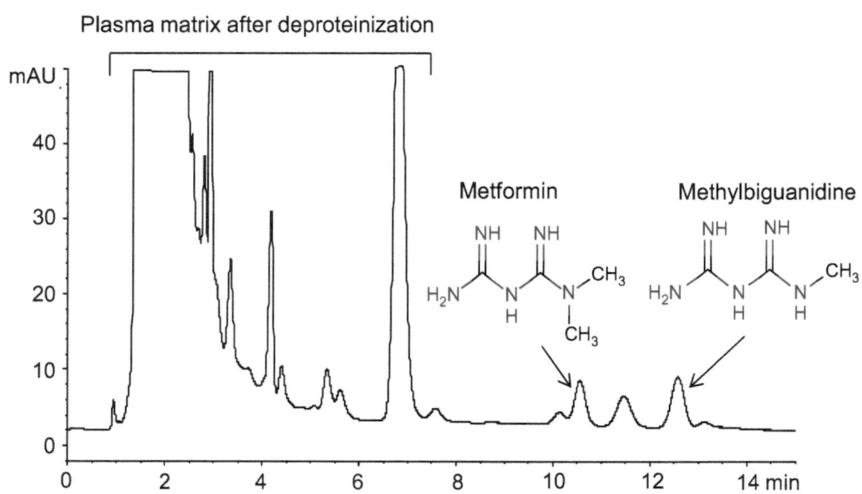

FIGURE 5.5.1 HPLC chromatogram of a plasma sample subjected to deproteinization with acetic anhydride and spiked with 1 μg/mL metformin and 1 μg/mL methylbiguanidine (detection in UV at 232 nm).

Role of the Amount of the Matrix

Separation of the matrix from the analytes follows similar rules as the separation of the analytes among themselves. One aspect that must be considered related to the separation of the matrix or matrix components from the analytes is related to the potential differences between the concentrations of the two. While the analytes may be in some cases present only in traces, the matrix may be present at much larger concentration. For the matrix at relatively low content in the sample, its presence may not affect the analysis. However, a larger concentration of the matrix must be taken into account related to the loading of the chromatographic column, and the distortion of the shape of chromatographic peaks (see Section 4.2 regarding the chromatographic sample loading). Also, in the case of selective detection (e.g., the use of mass spectrometry for detection), when only specific analyte characteristics are used for the measurements (e.g., a specific mass for MS detection, or a specific wavelength for UV detection), it is possible that the HPLC separation does not isolate the matrix (or a matrix component) from the analyte, and the matrix, although it is not seen, may still affect the detection. An exemplification of matrix influence on the analysis, when it is not separated chromatographically, of four standards of deuterated tobacco-specific nitrosamines (TSNAs) at concentrations of 1 ng/mL is given in Fig. 5.5.2. The analyzed compounds are D_4-nitrosonornicotine (NNN) (MRM transition 182.1 → 152.1), D_4-nitrosoanatabine (NAT) (MRM

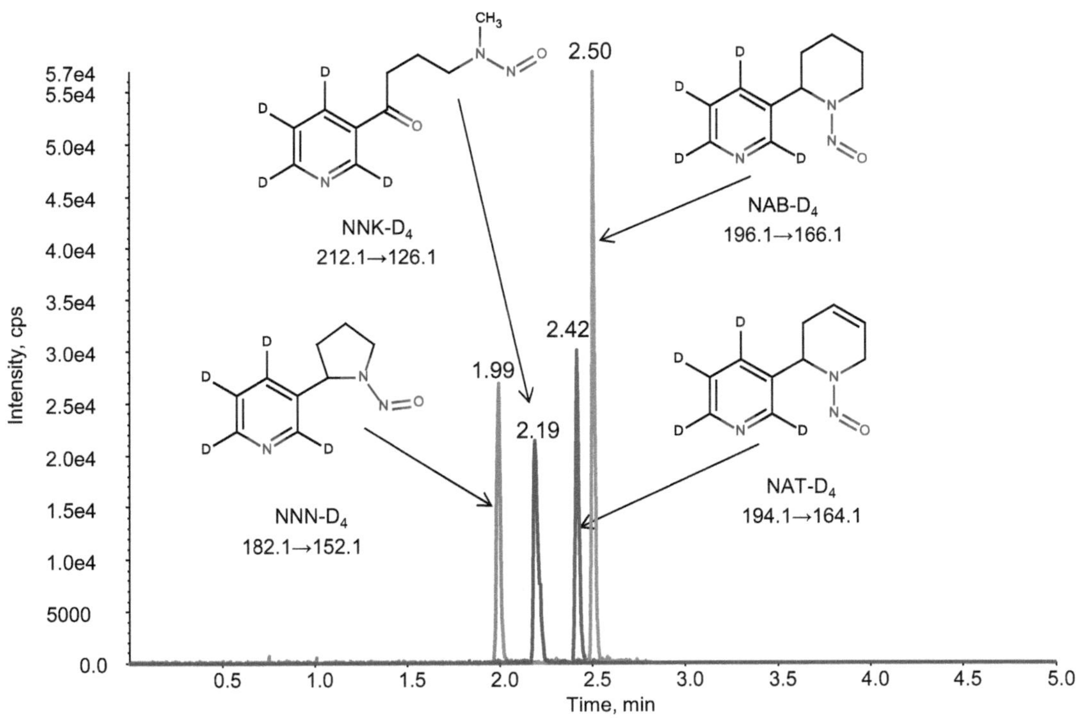

FIGURE 5.5.2 LC-MS/MS chromatogram performed in MRM positive mode for a solution containing 1 ng/mL NNN-D_4, NNK-D_4, NAB-D_4, and NAT-D_4.

transition 194.1 → 164.1), D_4-nitrosoanabasine (NAB) (MRM transition 196.1 → 166.1), and D_4-4-(methylnitrosamino)-1-(3-pyridyl)-1-butanone (NNK) (MRM transition 212.1 → 126.1) in a solution containing 1 ng/mL of each compound. The separation was performed on a Kinetex 1.7 μm EVO C18 column (100 × 2.1 mm). The elution was made using gradient with solution A aqueous 10 mM CH_3COONH_4, and solution B 0.1% CH_3COOH in acetonitrile [77]. The results from Fig. 5.5.2 demonstrate the high sensitivity of the LC-MS/MS technique in the absence of the interferences from the HPLC elution. For the same sample containing 5% nicotine, in addition to the TSNAs at 1 ng/mL level, the chromatogram is shown in Fig. 5.5.3. Nicotine is not well separated from the TSNAs in the chromatographic conditions used for the analysis, and the MS—MS conditions are not set for nicotine detection. The peaks for the analytes shown in Fig. 5.5.3 are either significantly decreased (as seen for NNN) or not detectable (for the other three analytes) in the presence of nicotine in the sample matrix.

The unseparated or incompletely separated matrix, although not acting as an obvious interfering element, may influence the analyte measurement as indicated in Section 2.3, due to ion suppression. For example, fluorescence can be affected by the unseparated matrix, although the matrix itself is not fluorescent. For this reason, the HPLC separation of the

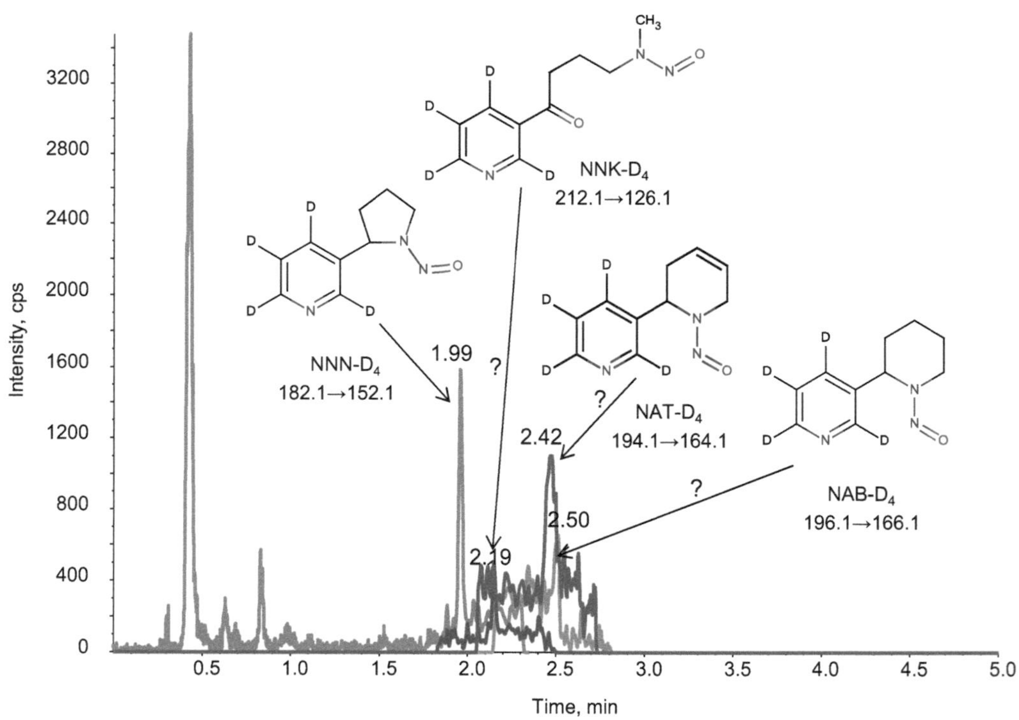

FIGURE 5.5.3 LC-MS/MS chromatogram performed in MRM positive mode for a solution containing 5% nicotine and 1 ng/mL NNN-D_4, NNK-D_4, NAB-D_4, and NAT-D_4, showing ion suppression due to incomplete separation of a matrix component (nicotine).

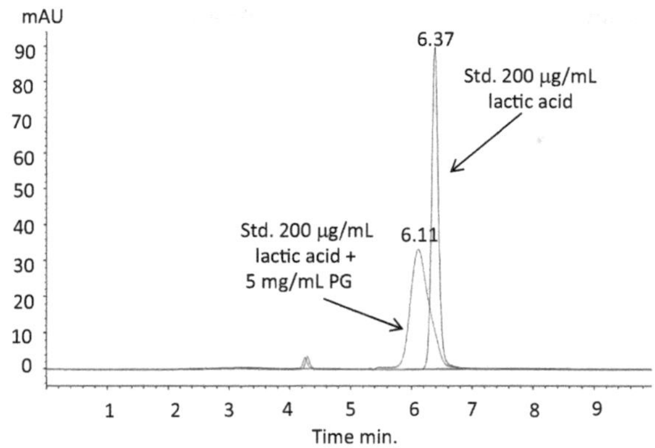

FIGURE 5.5.4 Chromatogram of a standard containing 200 μg/mL lactic acid on a Synergy Hydro RP column with mobile phase 20 mM phosphate buffer at pH 2.9 and the same sample in the presence of 5 mg/mL propylene glycol (PG) (UV detection at 215 nm).

matrix is frequently recommended and this separation is based on the same principles as that of the analytes.

Even when a separation takes place in the chromatographic column, the matrix can influence some HPLC peak shapes, similar to the influence of the sample solvent. The solvent of the sample, when it has weak eluting properties and is injected at larger volume (e.g., 30–50 μL) may leave all the analytes at the head of the chromatographic column before it is getting diluted with the mobile phase (see Section 14.1). This "focusing" of the analytes at the head of the column can improve the peak shape. A sample solvent with strong eluting properties (and at larger injection volume) may produce the reverse effect. The analytes can be eluted from the column head until the sample solvent is diluted with the mobile phase. This can generate distorted peaks. The same effect can be seen with a matrix component in large proportion present in the sample, and for large injection volumes, when the matrix component may act as a solvent until it is diluted with the mobile phase. As an example, Fig. 5.5.4 shows the chromatogram of a standard containing 200 μg/mL lactic acid on a Synergy Hydro RP column with mobile phase 20 mM phosphate buffer at pH 2.9 and the same sample in the presence of 5 mg/mL propylene glycol (PG).

5.6 REVIEW OF SAMPLE PROPERTIES WITH THE GOAL OF SELECTION OF A DETECTOR IN HPLC

A description of a number of detectors used in HPLC has been given in Section 4.1. These detectors have different characteristics regarding the following aspects: (1) applicability to specific physicochemical properties of the analyte, (2) selectivity/universal use, (3) restrictions regarding use of mobile phase, (4) sensitivity, and (5) other characteristics such as cost and ease of operation. The selection of a detector depends on a number of factors such as the purpose of analysis (qualitative, quantitative, or both), the type of sample or process, specific physicochemical properties of the analyte, the anticipated level of analyte, the quality of the HPLC separation. Also, the selection of a detector can be constrained by other factors (such as detector availability).

The quality of the HPLC separation is an important factor in selecting the detector. For a good separation, where all the potential interferences are eliminated from the peaks of the analytes, the detector to be selected can be a universal detector, and a selective detector is not required (although can be optionally used). If the separation does not separate well some of the sample components, the selectivity of the detector can be used for achieving the measurement. The quality of separation refers not only to the separation of other analytes, but also to the matrix background that may influence the detector reading (see Figs. 5.5.2 and 5.5.3). The mass spectrometric detector is capable of measuring only a specific ion for the analyte, or in case of LC-MS/MS working in MRM mode only a specific pair of ions (precursor ion → product ion) is detected such that very high selectivity is offered by the detector (multiple pairs can be selected for multiple analytes analysis).

The use of specific UV absorption, or specific fluorescence properties, can also be utilized to improve the selectivity of the analysis. Such detection procedures can be used successfully depending on the physicochemical properties of the analyte. On the other hand, for universal detectors (refractive index, ELSD, CAD), the HPLC separation must eliminate all the interferences.

The core analytical technique receives for analysis a processed sample in the case that a sample preparation step has been previously performed. This sample preparation step is important for the selection of the type of detector in several ways. One aspect is related to the cleaning of the matrix. If the matrix of the sample is simple and generates no interference, the sample preparation may not be necessary. Another role of sample preparation can be the chemical modification of the sample such that specific characteristics are changed to make it more amenable for the HPLC analysis. These chemical modifications may have the goal of enhancing the detection capability, for example by attaching chromophore or fluorophore groups to the analyte molecule [1]. Depending on the planned detection type and on the chemical structure of the analyte, a wide variety of derivatization procedures can be applied to the analytes (see, e.g., Refs. [1,78–81].)

Detector Selection for Qualitative or Quantitative Analysis

Qualitative analysis in HPLC is in general not diagnostic since no detector used in HPLC is able to provide universal information regarding the nature of the analyte. Mass spectrometry is the detection technique that comes the closest to a potential use in qualitative analysis (see Section 4.1). However, mass spectrometry coupled with HPLC is best characterized as highly selective such that, when searching for a specific compound, it is capable of providing reliable information on its presence or absence. Identification of an unknown compound by LC-MS can be done using unique characteristics of some compounds such as the value of m/z for the molecular ion, and specific fragments in the mass spectrum that can guide the compound identification. The use of high-resolution MS detection (e.g., using an Orbitrap instrument from Thermo Scientific) predicts the brute formula with very good accuracy. However, the brute formula is only a first step in a compound identification.

Other unique characteristics of a compound may be used for identification, such as a specific UV–Vis spectrum, or fluorescence characteristics. The broad bands in such spectra are a limiting factor in compound identification. The spectra can be useful when searching for a specific known compound or for verifying peak purity in HPLC.

Detector Selection Based on Specific Physicochemical Properties of the Analyte

Among the factors important in the detector selection are the specific physicochemical properties of the analyte. As indicated in Section 4.1 each type of detector is able to measure a specific physicochemical property including the following: (1) spectrophotometric UV–Vis absorption, (2) fluorescence, (3) chemiluminescence, (4) refractive index (RI), (5) mass spectrum (MS), (6) an electrochemical property, (7) evaporative light-scattering (ELS), and (8) other properties. The selection of a detector based on such properties is further discussed.

1) A convenient type of detection in HPLC is based on UV absorption. Such detection is usually highly reproducible and has a wide range of linear response to analyte concentration (see Eq. (2.2.3)). Also, many compounds are either colored or have strong absorption in UV. Such absorption is common for compounds containing aromatic rings, conjugated double bonds, or simply double bonds that include compounds containing functional groups such as −COOH, >CO, $CONH_2$. For this reason, the UV absorption spectra of the analytes should be evaluated for selecting the detection type. It is convenient to use UV–Vis detection when the compounds analyzed have strong absorption at wavelength higher than the absorption of the mobile phase components. When selecting the UV–Vis detection, it is recommended to verify if the ε_λ, the molar absorption (absorbance) coefficient (at the specific wavelength λ), is high enough for assuring sensitivity for the analytes. The absorption of the mobile phase (or of solvents composing mobile phase) is described by the "UV cutoff value" defined as the wavelength at which the absorbance A_λ of the solvent versus air, in a 1-cm cell, is equal to unity (see Chapter 13). The increase in absorption is common for many compounds at wavelength lower than 210–220 nm. For compounds with sole absorption bands in this range, the detection is utilized frequently, assuring a mobile phase with lower cutoff values and good chromatographic separation (e.g., for carbohydrates or aliphatic organic acids). However, since many compounds have strong absorption, the range 210–220 nm is not recommended for a selective detection. Unless well separated chromatographically, detection below 210–220 nm wavelength values is prone to interferences and possibly high background noise from absorption in this range of the mobile phase. Also, this range in the wavelength may impose restrictions to the mobile phase composition (which must have a lower cutoff value). For these reasons, if the compounds of interest have absorption bands at higher wavelengths, it is common to select them for detection [82]. For example, for the analysis of caffeine (spectrum shown in Fig. 2.2.1), the wavelength typically selected for detection is 274 nm, although below 210 nm the absorption of the compound is stronger.

2) Fluorescence detection requires fluorescent properties from the analyte molecule. This type of detection typically leads to very good sensitivity, and for already fluorescent compounds is highly recommended. However, fluorescence is not a very common property and derivatization techniques are used for attaching on the analyte molecule specific moieties that bring fluorescent properties (see, e.g., Ref. [1]). When fluorescence detection is considered and the analytes are not fluorescent, a decision should be made if a derivatization is preferred, or a different detection procedure is more appropriate. In cases when the analyte

is present in traces, it is frequently preferred to use derivatization because of the high sensitivity of fluorescence. In such cases, other aspects of the problem also must be considered, such as the availability of an adequate derivatization technique as well as the reproducibility of the derivatization. The nature of sample matrix is also important when choosing fluorescence for detection, since fluorescence can be affected by other components eluting together with the analytes.

3) Chemiluminescence as a detection technique can be selected after derivatization of the analyte (attaching a luminol moiety), and being the result of a chemical reaction, requires the addition of an oxidation reagent. Derivatization of the analyte requires the presence of reactive groups such as amino. One such reaction uses, for example, 6-isothiocyano-2, 3-dihydrobenzo[g]phthalazine-1,4-dione that reacts with the amino group. The resulting compound is oxidized with reagents such as H_2O_2, or $KHSO_5$. The addition of the oxidizing reagent can be done before the chromatographic column or postcolumn. Chemiluminescence techniques are reported to achieve very high sensitivity [83]. However, such procedures typically require a number of additional steps to the analysis (derivatization, addition of oxidizing reagent, etc.)

4) A common detector used in HPLC is based on the difference in refractive index (RI) of the eluate between mobile phase and mobile phase containing an analyte. This detector provides near universal detection and does not require special properties from the analytes. When other detection techniques are not possible, RI can be used with the condition to provide sufficient sensitivity for the analytes and to have an efficient separation.

5) Mass spectrometry became a common detection technique in HPLC. This technique is applied to a wide variety of molecules including small molecules as well as peptides. Due to the possibility to form molecules with more than one electric charge, and because the mass spectrometer separates the molecules based on their m/z values, relatively large molecules can be detected by MS. The instruments may be of LC-MS type, or LC-MS/MS type, and can have ESI or APCI ionization sources (see Section 4.1). The success in the utilization of mass spectrometry in HPLC detection is highly dependent on the nature of the analyte. Compounds that can form positive ions (e.g., containing nitrogen atoms in the molecule) typically can be detected at very low concentrations (see, e.g., in Section 5.5 the example of TSNAs shown in Fig. 5.5.2). Low detection limits (LODs) can also be obtained in negative ionization mode, depending on the nature of the compound, but usually not at sensitivities as low as in positive ionization mode. Stronger gas-phase basicity ΔG^0_{GPB} or gas-phase acidity ΔG^0_{GPA} (see Section 2.3) for the analytes dictates the choice of positive or negative ionization type in LC-MS. The values for ΔG^0_{GPB} or ΔG^0_{GPA}, if known, can be useful in the prediction of ionization and fragmentation under CI-like conditions. Also, the differences between the values of gas-phase free enthalpy for acidity versus basicity can explain why the formation of positive ions takes place more easily than that of negative ions in LC-MS, and why the sensitivity in positive ionization mode is frequently more sensitive than the sensitivity in negative ionization mode. The difference in the polarity between the solvent and the analyte molecules allows the selection of such conditions for the MS that the formation of ions takes place mainly from the analyte and less from the solvent [84].

The applications of LC-MS and LC-MS/MS systems are geared mainly toward quantitative measurements, where exceptionally good sensitivity can be achieved for numerous compounds. The selectivity of these techniques (in particular of LC-MS/MS) allows analyses where the separation is not perfect, and the interference is eliminated by selecting specific ions for measurement that are different between the two coeluting analytes.

6) Electrochemical detection is less common in HPLC as compared to UV, RI, or MS methods of detection. For amperometric methods of analysis, the compounds must have an oxidation potential below that of the constituent compounds of the mobile phase. Also, problems with the "poisoning" of the electrode surface following the electrochemical reaction must be addressed, as it is done in pulse amperometric detection (see Section 4.1). Conductometric measurements are typically applied in ion chromatography, and this type of detection is very successful in measuring ions in solution after the use of suppressors for eliminating the ions from the background.

7) Various molecules can be measured using ELSD and CAD. Except for volatile molecules, ELSD is a universal detector. The response in these techniques is basically independent of the chemical structure of the analyte. Since the detection with ELSD or CAD does not require for the analyte molecule to have chromophore groups, fluorophore groups, or to be able to ionize readily, these techniques have universal applicability. Also, while ELSD is not highly sensitive, CAD is reported as having very good sensitivity for a wide range of molecules.

Role of Analyte Concentration in the Selection of Detection in HPLC

Quantitative analytical methods are designed to determine with good accuracy and precision the level of specific analytes in a sample. Before performing the analysis, the level of the analyte is unknown, but it is common that a range of analyte concentrations can be anticipated. This may classify the analyte concentration as ultratraces, traces, low level, average level, or major constituent. For some samples, such information is not available, and for some other samples, details regarding specific ranges of concentration may be known. Depending on the analyte concentration, further selection of a detector can be made for the HPLC analysis. This selection is conditioned by the fact that the physicochemical properties of the analyte allow the use of the detector (e.g., the analyte has chromophore groups and UV can be the detection procedure).

The selection of the detector in HPLC considering various aspects of the detection must also include the consideration of the level of analyte expected to be analyzed. For analytes that are at relatively high levels in the processed sample, typically at concentrations above 50 μg/mL, detectors such as RI or ELSD can be used (as well as any other detector with higher sensitivity). For UV detection, depending on the analyte molar absorption (absorptivity) coefficient, concentrations as low as 0.2–0.5 μg/mL of the analyte can be measured (see Table 4.1.2 for detector sensitivity). Fluorescence detection can be extremely sensitive and levels in the range of 0.5 ng/mL analyte may be amenable for analysis. The sensitivity of mass spectrometric detection varies significantly from compound to compound, but some analytes can be measured at levels of 2–5 pg/mL in the injected solution.

Although the main problems in the analysis is typically the achievement of high sensitivity, too high a concentration of the analyte in the samples may also be a problem since at higher concentrations the detector response may deviate from linearity (with a lowering of the response). Although this problem can be solved by dilution for certain analysis, it starts to be a problem when analytes that are major components are attempted to be analyzed at the same time as trace analytes. In such cases, some analytes would benefit from higher sensitivity and sample concentration, while other analytes require dilution. In such cases, detectors with a wide dynamic range of response are very useful (see Table 4.1.3 for detector linear ranges).

References

[1] S.C. Moldoveanu, V. David, Sample Preparation in Chromatography, Elsevier, Amsterdam, 2002.
[2] S.C. Moldoveanu, V. David, Modern Sample Preparation for Chromatography, Elsevier, Amsterdam, 2015.
[3] R.E. Majors, Sample Preparation Fundamentals for Chromatography, Agilent Technologies, Wilmington, 2014.
[4] J. Pawliszyn (Ed.), Comprehensive Sampling and Sample Preparation, Elsevier, Amsterdam, 2012.
[5] http://www.chemaxon.com.
[6] A.L. McClellan, Tables of Experimental Dipole Moments, W. H. Freeman and Co., San Francisco, 1963.
[7] C. Hansch, A. Leo, D. Hoekman, Exploring QSAR, Hydrophobic, Electronic and Steric Constants, ACS, Washington, 1995.
[8] http://www.epa.gov/oppt/exposure/pubs/episuite.htm.
[9] http://www.daylight.com/.
[10] S.G. Machatha, S.H. Yalkowsky, Comparison of the octanol/water partition coefficients calculated by ClogP, ACDlogP and KowWin to experimentally determined values, Int. J. Pharm. 294 (2005) 185–192.
[11] V.N. Viswanadhan, A.K. Ghose, G.R. Revankar, R.K. Robins, Atomic physicochemical parameters for three dimensional structure direct quantitative structure-activity relationships. 4. Additional parameters for hydrophobic and dispersive interactions and their application for an automated superposition of certain naturally occurring nucleoside antibiotics, J. Chem. Inf. Comput. Sci. 29 (1989) 162–172.
[12] S.C. Moldoveanu, V. David, Dependence of the distribution constant in liquid–liquid partition equilibria on the van der Waals molecular surface area, J. Sep. Sci. 36 (2013) 2963–2978.
[13] D. Erős, I. Kövesdi, L. Őrfi, K. Takács-Novák, G. Acsády, G. Kéri, Reliability of log P predictions based on calculated molecular descriptors. A critical review, Curr. Med. Chem. 9 (2002) 1819–1829.
[14] W.J. Moore, Physical Chemistry, second ed., Prentice-Hall, Inc., Englewood Cliffs, 1955.
[15] J.H. Hildebrand, R.I. Scott, The Solubility of Non-electrolytes, Dover Pub., New York, 1964.
[16] L.R. Snyder, The role of the mobile phase in liquid chromatography, in: J.J. Kirkland (Ed.), Modern Practice of Liquid Chromatography, Wiley-Interscience, New York, 1971.
[17] G. Scatchard, Equilibria in non-electrolyte solutions in relation to the vapor pressures and densities of the components, Chem. Rev. 8 (1931) 321–333.
[18] R.A. Robinson, R.H. Stokes, Electrolyte Solutions, Dover Pub., Mineola, 2002.
[19] A. Bondi, Van der Waals volumes and radii, J. Phys. Chem. 68 (1964) 441–451.
[20] http://www.webelements.com/periodicity/van_der_waals_radius/.
[21] M. Ptitejean, On the analytical calculation of van der Waals surfaces and volumes: some numerical aspects, J. Comput. Chem. 15 (1994) 507–523.
[22] O. Sinanoğlu, Intermolecular forces in liquids, in: J.O. Hirschfelder (Ed.), Advances in Chemical Physics, vol. 12, J. Wiley, New York, 1967, pp. 283–326.
[23] T. Halicioğlu, O. Sinanoğlu, Solvent effects on cis-trans azobenzene isomerization; a detailed application of a theory of solvent effects on molecular association, Ann. N.Y. Acad. Sci. 158 (1974) 308–317.
[24] C. Horvath, W. Melander, I. Molnar, Solvophobic interactions in liquid chromatography with nonpolar stationary phases, J. Chromatogr. 125 (1976) 129–156.
[25] S. Moldoveanu, V. David, Estimation of the phase ratio in reversed-phase high-performance liquid chromatography, J. Chromatogr. A 1381 (2015) 194–201.
[26] S. Moldoveanu, A. Savin, Aplicatii in Chimie ale Metodelor Semiempirice de Orbitali Moleculari, Ed. Academiei, Bucuresti, 1980.
[27] S. Moldoveanu, Aplicatile Teoriei Grupurilor in Chimie, Ed. Stiintifica si Enciclopedica, Bucaresti, 1975.
[28] G. Popa, S. Moldoveanu, Reactivii Organici in Chimia Analitica, Ed. Academiei, Bucuresti, 1976.
[29] E. Pretsch, P. Bühlmann, M. Badertscher, Structure Determination of Organic Compounds, Springer-Verlag, Berlin, 2009.

[30] R.C. West, Handbook of Data on Organic Compounds, CRC Press, Boca Raton, 1985.
[31] http://www.colby.edu/chemistry/cmp/cmp.html.
[32] A.P. Demchenko (Ed.), Advanced Fluorescence Reporters in Chemistry and Biology, Sringer, Heidelberg, 2010.
[33] T. Jonsson, , reportPeroxyoxalate Chemiluminescence for Miniaturized Analytical Flow Systems,(Ph.D. thesis) Umeå University, Umeå, Sweden.
[34] http://www.engineeringtoolbox.com/refractive-index-d_1264.html.
[35] W.J. Lyman, W.F. Reehl, D.H. Rosenblatt, Handbook of Chemical Property Estimation Methods, ACS, Washington, 1990.
[36] M. Reinhard, A. Drefahl, Handbook for Estimating Physicochemical Properties of Organic Compounds, J. Wiley, New York, 1999.
[37] B.L.M. van Baar, Ionization methods in LC-MS and LC-MS-MS (TSP), APCI, ESP, and cf-FAB, in: D. Barcelo (Ed.), Application of LC-MS in Environmental Chemistry, Journal of Chromatography Library Series, vol. 59, Elsevier, Amsterdam, 1996, pp. 71–126.
[38] B.E. Conway, Electrochemical Data, Elsevier, Amsterdam, 1952.
[39] G. Milazzo, S. Caroli, V.K. Sharma, Tables of Standard Electrode Potentials, Wiley, Chichester, 1978.
[40] A.J. Bard, R. Parsons, J. Jordan, Standard Potentials in Aqueous Solutions, Marcel Dekker, New York, 1985.
[41] W.M. Haynes (Ed.), Handbook of Chemistry and Physics, ninety sixth ed., CRC Press, Boca Raton, 2015.
[42] S. Bayne, M. Carlin, Forensic Applications of High Performance Liquid Chromatography, CRC Press, Boca Raton, 2010.
[43] W.J. Lough, I.W. Wainer, C.M. Riley (Eds.), Pharmaceutical and Biomedical Applications of Liquid Chromatography, Elsevier Science Ltd, Amsterdam, 2013.
[44] T. Cserháti, Liquid Chromatography of Natural Pigments and Synthetic Dyes, Elsevier, Amsterdam, 2006.
[45] H. Mascher, HPLC Methods for Clinical Pharmaceutical Analysis, John Wiley & Sons, Hoboken, 2012.
[46] D. Jain, P.K. Basniwal, Forced degradation and impurity profiling: recent trends in analytical perspectives, J. Pharm. Biomed. Anal. 86 (2013) 11–35.
[47] D. Bartos, S. Görög, Recent advances in the impurity profiling of drugs, Curr. Pharm. Anal. 4 (2008) 215–230.
[48] R.J. Smith, M.L. Webb (Eds.), Analysis of Drug Impurities, Blackwell Publishing, Oxford, UK, 2007.
[49] M.V. Narendra Kumar Talluri, Impurity Profiling of Drugs and Pharmaceuticals, Lambert Academic Publishing, 2011.
[50] Y. Kazakevich, R. LoBruto, HPLC for Pharmaceutical Scientists, Wiley, Hoboken, 2007.
[51] J. Bernal, J.A. Mendiola, E. Ibáñez, A. Cifuentes, Advanced analysis of nutraceuticals, J. Pharm. Biomed. Anal. 55 (2011) 758–774.
[52] M. Urh, D. Simpson, K. Zhao, Affinity Chromatography: General Methods, in Methods in Enzymology, vol. 463, Elsevier, Amsterdam, 2009.
[53] S.S. Nielsen (Ed.), Food Analysis, Springer, New York, 2010.
[54] S. Ötles (Ed.), Handbook of Food Analysis Instruments, CRC Press, Boca Raton, 2009.
[55] A.K. Haghi, E. Carvajal-Millan, Food composition and analysis: methods and strategies, CRC Press, Boca Raton, 2014.
[56] S. Moldoveanu, W. Scott, J. Zhu, Analysis of small carbohydrates in several bioactive botanicals by gas chromatography with mass spectrometry and liquid chromatography with tandem mass spectrometry, J. Sep. Sci. 38 (2015) (2015) 3677–3686.
[57] S.C. Moldoveanu, Profiling of lipids from fruit and seed extracts, in: Su Chen (Ed.), Lipidomics: Sea Food, Marine Based Dietary Supplement, Fruit and Seed, Transworld Res. Network, Kerala, India, 2012.
[58] N.T. Faithfull, Methods in Agricultural Chemical Analysis. A Practical Handbook, CABI Pub., New York, 2002.
[59] P. Kusch, Identification of synthetic polymers and copolymers by analytical pyrolysis-gas chromatography/mass spectrometry, J. Chem. Educ. 91 (2014) 1725–1728.
[60] P.A. D'Agostino, J.R. Hancock, C.L. Chenier, Mass spectrometric analysis of chemical warfare agents and their degradation products in soil and synthetic samples, Eur. J. Mass Spectrom. 9 (2003) 609–618.
[61] R. Marsil, Flavor, Fragrance, and Odor Analysis, second ed., CRC Press, Boca Raton, 2012.
[62] G. Czichocki, H. Fiedler, K. Haage, H. Much, S. Weidner, Characterization of alkyl polyglycosides by both reversed-phase and normal-phase modes of high-performance liquid chromatography, J. Chromatogr. A 943 (2002) 241–250.

[63] P.A. Lara-Martín, E. Gonzalez-Mazo, B.J. Brownawell, Multi-residue method for the analysis of synthetic surfactants and their degradation metabolites in aquatic systems by liquid chromatography-time-of-flight-mass spectrometry, J. Chromatogr. A 1218 (2011) 4799–4807.
[64] A. Rodgman, T.A. Perfetti, The Chemical Components of Tobacco and Tobacco Smoke, CRC Press, Boca Raton, 2013.
[65] Food and Drug Administration (FDA), Guidance for Industry: Reporting Harmful and Potentially Harmful Constituents in Tobacco Products and Tobacco Smoke under Section 904(a)(3) of the Federal Food, Drug, and Cosmetic Act. Draft guidance, 2012. At: http://www.fda.gov/downloads/TobaccoProducts/GuidanceComplianceRegulatoryInformation/UCM297828.pdf.
[66] S.C. Moldoveanu, Analytical Pyrolysis of Natural Organic Polymers, Elsevier, Amsterdam, 1998.
[67] S.C. Moldoveanu, Analytical Pyrolysis of Synthetic Organic Polymers, Elsevier, Amsterdam, 2005.
[68] B. Wittmann-Liebold (Ed.), Methods in Protein Sequence Analysis, Springer, Berlin, 1989.
[69] Protein sequencing protocols, in: B.J. Smith (Ed.), Methods in Molecular Biology, vol. 211, Humana Press, Totowa, 2003.
[70] S. Sechi, Quantitative proteomics by mass spectrometry, in: Methods in Molecular Biology, vol. 359, Humana Press, Totowa, 2007.
[71] R.J. Simpson, P.D. Adams, E.A. Golemis (Eds.), Basic Methods in Protein Purification and Analysis: A Laboratory Manual, CSH Press, Cold Spring Harbor, 2009.
[72] A.S. Vila, A.I. Castellote-Bargalló, M.R. Palmero-Seuma, M.C. López-Sabater, High-performance liquid chromatography with evaporative light-scattering detection for the determination of phospholipid classes in human milk, infant formulas and phospholipid sources of long-chain polyunsaturated fatty acids, J. Chromatogr. A 1008 (2003) 73–80.
[73] Y. Iwasaki, A. Masayama, A. Mori, C. Ikeda, H. Nakano, Composition analysis of positional isomers of phosphatidylinositol by high-performance liquid chromatography, J. Chromatogr. A 1216 (2009) 6077–6080.
[74] N. El Tayar, H. van de Waterbeemd, B. Testa, The prediction of substituent interactions in the lipophilicity of disubstituted benzenes using RP-HPLC, Quant. Struct. Act. Relat. 4 (1985) 69–77.
[75] E.P. Kadar, C.E. Wujcik, O. Kavetskaia, Property Log D to hydrophilic interaction chromatography in the bioanalytical laboratory, in: P. Wang, W. He (Eds.), Hydrophilic Interaction Liquid Chromatography (HILIC) and Advanced Applications, CRC Press, Boca Raton, 2011.
[76] V. David, C. Barcutean, I. Sora, A. Medvedovici, Determination of metformin in plasma samples by HPLC-DAD based on plasma derivatization and precipitation with acetic anhydride, Rev. Roum. Chim. 50 (2005) 269–276.
[77] J. Zhu, N. Qian, S. Jones, S. Moldoveanu, A versatile method for the analysis of TSNAs in tobacco products and cigarette smoke by LC-MS-MS, in: 69th Tob. Sci. Res. Conf., Poster 73, Sept 20–23, Naples, 2015.
[78] G. Lunn, L.C. Hellwig, Handbook of Derivatization Reactions for HPLC, J. Wiley, New York, 1998.
[79] T. Toyo'oka (Ed.), Modern Derivatization Methods for Separation Sciences, J. Wiley, Chichester, 1999.
[80] K. Blau, J. Halket (Eds.), Handbook of Derivatives for Chromatography, J. Wiley, Chichester, 1993.
[81] D.R. Knapp, Handbook of Analytical Derivatization Reactions, J. Wiley, New York, 1979.
[82] T. Galaon, M. Radulescu, V. David, A. Medvedovici, Injection of a non-miscible diluent in ionic liquid/ion pair LC for the assay of active ingredients in a combination formulated as injectable solution, Cent. Eur. J. Chem. 10 (2012) 1360–1368.
[83] J. Woltman, W.R. Even, E. Sahlin, S.G. Weber, Chromatographic detection of nitroaromatic and nitramine compounds by electrochemical reduction combined with photoluminescence following electron transfer, Anal. Chem. 72 (2000) 4928–4933.
[84] M.S. Lee (Ed.), Mass Spectrometry Handbook, John Wiley & Sons, Hoboken, 2012.

CHAPTER

6

General Aspects Regarding the HPLC Analytical Column

6.1 CONSTRUCTION OF THE HIGH-PERFORMANCE LIQUID CHROMATOGRAPHY (HPLC) COLUMN

The chromatographic column is made from a tube (cylinder) filled with the stationary phase. Various physical dimensions of the tube hosting the stationary phase are available, and an even larger variety of stationary phases that are used in the column have been produced. Progress is continuously being made in the construction of a chromatographic column. In the past 5–10 years significant advances were made regarding the development of particles with diameter below 2 μm, and superficially porous particles are now very common. Significant progress was also made in the chemistry of the stationary phases, with columns being more stable in a wider pH range, and capable of being used in a variety of solvents (including 100% water) [1]. Several aspects of the construction of a chromatographic column are discussed in more detail in this section.

External Body of the Column

The external body of the column (empty column) is in the form of a tube made from stainless steel or a strong polymer [e.g., polyetheretherketone (PEEK)]. At the two ends of the column are special frits that keep the stationary phase from moving from inside the body of the column, and also fittings that allow the connection of the column with the narrow-bore tubing connecting the column to the pumps and to the detector (see Section 4.1). The geometry and the design of the frits may influence peak efficiency [2].

The physical dimensions of common analytical chromatographic columns vary, and values for length (internal) L can be between 30 and 250 mm (common length 50, 100, 150, 250 mm), and internal diameters (i.d.) d for usual columns can be between 1 and 10 mm (e.g., 2.1, 3.0, or 4.6 mm). Other dimensions are possible, particularly when the column is designed for special tasks. The newer columns have the tendency of being shorter and narrower, as the solid particles that form the stationary phase are made smaller. The physical shape of the body of the chromatographic column may have some effect on the separation. At the column wall, the packing material has a different distribution compared to the interior of the column, and

Selection of the HPLC Method in Chemical Analysis
http://dx.doi.org/10.1016/B978-0-12-803684-6.00006-8

the migration rate of the sample band in the region of the wall is different from the interior. The mobile phase flow at the walls is also different from that in the middle of the column. Such effects were studied in detail and reported in the literature [3–6].

Special cartridges (microfluidic chips) are also available as containers for the stationary phase. Based on the internal diameter of the analytical column, they are sometimes classified in the literature as (1) standard (3.0–4.6 mm i.d.), (2) minibore (2.0–3.0 mm i.d.), (3) microbore (0.5–2.0 mm i.d.), (4) capillary (0.2–0.5 mm i.d.), and (5) nanoscale (0.05–0.2 mm i.d.). Miniaturized sample introduction techniques, pumps, and specialized detectors have been developed for microcolumn LC [7]. Larger columns are used for semipreparative and preparative purposes. The empty volume of the column can be easily calculated as the volume of a cylinder $V = (\pi/4)d^2L$, and for analytical columns V ranges between 0.02 mL and 20 mL. The column is filled with the stationary phase, and a fraction of the empty column volume is filled with the mobile phase. This volume will give the dead (void) volume of the column, discussed in Section 4.2. Various dimensions of columns, empty column volumes, and typical void volumes are indicated in Table 4.2.1.

Packing of the Chromatographic Column

Most common stationary phases in HPLC consist of small particles. These particles are placed in the body of the column making the *particle packed columns*. Monolithic chromatographic columns are also manufactured, in which the stationary phase consists of a porous solid rod [8]. However, a much larger number of commercial columns are made using particles that are packed in the body of the column. A compact and uniform packing of the chromatographic column with the particles is important since it affects the column plate number N. When using high pressures with a specific column, the stationary phase should not change its volume. If the volume of the stationary phase shrinks, a void volume can be formed at the head of the column, affecting significantly the column plate number. A denser bed of the stationary phase may also affect the backpressure under which the column must be operated, affecting the chromatographic conditions. Also, for certain types of stationary phases, the separation itself can be drastically degraded if the stationary phase collapses under inappropriate elution conditions. Changes in the packing of the stationary phase or its degradation may occur in time, and the performance of the chromatographic column degrades. The reproducibility of the results of the analyses becomes a problem in such situations.

Different procedures of packing the columns with the stationary phase are used, depending on the stationary phase. Most commonly, the stationary phase is introduced in the column as a slurry in a specially selected liquid that allows the formation of a homogeneous suspension of the stationary phase and hinders the particle aggregation. Slurry packing involves the use of high pressure (usually 50% higher than the maximum pressure at which the column will be used) to push a dilute slurry of stationary phase through the column. Based on the packing procedures, it is common for most columns to have a required direction for the mobile phase flow.

Physical Characteristics of the Solid Supports for the Packed Columns

The most common material used as solid support for the particles in packed columns is porous hydrated silica ($SiO_2 \times H_2O$). Other hydrated oxides can be used as solid support for the stationary phase, such as hydrated zirconia ($ZrO_2 \times H_2O$), or hydrated alumina ($Al_2O_3 \times H_2O$), but their utilization is rather limited. Such materials offer a large surface covered with reactive groups (silanol or Si—OH in the case of silica, or Me—OH for other supports). At the same time these materials have a high rigidity and resilience to crashing and do not change their volume when placed in different solvents. The numerous reactive groups (Si—OH or Me—OH) on the surface allow derivatization for covering of the solid support with the active bonded stationary phase. This is done following a reaction that attaches organic moieties on the solid surface.

The physical characteristics of the particle's solid support include the following: (1) type of particles, (2) the dimension of particles, (3) shape of the particles, (4) the dimension distribution of the particles, (5) surface area of the particles, (6) pore size of the particles, (7) mechanical rigidity of the support, and (8) other solid support characteristics.

1) Three main types of particles are porous (fully porous), core—shell (superficially porous), and pellicular (nonporous). Porous particles have a porous structure for the entire particle. Core—shell particles (also known as fused-core or superficially porous microspheres) have a solid nonporous core surrounded by a porous outer shell. For particles of about 3 μm diameter, the porous shell may be 0.3—0.5 μm in depth. Pellicular particles are solid spheres covered with a thin layer of stationary phase. The type of particles plays an important role in the stationary-phase characteristic, mainly by affecting the diffusion process in the column. The eddy diffusion, for example, plays an important role in peak broadening and therefore in the decrease of theoretical plate number N (see expression (4.2.45)) for the same column length. In core—shell particles less eddy diffusion is noticed, since the path length for the separated molecules is shorter than in fully porous particles. In such a way, the number of theoretical plates for columns with core—shell particles is higher than for columns with fully porous particles. Pellicular particles typically offer only a small amount of active stationary phase and their use is limited to very small (diluted) samples.

2) The dimension of stationary-phase particles is an important parameter related to the column theoretical plate number and the backpressure of the column. The plate height H for packed columns is roughly proportional to the particle diameter, whereas the backpressure is inversely proportional to particle diameter. Usually, a linear increase in column efficiency due to the use of smaller particles is accompanied by a quadratic increase in column backpressure. Columns typically have particles with diameters of 5, 3, 2.1, or 1.7 μm. Other values of particle dimensions are available but less common, such as 10 μm particles. Larger particles, offering larger differences in the possible path length of the molecules when they penetrate the stationary phase lead to broader peaks (lower N/m). Particle size also influences the required pressure for the mobile phase to pass through the column, with larger particles generating lower backpressure for the column. In the past, when the pressure generated by the pumps was more limited, particles with a diameter around 5 μm were the most common, and the back pressure due to the column was not higher than about 240 bar (3500 psi), this depending on the solvent, column length, and column diameter. In modern HPLC the tendency is to reduce the particle size (3 μm, 1.7 μm, or even lower), these columns requiring higher working pressure up to 600 bar (9000 psi) or even higher. For a flow rate of 1 mL/min, columns with particles around 2 μm and lower may generate pressures up to 15,000 psi, and typically they are used with lower flow rates and narrow columns [9]. The decrease in particle size is also related to other properties of the column such as column lifetime. Particle size

also influences the *interstitial volume*, which is the space between particles. Interstitial volume also contributes to the peak broadening in HPLC. This volume is typically around 70% of the total internal volume of the column.

3) The shape of particles can be irregular or spherical. Significant effort has been involved in generating particles as close as possible to spherical form [10–12]. For spherical particles, it is possible to have a much more homogeneous stationary phase and with less variation in the particle dimensions (less than 1–2% difference in diameter). Particle size distribution is typically described by the parameter d_{90}/d_{10}. The value d_X refers to a cumulative distribution indicating the diameter for which X% from the total number (amount) of particles are below or at that diameter. Values for d_{90}/d_{10} lower than 1.2–1.3 indicate very good homogeneity. Values closer to $d_{90}/d_{10} = 1$ contribute to a higher theoretical plate number N (per column length).

4) The uniformity of the dimensions of the particles is also an important physical characteristic for the solid support. A mixture of smaller and larger particles is avoided for the stationary phase, and this is also related to the intention to limit the eddy diffusion. Large particles mixed with small particles will lead to larger differences in the path within the solid particles of different molecules and consequently to peak broadening. The quality of the column regarding peak broadening is related to particle homogeneity.

5) Surface area of the stationary phase is a very important parameter regarding the column. Since the active stationary phase is distributed on the surface of the solid support, a larger surface area is related to a larger amount of stationary phase for the same total volume of the space filled with mobile phase. For silica particles, for example, surface area varies between 200 m^2/g for low surface area particles and 300 m^2/g for high surface area particles. The larger surface area of the stationary phase is therefore related to a larger phase ratio Ψ. Larger values for phase ratio are related to larger retention time as shown by the formula:

$$t_R(X) = t_0[1 + K(X)\Psi] \tag{6.1.1}$$

Also, larger capacity factors k' and larger resolutions R are obtained when Ψ is larger (see expressions (4.2.17) and (4.2.63)).

6) The pore size (diameter) of the porous materials is also an important characteristic of the solid support. There are several classifications of pore size, but basically they are indicated as small (below 60 Å), medium (in the range 60–150 Å), and large (of about 300 Å or larger). The most common type used in HPLC is medium-pore phases used for the separation of nonpolymeric molecules and large, typically used for the separation of polymers (such as proteins). The pore size is important for being selected in agreement with the type of analyte to be separated on a column. For the separation of large molecules such as proteins, solid supports with large pores (300 Å or larger) are necessary in order to allow the intimate contact between the analyte and stationary phase. Pore size also plays an important role during silica derivatization through the differences in the reactivity of the −OH groups in small pores and in large pores, and also through steric effects. The pore volume is another solid support characteristic that is related to the pore size diameter.

7) Structural rigidity of the solid support is an important parameter to be considered on a column. Inorganic supports such as hydrated silica or hydrated zirconia withstand high pressures with values between 9000 and 15,000 psi (600–1000 bar) without changes, depending also on the particle dimension. Specific high-strength silica materials (HSS) are used in the

construction of some stationary phases (HSS technology). Other materials, such as organic polymers, are much more sensitive to higher pressure, and specification of the columns must indicate the maximum acceptable pressure. Rigid organic polymeric particles typically withstand pressure up to about 5000 psi, and soft gel columns must be used at even lower backpressures. Besides rigidity to high column backpressures, structural rigidity also refers to the lack of volume variation when the solvent present in the mobile phase is changed. Inorganic supports for stationary phase keep a constant volume in different solvents, while organic polymeric phases swell and shrink depending on the mobile phase solvents, these being a considerable disadvantage for the polymeric phases.

8) Other aspects related to the solid support include the proper packing of the material in the body of the chromatographic column, and the absence of unintended fine particles.

Chemical Characteristics of the Solid Supports for the Packed Columns

The chemical characteristics of the solid support include the following: (1) reactivity of the binding groups (Si–OH, Me–OH), (2) chemical resistance of the support to the mobile phase characteristics (e.g., pH), (3) chemical purity of the solid support, (4) chemical characteristics for the solid support that also contains in its structure the functional groups, and (5) other chemical properties. The chemical aspects are further discussed in this section for each specific solid support.

1) The reactive groups on the solid support play an important role in generating the active stationary phase. These groups are characterized by their nature and acidic/basic character and related to that is their reactivity. Most inorganic supports have –OH groups as reactive sites for binding the stationary phase. However, hydride-based silica is also used as stationary support. The density of the active groups versus that of backbone structure (e.g., –O–Si–O– groups) is another important chemical characteristic.

Porous polymers that are used as stationary phases typically act as both support and active phase. Porous polymers are used frequently as stationary phases for size-exclusion chromatography and for ion-exchange chromatography. For size-exclusion chromatography, the polymers do not need specific functionalities, and the main characteristic of these phases is their tridimensional structure. For ion exchange purposes, ionic groups are covalently bonded in the body of the polymer. Some polymers are also used for stationary phase in reversed-phase chromatography, but they are less resilient to high pressure than phases with inorganic supports.

2) The solid support is affected by the acid or basic character of the mobile phase, as well as the potential nature of solvents making the mobile phase. For example, hydrated silica is resistant to acid and bases only in the range of pH between 2 and 8, although several modifications of the silica were made allowing the extension of this range. Mobile phases with a lower or higher pH will lead to damage of the chromatographic column. Other supports, such as those based on organic polymers, are resistant to a wider range of pH, but they may have problems with incompatibility with specific solvents.

3) The purity of the solid support is related mainly to the elimination from the hydrated SiO_2, or other support material of traces of transitional metals such as iron. The presence of transitional ions, for example in silica, may lead to various interactions with the analytes

through the *d* electrons of the metal that will superpose over the intended interactions (e.g., hydrophobic interactions with the active stationary phase). The additional interactions lead to peak broadening and tailing.

4) Not all solid supports are used exclusively for binding an active stationary phase by derivatization. Some stationary phases already contain in their structure the functional groups that act as a true stationary phase (such as sulfonic groups in a polymer to be used as ion exchange stationary phase).

5) Other chemical properties are related, for example, to other types of interactions with the analytes besides the one intended by the user. For example, for silica supports, many silanol groups remain on the silica support surface even after derivatization and *end-capping* (reaction of free silanol groups with blocking reagents). These silanol groups may be more or less acidic and may have interactions with the analytes. Also, in the case of other hydrated oxides (such as hydrated ZrO_2), zirconium atoms from the solid backbone may interact with the analytes through its electrons from the *d* orbital.

Besides the reaction of the reactive groups with specific reagents for producing a bonded stationary phase, other procedures can be utilized to generate an active stationary phase. These procedures include coating or immobilizing a polymeric material on the solid support that is made, e.g., from silica. The goal was to take advantage of the porosity of the material that has a large surface area, mechanical strength, and pores of appropriate dimensions, and at the same time to have a stationary phase of a different nature than the solid support for acting as a stationary phase.

Silica as Solid Support for the Stationary Phase

Silica-based stationary phases are made using hydrated porous silica particles obtained from a chemical reaction that generates silicic acids (typically a controlled hydrolysis of an alkoxysilane or of a silicic acid salt). In extremely diluted solutions several simple silicic acids were identified, such as metasilicic acid (H_2SiO_3), ortosilicic acid (H_4SiO_4), disilicic acid ($H_2Si_2O_5$), and pyrosilicic acid ($H_6Si_2O_7$). All these are very weak acids (e.g., ortosilicic acid has $pK_1 = 9.84$, $pK_2 = 13.2$). In more concentrated solutions, polysilicic acids with the general formula $[SiO_x(OH)_{4-2x}]_n$ are rapidly formed from the condensation reaction between the silanol groups. This process gives rise to tridimensional structures (precipitates in gel form) containing siloxane (Si–O–Si) groups. The reaction of formation of silicic acid may start with acidic hydrolysis of salts such as Na_2SiO_3, Na_4SiO_4, and K_4SiO_4, or with the hydrolysis in the presence of an acid or a base (as catalysts) of several silicon compounds like SiH_4, $SiCl_4$, or of alkoxysilanes ($Si(OR)_4$ where $R = CH_3$, C_2H_5, C_3H_7, etc.). Tetraethyl orthosilicate $Si(OC_2H_5)_4$ is a common alkoxysilane used for the preparation of silica gels by hydrolysis. The following reactions schematically describe the formation of polysilicic acids from an alkoxysilane:

$$Si(OR)_4 + 4H_2O + H^+ \rightarrow Si(OH)_4 + 4ROH + H^+ \quad (6.1.2)$$

$$nSi(OH)_4 \rightarrow Si_n(O)_x(OH)_{4n-2x} + xH_2O \quad (6.1.3)$$

When the polysilicic acid is obtained by hydrolysis of alkoxysilanes, part of the alkoxy groups may remain attached to the silicon atom and further participate in the elimination reaction to form siloxane bonds. A partial hydrolysis of an alkoxysilane can be described schematically by the reaction:

$$Si(OR)_4 + 3H_2O + H^+ \rightarrow Si(OR)(OH)_3 + 3ROH + H^+ \quad (6.1.4)$$

Polysilicic acids initially formed by hydrolysis are initially present in solution form (sol form) and a large number of water molecules are bound through hydrogen bonds or mechanically retained in the structure. The choice of the reactions and of the conditions used to obtain silica gels varies considerably depending on the intended use of a specific material [13,14]. As the molecules of polysilicic acid increase in size, the sol particles grow and further change in a system containing both a liquid phase and a solid phase with a morphology ranging from discrete particles to a continuous polymeric network filled with a large number of pores of different dimensions. This material generates a gel (hydrogel or alcogel depending on the groups still left unreacted) where most of the structures are crosslinked. The distinction between a material in sol form and that of a gel is not precise, and as the amount of water in the structure decreases and the polymeric network increases, the gel form becomes more obvious. The two processes taking place during the gel formation are the growth of sol particles and the reticulation of the gel. These processes are schematically pictured in Fig. 6.1.1.

Sol particles can grow isolated from each other, having forms close to small spheres. This capability allows the formation of particles with a controlled shape and size, amenable to be used for chromatography. The control of the size of sol particles and the elimination of water with the increase in the crosslinking of the polymeric structure can be achieved by selecting the concentration of reagents that form the polysilicic acid and also by controlling other parameters such as temperature, pH, gelling time, addition of electrolytes for flocculation, addition of detergents, and specific solvents. The process is very complex, and as a function of selected parameters the growth of sol particles, or of the reticulation can be favored, such that the properties of the final material may be different. The structure of the gel continues to change in a process known as aging. Aging affects the properties of the final material by several mechanisms known as polycondensation (further reactions of silanol groups to

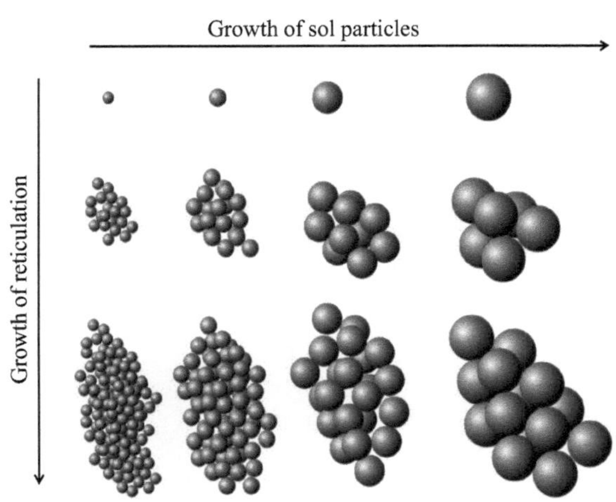

FIGURE 6.1.1 Schematic diagram of the two processes taking place during the gel formation: the growth of sol particles and the reticulation of the gel.

form siloxane bonds), syneresis or pore narrowing (the shrinkage of the gel network with the elimination of liquid from the pores), and coursing (the dissolution of small particles and the growth of larger ones).

Superficially porous particles are typically manufactured by adding solid silica cores during the sol formation. The hydrated silica grows on the cores and further changes in a system containing a continuous polymeric network filled with a large number of pores of different dimensions. The pore sizes are mainly controlled by the size of the silica nanoparticles, and the tortuous pore channel geometry is determined by how the nanoparticles randomly aggregate. A new process indicated as pseudomorphic transformation has been recently reported [15]. In this process the superficially porous particles are made with a narrower particle size distribution, thinner porous layer, high surface area, and nontortuous pore channels oriented normal to the particle surface.

Following the gelling and aging process, the hydrogel (or alcogel) is typically washed and dried (converted into a xerogel) to obtain a solid material. A silica gel, as first precipitated from water, may contain up to 300 moles of H_2O to 1 mole of SiO_2. The drying process consists of three stages. In the first stage the gel loses water, and its volume decreases with the volume of water that was evaporated. In this stage, the pore volume decreases, and the stiffness of the dried gel increases. In stage two, the liquid from the pores of the gel is eliminated by migration to the surface of the material where it is evaporated. In stage three, the liquid escapes from the pores directly by evaporation. When dried below 150°C, surfaces containing a large number of silanol groups (≡Si–OH) are still present in the gel. When heated at 300–1000°C, the surface covered with silanol groups dehydroxylates to form siloxane (Si–O–Si) surfaces. The final properties of the dried material, such as density (porosity), hardness, active surface area, pore volume, and pore size distribution depend not only on the initial hydrogel structure but also on the rate and temperature of drying. The number of silanol groups (silanol density) on the material active surface is determined by the same parameters. The xerogel pores contain some water that is retained by strong adsorption forces. The silanol groups have weak acidic properties, which are slightly different from group to group depending on whether the groups are isolated, vicinal, or geminal. Schematic formulas for different types of silanols are shown in Fig. 6.1.2.

Silanol groups present at the surface of the silica particle play a major role in the use of silica gel as a solid support for the stationary phase in chromatography. Bare amorphous silica can be used in direct-phase chromatography where a layer of water molecules adsorbed on the solid surface acts as a stationary phase. The silanol groups are active in the separation

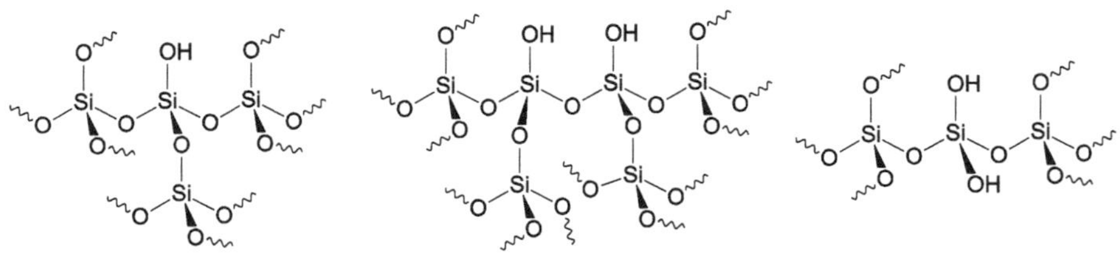

FIGURE 6.1.2 Schematic formulas for isolated, vicinal, and germinal silanols (the bond ∼∼ is not ∼∼H).

process, influencing this layer of adsorbed water. Direct adsorption on the silanol groups is also postulated as a potential separation mechanism. However, the main utilization of silica xerogels is related to the role of silanol groups as the place where other structural groups are introduced by derivatization in order to obtain modified silica for reversed-phase, chiral, hydrophilic, ionic exchange, and other type of stationary phases. The main property used for these derivatizations is the acidic character of −OH groups, which allows them to react with different derivatization reagents. Different silica gels have various surface characteristics regarding the density of silanol groups, and these can be derivatized to get functionalities attached to the surface. In general, a higher concentration of reactive hydroxyl groups is present in smaller-pore silicas, typically with a pore diameter less than 50 Å, whereas less reactive hydroxyls predominate on larger-pore silicas having pore diameters greater than 150 Å.

The pores considered for differentiating silicas are sometimes indicated as *mesopores*. Larger pores of 10−20 μm forming channels (through pores) with larger dimensions can also be present in silicas but they are in a small proportion. A common classification of silica gels is based on their average pore diameter and this differentiates type A (fine pores), type B (average pores), and type C (large pores). The main characteristics of these silicas are given in Table 6.1.1.

Silicas with characteristic values outside those indicated in Table 6.1.1 are also utilized in practice, and they still can be classified in the same types depending on the closeness of their characteristics to those indicated in the table. Type B (pore size) silicas are the most commonly used in HPLC, but type A and type C silicas are also used for specific purposes. The silicas for chromatographic use are available under various tradenames (Davisil, Astrosil, Lichrospher, Kromasil, etc.).

Other classifications of commercially available silica gels (e.g., Davisil grades) differentiate silicas based on both their pore size (22, 30, 60, and 150 Å) as well as on the size of particles [16].

Another important characteristic of silica solid supports is their purity. As previously indicated, traces of metallic ions, such as Al^{3+}, Fe^{3+}, and Ni^{2+}, are undesirable in the hydrated silica. To avoide the impurity problem, very high-purity silica has been produced, with a less acidic surface and a more homogeneous distribution of silanol groups. This type of silica is indicated as *type B*, while lower-purity silica that has a higher metal content and more acidity is termed *type A* silica.

The pH range where typical silica is stable is between pH = 2 and 8. The pH values of the mobile phase outside this range damage common stationary phases with silica support. For this reason, a number of attempts were made to enlarge the range of pH stability. One such attempt is based on the use of hybrid organic−inorganic materials still having siloxane group

TABLE 6.1.1 Typical Classification of Silica Gels Based on Their Pore Size

Property	Fine Pores	Average Pores (Common in HPLC)	Large Pore
Aspect	Transparent	Semitransparent	Milky
Average pore diameter (nm)	2.0−3.0	4.0−10.0	80−125
Pore volume (mL/g)	0.35−0.45	0.6−0.9	0.7−1.0
Surface area (m^2/g)	650−800	450−700	300−400

structures but also containing organic groups that provide some protection to the access of mobile phase (excess of OH^- or H^+ ions) to the backbone of the stationary phase. The result is a silica-base material that remains amenable to the derivatization necessary to make a bonded phase, but also having attached some organic fragments. One such material is obtained using the following type of reaction:

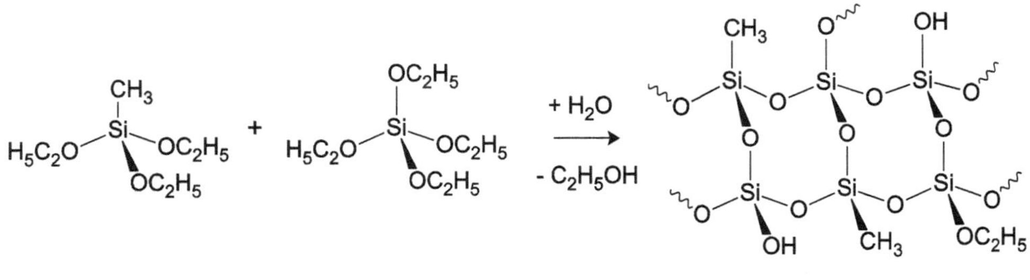

(6.1.5)

Materials with even better stability than those having CH_3 groups in the silica were obtained using ethylene bridged structures (indicated as BEH technology by Waters or TWIN technology or EVO by Phenomenex) [17] or propylene bridged materials. This type of silica-based support is prepared in two steps. In the first step, a polyethoxy-siloxane oil phase is prepared from hydrolytic condensation in acidic conditions of bis(triethoxysilyl)ethane and tetraethoxysilane using a small amount of water. Highly spherical porous hybrid particles are then prepared by further condensation under alkaline conditions in an oil-in-water emulsion. The first step of these reactions can be schematically written as follows:

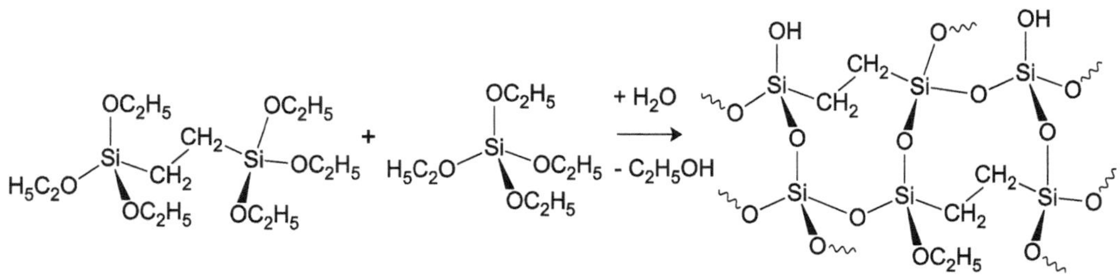

(6.1.6)

The resulting polyethoxysilane generates after further hydrolysis the desired material that can be derivatized to form the bonded phase. Once obtained, the silica-based organic/inorganic material is subjected to surface-ripening steps similar to those used for silica. Significant improvements at both low pH and high pH (up to pH = 12.0) were reported for stationary phases obtained with organic/inorganic phases. Hydrophobic columns (C18 or C8) obtained by the derivatization of silica with ethylene-bridged phases also have better behavior regarding wettability.

The combination of inorganic support with organic bound groups capable of further polymerization can also be achieved by reacting the silica surface with reagents such as vinyltriethoxysilane or methacryloxypropyltriethoxysilane. In this case, a reactive group capable of further polymerization is attached to the silica surface, allowing a better shielding of the possible free silanol groups from the silica surface. Other procedures for covering the silica-base material with an organic/inorganic layer are utilized for providing support materials with wide range of pH stability (e.g., in Kromasil Eternity columns that have an extended pH range up to 12).

Other modifications of the silica supports were reported, including a controlled surface charge procedure (CSH technology) [18]. This technology takes advantage of the fact that the silica surface is usually slightly negatively charged due to the dissociation of silanols. This charge can be neutralized by adding specific reagents such that the surface reactivity is decreased.

The silanol groups from the silica surface play an essential role in making the silica support amenable for binding the stationary phase. The amount of silanol groups on the silica surface varies depending on silica preparation, but values in the range 7–8 μmol/m^2 are common. Related to the amount of silica, the amount of silanol groups δ_{OH} (in mM/g) is in the range of 3–7 mM OH groups per g of SiO_2. The number of OH groups per unit mass of silica depends linearly on surface area of the material following an expression of the following form [19]:

$$\alpha_{OH} = 602.214 \frac{\delta_{OH}}{S_{surf}} \tag{6.1.7}$$

In expression (6.1.7), S_{surf} is the surface in m^2/g and α_{OH} is the number of OH groups per unit surface area (nm^2). The value of α_{OH} was found to vary between 4.1 and 5.6 OH groups per nm^2 (with an average of 4.9 OH groups per nm^2).

The binding of desired functionalities to the silica surface is usually performed using derivatization reactions, and the derivatization can be done using a variety of reagents, depending on the nature of the desired type of stationary phase. After binding the desired moieties on the silica surface, a large number of unreacted –OH groups (silanol groups) still remain present on the silica surface. These groups may not be desirable, for example, when a totally hydrophobic stationary phase is desired. For this reason, several procedures were developed to react with the remaining active silanol groups after the initial derivatization. The most common procedure to "block" the remaining silanol groups is the reaction indicated as end-capping. The end-capping is achieved by blocking as many as possible from the remaining –OH groups with small hydrophobic moieties such as $-Si(CH_3)_3$. The attachment of large fragments, such as octadecyl, to the silanol groups has a limited yield due to steric hindrances, while the attachment of small groups, such as $-CH_3$, is more efficient.

The presence of silanol groups on the surface of silica-based stationary phases even after derivatization and end-capping prompted continuous effort to generate new stationary phases that do not have this problem. One possibility to solve this problem is the use of hydride silica as support for further derivatization [20]. Silica hydride support material is sometimes indicated as type C silica. The basic chemical reaction for this process is the following:

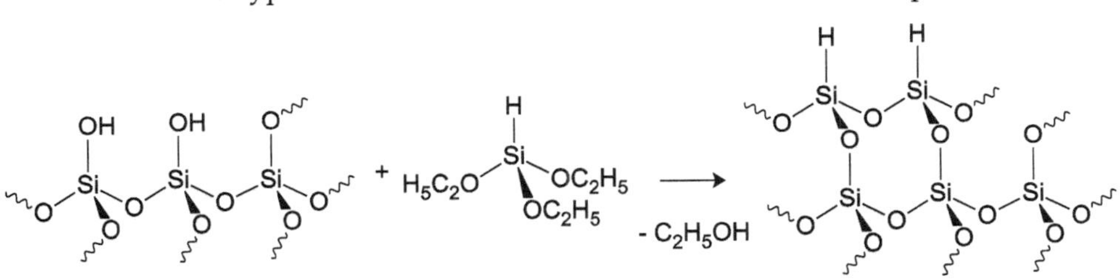

(6.1.8)

Other procedures for the synthesis of silica hydride materials are known, such as reduction with $LiAlH_4$ of silica having chlorine bonded groups [20].

The hydride silica can be further derivatized with specific moieties that act as active stationary phase, or can be used as is, as a stationary phase for certain applications. Silica surface covered with Si–H groups is sometimes indicated as *type C* silica [21], and should not be confused with pore size type C. Hydrated silica is also the most common material used for the monolithic columns [22].

Silica-Based Monolithic Chromatographic Columns

Monolithic stationary phases are made from a single piece (rod) of a solid porous material. Modified silica rods are probably the most common type of monolithic columns, but organic polymers are also used as monoliths. The monolithic rods are prepared by a polymerization or polycondensation process, either in situ in a column tube, such as in glass tubes or fused silica capillaries, or in a column mold. In situ preparation of monolithic columns has the advantage that no further encapsulation of the porous monolith in a tube resistant to the pressure of the solvent is needed. However, this approach is not compatible with monolithic silica columns having larger diameters (4.6 mm or more) due to the shrinkage that occurs during the sol–gel preparation process. When the monolithic silica rod is made in a mold, it has to be further clad with a suitable material such as PEEK, to which the column end fittings can be attached for use in the HPLC process [23].

For the production of silica-based monoliths, the basic sol–gel process is similar to the preparation of solid supports for porous silica materials. The process is conducted as a sequential hydrolysis followed by a polycondensation of silane derivatives. Similarly to the preparation of porous silica particles, either hydrogels or alcogels can be obtained. Used for this purpose are compounds such as tetraethoxysilane (TEOS), or tetramethoxysilane (TMOS) in aqueous acid or basic medium with an appropriate porogenic solvent [e.g., polyethyleneglycol (PEG)]. The hydrolysis conditions and the choice of the porogenic solvent are essential for obtaining mesopores and macropores [24,25]. Monoliths having the surface covered with the desired functionalities can be directly obtained by the hydrolysis of the appropriate compounds such as trimethoxyoctylsilane or trimethoxyoctadecylsilane.

Monoliths have a porous structure characterized by mesopores (pores between 2 and 50 nm in diameter) and macropores (about 4000 to 20,000 nm in diameter), with silica skeleton of approximately 1–2 μm thick, and a void volume of almost 80% of the entire column volume. These pores provide monoliths with high permeability, a large number of channels (the macropores are typically interconnected, the reason for which they are also indicated as through pores), and a high surface area (generated mainly by mesopores). The backbone of the monolithic column can easily be chemically derivatized for specific applications. Monolithic columns show an efficiency equivalent to 3–5 μm i.d. silica particles, but with a 30–40% lower pressure drop. The flow through monolithic channels is closer to laminar, and thus they allow less eddy diffusion. The transport of an analyte through the monolithic bed is based mainly on perfusion, and in the monolithic media with a low proportion of mesopores, the analyte diffusion into and out of the pores does not significantly contribute to the band broadening. Due to the reduced diffusion in solute transfer through the monolithic bed, the peak efficiency does not decrease as sharply as in particle-packed columns when increasing the linear flow rate of the mobile phase. This allows the application of higher flow rates and shorter analysis time, without a significant loss of plate numbers [26–28].

Monoliths having the surface covered with the desired functionalities can be obtained not only by derivatization of a silica monolith, but also by the hydrolysis of the appropriate compounds containing a desired group, such as trimethoxyoctylsilane or trimethoxyoctadecylsilane.

Core–Shell Particles in Packed Columns

A number of procedures are used to obtain core–shell particles. The core and the shell can be made of different materials although silica is the most commonly used for both the core and the shell, with the difference that the core is made of fused silica while the shell is made from porous silica. The core is usually a single sphere but it can be an aggregation of several small spheres. The size of the core particle, the shell thickness and the porosity in the shell are tuned in order to suit different types of chromatographic applications [29]. The particles can be synthesized by a two-step or by a multiple-step process. The core particles are synthesized first (e.g., as uniformly sized fused silica microspheres) and the shell is then added based on an electrostatic attraction procedure [30]. The thickness of the shell can be controlled by the number of deposition cycles that are applied to the core microspheres [29].

The porosity of the shell can vary, and medium-size porosity (80–100 Å) utilized for the separation of nonpolymeric molecules [31] as well as larger-size porosity (150–300 Å) used for separating larger molecules such as peptides and small proteins ($Mw < 15$ kDa) are available [32]. Larger superficially porous particles with a pore size of 400 Å or more allow large molecules ($Mw < 500$ kDa) unrestricted access to the phase and can be selected for protein separations [33].

Derivatization of Silica Solid Support

A variety of stationary phases with a range of polarities can be obtained by the derivatization of a hydrated silica solid support. The widespread use of silica as a solid support was previously described as being explained by the reactivity of silanol groups (–OH), the mechanical resilience of silica, the possibility to obtain silica with a high porosity (large surface area), the possibility to control the shape and dimensions of silica particles, and the relative neutrality of the silicon atoms. A number of aspects related to the derivatization of silica support are important in a column selection including the following: (1) the type of bonded phase, (2) the type of reagent used for derivatization, (3) the vertical or horizontal type of derivatization, (4) the extent of derivatization, (5) end-capping, (6) other derivatization possibilities, and (7) special derivatization reactions.

1) The type of bonded phase refers to the actual fragments that are bonded to the silica backbone. The bonded phase obtained by derivatization can have a hydrophobic character, a polar character, even an ionic character, or a mixed-type character (e.g., polar and nonpolar, or nonpolar and ionic). Various types of bonded phases are further discussed in more detail

in Sections 7.4, 8.3, and 9.3. Groups such as octyl (C8) and octadecyl (C18) are very common for making hydrophobic phases, although other groups with hydrophobic character, such as phenyl, may be attached to the silica backbone. For partially polar phases but still with some hydrophobic character, groups such as cyanopropyl or hydrophobic chains with polar embedded groups (e.g., −O−) can be synthesized. Groups with a polar character or ionic character can also be bonded to the silica backbone typically using a short hydrophobic "handle" consisting, for example, of two or three aliphatic carbon chains. Also, reactive groups can also be connected to the silica backbone through a short hydrophobic handle, making a material that can be further derivatized in another step. Among the groups utilized for second connection to the derivatized silica are the following: glycidyl, amino ($-NH_2$), azide ($-N_3$), isothiocyanate (−NCS).

2) Numerous types of reagents were used for hydrated silica derivatization. One group of reagents commonly used for derivatization consists of monofunctional silanization reagents (e.g., a chlorodimethylalkylsilane). This type of reagent allows the attachment of a silyl ligand R (such as dimethyloctadecylsilyl) to the silica backbone. The reaction is schematically described as follows:

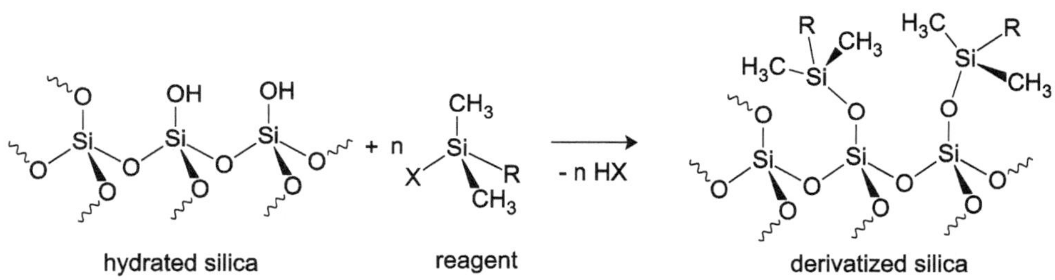

(6.1.9)

The reactive substituent X can be Cl, but also OCH_3, OC_2H_5, or other groups such as NH_2. The substituent R is the one determining the type of bonded phase (C8, C18, amino, cyano, and many others). The derivatization of silica surface with a monofunctional silanization reagent is typically indicated as *monomeric functionalization*, and this kind of derivatization leads to *brush* phases. Monomeric functionalization attaches a monomolecular layer of silyl ligands that form the active stationary phase on the silica surface [34]. When the derivatization reagent is a chlorosilane, the presence in the reaction medium of organic bases (pyridine or triethylamine) is necessary to remove HCl formed as byproduct. When alkoxydimethyl-R-silane is used as a silanization reagent, toluene is a proper solvent for the reaction. In this case, the alcohol formed as reaction byproduct should be removed from the surface, because it can subsequently react with silanol groups leading to alkoxy derivatized silica surface, which is unstable toward aqueous mobile phases. Monomeric functionalization (brush phases) leaves a considerable number of unreacted silanol groups on the solid support.

Another possibility of derivatization, which has the capability to involve in the reaction two or even three silanols from the silica surface, is the use of *di- or trifunctional reagents*. The hindrance in the reactions of Si−OH with reagents having larger R groups may be more frequent for silanols that are close to each other, such as vicinal ones. In monomeric functionalization a large number of such silanols remain underivatized. The reagents having two or three reactive functional groups such as ethoxy or chloro have a higher capability to react with vicinal silanols, as shown in the following schematic reaction:

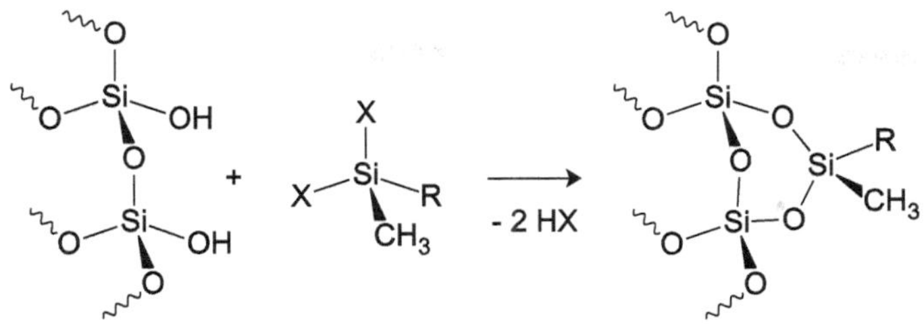

(6.1.10)

For trifunctional reagents the possibility to have all three reactive substituents connected to the silica surface is also possible. However, both difunctional and trifunctional reagents may react at only one point with the silica surface, and the attached silyl ligands still have reactive groups such that the treated surface retains additional capability for further reactions. A schematic description of a derivatization that leads to a surface containing reactive groups is shown below for a trifunctional triethoxysilane:

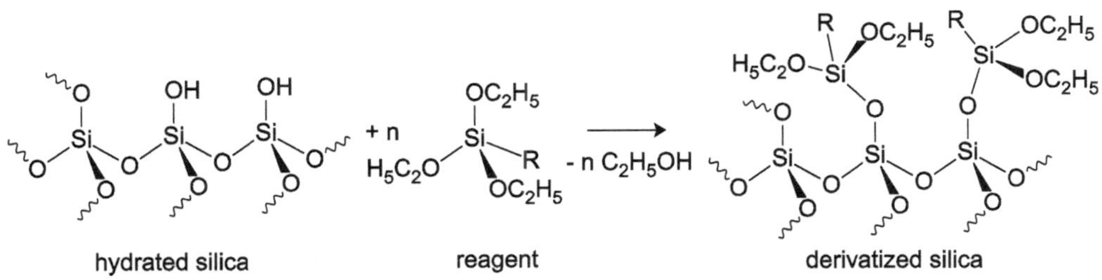

(6.1.11)

In Eq. (6.1.11) it is shown that the resulting derivatized silica still contains a number of reactive groups $-OC_2H_5$.

3) The remaining reactive groups on the derivatized silica with di- or trifunctional reagents can be used for generating one layer of bonded phase or more layers (up to five), depending on derivatization conditions. The two types of derivatization are indicated as *horizontal polymerization* and *vertical polymerization* [35]. Horizontal derivatization is sometimes indicated as generating an *oligomeric* phase, while vertical polymerization generates a *bulk* phase.

In horizontal polymerization, the reactive groups of di- or trifunctional reagent are intended to react completely with the silica surface. For this purpose, the reaction medium is kept anhydrous, and because long substituents *R* on the reagent will lead to steric hindrance, a mixture of *R* groups including a desired one (e.g., octyl) and a short one (e.g., propyl) are used for derivatization. Also, higher temperatures may be used for increasing the yield of derivatization. The result is schematically shown in Fig. 6.1.3.

If water is present in the derivatization medium, a different type of derivatization takes place, indicated as *vertical polymerization*. Vertical polymerization generates a *bulk* phase and takes place with a three-step sequence. In the first step the di- or trifunctional reagent reacts with silica to link trialkylethoxysilyl or dialkylchlorosilyl groups as shown in Eq. (6.1.11). In

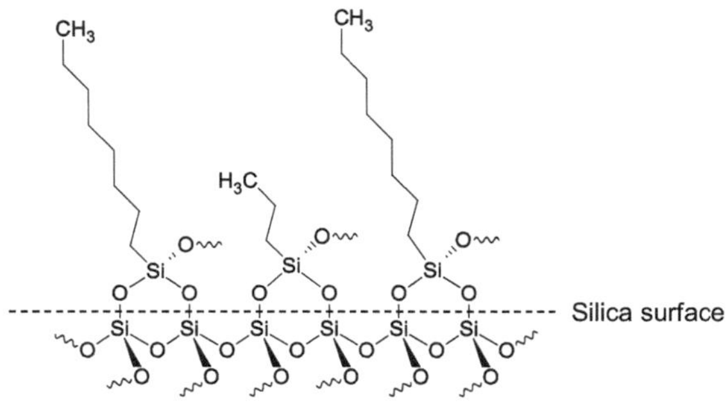

FIGURE 6.1.3 Schematic illustration of a horizontal polymerization showing octyl and propyl groups on silica surface.

the second step, the product is hydrolyzed in order to substitute the chlorine atom or the ethoxy groups with Si–OH groups as shown in the following reaction (for ethoxy groups):

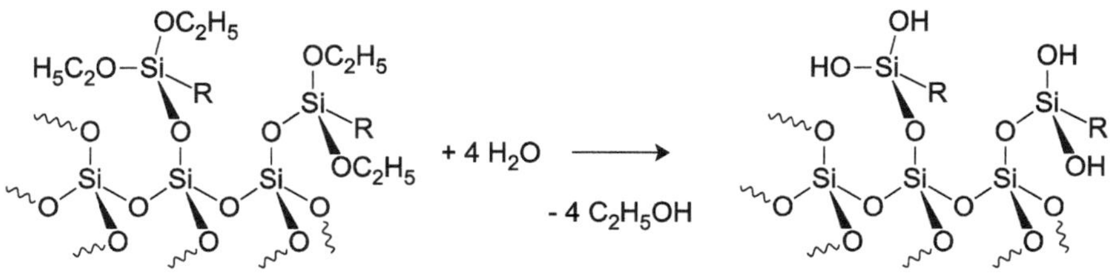

(6.1.12)

The third step is subjecting the hydrolyzed surface to another derivatization reaction with difunctional or trifunctional silane as shown in the following reaction:

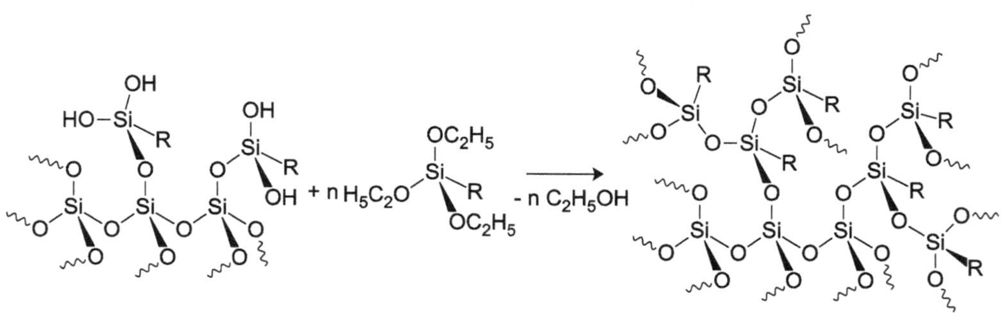

(6.1.13)

This sequence is repeated several times (8–10 times). The result is a thick layer of the bonded phase on the silica surface. The alkyl chains form an ordered packed bonded phase on the silica surface. A schematic illustration of the structure of a vertical polymerization process is shown in Fig. 6.1.4.

New stationary phases indicated as polar embedded can offer a more complex interaction with the analytes, and they can be synthesized using silanes having the polar group in the chain of one of the substituents. For example, a phase containing an ether group in a

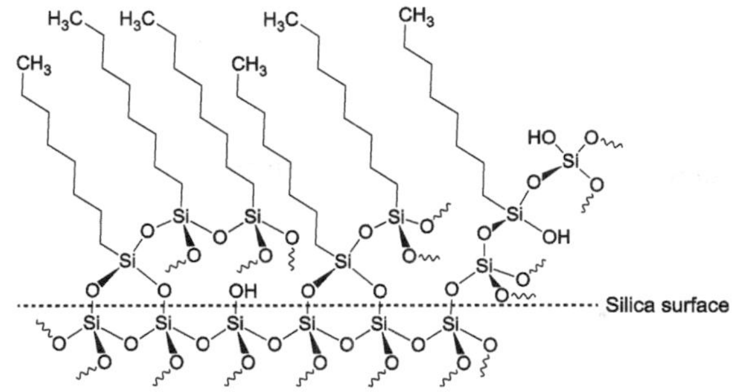

FIGURE 6.1.4 Schematic illustration of a vertical polymerization showing the octyl groups on silica.

hydrocarbon chain can be obtained by reaction of hydrated silica with triethoxy(propoxyoctyl)silane. Similarly, groups such as amide or carbamate can be imbedded in a hydrophobic phase. Such phases can also be obtained using two derivatization steps. For example, the hydrophobic phases that have an amide embedded group can be obtained by an initial derivatization of hydrated silica with a silane containing a propylamino group, followed by a secondary derivatization with an acid chloride. Such a reaction is schematically indicated below.

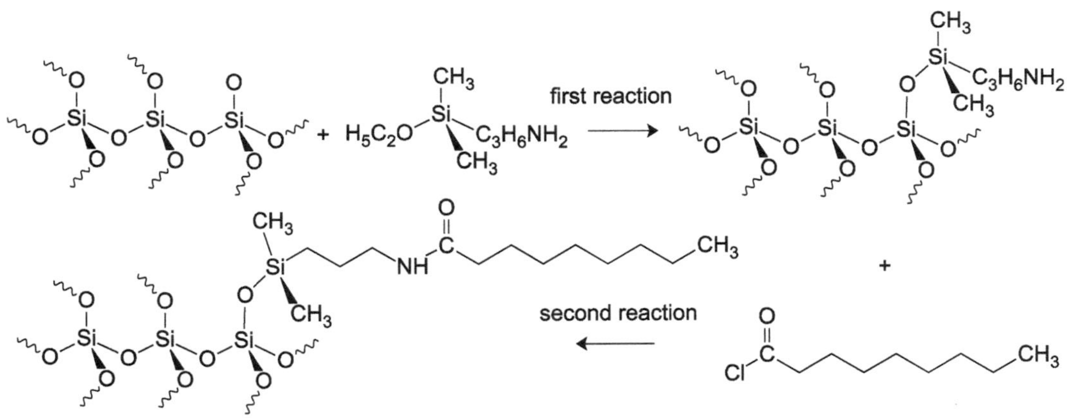

(6.1.14)

Other reactions with a silane reagent containing a short hydrocarbon "handle" such as propyl connected to a reactive group followed by a second reaction were used to produce a variety of bonded phases (see, e.g., Ref. [16]). Besides double reactions, polar embedded phases can also be obtained by single-step derivatization with appropriate reagents. Such reagents are silanes with a polar embedded functionality in a hydrophobic chain.

4) The extent of derivatization is an important parameter for column selection. Silica offers a complex surface containing macropores, mezopores, and also micropores. Pore surface area accessibility is key to effective derivatization of silica. Several parameters are utilized to describe the derivatization extent. One such parameter applied for hydrophobic phases (e.g., C8 or C18) is the carbon load (C%) that can be determined from elemental analysis of

the phase. Values of carbon load can vary between about 5% up to 25%. The carbon load is related to the number of micromoles of silyl ligands attached to the OH groups, which also characterize the extent of derivatization and is obtained with the following formula:

$$\delta_{\text{ligand}} = \frac{10^6 C\%}{(C_{\text{sil}} - C\%) Mw_{\text{sil}} S_{\text{surf}}} \tag{6.1.15}$$

In expression (6.1.15), δ_{ligand} is the amount in μmol of silyl ligands per m^2, $C\%$ is the carbon load, C_{sil} is the mole % of carbon in the silyl ligand, Mw_{sil} is the molecular weight of the silyl ligand, and S_{surf} is the surface area of the silica [36]. The value of δ_{ligand} for a ligand such as dimethyloctadecylsilyl is $\delta_{\text{ligand}} \approx 5.6\ \mu\text{mol/m}^2$ (for $C\% = 20\%$, $C_{\text{sil}} = 77.4\%$, $Mw_{\text{sil}} = 310$, and $S_{\text{surf}} = 200\ \text{m}^2/\text{g}$). The amount of silanol groups of 7–8 μmol OH/m^2 is not very different from that of the ligand amount. However, the value of 5.6 $\mu\text{mol/m}^2$ corresponds to a high $C\%$ load typical for stationary phases obtained for vertical polymerization. Because in vertical polymerization up to five alkylsilane ligands can connect to each other and not to the silica backbone, it results that a considerable number of silanol groups remain underivatized.

5) Since in both monomeric functionalized silica as well as in horizontal and vertical derivatized surfaces with di- and trifunctional reagents a significant number of free silanol groups remain on the silica backbone, attempts were made to reduce this number. Silanols affect column properties by generating polar interactions with the analytes. The free silanol groups can participate in the retention mechanism, generating tailing for compounds having besides hydrophobic moieties some polar groups (for example amino compounds). These silanol groups may be intended to exist in particular types of columns, but, frequently, a further derivatization is practiced to reduce the silanol number. This is performed by a separate procedure of blocking residual silanols by a subsequent derivatization with small groups such as trimethylsilyl to avoid steric hindrance. This is performed using as reagent trimethylchlorosilane or hexamethyldisilazane. Repeated end-capping operations are not uncommon. Besides trimethylchlorosilane or hexamethyldisilazane, difunctional derivatization reagents with small R groups are also used for end-capping. Such reactions are typically performed at elevated temperatures to achieve a derivatization as complete as possible [37–39]. By end-capping the carbon load $C\%$ of packing material does not significantly change, and thus its hydrophobic character for the reversed-phase phases is kept almost constant. End-capping with small polar groups is also possible for generating a "polar end-capped" phase, which offers additional interactions with the analytes.

6) Besides the reaction of hydrated silica with silanization reagents, such as chlorodimethylalkyl silane, other reactions were used for binding the desired moiety to the silica surface. For example, the silane groups on the silica surface can react with thionyl chloride and be replaced with chloride groups. The chloride groups can further react with a Grignard reagent to attach to the silica the desired moiety. Also, silica hydride can be used as solid support undergoing further reaction with an unsaturated hydrocarbon (containing either a double or a triple bond) in the presence of a catalyst. The resulting bonded phases do not have unreacted silanols that can produce undesirable interactions with the analytes.

7) Special derivatization reactions are necessary for generating stationary phases with particular applications. Among these derivatizations are those designed for producing columns for chiral separations, or those used in affinity and immunoaffinity chromatography.

Some of these columns have a porous silica support, and others have an organic polymeric support. For silica support special stationary phases, the typical procedure is to derivatize silica with a reagent having a short hydrocarbon handle ending with a reactive group such as NH_2, CHO, CN. Once bonded to the silica support, the reactive group is further utilized to bind the desired functionalities that act as "bonded phase." Specific examples of such phases are further discussed in Section 7.1.

Stationary Phases With Inorganic Support Different From Hydrated Silica

Besides hydrated silica, other hydrated oxides with a large surface may be used as support for the stationary phase, and can be derivatized to generate a bonded phase. One such material is hydrated zirconia ($ZrO_2 \times H_2O$). Some challenges in using hydrated zirconia are related to the fact that the OH groups bound to zirconium atoms are much less acidic as compared to the silanol groups (although still with the capability to act as a Brönsted acid). The lower acidity of zirconia makes this phase resilient to a wider pH range, between $pH = 1$ to about $pH = 10$, but it is less reactive as compared to hydrated silica. In addition, the zirconia surface may act as a Brönsted base, and also as a Lewis acid due to the *d* electrons of zirconium atoms. This may generate additional interactions with various ligands and broadening of the chromatographic peaks. Other hydrated oxides considered for solid support were hydrated Al_2O_3, hydrated ThO_2, and hydrated CeO_2. However, their use is not common.

Organic Polymers as Stationary Phases

Organic polymers are used in various forms for obtaining stationary phases in HPLC. The polymers can be used as small particles, similar to those obtained from hydrated silica, or can be made in monoliths. Also, some polymers can be used as stationary phase as synthesized, or can be used as a solid support that is further derivatized for obtaining the active stationary phase. For some specific types of chromatography, such as size exclusion or ion exchange, the use of organic polymers as stationary phase is very common. For other types of HPLC, such as reversed phase, the polymers as stationary phase play a less important role.

Perfusion particles of polymers usually contain very large pores (e.g., 600–1000 nm diameter) connected to a network of smaller pores (30–100 nm diameter). The flow of mobile phase through the column filled with perfusion particles takes place through large pores, with a diffusion process through the smaller pores. The size of the polymer particles is important to control (similar to the case of silica particles). This can be achieved using monosized polymer seed particles [40].

One of the most common polymeric supports used as a stationary phase is polystyrene crosslinked with divinylbenzene (PS-DVB). This material is obtained by suspension polymerization using a two-phase organic/aqueous system. The crosslinking polymerization is performed in the presence of inert diluents that are miscible with the starting monomers but must not dissolve in the aqueous phase. Submicrometer particles (microbeads) form as the styrene-divinylbenzene polymerizes and precipitates out of solution. The formed microbeads fuse together to form macroporous particles. Initially a network of microporosity may be present in the microbeads, and polymerization conditions must be controlled to minimize this type of porosity because it results in a soft polymer that has a poor mechanical strength and high propensity of swelling. After the crosslinked PS-DVB porous particles are formed, any residual reactants, diluents, and surfactants must be removed by thorough washing. Other polymers such as methacrylates and polyvinyl alcohol are also used as stationary phases. These organic polymers are used for both the preparation of particles, and also for the preparation of monolithic columns.

Stationary phases that are organic polymers and contain specific functionalities can be obtained either by derivatization or directly by the polymerization of a monomer having already attached the desired functionality. Stationary phases having C18, NH_2, or CN groups are commercially available for RP-HPLC, and those having COOH, SO_3H, NH_2, and NR_3^+ are commercially available for ion exchange chromatography.

A particular direction where effort was made to develop polymeric stationary phases is that of polymeric monoliths [41,42]. Monoliths were obtained from a variety of monomers, using a crosslinker or not. Initial monoliths were simply obtained from styrene and divinylbenzene that were copolymerized in a porogenic solvent and in the presence of a radical initiator. A new convenient procedure starts with glycidyl methacrylate, which is polymerized by UV, thermal, or γ-radiation initiation, in the presence of a crosslinker such as ethylenedimethacrylate, and using a porogenic solvent and an initiator (e.g. 2,2′-azobis-isobutyronitrile).

A special type of polymeric stationary phase is that coated or bonded to silica. The large surface area and mechanical strength of silica make this material better as support than organic polymers. However, the property of the polymer may be desired in some applications such as the resistance to a wider pH range than silica, or a specific chiral property such as that of cellulose or amylose. In such cases, the polymer may be simply coated on the silica surface or bonded through covalent bonds using, for example, amino or glycidyl groups present on a prederivatized silica. In such processes, preserving the silica porosity is important [43].

Study of Physicochemical Characteristics of a Stationary Phase

A prepared stationary phase can be the subject of an array of analyses for assessing its quality. These analyses may address the physical properties of stationary phase such as specific surface area, size and shape of the particles, volume and size distribution of the pores, etc. Also it may address the surface chemistry of the packing materials such as nature of bonded phase, its density on the surface, the concentration, conformation, and mobility of the organic attached groups, the content in specific elements in the bonded phase (e.g., carbon load %, nitrogen load %), metallic impurities in the stationary phase rigid support, etc. The applied analyses include a variety of techniques such as various absorption studies, elemental analysis by combustion, inductively coupled plasma–atomic emission spectrometry (ICP-AES), scanning electron microscopy (SEM), infrared spectroscopy (IR), diffuse reflectance infrared Fourier transform spectrometry (DRIFTS), nuclear magnetic resonance (NMR), fluorescence spectroscopy, and thermogravimetric analysis (TGA). Dedicated literature describes in detail many procedures for stationary phase control and evaluation (see, e.g., Refs. [13,16,44–51]).

6.2 COLUMN PROPERTIES AFFECTING SEPARATION

The presentation about HPLC column construction (addressed in Section 6.1) provides some information regarding the physical and chemical properties of the chromatographic column. These properties are key factors regarding column performance. For this reason, the same properties are the basis for the classification of HPLC columns in specific types. A summary description of the aspects involved in column construction, which is important at the same time for the column performance, is schematically shown in Fig. 6.2.1.

Because of the importance of the chromatographic column in HPLC analysis, different classes of columns are discussed in more detail in this book. This section presents only a summary regarding stationary phase and column classification based on their physical and chemical properties.

Dimensions of the Column Body Affecting Separation

The dimensions of the chromatographic column (length and internal diameter) may vary considerably. Related to their dimensions the columns are placed in several groups as indicated in Table 6.2.1.

A discussion about preparative and semipreparative columns is beyond the purpose of the present book. Micro- and nano-LC-capillary columns are utilized for analytical purposes, but specific instrumentation is necessary for using them, such as syringe pumps and dedicated small-volume cell detectors. Specific issues are related to such columns, and they are not very common in routine analytical HPLC practice. For these reasons, the main interest in

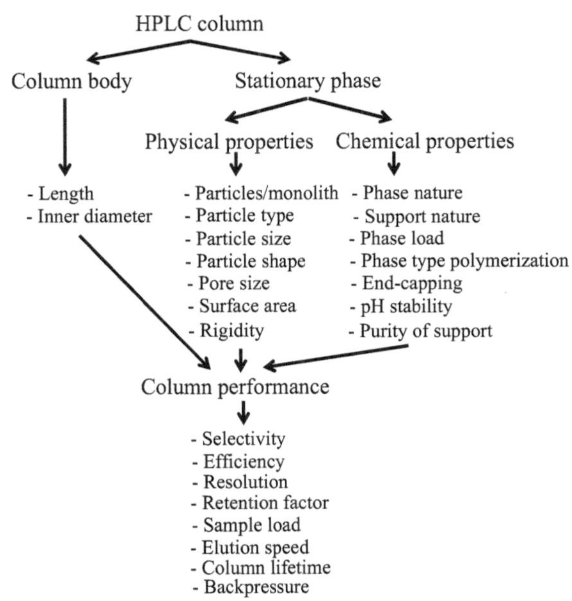

FIGURE 6.2.1 Aspects of column construction and their impact on column performance.

TABLE 6.2.1 Classification of HPLC Columns Based on Their Dimensions

Type	Inner Diameter (mm)	Length (mm)	Typical Flow Rate (mL/min)	Sample Loading
Preparative (not industrial)	>25	300, larger	>20	>25 mg
Semipreparative	10	250, larger	5–10	10–20 mg
Analytical conventional	3, 4.6	50, 100, 150, 250	0.5–2	50–200 μg
Analytical narrowbore	2, 2.1	50, 100, 150, 250	0.2–0.5	20–100 μg
Analytical microbore	1, 1.7	50, 100	0.05–0.1	<5 μg
Micro LC-capillary	<0.5	50,100	1–10 μL/min	1 μg
Nano LC-capillary	<0.1	50	<1 μL/min	<0.1 μg

HPLC practice is in the conventional, narrowbore, and microbore analytical columns, which will be further discussed.

Column length and diameter are important parameters related to several chromatographic characteristics of a chromatographic separation. Among the chromatographic parameters affected by the column dimensions are the following: (1) retention time t_R, (2) separation run time, (3) column efficiency (number of theoretical plates N), (4) column resolution R, (5) maximum sample load in the column, (6) column backpressure, and (7) column stability in time.

1) The retention time t_R is affected by column dimensions. The dead time t_0 for a chromatographic column is given by expression (4.2.3), which shows that t_0 is directly proportional with the column length L. The linear flow rate u in the chromatographic column is related to the volumetric flow rate U by expression (4.2.1), where parameter A_c is the area of the channel in which the flow takes place and $A_c = \varepsilon^*(\pi/4)d^2$ (d is the column inner diameter and ε^* is a constant depending on column porosity, e.g., $\varepsilon^* \approx 0.64 \times 10^{-3}$). Therefore, the following expression can be written:

$$u = \frac{4U}{\varepsilon^* \pi d^2} \tag{6.2.1}$$

Since U is the value set by the operator of the HPLC system, the following expression gives the dependence of t_0 on L, d, and the set volumetric flow rate:

$$t_0 = \frac{\varepsilon^* \pi d^2 L}{4U} \tag{6.2.2}$$

Expression (6.2.2) indicates that the column dead time is directly proportional with the square of the column diameter and the column length. The retention time $t_R(X)$ for a compound X is related to t_0 by expression (6.1.1), and therefore the following expression can be written:

$$t_R(X) = \frac{\varepsilon^* \pi d^2 L}{4U}[1 + K(X)\Psi] \tag{6.2.3}$$

Expression (6.2.3) shows how the retention time t_R of a compound changes when different column dimensions (d and L) are selected for the same flow rate U. (Other parameters in expression (6.2.3) are ε^* that depends on physical properties of the stationary phase, $K(X)$ that depends on the analyte and the chemical nature of the stationary and mobile phase, and Ψ that depends mainly on the stationary phase structure.) When no change is made to an HPLC separation except the column dimensions, expression (6.2.3) shows that the retention time of any analyte is directly proportional with the length and with the square of the internal diameter of the column.

In theory, the retention factor $k' = K(X)\Psi$ is not affected by the physical dimensions of the column because both t_0 and t_R are changing simultaneously (see expression (4.2.16)). However, the change in the column dimensions may affect (although not very much) the value of Ψ by small changes in the void spaces inside the column and slight increase in the volume V_0 of the mobile phase. Overall, the separations can be slightly better with a column with a larger diameter because of a lower linear flow u for the same U and a value closer to the optimum for the theoretical plate number of the column (see expression (4.2.50)). However, on columns with larger diameter, longer run times are generated and the peaks are broadening. Assuming broadening exclusively caused by diffusion (which is not the real case) based on Eqs. (4.2.24) and (4.2.35), the peak broadening (in time) W_b is given by the formula:

$$W_b = 4\,\sigma_t = 4\frac{\sqrt{2Dt_R}}{L}t_R \tag{6.2.4}$$

Expression (6.2.4) shows that the peaks are broadening for longer retention times. Since the resolution of the separation depends on N and not on W_b, and the number of theoretical plates N of the column remains the same if L is not changed, the resolution is not modified by using a wider column. However, broad peak integration may be performed with less accuracy, and method sensitivity may be affected (see Fig. 6.3.1 with smaller peak height for broader peaks at the same sample amount). For this reason, if other requirements are satisfied, a narrower column may offer some advantages.

2) The run time of the chromatogram, which is a time slightly longer than the retention time of the last peak in the chromatogram, is directly related to expression (6.2.3). The expression shows that for achieving shorter separation run times, shorter and narrower columns are recommended.

3) Column efficiency is also related to physical dimensions of a column. Expression (4.2.39) shows that the theoretical plate number N is directly proportional with the column length L. Therefore, longer columns have a larger number of theoretical plates. The height equivalent to a theoretical plate H is not affected by the column length. However, since the linear flow rate u changes with the column diameter the optimum volumetric flow rate U_{opt}, for which the minimum H value is obtained, depends on d. This value $u_{\text{opt}} = \sqrt{B/C}$ (where B and C are parameters in van Deemter Eq. (4.2.50)) will become for the volumetric flow rate U the following:

$$U_{\text{opt}} = \frac{\varepsilon^* \pi d^2}{4}\sqrt{B/C} \tag{6.2.5}$$

Expression (6.2.5) shows that changes in d will affect the optimum flow rate for obtaining a minimum H. For maintaining the same linear flow rate u in two columns, when the column diameter d_1 is changed to d_2, the following expression for the volumetric flow rates can be obtained from expression (6.2.1):

$$U_2 = \frac{d_2^2}{d_1^2} U_1 \tag{6.2.6}$$

For narrower columns (smaller d_2 than d_1), a lower volumetric flow rate U_2 is necessary in order to achieve the same linear flow rate u in the two columns. Since the height of theoretical plate H is dependent on u (see van Deemter Eq. (4.2.50)), the efficiency of the separation can be affected when changing columns with different diameters, even if all the other parameters are kept identical but the flow rate is not adjusted properly.

4) The resolution of the chromatographic column is also affected by column dimensions. Expression (4.2.61b) shows that the resolution (as measured for a compound X relative to previous peak) depends on a number of factors (nature of column stationary phase, nature of compounds to be separated, etc.) and also on the square root of the number of column theoretical plates N. Since N is directly proportional with the column length L, for the same compounds to be separated and identical columns except for their length, the variation of R can be described by the formula:

$$R = \text{Const.} L^{1/2} \tag{6.2.7}$$

Expression (6.2.7) indicates that longer columns offers better resolution, but only proportional to the square root of their length.

5) The sample load of the column refers to two different parameters: the maximum injection volume of the sample, and the maximum amount of sample to be injected. Formulas for the maximum volume of sample were developed (see expression (4.2.88)). These formulas show that larger columns allow larger injection volumes. Both column diameter and length influence the maximum injection volume, but the use of columns with larger diameter has more impact on volume load. Using the rough estimation for the theoretical plate height of a column as $H \approx 2d_p$ (d_p is diameter of particle) and using formula (4.2.39) for N, expression (4.2.88) becomes:

$$V_{\text{inj}} < 6 \times 10^{-4} d_p^{1/2} d^2 L^{1/2} (1 + k') \tag{6.2.8}$$

Expression (6.2.8) shows that V_{inj} depends on the second power of column diameter d and only on the square root of L. Columns with larger diameters can accommodate larger sample volumes. Too large an amount of sample injected in the chromatographic column may affect the peak shape, usually generating fronting (either steep front of the peak and shortened retention time, or peak less steep than the rear relative to the baseline due to nonideal equilibria in the chromatographic process).

The amount of sample injected is also limited by the amount of stationary phase but also by the nature of the stationary phase through its capacity to retain the solute. In analytical HPLC, the sample load is preferred to be low, but the load depends on the detector sensitivity, which limits the minimum amount of analyte that can be injected. Approximate formulas (see expression (4.2.89)) were developed for the estimation of maximum column load depending on column dimensions but also on other column and analyte characteristics (through the constant A). The problem of the need of a larger sample load occurs when a very low level of analyte is present in the sample. The analyte may sometimes be mixed with a large amount of matrix ("dirty" matrix) that must be separated. In such cases, a larger sample must be injected in the column to achieve an acceptable detection of the analyte, but at the same time the large amount of matrix must be accommodated by the column. For this reason, columns with more stationary phase (with larger diameter and length) are selected for "dirty" samples.

6) The column backpressure is also affected by the column dimensions. The acceptable value for the backpressure of a column when working in typical conditions is a very important parameter. This parameter is limited by the maximum possible pressure generated by the HPLC pumps, and also by the structure of the stationary phase, which may start collapsing above a specific pressure. The value of a column backpressure is determined by the particle size, the flow rate, the nature of the mobile phase, temperature, and the column dimension. A formula that governs backpressure in a packed bed column (difference between the pressure at the column inlet and that at the outlet) is the following (Darcy equation):

$$\Delta p = \frac{\eta u \phi_r L}{d_p^2} \tag{6.2.9}$$

In expression (6.2.9) η is the mobile phase viscosity, L is column length, d_p is the diameter of the particles in the bed, and ϕ_r is a column flow resistance factor. The replacement of linear flow rate u with the volumetric flow rate U leads to the equivalent expression:

$$\Delta p = \frac{4\eta U \phi_r L}{\varepsilon^* \pi d^2 d_p^2} \tag{6.2.10}$$

Expression (6.2.10) indicates the proportional increase in the backpressure of the column with the column length L. The backpressure also increases proportional to the decrease in the square of column diameter d. Also, backpressure depends on the mobile phase viscosity η, and since the viscosity of liquids typically decreases as the temperature increases, a higher temperature of the mobile phase may be used to lower the backpressure of the column.

7) The column stability in time can be estimated by the number of injections n_{inj} that can be made on a column with unaltered results for the chromatography. This number of injections depends on numerous factors besides the column dimension. The type of column, the nature of mobile phase, the type of sample, the nature of the matrix, the degree of clean-up performed on the sample before injection, the injection volume, and the sample concentration are factors influencing n_{inj}. In order to protect the columns from undesirable materials that may reach the stationary phase, filters and precolumns are commonly utilized in the sample path. The efficiency of the precolumn and the frequency with which this precolumn is changed also influence the number of injections n_{inj}. The only practical observation related to column dimensions is that the columns with larger amounts of stationary phase are in general more resilient to various factors that affect the column stability in time.

Physical Properties of Stationary Phase Affecting Separation

Chromatographic columns can be classified in specific types based on physical properties of their stationary phase. The selection of a chromatographic column from one of these types is very important for the success of an HPLC analysis since the physical properties of the stationary phase are closely related to column performance. Table 6.2.2 summarizes a classification of columns based on stationary phase physical properties.

From the classification described in Table 6.2.2, some stationary phase types are very common in HPLC columns. Other types, such as pellicular particles are not so frequently utilized. Also, the monolithic phases, although with a considerable number of applications, are much less common than phases made from particles packed in the chromatographic column. For this reason, the main discussion will be related to packed chromatographic columns. Also, different types of chromatography depend differently on physical properties of the stationary phase. For reversed phase (RP), HILIC, and ion exchange, the physical properties of stationary phase are important, but the chemical nature of the phase is the main component assuring the separation. For size-exclusion chromatography (SEC) the separation is based solely on the physical nature of the stationary phase, and therefore its role is even more important than in other chromatography types.

Some of the aspects connecting physical properties of the stationary phase with column performance were already indicated in Section 6.1 where the construction of an HPLC column is presented. It was shown that physical properties of the stationary phase are important parameters related to most chromatographic characteristics of a chromatographic separation. Among the chromatographic parameters affected by the stationary phase physical properties are the following: (1) retention time t_R, (2) retention factor k', (3) separation run time, (4)

TABLE 6.2.2 Classification of HPLC Columns Based on Stationary Phase Physical Properties

Stationary Phase Physical Type	Property
Phase consisting of packed particles	Stationary phase made from small particles
Monolithic phases	Stationary phase made of a porous solid rod
Porous particle	Phases made with porous particles frequently based on silica
Core–shell particles	Solid nonporous core surrounded by a porous outer shell
Pellicular particles	Solid spheres covered with a thin layer of stationary phase
Standard-diameter particles	3–5 μm diameter
Small particles	1.7–2.1 μm diameter
Uniform particles	Uniform spherical particles d_{90}/d_{10} close to 1 (<1.7–1.8)
Irregular particles	Irregular shape
Small/medium pore size particles	60–100 Å pore size
Large pore size particles	~300 Å pore size
Low/medium surface area particles	150–200 m^2/g surface area
Large surface area particles	~300 m^2/g surface area
Rigid phases particles	Rigid particles (withstand 600 to 1000 bar pressure)
Soft phases particles	Particles collapse above 300–400 bar (or even less)

column efficiency (value for N), (5) column resolution R, (6) maximum sample load in the column, (7) column backpressure, (8) results reproducibility, (9) column stability in time, and (10) utilization for the separation of small or large molecules.

1) The retention time is affected by several physical characteristics of stationary phase. Expression (6.1.1) shows that t_R depends on phase ration Ψ (defined by expression (4.2.15)) and larger values for phase ratio are related to larger retention times. Phase ratio can be different between porous and core–shell columns, by having a different amount of active stationary phase in the column. Also, larger surface area of the stationary phase may lead to a larger value for Ψ. Also, larger pores lead to smaller surface area of the stationary and therefore to lower Ψ. The columns with larger Ψ show larger retention times for the analytes. As a result, columns with fully porous particles and those with the stationary phase with a larger surface area, 200 to 300 m^2/g, typically have longer retention times for the analytes. However, core–shell particles also have a surface area around 150 m^2/g, such that the differences in the retention times are not necessarily very different. A direct comparison of columns regarding retention time capability is difficult to make since this characteristic depends on many other parameters.

An effect on retention time may also result from a combination column type plus flow rate in the column. Monolithic columns allow a reduction in elution time up to nine times due to the capability to use higher flow rates without having problems with the column backpressure. Due to a rapid mass transfer of the solutes between the bonded phase and the mobile phase, the decrease in the number of theoretical plates at higher flow rates, as predicted by van Deemter equation (see Eq. (4.2.50)), is not as intense for monolithic columns.

2) The value of retention factor k' is strongly influenced by the physical properties of the stationary phase. The value of Ψ is determined by several factors, including some related to the chemical properties of the stationary phase (phase load, for example, which may also be considered a physical property). The stationary phase surface area is directly related to the value for retention factor. Stationary phase is distributed on the surface of the solid support, and a larger surface area leads to a larger value for the volume V_{st} of stationary phase (see expression (4.2.15)). The same parameter affecting t_R also affects k' since t_0 is much less affected by the stationary phase surface area. The pore size of the silica support is related to the surface area and therefore to the retention factor of the stationary phase. Larger pores are associated with a smaller surface area of the solid support and consequently to smaller retention factor.

3) The longer retention times for the analytes result in longer run time for the chromatogram, and such longer times are typical for columns with porous particles or having larger surface area for the solid support of stationary phase.

4) Column efficiency (as expressed by theoretical plate height H or number of theoretical plates N) is a parameter highly dependent on physical characteristics of stationary phase, but also on some external factors such as how well the column is packed (with no void spaces). Expression (4.2.51) shows the linear dependence of H on particle size of packed chromatographic column. However, the formula does not take into account the shape of particles, which also influences significantly the value of H (and N). For core–shell particles, Van Deemter Eq. (4.2.50) is applicable, but expression (4.2.51) does not give a direct account for the dependence of H on total core–shell particle diameter d_p [52,53]. For typical packed columns, the efficiency per meter N for the column can be between 20,000 and 150,000, and for

core–shell columns it can be as high as 300,000. Some other stationary-phase physical properties influence the column efficiency, and there are differences between packed and monolithic columns. The variation of theoretical plate height H (as a function of the flow rate) for a monolith column versus several packed columns is shown in Fig. 6.2.2 [16]. The same figure shows the variation of H with the value for the flow rate for a C18 packed column having irregular particles, spherical particles of 5 μm diameter, or spherical particles of 3 μm diameter. The effect of particle shape and size on column efficiency indicates that irregular particles typically have a lower value for N (larger H) while spherical particles have higher values for N. As expected, smaller diameters for spherical particles lead to higher N values (lower H).

For many columns, the number of theoretical plates N (for a specified test compound) is provided by the manufacturer. An estimation of N for common porous particles columns can be obtained using the expression:

$$N \approx \frac{1000\,L}{Ct.\,d_p} \tag{6.2.11}$$

In expression (6.2.11), $Ct.$ is taken as 2, 2.5, or even 3.

5) Column resolution is an important parameter depending significantly on the physical properties of the stationary phase. From expression (4.2.63) of the resolution R, it can be seen that both Ψ and N are present in the formula. Since these two parameters are influenced by the surface area of the solid support of stationary phase, as well as by particle size and shape, the result is that R depends in more than one way on the physical properties of the stationary phase. In formula (4.2.63), another parameter is the constant K for the analytes in the equilibrium between the mobile phase and the stationary phase. This constant depends mainly on the chemical composition of the stationary phase.

6) Maximum sample load in the column, regarding both the volume and the amount of sample, depends on column dimensions (L and d) and also on k' and N (see expressions (4.2.88) and (4.2.89)). These two parameters depend on the stationary phase physical characteristics. In both expressions (4.2.88) and (4.2.89) the larger values for N lead to lower acceptable injected volume and mass of the analyte. This result is caused by the fact that the peak shape of columns with

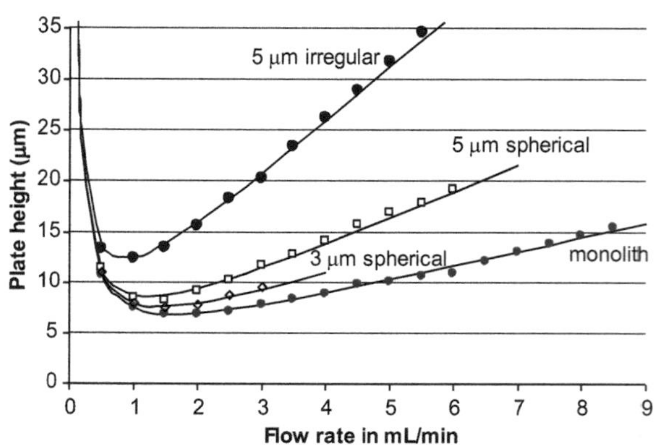

FIGURE 6.2.2 Variation in theoretical plate height H for different types of stationary phases (irregular particles 5 μm, spherical particles 5 μm, spherical particles 3 μm, and monolithic) exemplified for a C18 type stationary phase [16].

high efficiency is narrower and more easily affected by large injections. However, peak shape is affected by a number of additional factors, and sample loads larger than recommended are frequently used when improvements in sensitivity are necessary.

7) Column backpressure is affected by physical characteristics of stationary phase, as indicated by expression (6.2.10). Columns with larger particles (larger d_p) develop lower backpressure. Other aspects of column construction (through the value of ϕ_r) affect backpressure. However, UPLC systems with pumps that have the capability to produce higher pressures allow the use of columns with smaller particles for obtaining good column efficiency in spite of incurring higher backpressures. In order to withstand higher backpressures, the stationary phase of the column needs to have appropriate rigidity. If the stationary phase starts collapsing, an undesired space can be formed at the inlet side of the column, which can produce significant peak broadening. For softer stationary phases or columns made from organic polymers, high pressures must be avoided (see Section 6.1).

8) The short-term reproducibility of the results in HPLC may be affected by various problems related to instrumentation (e.g., pumps and injector), but the main factor can be considered to be the stability of the column. Protection of the column from excessive backpressure (e.g., by using too high flow rates), from precipitation of solids (e.g., of salts during gradient that changes the mobile phase to a higher organic content), from dewetting (use of high water content in the mobile phase in case of some hydrophobic columns), and from the use of mobile phase with pH outside column stability will ensure short-term stability.

9) The stability of the chromatographic column for longer periods of time should consider two aspects: (1) reproducibility of the column retention properties from run to run and (2) stability in time of the column when properly protected. Some of these properties are related to the chemistry of stationary phase, and some to the procedure of column packing. For resilience to high pressures, the stationary phase rigidity is very important to avoid the formation in time of a dead volume at the front end of the column. This dead volume can be caused by using high pressures, and once formed leads to significant peak broadening. For use in recommended pressure limits, and protection from "dirty matrix" that would not elute from the front end of the column, it is difficult to relate time stability to a specific stationary-phase physical characteristic. Columns with larger particles seem to be more resilient in time than those with small particles. However, since columns with smaller particles have higher values for N, and typically generate narrower peaks, it is possible that the change of column quality is more obvious for such columns, but in fact the peak broadening takes place equally regardless of the particle size. The column packing procedure is important for not allowing the formation in time of a void volume at the front of the column, which reduces the column efficiency (as expressed by N).

The stability in time of the column is also related to avoiding extreme conditions of use for the column such as very high pressure, pH values for the mobile phase exceeding the limits of the recommended range, solvents not recommended to be used as mobile phase with the specific column, and precipitation of buffers in the column.

10) For size-exclusion chromatography (SEC) the pore and channel sizes in the stationary phase play the main role in the separation mechanism. However, even for other types of chromatography, such as RP-HPLC, the pore size is an important parameter regarding the utilization of the column for the separation of small or large molecules. For allowing the interaction of the analyte molecule with the stationary phase from inside the pores of the solid

support, the molecule must be able to penetrate the pores. For this reason, depending on the application, stationary phases with medium pores (of about 100 Å) are typically used for the separation of small and medium-sized molecules, and stationary phases with large pores (of about 300 Å) are used for the separation of large molecules, such as proteins.

Chemical Properties of Stationary Phase Affecting Separation

The chemical properties of the stationary phase represent an essential factor assuring the chromatographic separation. In this section, only an overall view on the role of chemical properties of stationary phase is presented and the main focus is dedicated to bonded phases utilized in RP-HPLC and in HILIC. Details about stationary-phase chemical properties are further discussed in more detail elsewhere in this book (see Chapters 7–12). This detailed discussion will also include the stationary phases used in ion exchange and size exclusion. Besides the chemical nature of the main active phase, the discussion on the chemical nature of the stationary phase must consider the polymerization type (see Section 6.1), the load of bonded phase, other effects such as end-capping and abundance of silanol groups (for silica support), the chemical nature of the solid support (silica, zirconia, polymeric). Horizontal and vertical polymerizations, for example, lead to differences in the stationary-phase characteristics even if the basic active groups (e.g., C8 or C18) are the same. Horizontal polymerization offers a lower phase load and consequently a lower Ψ value for the phase. However, the number of free silanol groups is reduced compared to vertical polymerization because the surface coverage with the bonded phase is more homogeneous. In vertical polymerization a larger bonding layer is attached to the solid support, leading to a larger volume of stationary phase (V_{st}), but with a less homogeneous coverage of the silica backbone. The phase load on the solid support is therefore related to the bonded phase nature. The phase load by itself, for example for C8 or C18 columns described by the carbon load %, also plays an important role in column characterization. A summary classification of columns based on the chemical nature of their stationary phase (not specifying horizontal or vertical polymerization) is given in Table 6.2.3.

Among the chromatographic parameters affected by the chemistry of stationary phase are the following: (1) selectivity, (2) retention factor, (3) retention time t_R, (4) separation run time, (5) column efficiency (number of theoretical plates N), (6) column resolution R, (7) maximum sample load in the column, (8) column backpressure, (9) column stability in time, and (10) low bleed.

1) Selectivity α is one of the main properties of the chromatographic column and an essential parameter for compound separation. Expression (4.2.55) shows that selectivity is determined by the ratio of constants governing the equilibrium of the analytes between the mobile and stationary phases. This equilibrium depends on the chemical nature of the analytes, of the stationary phase, and of the mobile phase. For this reason the selection of a specific chemical nature for the stationary phase is a crucial component in achieving a desired separation. The choice of the chemical nature of the stationary phase is strictly correlated to the type of chromatography that should be used for a specific HPLC method. The classification of HPLC types such as reversed phase, normal phase, HILIC, ion exchange can be considered as based on stationary phase nature. Since the support for the stationary phase can also participate with interactions toward the analytes, the nature of this support must also be taken into consideration when selecting a specific column. Details regarding the selection of specific columns are further presented in this book for different chromatography types (see Chapters 7–12).

TABLE 6.2.3 List of Several Types of Stationary Phases

Types of Stationary Phase	Phase Nature (Examples)	Nature of Support	End-Capping
Hydrophobic, bonded	C4, C8, C18, other n-alkyl (e.g., C30), mixed alkyl, phenyl, diphenyl, mixed alkyl-aryl, cyclohexyl, fullerene, carbon (graphitized).	Silica	Yes (hydrophobic)
Hydrophobic, bonded	C8, C18, phenyl	Silica ethylene bridged	Yes (hydrophobic)
Hydrophobic	Polystyrene-divinylbenzene (PS-DVB), PS-DVB-methacrylate, DVB/methacrylate, polymethacrylate	Polymeric	Not needed
Hydrophobic with polar group embedded, bonded	C18-carbamate, C8-carbamate, C8-amide, C14-amide, C18-amide, C16-sulfonamide, phenyl ether	Silica	Yes (hydrophobic)
Hydrophobic, polar end-capped, bonded	C18 polar end-capped	Silica	Yes (polar)
Hydrophobic with polar group embedded, polar end-capped, bonded	Ether linked phenyl	Silica	Yes (polar)
Hydrophobic, bonded	C8, C18	Silica	No
Hydrophobic fluorinated, bonded	Pentafluorophenyl, perfluoroalkyl	Silica	?
Hydrophobic, bonded	C8, C18	Zirconia	?
Weakly polar, bonded	Cyano-propyl, pentafluorophenyl	Silica	Yes
Normal phase	Bare silica	Silica	No
HILIC, bonded	Amino, diol, zwitterionic, amide, urea, polyhydroxyl, diol with ether embedded	Silica	Yes/No
HILIC, bonded	Diol, amide	Silica ethylene bridged	No
HILIC	Polyamine	Polymeric	Not needed
Cation exchange, bonded	Sulfonic, carboxyl, phosphate	Silica	No
Cation exchange	Sulfonic, carboxyl	Polymeric	Not needed
Anion exchange, bonded	Ammonium quaternary $-NR_3X$ ($R = CH_3$, C_2H_5, $X = Cl$, OH), amino	Silica	No
Anion exchange	Ammonium quaternary $-NR_3X$ ($R = CH_3$, C_2H_5, $X = Cl$, OH), amino	Polymeric	Not needed
Anion exchange latex agglomerated	Ammonium quaternary $-NR_3X$ ($R = CH_3$, C_2H_5, $X = Cl$, OH), amino	Polymeric	Not needed

(*Continued*)

TABLE 6.2.3 List of Several Types of Stationary Phases—cont'd

Types of Stationary Phase	Phase Nature (Examples)	Nature of Support	End-Capping
Ion-moderated	Sulfonic, carboxyl, Na^+, K^+, Ca^{2+} Pb^{2+}	Polymeric	Not needed
Ligand exchange	Various	Various	?
Gel filtration	Silica gel	Silica	Not needed
Gel filtration	Polystyrene/divinylbenzene, vinyl alcohol copolymers, polyamide, poly(hydroxymethacrylate)	Polymeric	Not needed
Gel permeation	Silica gel	Silica	Not needed
Gel permeation	Polystyrene/divinylbenzene, vinyl alcohol copolymers, polyamide, poly(hydroxymethacrylate)	Polymeric	Not needed
Displacement	Special	Special	Special
Bioaffinity, affinity	Special	Various	Special
Chiral, bonded	Chiral substituents such as (S)-valine, other bonded chiral selectors on silica	Silica	?
Chiral	Dextrans, cellulose, proteins	Polymeric	Not needed
Other	Hydride	Silica	Not needed
Mixed mode	Various	?	?

? indicates no information available.

2) Retention factor k' (for a compound X) is another important parameter determined by the stationary phase composition. This parameter is affected through both $K(X)$ and Ψ for the specific stationary phase, as shown by expression (4.2.17). While $K(X)$ strongly depends on both stationary and mobile phase, Ψ is mainly dependent on the stationary phase (although the mobile phase also plays a role in the value of Ψ; see, e.g., Ref. [54]). Higher k' improves resolution R and this would suggest that higher k' values are desirable. However, large k' are related to long t_R (see expression (4.2.18)) and consequently to long chromatographic runs (which are not highly desirable). Therefore, the selection of the chromatographic stationary phase must be done by compromising one aspect or the other of the separation.

3) Retention time is strongly affected by the chemistry of the stationary phase, as previously described related to retention factor k'.

4) Chromatographic run time is determined by the longest t_R in the chromatogram and therefore strongly depends on the chemical nature of the stationary phase.

5) The column efficiency (expressed by N) is affected by the chemical composition of the stationary phase, but the dependence is more complex. The rate of transfer of solute into and out of the stationary phase depends on the depth of a bonded stationary phase (see expression (4.2.47)), but this is not the only source of peak broadening in HPLC [55,56]. The existence of multiple mechanisms of retention that may act differently on molecules, also plays an important role on peak broadening (e.g., by producing tailing).

6) The chemical nature of the stationary phase is a key factor in determining the column resolution R. Formulas (4.2.61a), (4.2.61b), and (4.2.63) can be used to understand the dependence of R on the chemistry of the stationary phase. Selectivity α, retention factor k', and the number of theoretical plates N are all affected by the chemical composition of stationary phase. As a result, the selection of a stationary phase/type of column for a specific separation must consider the way all these parameters (α, k', and N) are affected by the stationary phase chemical composition (nature of ligand, polymerization type, load of bonded phase, and other effects such as end-capping).

7) The maximum sample load in the column (volume of injection and mass of the sample) also depends on the stationary phase chemical nature. In formula (4.2.88) and (4.2.89) retention factor k' and the number of theoretical plates N appear explicitly. However, the maximum sample load is affected by the separation mechanism on the column.

8) Column backpressure is affected by the chemistry of the stationary phase through the column flow resistance factor ϕ (see expression (6.2.10)). General rules regarding this dependence are more difficult to establish.

9) Column stability is another factor significantly affected by the stationary phase chemical composition. The stability includes the following: (1) reproducibility of the column retention properties from run to run, (2) resilience of the column to pH values of the mobile phase close to recommended limits, (3) stability of the column to different solvents, and (4) resilience of the column to heating (within a limited temperature range). For properly protected columns, the time stability depends on many characteristics, and it is not always related to a specific nature of the stationary phase. Regarding the resilience of the column to extreme conditions of use such as pH outside the range 2–7, the column chemical properties are very important. Also, the stability of the column to various solvents, including 100% water, depends on the column chemical composition. This chemical composition includes the chemical structure of the solid support, the bonded phase (or other column active phases), the end-capping, as well as other treatments of the stationary phase. For example, the structure of the solid support, such as silica, ethylene bridge silica, polymeric substrate, hydrated zirconia, offers different pH stabilities, with silica the least resilient and polymeric support the most resilient. End-capping and other procedures for silica surface treatment such as controlled surface charge procedures (CSH technology) are also determining the resilience of the column to a wider pH range for the column. The nature of the bonded phase is also important for the type of solvent that can be used. Some hydrophobic columns, for example, cannot be used with 100% water in the mobile phase since a dewetting/phase collapse may occur.

10) The development of new detectors with high sensitivity, such as MS and MS/MS detectors, requires that the columns do not generate any bleeding materials, even at very low levels. Better derivatization procedures assure that columns do not generate any extraneous compounds in the eluting mobile phase.

Characterization of Stationary Phase Polarity With Octanol/Water Distribution Constant

Stationary phase polarity is a concept much more complicated to define in comparison to polarity of an analyte or even of a mobile phase. It is common for a stationary phase, typically made with a specific bonded material, to not offer a uniform surface. For example, in the case of a derivatized silica, the surface contains the groups from the bonded moieties, the groups from the end-capping process, the underivatized groups from the solid support (silanols), and also the groups of the solid support. For this reason, the phase characterization related to polarity of a stationary phase cannot be well represented by a parameter like log K_{ow}. Polarity of stationary phases is best described using a set of parameters obtained from the column behavior toward a set of "test" compounds and in a standard mobile phase. For example, the hydrophobic character of an RP column is well described by the retention factor for ethylbenzene $\left(\log k'_{EB}\right)$. The subject of stationary phase and column characterization is presented in more detail in Chapters 7–12.

In an attempt to have a simpler procedure for characterizing the polarity of a stationary phase, one possibility is to use a simulation of the phase with a model small molecule. Stationary phases can be seen as having a repetitive structure of a monomeric unit that forms the bonded material on silica surface. For example, a monomeric unit for a bonded phase on silica can be considered to be a short siloxane chain on which are bonded some silanol groups, a "functional" moiety (such as a C18 chain), and optionally some trimethylsilyl (TMS) groups that simulate end-capping. Such a monomeric unit can be small enough that its polarity can be estimated using the concept of a model octanol/water partition coefficient.

One advantage toward the qualitative estimation of a K_{ow}^{model} value for a stationary phase is offered by the existence of a simple procedure for the calculation of the values for log K_{ow} based on fragment additivity methodology [57]. In this procedure, the values of log K_{ow} for simple molecules can be obtained (as an approximation) using the following formula:

$$\log K_{ow} = \sum n f_n \tag{6.2.12}$$

In expression (6.2.12), f_n is a parameter associated with each molecular fragment (e.g., $-CH_2-$, $-O-$, $-NH-$, $-CONH-$, or larger fragments), and n is the number of occurrences of the fragments (in fact, the formula for the calculation of log K_{ow} is more complex, including a sum for the inclusion of different types of bonds) (see, e.g., Ref. [58]). Also, the value for f_n depends on which atom or cycle the fragment is connected to.

Based on the repetitive structure of a stationary phase, formula (6.2.12) can be used to model a molecule for which the log K_{ow} value can be easily estimated. The value of log K_{ow} for the model molecule can be used to obtain some information about the importance of certain groups in the structure of a stationary phase. For a bonded phase on silica, the number of types of functionalities is rather limited, and this provides a simple procedure to evaluate the polarity of certain desired models. For this purpose, an initial simple model can be selected. Such an initial model for a silica with an alkyl (propyl) group as bonded phase can be the one shown in Fig. 6.2.3. The selected small molecule shown in Fig. 6.2.3 to model a propyl bonded phase has log $K_{ow} = -0.26$. This small molecule can be modified in various ways to generate units that can be assumed to represent the model unit structure of different types of stationary phases. The values for log K_{ow} of any model can be immediately obtained using the fragment additivity (expression (6.2.12)). Values for various fragments f_n that can be added to modify the monomer structure are given in Table 6.2.4.

The values from Table 6.2.4 for different fragments can be used showing, for example, the importance of blocking an OH with a TMS group (the replacement of a silanol H with TMS increases log K_{ow}^{model} value with 2.18). Also, comparison regarding the change in polarity by

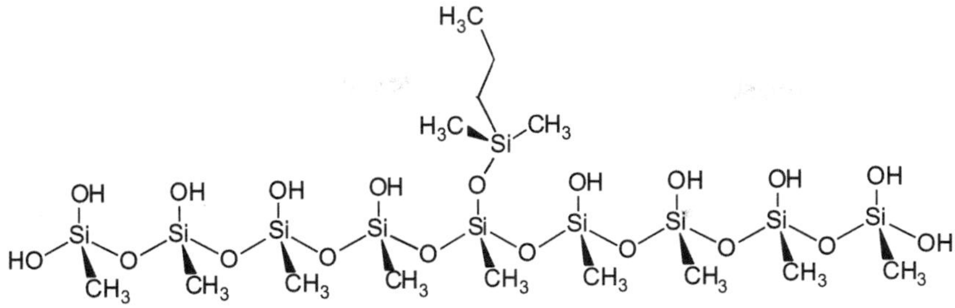

FIGURE 6.2.3 Model small molecule representing the unit for an alkyl (propyl) bonded phase on silica.

the attachment of specific groups to a hydrophobic chain can be based on such values. As indicated in Table 6.2.4 one additional free silanol group can reduce the log K_{ow}^{model} value more than two CH_2 groups added to the bonded phase. This observation indicates the importance of free silanols on the silica support for the bonded phase. A higher density of the bonded phase would increase the value of log K_{ow}^{model} considerably. Values from Table 6.2.4 also show how imbedded groups decrease hydrophobicity of the aliphatic chain. The addition of polar groups at the end of an alkyl chain also shows that the polarity increases in the order $-CN < -OH < -NH_2$ (not including the capability of NH_2 to be protonated). Addition of an NH_3^+ group to an alkyl chain produces a significant decrease in log K_{ow}^{model} (−4.27) indicating the role of protonation in the change of hydrophobicity. However, the polarity generated by the presence of two opposite ionic groups (e.g., $-SO_3^-$ and $-N(CH_3)_3^+$) cannot be estimated based on the values from Table 6.2.4, because the effect of two charges is counterbalanced by the proximity of the two ions of opposite charges. For example, in an HILIC zwitterionic phase, with the group $-N(CH_3)_2^+-(CH_2)_3-SO_3^-$, the total contribution to log K_{ow}^{model} is −3.34, while the calculated one should have a larger negative value. Also, for the groups that can be both neutral and anionic, such as amine or urea, the polarity of the phase depends on the mobile phase pH, and log K_{ow}^{model} can vary between the neutral and the charged value.

Unfortunately, the previous model does not have any quantitative significance. For example, the addition of five $-CH_2-$ groups to the model (to simulate an octyl group) will generate a log $K_{ow}^{model} = -0.26 + 5 \times 0.396 = 1.72$, and the simulation of an octadecyl group (by adding 15 CH_2 groups to the initial model molecule) generates a log $K_{ow}^{model} = -0.26 + 15 \times 0.396 = 5.68$. The significant difference in the calculated log K_{ow}^{model} values for an octyl as compared to an octadecyl fragment would suggest a significant difference in the polarity of the two types of stationary phases for reversed-phase chromatography. In reality, as described by log k' for hydrophobic compounds such as ethylbenzene, the differences in the retention properties of the two types of columns (C8 and C18) are not consistently different. Fig. 6.2.4 shows the variation of hydrophobic character of several columns with different number of carbons in the stationary phase as described by the value of log k' for ethylbenzene (log k'_{EB}) in a mobile phase acetonitrile/aqueous 60 mM phosphate buffer 50/50 v/v at pH = 2.8) [59]. From Fig. 6.2.4 it can be seen that hydrophobicity as expressed by log k'_{EB} as an average increases with the number of carbons (as expected from log K_{ow}^{model}), but a large variation in the hydrophobic character is seen from column to column with the stationary phase having the same number of carbons.

TABLE 6.2.4 Various Values for f_n Parameter for the Calculation of log K_{ow} of Model Molecules for Stationary Phase

Fragment	Connection	f_n
$-OSi(CH_3)_2OH$	Replace $-CH_3$ in siloxane chain with $-OSi(CH_3)_2OH$	−0.85
$-OSi(CH_3)(OH)_2$	Replace terminal −OH in siloxane chain with $-OSi(CH_3)(OH)_2$	−0.2
−OH	Replace nonterminal $-CH_3$ in siloxane chain with −OH	−1.05
$-CH_3$	Replace nonterminal −OH in siloxane chain with $-CH_3$	+0.4
$-CH_3$	Replace terminal −OH in siloxane chain with $-CH_3$	+0.4
$-CH_2-$	Insert additional $-CH_2-$ in the alkyl chain	+0.396
$-Si(CH_3)_3$ (TMS)	Replace −H from silanol with TMS	+2.18
Phenyl	Attach phenyl to the end CH_3 group of alkyl chain	+1.61
−O−	Replace $-CH_2-$ with −O− in the alkyl chain	−1.68
−CONH−	Replace $-CH_2-CH_2-$ in the alkyl chain with −CONH−	−2.68
$-CONH_2$	Attach $-CONH_2$ to the end CH_3 group of alkyl chain	−1.7
-NHCONH-	Replace $-CH_2-CH_2-CH_2-$ in the alkyl chain with −NHCONH−	−3.17
−C≡N	Attach −CN to the end CH_3 group of alkyl chain	−1.0
−OH	Attach −OH to the end CH_3 group of alkyl chain	−1.44
Diol	Attach −OH terminal and −OH vicinal	−2.51
$-NH_2$	Attach $-NH_2$ to the end CH_3 group of alkyl chain	−1.49
$-NH_3^+$	Attach $-NH_3^+$ to the end CH_3 group of alkyl chain	−4.27
$-NHCONH_2$	Attach $-NHCONH_2$ to the end CH_3 group of alkyl chain	−1.93
$-NHCONH_3^+$	Attach $-NHCONH_3^+$ to the end CH_3 group of alkyl chain	−5.21
$-NH_3^+$	Attach $-NH_3^+$ to the end CH_3 group of alkyl chain	−4.27
$-N(CH_3)_3^+$	Attach $-N(CH_3)_3^+$ to the end CH_3 group of alkyl chain	−5.35
$-N(CH_3)_2^+-$	Replace $-CH_2-$ with $-N(CH_3)_2^+-$ in the alkyl chain	−6.20
$-SO_3^-$	Attach $-SO_3^-$ to the end CH_3 group of alkyl chain	−3.28

Also, log k'_{EB} is not always higher as the number of carbons in the bonded chain is higher, as would indicate the calculated value for log K_{ow}^{model} based on the model previously described. This result can be easily explained considering the complexity of the solid surface of the stationary phase and the simplicity of the model of the small unit molecule used to describe the phase. A higher number of free silanol groups can easily modify the log K_{ow}^{model} value, and the result will be that a C8 column with less silanols is more hydrophobic than a C18 column with more silanols. As further shown in Section 7.1.1, the carbon load of the stationary phase is an important parameter related to its hydrophobic character as predicted by the simple

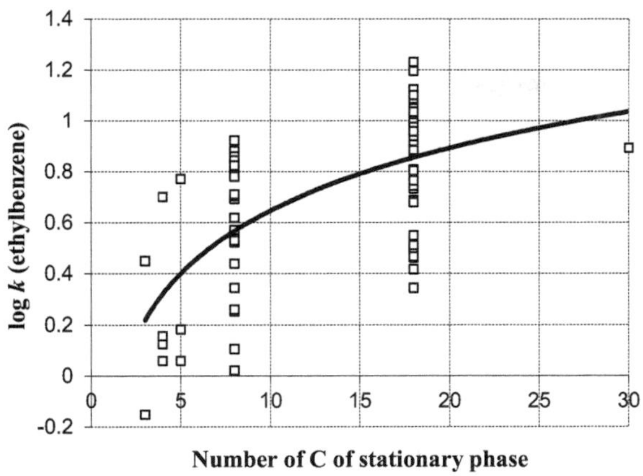

FIGURE 6.2.4 Variation of hydrophobic character of several columns with different number of carbons in the stationary phase as described by the value of log k'_{EB}. (mobile phase acetonitrile/aqueous 60 mM phosphate buffer 50/50 v/v).

model of log K_{ow}. In spite of obvious discrepancies between the quantitative predictions of the model and experimental results, the polarity of stationary phase can still be estimated in terms of contribution to log K_{ow}^{model} value for different fragments.

USP Classification of Chromatographic Columns

A standard classification of stationary phases has been proposed by US Pharmacopeia (USP) [60]. This classification uses codes indicated as L1, L2, L3, etc., and is shown in Appendix 1. The table also includes several examples of columns for each category [61]. The UPS classification offers a view on complexity of columns used in HPLC. However, even this classification is not always appropriate for all column classifications. For example, core–shell type columns, such as Kinetex XB-C18, Poroshell 120 SB-C18, or certain zirconia based columns, such as ZirChrom-EZ, are not easily classifiable using the criteria shown in Appendix 1.

6.3 GENERAL RULES FOR COLUMN SELECTION, UTILIZATION, AND CARE

The discussion about column selection will continue in more detail in Chapters 7–12, and only some general rules are presented in this section. Also, some recommendations regarding the best practices for the utilization of a chromatographic column, and the care for the column are further discussed.

Parameters to Consider for Column Selection

Selection of a chromatographic column should be made in such a manner that would lead to good results for the projected chemical analysis. The selection is guided by several general rules. However, external limitations in the selection are frequently encountered, such as the column availability, the capability of the HPLC system existent in the laboratory to generate the backpressure required by a certain column, and the time available to verify that the selected column is the optimum. New materials and better columns are continuously developed and become commercially available, such that it is very likely that a better column than the first selected one may exist. For this reason, besides the requirement that the column is fully adequate, the need for further optimization is optional. In developing an HPLC method, the selection of the mobile phase is also very important and only a combination of an adequate column plus optimum mobile phase composition (frequently using gradient) can lead to a successful HPLC method. While the recommendation about the HPLC column is to be "fully adequate," the optimization is always recommended for the mobile phase composition (and gradient). Mobile phase can be more easily modified, while the column once selected and installed requires more effort (manual and possibly financial) to be changed.

The main criteria for a fully adequate column are the following: () good selectivity, (2) good resolution, (3) good efficiency, (4) adequate retention factor k', (5) lack of peak asymmetry and tailing, (6) reproducible separation from run to run, and (7) properties to be fulfilled depending on operating requirements. These criteria are interrelated and sometimes a compromise must be made for achieving adequate results.

1) Column selectivity α describes the capability of the column to separate the analytes among themselves and from the matrix components. Parameter α depends on equilibrium constants $K(X)$ and $K(Y)$ for the compounds X and Y to be separated. Good selectivity requires a correct selection of the type of HPLC, and of the chemical nature of the stationary phase. For example, RP-HPLC must be selected for compounds with hydrophobic moieties, HILIC for polar compounds with small hydrophobic moieties, and ion chromatography for ionic compounds. For compounds with both polar groups and hydrophobic moieties, RP columns with polar embedded groups or polar end-capped can be selected. For special separations, even mixed-mode stationary phases can be used. The selectivity depends on the analytes, the mobile phase and its potential modification when using gradient, and the stationary phase. Other factors such as temperature of the column may also influence selectivity [62]. A more detailed discussion regarding the choice of a column for best selectivity can be found in Chapters 7–12.

In some cases, selective detection, such as mass spectrometry, is able to measure independently even overlapping analytes, and a good peak separation is less necessary. However, separation is frequently required even for selective detection when matrix interference may affect the quantitative results. Selectivity also affects column resolution R as shown by expression (4.2.61a) and (4.2.61b), and as illustrated in Figure 4.2.8.

2) Resolution R is the key chromatographic parameter for a good analyte separation, this parameter depending on selectivity α, retention factor k', and efficiency N (see expressions (4.2.61a) and (4.2.61b)). All three parameters α, k', and N depend on the column selection (and mobile phase composition). The influence of various column and stationary-phase characteristics on these three parameters have been discussed in Section 6.2. Higher resolution is achieved when all these parameters are higher. However, limitations can be encountered with too high values for k' and N, as further discussed.

3) Good efficiency N for the chromatographic column is also very important for a successful HPLC method. Peak broadening in columns with low N affects both the resolution R and also the integration accuracy of the chromatographic peaks. Electronic peak detection is

typically done based on the change in the slope of the chromatographic trace (traces and tangent at the start of the peak are illustrated in Fig. 6.3.1). This slope increases much faster along a narrow peak than for a broad peak, such that the peak detection and its (automatic) integration are done more accurately for narrow peaks.

For the detection and quantitation of low-level compounds, the peak broadening is also undesirable. Using a rough approximation for the peak shape as an isosceles triangle, it can be noticed that two peaks with the same area (same sample amount) have their height inversely proportional with the peak width (area $= {}^{1}/_{2}$ height $\times$ base). A peak x times narrower is x times higher (has x times higher signal), which corresponds to x times better limit of quantitation and limit of detection. For this reason, the use of shorter and narrower columns, or core–shell columns that generate narrower peaks is always recommended, when the separation, sample load, and backpressure allow their use [63].

Continuous effort is made to obtain columns with high values for N, for example by using core–shell particles and with very uniform spherical shape. Also, higher N are achieved for packed column by a decrease in the particle diameter d_p, and an increase in the column length L. Both these parameters are associated with an increase in column backpressure (see expression (6.2.8)). For this reason, UPLC instrumentation is capable of developing higher pressures (see Section 4.1) such that columns with small particles and high N can be used. The increase in the column length L is also associated with longer retention times (see expression (6.2.3)), which may not be desirable. A higher flow rate U would decrease the chromatographic run time, but higher flows may have a number of adverse effects such as a decrease in N by not having values for u close to the optimum recommended by van Deemter Eq. (4.2.50), and increase in column backpressure (see expression (6.2.8)). Also, large flow rates may not be compatible with certain detectors such as mass spectrometers.

4) Adequate retention factors k' (for sample components) are very important in selecting a chromatographic column. Since higher k' values increase resolution, they are preferable as long as they are not too high (e.g., higher than 10) and generate too long chromatographic run times. The values for k' are analyte X-dependent through the constant $K(X)$ and depend on the column construction through the phase ratio Ψ. Correct selections of the nature of the stationary and mobile phase are critical for the high value of $K(X)$. The stationary phase physical and chemical structures determine the values for Ψ (although mobile phase composition

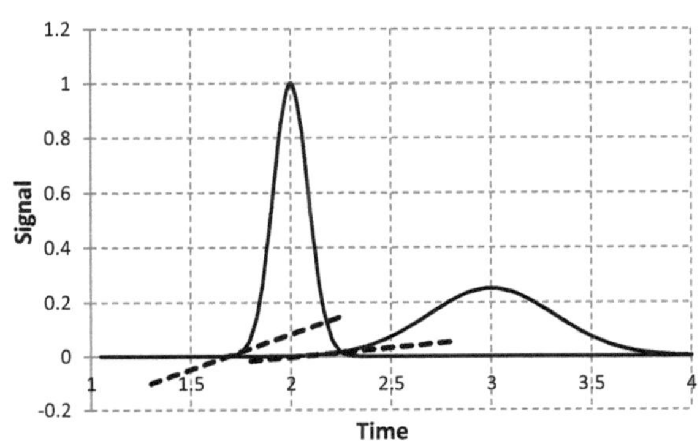

FIGURE 6.3.1 Difference in the slope of the trace at the start of a narrow and a broad peak.

also affects Ψ [64]). Phases with a high surface and a high load of stationary phase (e.g., bonded phase on silica) have higher Ψ values.

High values for k' are associated with long retention times (see expression (4.2.19)), and long run times for the chromatogram are not desirable. Besides the lower productivity (fewer runs in the same interval of time), long retention times are also associated with peak broadening (see expression (4.2.44)). However, large k' values in one mobile phase composition can be lowered by properly adjusting this composition, for example in the case of RP-HPLC by increasing the content of organic component. This is frequently achieved in gradient elution.

5) Lack of peak asymmetry and tailing (the column should have $As(X) \approx 1$, see expression (4.2.52)) is another property to consider for column selection. The tailing problem appears in particular for basic compounds (e.g., amines) and it is more obvious for columns displaying more types of interactions, such as hydrophobic and polar. For columns with other functional groups tailing is typically less intense. The metal impurities in the silica also play a role in tailing, and columns with ultra-pure silica backbone show less tailing than columns made with lower-purity silica (*type A*).

6) Reproducible separation from run to run is another important column characteristic. Short-term reproducibility in time of column properties is related with the type of stationary phase (e.g., bare silica columns may show variability in retention properties due to the different amount of water adsorbed on its surface). Long-term reproducibility depends on column construction (packing consistency) and also on the care of the column. Regarding backpressure this depends on a number of factors and the selection of the column should be made to fit the capability of the pumping system of the HPLC at a selected flow rate U.

7) Some other column properties necessary to be fulfilled depend on operating requirements. Among these properties should be listed the following: (1) resistant to a wide pH range, (2) resistant to a variety of solvents, (3) not producing excessive backpressure, (4) resilient to some heating, and (5) low bleed. The column selection must be made in agreement with manufacturer specifications, and operating requirements should be selected to meet these specifications. For example, the pH range of the utilized mobile phase should not exceed the recommended pH range for column utilization. Also, some solvents must be avoided if they are listed as not permitted for a specific column. One specific case on some RP type columns is that of 100% water, which must be avoided in order to avoid "dewetting/phase collapse" (see Section 7.2). Regarding backpressure of a column, this depends on particle diameter d_p in packed columns, on column length L, column diameter d, and on operating parameters flow rate U and viscosity η of the used mobile phase (at the operating temperature). The maximum backpressure recommended for a column should not be exceeded since it may produce crushing of some particles and formation of fine material that further increases the backpressure and make the column unoperable. On the other hand, it is possible that the backpressure of the column is too high to be overcome by the available HPLC system, and the flow of mobile phase may be shut-off by the pumping system. Stopping of the flow may be produced in the middle of a chromatographic run when the viscosity of the mobile phase increases due to gradient operation. The potential of column bleed must also be considered when selecting a column. Column bleed can interfere with analytical results, and particularly with some sensitive detectors such as LC-MS or CAD it is important that columns have the bonded phases stable enough to avoid bleeding. Experimental studies showed that the silica-based C-18 stationary phase gave the highest bleed,

and the carbon clad titanium dioxide column the lowest bleed [65]. All these parameters must be considered in order to avoid excessive backpressure.

The heating of the column during the HPLC run may be recommended in some methods for various reasons, some related to the separation and possibly for diminishing solvent viscosity. However, as indicated by expression (4.4.9) the typical effect of temperature increase is a decrease in retention factor k'. Nevertheless, for some specific HPLC separations, an increase in temperature is recommended [66].

Summary of Criteria for Column Selection

As previously described, the selection of an appropriate HPLC column must consider a number of characteristics, some related to the chemical and some to physical properties of the phase, and of the physical characteristics of the column body. A summary of these characteristics is listed in Table 6.3.1.

TABLE 6.3.1 Characteristics Used for Selecting a Good HPLC Column/Stationary Phase

No.	Criterion	Level
1	Efficiency and ability to obtain separation of sample analytes, between themselves and from the matrix components	Vital
2	Large values for theoretical plate number N	Critical
3	Adequate values for retention factor k'	Critical
4	Achieve separation within relatively low t_R	Critical
5	Acceptable values for the backpressure and the capability to maintain this backpressure constant in time when identical conditions of a separation are used	Critical
6	Low skew/low asymmetry/low tailing	Critical
7	No bleed of the stationary phase	Important
8	Stability to extreme operating conditions such as low pH, high pH	Important
9	Stability to a range of solvents (e.g., high water content for hydrophobic columns)	Important
10	Capability to maintain identical characteristics for a large number of injections	Important
11	Stability under varying operating conditions	Desirable
12	Ability to tolerate high pressure and a rapid change in pressure	Desirable
13	Reproducibility of the separation from column to column of the same type but from different batches	Desirable
14	Capability to achieve analyte separation within a wide range of experimental conditions (have versatility)	Desirable
15	Availability in different formats that fit a wide range of HPLC instruments, and ease of transferability from format to format of separation conditions	Desirable
16	Rapid regeneration capability	Desirable
17	Low price and extensive information regarding the column characteristics	Desirable

Selection of Columns for Orthogonal Separations

Most HPLC analyses require the use of a single column, and if a simpler possibility to solve an analytical problem exists, the use of complex instrumental setups should be avoided. However, some difficult separations may be solved by using two different columns. This can be, for example, the case of separation of complex samples, of separations that allow the differentiation of most compounds but not of all sample components, of separation coupled with optical isomers analysis, or the case of bidimensional separations. The HPLC separations performed on different types of columns and optionally using different solvents that lead to a different separation are labeled as orthogonal. In orthogonal separations, the types of selected columns must be different, such that the retention times of a set of compounds on one column should not correlate with the retention times from the other column. Special test mixtures can be used to verify column orthogonality. Although different, the orthogonal columns must have specific compatibility in order to be able to be used together in an instrumental setup that can be installation in series or as a bidimensional system [67,68].

Column Protection and Storing

Over the lifetime of the column some degradation of the stationary phase occurs. Column aging is usually accompanied by rising column backpressure, gradual shortening of the retention times and more peak tailing. Column lifetime of 500–2000 injections should be considered as satisfactory. Column lifetime is increased if some column protection measures are taken. Standard requirements for column protection include the following: (1) the use of a mobile phase with pH within manufacturer specifications, (2) injection of clean samples, avoiding the injection of samples with a "dirty" matrix (or unfiltered sample), (3) avoidance of high concentrations of buffers in the mobile phase that can precipitate in the column, (4) use only miscible solvents as mobile phase components (including during gradient), (5) use only solvents recommended by manufacturer (e.g., avoiding 100% water in some columns), (6) avoidance of excessive backpressures that may crush the stationary phase, (7) avoidance of sudden pressure changes, (8) avoidance of excessive temperatures outside the range recommended by manufacturer (typically below 60°C), (9) not reversing the flow in the column, and (10) avoidance of strong mechanical shocks.

Among the standard requirements for column protection, the characteristics of the mobile phase are in particular very important. The use of mobile phases with a pH outside the recommended range (e.g. lower than 2 or higher than 7) can easily destroy a chromatographic column. When using gradients, the solvents used as mobile phase must be miscible. Also switching from one solvent to another should be done considering solvent miscibility. The solvent composition must be kept within the limits recommended by the vendor. For example, the water content used with an RP column should not exceed the recommended limit in order to avoid phase collapse by dewetting. The addition of salts for the modification of pH or ionic strength must be done at the lowest possible concentration, and salt precipitation in the column will destroy it.

Another set of requirements is related to the injected sample. It is preferable that the sample is filtered, and sample preparation procedures are recommended to clean the sample matrix. The sample solvent must be selected to be miscible with the mobile phase, or even identical to the mobile phase, if possible. Also, the samples injected in the column should be (as much as possible) completely eluted. It is recommended for this purpose to use mobile phase gradients with high content of strong eluting solvents (e.g., acetonitrile for RP columns) toward the end of the chromatographic run, to assure all the sample components are eluted from the column. Noneluting sample components accumulating in HPLC columns can alter the chromatographic properties of a column. Also, particles (with $d > 2$ μm) suspended in the mobile phases or in the injected samples, not retained in filters before the chromatographic column, accumulate on the frit at the column inlet and can disrupt the uniform flow of samples onto the packing bed. These suspended particles in the mobile phase that enter the column can obstruct spaces in the packing bed and lead to an increase in the backpressure.

Although utilized with proper care, some of the column qualities, in particular the capability to maintain identical separation characteristics for a large number of injections, may deteriorate in time. This problem can be improved with the use of guard columns or guard cartridges. Even if the frits that are used in the path of mobile phase eliminate solid contaminants that may plug the analytical column, when the column is used for a longer period of time, certain strongly retained materials can accumulate on the column, be never eluted, and dramatically reduce column lifetime. By modifying the surface of the stationary phase, these retained materials can cause shifts in peak retention, loss of resolution and efficiency, as well as degradation of peak shape. Even when common solvents are used, some dissolution of silica (for silica-based columns) may occur, leading to the formation of a void volume at the head of the column and subsequent reduced column resolution. A common procedure to protect the analytical column from these problems is to use a guard column. Guard columns are short (or very short) columns or cartridges (1–2 mm thick), packed with a stationary phase having similar bonded phase as the analytical column. The guard columns have a very small dead volume and do not affect the column performance. At the same time, the guard column retains the noneluting contaminants and provides saturation of the mobile phase with silica by "bleeding" silica into the mobile phase instead of from the analytical column. Guard columns can substantially increase column lifetime and improve the quality of chromatographic results. The guard column (or cartridge) must be replaced from time to time, before the analytical column starts to degrade. The replacement time interval is determined by the specific experimental conditions and can be done either at a fixed schedule (such as 100 injections) or when signs such as an increase in the backpressure or some loss of resolution start to be noticed. A better way to determine the optimum time to replace a guard column for a certain number of sample and specific mobile phase conditions is by monitoring chromatographic performances, such as plate number (N), retention time (t_R) and resolution (R), or system backpressure. Significant modifications of any of these parameters in the range of 5–10% may indicate that replacing the guard column is necessary.

After a period of usage some column degradation may be noticed, with broader peaks, lower retention times, decreased separation capability, and increased column backpressure. When any of these problems appear, a column cleaning may be useful. The cleaning is typically done by rinsing the column (e.g., 10–20 column volumes) with a specific solvent or sequence of solvents at one-fifth to one-half the typical flow rate. For example, for RP columns, a set of solvents covering a wide range of polarity is recommended, starting with 95:5 water/acetonitrile v/v, followed by tetrahydrofuran (THF), followed by 95:5 acetonitrile/water v/v. Cleaning of columns used for protein separations may be washed with

0.1% trifluoroacetic acid (TFA) in water followed by 0.1% TFA in acetonitrile/isopropanol $^{1}/_{2}$ v/v, followed by rinsing with a common mobile phase (columns should not be stored in THF). Other specific cleaning procedures may be recommended by the column manufacturer. In general, it is not recommended to use a flow in the opposite direction than recommended with the hope to diminish a backpressure increase.

Before the use of a chromatographic column, it should be initially conditioned with the mobile phase. New columns are flushed with the mobile phase with 5–10 mL solvent (or smaller volumes for UPLC columns), and before any chromatographic run it should be assured that the column is properly conditioned (see Section 14.1). However, for storing the columns should be kept filled with specific solvents. Commercially available chromatographic columns are typically received filled with a solvent that is specified by the vendor, and their storage is recommended to be done in a similar solvent as received. The columns for reversed phase are usually received filled with acetonitrile/water 65:/35 v/v or methanol/water 50:50. Silica-based ion exchange columns are received filled with methanol or methanol/water; normal-phase columns are received filled with isopropanol or hexane. Other columns such as those used for ion exclusion can be received filled with water. Buffers are usually avoided as storage solvent, although some columns are received filled with a specific solvent with buffer, such as HILIC columns that can be received in acetonitrile/water (e.g. 90:10 v/v) that contains 100 mM of $HCOONH_4$. Before storing, the column must be cleaned from the presence of buffers or other mobile-phase components such as acids or bases. It is recommended to store the columns in the same solvent as received from the manufacturer. For example, RP-type columns can be stored in a mixture of 65% acetonitrile and 35% water. For normal-phase columns, isopropanol is typically recommended as storing solvent. Ion exchange columns are usually stored in methanol or methanol/water. For size-exclusion columns, water with 0.05% NaN_3 (to avoid bacteria growth) or with 10% methanol is typically recommended.

References

[1] D.S. Bell, New chromatography columns and accessories for 2016, LC/GC North America 34 (2016) 242–252.

[2] W. Sun, D. Hu, X. Mei, X. Zhou, Influence of ring frit structure on column efficiency, Chromatographia 78 (2015) 467–472.

[3] W. Guan, F. Xu, F. Tan, Y. Guan, Study of conical columns with 10° opening angle for preparative liquid chromatography, J. Chromatogr. A 1108 (2006) 218–224.

[4] R.A. Shalliker, B.S. Broyles, G. Guiochon, Physical evidence of two wall effects in liquid chromatography, J. Chromatogr. A 888 (2000) 1–12.

[5] J.H. Koh, G. Guiochon, Effect of the column length on the characteristics of the packed bed and the column efficiency in a dynamic axial compression column, J. Chromatogr. A 796 (1998) 41–57.

[6] D. Hu, P. Shi, X. Zhou, Influence of radial heterogeneity upon columns of different lengths, J. Liq. Chromatogr. Rel. Technol. 38 (2015) 948–954.

[7] J.P.C. Vissers, Recent developments in microcolumn liquid chromatography, J. Chromatogr. A 856 (1999) 117–143.

[8] A. Ghanem, T. Ikegami, Recent advances in silica-based monoliths: preparations, characterizations and applications, J. Sep. Sci. 34 (2011) 1945–1957.

[9] D. Cabooter, J. Billen, H. Terryn, F. Lynen, P. Sandra, G. Desmet, Detailed characterisation of the flow resistance of commercial sub-2 μm reversed-phase columns, J. Chromatogr. A 1178 (2008) 108–117.

[10] S.M. Chang, M. Lee, W.-S. Kim, Preparation of large monodispersed spherical silica particles using seed particle growth, J. Colloid Interf. Sci. 286 (2005) 536–542.

[11] S.B. Yoon, J.-Y. Kim, J.H. Kim, Y.J. Park, K.R. Yoon, S.-K. Park, J.-S. Yu, Synthesis of monodisperse spherical silica particles with solid core and mesoporous shell: mesopore channels perpendicular to the surface, J. Mater. Chem. 17 (2007) 1758–1761.

[12] F. Iskandar, Mikrajuddin, K. Okuyama, In situ production of spherical silica particles containing self-organized mesopores, Nano Lett. 1 (2001) 231–234.

[13] E.F. Vansant, P. van der Voort, K.C. Vrancken, Characterization and Chemical Modification of the Silica Surface, Elsevier, Amsterdam, 1995.

[14] A. Berthed, Silica: backbone material of liquid chromatographic column packings, J. Chromatogr. A 549 (1991) 1–28.

[15] T.-C. Wei, A. Mack, W. Chen, J. Liu, M. Dittmann, X. Wang, W.E. Barber, Synthesis, characterization, and evaluation of a superficially porous particle with unique, elongated pore channels normal to the surface, J. Chromatog. A 1440 (2016) 55–65.

[16] S.C. Moldoveanu, V. David, Essentials in Modern HPLC Separations, Elsevier, Amsterdam, 2013.

[17] K.D. Wyndham, J.E. O'Gara, T.H. Walter, K.H. Glose, N.L. Lawrence, B.A. Alden, G.S. Izzo, C.J. Hudalla, P.C. Iraneta, Characterization and evaluation of C18 HPLC stationary phases based on ethyl-bridged hybrid organic/inorganic particles, Anal. Chem. 75 (2003) 6781–6788.

[18] P.C. Iraneta, K.D. Wyndham, D.R. McCabe, T.H. Walter, The Evolution in LC Column Performance, Waters Corp, Milford, 2010.

[19] L.T. Zhuravlev, The surface chemistry of amorphous silica. Zhuravlev model,, Colloids and Sufaces 173 (2000) 1–38.

[20] J.E. Gomez, J.E. Sandoval, New approaches to prepare hydride silica, Anal. Chem. 82 (2010) 7444–7451.

[21] http://microsolvtech.com/hplc/typec_about3.asp.

[22] S. El Deeb, Monolithic silica for fast HPLC: current success and promising future, Chromatographia 74 (2011) 681–691.

[23] K. Cabrera, Applications of silica-based monolithic HPLC columns, J. Sep. Sci. 27 (2004) 843–852.

[24] A.M. Siouffi, Silica gel-based monoliths prepared by the sol-gel method: facts and figures, J. Chromatogr. A 1000 (2003) 801–818.

[25] T. Ellingsen, O. Aune, J. Ugelstad, S. Hagen, Monosized stationary phases for chromatography, J. Chromatogr. A 535 (1990) 147–161.

[26] G. Guiochon, Monolithic columns in high-performance liquid chromatography, J. Chromatogr. A 1168 (2007) 101–168.

[27] B. Chankvetadze, Monolithic chiral stationary phases for liquid-phase enantioseparation techniques, J. Sep. Sci. 33 (2010) 305–314.

[28] H. Zou, X. Huang, M. Ye, Q. Luo, Monolithic stationary phases for liquid chromatography and capillary electrochromatography, J. Chromatogr. A 954 (2002) 5–32.

[29] R. Hayes, A. Ahmed, T. Edge, H. Zhang, Core-shell particles: preparation, fundamentals and applications in high performance liquid chromatography, J. Chromatogr. A 1357 (2014) 36–52.

[30] R.G. Chaudhuri, S. Paria, Core/shell nanoparticles: classes, properties, synthesis mechanisms, characterization, and applications, Chem. Rev. 112 (2012) 2373–2433.

[31] J.J. DeStefano, S.A. Schuster, J.M. Lawhorn, J.J. Kirkland, Performance characteristics of new superficially porous particles, J. Chromatogr. A 1258 (2012) 76–83.

[32] S.A. Schuster, B.E. Boyes, B.M. Wagner, J.J. Kirkland, Fast high performance liquid chromatography separations for proteomic applications using Fused-Core® silica particles, J. Chromatogr. A 1228 (2012) 232–241.

[33] S.A. Schuster, B.M. Wagner, B.E. Boyes, J.J. Kirkland, Optimized superficially porous particles for protein separations, J. Chromatogr. A 1315 (2013) 118–126.

[34] J.J. Kirkland, P.C. Yates, Chromatographic Packing with Chemically Bonded Organic Stationary Phases, United States Patent, 3,795,313, 1974.

[35] J.J. Kirkland, Development of some stationary phases for reversed-phase high-performance liquid chromatography, J. Chromatogr. A 1060 (2004) 9–21.

[36] U.D. Neue, E. Serowik, P. Iraneta, B.A. Alden, T.H. Walter, Universal procedure for the assessment of the reproducibility and the classification of silica-based reversed phase packings. I. Assessment of the reproducibility of reversed-phase packings, J. Chromatogr. A 849 (1999) 87–100.

[37] M. Verzele, M. de Connink, C. Dewaele, On-column endcapping and derivatization in reverse-phase high performance liquid chromatography, Chromatographia 19 (1984) 443–447.

[38] Y. Sudo, Optimization of end-capping of octadecyl-silylated silica gels by high-temperature silylation, J. Chromatogr. A 757 (1997) 21–28.
[39] Z. Jiang, R.P. Fisk, J.E. O'Gara, J.E. Walter, K.D. Wyndham, U.S Patent Application No. 09/924,399, 2001.
[40] L.-I. Kulin, P. Flodin, T. Ellingsen, J. Ugelstad, Monosized polymer particles in size-exclusion chromatography: I. Toluene as solvent, J. Chromatogr. A 514 (1990) 1–9.
[41] S. Hjertén, J.-L. Liao, R. Zhang, High-performance liquid chromatography on continuous polymer beds, J. Chromatogr. 473 (1989) 273–275.
[42] Q.C. Wang, F. Švec, J.M.J. Fréchet, Reversed-phase chromatography of small molecules and peptides on a continuous rod of macroporous poly (styrene-co-divinylbenzene), J. Chromatogr. A 669 (1994) 230–235.
[43] S. Hayrapetyan, U.D. Neue, H.G. Khachatryan, Changes in the pore character of silicas caused by surface bonding and polymer coating, J. Sep. Sci. 29 (2006) 810–819.
[44] S. Kowalska, K. Krupczyńska, B. Buszewski, Some remarks on characterization and application of stationary phases for RP-HPLC determination of biologically important compounds, Biomed. Chromatogr. 20 (2006) 4–22.
[45] G. Rippel, E. Alattyani, L. Szepesy, Characterization of stationary phases used in reversed-phase and hydrophobic interaction chromatography, J. Chromatogr. A 668 (1994) 301–312.
[46] C.A. Doyle, J.G. Dorsey, Reversed-phase HPLC preparation and characterization of reversed-phase stationary phases, in: E. Katz, R. Eksteen, P. Schoenmakers, N. Miller (Eds.), Handbook of HPLC, Marcel Dekker, New York, 1998, pp. 293–323.
[47] S. Lowell, J.E. Shields, M.A. Thomas, M. Thomas, Characterization of Porous Solids and Powders: Surface Area, Pore Size, and Density, Kluwer Academic Publishers, Dordrecht, 2004.
[48] L. Hagel, M. Östberg, T. Andersson, Apparent pore size distributions of chromatography media, J. Chromatogr. A 743 (1996) 33–42.
[49] H. Guan-Sajonz, G. Guiochon, Study of physico-chemical properties of some packing materials: I Measurements of the external prosity of packed columns by inverse size exclusion chromatography, J. Chromatogr. A 73 (1996) 27–40.
[50] H. Guan-Sajonz, G. Guiochon, E. Davis, K. Gulakowski, D.W. Smith, Study of the physico-chemical properties of some packing materials: III Pore size and surface area distribution, J. Chromatogr. A 773 (1997) 33–51.
[51] B. Buszewski, S. Bocian, G. Rychlicki, Investigation of silanol activity on the modified silica surfaces using microcalorimetric measurements, J. Sep. Sci. 34 (2011) 773–779.
[52] F. Gritti, I. Leonardis, J. Abia, G. Guiochon, Physical properties and structure of fine core–shell particles used as packing materials for chromatography: relationships between particle characteristics and column performance, J. Chromatogr. A 1217 (2010) 3819–3843.
[53] E. Oláh, S. Fekete, J. Fekete, K. Ganzler, Comparative study of new shell-type, sub-2m fully porous and monolith stationary phases, focusing on mass-transfer resistance, J. Chromatogr. A 1217 (2010) 3642–3653.
[54] E. Caiali, V. David, H.Y. Aboul-Enein, S.C. Moldoveanu, Evaluation of the phase ratio for three C18 high performance liquid chromatographic columns, J. Chromatogr. A 1435 (2016) 85–91.
[55] F. Gritti, Quantification of individual mass transfer phenomena in liquid chromatography for further improvement of column efficiency, LC/GC North America 32 (2014) 928–940.
[56] F. Gritti, G. Guiochon, Comparison between the intra-particle diffusivity in the hydrophilic interaction chromatography and reversed phase liquid chromatography modes. Impact on the column efficiency, J. Chromatogr. A 1297 (2013) 85–95.
[57] C. Hansch, A. Leo, D. Hoekman, Exploring QSAR, Hydrophobic, Electronic and Steric Constants, ACS, Washington, 1995.
[58] S.C. Moldoveanu, V. David, Sample Preparation in Chromatography, Elsevier, Amsterdam, 2002.
[59] P.W. Carr, J.W. Dolan, U.D. Neue, L.R. Snyder, Contribution to reverse phase column selectivity. I. Steric interaction, J. Chromatogr. A 1218 (2011) 1724–1742.
[60] B. Bidlingmeyer, C.C. Chan, P. Fastino, R. Henry, P. Koerner, A.T. Maule, M.R.C. Marques, U. Neue, L. Ng, H. Pappa, L. Sander, C. Santasania, L. Snyder, T. Woznyak, HPLC column classification, Pharmacopeial Forum 31 (2005) 637–645.
[61] M. Marques (Ed.), USP Chromatographic Columns, U. S. Pharmacopeia, Rockville, 2009–2010.
[62] E.L. Zhu, J.W. Dolan, L.R. Snyder, Combined use of temperature and solvent strength in reversed phase gradient elution II. Comparing selectivity for different samples and systems, J. Chromatogr. A 756 (1966) 41–50.

[63] E.G. Franklin, T.W. Kahler, D.S. bell, Effect of column inner diameter and packed bad heterogeneities on chromatographic performance, LCGC North America 34 (2016) 388–399.

[64] S.C. Moldoveanu, V. David, Estimation of the phase ratio in reversed-phase high-performance liquid chromatography, J. Chromatogr. A 1381 (2015) 194–201.

[65] T. Teutenberg, J. Tuerk, M. Holzhauser, T.K. Kiffmeyer, Evaluation of column bleed by using an ultraviolet and a charged aerosol detector coupled to a high-temperature liquid chromatographic system, J. Chromatogr. A 1119 (2006) 197–201.

[66] V.R. Meyer, Practical High-Performance Liquid Chromatography, fifth ed., Wiley, New York, 2010.

[67] J. Pellett, P. Lukulay, Y. Mao, W. Bowen, R. Reed, M. Ma, R.C. Munger, J.W. Dolan, L. Wrisley, K. Medwid, N.P. Toltl, C.C. Chan, M. Skibic, K. Biswas, K.A. Wells, L.R. Snyder, Orthogonal separations for reversed-phase liquid chromatography, J. Chromatogr. A 1101 (2006) 122–135.

[68] E. Van Gyseghem, E. Deconinck, M. Dumarey, M. Jimidar, R. Sneyers, D. Redlich, E. Verhoeven, W. Peys, M. De Smet, Y. Vander Heyden, Evaluation of orthogonal/dissimilar RP-HPLC systems sets for their suitability as method-development starting points for drug impurity profiles, J. Chromatogr. Sci. 46 (2008) 793–803.

CHAPTER

7

RP-HPLC Analytical Columns

7.1 TYPES OF REVERSED PHASES AND THEIR PREPARATION

Hydrophobic phase columns (also indicated as RP columns) have very wide utilization in HPLC. These columns are used in RP-HPLC, which is the most common HPLC type, and also in ion-pair chromatography, hydrophobic interaction chromatography, and nonaqueous reversed-phase chromatography (NARP). Substantial progress has been made for the synthesis of new hydrophobic stationary phases. The novelties are related to both physical and chemical properties of the phase such as the particle size dimensions and homogeneity, resilience to a wider pH range, utilization of core—shell type, better end-capping, as well as introduction of new hydrophobic phases. Among hydrophobic phases, those having a C18 bonded phase are the most commonly used, followed by the C8 phase. Significant improvements were made, in particular regarding the columns with C18 and C8 phases. The main types of RP phases and some specific aspects of their synthesis are discussed in this section.

Main Types of Stationary Phases Used in RP-HPLC

Several types of stationary phases were already listed in Table 6.2.3, and some such phases are indicated as having a hydrophobic character. This classification is based on the chemistry of the stationary phase. A more detailed classification of hydrophobic phases is given in Table 7.1.1.

Preparation of RP Stationary Phases

Preparation of various types of stationary phases was presented in Section 6.1, and most RP stationary phases are prepared using those procedures. Only the restriction for the bonded phase to have a hydrophobic character makes the preparation procedures specific for the RP type. It was indicated that the most common support for bonded phases is porous silica. Silica purity is important for column quality and *type B* silica purity is commonly used. Also, besides phases bonded on porous silica, core—shell particles are successfully utilized as support for hydrophobic phases. Monolithic silica-based materials with bonded hydrophobic phases are also used in practice. For these phases, the bonded phase can be the same as the one for fully porous particles, some differences in column behavior (higher theoretical plate

Selection of the HPLC Method in Chemical Analysis
http://dx.doi.org/10.1016/B978-0-12-803684-6.00007-X

TABLE 7.1.1 Several Types of Hydrophobic Stationary Phases

No.	Type of Phase	Phase	Interactions
1	Common n-alkyl on silica end-capped	C8, C18	Strong hydrophobic
2	Special n-alkyl on silica end-capped	C2, C4, C12, C14, C20, C22, C27, C30	Medium or strong hydrophobic
3	n-Alkyl on organic/silica end-capped (BEH type)	C8, C18	Strong hydrophobic
4	Core–shell n-alkyl on silica end-capped	C8, C18, phenyl	Strong and medium hydrophobic
5	n-Alkyl on silica not end-capped	C8, C12, C14, C18	Strong hydrophobic and weak polar
6	n-Alkyl on silica polar end-capped	C8, C18	Strong hydrophobic and medium polar
7	Polar embedded on silica end-capped	C8, C18, etc. with-O-, -CONH-, urea, etc.	Strong hydrophobic and weak polar
8	Polar embedded on silica polar end-capped	C8, C18, etc. with-O-, -CONH-, urea, etc.	Strong hydrophobic and polar
9	Cyclic alkyl on silica	Cyclohexyl, n-C6 linked cyclohexyl	Medium hydrophobic
10	Mixed alkyl on silica	C18-short alkyl, C4-C18	Medium or strong hydrophobic
11	Aryl on silica	Phenyl, diphenyl, C2 linked phenyl, C6 linked phenyl	Medium hydrophobic and strong $\pi-\pi$ aromatic
12	Mixed alkyl-aryl on silica	C18-phenyl, C6-phenyl, cyclopropyl-phenyl, C14-phenyl	Strong hydrophobic and medium $\pi-\pi$ aromatic
13	Cyano on silica	Cyanopropyl, phenylcyanopropyl	Medium hydrophobic and strong polar
14	Fluorinated phases on silica	Pentafluorophenyl, perfluoroalkyl	Medium hydrophobic
15	Other on silica	Fulerene, cholesterol	Strong hydrophobic
16	Silicon hydride	C8, C18 bonded, silicon hydride partially graphitized	Special
17	Monolith silica n-alkyl bonded	C8, C18	Strong hydrophobic
18	Graphitized carbon	Graphitized carbon	Strong hydrophobic
19	Polymeric block	Polystyrene/divinylbenzene, ethylvinylbenzene/divinylbenzene	Strong hydrophobic
20	Phases bonded on organic polymers	C18 bonded on polymethacrylate, phenyl bonded on polymethacrylate, C18 bonded on divinylbenzene, pentafluorophenyl bonded on divinylbenzene	Medium or strong hydrophobic
21	Phases bonded on zirconia	C8, C18, etc.	Strong hydrophobic

number for core–shell particles or lower backpressure for monolithic columns) being caused by the special silica support. Besides the phases with the hydrophobic groups bound directly to a silica substrate, hydrophobic phases are also generated by bonding groups (such as C8 or C18) on organic/inorganic silica particles (such as ethylene-bridged hybrids indicated as BEH phases, XBridge phases, etc., depending on the name given by the manufacturer).

As described in Section 6.1, several characteristics must be considered regarding silica bonded phases: (1) type of bonded phase, (2) type of reagent used for derivatization (mono-, bi-, or trifunctional) and the associated type of derivatization (brush, horizontal polymerization, vertical polymerization), (3) extent of derivatization (leading to different carbon loads), and (4) end-capping.

1) The most common hydrophobic bonded phases are C8 and C18. Among other stationary phases listed in Table 7.1.1 are short alkyl phases (C2, C4) and long-chain phases such as C27 or C30. Such phases are used occasionally for special applications. Also considered as hydrophobic stationary phases, but having special properties, are the fluorinated phases. Phenyl phases are hydrophobic but also able to develop special $\pi-\pi$ interactions. Intermediate between hydrophobic and polar phases are the cyanopropyl type. Cyano groups can also be ending longer hydrophobic chains such as C8 or even longer. In such cases, a larger hydrophobic moiety is placed in the stationary phase structure, in addition to the CN group.

Two special groups of phases are those with polar embedded groups in a hydrophobic chain, and those with polar end-capping, displaying both hydrophobic character and some polar character. Such columns are very useful for the separation of compounds with small hydrophobic moieties that typically require a high content of water in the mobile phase. Some such columns can be used with 100% water mobile phase, which is not recommended for other columns that can develop "dewetting/phase collapse." The columns with polar embedded groups may also have end-capping with hydrophobic or with polar small groups. A number of groups that can be used as polar embedded moieties in the bonded phase of hydrophobic columns are indicated in Fig. 7.1.1.

Among advantages that may come from using the polar embedded phases include better capability to separate compounds with small hydrophobic moieties and strong polar groups, a wider pH range of stability, and less effect on the separation of the free silanol groups. The covering or encapsulating of the silica surface with polymeric materials is also used in some phases to reduce the activity of silanols. The results were phases showing minimal interaction and no or little tailing with strongly basic compounds. These phases are sometimes indicated as "base deactivated."

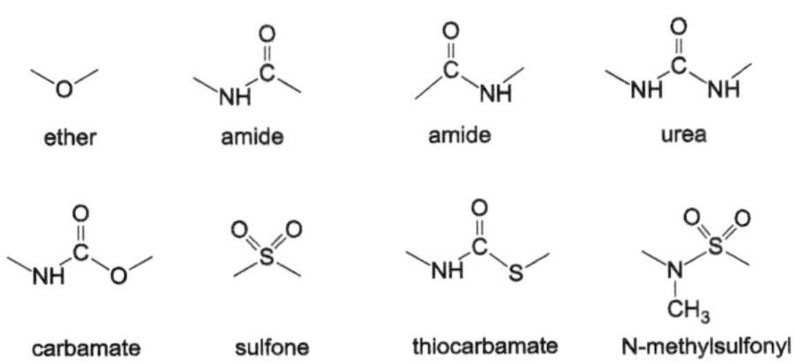

FIGURE 7.1.1 Some embedded polar groups used in hydrophobic stationary phases.

Polar embedded phases having two zones of polarity or an extended carbon linker (e.g., five or six atoms to the silica) are also known [1]. These phases reduce the possibility of interactions between the main polar group and silanol from silica surface. Bidentate embedded stationary phases were also reported in the literature [2]. The additional end-capping of polar embedded stationary phases can be done in the same manner as for other common hydrophobic phases.

2) The type of reagent used for derivatization (mono-, bi- or trifunctional) and the associated type of derivatization (brush, horizontal polymerization, vertical polymerization) has been discussed in Section 6.1. The type of reagent and type of derivatization influences the carbon load. The highest carbon load occurs for trifunctional reagents with vertical polymerization. However, the other types of derivatization can be used for special purposes (e.g., separation of proteins where adsorption has an important role in the retention process and the carbon load is not critical). The type of reagent is also related to the surface coverage and the proportion of free silanols.

3) The extent of derivatization that is related to carbon load and proportion of free silanols. Silicas may have different proportions of free silanols (from the preparation procedure), silanols with different reactivities, different available surface areas, and also different pore sizes. The extent of derivatization indicated by the ratio of free silanols versus derivatized silanols $\delta_{OH}/\delta_{ligand}$ (see Eq. 6.1.5) depends on all these parameters. The extent of derivatization may be high although the carbon load may not be necessarily very high if the initial silica has lower δ_{OH} values. In column selection all these parameters must be known and evaluated, having different effects on separation.

4) End-capping of the silica surface related to the treatment of free silanols remaining on the silica surface. The columns may be made with different types of silica, and with different end-capping. The silanol coverage is related to the phase resistance to the pH of the mobile phase, as well as the intensity of other interactions exhibited by the phase. The common pH range of stability for silica-based stationary phases is between 2 and 7 (or 8). This range is not entirely satisfactory in some applications. Basic pH is able to deteriorate the silica structure, and pH below 2 produces hydrolysis of the fragments that make the active phase, while pH higher than 8 leads to the dissolution of silica. Silica-based stationary phases resistant to a wider pH range are highly desirable, to allow the use of more acidic or more basic mobile phases. Lowering the pH of the aqueous component in mobile phase to very acidic values, close to 1, may improve some of the retention parameters (retention time, peak shape, or selectivity), in particular for analytes having $pK_a < 2$. One reason for this effect is that at low pH values the interactions of residual silanols with amino-containing solutes are suppressed and thus peak tailing is no longer observable. However, such low pH values may not be acceptable for the stationary phase stability. The end-capping of silanol groups, with the purpose of reducing secondary interactions with the analytes, also has some effect on extending the acceptable pH range for the mobile phase. This effect is due to the blocking of the access of H^+ or OH^- ions toward the silica.

Polymeric materials such as poly(styrene-divinylbenzene) (PS–DVB) with or with no phase attached to the polymer base can also be used as reversed-phase media (these being more resistant to high-pH mobile phases), but they typically have lower column efficiencies than those of silica gel-based packings. Polymeric stationary phases can be used as particles or as monoliths [3]. Special other phases such as graphitized carbon, silicon hydride, and zirconia based are also known and have hydrophobic character.

7.2 PROPERTIES OF REVERSED-PHASE HPLC COLUMNS

In Chapter 6 general comments were made regarding column properties, including properties of reversed-phase columns. Several types of columns can be classified as usable in reversed phase, and their USP (United States Pharmacopeia) classification can be found in Appendix 1. Besides the stationary phase, it must be emphasized that in RP-HPLC the mobile phase composition (which is more polar than the stationary phase) has a dominant importance, and the selection of a specific mobile phase is in many cases critical for a successful separation. Before any chromatographic run it should be assured that the column is conditioned with the mobile phase. The "strength" of the mobile phase containing usually a mixture of solvents (solvent strength changing in the order: acetonitrile > isopropanol > ethanol > methanol > water) is very important for the separation. For this reason, the selection of the column in RP-HPLC must always consider the behavior of the column toward the mobile phase, besides the intrinsic column characteristics. This behavior refers to the range of pH stability [4], the types of acceptable solvents on the column, wettability with water (since in common RP-HPLC the mobile phase always contains a certain proportion of water) [5], and temperature stability [6]. Some physical and chemical properties of reversed-phase columns are further discussed.

Basic Physical Properties of Hydrophobic Stationary Phases and Columns

The main physical properties of stationary phases and RP-HPLC columns are indicated in Table 7.2.1.

Several column dimensions are usually commercially available for the same phase. Regarding the other physical properties, they depend on the procedure for the synthesis of silica support, and they can vary considerably from one phase type to another. A wide variety of technologies are used in silica-base manufacturing. Some examples of physical (and chemical) properties of various hydrophobic columns are given in Appendix 2a. The

TABLE 7.2.1 Basic Physical Properties of Columns and Stationary Phases With Hydrophobic Character

Property	Common Range
Column internal diameter (i.d.)	2 mm–10 mm (common i.d. in mm 2.1, 3.0, 4.6)
Column length	50 mm–300 mm (or longer) (common length in mm 150, 250)
Particle type	Porous, core–shell, pellicular, monolithic
Particle shape (for silica)	Irregular or spherical
Particle size	1.7 μm–10 μm
Particle size distribution d_{90}/d_{10}	$d_{90}/d_{10} < 1.7$ ($d_{90}/d_{10} \approx 1.1$ indicates very homogeneous particles)
Surface area (for silica)	50–500 m^2/g
Particle size distribution	Narrow (1–2% size variation) to larger distributions
Pore size	50–4000 Å
Theoretical plate number (per m)	30,000–120,000 for porous particles, higher for core–shell

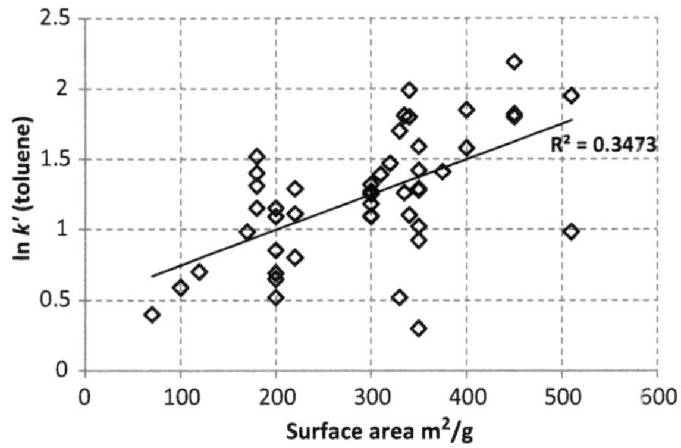

FIGURE 7.2.1 Dependence of ln k' on surface area of the stationary phase for different C18 columns (listed in Appendix 2a) for toluene as a test compound and a mobile phase 80% CH_3OH and 20% aqueous buffer at pH = 6.

influence of column and stationary phase physical properties on different aspects of separation was discussed in Sections 6.2 and 6.3.

Among the physical properties, the surface area of the silica support is correlated to the retention capability of the HPLC column. This correlation is not very strong, as shown in Fig. 7.2.1 for the set of C18 columns listed in Appendix 2a, and using as a test compound toluene in a mobile phase 80% CH_3OH and 20% aqueous buffer at pH = 6. The retention is characterized by log k' of the test compound toluene.

Basic Chemical Properties of Hydrophobic Stationary Phases

The basic chemical properties of stationary phases and RP-HPLC columns are indicated in Table 7.2.2.

TABLE 7.2.2 Basic Chemical Properties of Phases With Hydrophobic Character

Property	Common Range
Silica purity (for silica)	High to medium (type B or type A)
Metal activity from silica	Very low to high
Phase type (hydrophobic)	Bonded on silica, bonded on organic/silica (BEH type), graphitized carbon, polymeric, bonded on polymer
Reagent type	Monofunctional, bifunctional, trifunctional
Polymerization type (for bi- and trifunctional reagents)	Horizontal, vertical
Carbon load (for silica)	5–25%
Coverage (for silica)	Depending on phase, 2.0 μmol/m^2 to 4.5 μmol/m^2
End-capping (for silica)	Yes or no, nonpolar or polar
Silanol activity (for silica)	Very low to high
pH resilience	From range 2–7 to extended range
Wetting characteristics	Poor to good
Retention factor	Measured for a hydrophobic test compound such as ethylbenzene (k' in the range 1–10)

The importance of various chemical properties of phases was discussed in Sections 6.2 and 6.3. For hydrophobic stationary phases, several specific properties are essential for a successful separation. Each of these properties is further discussed in some detail.

Hydrophobicity

The hydrophobic character of the stationary phase is an essential property in RP-HPLC. Although in RP-HPLC the hydrophobicity of the stationary phase plays a secondary role to the polar character of the mobile phase that in the retention process "rejects" the analyte molecules, the role of solid phase hydrophobicity is still strongly manifested in the separation. A stationary phase with very low polarity is still more accessible to hydrophobic molecules compared to a less hydrophobic stationary phase. This explains why not all hydrophobic stationary phases respond equally, even to analyte molecules having no polar groups. Also, since most molecules separated by RP-HPLC also have polar groups, a stationary phase with significantly reduced polar interactions will act differently toward analytes as compared to less hydrophobic stationary phases. For this reason, stationary-phase hydrophobicity has been characterized by different procedures (see also [7]). Hydrophobicity of the stationary phase is related to several properties. For silica-based phases, these properties include the following: (1) the nature of the bonded substituent (C8, C18, phenyl, etc.), (2) the carbon content (carbon load $C\%$), (3) type of reagent used for derivatization (mono-, di-, or trifunctional) and reaction conditions (vertical or horizontal polymerization), (4) the density of surface coverage with the bonded phase, (5) the lack of silanol activity, (6) the purity of silica, (7) silica support specific chemistry, and (8) other stationary-phase construction characteristics such as embedded polar groups. Some examples of common physical and chemical properties of various hydrophobic type columns are given in Appendix 2a.

1) Several types of stationary phases differentiated based on the bonded substituent were listed in Table 7.1.1. The nature of bonded substituents is related to the hydrophobicity of the stationary phase. The model log K_{ow}^{model} described in Section 6.2 (see Table 6.2.4) gives some indication about phase hydrophobicity. Several other parameters are used for describing hydrophobicity. One important parameter related to hydrophobicity is log k' measured for hydrophobic compounds such as toluene, ethylbenzene, butylbenzene, and acenaphthene. Larger log k' values for such compounds indicate a more hydrophobic phase. Further discussion about hydrophobicity and its characterization is given in this chapter.

2) The carbon load (as $C\%$) offers a global characterization of hydrophobic phases. In general, a higher carbon load (which can reach up to 22–24%) is an indication of a more hydrophobic character. However, with new more modern phases that have polar embedded groups, organic/inorganic solid support, various types of end-capping, it is even more difficult to predict hydrophobicity from the stationary-phase construction and carbon load. For several C18 columns listed in Appendix 2a, the dependence of $\ln k'$ on $C\%$ is shown in Fig. 7.2.2, using toluene as test compound and a mobile phase 80% CH_3OH and 20% aqueous buffer at $pH = 6$.

3) The type of reagent used for derivatization (mono-, di-, or trifunctional) and reaction conditions (vertical or horizontal polymerization) influences column characteristics as described in Sections 6.1 and 7.1.

4) The density of surface coverage, also discussed in Section 6.2, is related to various parameters of column construction including carbon load but also the accessibility of the analytes to the active bonded phase. Depending on the density of silanol groups on silica

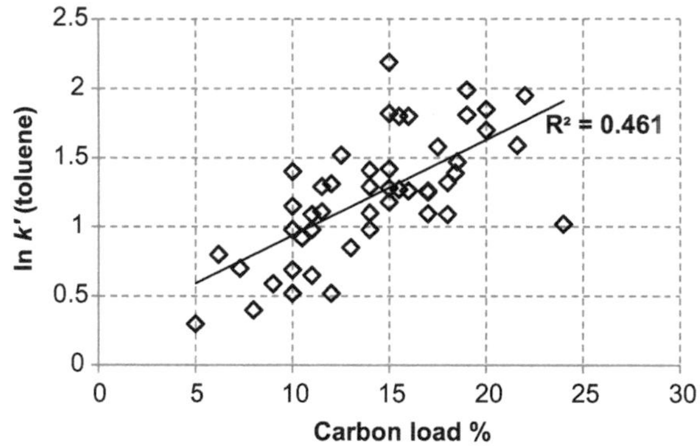

FIGURE 7.2.2 Dependence of ln k' on carbon load for different C18 columns (listed in Appendix 2a) for toluene in a mobile phase 80% CH_3OH and 20% aqueous buffer at pH = 6.

surface (see expressions 6.1.7 and 6.1.15), as well as the ratio of silanol density δ_{OH} and that of the ligand δ_{ligand}, different column characteristics can be achieved.

5) The lack of silanol activity achieved by a number of procedures such as end-capping is an important column characteristic. However, some columns intentionally have some silanol activity for enhancing the interaction with polar compounds. A discussion on end-capping is further presented in this section.

6) The purity of silica has been discussed in Section 6.1. Metal impurities are capable of interacting through coordinative bonding, in particular with amines. However, pure silica (*type B*) is now very common in stationary-phase construction.

7) The silica support specific chemistry refers to the various ways the hydrated silica has been prepared, and the use of organic/inorganic silica such as ethylene-bridged (BEH)-type materials. The base silica can be made with a larger or a smaller number of silanols (depending on the dehydration process as discussed in Section 6.1), which may lead to different silica reactivity. As a result, all silica materials (including ethylene-bridged-type products) are sensitive to the pH of the mobile phase, but with different degrees of stability. This stability may be limited to the range of pH from 2 to 7, or the range can be wider depending on silanol reactivity and protection. Ethylene bridged products are resilient to a wider pH range (e.g., 1–12) as compared to common silica-based columns.

8) Other stationary-phase construction characteristics are used for modification of the chemical properties of the stationary phase. These include the use of polar embedded groups, as well as other procedures such as CSH technology, some previously discussed in Section 6.1.

End-Capping and Silanol Activity

The advantages offered by the porous silica support with a bonded phase (in porous or core–shell particles) as compared with other supports such as zirconia or polymeric materials led to an overwhelming use of silica-based columns in RP-HPLC. A key problem with silica-based materials is related to the presence of residual silanol groups after the derivatization with the intended bonded phase. This problem remains even for BEH-type phases, although they offer some advantages regarding better resilience to a wider pH range. A number of procedures have been used for the reduction of the effect of residual silanols. Among these strategies the most common is end-capping using trimethylsilyl groups. This procedure was previously discussed in Section 6.1. Columns "double end-capped" are also reported as providing further coverage of silanol groups (e.g., Eclipse XDB). Repeated end-capping operations are not uncommon. Other procedures were utilized to shield the interaction of silanol groups with the analytes, such as the use of bidentate or tridentate reagents for attaching the desired group on silica surface [8]. Also, horizontal polymerization was shown to offer better protection of silanol surface interactions [9]. The introduction of a small percentage of groups displaying positive charges on silica surface (charged surface hybrid [CSH]) has also been used for reducing peak tailing for basic compounds by diminishing the interaction with the silanol groups. The free silanols may be present inside some small pores, and reaching them with the derivatization reagent may pose hindrance problems. However, even these free silanols present in very small pores remain accessible to small solvent molecules such as water that can form "water wires" as schematically shown in Fig. 7.2.3. The silanols from very small pores may still affect the hydrophobicity and interact with polar molecules through the water wires.

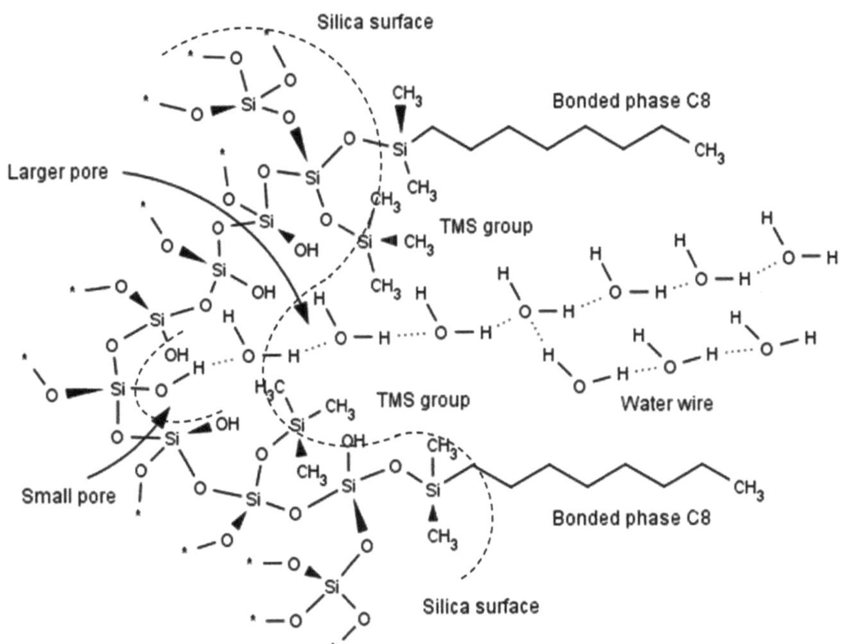

FIGURE 7.2.3 Schematic diagram showing a derivatized silica surface with C8 groups, TMS groups in a larger pore, and underivatized silanol from a small pore forming a water wire.

Besides potentially unwanted interactions with the analytes, the presence of free silanols is undesirable since they contribute to the stationary phase instability outside the common pH range of 2–7. Stationary phases with a large number of silanol groups are less stable in mobile phases with pH higher than 7 or lower than 2. Basic mobile phases tend to dissolve silica and the access of acids to the hydrated silica support leads to instability at pH lower than 2. For this reason, significant effort has been made to increase the range of pH stability of various stationary phases.

Enhancing the pH and Salt Stability of the Stationary Phase

The stability of bonded phases is influenced mainly by the pH, the nature of different salts in the buffers used for the mobile phase, and temperature. The degradation of different bonded phases is usually a result of silica support dissolution (at high pH), or hydrolysis of covalently bound silane ligands (at low pH). For enhancing the resilience of a stationary phase to pH values higher than 7–8 or lower than 2, and to make it more resilient to the use of buffers containing salts (in particular phosphates), a number of strategies were employed. Such strategies are usually based on the reduction of free silanol number or blocking the access to the free silanols. This has been achieved in various ways, some previously discussed, such as the reduction of the number of free silanol groups by exposing the initial hydrated silica to higher temperatures for the formation of silica gels. It has been shown that silicas made by the sol–gel process can be more resistant to dissolution than supports made by a silicate-gel (xerogel) process.

A common method for increasing the stability of the silica-based stationary phase is end-capping with double or multiple end-capping steps, which is superior to single end-capping [4,8]. Protection of stationary phases below pH = 2 relies on a special derivatization approach by which protecting groups are present close to the silica surface [10]. By steric hindrance, these groups provide a kind of shield for the area where the active phase is attached, protecting it from an acid attack from the mobile phase. This type of protection is achieved by performing end-capping with voluminous groups, such as isopropyl or isobutyl. The larger volumes of isopropyl and isobutyl groups provide some protection to the bond connecting the silica surface with the group acting as a stationary phase. This type of technology is sometimes indicated as StableBond or SB with the silica base protected by end-capping silanols with bulky groups.

Other procedures to enhance stability include the use of ethylene-bridged-type organic/inorganic silica (in BEH-type columns), the use of CSH technology, the use of extra dense bonding of the phase (XDB technology). Also, shielding of the silanol groups from the silica surface can be achieved by using silica derivatization with trifunctional reagents as previously shown in Eq. (6.1.13) (see Section 6.1). The resulting mass of vertical polymerized bonded phase may contain some silanol groups that are not directly bonded to the silica gel surface. This is achieved when the derivatization of silica is performed in special conditions (with a higher proportion of water). The silanol groups not connected to the silica gel surface may have lower acidity and have a protection effect to the attack of strong bases or acids.

Another procedure to protect the silica surface is the use of very long hydrophobic alkyl chains (C27, C30) for derivatization. Because of the higher degree of surface shielding, long-chain stationary phases are stable over a wider pH range than C8 or C18 bonded phases.

Besides end-capping with small hydrophobic groups such as trimethylsilyl or with bulky fragments, end-capping with small fragments containing polar groups was found to protect the silica structure and extend the range of pH where the columns can be utilized (sometimes indicated as base deactivation technology or BD). This procedure is also utilized for generating phases with some polar character in addition to the hydrophobic character of long alkyl groups. One such column (Synergi 4u Hydro-RP 80A) was used, for example, for the separation of amino acids using ion pair chromatography [11]. The polar-endcapping groups are usually amino or hydroxy, bonded to a propyl chain. These polar-endcapped stationary phases possess hydrophobicity similar to nonendcapped ones, and they display enhanced hydrogen bond type interactions, but the acidity of the polar groups is reduced and it is better controlled compared to that of silanols. Polar end-capping makes the phases more compatible with highly aqueous mobile phases, and this type of column has useful specific applications. For example, the use of a polar end-capped C18 phase can retain more polar water-soluble compounds such as water-soluble vitamins, being a better choice for their separation.

Stationary phases with polar embedded groups (e.g., −O−, −CONH−) seem to have a better resilience to extended range of mobile-phase pH. This can be explained by the disruption of water wires and by hindering the penetration of ionized acids or bases toward the silica backbone.

Depending on the specific application, the selection of a column resilient to a wider pH range may be or may not be necessary. The selection of a fully adequate column should be done based on analysis necessities. In column manufacturing, the utilization of a certain technology may be geared toward enhancing a specific property, but it can be at the same time somewhat detrimental to another property. For example, a larger number of silanol groups in the initial silica backbone will be preferable for achieving a higher degree of derivatization, but at the same time may diminish the resilience of the stationary phase to an extended pH range. The result is the availability of a variety of columns with many different characteristics. Some examples of stability of different columns at the pH of the mobile phase are given in Table 7.2.3 [12].

A different path for enhancing pH stability is the use of organic polymers for the stationary phase. Although typically much more resilient to an extended pH range than silica-based columns, polymeric columns have other disadvantages (see Section 6.1).

Phase Ratio

Phase ratio Ψ is an important characteristic of chromatographic columns defined by expression (4.2.15). Phase ratio is directly related to retention factor k' as shown by expression (4.2.17). Through k' other important parameters in HPLC depend on Ψ, among these being column resolution R (see expression 4.2.63). Also, the evaluation of thermodynamic parameters of an HPLC separation from van't Hoff plots (entropy changes evaluation) requires

TABLE 7.2.3 Stability of Several Columns to Mobile Phase pH [12]

Column	Manufacturer	pH 2	pH 3	pH 6	pH 7	pH 7	pH 8	pH 9	pH 10	pH 11
Medium		TFA	Phosph.	Phosph.	Acetat.	Phosph.	Bicarb.	Ammon./ Bicarb.	Ammon.	Triphosph.
XBridge C18	Waters	1	?	?	?	1	1	1	1	2
XTerra MS C18	Waters	1	?	3	1	3	1	1	?	?
XBridge phenyl	Waters	1	?	?	1	2	3	3	3	?
Luna C18	Phenomenex	1	?	1	2	2	2	2	?	?
Zorbax Bonus RP C18	Agilent	1	1	2	2	3	3	3	?	?
Sunfire C18	Waters	1	1	3	3	3	?	?	?	?
Zorbax SB C8	Agilent	1	?	?	?	1	2	?	?	?

Acetat., acetate buffer; *ammon.*, ammonia; *bicarb.*, bicarbonate buffer; *phosph.*, phosphate buffer; *TFA*, trifluoroacetic acid; *tri-phosph.*, sodium triphosphate.
Note: 1 = stable to more than 500 injections, 2 = stable to more than 200 but less than 500 injections, 3 = stable only to less than 200 injections, ? = no data available.

the value of phase ratio (see expression 4.4.10). In Section 4.2 several procedures for the evaluation of t_0 and of $V_0 (V_0 = t_0 U)$ were indicated. For the measurement of V_{st}, the following formula can be utilized [13]:

$$V_{st} = \frac{C\% \; Mw W_p}{1201.1 \rho n_C} \tag{7.2.1}$$

In expression (7.2.1), $C\%$ is the carbon load (g carbon per 100 g silica), Mw is the molecular weight of the silane used for preparation of the stationary phase, W_p is the weight (g) of the packing, ρ is the density of the bonded alkyl groups (g/cm^3), and n_C is the number of carbon atoms per silane molecule. The use of this formula is straightforward when specific parameters are known for the utilized column. The carbon load is frequently reported by the column manufacturer, the density of the bonded alkyl groups is around $0.86\ g/cm^3$ [14], the weight of the stationary phase can be estimated from column dimensions, and the number of carbons per silane molecule is typically known (except for some proprietary columns). Only the molecular weight of the silane used for preparation of the stationary phase requires some specific information. The calculated V_{st} using expression (7.2.1) should be considered a "constructional" value since this value does not account for the potentially immobilized solvent component on the stationary phase and therefore for the variation of the value of Ψ with the change in the mobile phase composition. The difficulties in evaluating Ψ are associated with a considerable number of theoretical and experimental studies on the subject [15–17].

An alternative procedure for the evaluation of an "effective" phase ratio Ψ is based on the correlation that exists between the retention factor k' ($\log k'$) and the octanol/water partition coefficient K_{ow} ($\log K_{ow}$) for hydrocarbons in a homologous series [18]. In Section 5.4 it was shown that based on solvophobic theory a relation between $\log K_{ow}$ and $\log k'$ has the form of expression (5.4.2). However, from Eq. (5.1.31) for hydrocarbons all $b' = 0$, which leads to $b_1 = 0$ and $b_2 = \log \Psi$ in expression (5.4.1). As a result the system 5.4.1 takes the form (for hydrocarbons):

$$\begin{cases} \log K_{ow}(X) = a_1 \mathcal{A}(X) \\ \log k'(X) = a_2 \mathcal{A}(X) + \log \Psi \end{cases} \tag{7.2.2}$$

The elimination of parameter $\mathcal{A}(X)$ (van der Waals surface area) between the two relations 7.2.2, leads to the following formula:

$$\log k'(X) = a \log K_{ow}(X) + \log \Psi \tag{7.2.3}$$

where X denotes the different hydrocarbons and $a = a_2/a_1$. From the linear plots of $\log k'(X)$ versus $\log K_{ow}(X)$ the values for a and $\log \Psi$ can be obtained as the slope of the line (a) and intercept ($\log \Psi$). An example is given in Fig. 7.2.4 for the evaluation of Ψ for a Luna C18(2) column, 150 × 4.6 mm with 5 μm particles (from Phenomenex, Torrance, CA, USA), with the mobile phase water/organic solvent 30/70 (v/v) where the organic solvent is either methanol (MeOH) or acetonitrile (AcCN) and the analytes are benzene, toluene, ethylbenzene, propylbenzene, and butylbenzene [19]. The results from Fig. 7.2.4 indicate that $\log \Psi = -0.6214$ ($\Psi = 0.2391$) for mobile phase water/methanol, and $\log \Psi = -0.4806$ ($\Psi = 0.3307$) for mobile phase water/acetonitrile. Results for $\log \Psi$ for a series of columns obtained by the previously described procedure, with the mobile phase consisting of 65/35 acetonitrile/aqueous buffer 20 mM $KH_2PO_4{-}K_2HPO_4$ at pH 7 are indicated in Table 7.2.4 [18].

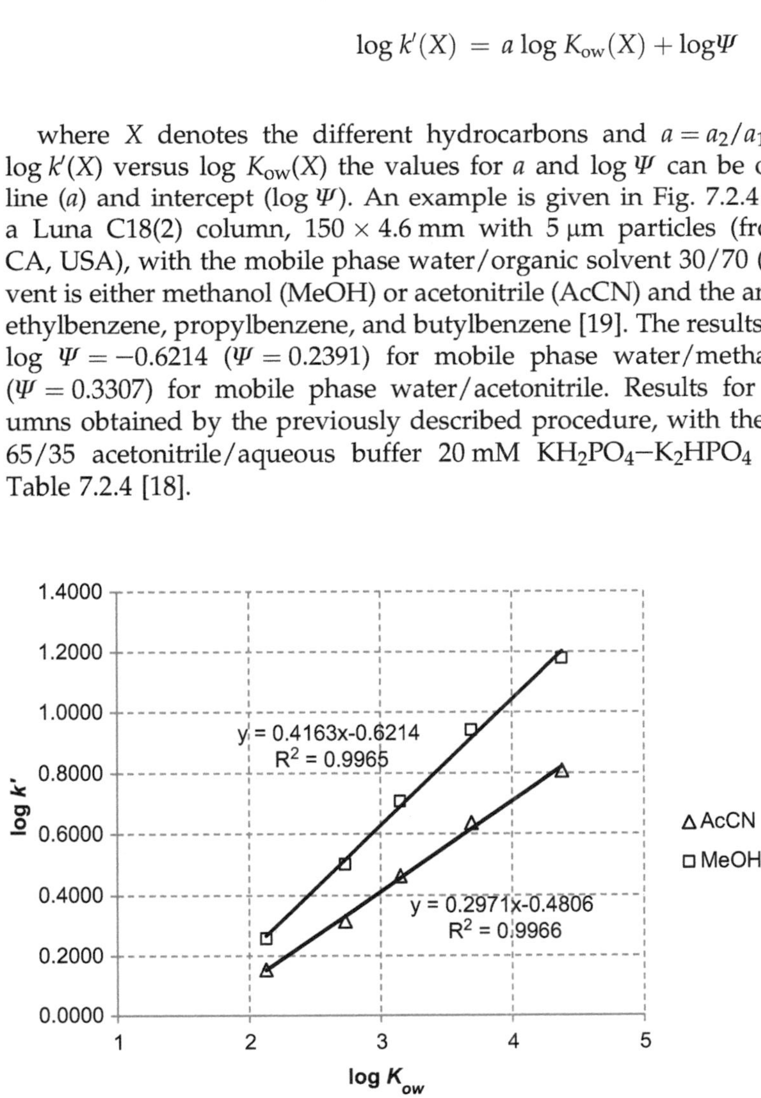

FIGURE 7.2.4 Determination of Ψ for a Luna C18(2) column with the mobile phase water/organic solvent 30/70 (v/v) where the organic solvent is either methanol (MeOH) or acetonitrile (AcCN) and the analytes are benzene, toluene, ethylbenzene, propylbenzene, and butylbenzene.

TABLE 7.2.4 Calculated Values for log Ψ for Several Reversed-Phase Columns Including Conventional C18, Polar-Embedded, and Polar-Endcapped

Column	log Ψ	Ψ
Luna5 um C18(2)	−0.523	0.300
Inertsil ODS(3)	−0.504	0.313
Zorbax XDB C18	−0.642	0.228
Symmetry C18	−0.559	0.276
Hypurity Elite C18	−0.834	0.146
Zorbax SB C18	−0.679	0.209
Prodigy ODS(3)	−0.558	0.277
Keystone Aquasil	−0.594	0.255
Aqua C18	−0.544	0.285
YMC Hydrosphere	−0.611	0.245
YMC ODS-Aq	−0.547	0.284
Prontosil C18 AQ	−0.610	0.246
Metasil AQ	−0.440	0.363
Synergi Hydro RP	−0.463	0.345
Zorbax Bonus-RP	−0.652	0.223
Polaris C18-A	−0.748	0.179
Discovery RP-Amide	−0.740	0.182
Symmetry Shield RP18	−0.491	0.323
Polaris Amide C18	−0.462	0.345

Equilibrium Constant K(X)

Very little information is available about the equilibrium constant $K(X)$ for the analyte X between the mobile and stationary phases. This constant depends on the analyte chemical structure, and also on the mobile phase and the stationary phase. Consequently, even for a given compound, no unique value exists for $K(X)$. However, the value for $K(X)$ determines the retention factor based on expression (4.2.17) ($k' = K\Psi$), and separation between two compounds based on expression (4.2.54) ($\alpha(X,Y) = K(X)/K(Y)$), such that information about $K(X)$ is important for understanding the separation.

Writing Eq. (4.2.17) in the form:

$$\log k'(X) = \log K(X) + \log\Psi \tag{7.2.4}$$

it can be seen that it is possible to obtain the value for $K(X)$ if $\log k'(X)$ and $\log \Psi$ are known for a specific stationary and mobile phase.

When X is a hydrocarbon, the comparison of expression (7.2.4) with expression (7.2.3), allows the calculation of equilibrium constant from the formula:

$$\log K(X) = a \log K_{ow}(X) \tag{7.2.5}$$

The values for a in Eq. (7.2.5) can be obtained as the slope of the graphs shown in Fig. 7.2.4, and those for $\log K_{ow}(X)$ are available in the literature [20]. For three chromatographic columns the values of a plotted as a function of organic solvent % for two solvent systems, methanol/water and acetonitrile/water, are shown in Fig. 7.2.5. The compounds X are benzene, toluene, ethylbenzene, propylbenzene, and butylbenzene, and the chromatographic columns are Gemini C18 and Luna C18, both with 5-μm particles, and Chromolith Performance RP-18 (monolithic) [19]. The results from Fig. 7.2.5 show that for the same mobile phase (nature and composition), the value for coefficient a does not differ significantly from C18 column to C18 column. The results indicate that for a separation where only hydrophobic interactions occur (as expected for hydrocarbons on a C18 column), the value for $K(X)$ does not change much when the chemistry of the stationary phase is of the same type (C18 in this case). The nature and the composition of the mobile phase on the other hand, are important factors affecting $K(X)$, as expected.

Wettability

Wettability is the ability of a liquid to maintain contact with a solid surface, and it is controlled by the balance between the intermolecular interactions of adhesive type (liquid to surface) and cohesive type (liquid to liquid). In RP-HPLC, the wettability problem occurs between the hydrophobic stationary phase and the mobile phase that has a high water

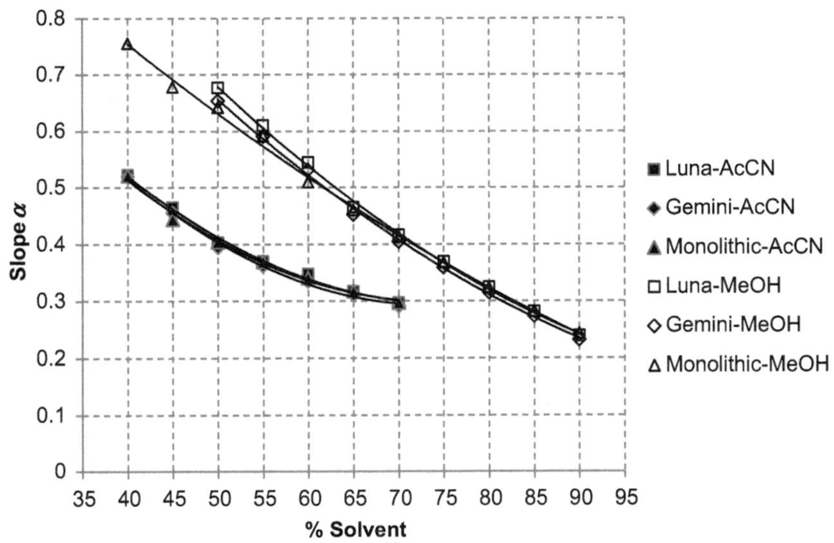

FIGURE 7.2.5 Variation of parameter a from Eq. (7.2.5) (measured using hydrocarbons between benzene and butylbenzene) for three RP-HPLC columns and two solvent systems [19].

content. The wettability can be estimated by measuring the maximum water concentrations in common organic solvents (methanol, acetonitrile, and isopropanol) that allow the phase to remain wetted. This can be checked by taking phase in bulk form and verifying the content of water in a solution when the phase does not float to the surface of the liquid. Hydrophobic columns that have as bonded phase CN or phenyl have less problems with wettability, while this problem may occur for C8 and C18 columns when no specific technology is utilized to improve wettability. All the brush phases, since they have a larger number of underivatized silanols, were shown to tolerate higher concentrations of water in the mobile phase than the bulk phases. It was also proved that the more dispersive and less polar is the solvent, the more water the solvent can contain before any of the phases become unwettable. Stationary phases that are incompletely derivatized can be wetted even by pure water. This can be explained by the high content of residual silanol on their surfaces, which allows the polar interactions with water molecules. Better wetting properties are also displayed by phases end-capped with polar groups, or containing embedded polar groups in a hydrophobic bonded phase. C18 phases are wetted by a lower content of water in organic modifier as compared to shorter-chain phases such as C2, which becomes wetted with solvents with a higher content of water (for a similar derivatization degree) [21].

Dewetting of the hydrophobic phases is possible when the concentration of water in the mobile phase is higher than 80–90%. It is manifested by the loss of retention capability of the stationary phase, retention irreproducibility, increased tailing, and inability of the column to regenerate when returned to a high concentration of organic component. For separation of compounds with small hydrophobic moieties it is sometimes necessary to use a high content of water in the mobile phase. For such separations, it is recommended to use columns having good tolerance for a high level of water, such as polar end-capped columns and hydrophilic end-capped columns, columns with polar embedded groups, or a combination of structures [22]. Also, the columns with ethylene-bridged technology have better resilience to mobile phases with high water concentration. For example, the EVO-type columns from Phenomenex are recommended for use even at 100% water in the mobile phase.

The study of this process [23] showed that two potential mechanisms are responsible for dewetting. One mechanism is related to the disturbance in the conformation of the alkyl chains in the presence of high water content in the mobile phase and is indicated as phase collapse. This explanation starts with the observation that short-chain stationary phases such as C4 and very long stationary phases such as C30 are less prone to dewetting. The short chains on the silica surface are not very well oriented, and the lack of effect from the water is expected since no specific conformation is to be disturbed. The very long-chain phases are in a strongly bundled form, and the water is not able to disturb easily the phase conformation, which may explain the resilience of this type of phase to higher water concentrations [24]. The interaction forces between the alkyl chains in C8 or C18 phases are easier to disturb. The use of melting point as a measure for interchain interactions indicates that $C_{18}H_{38}$ melts at 27–28°C and it is easier disturbed, while $C_{30}H_{62}$ melts at 64–67°C and forms more stable bundles. It can be assumed that water perturbs the aligned conformation of the C8 or C18 phases, reducing their capability of retaining the analyte. However, other studies [25] indicated that the alkyl bonded phases are always in the most compact conformation, regardless of the concentration of the organic modifier in the mobile phase, and the change in conformation due to the water is not the main cause of the dewetting. This suggests that dewetting is very likely caused by the exclusion of the aqueous mobile phase from the pores covered with a hydrophobic material and the inability of the mobile phase to reenter the pores. Since most of the stationary-phase surface is inside the pores, the exclusion of the mobile phase from the pores is the reason for the reduced active phase. A dry sorbent requires a specific pressure for the solvent to reenter the pores. This pressure can be estimated with the formula:

$$\Delta p = -\frac{4\gamma' \cos \theta}{d_c} \qquad (7.2.6)$$

where Δp is the pressure required for the liquid to enter the pore, γ' is the surface tension of the liquid, d_c is the effective pore (or capillary) diameter, and θ is the contact angle between water and air on the adsorbent surface. For water, $\gamma' = 72.8$ dynes/cm, $\theta \approx 110.6$ (for a hydrophobic surface), and $d_c \approx 150$ Å, generating a pressure of about 150 bar (in an experiment using a C8 column and 0.1% CH_3COOH in water, a pressure of 270 bar was necessary to rewet the column [26]). The mechanism based on inability of the mobile phase to enter the pores is in agreement with several experimental findings and can easily explain why short-chain alkyl phases are less prone to dewetting, their "exclusion" of water being much lower than that of C8 or C18 phases. The explanation for C30-type phases may be related to the fact that such phases are less densely packed, although the carbon content (C%) is high. However, it is more likely that the dewetting is a complex process where the exclusion of the highly aqueous mobile phase from the pores of the stationary phase plays an important role in the loss of separation capability of the phase, but changes in the conformation of the alkyl chains from the silica surface under the influence of water are also likely to occur and to play a role in the modification of phase properties.

Dewetting/phase collapse can be reversible (not always 100%) by regenerating the column with a partially organic aqueous mobile phase. The process of regenerating the phase can require hours of flushing the column with partially organic phase (e.g., 60% CH_3CN and 40% H_2O).

7.3 PARAMETERS USED FOR THE CHARACTERIZATION OF REVERSED-PHASE HPLC COLUMNS

Characterization of hydrophobic stationary phases and columns is a subject of considerable interest due to the importance of selecting the appropriate column for a successful separation, and due to the wide utilization of RP-HPLC. A number of chromatographic parameters, previously discussed in Section 4.2, such as retention capability (values for k'), selectivity α, efficiency (values of N), and resolution R depend on the analyte, the mobile phase, and also on the stationary phase. For this reason, a considerable number of studies were dedicated to stationary-phase characterization [27–31]. Three types of column characterization are basically utilized: (1) characterization based on physicochemical column characteristics (column efficiency, pH range of utilization, pore size, carbon load, reagent type used for derivatization), (2) tests for verifying specific properties using key analytes, and (3) fitting to specific retention models using multilinear regression.

The characterization based on physicochemical column properties has been discussed in part in Section 7.2, and is further commented on in this section. Other aspects related to the stationary-phase characterization in RP-HPLC using specific tests or fitting models are presented in this section.

Efficiency of Columns (Theoretical Plate Number)

Efficiency is a general characteristic of chromatographic columns, and the subject has been presented in Sections 4.2, 6.2, and 6.3. Theoretical plate number N depends on the type of particles in the column (fully porous, core–shell, monolithic), particle diameter, and other column construction (e.g., packing uniformity). Also, as explained in Section 4.2, N depends on the flow rate in the column following the van Deemter equation. Direct measurement of N for the column can also be performed using a test compound such as toluene. Some values of N (per m) for various hydrophobic columns are listed in Appendix 2a.

Retention Capability of Hydrophobic Columns

The retention capability of an analytical column is characterized by the retention factor k' (see expression 4.2.17). Related to retention factor are other column characteristics, including retention time t_R, and resolution R. The value for k' ($\log k'$) depends on the column but also on the compound nature as well as on the elution conditions such as mobile phase composition, temperature, pH, and other aspects related to mobile phase (see Chapter 13). For this reason, the description of column retention capability using the value of $\log k'$ must be done linked to a specified mobile phase and a selected analyte. Hydrophobicity, which is the main character of columns used in RP-HPLC is frequently the subject of column characterization. Hydrophobicity can be characterized by the value for $\log k'$ for a compound with no polar groups. This type of compound has only hydrophobic interactions with the stationary phase, and no other additional interaction types. Among test compounds for hydrophobicity are ethylbenzene (EB), toluene, butylbenzene, and acenapththene. Also, for developing standard conditions for hydrophobicity characterization, the mobile phase should always be specified and selected to represent a common separation condition in RP-HPLC when providing the $\log k'$ values.

The nature of the stationary phase influences the value of $\log k'$ through both the value of Ψ and that of $K(X)$ the equilibrium constant for the analyte between the stationary and mobile phase. For ethylbenzene as a test analyte X, and the mobile phase 50/50 (v/v) acetonitrile/aqueous buffer 60 mM phosphate at pH 2.8, the values for $\log k'$ for various types of columns are shown in Fig. 7.3.1. The data from Fig. 7.3.1 are informative regarding the expected values of $\log k'$ depending on the column bonded phase. One first observation is the relatively large spread of $\log k'$ for columns with the same type of phase. As expected, the values for $\log k'$ are the lowest for columns with a bonded phase containing the CN group (these columns make the transition toward polar ones). As an average, the increase in the bonded hydrocarbon length (from C4 to C18) shows an increase in the $\log k'$ values. Two columns having very long hydrocarbon moiety (C30) do not show an increase in the $\log k'$ value compared to C18 or even C8 columns. This happens, very likely, because the surface coverage with the bonded phase is not high in C30 columns, due to the high volume of this chain. Phenyl columns have $\log k'$ not very different from C8-type columns. The same remark is valid for fluorinated columns.

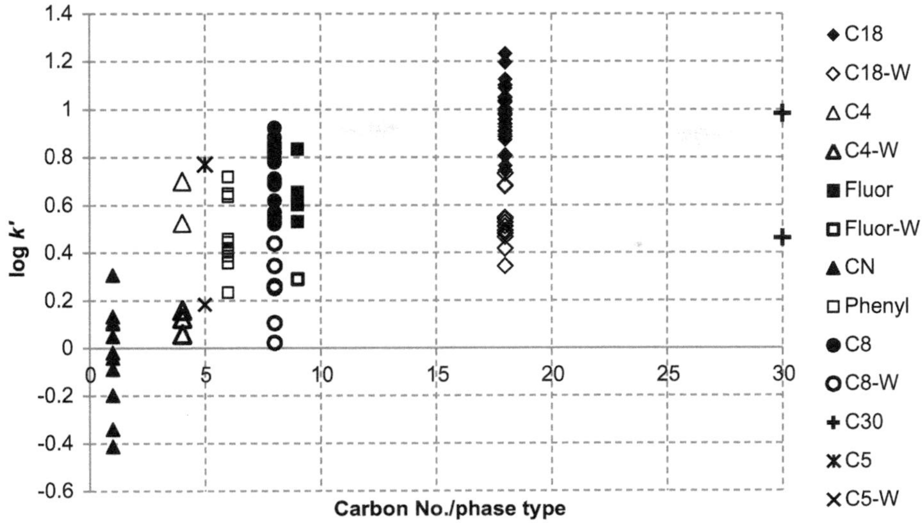

FIGURE 7.3.1 Values for log k' for various types of columns for ethylbenzene as a test analyte, and the mobile phase 50/50 (v/v) acetonitrile/aqueous buffer 60 mM phosphate at pH 2.8 (W indicates wide pore type stationary phase).

Another observation from the log k' values displayed in Fig. 7.3.1 is that columns with wide pores have lower log k' values as compared to small/middle-sized pore stationary phases. Larger pores of the solid support are typically associated with smaller surface area for the bonded phase support, and therefore a smaller Ψ value.

Some log k' values for toluene with a mobile phase consisting of 80% CH_3OH and 20% aqueous buffer 0.25 mM KH_2PO_4 at pH = 6 are indicated in Appendix 2a. More log k' values for acenaphthene in a mobile phase 65% CH_3OH/35% aqueous buffer 20 mM K_2HPO_4/KH_2PO_4 at pH = 7.0 (v/v) are given in Appendix 2b, and a number of log k' values for ethylbenzene in a mobile phase 50/50 acetonitrile/aqueous 60 mM phosphate buffer with pH = 2.8 are given in Appendix 2d.

Formulas for the Prediction of log k'

The evaluation of k' besides the experimental measuring should be done with sufficient accuracy for being useful for practical purposes. The evaluation of k' is a challenging goal, and the results provided in most cases are only approximations. The values for k' (log k') depend on too many factors related to compound, mobile-phase, and stationary-phase properties such that any calculation would require a number of parameters that are either not available or not precisely known [32].

One attempt to evaluate log k' for RP-HPLC separations was based on the energetic details of the process as explained by the solvophobic theory (see Section 4.3) [33]. A detailed description of the procedure for the evaluation of k' using solvophobic theory can be found in the literature [7,34] and is summarized in Section 4.3. The theory allows the evaluation of the equilibrium constant $K(X)$ for an analyte X in a partition process between a mobile phase L, and an immobilized phase S. The theory indicates that the formula for the retention factor k' can be obtained from an expression of the form:

$$\log k'(X) = a\mathcal{A}(X) - b + \Psi \tag{7.3.1}$$

where parameter a depends only on the nature of the stationary and mobile phase, b depends on polar functionality present in the molecule X, and Ψ is the phase ratio (see Section 4.3). The formula has been applied for the direct calculation of octanol/water partition coefficients [35], but the necessary parameters for other systems of the type mobile phase/stationary phase are not readily available.

A different approach to estimate $\log k'$ is based on linear solvation energy relationships theory [36]. This theory uses empirical correlations between different parameters characterizing the mobile phase mo, the stationary phase st, and the molecule X, such as Hildebrand solubility parameters and Kamlet–Taft solvatochromic parameters (see Section 13.1 [37,38]). This approach generates for the estimation of retention factor k' the following expression [39]:

$$\begin{aligned}\log k'(X) = {} & (\log k')_0 + m(\delta^2_{mo} - \delta^2_{st})V_X/100 + s(\boldsymbol{\pi}^*_{mo} - \boldsymbol{\pi}^*_{st})\boldsymbol{\pi}^*_X \\ & + a(\boldsymbol{\beta}_{mo} - \boldsymbol{\beta}_{st})\alpha_X + b(\boldsymbol{\alpha}_{mo} - \boldsymbol{\alpha}_{st})\boldsymbol{\beta}_X\end{aligned} \tag{7.3.2}$$

In Eq. (7.3.2) $(\log k')_0$ is a reference retention factor value, m, s, a, and b are the coefficients of the correlation, $V_X/100$ is the normalized molar volume of the solute, δ^2 is the Hildebrand solubility parameter, and $\boldsymbol{\pi}^*$, $\boldsymbol{\alpha}$ and $\boldsymbol{\beta}$ are the Kamlet–Taft solvatochromic parameters. Eq. (7.3.2) is usually simplified in different ways. One simplification is considering a system with a fixed mobile-phase composition and a fixed stationary phase, allowing the correlation of the retentions of different solutes. Another simplification considers one specific solute and stationary phase and considers only variations in the mobile-phase composition. This type of simplification allows comparison of results for different mobile-phase compositions (see Section 13.1).

Other expressions for the calculation of $\log k'$ are used for the comparison of different columns using the same analyte and the same mobile phase, and not for generating values of $\log k'$ to be used in the prediction of compound retention. For example, a suggested formula for $\log k'$ and based on column characteristics is the following:

$$\log k'(X) = a + bn + cn^2 + dd_{\mathrm{p}} + ed^2_{\mathrm{p}} + fc_{\mathrm{L}} + gc^2_{\mathrm{L}} + hE_{\mathrm{c}} \tag{7.3.3}$$

where n is the number of carbons in the alkyl chain of the ligand (e.g., 8 for C8, 18 for C18), d_{p} is the pore diameter of the particles of stationary phase (related to the surface area of the phase), c_{L} is the density of the active phase (ligand density), and E_{c} is a parameter describing the end-capping ($E_{\mathrm{c}} = 1$ for end-capped column and $E_{\mathrm{c}} = 0$ otherwise). Coefficients a, b, c, d, e, f, g, and h can be obtained from the regression fit from different columns for the same analyte and mobile phase [40].

Other formulas for the estimation of $\log k'$ are based on various correlations. For example, the correlation of $\log k'$ with octanol/water partition coefficient $\log K_{\mathrm{ow}}$ (for the same analytes) has been discussed in Section 5.4 and is given by Eq. (5.4.3). The prediction of $\log k'$ for the same compound and on the same stationary phase, with the change in the proportion of the same mobile phase constituents is given by Eq. (4.3.13). Prediction of $\log k'$ for the same compound and mobile phase but for different chromatographic columns is given by the following formula:

$$\log k'_{col.1} = a + b \log k'_{col.2} \tag{7.3.4}$$

In expression (7.3.4) a and b are constants depending on the two columns (col.1 and col. 2) and the mobile phase composition. All these correlations are useful for obtaining guidance regarding an expected value for log k', but usually not for an accurate value. Also, each formula requires values for correlation coefficients obtained from a large set of experimental data. Other estimations for log k' were reported in the literature, such as that obtained from the correlation with the values of van der Waals molecular area $\mathcal{A}_X$ of the analyte [18].

General Selectivity and Methylene Selectivity of Hydrophobic Columns

The capability of separating the analytes on a specific column (stationary phase) is a key factor for a successful HPLC analysis (see expression 4.2.55). For this reason, it would be very useful to describe with a unique parameter which column is better for the separation. However, because the chromatographic separation depends not only on the stationary phase but also on the composition of the mobile phase and on chemical nature of the analytes to be separated, the characterization of a stationary phase using selectivity α as a single parameter is not possible (see, e.g., [41]). However, it is possible for all the columns to be compared using a specific set of test compounds and a specific mobile phase. The use of such a set of identical selected conditions can provide values for a "standard" parameter $\alpha(X/Y)$ that can be used for the comparison of column selectivity related to a specific class of compounds. With such a parameter can be characterized: (1) hydrophobic character and (2) other types of interaction common for hydrophobic phases.

1) Methylene selectivity $\alpha(CH_2)$ is a parameter developed to characterize only the hydrophobic character of a stationary phase, and its capability to separate nonpolar molecules [42–47]. The parameter is defined as the selectivity α for two compounds, typically hydrocarbons, which are different by a CH_2 group (one example being the pair toluene and ethylbenzene). The formula for $\alpha(CH_2)$ is:

$$\alpha(CH_2) = \frac{k'(R - CH_2 - R)}{k'(R - R)} \tag{7.3.5}$$

The value of $\alpha(CH_2)$ depends, as expected, on the stationary phase and on the selected mobile phase. For the measurement of methylene selectivity the mobile-phase composition is not specifically recommended, but must be the same for the evaluated compounds.

The value of $\alpha(CH_2)$ is basically constant in a homologous series like the one from benzene to amylbenzene (for a given column and mobile phase). This is a result of the linear dependence that exists between log k' and the number of CH_2 units in a test homologous series. The linearity is expected, considering the linear dependence of the logarithm of retention factor log k' on log K_{ow} (see expression 5.4.3) and the fact that log K_{ow} is an additive parameter (expression 6.2.12) shows that the addition of a fragment f_n to a molecule produces a linear increment of log K_{ow}]. The linearity of the dependence of log k' on the number of CH_2 units was verified for various columns [45] and for various solvents [46]. Based on this observation, the values for log $\alpha(CH_2)$ can be obtained from the slope β of the trendline of the graph of log k' vs. the number of methylene groups in an alkylbenzene by the formula (see Fig. 7.3.2):

$$\log \alpha(CH_2) = \log \frac{k'(R - CH_2 - R)}{k'(R - R)} = \log k'(R - CH_2 - R) - \log k'(R - R) = \tan \beta \quad (7.3.6)$$

As an example, in Fig. 7.3.2 is illustrated the calculation of log $\alpha(CH_2)$ for a C18 monolithic column (Phenomenex Chromolith Performance RP-18) with the mobile phase 60% methanol 40% water (v/v) using the slope of the graph log k' vs. the number of methylene groups in an alkylbenzene. For the indicated phase system, log $\alpha(CH_2) = 0.2794$.

The values of $\alpha(CH_2)$ are different for different mobile phase compositions. The variation of $\alpha(CH_2)$ with the mobile phase composition for several columns is shown in Fig. 7.3.3. Five columns were evaluated: LiChrospher 100 RP-18 (LiChr.C18), Purospher RP-18 (Puro.C18), Symmetry-Shield RP18 (Symm.C18), LiChrospher 100 RP-8 (LiChr.C8), and Symmetry-Shield RP8 (Symm.C8). The evaluation used two different mobile phase systems, acetonitrile/water (ACN) and methanol/water (MeOH), at different concentrations [41]. All columns had 5 μm particle size, the column dimensions were 125 X 4 mm (LiChr. and Puro.), 150 × 3.9 mm (Symm), and except for LiChr.C8 the columns were end-capped.

When a column is selected for an application requiring the separation of hydrophobic compounds, the preferred characteristic should be a high $\alpha(CH_2)$ for obtaining a better separation. However, for compounds with polar groups and hydrophobic moieties, methylene selectivity may not be the only indicator for the best separation capability.

2) Since hydrophobic columns are frequently used for the separation of compounds that also contain polar or even ionic groups in addition to hydrophobic moieties, it is useful to have some test conditions in which besides hydrophobicity it is possible to assess polar interactions, π–π interactions, as well as steric selectivity of the stationary phase. For this purpose, specific pairs of test compounds have been recommended, as well as specific mobile phase compositions that provide α values relevant for separation.

A variety of test compounds and mobile phases can be selected as test conditions for providing α values capable of providing information regarding how strong the hydrophobic

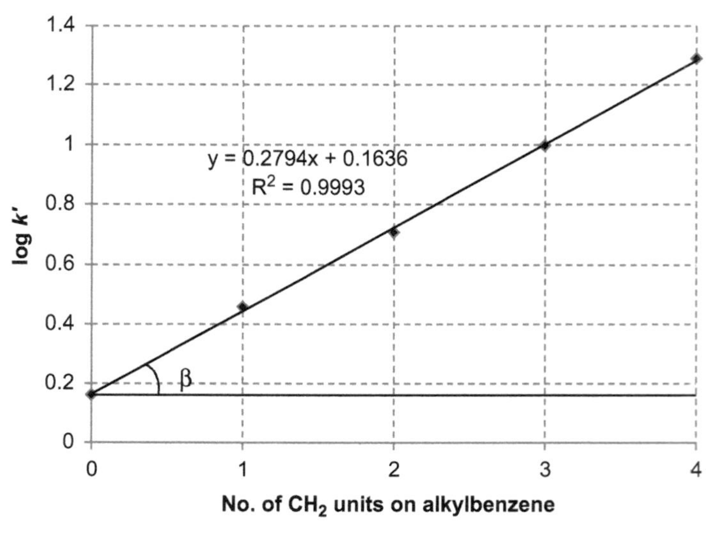

FIGURE 7.3.2 Calculation of log $\alpha(CH_2)$ for a C18 monolithic column (Phenomenex Chromolith Performance RP-18) with the mobile phase 60% methanol 40% water (v/v) for the series benzene to butylbenzene (log $\alpha(CH_2) = 0.2794$).

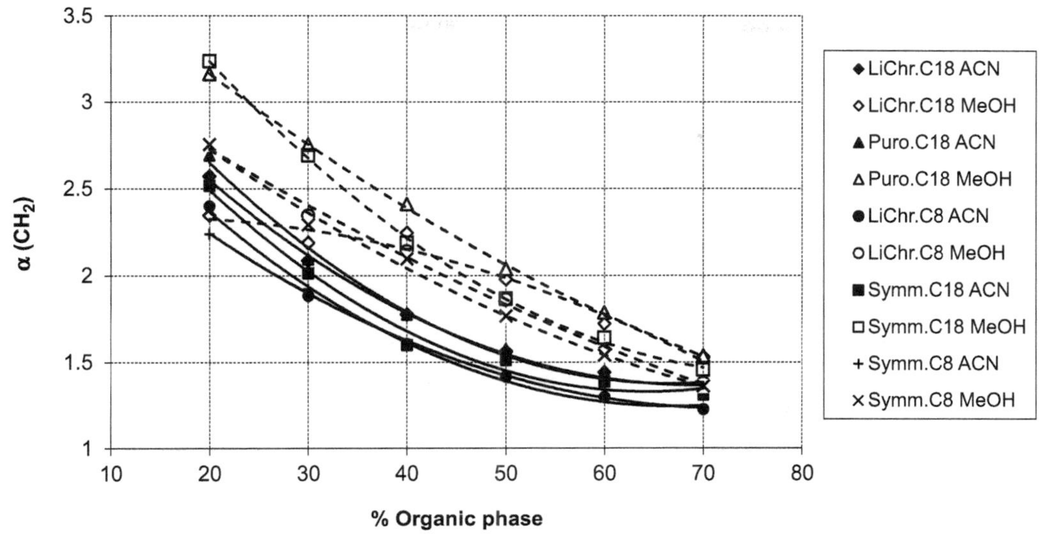

FIGURE 7.3.3 Variation of $\alpha(CH_2)$ for five RP columns and two mobile phase systems (acetonitrile/water and methanol/water) at different concentrations of the organic phase [41] (see also [7]).

interactions are and in addition other types of interactions such as polar, $\pi-\pi$, hydrogen bonding, and steric. Such interactions are in fact present in all hydrophobic columns. However, depending on the analytes and mobile phase such interactions are more or less manifested for a specific column. In order to put in evidence as well as possible multiple types of interactions in a hydrophobic column, some specific test compounds were recommended, as well as specific mobile phase compositions. One such test pair of compounds is amitriptyline and acenaphthene (shown in Fig. 7.3.4).

As an example of selectivity comparison, the values for log α for the separation of amitryptiline/acenaphthene with the mobile phase 65% CH_3OH/35% aqueous buffer 20 mM K_2HPO_4/KH_2PO_4 at pH = 7.0 (v/v) are given in Appendix 2b. The appendix also indicates the values for log k' for acenaphthene in the same mobile phase. A plot of the values for log k' for acenaphthene (on the Y axis) and for log α for the pair amitriptyline/acenaphthene (on the X axis) for the columns listed in Appendix 2b is shown in Fig. 7.3.5 (column number can be found in Appendix 2b).

Depending on the type of analytes to be separated, the selected column should have a larger log k' or a larger log α. For compounds with a strong hydrophobic character, the separation typically requires a strongly hydrophobic column, therefore a column with large log k' is needed. For the separation of compounds with higher polarity, the separation works better when a column with a larger log $\alpha_{\text{amitriptyline/acenaphthene}}$ is selected. However, the

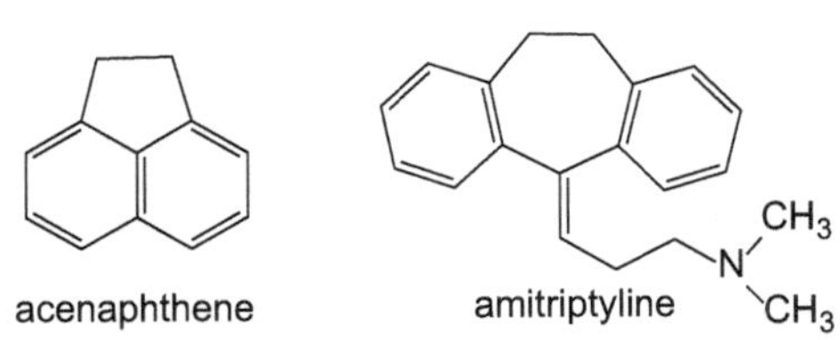

FIGURE 7.3.4 Chemical structures for acenaphthene and amitriptyline.

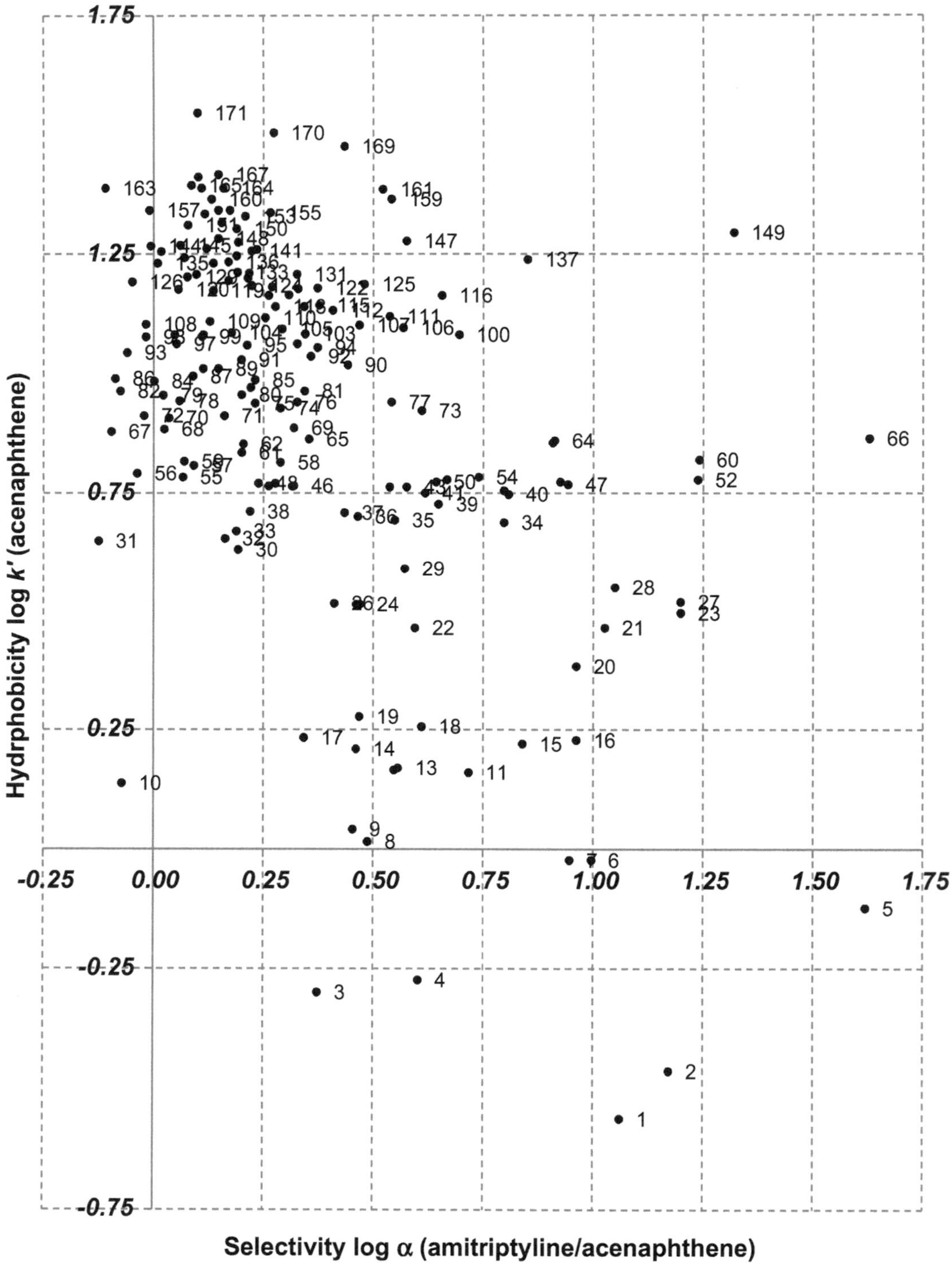

FIGURE 7.3.5 Plot of log k' vs. log α for a number of hydrophobic columns. The values of log k' for acenaphthene and the values for log α for the pair amitriptyline/acenaphthene for the columns listed in Appendix 2b [48] (column numbers corresponding to the table are also indicated in the figure). Mobile phase 65% CH_3OH/35% aqueous buffer 20 mM K_2HPO_4/KH_2PO_4 at pH = 7.0 (v/v).

mobile phase characteristics acceptable for the selected column should always be included in the selection decision. Other tests are designed to describe selectivity α, for specific compounds, but in addition other column properties will be further discussed.

Peak Asymmetry for RP-HPLC Columns

Peak asymmetry in HPLC is dependent on the analyzed compound, mobile phase composition, and also on the stationary phase of the chromatographic column. To the peak asymmetry in RP-HPLC the main contributors are the interactions of the analyte with the solid phase that do not affect all the analyte molecules equally. This is caused by the analyte having more than one molecular form in the mobile phase, and/or by the stationary phase interacting selectively with only a part of the analyte molecules. An illustration of this process is schematically indicated in Fig. 7.3.6.

For RP-HPLC columns, in order to put in evidence the column potential to generate asymmetric peaks, it is necessary to use specific test compounds. A column showing significant asymmetry for such compounds may show no asymmetry, for example, for a hydrocarbon. Two common test compounds used to put in evidence asymmetry are quinizarin and amitriptyline. Quinizarin is typically used for the evaluation of metal activity in a column, and amitriptyline, which is a basic compound ($pK_a = 9.4$), is used for the characterization of silanol activity based on its peak shape. Since quinizarin is a metal chelating reagent and its retention behavior is indicative of the presence or absence of metal ions in the stationary phase, low metal activity is characterized by symmetric peaks of quinizarin, while high metal activity by strong tailing. Amitriptyline has besides the hydrophobic interaction with the stationary phase, an additional interaction with the silanol groups. Peak asymmetry generated by quinizarin and amitriptyline is not correlated, since each compound has a different type of additional interaction with the stationary phase. Values for *As* measured for amitriptyline and for quinizarin for several common columns, in a mobile phase 80% CH_3OH and 20% aqueous potassium phosphate buffer with conc. 5 mM and with pH = 7 are shown in Table 7.3.1 [49].

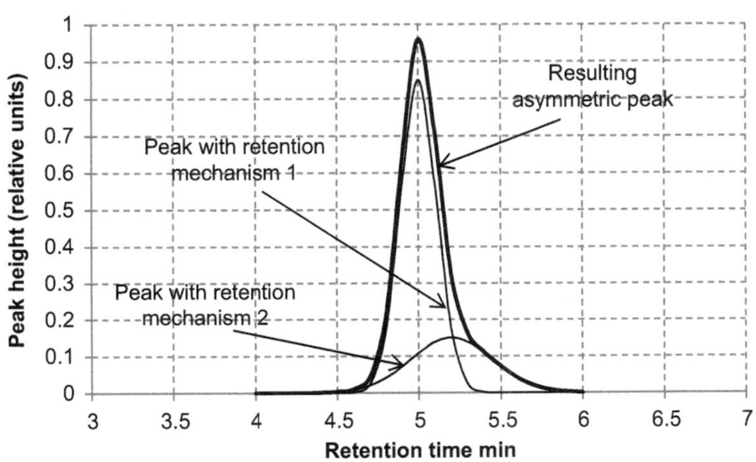

FIGURE 7.3.6 Illustration of the formation of asymmetrical chromatographic peaks as a result of two mechanisms of retention that do not affect equally all molecules of the analyte.

TABLE 7.3.1 Values for Asymmetry *As* for Various Columns as Measured for Amitriptyline and for Quinizarin [49]

	Column	*As* Amitriptyline	*As* Quinizarin
1	ACE C18	1.03	1.07
2	Xterra MS C18	1.12	1.12
3	Intersil ODS3	1.26	1.24
4	Luna C18(2)	1.52	1.46
5	Hypersil Elite C18	1.61	1.09
6	Discovery C18	1.78	1.37
7	Waters Spherisorb ODS1	1.88	2.27
8	Intersil ODS2	1.93	1.27
9	Symmetry C18	2.07	1.31
10	Hypersil BDS-C18	2.21	1.26
11	Partisil ODS2	2.74	3.07
12	Zorbax XDB-C18	3.05	1.07
13	Novapak C18	3.05	2.42
14	Zorbax SB-C18	3.07	1.43
15	Nucleosil C18	3.19	2.41
16	Vidac Selectapore 300M	6.50	1.34
17	Cosmosil C18rAR-II	8.30	1.16
18	μ Bondapack C18	8.99	2.56

Various Other Parameters and Tests for RP-HPLC Column Characterization

Some other column characterization procedures have been developed and they serve for column comparison and selection [45,50,51]. One such procedure (USP approach) uses a standard mixture of compounds known as Standard Reference Material (SRM) 870 and some information from the column manufacturer for column characterization. The SRM 870 material contains uracil (27.1 μg/g), toluene (1430 μg/g), ethylbenzene (1730 μg/g), quinizarin (90.8 μg/g), and amitriptyline hydrochloride (2740 μg/g) in a methanol solution. The test solution is used for the evaluation of several parameters. These parameters are the following: H_y given as capacity factor k' for ethylbenzene, chelating tailing factor *CTF* for quinizarin, capacity factor k' for amitriptyline *CFA*, tailing factor for amitriptyline *TFA*, and the bonding density *BD* in μmol/m^2 of the stationary phase [52]. The test is performed using as mobile phase composition 80% methanol and 20% aqueous phosphate buffer of pH = 7. The values for the parameters H_y, *CTF*, *CFA*, *TFA* and *BD* are reported for about 117 C18 columns (USP approach), and are listed in Appendix 2c [53]. Besides H_y, *CTF*, *CFA*, *TFA*, and *BD* the USP test mixture allows the evaluation of methylene selectivity as $\alpha(CH_2) = k'_{ethylbenzene}/k'_{toluene}$.

A number of other tests used for different characterizations of hydrophobic columns are given in Table 7.3.2.

Some other tests were reported in the literature [51,73,74].

The results generated with different tests are important for column characterization. As an example, such results are given in Table 7.3.3 for several conventional C18 columns, polar end-capped columns, and polar embedded columns [45] using *Test 17* from Table 7.3.2.

TABLE 7.3.2 Various Tests for the Characterization of Hydrophobic Columns

No.	Test Conditions	Parameters Measured	Ref
1	Test analytes: (1) uracil (t_0), (2) thiourea (t_0), (3) amylbenzene, (4) butylbenzene, (5) triphenylene, (6) *o*-terphenylene, (7) caffeine, (8) phenol, (9) benzylamine Eluent: MeOH–H_2O 80:20 v/v Eluent: MeOH–H_2O 30:70 v/v Eluent: MeOH-0.02 m phosph. buffer (pH = 7.6) 30:70 v/v Eluent: MeOH-0.02 m phosph. buffer (pH = 2.7) 30:70 v/v	• Hydrophobicity $\alpha(CH_2) = k'(3)/k'(4)$ • Alkyl ligands amount = $k'(3)$ • Steric selectivity = $k'(5)/k'(6)$ • Hydrogen bond capacity = $k'(7)/k'(8)$ • Ion exchange capacity (IEC) for pH > 7 = $k'(9)/k'(8)$ • IEC for pH < 3 = $k'(9)/k'(8)$	[54]
2	Test analytes: (1) thiourea (t_0), (2) aniline, (3) phenol, (4) *N,N*-dimethylaniline, (5) *p*-ethylaniline, (6) toluene, (7) ethylbenzene, (8) *m*-toluidine Eluent: MeOH–H_2O 55:45 v/v; $T = 40$ °C. $\lambda = 254$ nm	• Hydrophobicity $\alpha(CH_2) = k'(7)/k'(6)$; • Silanol activity $As(4)$, $As(5)$, $As(8)$ • Retention factors for basic solutes	[55]
3	Test analytes: (1) uracil (t_0), (2) triphenylene, (3) *o*-terphenyl Eluent: MeOH–H_2O 55:45 v/v; $T = 40$ °C. $\lambda = 254$ nm	• Shape selectivity = $k'(2)/k'(3)$;	[56]
4	Test analytes: (1) uracil (t_0), (2) benzene, (3) anthracene Eluent: AcCN–H_2O 65:35 v/v; $T = 40$ °C. $\lambda = 254$ nm	• Hydrophobicity = $k'(3)/k'(2)$	[57]
5	Test analytes: (1) uracil (t_0), (2) anthracene, (3) *N,N*-diethyl-*m*-toluamide, (4) nitrobenzene, (5) benzene, (6) toluene Eluent: 100% AcCN, $T = 40$°C	• Silanol activity = $k'(3)/k'(2)$ • Silanol index $k'(4)$ • Hydrophobicity $\alpha(CH_2) = k'(6)/k'(5)$ • Theoretical plate N from $W_b(6)$	[57]
6	Test analytes: (1) nortriptyline, (2) nicotine, (3) amphetamine, (4) pyridine, (5) benzylamine, (6) codeine, (7) quinine, (8) procainamide, (9) diphenhydramine Eluent: AcCN, MeOH, THF at pH 3.0 and neutral pH	• Retention factors • Asymmetry factors • Column efficiency	[58,59]
7	Test analytes: (1) naphthalene, (2) acetonilidine, (3) phenol, (4) benzonitrile Eluent: MeOH–H_2O and AcCN–H_2O	• Various characterization of columns for RP HPLC	[60]
8	Test analytes: (1) amitriptyline, (2) naphthalene, (3) acetylnaphthalene, (4) dipropylphthalate, (5) butylparabene Eluent: neutral pH, 35% 20 mm K_2HPO_4–KH_2PO_4 buffer pH = 7.0% and 65% MeOH, acidic pH, 20% 50 mm H_3PO_4–KH_2PO_4 buffer pH = 3.0% and 80% AcCN	• Estimation of the surface silanols • Markers of packing • Hydrophobicity • Determination of the packing polar selectivity	[28,29,61]

(Continued)

TABLE 7.3.2 Various Tests for the Characterization of Hydrophobic Columns—cont'd

No.	Test Conditions	Parameters Measured	Ref
9	Test analytes: (1) aniline, (2) phenol, (3) benzene, (4) toluene	• Hydrophobicity $(k'(3) + k'(4))/2$ • Silanophobicity (silanol index)	[62,63]
10	Test analytes: 1. benzo[a]pyrene (BaP), 2. phenantro[3,4-c]-phenantrene (PhPh), 3. tetrabenznaphthalene (TBN) Eluent: AcCN–H_2O 85/15, flow rate 2 mL/min. $\lambda = 254$ nm	• Column selectivity: • Monomer column $\alpha(3/1) > 1.7$ • Polymer column $\alpha(3/1) < 1$ • Oligomer column $1 < \alpha(3/1) < 1.7$	[64,65]
11	Test analytes: big molecules such as fullerenes Eluent: *n*-hexane	• Molecular shape recognition	[66]
12	Test analytes: macromolecules such as polymers	• Silanophilic interactions	[67,68]
13	Test analytes: (1) bronopol, (2) acetophenone, (3) benzene, (4) toluene Eluent: AcCN–H_2O 80:20	• Column classification according to selectivity to bronopol	[69]
14	Test: analytes: 1. bronopol Eluent: 100% water	• Column classification according to selectivity to bronopol	[69]
15	25 test analytes with different properties	• QSRR method based on structure–retention relationships	[70,71]
16	Test analytes: (1) naphthalene sulfonic acids Eluent: 0.4 m Na_2SO_4	• Surface coverage homogeneity • Silanol activity	[30,72]
17	Test analytes: (1) butylbenzene, (2) amylbenzene, (3) propylbenzene, (4) ethylbenzene, (5) toluene, (6) benzene, (7) caffeine, (8) phenol, (9) benzylamine, (10) *p*-hydroxybenzoic acid, (11) sorbic acid, (12) benzoic acid, (13) salicylic acid, (14) *p*-toluic acid, (15) triprolidine, (16) chlorpheniramine, (17) diphenyhydramine, (18) tricyclic antidepressants such as desipramine, amitriptyline Eluent: 65% AcCN, H_2O buffer 20 mM phosphate, pH = 7 Eluent: 15% AcCN, H_2O buffer 20 mM phosphate, pH = 7 Eluent: 30% AcCN, H_2O buffer 20 mM phosphate, pH = 7 Eluent: 30% AcCN, H_2O buffer 20 mM phosphate, pH = 2.5 Eluent: 75% AcCN, H_2O buffer 20 mM phosphate, pH = 2.5	• Hydrophobicity $k'(1)$, $k'(2)$ • Methylene selectivity $\alpha(CH_2)$ for benzene to amylbenzene • Hydrogen bonding capacity $\alpha = k'(7)/k'(8)$ • Silanol activity pH = 2.5 $\alpha = k'(9)/k'(8)$ • Silanol activity pH = 7 $\alpha = k'(9)/k'(8)$ • Retention acids $k'(i)$, $i = 10–14$ • Peak asymmetry $As(15)$, $As(16)$, $As(17)$ at pH = 2.5, $As(18)$ • Theoretical plate N from $W_b(5)$	[45]

Hydrophobic Subtraction Model for Selectivity Characterization

A more complete description of column separation capabilities is based on an empirical procedure known as the *hydrophobic subtraction model* [75–78]. This procedure is based on fitting to a specific retention model using multilinear regression. The procedure starts with

TABLE 7.3.3 Comparison on Several Test Parameters for Conventional, Polar End-Capped, and Polar Embedded Columns [45]

Column type	N	k'_{BB}	$\alpha(CH_2)$	$\alpha_{caffeine/phenol}$	$\alpha_{benzylamine/phenol\ pH = 2.5}$	$\alpha_{benzylamine/phenol\ pH = 7}$
	Plates m^{-1}	Hydro-phobicity	Methylene Selectivity	Hydrogen Bonding	Silanol Activity pH = 2.5	Silanol Activity pH = 7
CONVENTIONAL C18						
Luna 5 mm C18(2)	116,889	9.11	0.176	0.21	0.059	0.131
Inertsil ODS(3)	96,669	8.63	0.171	0.261	0.045	0.147
Zorbax XDB C18	95,997	8.57	0.187	0.213	0.088	0.134
Symmetry C18	91,515	9.23	0.181	0.224	0.049	0.147
Hypurity Elite C18	87,816	2.96	0.155	0.31	0.061	0.378
Zorbax SB C18	84,568	6.74	0.179	0.283	0.079	0.412
Prodigy ODS(3)	108,384	9.82	0.184	0.21	0.051	0.13
POLAR END-CAPPED						
Keystone Aquasil	109,457	4.41	0.147	0.725	0.131	1.659
Aqua C18	87,666	8.67	0.176	0.251	0.094	0.149
YMC Hydrosphere	109,430	6.25	0.167	0.27	0.055	0.14
YMC ODS-Aq	90,049	7.83	0.171	0.262	0.094	0.169
Prontosil C18AQ	92,800	7.04	0.173	0.303	0.079	0.254
Metasil AQ	88,846	7.77	0.158	0.217	0.073	0.162
Synergi Hydro-RP	113,187	10.07	0.174	0.24	0.063	0.203
POLAR EMBEDDED						
Zorbax Bonus-RP	110,156	3.86	0.147	0.201	0	0.167
Polaris C18-A	105,481	4.38	0.165	0.2	0.102	0.121
Discovery RP-Amide	115,577	3.34	0.15	0.166	0.05	0.097
SymmetryShield RP18	109,530	6.15	0.152	0.164	0.013	0.098
Supelco ABZ+	95,551	1.88	0.143	0.182	0	0.23
Experimental urea	54,458	6.25	0.157	0.133	0	0.8
Polaris Amide C18	87,744	5.11	0.139	0.123	0	0.08

The columns were evaluated using *Test 17* from Table 7.3.2 with parameters $k'_{BB} = k'_{butylbenzene}$, $\alpha_{caffeine/phenol} = \alpha(7,8)$, $\alpha_{benzylamine/phenol\ pH = 2.5} = \alpha(9,8)_{pH = 2.5}$, $\alpha_{benzylamine/phenol\ pH = 7} = \alpha(9,8)_{pH = 7.5}$.

expression (4.2.55) for the selectivity α, which indicates that α depends on the equilibrium constants for the two analytes (solutes) to be separated. When the mobile phase is kept unchanged, and selecting one of the solutes to be ethylbenzene (EB), the selectivity can be considered as dependent only on the chemical nature of the analytes (solutes) to be separated and the chromatographic column properties. Taking into consideration different types of interaction mechanisms for a separation on an RP-HPLC column, the following expression can be written for log α:

$$\log \alpha = \log\left(\frac{k'}{k'_{EB}}\right) = \eta' H^* - \sigma' S^* + \beta' A^* + \alpha' B^* + \kappa' C^* \tag{7.3.7}$$

The contribution of each term in expression (7.3.7) has a specific meaning, and $\eta' H^*$ accounts for column capability to separate the analyte from ethylbenzene based only on hydrophobic interactions, $\sigma' S^*$ accounts for steric interactions, $\beta' A^*$ accounts for hydrogen bonding between a basic solute and the acidic groups of the stationary phase, $\alpha' B^*$ accounts for hydrogen bonding between an acidic solute and basic groups of the stationary phase, and $\kappa' C^*$ accounts for cation exchange type and/or ion–ion interactions. Parameters η', σ', β', α', and κ' depend on solute properties (and mobile phase composition), and H^*, S^*, A^*, B^*, and C^* depend only on column properties. The values for H^*, S^*, A^*, B^*, and C^* can therefore be utilized for comparing different columns and consequently a possibility to establish when two different columns will have similar separation capabilities and can replace each other [79].

Regarding the interpretation of each type of interaction contributing to the column properties, the hydrophobic character, as well as the hydrogen bonding and ionic interactions described by the parameters H^*, A^*, B^*, C^*, respectively, were previously discussed (see Section 4.3). However, the steric interactions described for the column by parameter S^* were not discussed previously. This type of interaction can be noticed when molecules of different shape are retained by the chromatographic column. Differences in retention caused by steric differences can be evaluated using two aromatic hydrocarbons: one with a twisted structure and the other with a planar one. The two compounds should have practically identical hydrophobicity, but in the same time different volumes. The selectivity factor α for these compounds is a measure for the steric selectivity. For example, the selectivity factor can be evaluated for the following pairs: triphenylene/ortho-terphenyl [54], benzo[a]pyrene/tetrabenzonaphthalene [80], or benzoic acid/sorbic acid [81,82].

The values for both η', σ', β', α', κ' and H^*, S^*, A^*, B^*, C^* were obtained for a specific mobile phase that consisted of 50/50 acetonitrile/aqueous buffer containing 60 mM phosphate at pH = 2.8, or at pH = 7.0. The values for parameters η', σ', β', α', κ' and for parameters H^*, S^*, A^*, B^*, C^* were obtained initially based on a large number of measurements for the retention factors k' on a specific set of compounds and a set of columns. The procedure is well explained in the literature (see, e.g., [75]). The basic idea was to initially isolate a group of compounds that are separated exclusively based on hydrophobic interactions, for which it can be assumed that σ', β', α', and κ' are all zero. In such case expression (7.3.7) would be reduced to:

$$\log \alpha = \log\left(\frac{k'_{comp}}{k'_{EB}}\right) = \eta'_{comp} H^* \tag{7.3.8}$$

Expression (7.3.8) applied to two different columns will generate the expression:

$$\log \alpha_{col.2} = \left(\frac{H^*_{col.2}}{H^*_{col.1}}\right) \log \alpha_{col.1} \tag{7.3.9}$$

An arbitrary value $H^*_{col.1} = 1$ must be selected at this point. Using expression (7.3.9) the values for H^* for a number of columns can be obtained, using k' values for ethyl benzene and other compounds retained only based on hydrophobic interactions. For all these hydrophobic compounds the values η'_i can be obtained (where i indicates different compounds).

Using a large number of compounds with polar groups that have, besides hydrophobic interactions, other types of interactions with stationary phases, a deviation from expression (7.3.8) can be noticed. The values for these deviations can be written in the form:

$$\Delta_{i,col.j} = \log \alpha_{i,col.j} - \eta'' H^*_{col.j} \tag{7.3.10}$$

where η'' is the average of η'_i values. The test compounds for all tested columns are further grouped based on cross-correlations such that the compounds from each group manifest as much as possible one specific type of interaction and very weak interactions of other types. From averages of Δ values for each group, the values for S^*, A^*, B^*, and C^* were calculated. Once H^*, S^*, A^*, B^*, and C^* are established for each column, using multiple regression the values for η', σ', β', α', and κ' are generated for the whole set of test compounds, and further small adjustments to H^*, S^*, A^*, B^*, and C^* are made using multiple regression of log α vs. the values of η', σ', β', α', and κ'.

One addition to the previous summary description of establishing the values for H^*, S^*, A^*, B^*, and C^* is related to the dependence on pH of one group of interactions, namely that for cation exchange type and/or ion–ion interactions. The group of test compounds used for the determination of C^* are made of strong bases, and their retention (values for log $\alpha_{i,col.j}$ and therefore of $\Delta_{i,col.j}$) depends on the pH of the mobile phase. For this reason, C^* is indicated by two values, $C^*(2.8)$, which is the value obtained with the mobile phase at pH = 2.8, and $C^*(7.0)$, which is the value obtained with the mobile phase at pH = 7.0.

A considerable number of studies were later dedicated to the task of establishing H^*, S^*, A^*, B^*, and C^* for a variety of columns [40,83–91]. These parameters are similar but not identical to those used in the USP approach. For example, these parameters do not specifically indicate peak tailing, while the USP approach does. The values for the parameters H^*, S^*, A^*, B^*, and C^* are reported for about 674 different alkyl-silica columns (Product Quality Research Institute or PQRI approach), and are listed in Appendix 2d [53]. The parameters were finally measured for the mobile phase of 50/50 acetonitrile/aqueous buffer containing 60 mM phosphate, and using the following test compounds: thiourea, amitriptyline, 4-butylbenzoic acid, N,N-diethylacetamide, 5-phenyl-1-penthanol, ethylbenzene, N,N-dimethylacetamide, 5,5-diphenylhydanoin, toluene, nortriptyline, acetophenone, mefenamic acid, *p*-nitrophenol, anisole, 4-hexylaniline, *cis*/*trans* chalcone, benzonitrile, and berberine [52].

The hydrophobic subtraction model provides a useful procedure for the column comparison. A similarity parameter F between two columns can be calculated for each column pair by the formula:

$$F_{col.1-col.2} = \left[\left(H^*_{col.1} - H^*_{col.2}\right)^2 + \left(S^*_{col.1} - S^*_{col.2}\right)^2 + \left(A^*_{col.1} - A^*_{col.2}\right)^2 + \right. \\ \left. + \left(B^*_{col.1} - B^*_{col.2}\right)^2 + \left(C^*_{col.1} - C^*_{col.2}\right)^2\right]^{1/2} \tag{7.3.11}$$

Column comparison based on expression (7.3.11) can be done by accessing the USP site [53]. The utilization of expression (7.3.7) for the calculation of α for two specific compounds on a given column with known H^*, S^*, A^*, B^*, and C^* is limited by the need to know all the parameters η', σ', β', α', and κ' for both compounds. In comparison with the enormous number of chemicals, only a very limited number of compound parameters are reported in the literature (see, e.g., [7,78]).

Tests for the Evaluation of Aging of the Chromatographic Column

Stability of bonded phases can be evaluated periodically by measuring peak performances of test compounds. Also, for some columns the resilience to aging is reported after the column stability has been evaluated in an aging study where the column is subjected to different stress conditions with the aim of observing the performance modifications in time. One such test consists of continuously passing through the column a mobile phase containing acetonitrile and phosphate buffer with a specific pH and at a specified temperature (40–60°C), followed by column performance tests (for k', α, N). More detailed studies of column aging, including both evaluation of chromatographic parameter changes and material characterization with FTIR, ^{1}H-NMR, and solid state ^{13}C- and ^{29}Si-NMR of hydrolyzed bonded phases were also reported [92].

7.4 COMMON AND SPECIAL RP COLUMNS

A large variety of RP columns are commercially available (see, e.g., Appendix 2c), and the same type of stationary phase can be available in a number of formats: column dimensions, particle size, and particle type (fully porous or core shell). Although a distinction between common and less common columns is subjective, some RP-HPLC columns are more frequently reported as being used. These columns can be indicated as common. A number of more special types of columns are also available for RP-HPLC analyses. Comments regarding common and special types of RP columns are discussed in this section.

Common RP Columns

Good sources for finding commercially available columns, as well as for the guard columns recommended for column protection, are the vendor publications (for a list of several column vendors see, e.g., Appendix 2c). Published by the vendors are available numerous catalogs and websites (see, e.g., [93–97]). In these publications are reported the new columns that are continuously introduced onto the market, as well as many older columns that are still in use. A number of studies are published about the new trend in column technology [98–100].

Since common columns cover a wide range of types and manufacturers, only common characteristics of these columns are further commented on, and not particular columns. Older types of common columns were typically limited to C8, C18, phenyl, and CN types, with fully porous particles having 5 μm diameter. Some older columns did not have monodisperse particles (had irregular shape particles) and silica purity was not as high as *type B*. Several characteristics are emerging and now they are common for RP-HPLC columns: (1) use of pure silica (*type B*), (2) monodisperse spherical particle shape, (3) smaller particle size (3 μm, or even 1.7 μm diameter), (4) core–shell particles, (5) columns with the bonded phase generated by using trifunctional reagents, (6) columns with wider pH range utilization, (7) columns with wettability to 100% water, (8) polar embedded hydrophobic columns, (9) a wider range of hydrophobic phases capable of developing additional interactions beside hydrophobic, (10) columns with very low column bleed, (11) columns with longer utilization time (up to 1000–1500 injections without degradation), and (12) other improvements in column construction.

1) The utilization of *type* B silica of high purity has become common. Advantages of pure silica include the elimination of undesired metal–analyte types of interactions and better resilience to pH outside the range 2–7.

2) Technology for producing spherical monodisperse hydrated silica particles allows improvement of column efficiency, and these types of particles have become standard [101,102], while irregular particles of a range of dimensions are no longer common.

3) Smaller particles (3 μm, or even 1.7 μm diameter) that provide higher efficiency (higher N/m) are common. Although smaller particles generate higher backpressure, the newer UPLC instruments are capable of overcoming pressures as high as 1200 bar.

4) Core–shell particles have become frequently utilized [103]. For example, one series of core–shell columns is the Kinetex (from Phenomenex), available in C8, C18, XB-C18 (which has butyl side chains for protecting against silanol access), phenylhexyl, and pentafluorophenyl (PFP), with particle sizes of 2.6 μm and 1.7 μm [104,105]. Another series of core–shell columns is Ascentis Express (from Supelco) available in C18, phenyl-hexyl, C8, and pentafluorophenyl, and one more series is Accucore from Thermo Scientific that offers C18 and PFP columns as well as a column end-capped with polar groups (Accucore aQ). Agilent offers the Poroshell series of core–shell columns. The newer columns bring significant advantages such as higher efficiency (larger number of theoretical plate N values), and core–shell type columns using 1.7-μm particles can reach as much as $N/\mathrm{m} = 300{,}000$. The recommended sample load is typically lower for core–shell stationary phases compared to fully porous ones. However, the frequent use of more sensitive detection based on MS or MS/MS requires lower sample loads which is totally adequate for core–shell columns.

5) The use of trifunctional reagents for silica surface derivatization is now common. This type of reagent allows a more uniform derivatization of the silica surface, with fewer remaining underivatized silanol groups, higher phase stability, and higher carbon load.

6) Progress has been made in extending the pH range of utilization for silica-based columns. This is achieved by different procedures, one being the use of trifunctional derivatization reagents. However, significant advantages are offered by the use of organic/inorganic ethylene-bridged support (BEH type columns or with TWIN technology), which is now more frequently used. For example, X-Terra Shield C18 column from Waters is stable in the range of pH from 2 to 12, and Gemini-NX C18 from Phenomenex is stable in the range of pH from 1 to 12 (both using ethylene-bridged silica). Various other new technologies were introduced for extending the range of pH where the columns are stable. Among these are the CSH technology and end-capping with more voluminous groups (e.g., *tert*-butyl). CSH technology with particles incorporating a low level of surface charge also improves sample loadability and peak symmetry [106]. CSH technology can be applied to organic/inorganic ethylene-bridged phases bringing further improvements for pH stability of the column.

7) Wettability to 100% water in the mobile phase is another quality achieved. This is done using embedded polar groups and using polar end-capping such as in the Synergy Hydro-RP and Aqua type columns from Phenomenex, SunShell RP-AQUA from Nacalai, and Aquasil C18 from Restek.

8) Polar embedded hydrophobic columns are also common. These columns offer better wettability to high water content of mobile phase, and also additional interactions besides hydrophobic. Common commercially available columns contain embedded urea, amide, carbamate, or ether groups with C8, C18 hydrophobic chains, or with phenyl groups. It was demonstrated that the embedded polar groups do not necessarily decrease the column hydrophobicity, but they offer better wettability and capability to generate additional types of interactions with the analytes besides hydrophobic [45,107,108].

9) A variety of columns capable of producing additional interactions besides being strongly hydrophobic are also offered [109]. For example, fluorinated bonded phase columns have become more common. In these columns, a fluorinated fragment is attached to the silica surface with a short handle. Columns with bonded groups such as perfluorohexyl straight chain, perfluorohexyl branched chain, perfluorooctyl, perfluoropropyl, perfluorododecyl, pentafluorophenyl, pentafluorophenylalkyl, and pentafluorophenylpropyl are available. These phases cover a wide range for the k' values indicating that the fluorinated materials offer a wide range of hydrophobicity. By inspecting the values from Appendix 2c, the fluorinated phases tend to have high values for parameters $C^*(2.8)$ and $C^*(7.0)$ indicating propensity for ion–ion interactions in addition to the hydrophobic character. The fluorinated phases offer an alternative material for separations and were proven very useful in particular separations [105,110].

10) One other typical characteristic of newer columns is very low bleed, which is important for sensitive detectors such as MS and MS/MS. Progress made in silica surface derivatization with bi- and trifunctional reagents, and better end-capping techniques assure that the access of the mobile phase to the silica base is limited and the stability of the stationary phase is higher.

11) Also, newer columns display extended lifetimes with up to 1500 injections without significant degradation of the separation. This is a result of better derivatization techniques, higher silica purity, and use of additional techniques for protecting the access of mobile phase to the covalent connection between the silica surface and the bonded phase (see Section 6.1).

12) Other improvements to the column technology such as better and more uniform packing of the stationary phase in the body of the column, and special type of frits at the end of the column that allow a uniform distribution of mobile phase in the column body are common.

Special Types of Hydrophobic Phases

A number of more special columns are commercially available. Among these types of columns, the following can be mentioned: (1) columns with linear aliphatic bonded phase with an unusual number of carbons (C3, C4, C5, C27, C30), (2) columns with silica coated with a polymer, (3) monolithic columns, (4) silica hydride columns, (5) graphitic columns, (6) columns with zirconia support, and (7) organic polymer-type columns.

1) Besides the common C8 and C18 bonded phases, columns with bonded alkyl chains (on silica support) with different numbers of carbons were also commercialized. Two distinct

groups of such columns were made, one with a very short carbon chain, and the other with a long carbon chain. In the group with a very short carbon chain are C3 columns (L56 in USP classification), such as Zorbax StableBond 80A C3, Zorbax StableBond 300A C3, C4 columns (L26 in USP classification) such as Accucore 150-C4, ACE 5C4, Aeris WIDEPORE XB-C4, Bio-Basic 4, Biobond C4, Epic C4, Genesis C4 EC 120A, Ultra C4, ProntoSIL 300C4, and C5 columns such as Discovery BIO Wide pore C5, Jupiter 300C5, and Luna C5. These columns are designed to offer a lower hydrophobic interaction and some have wide pores for utilization in large molecule separations such as proteins.

The other group includes columns with an extra-long aliphatic chain, among these some having C27 chains such as Cogent UPHOLD C27 (from MicroSolv), CAPCELL CORE AQ (from Shiseido), and columns with C30 chains (L62 in UPS classification) such as Accucore C30, Develosil C30-UG-5, ProntoSIL 120-3-C30, ProntoSIL 200C30, ProntoSIL 300C30, and ProntoSIL 300C30 EC. The long alkyl chain stationary phases were found to be useful for the separation of small only partly hydrophobic molecules. These phases are more retentive for polar and nonpolar analytes than are most polar-embedded and even high-coverage C18 phases. Very long chain phases (e.g., C30) may offer greater pH stability than do C8 and C18 phases. They are also more resistant to phase collapse under high aqueous conditions than are C18 phases. This behavior may be explained by the resistance of long chains to conformation changes in the presence of water at the column temperature, or it may be related to a lower density of C30 chains on the silica surface for similar C% content as the C18 phases [88]. A column with cholesterol groups bonded on silica is also commercially available (Cogent UDC Cholesterol from MicroSolv).

2) Columns with different types of silica coating technology were developed with the goal of reducing free silanol activity. As an example, EnviroSep type phases (from Phenomenex) contain silica and polymer support and a hydrophobic bonded phase. Other columns are made using a variety of technologies. For example, Capcell Pak type columns (Shiseido Co. Ltd.) involve a surface coating of the silica-base support with a silicone polymer that shields the silanol groups, the bonded phase being connected to the coated surface.

3) Silica monoliths can also be considered special columns. The C18 bonded phase on monoliths is, however, a relatively common type, such as in Onyx Monolithic C18 or Chromolith Performance RP-18 columns, which are available in various formats. These columns have a typical dual porous structure with mesopores of about 130 Å and macropores of about 2 μm diameter. The nature of the bonded phase is similar to that for spherical particle C18 columns, and the silica surface is end-capped. This type of column allows a reduction in elution time up to nine times due to the capability to use higher flow rates without having problems with the column backpressure. Due to a rapid mass transfer of the solutes between the bonded phase and the mobile phase, the decrease in the number of theoretical plates at higher flow rates, as predicted by the van Deemter equation [see Eq. (2.2.8)], is not as intense. Shorter retention times also lead to better resolution [111].

4) Among other special stationary phases used in HPLC are those based on silica hydride. Bare silica hydride can be used for direct-phase separations (in organic normal-phase HPLC) or for aqueous normal-phase separations. Bonded phases on hydride silica (*type* C silica) with C18, C8, cholesterol groups, etc. are commercially available (e.g., Cogent Bidentate C18, Cogent Bidentate C8, Phenyl Hydride, Cogent UDC Cholesterol). Stationary phases with dimensions such as 4-μm particle diameter and pore size of 10 nm are available. These types of columns can be used in reversed-phase type conditions [112].

5) Porous graphitic carbon (PGC) can also be used as stationary phase. This material is obtained by decomposing organic matrices using silica as template. For example, the silica used as template is impregnated with a mixture of phenol and formaldehyde and then heated to 80–160°C to initiate the polycondensation. The characteristics of the silica gel as template material determine the size and porosity of the particles that will be obtained. The polymer is then pyrolyzed under inert atmosphere (N_2) at 1000°C. Thus, highly porous amorphous carbon is produced. This carbon corresponds to what is normally called carbon black. After this step the silica template is dissolved with a hot aqueous NaOH solution. The graphitization is realized through a thermal treatment at high temperature (about 2300°C) under inert atmosphere (Ar). This operation eliminates certain surface-attached functional groups, produces a structural rearrangement of C atoms, and removes the micropores. After cooling down to 1000°C, the replacement of argon with hydrogen can induce reactions between hydrogen and free radicals or functional groups still present at the carbon surface. By this deactivation procedure the PGC surface becomes more uniform. The result is porous graphitic carbon, which is now commercialized (e.g., trade name of Hypercarb) [113]. Unlike silica-based packing materials, carbon-based stationary phases have the advantages of being more resistant to hydrolysis, while the lack of swelling or shrinking makes them more useful than the polymeric materials. The efficiency of columns packed with these materials is comparable with modified silica-based columns but with higher $\alpha(CH_2)$ value compared to any C18 or C8 column [114,115]. The retention mechanism on PGC stationary phases seems to be based on adsorption, and these columns may have strong retention for both nonpolar analytes and polar compounds [7].

6) Hydrated zirconia can be used as a support for RP-bonded phases having the advantage of a wider pH stability compared to silica. The display of additional interactions besides hydrophobic of RP phases with zirconia directed efforts to obtain alternative supports still using zirconia resistence to high pH. One such alternative is the carbon-clad zirconia. This type of material can be produced either by high-temperature graphitization of an organic polymer covering the zirconia support or by chemical vapor deposition of carbon on zirconia. However, in practice, the surface of this type of stationary phase is not perfectly homogeneous, and a significant proportion of the surface of zirconia still remains uncovered by carbon, which creates highly acidic residual zirconia groups [116]. Another alternative uses the treatment of silica gel with zirconium or titanium tetrabutoxide followed by a hydrolysis step of the adsorbed layer of tetrabutoxide on the silica surface. This type of support is known as metalized silica. The bonded phases on metalized silica include poly(methyloctylsiloxane), poly(methyltetradecylsiloxane), poly(methyldecyl-(2–5%)-diphenylsiloxane), and poly(dimethylsiloxane) [117].

7) Organic polymer-based hydrophobic stationary phases are sometimes used in RP-HPLC, having the best resilience to a wide pH range for the mobile phase. Some polymeric phases can operate at pH values as low as 1 and as high as 13.5. Also, the polymeric substrate does not have a potential layer of unwanted polar groups as do the silica-base stationary phases. Polymeric materials typically show lower theoretical plate numbers N for the same dimensions as the silica-base particles, although phases with $N/\mathrm{m} = 80{,}000$ are available. Another disadvantage of polymeric supports is caused by the variation in the swelling of the polymeric particles when they are used in various solvents. This variation in swelling affects the volume occupied by the stationary phase in the column and can lead to the formation of void spaces and loss of efficiency. Organic polymeric materials are also less resilient to

high pressures, and this makes polymeric columns less adequate for the new developments toward UPLC. For these reasons, the use of polymeric columns, particularly for RP-HPLC, is limited. Some examples of RP polymeric phases are given in Table 7.4.1, where certain reported characteristics are also shown. The stationary phase from these columns may consist of the polymer itself or may have a specific bonded phase (such as C8, C18) on the polymeric support [118].

Besides stationary phases made with particles of rigid organic polymers, polymeric monoliths are also used as chromatographic media [119]. Polymeric phases with hydrophobic bonded groups such as C18 or pentafluorophenyl are commercially available (e.g., Jordi phases from MicroSolv). Among different polymeric materials used as stationary phase in HPLC, several molecular imprinted polymers (MIPs) were also available. The polymeric structures have been typically based on methacrylic acid and/or styrene crosslinked with

TABLE 7.4.1 Examples of Polymeric Columns and Their Main Characteristics

Name	Polymer	Size (μm)	Pore (Å)	N/m	pH Range	Functional Group	Max Press (psi)	Equivalent
Asahipak ODP	Polyvinyl alcohol	5	250	56,000	2–13	C18	2250	C18
Asahipak ODP40	Polyvinyl alcohol	4	250	68,000	2–13	C18	1950	C18
Asahipak C8P	Polyvinyl alcohol	5	250	45,000	2–13	C8	2250	C8
Asahipak C4P	Polyvinyl alcohol	5	250	40,000	2–13	C4	2250	C4
PolymerX	Styrene divinyl benzene	3, 5, 7	100	–	0–14	None	–	C18
Shodex ODP2 HP	Polyhydroxymethacrylate	5	40	80,000	2–12	None	2250	C18
Shodex DE	Polymethacrylate	4	100	70,000	2–12	None	2250	C18
Shodex DS	Polymethacrylate	3.5	100	70,000	2–13	None	3000	Mixed mode
Shodex RP18-413	Styrene divinyl benzene	3.5	100	80,000	1–13	None	3300	Mixed mode
Shodex RP18-613	Styrene divinyl benzene	3.5	100	80,000	2–13	None	3300	Mixed mode
Shodex RP18-415	Styrene divinyl benzene	6	430	36,000	2–13	None	3300	Mixed mode
Shodex NN	Polyhydroxymethylacrylate	10	100	40,000	2–12	None	2250	Mixed mode
Shodex JJ	Polyvinyl Alcohol	5	100	32,000	2–11	None	1400	Mixed mode
TSKgel Octadecyl-2PW	Styrene divinyl benzene	5	125	–	0–14	C18	–	C18
TSKgel Octadecyl-4PW	Styrene divinyl benzene	7, 13	500	–	0–14	C18	–	C18
TSKgel Phenyl-5PW	Styrene divinyl benzene	10, 13	1000	–	0–14	Phenyl	–	Phenyl
TSKgel Octadecyl-NPR	Styrene divinyl benzene	2.5	None	–	0–14	C18	–	C18

ethyleneglycol dimethacrylate. The monomers were polymerized in the presence of a template molecule (e.g., nortriptyline [120]). More information on monolithic stationary phases can be found in dedicated books [121].

7.5 SELECTION OF AN RP-HPLC COLUMN

The selection of the analytical column in an HPLC analysis is preceded by a number of decisions related to other aspects of the analysis. One important decision is related to the sample preparation step for the analysis. Depending on sample preparation, the processed sample may or may not have a simpler matrix and may or may not have the analytes in a higher concentration as compared to the raw sample. Also, following the sample preparation, the sample can be easily amenable for the HPLC analysis or can still pose significant challenges. However, the time/manpower consumed for sample preparation and the potential addition of sources of errors tend to contribute to the decrease in the utilization of sample preparation (see Section 1.2). The trend to simplify sample preparation increases the load of problems to be solved by the core HPLC analysis.

In selecting an HPLC column it is always very helpful to follow the results reported in the literature for the same type of compounds as the one to be analyzed. A large proportion of published analytical methods use C18 columns, with a smaller number using C8 columns, and significantly fewer methods using other columns. The popularity of C18 columns is explained by the fact that these columns are adequate for many types of analysis. However, a considerable variety of C18 columns (as well as of C8 columns) is available, the columns differing in physical construction, carbon load, range of pH resilience, wettability, etc. For this reason, even by limiting the choice to a C18 column, there are still many parameters that require a selection.

Since it is always easier to modify an already developed method than to develop a radically different one (or to reinvent the method), it is always advisable to search for reported analyses similar to the one intended to be implemented. It is common that small modifications to a reported method can make it adequate for a specific new task. In some instances, the columns recommended in the literature for a specific analysis are not available, and the PQRI or USB approaches offer the capability to select columns similar to the one initially recommended (see [53,122]). Besides general literature references, computer programs have been developed for guiding the selection and optimization of an HPLC separation using computer simulation [123]. Such a computer program (e.g., DryLab4 [124]) starts with input from two or more (up to 12) experimental chromatograms and allows virtual modification of various parameters such as mobile phase composition (isocratic or gradient), flow rate, and column dimensions. The program predicts retention times and optimum conditions for a separation. Also, the program can assist regarding the transferring of a method from one column to an equivalent one [125].

Before the column selection, other selections must be made for a core HPLC method. These selections include the following: (1) the type of HPLC chromatographic separation mechanism (RP, HILIC, IC, SEC, etc.), (2) the mobile phase composition, (3) the specific gradient or isocratic separation, (4) the flow rate, (5) the detector, (6) the injection volume, (7) the solvent for the injected sample, and (8) the column temperature. Only after such selections are made does the selection of the analytical column come into place. However, because for the development of an HPLC method several iterations can be involved, once a chromatographic column is selected, the whole decision process for performing the HPLC analysis may require reevaluation.

Since the change of a selected column that was proved to be inadequate represents some waste of resources (time, cost), it is preferable to select from the beginning a good column. The variety of columns commercially available on the market makes this selection a rather complicated task. However, the process is somewhat simplified by the fact that for many analyses not only a unique column can be fully adequate for a successful HPLC analysis. Similar results can be obtained with several different selections. Optimization of column selection between the fully adequate columns remains optional, depending on the goal of the analysis. In RP-HPLC, the mobile phase plays a very important role since the retention and elution process is governed mainly by hydrophobic interactions. In part because of this simple mechanism, the retention/elution process is fast, the peak shape in RP-HPLC is typically very good, and a considerable part of method optimization resides in the optimization of mobile phase composition and gradient conditions.

Column properties, their characterization, and criteria for column selection were already presented in Sections 6.2, 6.3, 7.2, 7.3, and 7.4. In this section, only a summary of the selection of an RP-HPLC analytical column is presented. The selection of the column must always be made in the context of analysis requirements such as goal of the analysis, sample complexity, relative concentration of analytes in the sample, sample size availability, number of samples to be analyzed, and requirements for time length of the chromatographic run. However, the RP columns frequently show similarities among themselves regarding the separation properties, and the replacement of one column with an equivalent one is not usually a problem. The differences reside mainly in the following: (1) the column efficiency (N/m), (2) resilience to a wider pH range, (3) wettability, (4) peak tailing in case of the separation of basic compounds, and (5) better separations when more difficult analytical tasks are encountered.

1) Column efficiency becomes critical when the analytes of interest have peaks in close proximity to other peaks. In such situations, when the modifications in the mobile phase are not capable of improving the peaks of interest separation, a column with better N may solve the problem. This can be achieved using, for example, columns with smaller particles, core–shell particles, longer columns. The selection depends on the backpressure sustained by the HPLC system (HPLC or UPLC), limitations in the length of run time, and other requirements of the chromatogram.

2) Columns resilient to a wider pH range are frequently necessary for the analysis of compounds capable of existing in more than one form at different pH values of the mobile phase (see Section 4.4). In such cases, the pH of the mobile phase is usually recommended to have a more extreme pH such that only one form of the analyte is present in solution (e.g., a high pH for the analysis of basic compounds capable to exist in forms like B and BH^+, or low pH for acidic compounds capable to exist in forms like AH and A^-). Depending on the pK_a (or pK_b) of the analytes, the pH of the mobile phase may need to be lower than 2 or higher than 7 or 8. In such cases, the column pH range of stability must be selected carefully. When a wider pH range for the mobile phase is necessary in a separation, the columns with silica technology should be preferred to those based on organic polymers. The ethylene bridge silica type columns offer the same mechanical characteristics as silica-based columns, and do not have the typical backpressure restrictions of organic polymeric columns.

3) In some separations it is useful to have a very high content of water in the mobile phase. This high content may be utilized only at the beginning of the chromatographic run when gradient elution is employed, or may be necessary for an extended range or even for the whole chromatographic separation. In such situations, the wettability of the chromatographic column is very important, since a mobile phase with 100% water may ruin some columns (see Section 7.2). Columns using polar end-capping, polar embedded groups, and those with ethylene bridge organic/inorganic silica support are usually more resilient to high water content.

4) In the analysis of some basic compounds (amines) a common problem is the peak tailing of the analytes of interest. Peak tailing is a common characteristic described by the manufacturers, and when necessary, columns with low tailing should be selected (usually made with silica of very high purity, or with special phase surface treatment).

5) When several alternatives for selecting an analytical column are available, it is recommended to start with the simplest and most straightforward choice. If the separation is possible using RP-HPLC or HILIC, RP-HPLC should be preferred. When the analytes have differences in the hydrophobic moiety, simple C18 or C8 columns are recommended. Phenyl-type columns typically offer a better separation when aromatic compounds should be separated. Cyano columns also offer better separation between compounds with lower polarity than those with higher polarity (see Fig. 7.3.5). If a simple hydrophobic column or a column with embedded polar groups or polar end-capped are the alternatives, and no restriction regarding wetting or better separation is present, the simple hydrophobic column is likely to offer a more robust analytical method. However, columns with embedded polar groups, polar end-capped, with phenyl or cyano groups may offer better separation characteristics, since they produce additional types of interactions that may help the separation. However, multiple types of interactions may not be necessary in all separations.

Selection of Physical Column Characteristics

The most frequently utilized columns in RP-HPLC are C8 or C18 columns of 100 or 150 mm length, 3 or 4.6 mm diameter, with uniform fully porous spherical particles of 3 or 5 μm diameter [126]. Many different physical parameters from those indicated can be selected, depending on the analysis requirements. For faster analyses, 50 mm length column can be selected, and for separation demanding higher N values, 250 mm column length may be necessary. For the analysis of cleaner samples, core–shell particles with 1.7 or 2.6 μm diameter can be better, such columns typically having a higher N. This choice depends also on the maximum backpressure that can be sustained by the HPLC system since smaller particles produce higher backpressure (see expression 6.2.10). Depending on the sample load, wider columns are recommended for more concentrated samples, but for low-volume injections, columns with 2.1 mm diameter are fully adequate. For trace analysis, columns with small dimensions are typically recommended, and core–shell columns offer advantages for high N at relatively short column length. Shorter columns and with a narrower diameter generate sharper peaks, with larger height (for the same peak area). This may significantly improve sensitivity, since signal to noise ratios (S/N) for the peaks are higher compared to S/N for the peaks of wider columns (see Section 6.3). For the analysis of large molecules it is recommended to select stationary phases with larger pores (200 Å or larger, depending on the molecule Mw).

Columns for the Analysis of Small Molecules With a Hydrophobic Moiety

The analysis of small molecules with average hydrophobic moiety should not pose difficult problems, unless the matrix is very complex and/or in large amount, or if the analysis is performed for ultra-trace analysis (see Section 5.4). Before selecting a column, it is always advisable to find the log K_{ow} values for the analytes. These values provide a good guidance regarding the choice of the type of RP column (see Figure 5.4.1). For compounds with average log K_{ow} values (0–4.5), C8 or C18 columns should be fully adequate. The pore size of the column should be medium (not with phases having large pores used for large molecules separations).

Any analysis may still encounter numerous problems. Among these can be listed the following: (1) sample contains basic compounds, (2) sample contains acidic compounds, (3) molecules have very low hydrophobic character, (4) molecules have very high hydrophobic character, (5) sample contains a mixture of compounds with a wide range of log K_{ow} values, (6) sample has a "dirty" matrix and has low analytes level, (7) similar compounds do not separate, and (8) other problems.

1) For the case of basic compounds, the pH of the mobile phase is necessary to be relatively high to keep the analyte from becoming ionized. In such cases, columns with higher resilience to pH above 7 or 8 must be selected (e.g., using ethylene-bridge technology). Also, columns with low tailing characteristics must be selected for basic compounds. Many amines have high propensity to form coordinative bonds with trace transitional metals (e.g., Fe^{2+}, Fe^{3+}, Ni^{2+}) and some columns show tailing of amines peaks (see USP test for tailing character in Section 7.3).

2) For acidic compounds, the columns must be resilient to low pH. In order to keep the organic acids in molecular form (not ionized), the mobile phase should have a low pH (2.5–3) and the column must be selected to be resilient to this pH.

3) Molecules with high polarity and low hydrophobic character (log K_{ow} −1.25 to 0.0) require a mobile phase with high water content or even 100% water, and a column providing high hydrophobicity and possibly able to develop additional types of interactions (with polar embedded groups or polar end-capping). Such columns should have excellent wetting capabilities (like those polar end-capped or with embedded polar groups), and at the same time a high carbon load (high C%) and possibly high surface area of silica support (see Figs. 7.2.1. and 7.2.2). As an example, the separation of several organic acids (acetic, malic, citric, quinic, pyruvic, lactic, fumaric), although they are hydrophilic compounds, can be done on a special RP-HPLC column (Synergi 4u Hydro-RP, 4.6 × 250 mm from Phenomenex, CA, USA) that has a strong hydrophobic character (C% = 19%, surface area = 475 m^2/g) and offers additional types of interactions besides hydrophobic, due to its polar end-capping. The separation requires no organic component in the mobile phase, which is an aqueous buffer at pH = 2.9 ($KH_2PO_4 + H_3PO_4$) (isocratic separation). For quinic acid, for example, at pH = 2.9, log $D_{ow} = -2.82$ and the separation can still be done using 100% aqueous mobile phase. Other acids in the series have higher log D_{ow}. For example, for fumaric acid log $D_{ow} = -0.16$ at pH = 2.9, but it is log $D_{ow} = -5.80$ at pH = 7.0. The separation for a set of standards of acids with concentrations around 100 μg/mL, except for pyruvic at 10 μg/mL and fumaric acid at 2 μg/mL, is shown in Fig. 7.5.1. The UV detection was performed at 215 nm. At higher pH than 3 of the mobile phase, the separation starts to deteriorate, and above pH 4.5–5.0 it is

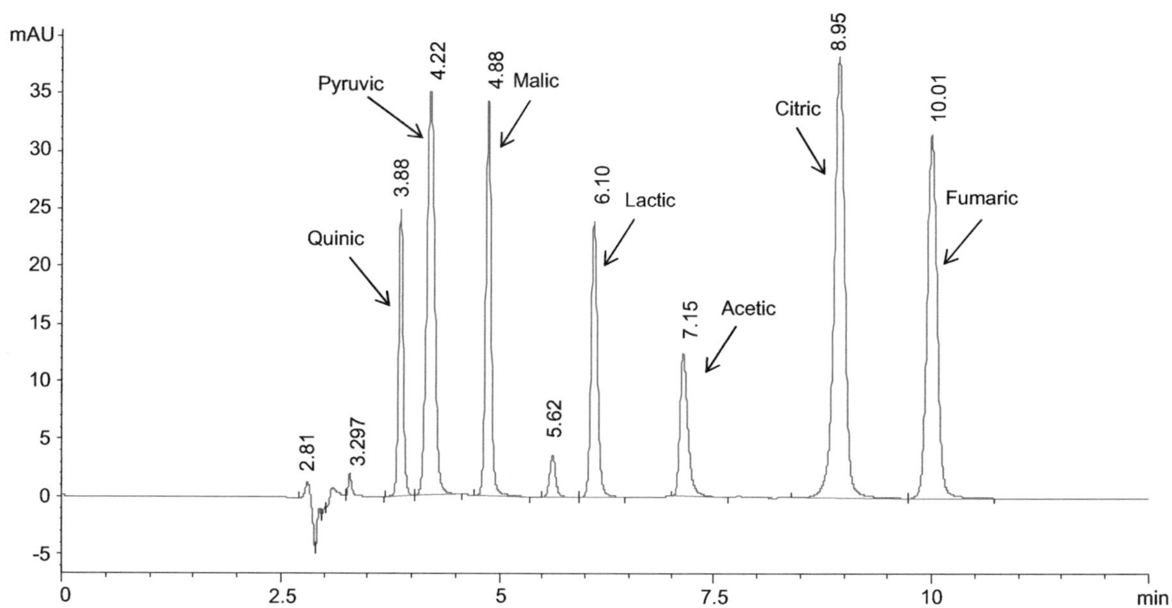

FIGURE 7.5.1 Separation for several acids on a Synergi Hydro-RP column at pH = 2.9 with UV detection at 215 nm. The concentrations were around 100 μg/mL except for pyruvic acid at 10 μg/mL and fumaric acid at 2 μg/mL.

no longer possible. The order of elution on the Synergi column is not exactly in the order of log D_{ow} values, due to the additional polar interactions on this column.

Some compounds with very high polarity cannot be well analyzed on RP-HPLC columns, and ion pairing chromatography (IPC) must be involved for utilization of RP separation (see Section 4.3). Some molecules with too low hydrophobicity can be separated on HILIC columns. For the selection of the RP column to be used in ion pairing, it must be assured that the stationary phase is resilient to the ion pairing agent (IPA) from the mobile phase (IPA can be a strong acid or base depending on the basic/acidic character of separated species).

4) Some molecules have high log K_{ow} indicating high hydrophobicity (above 4.5). For such compounds, the retention may be too strong on C18 or C8 columns and such compounds elute at long retention times even if the mobile phase has a high content of organic phase. In such cases, columns with a low C% load are used, and if separation of other compounds potentially present in the sample is acceptable, short columns are recommended. For compounds with very high hydrophobicity (log $K_{ow} > 5$), the alternative is to use nonaqueous reversed-phase chromatography (NARP) (see Section 3.4).

5) When the sample contains a mixture of compounds with very different log K_{ow} values, it is common to select a column capable of separating the compounds with low log K_{ow}, and then by using gradient to increase the organic component content in the mobile phase in order to elute all the analytes (and the matrix) from the column.

6) For samples that have a high content of matrix (and possibly with a complex composition), the HPLC core analysis has a more difficult task since it must separate the analytes and also the matrix components. When the matrix has basically a different chemical character from the analytes, it is possible to select the column and the mobile phase composition such that the matrix either elutes very close to the dead time of the column, or is eluted only when the mobile phase has a high organic content (is a "strong" eluent). Even when the analytes are well separated from the matrix components, the peak shape of the analytes may be affected (see Section 5.4). In some cases, a sample preparation step for cleanup is unavoidable. For samples with "dirty" matrix it is highly recommended to use precolumns for the protection of the analytical column, and to make as small injections as possible (depending on detector sensitivity). Also, for a sample with a "dirty" matrix it is recommended to use columns with a larger diameter (e.g., 4.6 mm) and larger particle size (e.g., 5 μm).

7) For the case of similar compounds that are not separated on one column, the first series of experiments should focus on changing the mobile phase composition and not the column. If this attempt fails, the selection of the second column should be done by trying the following: (1) select a column with a higher efficiency N (e.g., longer column or with smaller particles), (2) select a column that offers larger k' values for the compounds in the specific class, (3) select a column that offers additional types of interactions besides hydrophobic. Column comparison using the PQRI approach or USB approach should indicate if the second column has stronger hydrophobic character, stronger hydrogen bonding or ion exchange capabilities. Enantiomers cannot be separated on achiral stationary phases unless special mobile phases with chiral properties are used (see Section 13.5).

8) Several other problems may occur in an HPLC separation related to a selected column. One such problem can be related to the type of detector used in the method. For UV or RI detection, flows of 1.0 mL/min are common, but for MS or MS/MS detection such flows may be too high. In such situations, the flow may need to be limited to 0.2–0.5 mL/min,

and at these flows the retention time for some analytes may be too long as compared to the use of the same column at 1 mL/min flow. In such cases, an equivalent column should be selected, having narrower diameter (and possibly smaller particles). The equivalence for the separation can be obtained based on expression (6.2.6) (or using computer simulation programs such as DryLab4 [124]). Other problems may be related to the column having good separation, but with peak shape problems such as fronting or too much tailing. Fronting typically indicates either overloading of the column or nonideal equilibria in the chromatographic process. Overloading can be corrected by diminishing the sample volume, concentration, or both. For correcting the nonideal equilibrium in the chromatographic process, the column and/or the mobile phase must be changed.

Columns for the Analysis of Peptides and Proteins

Successful analysis of proteins and peptides can be performed on RP-HPLC columns [127]. Several aspects regarding the column (and mobile phase) selection are specific for this type of analysis. For example, for easy access to the pores of stationary phase, the pore size should be at least three times the hydrodynamic diameter of the molecule (often pores of 300 Å diameter or even larger are used). Different proteins may have a wide range of molecular shapes and also of polarities, and their retention on an RP column may vary considerably. The most common hydrophobic phases used for protein separations are C8 and C18. Also, columns that have lower hydrophobicity but additional interactions with polar groups from the protein molecule may be better suited for some separations. Phenyl and cyano columns may show some special selectivity and are used in specific applications where the differences in hydrophobicity between different proteins are not sufficient for their separation. The same is true for columns end-capped with polar groups or not end-capped and having a larger proportion of free silanols. However, some proteins with higher hydrophobicity can be separated more easily using short-chain bonded phases such as C1 to C3 since they may be retained too strongly on C8 or C18 columns.

One particularly important aspect in protein analysis using RP-HPLC is the selection of the pore size of the stationary phase (see, e.g., [128]). This parameter is related to the molecular weight of the analyte. The free access of the analyte to the bonded phase is very important since the retention on the stationary phase depends on the accessible surface area of the packing. The estimation of steric hindrance for a molecule at a pore entrance has been reported in the literature [129]. The theory indicates that for relatively small molecules, such as small peptides ($Mw < 1000$ Da), the stationary phases with 80−120 Å pore size are well suited. For larger molecules with Mw higher than 1000−2000 Da and up to 10,000, stationary phases with pore size around 300 Å are necessary. The larger pores are necessary to allow the interaction of the analyte with the stationary phase and to avoid size exclusion effects. For large proteins (Mw 10,000−100,000 Da or higher) even larger pores are necessary, up to 1000 Å. Larger pores are typically associated with smaller surface areas for the phase, and as shown in Appendix 2c and Fig. 7.3.1, the columns with wider pores (300 Å) have typically lower k' for hydrophobic molecules (e.g., ethylbenzene), but in the case of proteins they offer better separation [130].

Besides large pores utilization for protein analysis, another alternative is the use of nonporous C18 columns (e.g., Presto FF-C18 column from Imtakt). Such columns have lower loading capacity, but large proteins (of about 150 kDa) were separated on such columns [131].

Since the interaction of large molecules with the hydrophobic bonded phase is probably mainly based on an adsorption process (not on distribution), monomeric type bonded phases or phases obtained by horizontal polymerization display a similar capacity factor as vertical polymeric bonded phases (see Section 6.1). Monomeric bonded phases and those with horizontal polymerization typically show better reproducibility in protein/peptide separations. For this reason, monomeric phases are preferred to phases obtained with vertical polymerization, although these may have a larger carbon load.

The diffusion of proteins in the phases is in general slower than that of small molecules (see Eqs. 4.2.46–4.2.48). When the diffusion coefficients of the analyte in the mobile phase D and in the stationary phase D_s are small, several terms contributing to the plate height (H_C, H_T, H_S) are large, and for this reason the resulting peaks of the proteins may be significantly wider than for small molecules. As shown from expression (4.2.49), a lower flow rate u in the column has the effect of diminishing plate height components H_C, H_T, and H_S, but longer separation times result in this case. For this reason, columns with a high theoretical plate number N (as measured for a test small molecule compound) are preferred for protein separation to compensate for peak broadening inherent to large molecules (although the N value for the protein is much smaller than obtained with the test compound). Columns with small particles or with core–shell that have high N values can be useful for protein separation. The flow rate is typically maintained at a constant value in the range 0.5–2 mL/min. Some problems in protein separation, such as peak tailing are corrected by increasing the acid content in the mobile phase. In this way, the silanol activity of the column is reduced and the tailing of basic compounds is diminished. The acids used in the mobile phase for proteins and peptides, in particular TFA, may act as an ion pair to the proteins with basic character ($pI > 7$). The ion pair formed with TFA has a stronger retention (larger log k') than expected for the free compound. Protein degradation during separation must be avoided, in particular when the physiological property of the protein must be preserved.

New Developments

The column technology is continuously evolving, and various new columns are available on the market every year. The new trends in column development include sub-2 μm particle sizes, utilization of core–shell technology, numerous improvements in pH range stability, excellent wettability, new improvements in stationary phase chemistry such as addition of polar groups in hydrophobic columns (polar enhanced phases), excellent reproducibility in time, development of additional phases besides C18 and C8, and utilization of trifunctional reagents for silica derivatization to ensure good silanol coverage. A few new columns more recently commercially available are listed in Appendix 2e [132].

References

[1] X. Liu, A. Bordunov, M. Tracey, R. Slingsby, N. Avdolovic, C. Pohl, Development of polar embedded stationary phase with unique properties, J. Chromatogr. A 1119 (2006) 120–127.

[2] X. Liu, A.V. Bordunov, C.A. Pohl, Preparation and evaluation of a hydrolytically stable amide-embedded stationary phase, J. Chromatogr. A 1119 (2006) 128–134.

[3] A.M. Faria, C.H. Collins, I.C.S.F. Jardim, State-of-the-art in immobilized polymer stationary phases for high-performance liquid chromatography, J. Braz. Chem. Soc. 20 (2009) 1385–1398.

[4] J.J. Kirkland, J.W. Henderson, J.J. DeStefano, M.A. van Straten, H.A. Claessens, Stability of silica-based, endcapped columns with pH 7 and 11 mobile phases for reversed-phase high-performance liquid chromatography, J. Chromatogr. A 762 (1997) 97–112.
[5] S.W. Pettersson, B.S. Persson, M. Nyström, General method allowing the use of 100% aqueous loading conditions in reversed-phase liquid chromatography, J. Chromatogr. B 803 (2004) 159–165.
[6] H.A. Claessens, M.A. van Straten, Review on the chemical and thermal stability of stationary phases for reversed-phase liquid chromatography, J. Chromatogr. A 1060 (2004) 23–41.
[7] S.C. Moldoveanu, V. David, Essentials in Modern HPLC Separations, Elsevier, Amsterdam, 2013.
[8] J.J. Kirkland, M.A. van Straten, H.A. Claessens, Reverse-phase high-performance liquid chromatography of basic compounds at pH 11 with silica-based column packings, J. Chromatogr. A 797 (1998) 111–120.
[9] M.J. Wirth, H.O. Fatunmbi, Horizontal polymerization of mixed trifunctional silanes on silica. 2. Application to chromatographic silica gel, Anal. Chem. 65 (1993) 822–826.
[10] J.J. Kirkland, J.L. Glajch, R.D. Farlee, Synthesis and characterization of highly stable bonded phases for high-performance liquid chromatography column packages, Anal. Chem. 61 (1989) 2–11.
[11] S.C. Moldoveanu, J. Zhu, N. Qian, Free amino acids analysis by liquid chromatography with tandem mass spectrometry in several botanicals with antioxidant character, J. Sep. Sci. 38 (2015) 2208–2222.
[12] C. Ye, G. Terfloth, Y. Li, A. Kord, A systematic stability evaluation of analytical RP-HPLC columns, J. Pharm. Biomed. Anal. 50 (2009) 426–431.
[13] K.B. Sentell, J.G. Dorsey, On the calculation of the stationary phase volume in reversed phase chromatography, J. Liq. Chromatogr. 11 (1988) 1875–1885.
[14] W. Cheng, Differential density method for determination of carbon load on chromatographic packings, Anal. Chem. 57 (1985) 2409–2412.
[15] K.S. Yun, C. Zhu, J.F. Parcher, Theoretical relationships between the void volume, mobile phase volume, retention volume, adsorption, and Gibbs free energy in chromatographic processes, Anal. Chem. 67 (1995) 613–619.
[16] M. Wang, J. Mallette, J.F. Parcher, Comparison of void volume, mobile phase volume and accessible volume determined from retention data for oligomers in reversed-phase liquid chromatographic systems, J. Chromatogr. A 1218 (2011) 2995–3001.
[17] F. Gritti, G. Guiochon, Adsorption mechanism in reversed-phase liquid chromatography. Effect of the surface coverage of a monomeric C18-silica stationary phase, J. Chromatogr. A 1115 (2006) 142–163.
[18] S. Moldoveanu, V. David, Estimation of the phase ratio in reversed-phase high-performance liquid chromatography, J. Chromatogr. A 1381 (2015) 194–201.
[19] E. Caiali, V. David, H.Y. Aboul-Enein, S.C. Moldoveanu, Evaluation of the phase ratio for three C18 high performance liquid chromatographic columns, J. Chromatogr. A 1435 (2016) 85–91.
[20] http://www.epa.gov/oppt/exposure/pubs/episuite.htm.
[21] R.P.W. Scott, P. Kucera, Examination of five commercially available liquid chromatographic reversed phases (including the nature of the solute-solvent-stationary phase interactions associated with them), J. Chromatogr. A 142 (1977) 213–232.
[22] R.E. Majors, M. Przybyciel, Columns for reverse-phase LC separations in highly aqueous mobile phases, LC-GC North America 20 (2002) 584–593.
[23] I. Rustamov, T. Farcas, F. Ahmed, F. Chan, R. LoBrutto, H.M. McNair, Y.V. Kazakevich, Geometry of chemically modified silica, J. Chromatogr. A 913 (2001) 49–63.
[24] M. Pursch, S. Strohschein, H. Handel, K. Albert, Temperature dependent behavior of C30 interphases. A solid-state NMR and LC-NMR study, Anal. Chem. 68 (1996) 386–393.
[25] Y.V. Kazakevich, R. LoBrutto, F. Chan, T. Patel, Interpretation of the excess adsorption isotherms of organic eluent components on the surface of reversed-phase adsorbents: effect on the analyte retention, J. Chromatogr. A 913 (2001) 75–87.
[26] M. Przybyciel, R.E. Major, Phase collapse in reversed-phase LC, LC-GC Europe (October 2002) 2–5.
[27] U.D. Neue, HPLC Columns, Theory, Technology, and Practice, Wiley Blackwell, New York, 1997.
[28] U.D. Neue, E. Serowik, P. Iraneta, B.A. Alden, T.H. Walter, Universal procedure for the assessment of the reproducibility and the classification of silica-based reversed phase packings. I. Assessment of the reproducibility of reversed-phase packings, J. Chromatogr. A 849 (1999) 87–100.
[29] U.D. Neue, B.A. Alden, T.H. Walter, Universal procedure for the assessment of the reproducibility and the classification of silica-based reverse phase packings. II. Classification of reverse phase packings, J. Chromatogr. A 849 (1999) 101–116.

[30] K. Krupczyńska, B. Buszewski, Characterizing HPLC stationary phases, Anal. Chem. 76 (2004) 226 A–234 A.
[31] S. Kowalska, K. Krupczyńska, B. Buszewski, Some remarks on characterization and application of stationary phases for RP-HPLC determination of biologically important compounds, Biomed. Chromatogr. 20 (2006) 4–22.
[32] M.H. Abraham, A. Ibrahim, A.M. Zissimos, Determination of sets of solute descriptors from chromatographic measurements, J. Chromatogr. A 1037 (2004) 29–47.
[33] O. Sinanoğlu, in: B. Pullman (Ed.), Molecular Associations in Biology, Academic Press, New York, 1968, pp. 427–445.
[34] C. Horvath, W. Melander, I. Molnar, Solvophobic interactions in liquid chromatography with nonpolar stationary phases, J. Chromatogr. 125 (1976) 129–156.
[35] S.C. Moldoveanu, V. David, Dependence of the distribution constant in liquid–liquid partition equilibria on the van der Waals molecular surface area, J. Sep. Sci. 36 (2013) 2963–2978.
[36] W.J. Cheong, P.W. Carr, Limitations of all empirical single-parameter solvent strength scales in reversed-phase liquid chromatography, Anal. Chem. 61 (1989) 1524–1529.
[37] R.W. Taft, M.J. Kamlet, The solvatochromic comparison method 2. The alpha scale of solvent hydrogen-bond donor (HBD) acidities, J. Amer. Chem. Soc. 98 (1976) 2886–2894.
[38] M.J. Kamlet, R.W. Taft, The solvatochromic comparison method 1. The beta scale of solvent hydrogen-bond acceptor (HBa) basicities, J. Amer. Chem. Soc. 98 (1976) 377–383.
[39] M. Rosés, E. Bosch, Linear solvation energy relationship in reversed-phase liquid chromatography. Prediction of retention from a single solvent and a single solute parameter, Anal. Chim. Acta 274 (1993) 147–162.
[40] L.R. Snyder, A. Maule, A. Heebsch, R. Cuellar, S. Paulson, J. Carrano, L. Wrisley, C.C. Chan, N. Pearson, J.W. Dolan, J. Gilroy, A fast, convenient and rugged procedure for characterizing the selectivity of alkyl-silica columns, J. Chromatogr. A 1057 (2004) 49–57.
[41] A. Sándy, L. Szepesy, Evaluation and modulation of selectivity in reversed-phase high-performance liquid chromatography, J. Chromatogr. A 845 (1999) 113–131.
[42] K.B. Sentell, J.G. Dorsey, Retention mechanisms in reversed-phase chromatography: stationary phase bonding density and solute selectivity, J. Chromatogr. 461 (1989) 193–207.
[43] P.W. Carr, J. Li, A.J. Dallas, D.I. Eikens, L.C. Tan, Revisionist look at solvophobic driving forces in reversed-phase liquid chromatography, J. Chromatogr. A 656 (1993) 113–133.
[44] K.B. Sentell, N.I. Ryan, A.N. Henderson, Temperature and solvation effects on homologous series selectivity in reversed phase liquid chromatography, Anal. Chim. Acta 307 (1995) 203–215.
[45] J. Layne, Characterization and comparison of the chromatographic performance of conventional polar-embedded and polar-endcapped reverse phase liquid chromatography stationary phases, J. Chromatogr. A 957 (2002) 149–164.
[46] A.M. Kristulović, H. Colin, A. Tchapla, G. Guiochon, Effects of the bonded alkyl chain length on methylene selectivity in reversed phase chromatography, Chromatographia 17 (1983) 228–230.
[47] R.K. Gilpin, M. Jaroniec, S. Lin, Dependence of the methylene selectivity on the composition of hydro-organic eluents for reversed-phase liquid chromatographic systems with alkyl bonded phases, Chromatographia 30 (1990) 393–399.
[48] http://www.waters.com.
[49] Comparison Guide to C18 Reversed Phase HPLC Columns, MAC-MOD Analytical, Chadds Ford, 2008.
[50] P.A. Bristow, J.H. Knox, Standardization of test conditions for high performance liquid chromatography columns, Chromatographia 10 (1977) 279–289.
[51] E. Lesellier, C. West, Description and comparison of chromatographic tests and chemometric methods for packed column classification, J. Chromatogr. A 1158 (2007) 329–360.
[52] B. Bidlingmeyer, C.C. Chan, P. Fastino, R. Henry, P. Koerner, A.T. Maule, M.R.C. Marques, U. Neue, L. Ng, H. Pappa, L. Sander, C. Santasania, L. Snyder, T. Woznyak, HPLC column classification, Pharmacopeial Forum 31 (2005) 637–645.
[53] http://apps.usp.org/app/USPNF/columnsDB.html.
[54] K. Kimata, K. Iwaguchi, S. Onishi, K. Jinno, R. Eksteen, K. Hosoya, M. Araki, N. Tanaka, Chromatographic characterization of silica C18 packing materials. Correlation between a preparation method and retention behavior of stationary phase, J. Chromatogr. Sci. 27 (1989) 721–728.
[55] H. Engelhardt, M. Jungheim, Comparison and characterization of reversed phases, Chromatographia 29 (1990) 59–68.

[56] H. Engelhardt, M. Arangio, T. Lobert, A chromatographic test procedure for reversed-phase HPLC column evaluation, LC/GC 17 (1997) 856–865.

[57] M.J. Walters, Classification of octadecyl-bonded liquid chromatography columns, J. Assoc. Off. Anal. Chem. 70 (1987) 465–469.

[58] D.V. McCalley, Effect of temperature and flow-rate on analysis of basic compounds in high-performance liquid chromatography using a reversed-phase column, J. Chromatogr. A 902 (2000) 311–321.

[59] D.V. McCalley, Selection of suitable stationary phases and optimum conditions for the application in the separation of basic compounds by reversed-phase HPLC, J. Sep. Sci. 26 (2003) 187–200.

[60] L. Rohrschneider, Characterization of stationary phases by retention data and solvation parameters, J. Sep. Sci. 24 (2001) 3–9.

[61] U.D. Neue, K. Van Tran, P.C. Iraneta, B.A. Alden, Characterization of HPLC packings, J. Sep. Sci. 26 (2003) 174–186.

[62] S.V. Galushko, Calculation of retention and selectivity in reversed phase liquid chromatography, J. Chromatogr. 552 (1991) 91–102.

[63] S.V. Galushko, The calculation of retention and selectivity in reversed phase liquid chromatography, Chromatographia 36 (1993) 39–42.

[64] L.C. Sander, S.A. Wise, Recent advanced in bonded phases for liquid chromatography, Crit. Rev. Anal. Chem. 18 (1987) 299–417.

[65] L.C. Sander, S.A. Wise, Influence of stationary phase chemistry of shape recognition in liquid chromatography, Anal. Chem. 67 (1995) 3284–3292.

[66] K. Jinno (Ed.), Chromatographic Separation Based on Molecular Recognition, Wiley-VCH, New York, 1997.

[67] D. Berek, I. Novák, Structural inhomogeneities in wide-pore silica gels, J. Chromatogr. A 665 (1994) 33–36.

[68] D. Berek, J. Tarbajovska, Evaluation of high-performance liquid chromatography column retentivity using macromolecular probes: II. Silanophilic interactivity traced by highly polar polymers, J. Chromatogr. A 976 (2002) 27–37.

[69] B. Buszewski, I. Cendrowska, K. Krupczyńska, R.M. Gadzala-Kopciuch, Bronopol as an ingredient of a new test mixture for evaluation of HPLC columns, J. Liq. Chromatogr. Rel. Technol. 26 (2003) 737–750.

[70] R. Kaliszan, Quantitative Structure–chromatographic Retention Relationship, Wiley, New York, 1987.

[71] B. Buszewski, R.M. Gadzala-Kopciuch, M. Markuszewski, R. Kaliszan, Chemically bonded silica stationary phases: synthesis, physicochemical characterization and molecular mechanism of reversed-phase HPLC retention, Anal. Chem. 69 (1997) 3277–3284.

[72] P. Jandera, J. Fischer, V. Staněk, M. Kučerová, P. Zvoníček, Separation of aromatic sulphonic acid dye intermediates by high-performance liquid chromatography and capillary zone electrophoresis, J. Chromatogr. A 738 (1996) 201–213.

[73] E. Cruz, M.R. Euerby, C.M. Johnson, C.A. Hackett, Chromatographic classification of commercially available reverse-phase HPLC columns, Chromatographia 44 (1997) 151–161.

[74] A. Sándi, A. Bede, L. Szepesy, G. Rippel, Characterization of different RP-HPLC columns by a gradient elution technique, Chromatographia 45 (1997) 206–214.

[75] N.S. Wilson, M.D. Nelson, J.W. Dolan, L.R. Snyder, R.G. Wolcott, P.W. Carr, Column selectivity in reversed-phase liquid chromatography: I. A general quantitative relationship, J. Chromatogr. A 961 (2002) 171–193.

[76] N.S. Wilson, M.D. Nelson, J.W. Dolan, L.R. Snyder, P.W. Carr, Column selectivity in reversed phase liquid chromatography: II. Effect of a change in conditions, J. Chromatogr. A 961 (2002) 195–215.

[77] N.S. Wilson, M.D. Nelson, J.W. Dolan, L.R. Snyder, P.W. Carr, L.C. Sander, Column selectivity in reversed-phase liquid chromatography: III. The physico-chemical basis of selectivity, J. Chromatogr. A 961 (2002) 217–236.

[78] J.J. Gilroy, J.W. Dolan, L.R. Snyder, Column selectivity in reversed-phase liquid chromatography: IV. Type-B alkyl-silica columns, J. Chromatogr. A 1000 (2003) 757–778.

[79] J.W. Dolan, A. Maule, L. Wrisley, C.C. Chan, M. Angod, C. Lunte, R. Krisko, J. Winston, B. Homeierand, D.M. McCalley, L.R. Snyder, Choosing an equivalent replacement column for a reversed-phase liquid chromatographic assay procedure, J. Chromatogr. A 1057 (2004) 59–74.

[80] L.C. Sander, S.A. Wise, Determination of column selectivity toward polycyclic aromatic hydrocarbons, J. High Resol. Chromatogr. Commun. 11 (1988) 383–388.

[81] M.R. Euerby, P. Petersson, Chromatographic classification and comparison of commercially available reversed-phase liquid chromatographic columns containing polar embedded groups/amino endcappings using principal component analysis, J. Chromatogr. A 1088 (2005) 1–15.

[82] A.U. Kulikov, M.N. Galat, Comparison of C18 silica bonded phases selectivity in micellar liquid chromatography, J. Sep. Sci. 32 (2009) 1340–1350.
[83] J.J. Gilroy, J.W. Dolan, L.R. Snyder, Column selectivity in reversed-phase liquid chromatography: V. Higher metal content (type-A) alkyl-silica columns, J. Chromatogr. A 1026 (2004) 77–89.
[84] N.S. Wilson, J. Gilroy, J.W. Dolan, L.R. Snyder, Column selectivity in reversed-phase liquid chromatography: VI. Columns with embedded or end-capping polar groups, J. Chromatogr. A 1026 (2004) 91–100.
[85] L.R. Snyder, J.W. Dolan, P.W. Carr, The hydrophobic-subtraction model of reversed-phase column selectivity,, J. Chromatogr. A 1060 (2004) 77–116.
[86] D.H. Marchand, K. Croes, J.W. Dolan, L.R. Snyder, Column selectivity in reversed-phase liquid chromatography: VII. Cyanopropyl columns, J. Chromatogr. A 1062 (2005) 57–64.
[87] D.H. Marchand, K. Croes, J.W. Dolan, L.R. Snyder, R.A. Henry, K.M.R. Kallury, S. Waite, P.W. Carr, Column selectivity in reversed-phase liquid chromatography: VIII. Phenylalkyl and fluoro-substituted columns, J. Chromatogr. A 1062 (2005) 65–78.
[88] C.A. Rimmer, L.C. Sander, S.A. Wise, Selectivity of long chain stationary phases in reversed phase liquid chromatography, Anal. Bioanal. Chem. 382 (2005) 698–707.
[89] L.R. Snyder, J.W. Dolan, P.W. Carr, A new look at the selectivity of reversed-phase HPLC columns, Anal. Chem. 79 (2007) 3255–3261.
[90] P.W. Carr, J.W. Dolan, U.D. Neue, L.R. Snyder, Contributions to reversed-phase column selectivity. I. Steric interaction, J. Chromatogr. A 1218 (2011) 1724–1742.
[91] D.H. Marchand, P.W. Carr, D.V. McCalley, U.D. Neue, J.W. Dolan, L.R. Snyder, Contributions to reversed-phase column selectivity. II. Cation exchange, J. Chromatogr. A 1218 (2011) 7110–7129.
[92] H.A. Claessens, C.A. Cramr, J.W. de Haan, F.A.H. den Otter, L.J.M. van der Ven, P.J. Andree, G.J. de Jong, N. Lammers, J. Wijma, J. Zeeman, Ageing processes of alkyl bonded phases in HPLC; a chromatographic and spectroscopic approach, Chromatographia 20 (1985) 582–586.
[93] http://www.phenomenex.com/hplc-column.
[94] http://www.waters.com/waters/en_US/HPLC-Columns/nav.htm?cid=511505&locale=en_US.
[95] http://www.analytics-shop.com/gb/chromatography/hplc-columns/hplc-columns-by-manufacturer/agilent-hplc-columns.html.
[96] http://www.mac-mod.com.
[97] http://www.mz-at.de/resources/brochures/ESIcatalog.pdf.
[98] F. Gritti, G. Guiochon, The current revolution in column technology: how it began, where is it going? J. Chromatogr. A 1228 (2012) 2–19.
[99] R.E. Majors, Highlights of HPLC 2014, LCGC North America 32 (7) (2014) 1–11.
[100] R.E. Majors, Historical developments in HPLC and UHPLC column technology: the past 25 years, LCGC North America 33 (11) (2015) 1–15.
[101] S.M. Chang, M. Lee, W.-S. Kim, Preparation of large monodispersed spherical silica particles using seed particle growth, J. Colloid Interf. Sci. 286 (2005) 536–542.
[102] F. Iskandar, Mikrajuddin, K. Okuyama, In situ production of spherical silica particles containing self-organized mesopores, Nano Lett. 1 (2001) 231–234.
[103] S.B. Yoon, J.-Y. Kim, J.H. Kim, Y.J. Park, K.R. Yoon, S.-K. Park, J.-S. Yu, Synthesis of monodisperse spherical silica particles with solid core and mesoporous shell: mesopore channels perpendicular to the surface, J. Mater. Chem. 17 (2007) 1758–1761.
[104] F. Gritti, G. Guiochon, Performance of columns packed with the new shell Kinetex-C18 particles in gradient elution chromatography, J. Chromatogr. A 1217 (2010) 1604–1615.
[105] M. Przybyciel, Novel Phases for HPLC Separations, LCGC, LC Column Technology Supplement, April 2006, pp. 49–52.
[106] http://www.waters.com/waters/en_RO/CSH-%28Charged-Surface-Hybrid%29-Technology/.
[107] G.P. O'Sullivan, N.M. Scully, J.D. Glennon, Polar-embedded and polar-endcapped stationary phases for LC, Anal. Lett. 43 (2010) 1609–1629.
[108] J.E. O'Gara, D.P. Walsh, C.H. Phoebe Jr., B.A. Alden, E.S.P. Bouvier, P.C. Iraneta, M. Capparella, T.H. Walter, Embedded-polar group bonded phases for high performance liquid chromatography, LC-GC 19 (2001) 632–642.
[109] M. Przybyciel, Fluorinated HPLC phases-looking beyond C18 for reversed-phase HPLC, LCGC North America 19 (1) (2006) 1–8.

[110] E. Bacalum, T. Galaon, V. David, H.Y. Aboul-Enein, Retention behavior of some compounds containing polar functional groups on perfluorophenyl silica based stationary phase, Cromatographia 77 (2014) 543–552.

[111] D. Lubda, K. Cabrera, W. Kraas, C. Schaefer, D. Cunningham, New developments in the application of monolithic HPLC columns, LC-GC Europe (December 2001) 2–5.

[112] J.J. Pesek, M.T. Matyska, Hydride-based silica stationary phases for HPLC: fundamental properties and applications, J. Sep. Sci. 28 (2005) 1845–1854.

[113] J.H. Knox, P. Ross, Carbon-based packing materials for liquid chromatography, structure, performance and retention mechanism, in: P.R. Brown, E. Grushka (Eds.), Advances in Chromatography, 37, 1997, pp. 74–119.

[114] L. Pereira, Porous graphitic carbon as a stationary phase in HPLC: theory and applications, J. Liq. Chromatogr. Rel. Technol. 31 (2008) 1687–1731.

[115] H. Mockel, A. Braedikow, H. Melzer, G.A. Aced, A comparison of the retention of homologous series and other test solutes on an ODS column and a Hypercarb carbon column, J. Liq. Chromatogr. 14 (1991) 2477–2498.

[116] J. Nawrocki, M.P. Rigney, A. McCormick, P.W. Carr, Chemistry of zirconia and its use in chromatography, J. Chromatogr. A 657 (1993) 229–282.

[117] L.F.C. Melo, C.H. Collins, K.E. Collins, I.C.S.F. Jardim, Stability of high-performance liquid chromatography columns packed with poly(methyloctylsiloxane) sorbed and radiation-immobilized onto porous silica and zirconized silica, J. Chromatogr. A 869 (2000) 129–135.

[118] F. Švec, Organic polymer monoliths as stationary phases for capillary HPLC, J. Sep. Sci. 27 (2004) 1419–1430.

[119] F. Švec, J.M.J. Fréchet, Continuous rods of macroporous polymer as high-performance liquid chromatography separation media, Anal. Chem. 64 (1992) 820–822.

[120] P.T. Vallano, V.T. Remcho, Affinity screening by packed capillary high-performance liquid chromatography using molecular imprinted sorbents I. Demonstration of feasibility, J. Chromatogr. A 888 (2000) 23–34.

[121] P.G. Wang, Monolithic Chromatography and its Modern Applications, ILM Publications, Glendale, USA, 2010.

[122] J.W. Dolan, L.R. Snyder, The hydrophobic-subtraction model for reversed-phase liquid chromatography: A reprise, LC/GC North America 34 (9) (2016) 730–741.

[123] I. Molnar, Computerized design of separation strategies by reversed-phase liquid chromatography: development of DryLab software, J. Chromatogr. A 965 (2002) 175–194.

[124] http://molnar-institute.com/drylab.

[125] R. Kormány, J. Fekete, G. Davy, S. Fekete, Reliability of computer-assisted method transfer between several column dimensions packed with 1.3–5 μm core–shell particles and between various instruments, J. Pharm. Biomed. Anal. 94 (2014) 188–195.

[126] C.S. Young, R.J. Weigand, An efficient approach to column selection in HPLC method development, LC/GC North America 20 (May 2002) 464–473.

[127] I. Neverova, J.E. Van Eyk, Role of chromatographic techniques in proteomic analysis, J. Chromatogr. B 815 (2005) 51–63.

[128] http://www.sge.com/uploads/b8/4c/b84c77ffb452a93fe4d12d7401dfa60b/TA-0136-H.pdf.

[129] E.M. Renkin, Filtration, diffusion, and molecular sieving through porous cellulose membranes, J. Gen. Physiol. 38 (1954) 225.

[130] M. Stromqvist, Peptide mapping using combinations of size-exclusion chromatography, reversed-phase chromatography and capillary electrophoresis, J. Chromatogr. A 667 (1994) 304–310.

[131] K. Sakai-Kato, K. Nanjo, T. Yamaguchi, H. Okuda, T. Kawanishi, High performance liquid chromatography separation of monoclonal IgG2 isoforms on a column packed with nonporous particles, Anal. Methods 5 (2013) 5899–5902.

[132] D.S. Bell, New chromatography columns and accessories for 2016, LC/GC North America 34 (4) (2016) 242–252.

CHAPTER

8

Polar Analytical Columns

8.1 TYPES OF POLAR HPLC PHASES AND THEIR PREPARATION

Polar stationary phases are used in hydrophilic interaction chromatography (HILIC), normal-phase chromatography (NPC), and also in aqueous normal-phase chromatography (ANPC or ANP). USP (United States Pharmacopeia) classification of polar columns is given in Appendix 1. HILIC chromatography is typically performed on bonded phases containing terminal polar groups such as aminopropyl, diol, amide, peptide, and also on bare silica. Other groups bonded on silica, such as cyano (e.g., cyanopropyl) have some polarity, and they make a transition between reversed-phase materials and polar ones. The classification of chromatography as RP-HPLC or HILIC on phases containing bonded cyano terminal groups depends on the polarity of the mobile phase, and in RP-HPLC the mobile phase is always more polar than the stationary phase, while in HILIC it is less polar (although it contains some water). The separations are performed only after a chromatographic column is conditioned with the mobile phase. NPC has been practiced for a long period of time on bare silica, with a nonpolar organic solvent as the mobile phase. ANPC is performed on special stationary phases (see Section 3.4) such as silica hydride. A more detailed discussion about polar phase columns is presented in this chapter. This discussion will be focused mainly on HILIC type columns, which are the most frequently utilized columns having a polar stationary phase.

Main Types of Stationary Polar Phases

The HILIC columns can be classified into several main types: (1) bare silica, (2) neutral, (3) anion exchange, (4) cation exchange, (5) zwitterionic, and (6) mixed polar and hydrophobic. The base material for polar phases can be either silica or a polymer, but like reversed-phase columns, the majority of HILIC columns on the market are silica-based. A few HILIC columns are not silica-based, being made from one of the following polymers: poly(vinyl alcohol), polymethacrylate, or styrene-divinylbenzene. The column Chrometa EP-NH2, for example, is made from a hybrid material and has both silica and polymer structure, and HILIC columns with a zirconia base are also known.

The classification of a stationary phase as being polar is based on its chemistry, but in the case of phases with alkyl-cyano groups, this is also determined by the mobile phase composition (when the mobile phase is more polar than the alky-cyano phase the mechanism is of

Selection of the HPLC Method in Chemical Analysis
http://dx.doi.org/10.1016/B978-0-12-803684-6.00008-1

RP-type). A list of several types of polar phases is given in Table 8.1.1 [1]. Other types of stationary phases, some based on silica and others on polymers, have been reported in the literature [2,3].

One particular aspect of HILIC (and NPC) columns is related to the dependence of column property on the mobile phase used for the separation. While the nature of stationary phase in RP-HPLC is less influenced by the mobile phase, in HILIC the pH of the mobile phase controls the ionization state of the polar groups of the stationary phase (in addition to the state of the analyte ionization). The action of a stationary phase as anion or cation exchanger or only as a polar phase depends on mobile phase pH. Even the ionization of silanol groups on bare silica depends on the pH of the mobile phase. In addition, the water content adsorbed on the surface of the polar phase influences (in particular in NPC) the retention capability of the column.

Preparation of Polar Stationary Phases

Preparation of polar stationary phases uses many of the general procedures for solid phase preparation, already described in Section 6.1. Silica is frequently used as solid support for polar phases, and bare silica is also a common polar stationary phase. Derivatization of the silica surface with a reactive silane, such as (3-aminopropyl)trietoxysilane, will lead to the

TABLE 8.1.1 Several Types of Polar Stationary Phases Used in NPC and HILIC Chromatography

Phase	Type	Phase Structure
Silanol	Silica	Bare silica
Diol on silica	Neutral	Diol bonded on different alkyl handle (spacer)
Diol + ether on silica	Neutral	Diol, ether embedded and diol
Amide on silica	Neutral	Amide terminal, e.g., on propyl handle, polyamide
Other weak polar on silica	Neutral	Cyano (also used in RP mode)
Special polar groups on silica	Neutral	Cyclodextran; bonded carbohydrates
Polymers bonded on silica	Neutral	Polyvinyl alcohol
Imide on silica	Neutral	Poly(succinimide), other
Urea on silica	Anion exchange	Urea terminal (e.g. on propyl handle)
Amine on silica	Anion exchange	$-C_3H_6-NH_2$, $-N^+(CH_3)_3$, diethylamine, triazole, etc.
Polyethyleneimine	Anion exchange	$-(CH_2)_n-N^+(CH_3)_3$
Weak anionic polymeric	Anion exchange	Different polar groups on porous polymers
Weak cationic on silica	Cation exchange	Sulfonylethyl, etc.
Weak cationic on silica	Cation exchange	Nitrophenyl (very weak cationic)
Zwitterionic on silica	Zwitterionic	Amino-sulfonic, amino-carboxilic, poly(2-hydroxyethyl aspartamide), phosphorylcholine, etc.
Mixed mode	Mixed	Various types of polar groups and hydrophobic moieties

formation of a silica-based material that can be used itself as a polar stationary phase and also is capable of undergoing further reactions. The reaction of free silanols with 3-aminopropyltrietoxysilane is schematically shown below:

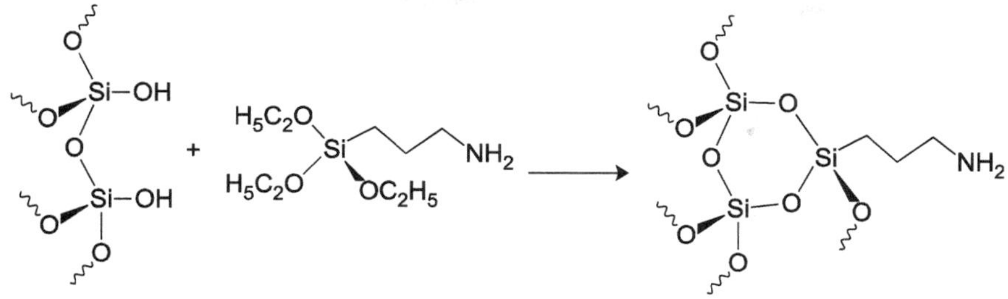

(8.1.1)

Formation of silica containing terminal NH_2 groups can be obtained using trifunctional silanes, bifunctional silanes such as γ-aminopropylmethyldiethoxysilane, or even monofunctional ones. Different groups can be attached using other reagents such as cyanopropyltriethoxysilane to form phases containing a CN terminal group, or dihydroxypropyltriethoxysilane to form phases with diol terminal groups.

In accordance with the diversity of stationary phases used in HILIC separations, various synthetic procedures were utilized for their preparation. For example, some polar phases can be generated in two step reactions, the first being the synthesis of silica derivatized with active groups. Such initial derivatizations can be performed for example with reagents such as γ-glycidoxypropyl-trimethoxysilane, 3-azidopropyltrimethoxysilane, or 3-aminopropyl-trietoxysilane [1]. For example, aminopropyl silica can be further treated with specific reagents such as polysuccinimide and form polymeric type phases bonded on silica (poly(succinimide)-silica). The reaction with polysuccinimide is schematically indicated below:

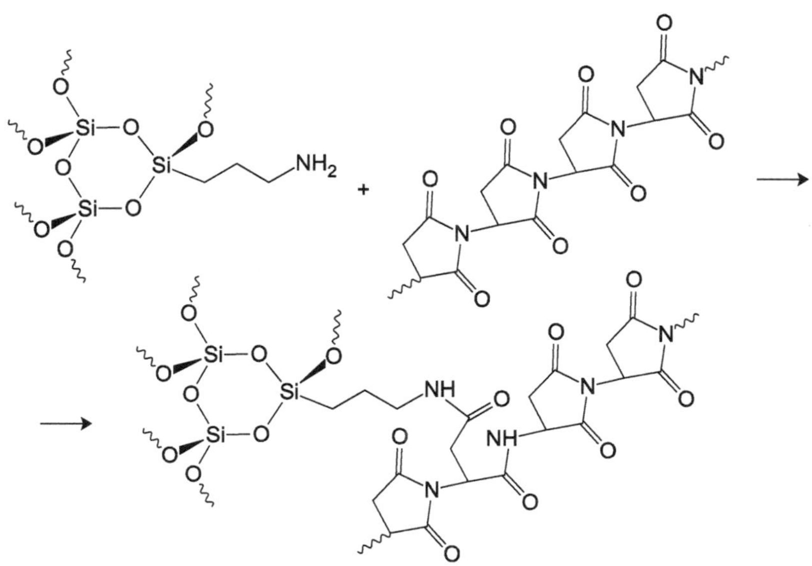

(8.1.2)

Only a fraction of the initial polysuccinimide rings are engaged in the bonding to silica and the rest of the rings remain intact and can be used as a polysuccinimide phase. A polysuccinimide phase is also susceptible to reaction with nucleophiles, which can be used to produce a variety of other functional silicas [4]. For example, the poly(succinimide)silica can undergo an alkaline hydrolysis in water to form a poly(aspartic acid) bonded phase. The material can also react with water in the presence of aminoethanol to form poly(2-hydroxyethyl)aspartamide bonded phase, and in the presence of aminoethansulfonic acid to form poly(2-sulfonylethyl)aspartamide bonded silica. The poly(sulfonylethyl)aspartamide silica phase displays a zwitterionic character.

Silica derivatized with 3-azidopropyltrimethoxysilane can react in a second step of phase preparation with different reagents having an ethynyl group attached to different fragments, following a 1,3-dipolar cycloaddition (Huisgen cycloaddition). The reaction is schematically shown below:

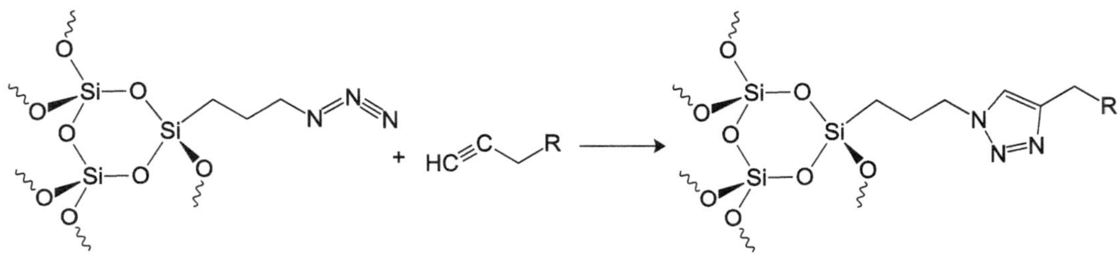

(8.1.3)

In reaction 8.1.3, the group R can be derived from β-cyclodextrin, galactose, lactose, as well as from other polar molecules. Due to their specific configurations, these bonded phases containing various carbohydrate moieties can be used for HILIC separations as well as for chiral separations of highly polar amino acids, glycopeptides, oligonucleotides, or other natural products, such as flavonoids [5]. Synthesis involving binding to silica of polar macrocycles was also described in the literature [6].

Some stationary phases used in HILIC are based on an organic polymer backbone such as Asahipak NH_2P, Astec apHera, COSMOSIL DEAE, Chromenta EP-NH2 (aminopropyl), and Jordi Amino. Also, several HILIC type materials based on organic polymers were prepared as monoliths. The polymer backbone of organic polymer monoliths can be based on styrene, acrylamide, or acrylate/methacrylate. Methacrylate-derived monoliths are frequently used since they have high chemical stability over a wide pH range, excellent mechanical resilience, flexibility for surface modification, and good reproducibility of the polymerization process [7]. For example, a porous zwitterionic monolith can be prepared by thermal copolymerization of N,N-dimethyl-N-methacryloxyethyl-N-(3-sulfopropyl)ammonium betaine and ethylene dimethacrylate inside a 100-μm inner dimension (i.d.) capillary [8].

The surface of the pores of a monolithic phase plays an important role in determining the selectivity in a separation [9]. For this reason, postpolymerization functionalization is a successful procedure to introduce functional groups on the pore surface of monoliths [10–12]. One common postpolymerization procedure is applied to polymers obtained starting with glycidyl methacrylate used as the reactive functional monomer. The resulting monolith can undergo further modifications for attaching the desired polar functionality, based on epoxy group reactivity. One advantage of postpolymerization functionalization is that the optimization of polymerization conditions for obtaining the desired monolith mechanical and flow-through porous properties is established for one polymer instead of reoptimization of the polymerization conditions from monolith to monolith.

8.2 PROPERTIES AND CHARACTERIZATION OF POLAR HPLC PHASES

Physical Properties of Polar Stationary Phases and Columns

The physical properties of polar stationary phases and columns are very similar to those for RP-type stationary phases. These include all the parameters listed in Table 7.2.1. Column body construction, as well as stationary phase physical properties, influence the separation on HILIC phases in the same way as they influence the separation in RP-HPLC and the discussions from Sections 6.2 and 6.3 remain valid for HILIC columns. For example, column efficiency (measured in N/m) depends on the type of particles in the column (fully porous, core–shell, monolithic), particle diameter, and other column construction (e.g., packing uniformity). Similar to RP phases, HILIC phases are available on porous particles, and core–shell particles. Also, similar to RP columns, HILIC columns having the support of organic/inorganic silica are available (e.g., XBridge HILIC from Waters). In the selection of the HILIC column, particle dimensions, particle surface area, particle size distribution, etc. are important parameters that should be considered for a successful separation.

Chemical Properties of Polar Stationary Phases

The nature of the stationary phase in HILIC plays a more important role than in RP-HPLC. The phases used in HILIC separation have basically three different properties: (1) polarity (including hydrogen bonding), (2) ion exchange character, and (3) hydrophobicity. Because of various mechanisms of interaction displayed by the HILIC columns, such as polar (including hydrogen bonding), ion exchange, hydrophobic, plus other types such as steric hindrance, the properties of the stationary phase in HILIC are more difficult to classify. In addition, the role of the mobile phase (its pH) may influence the nature of polar stationary phases.

1) Regarding the polarity, a guidance can be obtained from the values of $\log K_{ow}^{model}$ indicated in Table 6.2.4. It can be considered that polarity varies in the following order: bare silica ≈ tertiary amine > primary amine > amide > zwitterionic > imide > urea > diol > cyano. This order can change significantly depending on the silica-base structure, coverage of the silica support, and use of other support materials (polymeric or zirconia).

2) The ion exchange properties are not necessarily in the same order as polarity and, for example, amine columns display stronger ion exchange characteristics compared with bare silica or phases with diol, amide, or CN groups. Zwitterionic columns display ion exchange properties, but the two groups seem to suppress this character as a whole.

3) The phases with different structures have different polarities, but beside polarity the hydrophobicity of the spacer (handle) connecting the polar group with the silica-base plays an important role in the separation. For this reason, the carbon load C% of the HILIC columns should also be considered for column characterization. The spacer may be a propyl group, but also a longer group such as a hydrocarbon chain with 10 carbon atoms. It also may include phenyl groups, or even embedded polar groups such as ether. Because of the presence of a hydrophobic component in the HILIC retention mechanism, the characterization of retention cannot be done with the values of k' for compounds such as toluene or ethylbenzene. These compounds are strongly hydrophobic and will interact mainly based on hydrophobic interactions, and polar interactions will be ignored. The values for k' for series such as uracil–5-methyluracyl or uridine–methyluridine are typically used for the evaluation of retention.

Other chemical properties of the stationary phase show similarities with RP phases. These include the following: (1) silica purity, (2) metal activity from silica, (3) coverage of support with the bonded phase, (4) pH resilience, and (5) preparation procedure using mono-, bi-, or trifunctional reagents. Even end-capping and silanol activity are properties to be considered for HILIC columns (for NPC columns water adsorption on the stationary phase plays an important role). The nature of the bonded phase in HILIC is different from RP-HPLC (except for some cyano phases). Some HILIC columns and many other columns on which NPC is practiced have bare silica as a stationary phase. The silanol groups on silica may have various properties depending on the preparation procedure of hydrated silica material. On these materials the role of silanol groups is critical for the separation. However, for HILIC columns, the silanol activity may interfere with the polar interactions generated by the bonded polar groups (the interaction with polar solvents such as water is also affected by the polar bonded group). For this reason, some polar bonded stationary phases are also end-capped with small hydrophobic groups (e.g., TSKgel NH2-100 is end-capped with TMS groups).

The pH range of stability of HILIC columns is also similar to that for RP-HPLC columns. Standard columns made from bare silica or with a bonded polar phase are typically stable between pH 2–8. Special phases may have an extended pH range of stability, and this is indicated by the manufacturer for specific columns.

In HILIC, because the mobile phase always contains water, the wetting characteristic of the stationary phase is not an issue. For NPC, water adsorbed on the stationary phase plays an important role in the separation. However, because in NPC the mobile phase does not contain water, the mobile phase slowly removes the water from the stationary phase surface. This affects the separation and leads to lack of reproducibility in NPC separations. For this reason, it is important in NPC to keep the silica surface wet with water. This can be achieved either by adding a very small proportion of water in the mobile phase or by conditioning the stationary phase surface before use with a solvent containing water.

One additional characteristic of HILIC (and NPC) columns is the rate with which the column is equilibrated in a specific mobile phase. The HILIC separation process is rather complex, and the mobile phase plays an important role in the separation. The equilibration of the stationary phase with a specific mobile phase takes place by a complex diffusion and adsorption process. Since the water adsorbed on the stationary phase surface plays an important role in column properties, the HILIC (and NPC) columns require in general a longer time for equilibration as compared to RP-HPLC columns.

Retention and Separation Capability of Polar Columns

The chromatographic parameters characterizing the separation capability of the polar analytical columns are the same as for hydrophobic columns, and these parameters include the following: retention time t_R, retention factor k', number of theoretical plates N, selectivity α, resolution R, and peak asymmetry As. Retention factor k' in HILIC depends on several parameters which include the column properties, the nature of compounds to be separated, and elution conditions such as mobile phase composition, temperature, pH, and other aspects related to mobile phase. However, prediction of log k' is more difficult to make for HILIC than it is for RP-HPLC since in HILIC the retention and elution process is determined by more types of interactions (see Section 4.3). The variation of log k' for a compound in HILIC depends on the polarity of the mobile phase in an opposite way compared to RP-HPLC, and a higher content in water in the mobile phase for an HILIC separation leads to a faster elution of the solutes (see expression 4.3.44). The mobile phase is typically a mixture of two or more solvents with "strength" varying for HILIC in the following order: water > methanol > dimethylformamide > dioxane > ethanol > isopropanol > acetonitrile > acetone [13]. In HILIC, the pH of the mobile phase is also very important. The ionization of polar analytes containing groups such as $-COOH$, $-NH_2$, >NH depends strongly on the pH of the mobile phase, and their log D_{ow} changes by the modification of pH. For this reason, the use and therefore the selection of a specific HILIC column must always consider the mobile phase nature and pH to be utilized on that column. Different tests are used for the characterization of separation capability of HILIC columns, and some details about these tests are given in this section.

Parameters and Tests for HILIC Column Characterization

The interactions in HILIC comprise polar (including hydrogen bonding), hydrophobic, ion exchange (ionic), and steric effects [3,14]. For this reason, the characterization of HILIC columns needs to include a number of parameters selected for the characterization of specific properties. A set of such parameters may include the following: (1) k' for uridine or $k'(U)$, (2) efficiency, (3) methylene selectivity $\alpha(CH_2)_{HILIC}$, (4) hydroxy group selectivity $\alpha(OH)$, (5) isomer selectivity for diastereoisomers α_{dia}, (6) selectivity for position isomerism (regioisomers) α_{regio}, (7) molecular shape selectivity α_{shape}, (8) anion exchange selectivity α_{AX}, (9) cation exchange selectivity α_{CX}, (10) acidity character α_{tt}, and (11) peak asymmetry $As(X)$. As previously indicated, mobile phase characteristics and particularly the pH must also be kept at specific values when testing and evaluating HILIC columns. The pH of the mobile phase controls not only the analyte properties, but also the ionization state of the polar groups of the stationary phase.

1) Similar to the use of the retention factor k' for the characterization of an RP phase using a hydrophobic compound (such as toluene or ethylbenzene), the characterization of retention of HILIC columns is typically done using k' for a polar compound, and uridine (U) is one of the compounds frequently used for this purpose. The values of $k'(U)$ depend on the mobile phase composition, and it is common to use as a mobile phase for the measurement of $k'(U)$ acetonitrile/aqueous buffer 90/10, the buffer being 20 mM ammonium acetate at pH = 4.7. Since the values of k' depend on mobile phase composition, the k' values can be used only

for a relative comparison of columns, with the same mobile phase and test compound utilized.

2) The number of theoretical plates N (better indicated as N/m since the columns may have different lengths) describes column efficiency for HILIC similar to RP or other types of chromatography. However, the value for N is compound-dependent, and even for the same column different N values are obtained with different test compounds. The test compounds used in RP-HPLC are also different from those used in HILIC. Among the compounds used for the evaluation of N in HILIC are uridine (U) and 5-methyluridine (5MeU). The column efficiency in HILIC follows the same rules as for RP-HPLC, with smaller particles and core–shell particles leading to higher N (lower theoretical plate height H). However, typically in HILIC and in NPC the values for N are lower than in RP-HPLC (the peaks are usually broader). This can be explained by the existence of a combination of different retention mechanisms in HILIC, which influence in different ways the analyte exchange between the mobile and stationary phases [15].

3) Methylene selectivity $\alpha(CH_2)_{HILIC}$ (or simply $\alpha(CH_2)$) is a parameter that quantifies the hydrophobic character (as described in Section 7.3). HILIC separations also involve hydrophobic effects, and the disruption of the interactions between the solvent molecules produced by the analyte plays a noticeable role in the energy values involved in the separation (see Section 4.3). For HILIC columns the evaluation of methylene selectivity cannot be done using hydrophobic compounds such as those recommended for RP columns since they elute too early. Different compounds that are one CH_2 apart can be used for defining the $\alpha(CH_2)_{HILIC}$ value. A recommended comparison for the measurement of methylene selectivity in HILIC separations is the ratio of capacity factors for uridine (U) and 5-methyluridine (5MeU). The value for $\alpha(CH_2)_{HILIC}$ can also be indicated as α(U/5MeU). The mobile phase conditions recommended for such evaluation are acetonitrile/aqueous buffer 90/10, the buffer being 20 mM ammonium acetate at pH = 4.7 [14]. This methylene selectivity is given by the expression:

$$\alpha(CH_2)_{HILIC} = \frac{k'(U)}{k'(5MeU)} \tag{8.2.1}$$

The choice of nucleosides for obtaining a value for $\alpha(CH_2)_{HILIC}$ was made because these compounds are well retained on HILIC columns and the retention of uridine and 5-methyluridine is less affected by ion exchange effects.

4) Hydroxy group selectivity α(OH), also indicated as α(U/2dU), is a parameter that quantifies the hydrophilic character. A recommended comparison is the measurement of α(OH) as the ratio of capacity factors for uridine (U) and 2-deoxyuridine (2dU), the values for k' being obtained in the same mobile phase conditions as for $\alpha(CH_2)_{HILIC}$. The formula for α(OH) is the following:

$$\alpha(OH) = \frac{k'(U)}{k'(2dU)} \tag{8.2.2}$$

Other compounds differing by an OH group can be used for calculating α(OH) selectivity [16,17].

5) Isomer selectivity for diastereoisomers α_{dia}, also indicated as α(Vi/Ad) or α(V/A), is an important column characteristic and can be estimated based on the ratio of k' values for

vidarabine (Vi) and adenosine (Ad). The mobile phase for the separation is the same as for the measurement of $\alpha(CH_2)_{HILIC}$.

6) Selectivity for position isomerism α_{regio}, also indicated as α(2 dG/3 dG), can be estimated based on the ratio of k' values for 2′-deoxyguanosine (2 dG) and 3′-deoxyguanosine (3 dG) (and the same mobile phase as for the previous measurements).

7) Molecular shape selectivity α_{shape}, also indicated as α(NPαGlu/NPβGlu), can be measured based on the ratio of k' values for two diastereoisomers: 4-nitrophenyl-α-D-glucopyranoside (NPαGlu) and 4-nitrophenyl-β-D-glucopyranoside (NPβGlu) (and the same mobile phase as for the previous measurements).

8) Anion exchange selectivity α_{AX}, also indicated as α(SPTS/U), can be measured from the ratio of k' values for sodium *p*-toluenesulfonate (SPTS) and uridine (U).

9) Cation exchange selectivity α_{CX}, also indicated as α(TMPAC/U), can be measured from the ratio of k' values for N,N,N-trimethylphenylammonium chloride (TMPAC) and uridine (U).

10) The acidity character α_{tt}, also indicated as α(Tb/Tp) or α_{tt}(Tb/Tp), can be obtained from the ratio of k' values for theophylline (Tp) and theobromine (Tb). This parameter differentiates columns regarding their basic, neutral, or acidic character with $\alpha_{tt} < 1$ for basic, $\alpha_{tt} \approx 1$ for neutral, and $\alpha_{tt} > 1$ for acidic stationary phases.

11) Peak asymmetry $As(X)$ is another parameter used for HILIC column characterization (see expression 4.2.52). Peak asymmetry is compound-dependent, and among the test compounds for the estimation of As in HILIC separations are U and 5MeU.

Differently from the case of hydrophobic columns, where the characterization is performed with parameters H^*, S^*, A^*, B^*, $C(2.8)^*$, and $C(7.0)^*$ that were obtained using regression techniques, in the case of HILIC columns it was found more convenient to simply characterize the columns based on measured α parameters for specific pairs of compounds. The HILIC mechanism being more complex and less investigated than the mechanisms for hydrophobic columns, it would require further characterization of a multitude of columns for a similar approach to that used in RP-HPLC. Some examples of HILIC columns commercially available and the values for their characterization with the parameters previously described (in a mobile phase acetonitrile/aqueous buffer 90/10, the buffer being 20 mM ammonium acetate at pH = 4.7) are given in Table 8.2.1 [14].

The parameters for the same type of phase may vary significantly. Zwitterionic phases seem to have the larger α(OH) and $\alpha(CH_2)$. The bare silica phases typically have low k'(U), α(OH), and $\alpha(CH_2)$ and large α_{CX}, although the actual values can differ significantly.

It is common for the characterization of HILIC stationary phases to use radar plots including k'(U), α(OH), $\alpha(CH_2)$, α_{dia}(Vi/Ad), α_{regio}(2 dG/3 dG), α_{AX}, α_{CX}, and α_{tt}(Tb/Tp). The values for α_{shape} and for As are not usually included in the plots. Sometimes, specific values are normalized in order to bring closer to each other the ranges of the numbers representing each parameter. Useful presentation for the set of parameters describing an HILIC column is done using a radar plot. Several such radar plots for the columns ZIC-HILIC (5 μm) (plot A), Amide-80 (3 μm) (plot B), CYCLOBOND I (5 μm) (plot C), and NH2-MS (5 μm) (plot D) are shown in Fig. 8.2.1. These radar plots allow a rapid estimation of column characteristics and help in column selection for HILIC separations.

TABLE 8.2.1 List of Several Commercially Available HILIC Columns, and Their Characteristic Parameters [14]

Column	k'(U)	N(U)	α(OH)	$\alpha(CH_2)$	α_{dia}	α_{regio}	α_{shape}	α_{AX}	α_{CX}	α_{tt}	As(U)
ZIC-HILIC (5 μm)	2.11	40,000	2.03	1.67	1.50	1.11	1.14	0.33	1.57	1.18	1.34
ZIC-HILIC (3.5 μm)	2.10	83,333	2.07	1.71	1.51	1.12	1.14	0.27	1.64	1.20	1.26
Nucleodur HILIC (3 μm) (zwitterionic)	2.20	71,429	1.55	1.28	1.46	1.08	1.14	0.34	0.95	1.00	0.88
Amide-80 (5 μm)	3.30	38,462	1.67	1.27	1.29	1.08	1.18	0.19	1.00	1.39	1.37
Amide-80 (3 μm)	4.58	111,111	1.64	1.27	1.28	1.08	1.18	0.41	2.75	1.32	0.99
XBridge amide (3.5 μm)	2.55	83,333	1.70	1.29	1.30	1.07	1.16	0.47	1.20	1.38	1.42
PolySULFONYLETHYL (3 μm)	1.58	16,129	2.13	1.48	1.21	1.06	1.24	0.06	0.35	1.00	1.11
PolyHYDROXYETHYL (3 μm)	3.92	16,393	1.92	1.36	1.31	1.07	1.21	0.34	1.31	1.14	0.99
CYCLOBOND I (5 μm)	0.70	55,556	1.21	1.13	1.24	1.10	1.20	4.73	0.63	1.01	1.73
LiChrospher diol (5 μm)	1.50	58,824	1.36	1.15	1.32	1.06	1.17	0.63	1.16	1.04	0.98
Chromolith Si	0.31	83,333	1.00	1.12	1.16	1.11	1.31	0.09	8.21	1.22	1.00
HALO HILIC Si (2.7 μm)	0.64	125,000	1.08	1.16	1.18	1.13	1.29	0.64	29.03	1.26	1.47
COSMOSIL HILIC (5 μm) (Triazole)	1.60	83,333	1.6	1.14	1.36	1.03	1.13	0.80	0.49	0.89	1.11
Sugar-D (5 μm)	1.58	58,824	1.74	1.44	1.45	1.10	1.22	1.90	0.25	0.52	1.12
NH2-MS (5 μm)	2.44	83,333	1.88	1.3	1.36	1.07	1.20	0.82	0.28	0.54	1.04

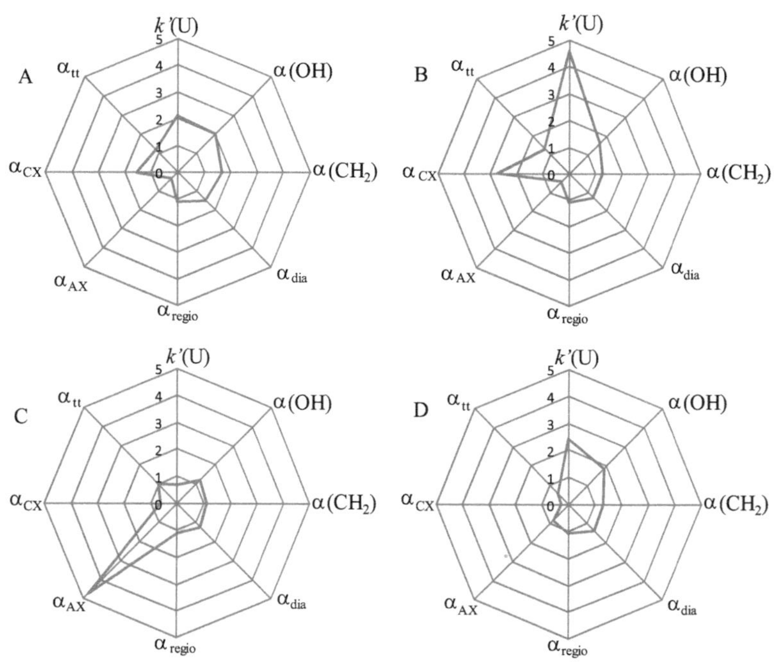

FIGURE 8.2.1 Radar plots for the columns ZIC-HILIC (5 μm) (plot A), Amide-80 (3 μm) (plot B), CYCLOBOND I (5 μm) (plot C), and NH2-MS (5 μm) (plot D).

8.3 COLUMNS WITH A POLAR STATIONARY PHASE

Polar columns can be classified in several groups depending on the chemical structure of the active stationary phase, as indicated in Table 8.1.1. The development of stationary phases with a bonded material containing polar groups generated a considerable expansion of applications using HILIC columns. The bonded phase typically uses silica as a substrate. The polarity of bonded-phase HILIC columns is lower than that of bare silica. However, their reproducibility and rapid regeneration when changing the mobile phase corrected some of the main problems of bare silica columns. HILIC columns are widely used for the separation of important classes of polar organic compounds such as carbohydrates, amino acids, and peptides [18]. Also, due to the use of mobile phases containing both water and an organic solvent, HILIC columns are frequently recommended to be selected when MS or MS/MS detection is utilized, since these solvents provide good ionization media and lead to good sensitivity. A variety of types of stationary phases can be used for HILIC separations. They include mainly columns made using porous particle materials (hydrated silica), but several are made using core–shell (fused core) support and even inorganic/organic (ethylene-bridged structures). In this section, each specific type of HILIC column is discussed in some detail.

Bare Silica Columns

Silica stationary phase can be used either in NPC mode or in HILIC mode. In NPC mode, the organic mobile phase has no water, but this does not imply the total absence of water from the silica surface. Anhydrous silica still contains a layer of water (possibly monomolecular) on its surface. In HILIC mode, silica is covered with a thicker layer of water, and a partition mechanism for the separation is more likely than on dry silica.

For both NPC and for HILIC applications, silica of high purity (*type B* purity) is commonly used. Silica phases based on *type B* purity specifically developed for work in HILIC mode were developed and are commercially available. Among these are the columns indicated in Appendix 3a. The properties of silica columns can vary significantly from one brand column to another. As shown in Table 8.2.1, the α_{CX} values for silica are typically high and the phases have an acidic character (α(Tb/Tp) > 1.0).

Neutral HILIC Stationary Phases With a Bonded Phase

Neutral HILIC stationary phases have the polar group amide, diol, cyano, imide, and also hydroxyl. These groups can be bonded to silica or other materials (cyclodextran, polyvinyl alcohol, polymethacrylate). These phases are less polar than bare silica. The columns containing cyano groups have the lowest polarity, and this type of column can also be used in RP-HPLC when the mobile phase is more polar than the stationary phase. For this reason several cyano columns were also discussed in Section 7.4. The polar groups bonded on silica are typically at the end of a hydrocarbon chain serving as a "handle" (see Section 8.1). The length of the aliphatic chain contributes to the hydrophobicity of the phase (see, e.g., Section 7.2). One other phase recommended for HILIC is a fluorinated stationary phase (Epic HILIC FL with undisclosed structure), although fluorinated phases are typically considered of RP type.

A common polar group in HILIC stationary phases is the amide. Amide polarity is higher than cyano (see Table 6.2.4 with $-CN$ contribution to log K_{ow}^{model} of -1, and $-CONH_2$ contribution of -1.7). Another common polar group is hydroxy. According to Table 6.2.4, one hydroxy group has a polarity contribution between cyano and amide, but a diol group brings an even stronger polar contribution (from Table 6.2.4 the $-OH$ contribution to log K_{ow}^{model} is -1.44, and the diol contribution is -2.51). Some diol phases may have a specific structure where ether groups are also present in the connecting chain (handle) to the silica surface (crosslinked diol). Polyethylene glycol bonded to silica is also used as a polar stationary phase. The retention mechanisms for neutral HILIC stationary phases consist of polar interactions and hydrogen bonding between the hydroxy or amide groups on the stationary media and the polar groups of the analytes. HILIC columns with amide groups were proved to be useful for sugar, amino acid, and peptide analyses. Dihydroxypropyl (diol) groups are able to form stronger hydrogen bonds compared to the amide groups, and these columns are used when stronger polar interactions are necessary for the separation.

More complicated structures that contain OH groups can be used as neutral HILIC stationary phases. Among these are phases with bonded cyclodextrin or bonded perhydroxyl-cucurbit [6] uril groups that contain numerous OH functionalities and can act as HILIC stationary phases [19].

Other materials such as those made from silica-bonded polysuccinimide [20] or polyhydroxyethyl-aspatamide can be used as neutral phases for HILIC. Phases containing a sulfur atom embedded in the chain bearing an OH group, such as mercaptoethanol silica and thioglycerol silica, were also reported [1]. Several neutral HILIC columns that are commercially available are listed in Appendix 3b. As indicated in Appendix 3b, organic polymers are also used as a support for phases with neutral polar groups, such as methacrylic polymers with 2,3-dihydroxypropyl groups [21], sorbitol bonded to a methacrylate polymer covering silica, etc.

Anion Exchange HILIC Type Stationary Phases

Anion exchange type stationary phases have the polar group amine or triazole. These phases are in fact weak anion exchange stationary phases that can be used in HILIC mode being applied for the separation of neutral molecules. The propyl amino type stationary phase has been in use for a long time and applied in particular for the separation of carbohydrates. Besides amino groups attached to an aliphatic hydrocarbon chain, the amine group can also be attached to an aromatic ring. Several commercially available HILIC columns with weak anion exchange properties are listed in Appendix 3c.

Cation Exchange HILIC Type Stationary Phases

Cation exchange HILIC are cation exchange columns that can be used in HILIC mode for the separation of neutral molecules. Among such columns some are weak cation exchangers and some strong cation exchangers (with sulfonic groups). However, these columns can be successfully used in HILIC mode. Some columns with cation exchanger properties are listed in Appendix 3d.

Zwitterionic Type HILIC Stationary Phases

Zwitterionic HILIC (Zic-HILIC) stationary phases contain both an anionic and a cationic group in their structure. Common groups in the bonded phase are quaternary amine (embedded) and sulfonic terminal, and quaternary amine (embedded) and carboxyl terminal. Some other zwitterionic structures have been made such as phases containing polypeptides bonded to silica. Among the commercially available HILIC columns with a zwitterionic structure, several are listed in Appendix 3e.

Zic-HILIC stationary phases can be divided into two types depending on the position of charged groups toward the stationary phase surface. In one type, the cation type group is placed toward the surface of the phase, and the anion group toward the silica base, the two groups being separated by a hydrophobic spacer. In the other type of phase, the anion type group is toward the surface of the phase, and the cation group toward the silica base. For example, the stationary phases Obelisc R and Obelisc N (see Appendix 3e), differ in the type and proximity of their charged groups as well as the hydrophobicity of their long chains. Obelisc R has the cationic group close to the silica surface separated from anionic group by a hydrophobic chain, while Obelisc N has an anionic group close to the surface separated from a cationic group by a hydrophilic chain. The two different charged structures interact differently with polar analytes and lead to different chromatographic results [22].

The functional group sulfonylalkylbetaine for ZIC-HILIC and ZIC-pHILIC and functional group phosphorylcholine for ZIC-cHILIC are schematically illustrated in Fig. 8.3.1.

The electrostatic interactions for ZIC-HILIC columns are diminished compared to those of individual ionic groups. Even the values of log K_{ow}^{model} for a model stationary phase with zwitterionic groups cannot be estimated based on the values from Table 6.2.4, because the effect of two charges is counterbalanced by the proximity of the two ions of opposite charges. For this reason, zwitterionic phases are successfully utilized for separations where the polar and ionic interactions are not very strong.

Stationary Phases With More Than One Type of Group (Mixed-Mode HILIC Phases)

The polar groups present in the bonded phases are connected with a "handle" to the support (usually silica), and this handle is typically formed from a hydrocarbon chain. A short chain such as propyl, although bringing some hydrophobic character to the phase, does not significantly affect the polar character of the stationary phase when polar groups are attached to the chain end. However, longer hydrophobic chains (such as C8 or longer) with polar groups attached at the chain end or embedded in the hydrophobic chain produce phases with a mixed mode of action. For example, some mixed-mode HILIC columns have

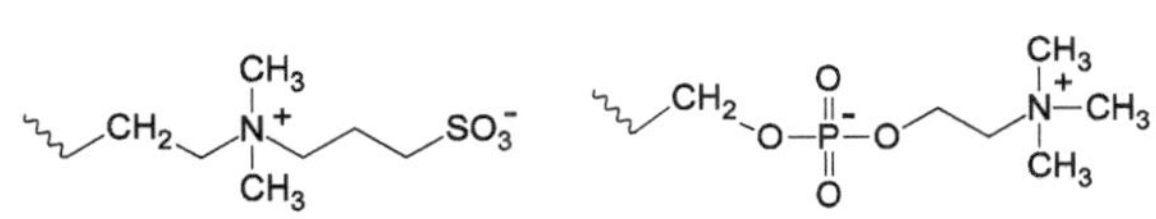

FIGURE 8.3.1 Schematic structure of functional group sulfonylalkylbetaine for ZIC-HILIC and ZIC-pHILIC and functional group phosphorylcholine for ZIC-cHILIC stationary phase.

bonded phases with nonpolar chains and groups of the type polar neutral, anion exchange, cation exchange, or zwitterionic. For example, Primesep N (Sielc) has embedded acidic groups (negatively charged, thus cation exchange) on a hydrocarbon chain bonded phase, Primesep AP (Sielc) has weak amino anion exchange groups and nonpolar moieties, and Obelisc N (Sielc) has both negatively and positively charged groups (zwitterionic) on the same long chains of bonded phase. The ion exchange groups give these columns enhanced selectivity in addition to RP character. For example, the phase in Acclaim Mixed Mode HILIC-1 column (Dionex/Thermo Scientific) contains a diol group and a long (C10) hydrocarbon spacer, the column Acclaim Mixed Mode WCX-1 contains a carboxyl group at the end of a C10 spacer (and can act as an ion exchange and RP phase), and Acclaim Mixed Mode WAX contains an amide group, a tertiary amine. Various other mixed mode phases were reported in the literature (see, e.g., [1,23].

Besides small particle phases, monolithic HILIC phases were also prepared. Besides bare silica Chromolith (see Appendix 3a), polymer-coated monolithic silica was used as HILIC stationary phase, the coating being performed using 3-diethylamino-2-hydroxypropylmethacrylate, *p*-styrenesulfonic acid, or other similar monomers. Also, monolithic organic polymers with polar functionalities such as poly(hydroxymethacrylate) were reported in the literature [24].

Silica Hydride-Based Phases

Silica hydride stationary phases contain Si-H groups on their surface and not Si-OH groups as do many silica-based columns (see Section 6.1 for the synthesis of hydride silica materials). A number of silica hydride (silica *type* C) phases are commercially available (e.g., from MicroSolv/Cogent) and can be used in HILIC mode. These columns include a bare hydride column (Diamond Hydride) as well as columns with a bonded phase containing polar groups such as butylamino (Cogent HPS Amino). The types of chromatography that can be practiced on bare hydride columns include ANP and HILIC [25]. Several commercially available HILIC columns with silica hydride structure are listed in Appendix 3f.

8.4 SELECTION OF A POLAR COLUMN

Selection of an HILIC column may represent a relatively challenging task. In HILIC (and NPC) the role of the stationary phase becomes more important, and for this reason the column selection is not very simple. The interaction of the analytes with the polar groups from the stationary phase plays a crucial role in the separation, affecting both the retention time, the resulting retention factor k', the peak shape and therefore the efficiency (N), and the selectivity α. Through these parameters resolution R is affected. Mobile phase composition and pH still remain in HILIC important factors related to separation. For example, the pH of the mobile phase may influence not only the form in which the analytes are (free molecules or ions), but also may influence the polarity of the stationary phase when groups such as $-NH_2$, $-N^+(CH_3)_2-$, $-N^+(CH_3)_3$, $-SO_3^-$ are present in the stationary phase. Also, the peak shape and efficiency in HILIC may be a problem, and typically they are not as good as in RP-HPLC. For these reasons, when a separation is still possible on an RP-HPLC column, this offers advantages. The development of RP-HPLC columns having polar embedded groups, hydrophilic end-capping and use of trifunctional reagents for the preparation of the stationary phase allows the extension of RP-HPLC separations to highly polar compounds, which usually require a very low content of organic component in the mobile phase (or even 100% aqueous mobile phase). In Figure 7.5.1 the separation of several organic acids with highly polar molecules can be seen on an RP-type column.

Another alternative to HILIC is the use of ion pair chromatography (IPC), which takes advantage of the excellent performances of RP-HPLC. For example, amino acids can be analyzed using an HILIC column (e.g., ZIC-pHILIC), but also using a separation on a C18 column (e.g., AcclaimTM RSLC PolarAdvantage II from Dionex) in the presence of heptafluorobutyric acid as IPA [26]. A considerable number of molecules that have the capability to display ionic character in specific pH conditions can be analyzed by IPC on RP-type columns.

An alternative to NPC for the separation of hydrophobic compounds is convergence chromatography (see Section 3.3). This technique is a new development in supercritical fluid chromatography (SFC)-type separations and has been successfully utilized in the analysis of lipids, fat-soluble vitamins, steroids, and nonionic surfactants.

Many compounds that are too polar to be separated on RP-HPLC such as sugars, oligosaccharides, polysaccharides, certain peptides, proteins, and nucleotides still must be separated on polar columns in HILIC mode. Also, in some applications, HILIC or NPC are necessary for the separation of certain isomers (achiral) since polar phases are more efficient than RP for isomer separations. As discussed in Section 4.3, the interactions between the stationary phase and the solute (analyte) in NPC and HILIC are stronger for more polar compounds. Since the polarity of the isomers (not chiral isomers) is frequently very different, for example because the polar groups are positioned in different parts of the molecule, the interactions with the stationary phase are different from isomer to isomer. This explains the efficiency of polar phases in isomer separation. Besides the polar group position in an analyte molecule, other effects such as steric hindrance and additional effects such as propensity to form hydrogen bonds or having electron donation or withdrawal interactions may affect the differences in retention on polar phases [13,27].

In some cases, polar columns are used in NPC-type separations. Among the NPC uses are the separations of highly hydrophobic compounds (for example, carotenoids or lipids) that require an organic mobile phase with no water or polar constituents for analyte dissolution. Such compounds are too strongly retained on RP columns. The separations can be performed on silica columns. There are several disadvantages to silica utilization in NPC mode. The main problem is related to the reproducibility of such separations. Since the silica is always covered with a layer of water, the immobilized material acting as stationary phase can be considered to be water. The variation in the amount of water retained on the silica produces variations in the retention times. The different levels of water can be caused by different levels of exposure of the stationary phase to water from solvents with different degrees of dryness or by the length of time of exposure of the column to a dry solvent that may dissolve some of the water forming the stationary phase. The sensitivity of column reproducibility to the water exposure is very high. Also, related to the water layer formation is the slow equilibration of the column when changing solvents and frequent irreproducibility in this equilibration process. Some compounds are retained irreversibly on bare silica. Although the structure of *type*

B silica is highly homogeneous, silica columns frequently produce tailing of the peaks, indicating variability in the type of interactions on the stationary-phase surface.

The problem of diminishing the water layer is not present (or it is significantly diminished) in HILIC applications, even when using silica columns. The layer of water on the silica surface is much thicker in HILIC than when working in NPC mode, and the separations are less susceptible to silica dryness caused by mobile phase removal of the water layer [28,29]. Some other disadvantages of the use of bare silica as stationary phase in NPC mode are diminished in HILIC, although peak tailing may still be a problem. Also, column equilibration in a specific mobile phase must be considered in all polar separations [30].

Selection of Physical Column Characteristics

Similar to RP-HPLC, the columns with an HILIC phase that are most frequently utilized have 100 or 150 mm length, and 2.7, 3, or 4.6 mm diameter, with uniform fully porous spherical particles of 3 or 5 μm diameter. However, the HILIC technique gained popularity more recently compared to RP-HPLC, and the utilization of columns with more recent developments such as smaller particle size and core–shell particles are relatively more common in HILIC. To achieve better efficiency, columns with smaller particles or core–shell particles are recommended. Smaller particles lead to higher backpressures in the HPLC system, and these pressures must be acceptable for the HPLC or UPLC pumping system. Also, phase rigidity and resilience to crushing must be considered when a column with small particles is selected. In the selection of an HILIC column for the analysis of large molecules (polymeric), similar to the case of RP-HPLC columns, it is recommended to select stationary phases with larger pores (200 E or larger, depending on the molecule *Mw*).

Selection of the Nature of Stationary Phase for the Column

One path for the selection of the nature of the stationary phase for an HILIC column is based on the results reported in the literature for a similar class of compounds as that of the analytes to be separated in the new method. Various written materials are available describing the methods practiced on HILIC columns [18,31,32].

The general rule for the selection of a phase for HILIC separation is that the more negative is the log D_{ow} value for an analyte, the greater should be stationary phase polarity required to retain it. The polarity of the phase is, however, difficult to estimate since it depends on the functionality but also on the silica structure, spacer length that brings hydrophobic character, end-capping of silica as well as the pH of the mobile phase. Using as a guide the results for log K_{ow}^{model} values from Table 6.2.4, the order of polarity for different phases is bare silica ≈ tertiary amine > primary amine > amide > zwitterionic > imide > urea > diol > cyano. However, this polarity can be changed depending on the mobile-phase pH, and in an acidic medium, the amine and urea groups can be much more polar [33]. Zwitterionic phases seem to have a balance between the charged groups, such that their apparent polarity is not very high. The value of K_{ow}^{model} offers only a modest guidance for choosing an HILIC column.

Useful information for selecting an HILIC column is provided by the set of parameters described in Section 8.2. The parameters $k'(U)$, $\alpha(OH)$, $\alpha(CH_2)$, $\alpha_{dia}(V/A)$, $\alpha_{regio}(2\,dG/3\,dG)$, α_{AX}, α_{CX}, and $\alpha_{tt}(Tb/Tp)$ are not always reported for a column, but, when available, these parameters offer good criteria for selecting the column. When such characterizing parameters are available, it should be noticed that they are somewhat independent from one another, and a higher value of $\alpha(OH)$, for example, does not necessarily indicate a low value for $\alpha(CH_2)$. The hydrophobicity $\alpha(CH_2)$ is related as expected to the carbon load of the column C%. When $\alpha(CH_2)$ is not known, the capacity of the column to accommodate hydrophobic compounds can be estimated based on C%.

For compounds with anionic character, a column with anionic character is useful for the separation, while one with high cationic character is not likely to provide good separation. The same is applicable for cations that are not well separated on columns with high α_{AX} values. Acidic and basic character of stationary phases can also be estimated based on $\alpha_{tt}(Tb/Tp)$ values. For mixtures of polar analytes with neutral, basic, and acidic characters, neutral HILIC columns are recommended for the separation. Other similarities between the column structure and analyte structure offer also guidance for column selection. For example, a number of amino acids (that have a zwitterionic character) can be separated on a Zic-pHILIC column and the separation may be favored by the interaction of amino acids with the zwitterionic stationary phase. Similarities in the phase structure and analyte structure are not always required. For example, glucose and other monosaccharides can be separated with good results on an amino-type HILIC phase and not on a diol phase. As an example, the separation of several small carbohydrates in the plant materials (fructose, glucose, mannose, sucrose, maltose, xylose, sorbitol, and myo-, chiro-, and scyllo-inositols) can be separated on a YMC-Pack Polyamine II column, followed by MS/MS detection [34]. The chromatogram of a standard mixture of small carbohydrates with concentrations between 25 and 50 μg/mL is shown in Fig. 8.4.1. The mobile phase was 75% CH_3CN and 25% water in isocratic conditions. The MS/MS detection allowed the measurement without interference of the overlapping peaks since they were detected using different ions. On the other hand, the diol phase can be used, for example, for the separation of nicotine-type alkaloids [35]. This phase can also be used for other separations.

Other Parameters Important in HILIC Column Selection

Similar to the case of RP columns, other column characteristics are important in the selection of an HILIC column. Among these can be listed the following: (1) resilience to a wider pH range, (2) low asymmetry, (3) low column bleed, (4) short equilibration times, and (5) reproducibility for a large number of injections. All these characteristics must be evaluated during the selection process and considered in the context of analysis requirements.

New Developments

The continuous development in column technology included the HILIC columns. New phases, implementation of core–shell technologies, and the decrease in particle size of the stationary phase, typical for reversed-phase columns were also used for HILIC columns. A few newly commercially available HILIC columns are listed in Appendix 3g [36].

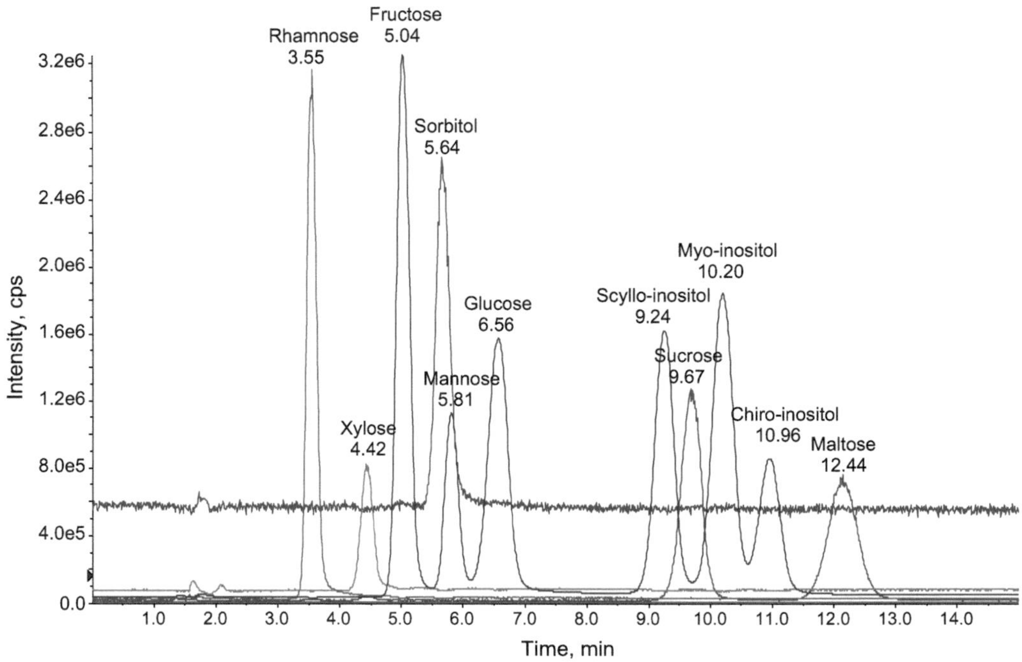

FIGURE 8.4.1 Chromatogram of a set of small carbohydrate standards with concentrations between 25 and 50 μg/mL on a YMC-Pack Polyamine II column with the mobile phase 75% CH_3CN and 25% water [34].

References

[1] P. Jandera, Stationary and mobile phases in hydrophilic interaction chromatography: a review, Anal. Chim. Acta 692 (2011) 1–25.

[2] W. Jiang, G. Fischer, Y. Girmay, K. Irgum, Zwitterionic stationary phase with covalently bonded phosphorylcholine type polymer grafts and its applicability to separation of peptides in the hydrophilic interaction liquid chromatography mode, J. Chromatogr. A 1127 (2006) 82–91.

[3] T. Ikegami, K. Tomomatsu, H. Takubo, K. Horie, N. Tanaka, Separation efficiencies in hydrophilic interaction chromatography, J. Chromatogr. A 1184 (2008) 474–503.

[4] P. Hemström, K. Irgum, Hydrophilic interaction chromatography, J. Sep. Sci. 29 (2006) 1784–1821.

[5] L. Moni, A. Ciogli, I.D. Acquarica, A. Dondoni, F. Gasparrini, A. Marra, Synthesis of sugar-based silica gels by copper-catalysed azide-alkyne cycloaddition via a single-step azido-activated silica intermediate and the use of the gels in hydrophilic interaction chromatography, Chem. Eur. J. 16 (2010) 5712–5722.

[6] S.-M. Liu, L. Xu, C.-T. Wu, Y.-Q. Feng, Preparation and characterization of perhydroxylcucurbit[6]uril bonded silica stationary phase for hydrophilic-interaction chromatography, Talanta 64 (2004) 929–934.

[7] M. Jonnada, R. Rathnasekara, Z. El Rassi, Recent advances in nonpolar and polar organic monoliths for HPLC and CEC, Electrophoresis 36 (2015) 76–100.

[8] Z. Jiang, N.W. Smith, P.D. Ferguson, M.R. Taylor, Hydrophilic interaction chromatography using methacrylate-based monolithic capillary column for the separation of polar analytes, Anal. Chem. 79 (2007) 1243–1250.

[9] I. Tijunelyte, J. Babinot, M. Guerrouache, G. Valincius, B. Carbonnier, Hydrophilic monolith with ethylene glycol-based grafts prepared via surface confined thiol-ene click photoaddition, Polymer 53 (2012) 29–36.

[10] V. Škeříková, J. Urban, Highly stable surface modification of hypercrosslinked monolithic capillary columns and their application in hydrophilic interaction chromatography, J. Sep. Sci. 36 (2013) 2806–2812.

[11] Y. Lv, Z. Lin, F. Svec, Hypercrosslinked large surface area porous polymer monoliths for hydrophilic interaction liquid chromatography of small molecules featuring zwitterionic functionalities attached to gold nanoparticles held in layered structure, Anal. Chem. 84 (2012) 8457–8460.

[12] Y. Lv, Z. Lin, F. Svec, "Thiol-ene" click chemistry: a facile and versatile route for the functionalization of porous polymer monoliths, Analyst 137 (2012) 4114–4118.
[13] B. Buszewski, S. Noga, Hydrophilic interaction liquid chromatography (HILIC) – a powerful separation technique, Anal. Bioanal. Chem. 402 (2012) 231–247.
[14] Y. Kawachi, T. Ikegami, H. Takubo, Y. Ikegami, M. Miyamoto, N. Tanaka, Chromatographic characterization of HILIC stationary phases: hydrophilicity, charge effects, structural selectivity, and separation efficiency, J. Chromatogr. A 1219 (2011) 5903–5919.
[15] F. Gritti, G. Guiochon, Comparison between the intra-particle diffusivity in the hydrophilic interaction chromatography and reversed phase liquid chromatography modes. Impact on the column efficiency, J. Chromatogr. A 1297 (2013) 85–95.
[16] Y. Guo, S. Gaiki, Retention and selectivity of stationary phases for hydrophilic interaction chromatography, J. Chromatogr. A 1218 (2011) 5920–5938.
[17] M. Dolci, Hydrophilic Interaction Liquid Chromatography: An Investigation into the Solvent and Column Selectivity, Chromatography Today, May/June, 2013.
[18] P.G. Wang, W. He (Eds.), Hydrophilic Interaction Chromatography (HILIC) and Advanced Applications, CRC Press, Boca Raton, 2011.
[19] L. Yu, X. Li, Z. Guo, X. Zhang, X. Liang, Hydrophilic interaction chromatography based enrichment of glycopeptides by using click maltose: a matrix with high selectivity and glycosylation heterogeneity coverage, Chemistry 15 (2009) 12618–12626.
[20] A.J. Alpert, Cation-exchange high performance liquid chromatography of proteins on poly(aspartic acid)-silica, J. Chromatogr. A 266 (1983) 23–37.
[21] M. Xu, D.S. Peterson, T. Rohr, F. Svec, J.M. Fréchet, Polar polymeric stationary phases for normal phase HPLC based on monodisperse macroporous poly(2,3-dihydroxypropyl methacrylate-co-ethylene dimethacrylate) beads, Anal. Chem. 75 (2003) 1011–1021.
[22] http://www.sielc.com/Products_Obelisc.html.
[23] M. Lämmerhofer, M. Richter, J. Wu, R. Nogueira, W. Bicker, W. Lindner, Mixed-mode ion-exchangers and their comparative chromatographic characterization in reversed-phase and hydrophilic interaction chromatography elution modes, J. Sep. Sci. 31 (2008) 2572–2588.
[24] C. Viklund, A. Sjögren, K. Irgum, Chromatographic interactions between proteins and sulfoalkylbetaine-based zwitterionic copolymers in fully aqueous low salt buffers, Anal. Chem. 73 (2001) 444–452.
[25] E.M. Borges, D.A. Volmer, Silica, hybrid silica, hydride silica and non-silica stationary phases for liquid chromatography. Part II: chemical and thermal stability, J. Chromatogr. Sci. 53 (2015) 1107–1122.
[26] S.C. Moldoveanu, J. Zhu, N. Qian, Free amino acids analysis by liquid chromatography with tandem mass spectrometry in several botanicals with antioxidant character, J. Sep. Sci. 38 (2015) 2208–2222.
[27] M.J. Kailemia, L.R. Ruhaak, C.B. Lebrilla, I.J. Amster, Oligosaccharide analysis by mass spectrometry: a review of recent developments, Anal. Chem. 86 (2014) 196–212.
[28] D.V. McCalley, U.D. Neue, Estimation of the extent of the water-rich layer associated with the silica surface in hydrophilic interaction chromatography, J. Chromatogr. A 1192 (2008) 225–229.
[29] N.P. Dinh, T. Jonsson, K. Irgum, Water uptake on polar stationary phases under conditions for hydrophilic interaction chromatography and its relation to solute retention, J. Chromatogr. A 1320 (2013) 33–47.
[30] M.B. Sorde, B.N. Poul, A.Y. Ghodke, O.G. Bhusnure, A review on hydrophilic interaction chromatography – a useful review, Int. Res. J. Pharm. App. Sci. 3 (2013) 137–142.
[31] http://www.thermoscientific.com/content/dam/tfs/ATG/CMD/cmd-documents/bro/bro/chrom/lc/col/TG-21003-HILIC-Separations-TG21003-EN.pdf.
[32] Comprehensive Guide to HILIC: Hydrophilic Interaction Chromatography, Waters Corp, Hoboken, 2014.
[33] G. Greco, T. Letzel, Main interactions and influences of the chromatographic parameters in HILIC separations, J. Chrom. Sci. 51 (2013) 684–693.
[34] S. Moldoveanu, W. Scott, J. Zhu, Analysis of small carbohydrates in several bioactive botanicals by gas chromatography with mass spectrometry and liquid chromatography with tandem mass spectrometry, J. Sep. Sci. 38 (2015) 3677–3686.
[35] H. Tanaka, X. Zhou, O. Masayoshi, Characterization of a novel diol column for high-performance liquid chromatography, J. Chromatogr. A 987 (2003) 119–125.
[36] D.S. Bell, New chromatography columns and accessories for 2016, LC/GC North America 34 (4) (2016) 242–252.

CHAPTER

9

Stationary Phases and Columns for Ion Exchange, Ion-Moderated, and Ligand Exchange Chromatography

9.1 TYPES OF PHASES AND THEIR PREPARATION

Types of Ion Exchange Phases

The stationary phases used in ion exchange chromatography (IC or IEC) are basically similar to other columns having an inert support that is carrying the functional groups, which in IC are ionic [1]. Most supports are either silica or organic polymers, but other materials are sometimes used for obtaining ion exchange phases (e.g., alumina, zeolites). Differently from reversed-phase high-performance liquid chromatography (RP-HPLC) or hydrophilic interaction chromatography (HILIC), the organic polymers are more frequently employed as support materials in IC. These materials have a better stability toward extreme pH conditions than silica, and a wider pH range is sometimes necessary in IC. While silica-based LC columns are usually stable at pH between 2 and 8, polymeric ion exchangers (resins) can be stable in a wider range of pH of the mobile phase [2]. The main types of ion exchange stationary phases related to their functionality are indicated in Table 9.1.1.

Regarding the ion-moderated phases (also known as ion exclusion phases), they are usually made from a cation exchange material in H^+ or metal form (e.g., Na^+, K^+, Ca^{2+}, Pb^{2+}), which is used in specific conditions for the separation of neutral species based on a selective partition of the analyte between the liquid inside the resin and the mobile phase. The retention inside the resin is based on weak polar interactions between the analyte and the stationary phase. Some ion-moderated phases may be considered as functioning based on an ligand exchange mechanism. A different type of phase from the same category is based on crown ether moieties. The crown ether groups can be present in an organic-based solid support or immobilized on silica. Crown ethers strongly bind certain cations, forming complexes. For example, 18-crown-6 ether has affinity for potassium ions, while 15-crown-5 has affinity for sodium. The crown ether can also display Donnan-type exclusion and act as an ion exclusion phase. Other neutral materials, such as polyethylene glycol, exhibit retention abilities of various anions and are used as the stationary phase for the separation of inorganic anions in IC [3].

Selection of the HPLC Method in Chemical Analysis
http://dx.doi.org/10.1016/B978-0-12-803684-6.00009-3

TABLE 9.1.1 Types of Functionalities in Ion Exchange Phases

Phase Character	Strength	Examples of Phase Structure
Cation exchange	Weak	$-COO^-$, $-C_6H_4-O^-$, $-AsO_3H^-$
Cation exchange	Medium	$-PO_3H^-$
Cation exchange	Strong	$-PO_4H^-$, $-SO_3^-$
Anion exchange	Weak	$-NH_3^+$, $-[NH_2(CH_3)]^+$
Anion exchange	Medium	$-[N(CH_3)_2(CH_2CH_2OH)]^+$
Anion exchange	Strong	$-[N(CH_3)_3]^+$, $-[N(C_2H_5)(CH_3)_2]^+$
Zwitterionic	–	$-N(CH_3)_2^+-(CH_2)_n-SO_3^-$ or $-CH(SO_3^-)-(CH_2)_n-N(CH_3)_3^+$
Amphoteric*	–	Various polymers

*The presence in a polymeric structure of both acidic and basic groups such as $-N(CH_3)_2^+-$ and $-SO_3^-$ is usually indicated as zwitterionic. An amphoteric group must be able to act as an acid in basic medium and as a base in acidic medium.

Ion-Moderated (Ion Exclusion) and Ligand Exchange Phases

Ion-moderated stationary phases (also indicated as ion exclusion phases) typically consist of a cation exchange stationary phase having covalently bonded groups such as $-SO_3^-$, with these groups having as counterion cations such as K^+, Na^+, Ag^+, Ca^{2+}, Pb^{2+} (in some cases H^+). Such columns are frequently utilized for the analysis of monosaccharides, oligosaccharides, hydroxy-organic acids, glycerol, sorbitol, and other analytes in an aqueous or almost aqueous mobile phase. The cation exchange material is usually an organic polymer-based styrene/divinylbenzene, and the specific metal form is maintained by having the ions at a low concentration in the mobile phase.

A special type of ion-moderated stationary phase is utilized in the separation of *cis/trans* lipids and fatty acids. This stationary phase contains Ag^+ ions typically in a silica-type column such as ChromSpher 5 Lipids [4]. The separation of *cis/trans* isomers is based on differences in the interaction of π bonds in the lipid or fatty acid with the silver ions. Such columns are used with nonpolar organic mobile phases [5].

Ligand exchange phases are ion exchange phases loaded with a transitional metal capable of forming coordinative bonds with the analyte. Special phases are used for immobilized metal affinity usually derived from iminodiacetic acid or tricarboxymethylethylenediamine. These phases are able to form very strong coordinative complexes with transitional metals, which are further complexing the analytes [6].

Stationary Phases with More than One Type of Group (Mixed-Mode Ion Exchange Phases)

Some ionic exchange phases have intentionally, besides ionizable groups, specific hydrophobic moieties that allow the phase to work in mixed mode, both reversed phase and ion exchange [7]. For example, Acclaim Mixed-Mode WCX-1 columns are packed with a silica-based stationary phase that incorporates both hydrophobic and weak cation exchange (WCX) properties. Unlike conventional RP material, the packing has a hydrophobic alkyl chain with a carboxylate terminus and offers the potential for a wide range of applications, depending on the pH of the mobile phase [8]. The Acclaim Mixed-Mode WCX-1 column offers multiple retention mechanisms, including RPLC, cation exchange, ion exclusion, and HILIC. These features make this column versatile in many applications that require different types of selectivity.

Columns containing a stationary phase that have both hydrophobic and weak anion exchange (WAX) properties are also known. For example, Acclaim Mixed-Mode WAX-1 columns are packed with a silica-based stationary phase that incorporates both hydrophobic and WAX properties. Retention of basic, neutral, and acidic molecules can be either independently or concurrently adjusted by changing ionic strength, pH, and organic solvents content in the mobile phase. In these columns, the presence of anion exchange functionality controls through electrostatic interactions the attraction of anions and repulsion of cations, while the alkyl chain controls the separations by hydrophobic interactions [9].

Mixed-mode ion exchange phases can be based on silica or may have a polymeric base. For example, the OmniPac PAX- and PCX-100 and -500 are latex-based columns having both hydrophobic and ion exchange character. The PAX anion exchange capacity is about 40 μEq per column and the PCX cation exchange capacity is approximately 120 μEq per column. The columns are 100% solvent-compatible and allow the separation of inorganic and organic anions, having acid and base compatibility over 0–14 pH range.

Summary of Procedures for the Synthesis of Ion Exchange Phases

The ion exchange stationary phases can be made using a variety of procedures. A classification of stationary phases used in IC based on their construction is indicated in Table 9.1.2.

Silica-based ion exchange phases include those with a bonded phase containing ionic groups, and also other types of phases generated, for example, by coating a porous silica material with a polymer with specific ionic groups. Bonded ionic groups are more common in these types of ionic phases. The synthesis of such materials can be done using one step derivatization of porous (hydrated) silica with appropriate reagents. For example, acidic

TABLE 9.1.2 Types of Construction for Ion Exchange Phases

Phase Support	Type of Phase Structure
Silica-based	Bonded ionic groups
Silica-based	Polymer-coated
Organic polymer	Synthesized with ionic functional groups
Organic polymer	Surface-functionalized
Organic polymer	Latex-agglomerated
Inorganic silicates	Zeolites
Inorganic oxides	Alumina, silica

groups such as $-SO_3^-$ can be attached by using, for silica derivatization, a silane containing ethylbenzenesulfonic groups. More commonly ionic phases are prepared in two-step reactions, one example of synthesis for an anion exchange phase being:

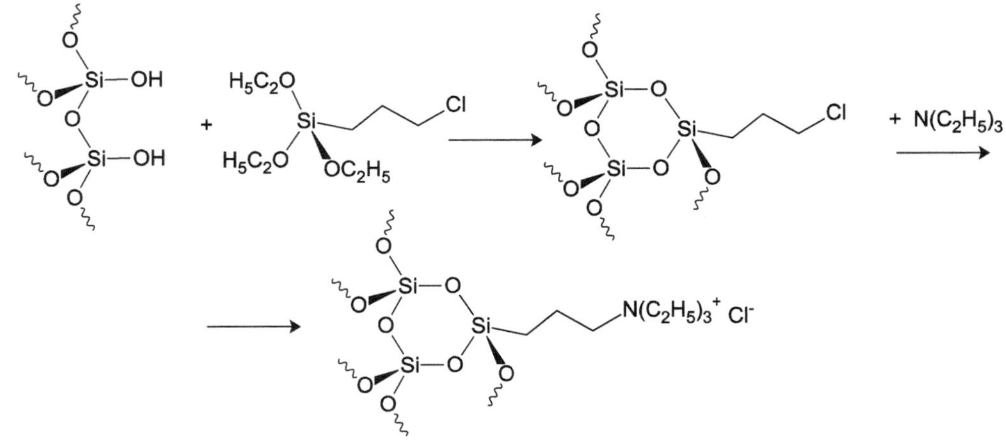

(9.1.1)

Other procedures were also described in the literature (see, e.g., [10]), such as reacting the silica surface with a silane containing an alkylthiol group followed by an oxidation of the thiol to a sulfonic group.

A large proportion of ion exchange stationary phases are based on organic polymers. These materials consist of an organic polymeric (resin) backbone containing covalently bound groups that are able to exist in ionic form. Among the common backbone polymers are the copolymers of styrene-divinylbenzene (PS-DVB), ethylvinylbenzene-divinylbenzene (EVB-DVB), polyvinyl copolymers such as those obtained from polyvinyl alcohol (PVA), and various polymethacrylates [11]. Materials with ion exchange properties can also be obtained based on cellulose, dextrans, and other similar materials. The presence of the ionic groups on these resins makes these resins act as polyelectrolytes. The main advantage of resin-based ion exchangers is their tolerance toward eluents with extreme pH values, between 0 and 14, in contrast to the silica-based stationary phases, whose pH limits are 2–7. This wide range of pH values allows the use of selectivity effects on multicharged or weakly ionizable solutes. The use of polymeric resins has pressure limitations because they have lower mechanical resilience. Macroporous resins with a high degree of crosslinking are relatively more rigid and stable and, although they have lower ion exchange capacity, are used in HPLC and can be included in longer columns at higher flow rates. Monolith columns were also made from polymers such as polyacrylamide [12].

Several procedures are used for obtaining polymers with ion exchange properties. One such procedure consists of the synthesis of the polymer already containing the desired ionic groups in the monomer. Another procedure is the derivatization of an already polycondensed or polymerized resin. Many other synthetic paths have been either used or only explored for producing ion exchange resins. Some resins are prepared to have more than one type of functional group, others are made to contain unique structures. Amphoteric ion exchangers with both basic and acidic groups, as well as ion exchangers with specific chelating properties, are also available for various applications.

Silver ion-impregnated HPLC columns are typically made using a silica column or an ion exchange column (such as Nucleosil 5 SA containing phenylsulfonic acid groups bonded chemically to silica) and treating it with a solution of a silver salt, such as $AgNO_3$ [13].

Latex-Agglomerated Ion Exchangers

The direct use of polymeric ion exchange materials as stationary phase for HPLC encounters several problems. One problem is the relatively low mechanical stability even at moderate pressure, and the other is the swelling and shrinkage of the phase during the ion exchange process. These problems are significantly alleviated using latex-agglomerated ion exchange stationary phases. In addition to better mechanical stability and reduced swelling and shrinkage, these phases also offer high efficiencies and a better ion exchange capacity (between 0.03 and 0.1 mEquiv/g).

Latex-agglomerated exchangers contain an internal core particle (or support) that contains on its surface ionic functionalities. On this support is attached a monolayer of small-diameter particles that carry functional groups consisting of bonded ions that are of the opposite charge to the functionality of the support. The groups of the outer particles have a double role, to attach the small particles to the core, and also to act as an ion exchanger for the ions in solution. The core particles are typically PS-DVB resin of moderate to high crosslinking, with a particle size in the range 5–25 μm. The outer microparticles consist of finely ground resin or monodisperse polymer (latex) with diameters up to 0.1 μm that have very high porosity, and are functionalized to contain an appropriate ion exchange group. This group determines the ion exchange properties of the composite particle. For example, an aminated latex produces agglomerated anion exchangers, although the core support contains groups such as SO_3^-. A sulfonated latex produces agglomerate cation exchangers although the core support has positive groups (see, e.g., [1,14]).

9.2 CHARACTERIZATION OF ION EXCHANGE PHASES

Besides common physical characteristics of any stationary phase in HPLC (such as particle size, particle uniformity, porosity, pore size, and type of active groups), ion exchange phases are also characterized by their ion capacity. The ion capacity of the ion exchanger is determined by the number of functional groups per unit weight of the stationary phase. The most commonly used units for ion capacity characterization are milliequivalents of charge per gram of dry packing and milliequivalents per mL of wet packing. Typical ion exchange capacity in IC is 10–100 mEquiv/g. The counter-ion present in the stationary phase must be indicated together with the ion capacity since it affects the degree of swelling of the packing and hence its volume. The ion exchange capacity of a stationary phase plays a significant role in determining the concentrations of competing ions used in the mobile phase for elution. Higher-capacity stationary phases generally require the use of more concentrated mobile phases. This is not a recommended feature in high-performance IC, since the use of the common conductometric detectors cannot function well with high salt concentrations. However, the lower ion exchange capacity limits the sample loading of the phase, which must be low for low ion capacity phases. For strong ion exchange phases, the ionization of the stationary phase does not depend on the pH of the mobile phase, and regardless of the pH, the loading capacity is the same. For weak ion exchangers the use of a pH of the mobile phase where the stationary phase is not ionized may considerably decrease the phase retention.

Another important characteristic of the IC columns is related to their hydrophobic character, which should be as low as possible. In many IC separations, besides the inorganic ions, other molecules are present, and the hydrophobic character of the analyte should not interfere with the separation. However, the ionic groups of the stationary phase are typically bonded to an organic fragment that is connected either to silica or to an organic polymer. This part of the ionic stationary phase may influence the separation of the analytes by undesired hydrophobic interactions. Phases with low or ultra-low hydrophobicity were manufactured, and this property is typically indicated for commercially available IC columns.

Besides ionic and hydrophobic interactions, some phases develop polar interactions with the analytes. The polarity of the stationary phase is not usually considered as detrimental. The polar character influences the retention of polar compounds, but the overall ion exchange character is dominant with both ionic and polar molecules. Polar character and ionic character do not act opposite to each other, and the polar interactions do not affect adversely the separations.

Other types of characterizations of ion exchange phases are related to their physical properties. Basically, the same physical characteristics are applicable to ion exchange columns as for other types of columns. Silica-based ion exchange phases usually have good mechanical resilience to the column backpressure, while for polymeric phases the backpressure limit is usually lower. Changes in the phase volume in different mobile phases, in particular for polymeric type phases, are undesirable, and phases with minor variation in volume are required for HPLC applications. Also, parameters such as thermodynamic exchange constant and the free energy of solute/phase interaction were used for phase characterization [15].

Ionic Capacity Measurement

Ion capacity of an ion exchange resin can be measured by titration with a strong acid or base solution, depending on the type of resin. However, this approach is rather tedious [16]. Another procedure is based on elemental analysis for the identification of specific functional groups (e.g., measurement of sulfur for the evaluation of the number of SO_3H groups). Ion exchange capacity can also be obtained using the measurements of transient pH change that occurs when a step is made in the ionic strength of the mobile phase with an unmodified pH [17]. This method uses two solutions of exactly the same pH, but at different ion strengths (different buffer concentrations) as mobile phase. The ion exchange column is first loaded with a specific ion in a high concentration of a buffer and, after being equilibrated, the pH of the eluent is measured. By step decreasing the buffer concentration, some of the attached ions on the ion exchanger are released in the solution. For example, a column loaded with $H_2PO_4^-$ ions in a higher concentration of the buffer in the mobile phase will release $H_2PO_4^-$ ions when the buffer concentration is suddenly decreased. The released $H_2PO_4^-$ ions produce a pH drop (although the initial pH of the concentrated buffer solution and diluted buffer solution are the same). After a specific amount of ions is released, the equilibrium is reestablished and the pH is reset. The duration of the pH drop transient time is proportional with ionic capacity of the column, allowing the evaluation of this capacity [18,19].

Solvent Compatibility

Compatibility of the stationary phase with different organic solvents from the mobile phase is an important phase characteristic. The presence of an organic solvent in the mobile phase is common in IC for assuring organic bases or acids solubility and affecting selectivity by changing the retention characteristics of the column [20]. Selectivity is affected by a complex process of modifying the solvation sphere of the ions, and by the strength of ionic analyte/solid phase interaction [21]. Silica-based ion exchange phases are less affected by the presence of an organic solvent in the predominantly aqueous mobile phase. However, ion exchanges based on organic polymers are prone to swelling even in the presence of a low level of organic solvent in an aqueous mobile phase. By increasing the polymer crosslinking, e.g., to as much as 50%, the swelling is significantly reduced in modern ion exchangers such that they are compatible with many solvents commonly used in IC. The swelling characteristics of the phase are important in the phase selection.

Phase Affinity for Specific Ions

The affinity of the chromatographic column for certain ions is an important parameter for column selection. The columns typically come loaded with a counter-ion that has low affinity for the column. For example, the anion exchange columns typically come loaded with OH^- or with CO_3^{2-} ions. As indicated in Section 4.1, IC frequently uses conductivity detectors with ion suppression capability. The ion suppression eliminates the OH^- from the mobile phase by changing it into H_2O, while an ion X^- is changed into HX that is dissociated and produces high conductivity (similar mechanism works for CO_3^{2-}). Both OH^- and CO_3^{2-} ions have low affinity for anion exchange columns, and they can be easily replaced by ions such as F^-, Cl^-, Br^-, I^-, NO_3^-, ClO_3^-, SO_4^{2-}, and PO_4^{3-}. However, the ion exchange constant $K_{X,OH}$ (see expression 4.3.48 for a cation exchange) depends on the nature of the resin. For an anion X^- and the OH^- as a counter-ion, expression (4.3.48) can be written in the form:

$$K_{X,OH} = \frac{C_{resin}(X^-)}{C(X^-)} \frac{C(OH^-)}{C_{resin}(OH^-)} \tag{9.2.1}$$

where C_{resin} is the concentration in the stationary phase and C is the concentration in the mobile phase for each ion. The equivalent of expression (4.3.51) for the retention factor k' in the case of an anion exchange where the counter ion in the resin is OH^- and the mobile phase uses a strong hydroxide such as KOH or NaOH is the following:

$$k'(X^-) = K_{X,OH} \frac{C_{resin}(OH^-)}{C(OH^-)} \Psi \tag{9.2.2}$$

Expression (9.2.2) shows that at low hydroxide concentration in the mobile phase ($C(OH^-)$) the retention factor k' for the ion X^- is high, and when the concentration of OH^- increases, k' decreases and the species X^- can be eluted from the column. The phase ratio Ψ in expression (9.2.2) can be expressed as V_{resin}/V_{mo} (see expression 4.2.15), and formula (9.2.2) will become:

$$k'(X^-) = K_{X,OH} \frac{C_{resin}(OH^-)}{C(OH^-)} \frac{V_{resin}}{V_{mo}} \tag{9.2.3}$$

In expression (9.2.3), IC column capacity can be defined as:

$$Q_{IC} = C_{resin}(OH^-)V_{resin} \tag{9.2.4}$$

The use of Q_{IC} in expression (9.2.3) leads to the following formula:

$$k'(X^-) = K_{X,OH}\frac{Q_{IC}}{C(OH^-)V_{mo}} \tag{9.2.5}$$

The values for $K_{X,OH}$ and for Q_{IC} for various columns are discussed in the literature [22,23] and reported by some column manufacturers [24]. For example, the values for $K_{X,OH}$ where X^- is F^-, Cl^-, Br^-, I^-, NO_3^-, ClO_3^-, SO_4^{2-}, or PO_4^{3-} are higher for stationary phases containing bonded trimethylamine or triethylamine in comparison to alkanolamine. This can be explained by the more hydrophilic character of alkanolamine sites, which become more hydrated than trimethylamine and triethylamine moieties and display different affinity for the ions. For several Dionex columns (see Appendix 4d) the values for $K_{Cl,OH}$ are in the order: AS10 << AS15 << AS19 < AS24 < AS18 < AS11-HC ≈ AS20 < AS16. For the same types of columns, $K_{NO_3,OH}$ is in the order: AS10 << AS15 << AS24 ≈ AS19 ≈ AS18 < AS11-HC < AS20 < AS16. The column capacity Q_{IC} for OH^- of these columns is in the order: AS24 < AS10 ≈ AS16 < AS15 < AS19 < AS18 < AS11-HC < AS20. The physical construction and chemical nature of the phase influence both $K_{X,OH}$ and Q_{IC} values.

Hydrophobicity of IC Columns

The hydrophobicity of the stationary phase for IC columns caused by various hydrophobic moieties in the phase construction (e.g., polymeric backbone) interferes in the ion exchange process and influences the retention and selectivity of IC columns. This characteristic can be detrimental in some separations, but also can be used to the advantage of certain separations. The hydrophobic contribution of a column can be evaluated by studying the retention of compounds that also have both ionic and hydrophobic moieties such as aliphatic carboxylic acids [25] or alkyl sulfonic acids [22].

9.3 COMMON COLUMNS WITH ION EXCHANGE PHASES

Cation Exchange Phases Based on Silica

Silica-based ion exchange columns are preferable for some of their qualities, in particular their mechanical resilience and small variation in the volume in different mobile phases. However, silica-based columns can be typically used within a narrower pH range. In IC, it is frequently necessary to use basic or acidic mobile phases, and the acceptable pH range of use may be a limitation for some applications. Some commercially available silica-based cation exchange columns are listed in Appendix 4a.

Attempts were made to use the advantages of silica support, such as the large surface area and lack of volume changes in different mobile phases or when the counterion is changing, and at the same time to extend the pH range utilization. One path to reach this goal is to make polymer-coated silica phases [26]. The silica surface can be coated using "prepolymers," which are oligomers synthesized in a separate step and then applied to the support material and immobilized. For example, a prepolymer can be made using a mixture of butadiene and maleic acid (PBDMA) [2,27].

Organic Polymeric Cation Exchange Phases

Organic polymeric cation exchange columns are common. Styrene/divinylbenzene copolymers are the most widely used support materials for obtaining this type of cation exchangers. Their resilience to a wide pH range is an advantage for IC applications. The ionic capacity is determined by the degree of sulfonation. Typical values for ionic capacity are 0.005–0.1 mEquiv/g, but some manufactured exchangers can reach higher values. Some cation exchange polymeric columns that are commercially available are listed in Appendix 4b.

Anion Exchange Phases Based on Silica

Silica-based IC stationary phases are characterized by higher chromatographic efficiencies and greater mechanical stabilities compared to organic polymers. They have better mechanical resilience and do not show swelling and shrinkage in different solvents or when the counterion is different. Also, silica-based columns can be used at temperatures up to 80 degree. However, silica-based columns typically have a narrower pH range of utilization (pH 2–8). The columns with up to $N/m \approx 20{,}000$ are common for phases with particles of 3 μm. The ion exchange capacities are in the range 0.1–0.3 mEquiv/g. A number of anion exchange columns with a silica backbone are commercially available. Some of these columns are listed in Appendix 4c.

Besides fully porous particles, pellicular phases consisting of much larger particles with a diameter of 25–40 μm and covered with 1–3-μm polymeric film are also known. The polymer can be, for example, lauryl methacrylate (Zipax SAX) with quaternary ammonium ion exchange functionality. The ion exchange capacity of pellicular materials is, however, lower, at about 0.01 mEquiv/g [2].

Organic Polymeric Anion Exchange Phases

Organic polymer anion exchange columns are very common. Similar to the case of cation exchange polymeric columns, the resilience of organic polymer-based phases to a wide pH range is an advantage for IC applications. Styrene/divinylbenzene copolymers, polymethacrylate, and other polyvinyl resins are common as support materials in the manufacturing process for polymer-based anion exchangers [2]. Some of the phases are made using latex technology (see Section 9.1). Besides typical anions, some columns are designed to work at very high pH and are capable of analyzing monosaccharides, disaccharides, specific oligosaccharides, and sialic acids [28]. Some anion exchange polymeric columns that are commercially available are listed in Appendix 4d.

The use of conductivity detection following ion suppression is commonly used for anion detection. The columns can be designed specifically for use with aqueous NaOH (or KOH) eluent, or with aqueous Na_2CO_3 eluent (see, e.g., [29]).

Capillary IC Columns

Besides columns with inner diameter (i.d.) of 4.6, 4.0, or 2 mm, several capillary IC columns are commercially available (see, e.g., [30]). For example, Thermo Scientific/Dionex has commercially available columns with 0.4 mm i.d. (indicated as capillary). These columns are packed with the same material as the equivalent analytical scale version, and produce separations similar to 4 mm i.d. columns (when appropriate flow is selected). The typical eluent flow rate in capillary columns must be selected only at about 1/100 from the flow rate of an analytical format. Due to the smaller cross-sectional area, the flow rate on a capillary format can be scaled down to 10 μL\min. The first advantage of such flow is the reduction of the eluent usage. Also, scaled down when using capillary columns must be the sample load. However, in IC the most common detector is the conductivity detector, which is a concentration-sensitive detector (see Section 4.1). Because of the difference in the separation flow rates (e.g., 100 times lower than in conventional columns), the volume passing through the conductivity detector at a specific time is 100 times smaller with the capillary system than with the standard system, and the concentrations of the separated analytes passing through the conductivity detector are 100 times higher. In this way, the capillary column provides increased mass sensitivity such that a much smaller sample can be injected with no loss or even with gain in sensitivity compared to standard columns and flows. One disadvantage of the use of capillary columns is the need for adequate instrumentation that may be different from the standard equipment. The dedicated equipment requires high-precision pumping at low flow rates, low-volume injectors, and low-volume conductivity cell for the IC detector (see Section 4.1) [31,32].

Ion-Moderated (Ion Exclusion) Phases

Most of the ion-moderated (ion exclusion) phases are based on a polymeric resin typically with covalently bonded acidic groups ($-SO_3^-$ or $-COO^-$) with counterions such as H^+, K^+, Na^+, Ag^+, Ca^{2+}, or Pb^{2+}. Some ion-moderated polymeric columns that are commercially available are listed in Appendix 4e. While most ion-moderated columns are used with an aqueous mobile phase, the lipid columns (containing Ag^+) are used with a nonaqueous mobile phase.

Other Phases

Columns with stationary phases containing a crown ether moiety are utilized in particular for chiral separations, when the crown ether has a specific chiral structure (see Chapter 10). Some ligand exchange columns are also used in chiral separations.

9.4 SELECTION OF AN ION EXCHANGE PHASE

Strong ion exchange stationary phases are completely ionized over a wide range of pH. The degree of dissociation of weak ion exchangers depends on the mobile phase pH. For this reason, the ion exchange behavior of strong ion exchange phases, within the typical range of pH operation, is not influenced by the pH of the mobile phase. The selection of a strong or

TABLE 9.4.1 Example of Recommended Applications for Several Thermo Scientific/Dionex Columns

Column Type	Common Dimension (mm)	μEquiv/col	Application With Cation Exchange
IonPac CS19-4 μm	250 × 4	2410	Common inorganic cations, amines
IonPac CS19	250 × 4	2410	Amines, morpholine, choline, etc.
IonPac CS18	250 × 2	290	Polar amines
IonPac CS17	250 × 4	1450	Amines
IonPac CS16	250 × 4	8400	Amines, ammonia, Li^+, Na^+
IonPac CS12A	250 × 4	2800	Common cations, Li^+, Na^+, Ca^{2+}, Mg^{2+}
IonPac CS5A	250 × 4	40	Hg^{2+}, transitional metals
Column Type	**Common Dimension (mm)**	**μEquiv/col**	**Application With Hydroxide Mobile Phase**
IonPac AS27	250 × 4	220	Halogen anions
IonPac AS26	250 × 4	250	Haloacetic acids
IonPac AS25	250 × 4	350	Various inorganic anions
IonPac AS24A	250 × 4	560	Haloacetic acids
IonPac AS24	250 × 2	140	Halogen anions, small organic acids
IonPac AS20	250 × 4	310	Various anions
IonPac AS19	250 × 4	240	Various inorganic anions
IonPac AS18	250 × 4	285	Various anions
IonPac AS17-C	250 × 4	30	Fast analysis of various inorganic anions
IonPac AS16	250 × 4	170	Various inorganic anions
IonPac AS15	250 × 4	225	Various anions, carbohydrates, organic acids
IonPac AS11	250 × 4	45	Various inorganic anions, organic acids
IonPac AS7	250 × 4	100	Cr(VI)
Column Type	**Common Dimension (mm)**	**μEquiv/col**	**Application With Carbonate Mobile Phase**
IonPac AS23	250 × 4	320	Inorganic anions
IonPac AS22	250 × 4	220	Inorganic anions, organic acids
IonPac AS14	250 × 4	65	Inorganic anions
IonPac AS9-HC	250 × 4	190	Inorganic anions
IonPac AS4A-SC	250 × 4	20	Fast analysis of inorganic anions

The μEquiv/col refers to the common column dimensions. Other formats are also available for the same column type.

weak ion exchange stationary phase is determined by the nature of the analytes and by the purpose of analysis. The use of weak ion exchangers allows a better modulation of the retention properties depending on the mobile phase.

The basic concept in the selection of ion exchange columns is simple: strong acids ($pK_a < 2$) are best separated on weak anion exchange columns, while the weak acids ($pK_a = 2–5$) are separated on a strong anion exchange column. Similarly, weak bases are well separated on a strong anion exchange column, and strong bases ($pK_a > 10$) on a weak anion exchange column. However, the utilization of ion exchange separations is much more complex, and various vendors developed specific applications for their columns. As an example, the recommended applications for several Thermo Scientific/Dionex columns are indicated in Table 9.4.1 (see, e.g., [33]). Separation in IC depends on, besides the column, other separation parameters such as mobile phase composition, temperature, and choice of gradient versus isocratic mode. Since in IC the detection is frequently performed using conductivity detection, the ion suppression is an important step, and a selection of the suppressor type and its parameters is required for a successful analysis. Specialized software guiding the column selections in IC is available in the literature [34].

Similar recommended applications can be found in the literature from other column manufacturers and for a variety of other columns. For example for Rezex-type columns working in ion-mediated type chromatography, the recommendations indicated in Table 9.4.2 can be found [35].

TABLE 9.4.2 Example of Applications for Rezex-Type Columns

Column	Part. Size (μm)	Max. Back Pressure (psi)	Applications
RAM-carbohydrate Ag^+ (8%)	8	1000	Monosaccharides
RCM-monosaccharide Ca^{+2} (8%)	8	1000	Monosaccharides and sugar alcohols
RHM-monosaccharide H^+ (8%)	8	1000	Monosaccharides, organic acids, fatty acids, alcohols, ketones, neutral compounds, and inorganic salts
RKP-potassium K^+ (8%)	8	1000	Glyphosate
RNM-carbohydrate Na^+ (8%)	8	1000	Sugars in matrices that contain a high level of sodium
ROA-organic acid H^+ (8%)	8	1000	Organic acids, carbohydrates, alcohols, fatty acids
RPM-monosaccharide Pb^{+2} (8%)	8	1000	Cellobiose, glucose, xylose, arabinose, mannose
RSO-oligosaccharides Ag^+ (4%)	8	1000	Oligosaccharides
RNO-oligosaccharides Na^+ (4%)	8	1000	Oligosaccharides
RFQ-fast acid H^+ (8%)	8	1000	Ethanol, acetic acid, glycerol, other alcohols
RCU-USP sugar alcohols Ca^{+2} (8%)	8	1000	Sugar alcohols such as sorbitol, mannitol

In addition to single-column IC, more complex IC separations are sometimes practiced. The columns used in comprehensive multidimensional ion chromatography (IC × IC) are frequently selected to have orthogonal selectivity [36].

References

[1] J. Weiss, Ion Chromatography, second ed., VCH, Weinheim, 1995.

[2] J. Weiss, D. Jensen, Modern stationary phases for ion chromatography, Anal. Bioanal. Chem. 375 (2003) 81–98.

[3] L.W. Lim, L. Rong, T. Takeuchi, Polyoxyethylene as the stationary phase in ion chromatography, Anal. Sci. 28 (2012) 205–213.

[4] A. Müller, K. Düsterloh, R. Ringseis, K. Eder, H. Steinhart, Development of an alternative eluent system for Ag^+-HPLC analysis of conjugated linoleic acid isomers, J. Sep. Sci. 29 (2006) 358–365.

[5] G. Dobson, W.W. Christie, B. Nikolova-Damyanova, Silver ion chromatography of lipids and fatty acids, J. Chromatogr. B Biomed. Sci. App. 671 (1995) 197–222.

[6] V.A. Davankov, Ligand exchange chromatography, in: M. Cooke, C.F. Poole (Eds.), Encyclopedia of Separation Science, Academic Press, New York, 2000, pp. 2369–2380.

[7] L. Wang, W. Wei, Z. Xia, X. Jie, Z.Z. Xia, Recent advances in materials for stationary phases of mixed-mode high-performance liquid chromatography, TrAC Trends Anal. Chem. 80 (2016) 495–506.

[8] X. Liu, C. Pohl, A weak cation-exchange, reversed-phase mixed-mode HPLC column and its applications, Am. Lab. 41 (2009) 26–29.

[9] http://www.dionex.com/en-us/products/columns/lc/mixed-mode/acclaim-mixed-mode-wax-1/lp-71737.html.

[10] L.M.L.A. Auler, C.R. Silva, K.E. Collins, C.H. Collins, New stationary phase for anion exchange chromatography, J. Chromatogr. A 1073 (2005) 147–153.

[11] B. Buszewski, M. Jaćkowska, S. Bocian, P. Kosobucki, B. Gawdzik, Functionalized polymeric stationary phases for ion chromatography, J. Sep. Sci. 34 (2011) 1–8.

[12] I.N. Savina, I.Y. Galaev, B. Mattiasson, Ion-exchange macroporous hydrophilic gel monolith with grafted polymer brushes, J. Molec. Recognit. 19 (2006) 313–321.

[13] S. Momchilova, B. Nikolova-Damyanova, Stationary phases for silver ion chromatography of lipids: preparation and properties, J. Sep. Sci. 26 (2003) 261–270.

[14] S.C. Moldoveanu, V. David, Essentials in Modern HPLC Separations, Elsevier, Amsterdam, 2013.

[15] A. Marton, Classification and characterization of stationary phases for liquid chromatography, Part II. Characterization of ion exchange chromatographic stationary phases, Pure Appl. Chem. 69 (1997) 1481–1487.

[16] F.G. Helfferich, Ion Exchange, McGraw-Hill, New York, 1962.

[17] N. Lendero, J. Vidič, P. Brne, V. Frankovič, A. Štrancar, A. Podgornik, Characterization of ion exchange stationary phases via pH transition profiles, J. Chromatogr. A 1185 (2008) 59–70.

[18] N. Lendero, J. Vidic, P. Brne, A. Podgornik, A. Strancar, Simple method for determining the amount of ion-exchange groups on chromatographic supports, J. Chromatogr. A 1065 (2005) 29–38.

[19] F.M. Plieva, I.N. Savina, S. Deraz, J. Andersson, I.Y. Galaev, B. Mattiasson, Characterization of supermacroporous monolithic polyacrylamide based matrices designed for chromatography of bioparticles, J. Chromatogr. B 807 (2004) 129–137.

[20] S. Rabin, J. Stillian, Practical aspects on the use of organic solvents in ion chromatography, J. Chromatogr. A 671 (1994) 63–71.

[21] J.S. Fritz, Factors affecting selectivity in ion chromatography, J. Chromatogr. A 1085 (2005) 8–17.

[22] C. Liang, C.A. Lucy, Characterization of ion chromatography columns based on hydrophobicity and hydroxide eluent strength, J. Chromatogr. A 1217 (2010) 8154–8160.

[23] R.W. Slingsby, C.A. Pohl, Anion-exchange selectivity in latex-based columns for ion chromatography, J. Chromatogr. 458 (1988) 241–253.

[24] http://www.dionex.com/en-us/products/columns/ic-rfic/hydroxide-selective-packed/ionpac-as10/lp-73255.html.

[25] M.C. Bruzzoniti, E. Mentasti, C.A. Pohl, J.M. Riviello, C. Sarzanini, Effect of ion-exchange site and eluent modifiers on the anion-exchange of carboxylic acids, J. Chromatogr. A 925 (2001) 99–108.

[26] P. Kolla, J. Köhler, G. Schomburg, Polymer-coated cation-exchange stationary phases on the basis of silica, Chromatographia 23 (1987) 465–472.

[27] L. Nair, R. Saari-Nordhaus, J.M. Anderson Jr., Ion chromatographic separation of transition metals on a polybutadiene maleic acid-coated stationary phase, J. Chromatogr. A 671 (1994) 43–49.
[28] http://www.dionex.com/en-us/products/columns/bio/carbohydrate/carbopac-pa1/lp-73364.html.
[29] http://www.dionex.com/en-us/products/columns/ic-rfic/carbonate-eluent-packed/ionpac-as4asc/lp-73217.html.
[30] http://www.dionex.com/en-us/products/ion-chromatography/ic-rfic-solutions/capillary-ic/columns/lp-88588.html.
[31] P. Kuban, P.K. Dasgupta, Capillary ion chromatography, J. Sep. Sci. 27 (2004) 1441–1457.
[32] B. Wouters, C. Bruggink, G. Desmet, Y. Agroskin, C.A. Pohl, S. Eeltink, Capillary ion chromatography at high pressure and temperature, Anal. Chem. 84 (2012) 7212–7217.
[33] http://www.dionex.com/en-us/webdocs/114317-ST-IC-Columns-Reference-Card-ST70223_E-12-13.pdf.
[34] http://www.dionex.com/en-us/webdocs/4572-Column_Selector_Documentation_v0.93_V17.pdf.
[35] https://www.phenomenex.com/Products/HPLCDetail/rezex.
[36] R.A. Shellie, É. Tyrrell, C.A. Pohl, P.R. Haddad, Column selection for comprehensive multidimensional ion chromatography, J. Sep. Sci. 31 (2008) 3287–3296.

CHAPTER

10

Stationary Phases and Columns for Chiral Chromatography

10.1 TYPES OF CHIRAL PHASES AND THEIR PREPARATION

Types of Chiral Phases

The separation of enantiomers typically requires a chiral stationary phase (CSP), although separations with a chiral agent in the mobile phase on achiral phases are also possible (see, e.g., [1]). Most chiral phases having the separation based on polar-type interactions are designed to be used in nonpolar solvents. Water being strongly polar may interfere with the polar interactions involved in chiral recognition, and a water excess in the mobile phase is typically undesirable for a successful chiral separation. However, some analytes, in particular biomolecules, may not be soluble in a mobile phase with very low polarity, and chiral phases that can still work in the presence of polar solvents such as methanol and an organic acid, base, or salt have been developed. The chiral phases are designed to work with analytes of different polarities. The type of utilization depends on both the nature of the stationary phase, and also on the mobile phase nature [2–4]. The types of utilization of chiral phases classify them in types similar to those used in nonchiral chromatography as follows [5] (see also Section 13.5).

1) Utilization similar to that practiced in normal-phase (NP) chromatography and applicable to the separation of many analytes, except some with very high polarity. The phases are used with nonpolar or weakly polar solvents as mobile phases, such as heptane with 20–30% addition of ethanol or isopropanol. This is the most common type of utilization of chiral phases. A disadvantage of the use of nonpolar mobile phase is related to the case of MS detection with an electrospray ion source (ESI). The nonpolar solvents utilized for mobile phase are typically not compatible with ESI. A nonpolar organic solvent content prevents the ionization of dissolved solute molecules, as the solvent itself does not contain charge carriers like protons. Atmospheric pressure chemical ionization (APCI) and atmospheric pressure photoionization (APPI) are still applicable, although the use of polar additives in the mobile phase, like formic acid, acetic acid, or ammonium acetate, enhances sensitivity [6].

2) Utilization similar to that practiced in reversed phase (RP) chromatography applicable to many analytes having hydrophobic moieties, but less amenable for analysis of strongly polar compounds. The phases are used with (nonaqueous) more polar mobile phases than the

Selection of the HPLC Method in Chemical Analysis
http://dx.doi.org/10.1016/B978-0-12-803684-6.00010-X

stationary phase. Among the mobile phase components can be methanol, acetonitrile, mixed with lower levels of nonpolar solvents, and with or without the addition of a salt such as ammonium acetate of ammonium formate. Water, if necessary, should not exceed 2–3% in the mobile phase [7].

3) Utilization applicable to polar molecules capable of forming strong hydrogen bonds (corresponding to hydrophilic interaction [HILIC] nonchiral chromatography). In such cases the phases are used with polar solvents such as methanol and acetonitrile with additives such as ammonium formate or low level of buffers such as acetic acid/triethanolamine or trifluoroacetic acid/triethanolamine, or specific chaotropic salts. However, if water is necessary in the mobile phase, it should not exceed a few percent.

4) Utilization applicable to ionizable compounds (corresponding to IC nonchiral chromatography). The phases are used with polar solvents such as methanol and acetonitrile with additives such as acetic acid/triethanolamine. Only low levels of water (2–5%) are tolerable in some separations.

Regarding the construction/chemical composition a variety of chiral phases were manufactured. All these phases have chiral centers, and as previously discussed in Section 4.3, at least three interaction sites with the analyte. Based on their chemical composition, the chiral phases can be placed in specific classes that include the following: (1) brush or "Pirkle" type, (2) cellulose-based, (3) cyclodextrin- or cyclofructan-based, (4) amylose-based, (5) crown ether-based, (6) macrocyclic antibiotic type, (7) protein-based, (8) ligand exchange type, and (9) chiral synthetic polymer type. Many chiral phases are placed on a silica gel surface in order to take advantage of the high porosity of this material. Some of the phases are just coated on silica, but some are connected with covalent bonds. Some organic polymers are also used to bind chiral phases. Numerous types of chiral phases are commercially available. In this section, some examples of chiral phases from each class are discussed [8,9].

Brush or "Pirkle" Chiral Phases

This type of stationary phase uses a silica support with the attached moieties capable of developing both donor and acceptor hydrogen bonds, and a group able to establish either π–donor, π–acceptor, or both π–donor and π–acceptor intermolecular bonds [10–12]. It is also required that the moiety has a stereospecific structure. Such structures are frequently obtained using an amino acid involved in an amide bond, and one or more aromatic moieties such as the "electron-poor" dinitrobenzene, which offers π–acceptor interactions or an "electron-rich" moiety that offers π–donor interactions [13,14]. Typical structures for Pirkle chiral phases are given in Fig. 10.1.1, and the bonding silane to silica can be either monofunctional or trifunctional (see Section 6.1).

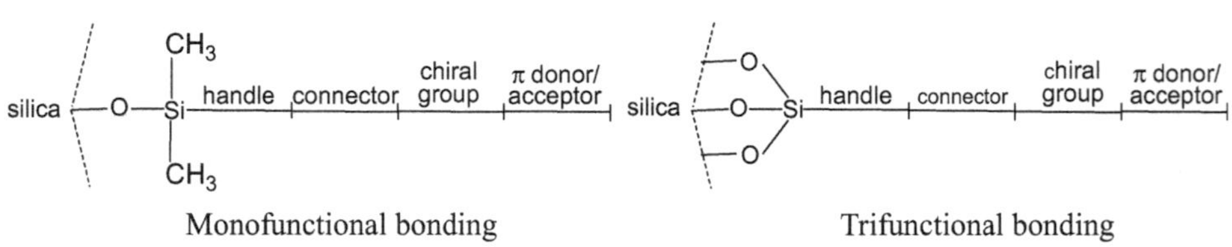

FIGURE 10.1.1 Typical structure of a chiral Pirkle phase.

An example of such a structure and the path followed for its synthesis are shown in Fig. 10.1.2. This synthesis typically starts with aminopropyl silica that is derivatized with an amino acid such as (S)-leucine, followed by derivatization with 3,5-dinitrobenzoic acid (e.g., as acid chloride). The three-point interactions with a structure such as the one shown in Fig. 10.1.2 are offered by the CO in the amide group, NH in the amide group, and the dinitrobenzene moiety.

Other amino acids can be used in the synthesis such as (R)-leucine, (S)- or (R)-phenylglycine, and (S)-proline [5,15–18]. Also, moieties different from π–acceptor 1,3-dinitrobenzene and capable of offering π–donor interactions have been synthesized. Besides the use of a propylamino handle such as in the aminopropylsilica for the connection to the silica the connection can be achieved using alkylglycidyl or alkylisocyanate handles.

Common utilized solvents for Pirkle phases are ethanol, isopropanol, hexane, and CH_2Cl_2. The phases can be used in water–alcohol mixtures, but the water content in the mobile phase must be very low. Water has the capability to form strong hydrogen bonds and interfere with the interaction between the analyte and stationary phase. Also, the pH of the mobile phase must be in a relatively narrow range (pH = 2.5–7.5). Another limitation of these phases is that they are usable mainly with aromatic compounds (or compounds with a system of π electrons), since the π–donor or π–acceptor phases should interact with a π–acceptor or π–donor analyte, respectively. For the analysis of compounds not having such groups, derivatization before separation with reagents introducing, e.g., a π–donor group such as naphthylamine can be performed. Some examples of Pirkle-type stationary phases are given in Appendix 5a. Other Pirkle phases manufactured by Sumika Chemical Analysis Service are listed in Appendix 5b. Various suppliers offer Pirkletype chiral chromatographic columns, including Advanced Separation Technologies (Astec) (three types), IRIS Tecnologies Inc.

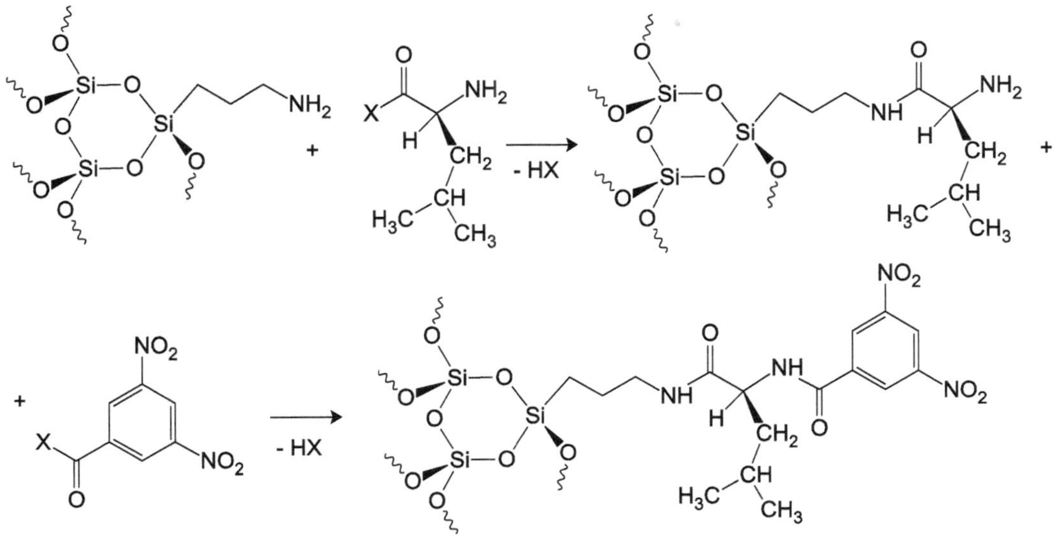

FIGURE 10.1.2 Schematic synthesis of a Pirkle chiral stationary phase (CSP) containing a propyl handle, an amide group connector, (S)-leucine as chiral group, and 1,3-dinitrobenzene moiety as π–acceptor group.

(six types), Machery-Nagel (two types), Merck (two types), Phenomenex (11 types), Regis Technologies (nine types), Sumika Chemical Analysis Service (11 types), and YMC (two types).

Cellulose Chiral Phases

The chiral structure of cellulose and its capability to be derivatized at the active -OH groups allowed the synthesis of a number of chiral stationary phases. Some of these phases were obtained by derivatizing microcrystalline cellulose with groups such as triacetate (to form microcrystalline cellulose triacetate [MCTA]), tribenzoate, trisphenylcarbamate, and tris(3,5-dimethylphenylcarbamate). In order to obtain a larger surface for the stationary phase, the derivatized cellulose material can be coated on a silica support [19,20]. The coating of silica can be achieved either by simple adsorption on silica surface (physically coated) of the derivatized cellulose using a proper solvent or by simultaneous forming of bonds between cellulose and silica layer. One possibility to anchor cellulose on silica is through glycidyl groups already attached to the silica surface by previous derivatization (see Section 6.1). Other procedures for binding cellulose to silica have been described in the literature [21]. The coating or binding must preserve the porous structure of silica. Typical derivatizations of the cellulose are performed as esters or as carbamates. Two examples of cellulose derivatization are given in Fig. 10.1.3. Other substituents, R, can be used for various phases such as methoxy, acetyl, hydroxypropyl ether, naphthenylcarbamate, 3,5-dinitrophenylcarbamate, 3,5-dimethylphenylcarbamate, and 2,6-dinitro-4-trifluoromethylphenyl ether [22]. Several commercially available chiral columns based on cellulose derivatives on silica support with different types of derivatization are listed in Appendix 5c. Coating of macroporous polymers such as poly(2-aminoethyl methacrylate-co-ethylenedimethacrylate) with derivatized cellulose was also reported in the literature.

Besides silica generally used as support for coating with chiral materials, the surface of zirconia previously sintered but not rehydroxylated (see Section 10.1), can also be utilized as support, providing a stable surface for depositing the chiral polymers, such as derivatized cellulose with tris(3,5-dimethylphenyl-carbamate) in proportions of 3–4% (w/w) [23,24].

The utilization of cellulose-based columns requires a normal mobile phase, such as hexane-ethanol, or hexane-isopropyl alcohol. The use of these columns with a mobile phase containing some water can also be done, but addition of a salt in the mobile phase, such as a perchlorate, is necessary to prevent the column degradation by the dissolution of the stationary active coating.

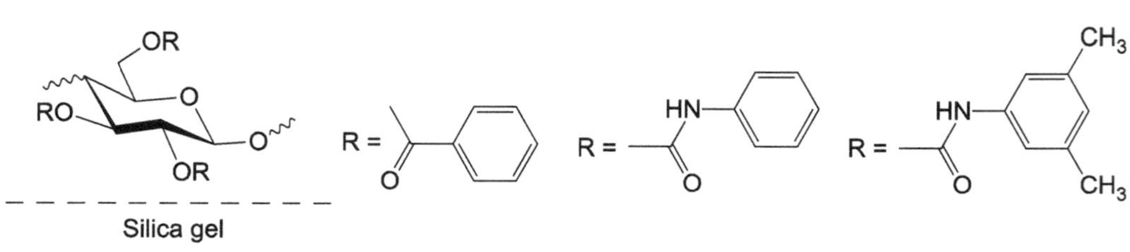

FIGURE 10.1.3 Examples of derivatized cellulose on a silica support.

Amylose Chiral Phases

Amylose is one of the components of starch and the polymer can assume a helical shape and can form inclusion compounds, such as with iodine, fatty acids, and aromatic compounds. Similarly to cellulose, it also offers the capability of hydrogen bonding and some hydrophobic interactions. Amylose can also be derivatized to generate chiral stationary phases, typically in the form of carbamates in the form of tris(3,5-dimethylphenyl-carbamate), tris[(S)-α-methylbenzyl carbamate], or tris(chloro-2-methylphenylcarbamate). Amylose or derivatized amylose is placed on the silica surface to achieve the high surface area of the phase [25]. Several commercially available chiral columns based on amylose derivatives on silica support with different types of derivatization are listed in Appendix 5d.

Cyclodextrin and Cyclofructan Chiral Phases

A number of chiral phases are based on cyclodextrins or on derivatized cyclodextrins [26]. These phases are placed on silica gel surfaces similarly to cellulose. Cyclodextrins are produced by the action of enzymes cyclodextrin glycosyltransferase and amylase on starch. The main types of cyclodextrins produced are α, β, and γ (with six, seven, and eight glucose residues, respectively), and they are generated by the enzymatic cleavage of starch polymeric chain and the formation of cyclic oligomers. Larger cyclodextrins are also known. Cyclodextrins offer a series of advantages to be used as chiral stationary phases. They can be anchored to silica (e.g., through reactions with glycidyl groups bonded to silica, as shown in Section 6.1), they offer a large number of chiral centers (30 for α, 35 for β, and 40 for γ-cyclodextrin), and also offer the possibility to be derivatized at their free OH groups with less polar substituents. As discussed in Section 4.3, the separation on the cyclodextrin type of phases is based on inclusion properties provided by their conical cavity of one part of the chiral molecule, and differences in the interaction with the chiral OH or OR groups of cyclodextrin with the enantiomer remaining substituents. The difference in the size of enantiomers is important for the choice of α, β, or γ cyclodextrin, and the molecular fit determines the range of analytes that can be separated. For inclusion, α-cyclodextrins will host single phenyl groups or naphthyl end-groups, β-cyclodextrins will accept naphthyl groups and heavily substituted phenyl groups, and γ-cyclodextrins are useful for bulky steroid-type molecules. Other interactions such as those involving the polar regions of an analyte and the surface hydroxyls of the stationary phase and the potential hydrophobic interactions in the cavity provide the other two or more points of interaction required for chiral recognition.

Besides unmodified cyclodextrins, different modified cyclodextrins have been developed, and they expand the nature of compounds that can be separated. The derivatives are formed by bonding various groups onto the surface hydroxyls of the cyclodextrin cavity. The derivatives include acetyl (analogous to acetylated cellulose), (S)-hydroxypropyl ether, (S) or (R)-naphthylethylcarbamate (analogous to naphthyl Pirkle-type columns), 3,5-dimethyl-phenylcarbamate, and *p*-toluoylester. These modified cyclodextrin phases have several advantages. They offer different polarities from simple cyclodextrins and are more stable to different mobile phases, allowing the use of a wider range of solvents. Cyclodextrins are used mainly in normal phase-type separations, with no or little water/buffers. It is possible to operate cyclodextrin columns in an alternative polar organic mode with the mobile phase consisting of acetonitrile with up to 10% methanol plus up to 0.5% acetic acid and/or 0.5% triethylamine, and even in reversed-phase conditions using acetonitrile/water mobile phase. Various suppliers offer chiral chromatographic columns based on derivatized cyclodextrins and resulting from α, β, or γ cyclodextrins. Among these suppliers are the following: Advanced Separation Technologies (Astec) (Cyclobond 12 types), Machery-Nagel (four types), Merck (two types), Phenomenex (cyclodextrin with carboxymethyl functionalities bonded to methacrylate polymer), Showa Denko (four types), Serva (three types), Thermo Hypersil (two types), and YMC (three types). Besides cyclodextrins or modified cyclodextrins bonded to silica (typically with the help of a spacer), derivatized cyclodexrin (e.g., carboxymethyl cyclodextrin) can also be bonded to organic polymers. Stationary phases based on cyclofructans are also developed in the same manner as those based on cyclodextrin. Several commercially available chiral columns based on cyclodextrins or cyclofructans on silica support with different types of derivatization are listed in Appendix 5e.

Crown Ether Chiral Phases

Crown ether-based chiral stationary phases are also used for enantiomer separations [27]. The crown ether can be bonded to the silica surface, for example by the derivatization (as a second step) of aminopropyl silica (e.g., Chirosil RCA(+)). The crown macrocyclic groups consist of a polyether with the formula $[-(CH_2)_n-O-]_m$ ($n = 2$ or 3 and $m = 5-10$). The crown ether must achieve asymmetry by incorporating additional groups such as binaphthyl, biphenanthryl, tartaric acid, and several carbohydrates. A model chemical reaction with the formation of a bonded chiral phase starting with aminopropyl silica and tartaric acid modified crown ether is shown in Fig. 10.1.4 [28]. Among the commercially available crown ethers used for chiral separations are Daicel Chemical Industries Ltd. (Crownpack CR(+) and

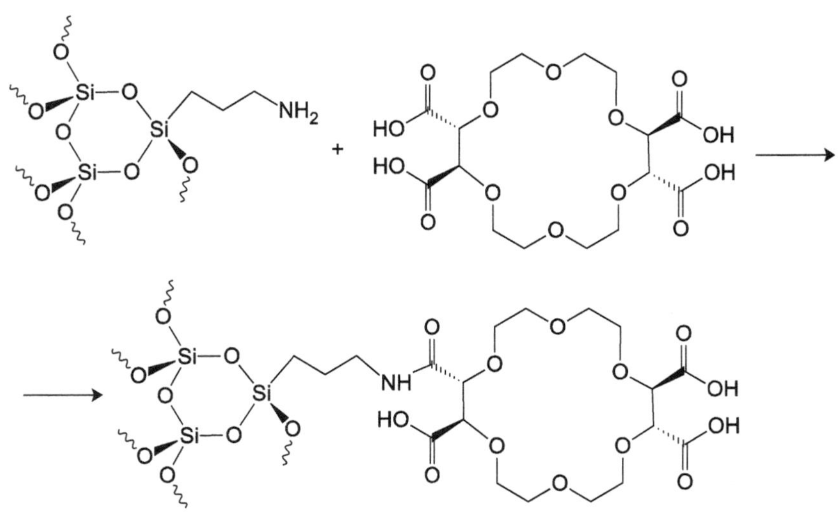

FIGURE 10.1.4 Schematic synthesis of a chiral stationary phase (CSP) based on a tartaric acid-modified crown ether.

Crownpack CR(−)), Regis Technologies (Chirosil RCA(+) and SCA(−)), and Phenomenex (Sumichiral OA-8000).

Macrocyclic Antibiotics and Glycopeptides

Another type of chiral stationary phase is based on macrocyclic-type antibiotics immobilized on silica [29,30]. The macrocyclic antibiotics contain numerous chiral centers, functionalities that allow bonding to silica (using a prederivatized silica with reactive groups) and capability to offer $\pi-\pi$ interactions, hydrogen bonding, inclusion/complexation, and ionic interactions. Several glycopeptide-type antibiotics have been used for making stationary phases, including the following: rifamycin(s), vancomycin (18 chiral centers), avoparcin, ristocetin, glycopeptide A-40,926 (MDL 62,476), and teicoplanin (23 chiral centers). Macrocyclic antibiotics used as bonded stationary phase also include a family of thiopeptides with the parent compound being thiostrepton (17 chiral centers). The antibiotics that are glycopeptides have dissociable groups (−OH of phenolic type, $-NH_2$, −COOH, and −OH from carbohydrate). Some of the groups are capable of existing in zwitterionic form, and are likely to play a major role in the association with analytes during the chiral recognition. Similar to other chiral columns, the macrocyclic antibiotic-type columns are used in normal phase but also in reversed-phase mode.

Protein Chiral Phases

Proteins bonded to silica were also used as stationary phases for chiral separations. Proteins contain a large number of chiral centers, they are known to interact with enantiomers to produce acceptable chiral selectivity, and can be bonded to silica, e.g., through one of their amino groups. Several types of proteins have been used so far for separation of enantiomers, such as bovine serum albumin (BSA), human serum albumin (HSA), α1-acid glycoprotein, other glycoproteins, ovomucoid, ovoglycoprotein, avidin with cellobiohydrolase, and others. Some of these phases can be used in either normal-phase or reversed-phase mode with solvents such as isopropanol, ethanol, or acetonitrile in mixture with aqueous buffers [31,32]. Several commercially available chiral columns based on proteins on silica support are listed in Appendix 5f.

Besides bonding directly to silica, proteins were also bonded to organic polymers such as acrylates and further coated on silica or silica functionalized with amino groups that provide a stronger bonding [33]. Another possibility to bond proteins to silica-based supports is to functionalize the silica surface to become an anion exchanger (e.g., quaternized polyvinylimidazole coated silica) and then to bond electrostatically the proteins to its surface [34].

Ligand Exchange Chiral Phases

Ligand exchange chromatographic columns contain chiral molecular units bonded on an organic polymer or on silica, the chiral units being capable of forming complexes with transitional metal cations such as Cu^{2+}, Ni^{2+}, or Zn^{2+}. Some analytes are capable of forming ternary complexes with the metal ions already bonded to the chiral moiety from the stationary phase. The enantiomer separation results from the difference between the stability constants of the ternary complexes. For example, (S)-proline attached to a solid phase forms a complex with Cu^{2+}. The ternary complex with an (S) or (R) analyte, such as leucine, will have different stability constants [35]. Examples of ligand exchange phases are Astec CLC-L and CLC-D containing a chiral bidentate ligand and recommended to work in the presence of Cu^{2+} in the mobile phase, Davankov LEC (from Regis), Nucleosil Chiral1, Chiralpak WH, WM and WE (from Capital HPLC Limited) [36], and Chirex 3126 that contains D-penicillamine bonded on silica through a C18 handle in Cu^{2+} form.

Chiral Synthetic Polymers

It is possible to synthesize chiral synthetic polymers that can be used as stationary phase for enantiomer separation. These polymers can be classified into three categories: addition polymers, condensation polymers, and crosslinked gels [37]. Such polymers can be bonded to silica in order to assure a high contact surface and mechanical rigidity. The polymer must contain chiral moieties to act as chiral selectors. One example is a polymer having diaminocyclohexylacrylamide chiral selectors with polymeric structure bonded to silica. Some other phases based on chiral polymers are commercially available (e.g., poly(diphenylethylenediamine-bis-acryloyl)) [38,39]. These stationary phases can be used in normal phase (NP) mode (e.g., hexane + isopropanol, or methylene chloride + ethanol or isopropanol as a mobile phase), and also in NP with polar-organic mobile phase consisting of acetonitrile, methanol, and a polar additive (trifluoroacetic acid, triethanolamine). Several commercially available chiral columns based on synthetic polymers are listed in Appendix 5g.

A number of chiral synthetic polymers were reported as synthesized, but they are not commercially available and were used only for research purposes. Examples include *trans*-9, 10-dihydro-9,10-ethanoanthracene-(11S,12S)-11,12-dicarboxylic acid bis-4-vinylphenylamide [40], and - N,N′-[(1R,2 R)-1,2-diphenyl-1,2-ethanediyl]bis-2- propenamide [41].

Among the polymers evaluated for potential use in chiral separations are also certain molecular imprinted polymers [42–44]. For example, 4-vinylpyridine can be used as a monomer with N-carbazolyl-L-aspartic acid as template for preparing a resin capable of separating N-carbobenzoxy-D,L-aspartic acids [45].

10.2 CHARACTERIZATION OF CHIRAL PHASES

Several characteristics of chiral phases and columns are common to those used in other chromatographic techniques. Besides the obvious characteristics, such as column dimensions, others are particle size of stationary phase, shape of particles, their mechanical resilience, range of pH stability, chemical stability in different solvents, and lack of bleeding. One important aspect related to chiral columns is related to their mode of utilization (NP, RP, HILIC, or IC like). Some columns are capable of being utilized in NP mode only, and others in more than one mode. For example, it is common for polysaccharide-type CSPs to be effective under both normal-phase conditions and also under reversed-phase conditions. The stability of the stationary phase in different solvents is also very important.

The main role of chiral phases is the efficient separation of enantiomers, which is characterized by enantioselectivity. Enantioselectivity follows the general definition of chromatographic selectivity, and can be characterized by the enantiomeric separation factor α (given by

Eq. 4.2.54 with $\alpha = k_2'/k_1'$, where k_1' and k_2' are the retention factors of the first and second enantiomers, respectively). It depends on which of the enantiomers forms the more stable association with the chiral selector from CSP and thus will be the more strongly retained species of the racemate. Enantioselectivity depends not only on the stationary phase, but also on the composition, pH, and temperature of the mobile phase [46].

It is rather difficult to predict the efficiency of enantiomer separations and, when separated, the elution order of each enatiomer is uncertain. For the identification of a specific enantiomer peak, isolated enantiomers should be injected on the chiral column and their retention time measured. When isolated enantiomers are not available, information about their retention may be available in databases regarding enantiomer separation. For example, two known commercial databases, namely ChirBase [47] and Chirsource [48], contain information for over 100,000 chiral separations [49,50]. A theoretical possibility to predict the elution order of enantiomers is offered by chemometric approaches based on quantitative structure–enantioselective retention relationships [51].

10.3 SELECTION OF A CHIRAL PHASE

Chiral separations occupy an important sector of HPLC-type analyses since many bioorganic molecules, as well as many pharmaceutical drugs, are chiral. The subject of chiral chromatography is covered in numerous publications including books (see, e.g., [52–56]), dedicated journals such as *Chirality* (Wiley), *Tetrahedron: Asymmetry* (Elsevier), numerous peer-reviewed publications, and companies' application notes (see, e.g., [57]). However, the selection of a column in chiral separation and the development of a successful HPLC procedure remain a rather difficult subject. The differences in the interaction forces between enantiomers are typically small, and specific phases are more suitable than others for a specific separation. In addition, the restriction of using low level or no water in the mobile phase limits the type of polar compounds that can be separated using chiral phases. For this reason, trial and error in chiral separations is not uncommon. Theoretical considerations such as QSAR modeling are sometimes used for helping in the chiral column selection [58].

Since chiral separations may be necessary only for a portion of the compounds present in a sample, and the separation of all the components in a sample may be difficult using the chiral phase, bidimensional separations of samples containing chiral compounds is frequently utilized. In such separations, the components of the sample are initially separated on a nonchiral column with the enantiomer analytes not separated between themselves. A heart-cut is typically taken for the region where the enantiomers elute, and the second separation is performed on a chiral column [59].

The Role of Column Selection in the Development of a Method for Chiral Separations

The selection of a chiral column for developing a separation is typically based on manufacturer recommendations. For the same task, it is possible to select several types of columns. Some recommendations for the use of Pirkle-type columns on different classes of compounds are given in Table 10.3.1 (see also Appendix 8).

TABLE 10.3.1 Some Recommendations for the Use of Pirkle-Type Columns

Chiral Selector	Compounds to Be Analyzed
(R)-Phenylglycine and 3,5-dinitrobenzoic acid amide linkage	Carboxylic acids, alcohols, esters, sulfoxides
(R)-Phenylglycine and 3,5-dinitroaniline urea linkage	Carboxylic acids, amino acids derivatives, compounds containing π-basic groups
(R)-1-Naphthylglycine and 3,5-dinitrobenzoic acid amide linkage	Carboxylic acids, alcohols, esters, nonsteroidal antiinflammatory agents
(S)-Valine and 3,5-dinitroaniline urea linkage	Carboxylic acids, amino acids derivatives, hydroxy acids
(S)-Tert-leucine and 3,5-dinitroaniline urea linkage	Carboxylic acids, amino acids derivatives, benzodiazapines
(S)-Valine and (R)-1-(α-naphthyl)ethylamine urea linkage	Π—Acceptor derivatives of amines, carboxylic acids and amino acids, esters and amides of these acids, alcohols
(S)-Proline and (R)-1-(α-naphthyl)ethylamine urea linkage	Amines, alcohols and amino acids, underivatized β-blockers, aromatic amines, pesticides
(S)-Tert-leucine and (S)-1-(α-naphthyl) ethylamine urea linkage	Esters, amino alcohols, underivatized β-blockers, aromatic amines, cyano alcohols, pesticides
(S)-Tert-leucine and (R)-1-(α-naphthyl) ethylamine urea linkage	Amines, amino alcohols, alcohols, underivatized β-blockers, aromatic amines, cyano alcohols, pesticides
(S)-Indoline-2-carboxylic acid and (R)-1-(α-naphthyl) ethylamine, urea linkage	Amines, amino alcohols, alcohols
3,5-Dinitrobenzoyl derivative of 1,2-diaminocyclohexyl bonded to silica by ether containing handle	Amides, alcohols, esters, ketones, acids, sulfoxides, phosphine oxides, heterocycles.
Dimethyl-N-3,5-dinitro-benzoyl-amino-2,2-dimethyl- 4-pentenyl phosphonate silica	β-Blockers, π—acceptor derivatives of pharmaceuticals
N-(1-Propylsilica-2-oxo-4-phenylazetidin-3-yl)-3,5-dinitrobenzamide	β-Blockers
N-3, 5-Dinitrobenzoyl-3-amino- 3-phenyl-2-(1,1-dimethylethyl)-propanoate bonded to silica by undecyl handle	Carboxylic acids, including nonsteroidal antiinflammatory agents.
3,5-Dinitrobenzoyl derivative of diphenylethylenediamine bonded to silica by tetradecanoic acid handle	Aryl carbinols, pharmaceuticals
1-N,1-N,3-N,3-N-Tetramethylidene-5-(1,2,3,4-tetrahydrophenanthren- 4-yl)carbamoyl]-benzene-1,3-diamine (propyl handle to silica)	Amides, epoxides, esters, ureas, carbamates, ethers, aziridines, phosphonates, aldehydes, ketones, carboxylic acids, alcohols, and nonsteroidal antiinflammatory drugs

A summary of other recommendations regarding the type of column to be used for a chiral separation is indicated in Table 10.3.2 [60]. The recommendations cover utilization mode (NP, RP, HILIC, or IC-like), types of analyte, and detection mode.

The success in developing a good chiral separation is also highly dependent on the selection of the mobile phase, as also it is in other types of chromatographic separations [61,62]. In the development of chiral methods of separation, the mobile phase must be selected according to the type of utilization (NP, RP, HILIC, or IC-like), and changes in the mobile phase composition are recommended for tuning a separation. These changes include modification in the ratio of nonpolar/polar solvent composition, addition of modifiers such as acid/base mixtures including the acids formic, acetic, trifluoroacetic, methanesulfonic, and the bases diethylamine, triethylamine, or salts such as ammonium acetate, ammonium formate. The content of water in the mobile phase (0–5%) can also be utilized for modifying the elution and separation.

TABLE 10.3.2 Suggestions for the Utilization of Different Types of Chiral Columns (Not Including Pirkle Type)

Analysis Type	Cellulose DMP	Cyclodextrin	Cyclofructan	Glycopeptide	Imobilized Protein	Ligand Exchange	Synthetic Polymer
Routine chiral column screening	3	3	3	3	2	1	2
Normal-phase HPLC	3	2	3	2	1	1	3
Reversed-phase HPLC	1	3	2	3	3	2	1
Hydrophilic interaction HPLC	1	2	2	3	1	1	1
Polar organic mode HPLC	3	2	3	3	1	1	3
Polar ionic mode HPLC	1	1	1	3	1	1	1
Polar/ionic analytes	2	2	2	3	3	1	1
Amino acids, peptides	1	2	2	3	1	3	2
Primary amines, chiral	1	1	3	1	1	1	1
Nonaromatic organic acids	1	2	1	2	1	3	1
Bioanalysis (drugs in biological fluids)	2	3	1	3	2	1	1
Mass spectrometry (LC/ESI)	2	3	3	3	3	1	1
Mass spectrometry (LC/APCI)	3	3	3	3	1	1	2

Note: 3 = recommended, 2 = possible use, 1 = not recommended.

References

[1] S.C. Moldoveanu, V. David, Essentials in Modern HPLC Separations, Elsevier, Amsterdam, 2013.

[2] A. Cavazzini, L. Pasti, A. Massi, N. Marchetti, F. Dondi, Recent applications in chiral high performance liquid chromatography: a review, Anal. Chim. Acta 706 (2011) 205–222.

[3] W.J. Lough, Classification of LC chiral stationary phases: Wainer Types I–V revisited, J. Chromatogr. B 968 (2014) 1–7.

[4] D.V. Armstrong, B. Zhang, Chiral stationary phases for HPLC, Anal. Chem. 73 (2001) 557A–561A.

[5] M. Lämmerhofer, Chiral recognition by enantioselective liquid chromatography: mechanism and modern chiral stationary phases, J. Chromatogr. A 1217 (2010) 814–856.

[6] http://www.sigmaaldrich.com/technical-documents/articles/analytical/brochures/chiral-hplc-separations.html#sthash.EJ3AR0T8.dpuf.

[7] I. Ali, K. Kumerer, H.Y. Aboul-Enein, Mechanistic principles in chiral separations using liquid chromatography and capillary electrophoresis, Chromatographia 63 (2006) 295–307.

[8] H.Y. Aboul-Enein, I. Ali (Eds.), Chiral Separations by Liquid Chromatography and Related Technologies, Marcel Dekker, New York, 2003.

[9] S. Ahuja (Ed.), Chiral Separation Methods for Pharmaceutical and Biotechnological Products, John Wiley & Sons, Hoboken, 2011.

[10] W.H. Pirkle, D.W. House, Chiral high-performance liquid chromatographic stationary phases. 1. Separation of the enantiomers of sulfoxides, amines, amino acids, alcohols, hydroxy acids, lactones, and mercaptans, J. Org. Chem. 44 (1979) 1957–1960.

[11] W.H. Pirkle, J.M. Finn, Chiral HPLC stationary phases 3. General resolution of arylalkylcarbinols, J. Org. Chem. 46 (1981) 2935–2938.

[12] W.H. Pirkle, C.J. Welch, An improved chiral stationary phase for the chromatographic separation of underivatized naproxen enantiomers, J. Liq. Chromatogr. 15 (1992) 1947–1955.

[13] W.H. Pirkle, T.C. Pochapsky, Considerations of chiral recognition relevant to the liquid chromatography separation of enantiomers, Chem. Rev. 89 (1989) 347–362.

[14] K. Kurata, S. Sakamoto, A. Dobashi, Chiral separation with novel (S)-biotin-bonded silica gel for liquid chromatography, J. Chromatogr. A 1068 (2005) 335–337.

[15] M. Lämmerhofer, P. Franco, W. Lindner, Quinine carbamate chiral stationary phases: systematic optimization of steric selector-select and binding increments and enantioselectivity by quantitative structure-enantioselectivity relationship studies, J. Sep. Sci. 29 (2006) 1486–1496.

[16] M. Ashtari, N.M. Cann, The docking of chiral analytes on proline-based chiral stationary phases: a molecular dynamics study of selectivity, J. Chromatogr. A 1409 (2015) 89–107.

[17] M. Kato, T. Fukushima, T. Santa, S. Nakashima, R. Nishioka, K. Imai, Preparation and evaluation of new Pirkle type chiral stationary phases with long alkyl chains for the separation of amino acid enantiomers derivatized with NBD-F, Analyst 123 (1998) 2877–2882.

[18] K.M. Kacprzak, W. Lindner, Novel Pirkle-type quinine 3,5-dinitrophenylcarbamate chiral stationary phase implementing click chemistry, J. Sep. Sci. 34 (2011) 1–6.

[19] W. Huang, Y. Xing, Y. Yu, S. Shang, J. Dai, Enhanced washing durability of hydrophobic coating on cellulose fabric using polycarboxylic acids, Appl. Surf. Sci. 257 (2011) 4443–4448.

[20] F. Ling, E. Brahmachary, M. Xu, F. Svec, J.M.J. Fréchet, Polymer-bound cellulose phenylcarbamate derivatives as chiral stationary phases for enantioselectve HPLC, J. Sep. Sci. 26 (2003) 1337–1346.

[21] X. Chen, H. Zou, Q. Zhang, J. Ni, Z. Zhang, Synthesis of chemically bonded cellulose trisphenylcarbamate chiral stationary phases for enantiomeric separation, J. Chromatogr. Sci. 40 (2002) 315–320.

[22] K. Tachibana, A. Ohnishi, Reversed-phase liquid chromatographic separation of enantiomers on polysaccharide type chiral stationary phases, J. Chromatogr. A 906 (2001) 127–201.

[23] C.B. Castells, P.W. Carr, Cellulose tris(3,5-dimethylphenylcarbamate)-coated zirconia as a chiral stationary phase for HPLC, Anal. Chem. 71 (1999) 3013–3021.

[24] I.W. Kim, J.K. Ryu, S.D. Ahn, J.H. Park, K.-P. Lee, J.J. Ryoo, M.H. Hyun, Y. Okamoto, C. Yamamoto, P.W. Carr, Comparison of chiral separation on amylose and cellulose tris(3,5-dimethylphenylcarbamate)-coated zirconia in HPLC, Bull. Korean Chem. Soc. 24 (2003) 239–242.

[25] K. Kalíková, M. Riesová, E. Tesařová, Recent chiral selectors for separation in HPLC and CE, Cent. Eur. J. Chem. 10 (2012) 450–471.

[26] T.J. Ward, D.W. Armstrong, Improved cyclodextrin chiral phases: a comparison and review, J. Liq. Chromatogr. 9 (1986) 407–414.

[27] M.H. Hyun, Development and application of crown ether-based HPLC chiral stationary phases, Bull. Korean Chem. Soc. 26 (2005) 1153–1163.

[28] M.-J. Paik, J.S. Kang, B.–S. Huang, J.R. Carey, W. Lee, Development and application of chiral crown ethers as selectors for chiral separation in high-performance liquid chromatography and nuclear magnetic resonance spectroscopy, J. Chromatogr. A 1274 (2013) 1–5.

[29] D.W. Armstrong, Y. Tang, S. Chen, Y. Zhou, C. Bagwill, J.R. Chen, Macrocyclic antibiotics as a new class of chiral selectors for liquid chromatography, Anal. Chem. 66 (1994) 1473–1484.

[30] I. D'Acquarica, F. Gasparrini, D. Misiti, M. Pierini, C. Villani, Hplc chiral phases containing macrocyclic antibiotics: practical aspects and recognition mechanism, in: E. Grushka, N. Grinberg (Eds.), Advances in Chromatography, vol. 46, CRC press, Boca Raton, 2007.

[31] J. Haginaka, Protein-based chiral stationary phases for high-performance liquid chromatography enantioseparations, J. Chromatogr. A 906 (2001) 253–273.

[32] J. Haginaka, Recent progress in protein-based chiral stationary phases for enantioseparations in liquid chromatography, J. Chromatogr. B 875 (2008) 12–19.

[33] M.C. Millot, N.L. Taleb, B. Sebille, Binding of human serum albumin to silica particles by means of polymers: a liquid chromatographic study of the selectivity of resulting chiral stationary phases, J. Chromatogr. B 768 (2002) 157–166.

[34] S.C. Jacobson, G. Guiochon, Enantiomeric separations using bovine serum albumin immobilized on ion-exchange stationary phases, Anal. Chem. 64 (1992) 1496–1498.

[35] V.A. Davankov, Ligand exchange chromatography, in: I.D. Wilson, E.R. Adlard, M. Cooke, C.F. Poole (Eds.), Encyclopedia of Separation Science, vol. 5, Academic Press (Elsevier), Amsterdam, 2000, pp. 2369–2380.

[36] http://www.capitalhplc.com/index.php/hplc-product/chiral-hplc-columns/pirkle-type-phases.

[37] T. Nakano, Optically active synthetic polymers as chiral stationary phases in HPLC, J. Chromatogr. A 906 (2001) 205–225.

[38] T. Payagala, E. Wanigasekara, D.W. Armstrong, Synthesis and chromatographic evaluation of new polymeric chiral stationary phases based on three (1S,2S)-(-)-1,2-diphenylethylenediamine derivatives in HPLC and SFC, Anal. Bioanal. Chem. 399 (2011) 2445–2461.

[39] K.-P. Lee, S.-H. Choi, S.-Y. Kim, T.-H. Kim, J.J. Ryoo, K. Ohta, J.-Y. Jin, T. Takeuchi, C. Fujimoto, Comparison of monomeric and polymeric chiral stationary phases, J. Chromatogr. A 987 (2003) 111–118.

[40] X. Han, C. Wang, L. He, T.E. Beesley, D.W. Armstrong, Preparation and evaluation of a new synthetic polymeric chiral stationary phase for HPLC based on the trans-9,10-dihydro-9,10-ethanoanthracene-(11S,12S)-11,12-dicarboxylic acid bis-4-vinylphenylamide monomer, Anal. Bioanal. Chem. 387 (2007) 2681–2697.

[41] X. Han, L. He, Q. Zhong, T.E. Beesley, D.W. Armstrong, A synthetic polymeric chiral stationary phase for liquid chromatography based on N, N′-[(1R,2R)-1,2-diphenyl-1,2-ethanediyl]bis-2-propenamide, Chromatographia 63 (2006) 13–23.

[42] K. Hosoya, N. Tanaka, Development of uniform sized, molecular-imprinted stationary phases for HPLC, in: R.A. Bartsch, M. Maeda (Eds.), Molecular and Ionic Recognition with Imprinted Polymers, vol. 703, ACS Ser, 1998.

[43] C.-Y. Hung, H.-H. Huang, C.-C. Hwang, Chiral separation of mandelic acid by HPLC using molecular imprinted polymers, Eclet. Quim. 30 (2005) 67–73.

[44] B. Sellergren, Imprinted chiral stationary phases in high-performance liquid chromatography, J. Chromatogr. A 906 (2001) 227–252.

[45] M. Kempe, K. Mosbach, L. Fischer, Chiral separation using molecularly imprinted heteroaromatic polymers, J. Mol. Recognit. 6 (1993) 25–29.

[46] V.A. Davankov, Analytical chiral separation methods (IUPAC Recommendations 1997), Pure Appl. Chem. 69 (1997) 1469–1474.

[47] http://chirbase.u-3mrs.fr/.

[48] http://chempedia.info/info/141240/.

[49] C. Roussel, P. Piras, A screening study of ChirBase molecular database to explore the expanded chiral pool derived from the application of chiral chromatography, J. Pharm. Biomed. Anal. 46 (2008) 839–847.

[50] C.F. Poole, The Essence of Chromatography, Elsevier, Amsterdam, 2003, p. 799.

[51] K. Heberger, Quantitative structure—(chromatographic) retention relationships, J. Chromatogr. A 1158 (2007) 273—305.
[52] T.E. Beesley, R.P. W Scott, Chiral Chromatography, J. Wiley & Sons Ltd, New York, 1998.
[53] H.Y. Aboul-Enein, I. Ali, Chiral Separations by Liquid Chromatography: Theory and Applications, Chromatographic Science, vol. 90, CRC Press, Boca Raton, 2003.
[54] S. Ahuja, Chiral Separations by Chromatography, ACS Publication, Washington, 2000.
[55] Y.V. Kazakevich, R. LoBrutto, HPLC for Pharmaceutical Scientists, Wiley-Interscience, New York, 2007.
[56] K. Busch, M. Busch (Eds.), Chiral Analysis, Elsevier, Amsterdam, 2006.
[57] T. Cleveland, Phenomenex, Application Note TN-1015.
[58] R. Sheridan, W. Schafer, P. Piras, K. Zawatzky, E.C. Sherer, C. Roussel, C.J. Welch, Toward structure-based predictive tools for the selection of chiral stationary phases for the chromatographic separation of enantiomers, J. Chromatogr. A, http://dx.doi.org/10.1016/j.chroma.2016.05.066.
[59] A. Medvedovici, F. Albu, C. Georgita, D.I. Sora, T. Galaon, S. Udrescu, V. David, Achiral— chiral LC/LC—FLD coupling for determination of carvedilol in plasma samples for bioequivalence purposes, J. Chromatogr. B 850 (2007) 327—335.
[60] www.sigmaaldrich.com/content/dam/sigma-aldrich/docs/Supelco/General_Information/1/T409107.pdf.
[61] G. Subramanian, Chiral Separation Techniques: A Practical Approach, Wiley VCH Verlag, Weinheim, 2007, p. 7.
[62] G. Gübitz, M.G. Schmid, Chiral Separations: Methods and Protocols, Humana Press, Totowa, 2004.

CHAPTER

11

Stationary Phases and Columns for Size Exclusion

11.1 TYPES OF PHASES AND THEIR PREPARATION

Size-exclusion chromatography (SEC) is a technique applied mainly for the separation of mixtures of organic polymers or for the separation between organic polymers and small molecules. SEC is also used for the estimation of molecular weight of polymers (see expression 4.3.62) [1]. Depending on the nature of the polymers, they may be soluble in an aqueous solvent with separation based on gel filtration (GFC) or soluble in organic solvents with separation based on gel permeation (GPC). Solubilization of certain polymers can be a difficult task, and for this purpose special solvents and sometimes elevated temperatures, must be used. The organic polymers can be either synthetic or natural, many natural polymers being of biological origin. The average molecular weight of polymers separated by SEC varies in a large range starting with *Mw* of oligomers formed from several monomeric units and being as high as $2 \cdot 10^7$ Daltons [2].

Size exclusion may be performed at higher pressures and be considered part of the HPLC family, but also can be performed at low pressure. More recently, some success has been obtained in the attempts to perform SEC in ultra-performance liquid chromatography (UPLC) mode, on particles below 2 μm diameter [3].

Stationary phases in SEC can be based on porous silica but also on other inorganic materials and very commonly on organic polymeric materials. Polymeric materials for making the stationary phase are more frequently used in SEC than in other HPLC procedures [2].

Silica-Based SEC Stationary Phases and Glass Phases

Silica-based stationary phases are common for size-exclusion separations, their advantages being mainly related to the good mechanical properties and the lack of swelling, such that their pore size is independent of the nature of mobile phase [4]. This typically is associated with good column efficiency. However, for SEC, polymer-based phases are more frequently utilized than silica-based ones, in particular due to the larger range of pore sizes offered by the polymeric materials (between 60 Å to 4000 Å). The progress made in obtaining good mechanical strength for newer organic polymers contributed to their utilization in HPLC.

Selection of the HPLC Method in Chemical Analysis
http://dx.doi.org/10.1016/B978-0-12-803684-6.00011-1

Silica-based phases have been used in gel permeation-GPC (with organic-type mobile phase) but more often in gel filtration-GFC (with aqueous-type mobile phase) modes. Pore size for SEC silica phases can be controlled similarly as for other silica-based phases (see Section 6.1), and usually pore sizes of 300 Å or larger are used in SEC.

Silica-based phases include bare silica as well as derivatized silica, similar to the construction of other bonded phases on silica. Bonded phase silica-based materials with large and controlled pore dimensions can be made to also have groups such as diol, diol on Zr-clad silica, glycol ether, polyether, and amide. Special polymer separation can be done on large-pore C18-type columns [5]. These phases frequently have a propyl anchor group to the silica material. The derivatization of silica for SEC separations is performed for altering the interaction of pore walls with the analytes. The goal in SEC is to eliminate as much as possible the enthalpic interactions (see Section 4.3) between the stationary phase and the analytes. These interactions interfere with the "pure" entropic separation and affect the elution order of the analytes. The selection of the structure of the stationary phase allows recommendations for which types of analytes are expected to have less enthalpic interactions. For example, diol phases seem to be the most successful in separating proteins since they have minimal enthalpic interactions with them.

Other procedures are also used to limit the enthalpic interactions. For example, the purity of the silica material used for making the packing is very important in order to avoid enthalpic interactions with the analytes. Some parameters of the mobile phase may contribute to minimization of enthalpic interactions between stationary phase and analyte, such as added salts and other modifiers, or a specific pH [6]. The silica-based phases should be operated within the recommended range of pH, which is typically from 2.5 to 7.5. The silica-based phases are not stable outside this range. Several commercially available silica-based packings are listed in Appendix 6a.

Synthesis of silica-based stationary phases for SEC columns uses similar procedures as those previously described for other silica-type phases. Controlled hydrolysis of an alkoxysilane with the purpose of generating a silica gel with the appropriate pore size is the first step in the phase preparation (see Section 6.1). The aging of the silica gel by controlling parameters such as temperature, pH, gelling time, addition of electrolytes for flocculation, addition of detergents, and specific solvents determines the growth of sol particles and of the reticulation to obtain the desired properties of the final material. Drying of the gel is also important for generating the desired silanol density. The silica gel can be used without modifications or can be derivatized, e.g., with diol groups.

Besides silica, porous glass has been used as a packing material for SEC columns [7]. The surface of the porous glass can be derivatized with diol groups, similar to that of porous silica [8].

Polymer-Based Phases Used in SEC

Polymeric-based SEC phases are more common than silica-based ones. Some phases are designed for GPC, a very common material used for this purpose being polystyrene-divinylbenzene (PS-DVB). This polymer is typically prepared by suspension polymerization (see Section 6.1), and depending on the proportion of crosslinking reagent (DVB), either soft gels are obtained (2–12% DVB), or rigid polymers are generated (more than 20% DVB) [9].

Other phases are used for GFC (aqueous SEC), examples of these including several soft gels based on dextran or agarose, more rigid polymers such as hydroxylated poly(methyl methacrylate) (HPMMA), and polyvinylalcohol (PVA) copolymers.

The porosity of the polymeric materials is controlled by the conditions in which the polymerization is carried out. When more than 20% or more of a nonpolymerizable compound is present, this compound dissolving the monomer but not the polymer, a product with a macroporous structure is obtained. This macroporous structure remains in the dry state of material. Porosity of the stationary phase is critical for the separation of compounds within a specific range of *Mw*. However, the pore size of the polymer is difficult to measure when the polymer is swollen, and it is usually assessed with calibrants of known *Mw*. Examples of calibrants are sets of polystyrenes with different known *Mw* for GPC separations and sets of polyethylene oxide, dextrans, and proteins, also with different known *Mw*. Polymeric packings with a narrow range of pore size have a separation capacity concentrated in a limited molecular weight range. Although the resolution of such columns is high, their use for SEC analyses is possible only for a narrow molecular weight distribution of the sample. In practice, SEC columns of different pore sizes are connected in series to provide a wider molecular weight separation range. This procedure of multiple columns is not very convenient, and unique columns in which the pore size distribution of the stationary phase is broadened by blending together two or more phases are available. Mixed-pore packings obtained by blending together several selected pore size materials can be made such that the column exhibits a linear calibration for analytes in a wide range of *Mw* [10,11].

The particle size of polymer-based packings is another parameter important for SEC column efficiency (height of theoretical plate). The packing materials must be as homogeneous as possible, and this is achieved by having particles of equal size and spherical shape, with uniform flow channels. Polymer particles with a highly mono-disperse particle size distribution can be produced by a two-step microsuspension method. This process is based on using monosized polymer seed particles mixed with the low-molecular-weight material during the polymerization. This process allows the preparation of monosized compact or macroporous particles of predetermined size in the range of 1−100 μm [12]. However, it is common that a separation by size of the synthesized polymer particles is still necessary for obtaining particles with a narrow size distribution. This is achieved by using sedimentation or centrifugation [13]. Information regarding the particle shape and size can be obtained by microscopic methods. Smaller particles offer improved resolution but result in higher operating pressures and can prove more difficult to pack.

Mechanical stability of the polymer is important for establishing the limit of flow rate operation. The maximum operating pressure of the packing should be below the compression point of the packing material. Since in both GPC and GFC it is possible that the mobile phase viscosity can be increased during a gradient separation, particular attention must be given to the potential changes in this parameter. SEC stationary phases may be sensitive to excessive backpressure that can collapse the phase and destroy the column.

The surface chemistry of the packing material is very important and must be selected such that any enthalpic contribution (interactions with the polymer surface) is minimal. For GFC, the packing materials are typically highly hydrophilic and should not possess charges. Supports for aqueous-phase size-exclusion chromatography may be produced by coating the particle surface with a hydrophilic crosslinked polymer. The enthalpic effects are difficult to completely eliminate, and salts (buffer) as well as organic modifiers are typically added to the mobile phase to diminish such interactions.

Chemical and temperature stability of the packing is related to its inertness to specific solvents, and typically the crosslinked polymers are capable of being used with a wide range of solvents. The temperature stability of the packing is also necessary since some polymeric analytes are soluble in the mobile phase only at elevated temperature. Packing materials capable to be used at temperatures in the range 50–120°C are common. Several phases used in GPC are listed in Appendix 6b, and some used in GFC are listed in Appendix 6c.

11.2 CHARACTERIZATION OF SIZE-EXCLUSION PHASES AND COLUMNS

The separation in SEC is influenced by specific characteristics of the packing materials. Several such characteristics of SEC columns/packing materials are described in Table 11.2.1 [14].

Column characterization should be made in association with the purpose of analysis and nature of the analytes [15]. The interaction between the SEC column and the analyte can be very different for different classes of macromolecules, and proteins, polymeric carbohydrates, and synthetic polymers may have very different behavior on the same type of column.

TABLE 11.2.1 Characteristics of SEC Columns/Packing Materials

Attribute	Variable	Importance	Common Values	Range
Porosity	Pore size. Pore volume	Range of MW	50–10^5 Å	50–10^5 Å
Pore size distribution	Range of pore size	Resolution		
Resolution	Column efficiency (N)	Separation efficiency	18,000	7000–25,000
Resolution	Particle size	Separation efficiency	5 μm, 7 μm	4–20 μm
Resolution	Particle shape	Separation efficiency	Spherical	Irregular—sphere
Column pressure	Particle size	Flow rate, analysis time	500 psi	600–2000 psi
Inertness	Column activity	Utility for MW evaluation	Inert	Some inertness
Recovery	Column activity	Recovery of the analyte	95%	Variable
Mechanical stability	Support type	Maximum working pressure	500, 2000 psi	Variable
Chemical stability	Support type. Phase type	Solvents, range of pH	Stable	Variable
Thermal stability	Support type.	Temperature of utilization	Up to 130°C	20–160°C
Load capacity	Amount of sample	Resolution	2–4 mg	2–4 mg

Porosity and Particle Size

Some of the characteristics indicated in Table 11.2.1 are relevant for any chromatographic column, such as resolution, column pressure, and chemical and thermal stability. Other characteristics are more relevant for SEC columns. One of these is the stationary-phase porosity, which is also important for other chromatography-type phases, but in SEC it plays a crucial role in the separation mechanism. Porosity determines the range of molecular weights (*Mw*) that can be covered by a specific column. The resolution of the columns regarding the separation of two analytes of different molecular weights depends on pore size and also on pore size distribution. Columns with a wider range of pore diameters are able to separate a wider range of molecular weights for the analytes, but typically the resolution of these columns is lower, as compared to columns with a narrow pore-size distribution. However, columns with a narrow range of pore sizes (narrow pore-size distribution) can be used only within a narrow range of molecular weights for the analytes.

Particle size diameter of the columns is also important for the resolution of the separation. For high column efficiency, packing materials in SEC should be fine, uniform, and spherical particles (3–20 μm), as in the common interactive HPLC. However, smaller particles (e.g., 3–5 μm), which lead to better resolution, also increase the column backpressure. In SEC, the columns have very frequently a limited value for the backpressure in order to avoid the collapse of the pores, which damages the column. For this reason, when columns with small particles are used, a decrease in the flow rate of the mobile phase is typically recommended (below 1 mL/min, e.g., 0.2 mL/min). The increase in the column length, which also enhances resolution, is associated with an increase in the column backpressure. The use of two or more columns in series for SEC, which is useful for covering the separations of a wider range of *Mw*, is also limited by the increase in the backpressure.

Inertness and Recovery

Two other characteristics are the inertness of the phase and the recovery (which is related to inertness). The inertness of the SEC column is important in particular when estimations of the *Mw* of polymers are done using SEC. When the column is not inert, other interactions besides size exclusion will influence the retention process, and the result for the *Mw* evaluation may be incorrect. From expressions 4.3.57 and 4.3.59, the following formula is obtained for the retention volume:

$$V_R(X) = V_{inter} + K_{SEC}^{pure}(X) K_{SEC}^{interaction}(X) V_{pores} \tag{11.2.1}$$

In expression 11.2.1, only $K_{SEC}^{pure}(X)$ depends on *Mw* of *X*. As shown by expression 4.3.58, $K_{SEC}^{pure}(X)$ depends only on ΔS^0 which depends only on *Mw*. On the other hand, $K_{SEC}^{interaction}(X)$ depends on specific (enthalpic) interactions of compound *X* with the stationary phase. Different compounds having different $K_{SEC}^{interaction}$, the dependence of retention volume V_R will not depend anymore only on *Mw*.

For preparative reasons in particular, the recovery of the analytes is another important characteristic of the SEC columns. It is common that proteins or other biologically active materials are separated (e.g., for purification) on SEC columns. In some instances it may be important that the material injected in the column is recovered and its biological activity is not altered. When the biological activity of the analyte must be preserved, the nature of the mobile phase also plays an important role in preserving biological activity. For such GFC separations, the mobile phase is either water or a dilute buffer.

11.3 SELECTION OF COLUMNS USED IN SEC

Selection Factors for SEC Columns

The main factors to be considered in the selection of the column (or of a set of columns) for SEC separations are (1) the purpose of separation, (2) the nature of the analytes to be separated, and (3) other analysis requirements. The purpose of separation in SEC may include the following: (1a) separation of small molecules from polymeric molecules, (1b) separation of polymeric molecules for analytical purposes, (1c) purification of polymeric molecules for small-scale preparative purposes, (1d) evaluation of molecular weight of polymers. Regarding the nature of the analytes to be separated the following aspects should be considered: (2a) solubility of the analytes in aqueous/polar solvents (for GFC application), (2b) solubility of the analytes in organic nonpolar solvents (for GPC applications), (2c) range of molecular weights expected for the analytes and matrix. Regarding other analysis requirements, these can be related to the following: (3a) general column properties, (3b) analysis time, (3c) use of specific equipment such as specific detectors. Some details about these factors are further discussed.

1a) Separation of small molecules from large molecules is easily achieved in SEC. For this reason, the selection of a column to separate large molecules from small molecules should not be a difficult task. The column selection must consider other aspects such as the nature of the polymers, which indicates if GFC or GPC is going to be utilized. However, one important parameter to consider when separating large molecules from small molecules is the lack of enthalpic interactions of the macromolecular components with the stationary-phase material (see Section 4.3). When strong interactions take place between the macromolecules and the stationary phase (interactions with enthalpic contribution are present), these macromolecules will be retained longer in the column, and they may elute in the region of small molecules (having larger V_R than expected).

1b) Separation of polymeric materials for analytical purposes is a common task for SEC analyses. In this type of separation the selection of the SEC column also must consider the use of GFC or GPC depending on sample nature. For many natural polymers such as polymeric carbohydrates, proteins, and polyethylene glycols, GFC is usually utilized [16–18]. For many synthetic polymers, the solubility in organic nonpolar solvents makes GPC more appropriate for separation [19]. For successful separations, it is very useful to have some information about the Mw range of the components of the sample that must be separated. For the separation of mixtures of molecules within a wide range of Mw values, more than a single SEC column can be used in a series. In such cases, the column with the highest porosity is used first. For example, Phenogel columns cover a wide range of porosities (between 10^2 and 10^6) and can be used up to four columns in series for achieving some separations. Since high backpressure can damage gel porosities, the maximum acceptable backpressure (up to 1500 psi) of the column should never be exceeded. A reduced flow rate of the mobile phase is sometimes necessary for maintaining the correct backpressure, especially when more than

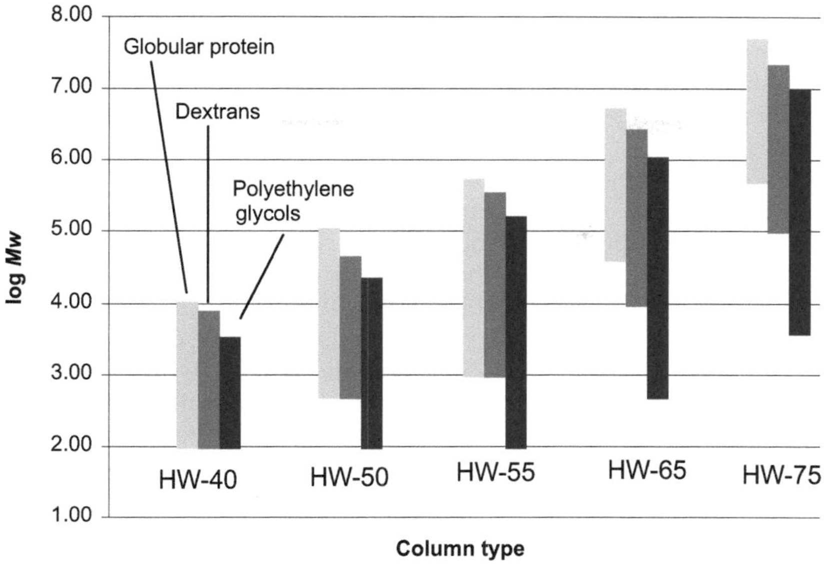

FIGURE 11.3.1 Types of Toyopearl columns recommended for use with polymers of different *Mw* and different chemical natures [21].

one column in series is utilized for the separation. Recommendations from column manufacturers indicate the solvent compatibility and the *Mw* range where the column is effective [20]. The range also depends on the molecular structure of the polymeric molecule. For example, the recommended use of Toyopearl columns (Tosoh Bioscience) [21] for different types of analytes and for different ranges of *Mw* is illustrated in Fig. 11.3.1. As indicated in Fig. 11.3.1, dextrans, globular proteins, and polyethylene glycols, having different molecular structures, are not covered in the same *Mw* range for a specific type of column. A variety of column formats are available for specific applications [22]. Particular advantages and limitations as well as typical applications for each type of column are reported by the suppliers.

1c) Some SEC separations are performed with the purpose of further utilization of the separated macromolecules such as proteins. In such cases, lack of denaturation and as complete as possible recovery from the HPLC column are very important [23]. Particular attention must be given to the nature of solvents, pH, and temperature of the mobile phase.

1d) When the purpose of SEC separation is the *Mw* determination of a polymer or polymer mixture, the selection of the column type is very important. The linearity of the correlation between retention time t_R (retention volume V_R) and molecular weight (in fact molecular hydrodynamic volume) strongly depends on the nature of the polymer and the selected column. A particular problem in the use of SEC for the measurement of *Mw* is related to the dimensions of the pores and pore size distribution of the phase. As previously indicated (see Section 11.2) the columns with a narrow pore-size distribution offer better resolution and allow a more precise measurement of *Mw* of analytes, but they can be applied only in a narrow range of *Mw* values. In order to achieve both higher resolution and separation for a wider range of *Mw*, SEC can be conducted using two to four columns of large dimensions (7.8 mm inner diameter [i.d.] × 300 mm) connected in series. Similar to separation for polymer analysis, the range of utilization of SEC columns is also related to the nature of the analytes that are separated. The measurement of *Mw* is typically performed using a calibration with standards of known *Mw* (standards are available from different suppliers (e.g., Ref. [24])). For relevant results, the calibration must be performed using standards with a similar molecular structure as the analytes.

2a) Solubility of the polymeric samples can be a challenging task, in particular because the sample concentration in SEC is typically higher than in other HPLC techniques (can be in the range of 1−3%). Some polymers are soluble only in special solvents (e.g., dissolution of cellulose may require ionic solvents [25]) or solvents such as water or diethylformamide but with additives such as LiOH and urea [26], and dissolution may affect the polymerization degree (and therefore the *Mw*) of the polymer. The selection of GFC or GPC is related to the polymer nature and the nature of the sample solvent.

2b) The use of GPC is common for many synthetic polymers with hydrophobic molecules. For these polymers, solubilization may also be a problem and some polymers are soluble only at temperatures of 110−130°C. For the analysis of such polymers, the columns (and the whole HPLC system) must be kept at elevated temperatures. Specific columns are recommended for GPC separations and maximum allowed temperatures are usually specified by the manufacturer (see Appendix 6).

3a) Some of the column general properties were already indicated, such as column efficiency, maximum column backpressure (limited by mechanical resilience of the phase), stability in specific solvents, and pH stability range (particularly for silica-based columns). Also, column loading capacity must be considered in SEC, since in SEC practice it is common to use rather concentrated samples for injection. The selection of the column for SEC separation must consider all those elements for a successful separation.

3b) The use of large column bunks (two to four columns in the order of decreasing pore size) achieves high resolution and accurate *Mw* measurement. However, it requires long analysis time and significant solvent consumption. Since there is increasing interest in developing high-speed SEC using a single column of small dimensions, special columns with higher phase resilience have been made commercially available (see Appendix 6). Help in reducing the run time was also obtained from the use of mass spectrometers as detectors in SEC. The higher sensitivity of the mass spectrometers allowed the use of shorter columns with lower loading capacity and better resolution where the use of lower flow rates overcame the backpressure problem [4].

Utilization of UPLC in SEC is another attractive alternative to solve the problem of large solvent volumes and long analysis time. However, the implementation of UPLC for the analysis of macromolecules poses several challenges, including the following: (1) development of packing materials with large pore diameters and pore volumes that are mechanically stable at ultrahigh pressure, (b) avoidance of high shear stress generated by the flow of polymer solutions, which may affect the conformation of the polymer chains, and (c) assurance of proper diffusion of the polymers in the pore of the stationary phase during the separation. Effort is continually made in solving such problems [5], and several phases recommended to be used in UPLC mode are commercially available (e.g., TSKgel UP-SW3000 from Tosoh Bioscience). Also, the possibility to perform fast and efficient two-dimensional analysis of polymers using ultrahigh-pressure liquid chromatography (UHPLC × UHPLC) was proved for the separations of some copolymers based on their chemical composition [27].

New Developments

New developments in SEC column technology included the use of core—shell phases and small particle size ones (sub 2-μm) [3]. A few newly commercially available SEC columns are listed in Appendix 6d [28]. The long separation times typical for SEC separation are significantly shortened on newer columns. Also, the decrease in particle size leads to an increase in column efficiency with better separations of biopolymers and more accurate results regarding evaluation of *Mw*.

References

[1] S. Mori, H.G. Barth, Size Exclusion Chromatography, Springer Verlag, Berlin, 1999.
[2] C.-S. Wu (Ed.), Handbook of Size Exclusion Chromatography, M. Dekker, New York, 1995.
[3] E.S.P. Bouvier, S.M. Koza, Advances in size-exclusion separations of proteins and polymers by UHPLC, Trends Anal. Chem. (TrAC) 63 (2014) 85—94.
[4] Y. Han, S.S. Lee, J.Y. Ying, Spherical siliceous mesocellular foam particles for high-speed size exclusion chromatography, Stud. Surf. Sci. Catal. 165 (2007) 829—832.
[5] E. Uliyanchenko, P.J. Schoenmakers, S. van der Wal, Fast and efficient size-based separations of polymers using ultra-high-pressure liquid chromatography, J. Chromatogr. A 1218 (2011) 1509—1518.
[6] P. Hong, S. Koza, E.S.P. Bouvier, Size-exclusion chromatography for the analysis of protein biotherapeutics and their aggregates, J. Liq. Chromatogr. Relat. Technol. 35 (2012) 2923—2950.
[7] P.L. Dubin, M.M. Tacklenburg, Size exclusion chromatography of strong polyelectrolytes on porous glass columns, Anal. Chem. 57 (1985) 275—279.
[8] S. Mori, M. Kato, High performance aqueous size-exclusion chromatography with diol-bonded porous glass packing material, J. Chromatogr. A 363 (1986) 217—222.
[9] C.W. Huck, G.K. Bonn, Poly(styrene-divinylbenzene) based media for liquid chromatography, Chem. Eng. Technol. 28 (2005) 1457—1472.
[10] L.K. Kostanski, D.M. Keller, A.E. Hamielec, Size-exclusion chromatography — a review of calibration methodologies, J. Biochem. Biophys. Methods 58 (2004) 159—186.
[11] B. Trathnigg, Size-exclusion chromatography of polymers, in: R.A. Meyers (Ed.), Encyclopedia of Analytical Chemistry, John Wiley & Sons Ltd, Chichester, 2000, pp. 8008—8034.
[12] T. Ellingsen, O. Aune, J. Ugelstad, S. Hagen, Monosized stationary phases for chromatography, J. Chromatogr. 535 (1990) 147—161.
[13] C.M. Cheng, F.J. Micale, J.W. Vanderhoff, M.S. El-Aasser, Pore structural studies of monodisperse porous polymer particles, J. Colloid Interface Sci. 150 (1992) 549—558.
[14] H.G. Barth, B.E. Boyes, C. Jackson, Size exclusion chromatography and related techniques, Anal. Chem. 70 (1998) 251R—278R.
[15] R.E. Majors, L.L. Loyd, Characterizing SEC columns for the investigation of higher-order monoclonal antibody aggregates, LC/GC North Am. 34 s4 (2016) 28—36.
[16] M. Gaborieau, P. Castignolles, Size-exclusion chromatography (SEC) of branched polymers and polysaccharides, Anal. Bioanal. Chem. 399 (2011) 1413—1423.
[17] E. Folta-Stogniew, K. Williams, Determination of molecular masses of proteins in solution: implementation of an HPLC size exclusion chromatography and laser light scattering service in a core laboratory, J. Biomol. Tech. 10 (1999) 51—63.
[18] M. Plante, T. Villasenor, C. Crafts, B. Bailey, I. Acworth, D. Clark, Analysis of Various Polyethylene Glycol (PEG) Species Including Simultaneous Free PEG and PEGylated Proteins with HPLC and Charged Aerosol Detection, Dionex, http://www.dionex.com/en-us/webdocs/88880-PO-EAS-Analysis-Various-PEG-Species-2Nov2010-LPN2672-01.pdf.
[19] D. Berek, Size exclusion chromatography-a blessing and a curse of science and technology of synthetic polymers, J. Sep. Sci. 33 (2010) 315—335.
[20] https://www.phenomenex.com/Products/HPLCDetail/phenogel.

[21] http://wolfson.huji.ac.il/purification/PDF/Gel_Filtration/TOSOH_GelFiltration.pdf.
[22] J.D. Caldwell, B.S. Cooke, M.K. Greer, High performance liquid chromatography-size exclusion chromatography for rapid analysis of total polar compounds in used frying oils, J. Am. Oil Chem. Soc. 88 (2011) 1669–1674.
[23] N. Aust, Application of size-exclusion chromatography to polymers of ultra-high molar mass, J. Biochem. Biophys. Meth. 56 (2003) 323–334.
[24] http://www.sigmaaldrich.com/materials-science/material-science-products.html?TablePage=9545551.
[25] S. Zhu, Y. Wu, Q. Chen, Z. Yu, C. Wang, S. Jin, Y. Ding, Gang Wu, Dissolution of cellulose with ionic liquids and its application: a mini-review, Green Chem. 8 (2006) 325–327.
[26] J. Cai, L. Zhang, Rapid dissolution of cellulose in LiOH/urea and NaOH/urea aqueous solutions, Macromol. Biosci. 5 (2005) 539–548.
[27] E. Uliyanchenko, P.J.C.H. Cools, S. van der Wal, P.J. Schoenmakers, Comprehensive two-dimensional ultrahigh-pressure liquid chromatography for separations of polymers, Anal. Chem. 84 (2012) 7802–7809.
[28] D.S. Bell, New chromatography columns and accessories for 2016, LC/GC North Am. 34 (4) (2016) 242–252.

CHAPTER

12

Stationary Phases and Columns for Immunoaffinity Type Separations

12.1 TYPES OF PHASES AND THEIR PREPARATION

Separation using immunoaffinity liquid chromatography (IAC) is a common technique for many analyses of biological samples. The technique can be applied in both low-pressure [1] and high-performance chromatography (HPLC) modes [2]. Because of the importance of this technique in medical and biological applications, a significant number of publications are available on this subject. Details about immunoaffinity chromatography are given in a number of publications (see, e.g., Refs. [3,4]). Because the applications of IAC are very diverse, and each application has its own particularities, it is not possible to provide general recommendations for the selection of a type of column to be used in IAC. In this chapter only a short summary regarding this technique is given.

Supports for Stationary Phases in Immunoaffinity Chromatography

Similar to other chromatographic stationary phases, the immunoaffinity phases have a support and an active component. For being used in HPLC, the support for IAC should be chemically and physically stable, with a good mechanical strength, and should have minimal nonspecific adsorption capability. Because IAC is used for large-molecule separation, the particle size and pore-size distribution of the support are very important parameters for assuring the access of the analyte to the active phase. The pore size for IAC is typically large, between 300 and 500 Å. The immobilization of ligands of interest should be such that high ligand accessibility is achieved while the phase properties are not affected. A wide variety of materials, including organic and inorganic polymers, have been used for the design of these solid-phase supports. Among the first materials used as support in IAC were natural polysaccharides such as agarose (agarose is a linear polymer made up of repeating units of agarobiose and it is one of the components of the natural polysaccharide agar), crosslinked agarose (sepharose), cellulose, and crosslinked dextran (sephacryl). Gels made from purified agarose have a relatively large pore size, making them useful for separation of large molecules. These materials are stable over a wide interval of pH (3−13) and are characterized

Selection of the HPLC Method in Chemical Analysis
http://dx.doi.org/10.1016/B978-0-12-803684-6.00012-3

by a high content of hydroxyl groups available for activation and derivatization. Their hydrophilic surface generally does not interact with proteins and exhibits only a low nonspecific adsorption. A major drawback of these materials is their poor mechanical strength and their swelling propensity in aqueous solvents. Other supports used in IAC for the stationary phase include silica and synthetic organic polymers such as methacrylates, polyacrylamide, and copolymers of polyacrylamide with other polyvinyl polymer. Although more resistant to pressure than polysaccharides, the organic supports still exhibit a lower pressure tolerance in comparison to inorganic materials such as silica. In addition, some of the synthetic polymers show swelling differences in the presence of organic solvents, a broader pore-size distribution, decreased ligand binding efficiency, and nonspecific interaction due to their hydrophobic character compared to the phases based on natural polysaccharides. The stationary phases can also be porous, nonporous [5], or monolithic [6].

Monolithic supports based on glycidyl methacrylate-co-ethylene dimethacrylate (GMA-EDMA) have been shown to have some advantages, in particular related to the lack of swelling and to the lower backpressure that eliminates the need of high mechanical resilience. Such columns are commercially available from BIA Separations (Slovenia) under the trade name of CIM disk monolithic columns. The main advantage of the GMA-EDMA support is the significant concentration of chemically reactive epoxy groups, which can be used for immobilization of ligands, e.g., containing amino groups [7].

Silica is also used as support in IAC mainly because of its excellent stability under pressure and because it can be easily derivatized in order to introduce different functional groups. However, silica stability is not good in alkaline pH values and dissolves significantly above pH of 8. In addition, nonspecific interactions can occur between silanol groups and the basic parts of the biomolecules. Surface modification of silica can be performed either by chemical modification or physical adsorption of ligands in order to minimize this nonspecific adsorption and to introduce a high density of functional groups [8].

The Active Phase in Immunoaffinity Chromatography

The active phase in immunoaffinity can be very diverse, depending on the nature of the material immobilized on the stationary phase. One important group of active phases is that of antibodies or antibody-related moieties bonded to the solid support [9,10]. Stationary phases containing an immobilized antibody, enzyme, or other compounds such as heparin [11], lectins, or nucleotides were successfully utilized in various chromatographic procedures, some as low-pressure liquid chromatography and others in HPLC mode.

The two main types of antibodies that are used in IAC are polyclonal antibodies and monoclonal antibodies. Polyclonal antibodies are produced from multiple cell lines within a living body and they can bind a variety of antigens and with a range of binding strengths (the binding takes place with the part of the antigen recognized by the immune system, known as epitope; the nature of antibodies—antigen interactions is very complex and the term binding is preferred instead of bonding that would designate the formation of a definite type of bond). Because monoclonal antibodies are generated from a single cell line, they bind to a single epitope with identical binding affinities. Recent advances have led to the production of artificial antibodies that have high binding affinities [12]. Before the antibodies can be used in IAC, some further purifications are often required. This purification may involve the use of ion-exchange chromatography [13], precipitation with ammonium or dextran sulfate, or isolation on a low-pressure liquid chromatography (LC) immunoaffinity column [14].

Antibodies can be immobilized onto supports by using a variety of techniques that range from covalent attachment to adsorption-based methods. Of these techniques, those that make use of covalent attachment are the most common, but even these methods can range from the use of random attachment via amino or carboxyl groups to more site-selective immobilization approaches that make use of modified carbohydrate residues or thiol groups. The ideal situation in any of these immobilization methods is to have antibodies attached to the support in a way that does not affect the activity of the binding sites or the accessibility to these sites of the target compounds [15]. The immobilization can be done through free amine groups by using supports that have been activated with agents such as *N,N′*-carbonyldiimidazole, cyanogen bromide, *N*-hydroxysuccinimide, or tresyl chloride/tosyl chloride. Antibodies can also be immobilized through amine groups using a support that has been treated to produce reactive epoxy or aldehyde groups on its surface [16]. Antibodies can be covalently linked to IAC supports through more site-selective methods. This can be achieved by utilizing the free sulfhydryl groups of the protein. These groups can be used for immobilization by using techniques such as the divinylsulfone, epoxy, iodoacetyl/bromoacetyl, maleimide, or tresyl chloride/tosyl chloride methods. Another procedure for the immobilization of antibodies is using antibodies that have been reacted with biotin and then are adsorbed to a support that contains immobilized avidin or streptavidin. The most popular biotinylation technique is to incubate antibodies with *N*-hydroxysuccinimide-*D*-biotin at pH 9. These linkages can resist many types of elution conditions without dissociation.

For silica-based stationary phases, a common material utilized as a starting material is silica pretreated with an aminosilane (e.g., γ-aminopropyltriethoxysilane, γ-aminopropylmethyldiethoxysilane, γ-aminopropyldimethylethoxysilane). The reaction of the propylamino silica surface with glutaraldehyde generates a reactive material on which an amino group of a protein (e.g., with lysine units in its structure) can be readily attached. The reactions are schematically shown in Fig. 12.1.1. Other similar procedures use succinic anhydride instead of glutaraldehyde as an intermediate step of derivatization. Some proteins that contain phenolic groups (e.g., with tyrosine units) can be bonded to the aminosilica using azo coupling reactions.

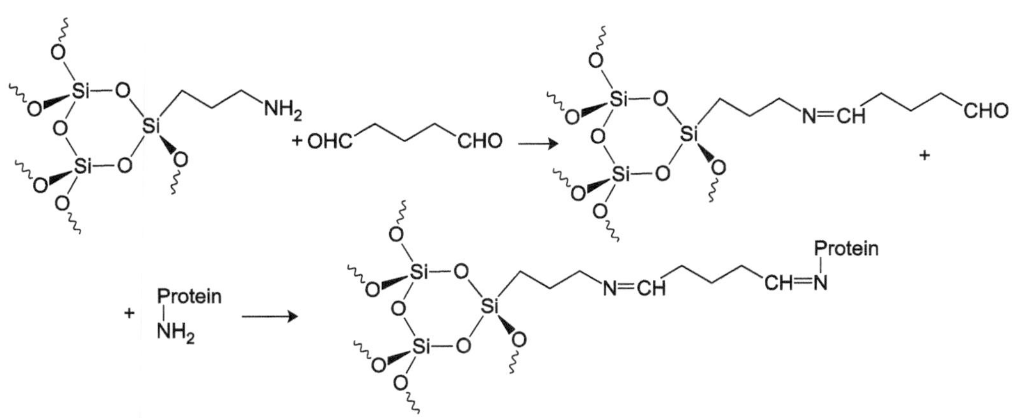

FIGURE 12.1.1 Schematic reaction showing the binding of a protein to amino silica.

For an organic support, the linking process is usually done using an activating reagent such as cyanogen bromide when the molecule to bond contains a free primary amine, sulfhydryl, or hydroxyl groups for attachment. On the activated support, proteins (or other molecules) with specific binding capability are further immobilized. These can be selected by the user or can be general-purpose immobilized compounds. Immobilized heparin, for example, acts with a specific binding site to retain certain proteins, lectin resins can be used for the separation of glycoproteins from other glycoconjugate molecules, and nucleotide resins are used for the separation of specific proteins. Heparin resins, lectin resins, and others are commercially available. Specific immunoproteins also can be bonded, for example, on agarose activated with cyanogen bromide or with other activation reagents such as 6-aminohexanoic acid, carbonyldiimidazole, and thiol [16]. These types of materials have a very high specificity for the specific antigen that generates the immunoprotein. Affinity resins containing immobilized sugars and sugar derivatives and resins with immobilized biotin or avidin are also available. Several activation reagents for agarose and crosslinked dextrans are indicated in Table 12.1.1. Other resins are used having immobilized ligands such as *p*-amino benzamidine (ABA-5PW), *m*-aminophenyl boronic acid (Boronate-5PW), and iminodiacetic acid (Chelate 5 PW), all from Sigma, and TSKgel Chelate-5PW from Tosoh Bioscience LLC. Numerous other immobilization procedures are reported in the literature [17–19].

TABLE 12.1.1 Reagents Used to Make Activated Resins Able to Bind Proteins

Activating Reagent	Linkage to Resin by	Reactive Group in the Ligand	Specificity of the Group to	Reaction Conditions	Bond Type to Ligand	Stability
6-Aminohexanoic acid	Isourea	Carboxyl	Amine, with carbodiimide coupler	pH 4.5–6.0	Amide	Good
6-Aminohexanoic acid N-hydroxy-succinimide ester	Isourea	Succinimidyl ester	Amine	pH 6.0–8.0	Amide	Good
Carbonyldiimidazole	Carbamate	Imidazolyl carbamate	Amine	pH 8.0–10.0	Carbamate	Good below pH 10
Cyanogen bromide	Ester	Cyanate	Amine	pH 8.0–9.5	Isourea	Moderate
Epoxy	Ether	Epoxy	SH > NH	pH 7–8 SH pH 8–11 NH_2	SH: Thioether, NH_2 amino ether	Very good
N-hydroxy-succinimide ester	Isourea	Succinimidyl ester	Amine	pH 6.0–8.0	Amide	Good
Periodate	Oxidizes agarose, saccharides	Aldehyde	Amine	pH 4.0–10.0	Reductive amination with $NaBH_3(CN)$	Very good
Thiol	Isourea	Disulfide	Sulfhydryl	pH 6.0–8.0	Disulfide	Good in nonreducing conditions

The immunoaffinity-type phases have excellent selectivity and work well in aqueous solutions, but each material must be developed for a specific analyte, the phases can be unstable toward organic solvents, and may be stable only in a narrow pH range. Some bioaffinity phases are used for HPLC separations [19–21], while others must be used at lower pressures (1–2 bar) since some materials impose considerable limitations regarding the maximum pressure that can be applied to the phase.

Besides bioaffinity (immunoaffinity) phases, affinity chromatography can be also used for small-molecule analysis, such as separation of amino acids on a column having ethylenediaminotetraacetic acid moiety bonded on a solid material, in the form of a transitional metal chelate capable of forming further chelates with the amino acids in solution. Such phases can be used, for example, in chiral separations (see Chapter 10).

References

[1] D.S. Hage, D.H. Thomas, M.S. Beck, Theory of a sequential addition competitive binding immunoassay based on high-performance immunoaffinity chromatography, Anal. Chem. 65 (1993) 1622–1630.

[2] Y.D. Clonis, Affinity chromatography matures as bioinformatic and combinatorial tools develop, J. Chromatogr. A 1101 (2006) 1–24.

[3] Affinity Chromatography, Principles and Methods, Amersham Pharmacia, Uppsala, 1981.

[4] A.C. Moser, D.S. Hage, Immunoaffinity chromatography: an introduction to applications and recent developments, Bioanalysis 2 (2010) 769–790.

[5] C.-M. Bo, B.-L. Gong, W.-Z. Hu, Preparation of immobilized metal affinity chromatographic packings based on monodisperse hydrophilic non-porous beads and their application, Chin. J. Chem. 26 (2008) 886–892.

[6] J. Sproβ, A. Sinz, Monolithic media for applications in affinity chromatography, J. Sep. Sci. 34 (2011) 1–16.

[7] T. Tennikova, A. Strancar, Short High-throughput Monolithic Layers for Bioaffinity Processing, LabPlus International, February/March 2002, pp. 1–3.

[8] A.E. Ivanov, L.V. Kozlov, B.B. Shojbonov, V.P. Zubov, V.K. Antonov, Inorganic supports coated with N-substituted polyacrylamides: application to biospecific chromatography of proteins, Biomed. Chromatogr. 5 (1991) 90–93.

[9] D.S. Hage, T.M. Phillips, Immunoaffinity chromatography, in: D.D. Hage (Ed.), Handbook of Affinity Chromatography, Taylor & Francis, New York, 2006.

[10] M.M. Rhemrev-Boom, M. Yates, M. Rudolph, M. Raedts, Immunoaffinity chromatography: a versatile tool for fast and selective purification, concentration, isolation and analysis, J. Pharm. Biomed. Anal. 24 (2001) 825–833.

[11] H. Sasaki, A. Hayashi, H. Kitagaki-Ogawa, I. Matsumoto, N. Seno, Improved method for the immobilization of heparin, J. Chromatogr. 400 (1987) 123–132.

[12] B.A.R. Williams, C.W. Diehnelt, P. Belcher, M. Greving, N.W. Woodbury, S.A. Johnston, J.C. Chaput, Creating protein affinity reagents by combining peptide ligands on synthetic DNA scaffolds, J. Am. Chem. Soc. 131 (2009) 17233–17241.

[13] M.D.P. Boyle, J.J. Langone, A simple procedure to use whole serum as a source of either IgG-or IgM-specific antibody, J. Immunol. Meth. 32 (1980) 51–58.

[14] M. Eliasson, R. Andersson, A. Olsson, H. Wigzell, M. Uhlen, Differential IgG-binding characteristics of staphylococcal protein A, streptococcal protein G, and a chimeric protein AG, J. Immunol. 142 (1989) 575–581.

[15] H. Kim, D.S. Hage, Immobilization methods for affinity chromatography, in: D.S. Hage (Ed.), Handbook of Affinity Chromatography, Taylor & Francis, New York, 2006.

[16] G.T. Hermanson, A.K. Mallia, P.K. Smith, Immobilized Affinity Ligand Techniques, Academic Press, Ney York, 1992.

[17] D.J. O'Shannessy, M. Wilchek, Immobilization of glycoconjugates by their oligosaccharides: use of hydrazido-derivatized matrices, Anal. Biochem. 191 (1990) 1–8.

[18] P.L. Domen, J.R. Nevens, A.K. Mallia, G.T. Hermanson, D.C. Klenk, Site-directed immobilization of proteins, J. Chromatogr. 510 (1990) 293–302.

[19] N. Jonker, J. Kool, H. Irth, W.M.A. Niessen, Recent developments in protein-ligand affinity mass spectrometry, Anal. Bioanal. Chem. 399 (2011) 2669–2681.
[20] D.J. Winzor, Determination of binding constants by affinity chromatography, J. Chromatogr. A 1037 (2004) 351–367.
[21] K.K.R. Tetala, T.A. van Beek, Bioaffinity chromatography on monolithic supports, J. Sep. Sci. 33 (2010) 422–438.

CHAPTER

13

Solvents, Buffers, and Additives Used in the Mobile Phase

13.1 CHARACTERIZATION OF LIQUIDS AS SOLVENTS

The mobile phase plays a role as important as the stationary phase in HPLC separations. Before separation, the mobile phase and stationary phase must be at equilibrium (the column is conditioned with the mobile phase). In the development of a new analytical method or in its adaptation from the literature, it is much simpler to modify the mobile phase than to replace a chromatographic column, and for this reason, it is very useful to understand the role of the mobile phase and how its changes can affect a separation. The mobile phase consists of one or a mixture of liquid solvents and also may contain various additives. The additives contribute to the modification of mobile phase pH, ionic strength, or other properties [e.g., elution characteristics in ion pairing chromatography (IPC) when an ion pairing reagent is added, or chiral selectivity when a chiral agent is added]. The mobile phase composition may be kept constant for the whole chromatographic run (in isocratic HPLC separations) or can be changed (in gradient type separations). Gradient HPLC is further discussed in Chapter 14.

A variety of solvents are used in HPLC, although three solvents are the most common: water, methanol, and acetonitrile. These solvents are used extensively in RP-HPLC and also in HILIC, which are the most common HPLC techniques, and in gel filtration (GFC). The utilization of other solvents in HPLC, although less frequent, may provide advantages which are sometimes necessary for specific separations. In particular the normal-phase (NP) separations and chiral separations that are practiced in NP mode may use a variety of solvents. Also, GPC may use various solvents for the separation of polymers.

The properties of liquids are also important as dissolving agents for the sample, since in HPLC the most common procedure for placing the analytes in the mobile phase is by injecting them in the form of a solution. The properties of solvents are presented in numerous publications [1–5], and in this section only some of these properties related to chromatographic separations are discussed.

Selection of the HPLC Method in Chemical Analysis
http://dx.doi.org/10.1016/B978-0-12-803684-6.00013-5

Miscibility of Solvents

Of particular interest for HPLC is the miscibility (solubility in any proportions) of one solvent in another. When using gradient HPLC, the pumps are delivering solvents in various proportions, and these solvents must be perfectly soluble one in another in the range of utilized composition. Solvents that can separate from each other when mixed should not be used as mobile phase. Precipitation of solids when the mobile phase components are mixed must always be avoided in selecting the mobile phase. The use of buffers is discussed in Section 13.4. For various solvents, the miscibility is typically based on experimental results. A chart showing various solvents' miscibility is given in Fig. 13.1.1. Quantitative data for solubility are also available in the literature. The solubilities in water of several solvents used in HPLC are given in Appendix 7b.

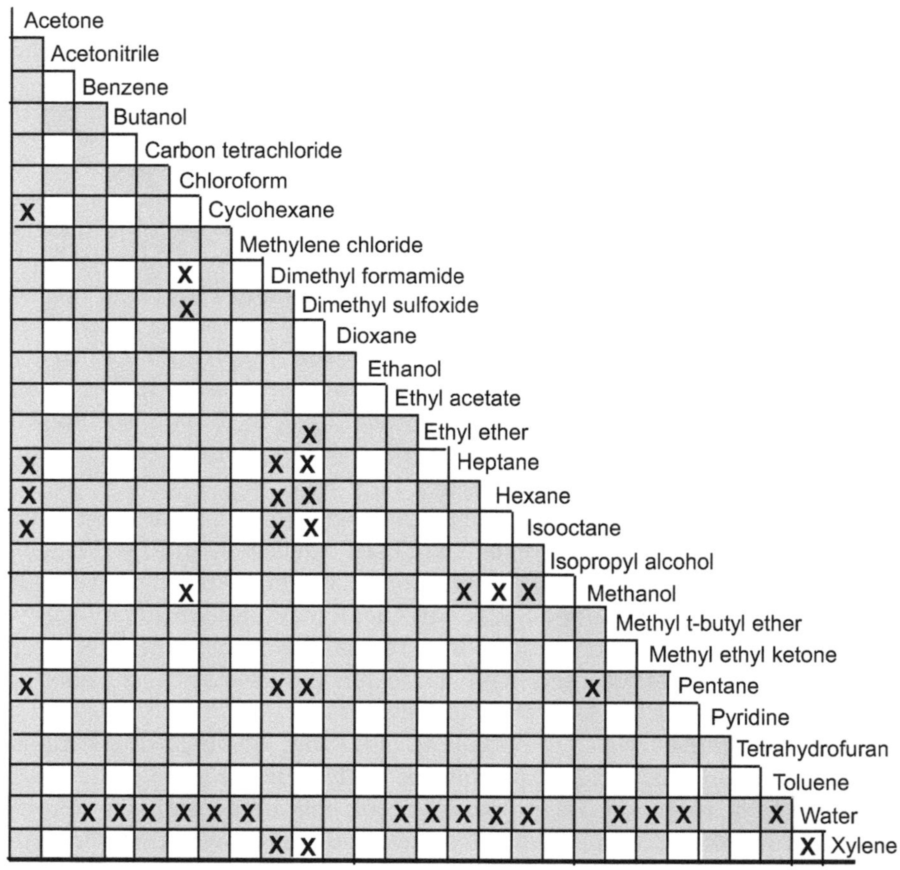

FIGURE 13.1.1 Miscibility chart of various solvents used in HPLC. **X** indicates solvents that are not miscible (gray or white squares do not signify differences in miscibility).

Characterization of Solvents with Hildebrand Solubility Parameter

Among the parameters used for solvent characterization, Hildebrand solubility parameter δ (see Section 5.1) is very useful for solvent characterization related to HPLC. This parameter is related to a thermodynamic property of the solvent, and it is determined from the vaporization energy and molar volume by expression (5.1.20) $\left(\delta_X^2 = \Delta E_{vap,X}/V_X\right)$. Vaporization energy $\Delta E_{vap,X}$ of a compound is determined by the internal molecular interactions, which can be of several types. For nonelectrolytes, such interactions result from van der Waals forces (forces between molecules that are not caused by ionic charges). More than one type of van der Waals forces act between molecules including dispersion forces, polar interactions, and hydrogen bonding interactions (see, e.g., Ref. [6]). As a result, δ is composed from different contributions. These partial δ values are indicated as δ_d (for dispersion van der Waals forces), δ_p (for polar van der Waals forces), δ_a (for proton acceptor hydrogen bonding), and δ_h (for proton donor hydrogen bonding). Values for δ and partial δ describing different types of interaction were reported in the literature and some of these values are listed in Appendix 7b (together with the compound molar volume). Because partial δ are obtained from extrathermodynamic considerations their sum is not equal to δ which is obtained based on thermodynamic considerations [3].

One application of Hildebrand solubility parameter δ is the evaluation of solubility of a compound in a specific solvent. For this purpose, formula (5.1.15) for solubility as discussed in Section 5.1 can be used. Based on formula (5.1.26) for ΔH_{mix}, the expression for solubility x_X is given by the expression:

$$\ln x_X = -\frac{V_X(\delta_X - \delta_S)^2}{RT} - \frac{T_{fus}\Delta S_{fus}}{R}\left(\frac{1}{T} - \frac{1}{T_{fus}}\right) \tag{13.1.1}$$

where the approximation $\Delta S_{fus} \approx 13$ cal/mol deg can be made. The limited availability of δ values for many analytes is a drawback regarding the use of expression (13.1.1). Otherwise, the formula can be successfully used for evaluating the solubility of nonelectrolytes.

The Hildebrand solubility parameter also offers a criterion for describing the *elution strength* of a solvent (see Section 4.3). For describing the elution process based on the value of δ, the assumption should be made that the separation is based on a pure partition process. This assumption is valid for a reversed-phase separation in which only hydrophobic interactions take place. However, since the theory of liquid–liquid partition is in many cases only an approximation of the real chromatographic processes, the use of parameter δ for the characterization of mobile phase and of the stationary phase (viewed as an immobilized liquid) must be done with caution.

A partition equilibrium of X between mobile and stationary phase is typically indicated as follows:

$$X_{mo} \leftrightarrow X_{st} \tag{13.1.2}$$

The equilibrium constant $K(X)$ for equilibrium (13.1.2) is given by expression (4.3.5), also written below:

$$K(X) = \frac{\gamma_{X,mo}}{\gamma_{X,st}}\exp\left(-\frac{\mu^0_{X,st} - \mu^0_{X,mo}}{RT}\right) \tag{13.1.3}$$

Since the standard chemical potentials $\mu^0_{X,st}$ and $\mu^0_{X,mo}$ are independent of solution concentration, the values for these potentials must be selected. One choice for the standard state is that of the pure liquid X at the working temperature and pressure. In this case, $\mu^0_{X,st} = \mu^0_{X,mo} = \mu_X^{\text{Liquid}}$. The other choice is that of the pure solid $\mu^0_{X,st} = \mu^0_{X,mo} = \mu_X^{\text{Solid}}$. For both these choices, the result is $\Delta\mu^0_X = 0$. Another convention is to choose the standard state to be that of the pure liquid only for a solvent, with all the other components having a "fictitious" state with the properties that pure X would have if its limiting low concentration properties in solution were to be retained in a pure substance. Selecting the choice of standard state $\mu^0_{X,st} = \mu^0_{X,mo}$, the expression of $K(X)$ is reduced to the following form:

$$K(X) = \frac{\gamma_{X,mo}}{\gamma_{X,st}} \tag{13.1.4}$$

The activity coefficients γ can be obtained based on formula (4.4.12), which expresses activity as a function of enthalpy of mixing of compound X in a dissolution phase. The enthalpy of mixing is given by formula (5.1.26) and from these two expressions the following formula can be obtained for $K(X)$:

$$K(X) = \exp\left\{V_X\left[(\delta_X - \delta_{mo})^2 - (\delta_X - \delta_{st})^2\right]\Big/RT\right\} \tag{13.1.5}$$

Expression (13.1.5) can be easily rearranged in the form:

$$K(X) = \exp\{[(\delta_{st} - \delta_{mo})V_X(2\delta_X - \delta_{st} - \delta_{mo}))]/RT\} \tag{13.1.6}$$

For the evaluation of the equilibrium constant (or of retention factor $k' = K\Psi$ when phase ratio Ψ is known) is it necessary to know the values of δ for the stationary phase, mobile phase, and analyte, which are seldom available. Also, even assuming that a pure liquid is adsorbed on a solid material and acts as stationary phase, the direct use of the values δ for this liquid in pure form may not correspond to the one of the liquid adsorbed on a solid surface. For this reason a true calculation is not usually feasible. Nevertheless, useful considerations can be made based on formula (13.1.6). With the assumption that the stationary phase is very nonpolar and the mobile phase (eluent) is polar, for RP-HPLC it can be concluded that $\delta_{st} < \delta_{mo}$ (e.g., the values in Appendix 7b show that nonpolar compounds have lower δ than polar ones), and the term $\delta_{st} - \delta_{mo}$ is always negative. Also, in order to be retained on a hydrophobic column, the solute must be rather hydrophobic and its δ_X value should be lower than δ_{mo} and closer to δ_{st}. As a result, $(2\delta_X - \delta_{st} - \delta_{mo})$ is also negative. Eq. (13.1.6) shows that $K(X)$ for a solute is larger when the difference $(\delta_{st} - \delta_{mo})$ is larger, in other words, when the solvent used as eluent has a large δ_{mo} value. This can be easily exemplified by considering a hydrophobic compound retained on a C18 column, which is not eluted by water. As δ_{mo} of the eluent decreases, $K(X)$ decreases. When $2\delta_X > \delta_{st} + \delta_{mo}$, the equilibrium is rapidly displaced toward having the analyte in the mobile phase (eluent). This process can be visualized in Fig. 13.1.2 with a hypothetical example of two analytes, X and Y, one with $\delta_X = 8.0$ (cal/cm^3)$^{1/2}$, the other with $\delta_Y = 8.5$ (cal/cm^3)$^{1/2}$, and with $V_X = V_Y = 100$ cm^3/mol. The analytes are supposed to be retained on a stationary phase with $\delta_{st} = 6.0$ (cal/cm^3)$^{1/2}$ and with the value of δ_{mo} of the solvent varying between 10 and 20 (cal/cm^3)$^{1/2}$. It can be seen that for $\delta_{mo} < 14$, neither compound is retained on the column (e.g., methanol will elute this compound). As δ_{mo} increases (e.g., for a water/methanol mixture), the value of K increases rapidly indicating retention on the stationary phase [6].

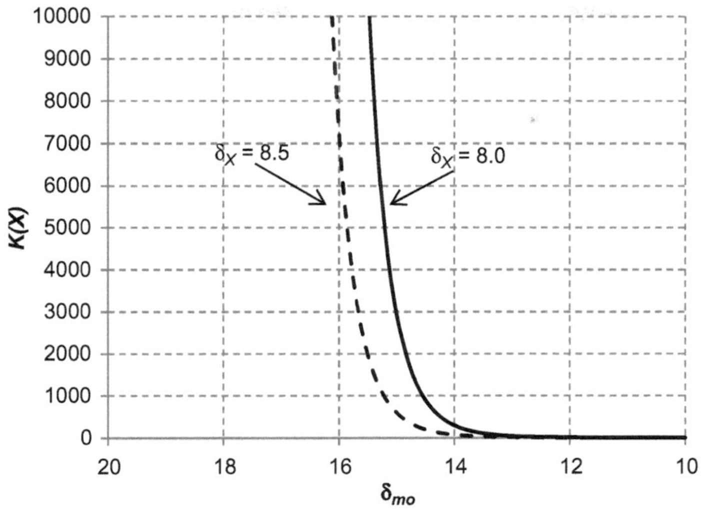

FIGURE 13.1.2 Exemplification of the variation of retention in an RP-HPLC separation as a function of δ_{mo}. (The value of equilibrium constant $K(X)$) as a function of the variation in δ_{mo} $(cal/cm^3)^{1/2}$ of the eluent for two analytes, one with $\delta_X = 8.0$ $(cal/cm^3)^{1/2}$ and the other with $\delta_Y = 8.5$ $(cal/cm^3)^{1/2}$, respectively, both with $V_X = 100$ cm^3/mol on a hypothetical stationary phase ($\delta_{st} = 6.0$ $(cal/cm^3)^{1/2}$).

Expression (13.1.6) can be related to the *strength* of an eluent. For RP-HPLC the relation shows that an eluent with lower δ, which is typical for hydrophobic solvents, leads to lower $K(X)$ values for the analytes, indicating a "stronger" eluent. In a solvent with larger δ, the hydrophobic interactions in the liquid are expelling more efficiently the analyte from the mobile phase placing it in the stationary phase (which is a liquid with lower δ). However, the value of δ cannot be used directly for deciding if a compound is eluted or not from the chromatographic column. For example, acetonitrile has $\delta = 11.8$, while the stationary phase has a lower value for δ (e.g., a hydrocarbon such as isooctane has $\delta = 7$). Nevertheless, acetonitrile is capable of eluting many compounds from a stationary phase with a C8 (or C18) bonded phase.

In gradient RP-HPLC, the initial polarity of the mobile phase is typically high (e.g., water has $\delta = 21$ $(cal/cm^3)^{1/2}$), such that all analytes are retained on the column (large $K(X)$). As the content in organic phase increases, δ_{mo} decreases and the $K(X)$ values decrease. The compound slightly less hydrophobic ($\delta = 8.5$) is first eluted ($K(X)$ for $\delta = 8.5$ is low enough earlier than $K(X)$ for $\delta = 8.0$ as shown in Fig. 13.1.2). As δ_{mo} continues to decrease, the $K(X)$ for the compound with $\delta = 8.0$ also decreases sufficiently that the compound is eluted. The steepness of the decrease of $K(X)$ with the decrease of δ_{mo} can be related to the good separation capability of RP-HPLC in gradient elution.

In HILIC, the stationary phase is more polar than the mobile phase and in expression (13.1.6) $\delta_{st} > \delta_{mo}$ (opposite than in RP-HPLC). In order to be retained on a polar column, the solute must be rather polar and its δ_X value should be higher than δ_{mo} and closer to δ_{st}. In conclusion, $(2\delta_X - \delta_{st} - \delta_{mo})$ is positive. Eq. (13.1.6) shows that $K(X)$ for a solute is larger when the difference $(\delta_{st} - \delta_{mo})$ is larger, in other words, when the solvent used as eluent has a small δ_{mo} value. This indicates that in HILIC the organic solvents are "weak" eluents, while polar solvents such as water are "strong" eluents. For this reason, parameter δ can be used for "strength" characterization, but this should be done depending on the type of separation.

Characterization of a liquid based on its solubility parameter δ provides useful information regarding its solvent characteristics, but its utilization in HPLC is still rather limited. The mobile phase is seldom made using a single solvent, and the parameter δ can only be approximated for solvent mixtures. An approach to estimate δ as to be unique to a compound is to disregard the true meaning of this parameter. The value for δ can be assigned to a "hypothetical" liquid with identical property as the mixture. The expression of δ for a mixture of solvents can be based on formula (4.2.69), which shows the variation of log k' for an RP-HPLC separation as a function of mobile phase composition ϕ. For a compound X, formula (4.2.69) for retention factor k' also holds for the equilibrium constant K and can be written in the form:

$$\log K(X) = \log K_w(X) - S(X)\phi \tag{13.1.7}$$

Notations previously used include parameter $S(X)$ specific for a compound X, and the volumetric fraction of organic component ϕ. Equilibrium constant $K_w(X)$ corresponds to retention factor $k'_w(X)$ in 100% water in the mobile phase. The use of expression (13.1.5) for $K(X)$ in formula (13.1.7) leads to the following relation for δ in a solvent mixture versus δ in water:

$$(\delta_X - \delta_{mo})^2 = (\delta_X - \delta_w)^2 - S'(X)\phi \tag{13.1.8}$$

where $S'(X)$ is given by the formula:

$$S'(X) = \frac{2.303\ RT}{V_X} S(X) \tag{13.1.9}$$

For relatively close values for δ_{mo} and δ_w, and assuming $2\delta_X\delta_{mo} \approx 2\delta_X\delta_w$, the following relation can be written:

$$\delta_{mo}^2 = \delta_w^2 - S'(X)\phi \tag{13.1.10}$$

Separation between two compounds X and Y can also be considered based on the differences in the values in their Hildebrand solubility parameters. The selectivity α is given by expression (4.2.55). The replacement in the formula for $\alpha(X,Y) = K(X)/K(Y)$ the expressions for $K(X)$ and $K(Y)$ given by formula (13.1.6), leads to the following expression:

$$\alpha(X, Y) = \exp\{(\delta_{st} - \delta_{mo})[V_X(2\delta_X - \delta_{st} - \delta_{mo}) - V_Y(2\delta_Y - \delta_{st} - \delta_{mo})]/RT\} \tag{13.1.11}$$

Expression (13.1.11) leads to the estimation of the value of α for a given separation (when the molar volumes V and parameters δ are known). The first factor in formula (13.1.11) shows that α increases as the difference between the δ parameters for the mobile phase and the stationary liquid phase increases ($\delta_{st} - \delta_{mo}$ is larger). The second factor shows that when the differences between $V_X\ \delta_X$ and $V_Y\ \delta_Y$ are larger, the value of α is also larger. These considerations are in accord with experimental observations that a stationary phase more different from the mobile phase will lead to a higher selectivity between the species to be separated. As expected, the more different are the analytes (different V and different δ), the better is the separation. The formula also shows that at lower temperatures, it can be expected to obtain a higher selectivity than at a higher temperature.

The interactions between the analyte, the mobile phase, and the stationary phase, not being completely described by the solubility parameter δ, it is important to consider in selecting a solvent the values of the other solubility parameters: δ_d (dispersion), δ_p (polar), δ_a (proton acceptor), and δ_h (proton donor) for both analyte and eluent. For example the comparison of acetonitrile (AcCN) and methanol (MeOH) using the values from Appendix 7a indicates that $\delta_{AcCN} = 11.8$ and $\delta_{MeOH} = 12.9$, which shows AcCN as being a stronger eluent in RP-HPLC. However, only δ_d (dispersion) and δ_p (polar) are larger for AcCN, while δ_a (proton acceptor) and δ_h (proton donor) are larger for MeOH. This indicates that differences between the two solvents can be expected to depend on the nature of the analytes to be separated. For example, the difference $\delta_X - \delta_{mo}$ is expected to be smaller for a compound containing OH groups, for mo = MeOH as compared to mo = AcCN. This indicates that for RP-HPLC although overall acetonitrile is a stronger eluent than methanol, for compounds capable of forming strong hydrogen bonds, methanol may also act as a strong eluent. Unfortunately, quantitative estimations of eluent strengths cannot be made based on δ and partial δ values, and the inspecting of such values offers only guidance in the selection of a solvent for the mobile phase.

Solvent Characterization Using Octanol/Water Partition Constant K_{ow}

Octanol/water partition coefficient K_{ow} is probably the most useful parameter for solvent characterization, not necessarily because it correlates the best with solvent properties, but because this parameter is readily available (see Section 5.1) [7,8] and can be calculated using computer programs if the experimental value is not known (e.g., MarvinSketch 5.4.0.1, ChemAxon Ltd. [9], EPI Suite [8]). Also, log K_{ow} values can be obtained using additive fragment methodology [10,11] and by other procedures. The use of octanol/water partition coefficient for the characterization of hydrophobicity (and polarity) of compounds has been already presented (see Sections 4.3 and 5.1), and in Section 5.4 it was shown that the selection between NARP, RP-HPLC, HILIC, and IC can be guided by the value of log K_{ow} of the analyte (see Fig. 5.4.1). For RP-HPLC, even a direct good correlation exists between log K_{ow} and log k' [12–14].

For pure solvents, K_{ow} (log K_{ow}) can be used for the estimation of their hydrophobic character as it is done for any other simple compound. The values of log K_{ow} for a number of solvents are given in Appendix 7c. These values were obtained using the MarvinSketch computer package from ChemAxon Ltd. [9]. A higher value for K_{ow} indicates a less polar compound and therefore the values for this parameter are growing in the opposite direction to δ. However, log K_{ow} and δ do not show a good (negative) correlation (for a set of 36 common solvents, the correlation between log K_{ow} and δ has $R = -0.65{,}575$). From the values of log K_{ow} of the solvent and based on their common use in the mobile phase in HPLC (e.g., in RP-HPLC), it can be estimated which solvent is potentially a stronger eluent, but the comparison of the values for log K_{ow} of the solvent and log K_{ow} for the analyte cannot be used for establishing the true elution capability. For example, even solvents with lower log K_{ow} are capable of eluting strongly retained compounds with significantly higher log K_{ow} values.

For RP-type separations, log K_{ow} of the solvent (log $K_{ow}(mo)$) can be used as a guidance regarding its "strength." For example, the strength of a solvent in RP-HPLC and the values for log $K_{ow}(mo)$ (see Appendix 7c for several common solvents) are in the order: n-hexane log $K_{ow} = 3.13 >$ diethyl ether log $K_{ow} = 0.84 >$ tert-butanol log $K_{ow} = 0.54 >$ tetrahydrofuran log $K_{ow} = 0.53 >$ isopropanol log $K_{ow} = 0.25 >$ acetone log $K_{ow} = 0.11 >$ dioxane log $K_{ow} = -0.09 >$ ethanol log $K_{ow} = -0.16 >$ acetonitrile log $K_{ow} = -0.17$ (experimental -0.34) $>$ methanol log $K_{ow} = -0.52$ (experimental -0.77) $>$ dimethylformamide log $K_{ow} = -0.63 >$ water log $K_{ow} = -0.65$ (experimental -1.38). Although for RP-HPLC the elution "strength" of these solvents should follow the order of their log K_{ow} values, the changes in log $k'(X)$ (decrease for higher log $K_{ow}(mo)$) for different solvents is also dependent on the nature of X. For this reason, a higher or a lower log K_{ow} of the mobile phase influences differently different compounds. Usually, larger k' values lead to better resolution, and this can be achieved by decreasing the strength of the mobile phase. However, the differences between solvents in influencing k' differently depending on the nature of the separated compounds, can be utilized for better separations, and this can be based on the use of weaker but in some cases even of stronger eluents.

One problem related to the use of log K_{ow} for the characterization of the hydrophobicity/polarity of the mobile phase in HPLC, is that the mobile phase is typically made using a mixture of solvents (and sometimes also contains additives). However, a hypothetical K_{ow}^{hyp} can be utilized for the description of the apparent hydrophobicity of a solvent mixture. Such a value can be taken as being the weighted (by content) average of the K_{ow} values of the participating solvents. For example, for a liquid mixture X containing a pure solvent X_1 plus another solvent X_2 at the volume fraction ϕ, a hypothetical $K_{ow}^{hyp}(X)$ can be defined by the expression:

$$\log K_{ow}^{hyp}(X) = \log K_{ow}(X_1) + S''\phi \tag{13.1.12}$$

where S'' is a constant specific for the solvent system and can be positive or negative. Assuming a linear variation of $K_{ow}^{hyp}(X)$ for different volume fractions of X_1 and X_2, formula (13.1.12) will give the expression:

$$\log K_{ow}^{hyp}(X) = (1-\phi)\log K_{ow}(X_1) + \phi \log K_{ow}(X_2) \tag{13.1.13}$$

Expression (13.1.13) indicates that for a binary solvent an estimation of polarity can be based on the proportion of the two solvents and individual log K_{ow} values. For example, water has log $K_{ow} = -1.38$, acetonitrile has log $K_{ow} = -0.34$, and methanol has log $K_{ow} = -0.77$. The values for log K_{ow}^{hyp} for any mixture acetonitrile + water or methanol + water can be roughly estimated from these values.

Solvent Characterization Based on Liquid–Gas Partition

One possibility of characterizing a solvent is based on the measurement of "how good" the solvent is for dissolving a volatile solute. Such a parameter is expected to be related to the dissolution energy for the compound X in the solvent S, and to the energy of vaporization of compound X. For the characterization of solvent S, a set of trial solutes X can be selected followed by the measurement of the distribution constant $K_S(X)$ between the solvent and headspace. Experimentally, the distribution constants can be measured by adding a volume of test compound in a given volume of solvent S placed in a closed vial of a specific volume. A protocol for this procedure reported in the literature [15] used a 5 μL mixture of ethanol, dioxane, and nitromethane in a chamber of 13.4 mL with 2 mL solvent S at 25°C. After equilibration, the concentration of each test compound X in the gas phase and in the liquid can be measured (e.g., by gas chromatography). The K_S value for each component in the mixture was calculated using the formula:

$$K_S(X) = \frac{C_S(X)}{C_g(X)} \tag{13.1.14}$$

The selection of the three test compounds, ethanol, dioxane, and nitromethane, was made on the assumption that they have different types of polar interactions with the molecules of tested solvent *S*. Dipole–dipole interactions are stronger with nitromethane, acidic–polar interactions are stronger with dioxane, and basic–polar interactions are stronger with ethanol.

The K_S values are further used in the calculation of a modified constant that is intended to eliminate the effect of the solvent molecular weight. The modification is done by the use of the solvent molar volume V_S (mL/mole) in the expression:

$$K'_S(X) = K_S(X)V_S \tag{13.1.15}$$

The values K'_S are then used to calculate the coefficients K''_S, which are obtained with further correction of K'_S values with the intention to correct for nonpolar (dispersive) interactions. This is done with the relation:

$$K''_S(X) = K'_S(X)/K'_v \tag{13.1.16}$$

where K'_v is the estimated K'_S value of *S*, an n-alkane whose molar volume is the same as that of the solute X. The values of K'_v are calculated using the expression:

$$\log K'_v = (V_X/163)\log K_S(\text{octane}) \tag{13.1.17}$$

In expression (13.1.17), V_X is the molar volume of the solute (ethanol, dioxane, or nitromethane), and K_S(octane) is the experimental distribution coefficient of n-octane in the evaluated solvent. The constants K''_S are further corrected to have zero value for n-hexane as a solvent. The resulting constants K''_S are used to measure the excess retention of a solute relative to an n-alkane of equivalent molar volume.

For any evaluated solvent *S*, an experimental polarity parameter also known as chromatographic strength P' is then defined by the sum:

$$P'_S = \log K''_S(\text{ethanol}) + \log K''_S(\text{dioxane}) + \log K''_S(\text{nitromethane}) \tag{13.1.18}$$

Larger values for P'_S indicate a polar solvent such as alcohol or water, and values close to zero show nonpolar solvents such as hexane and cyclohexane. Solvent polarity can be used in selecting solvents in LC separations.

The polarity parameter P' can be hypothetically extended to a solvent mixture. An approximation of this parameter for a mixture of solvents is given by the expression:

$$P'_{\text{mix}} = P'_1\phi_1 + P'_2\phi_2 + \ldots + P'_n\phi_n \tag{13.1.19}$$

where P'_1, P'_2, ...P'_n are the polarities of individual solvents, and ϕ_1, ϕ_2, ...ϕ_n are the volume fractions of each component of the mixture. Expression (13.1.19) is only an approximation of polarity, and nonlinear variations of parameters with the composition are common for solvent mixtures.

Parameter P'_S is not always sufficient for the characterization of solvent properties. The types of interactions that dominate solvent behavior can be quite different between solvents with the same P'. For example, a polar solvent and a solvent forming hydrogen bonds, although they may have identical P', may not act in the same manner toward different solutes. An additional separation parameter x_i was developed for solvent characterization, defined by the ratio:

$$x_i = \log\left[K''_S(i)/P'_S\right] \tag{13.1.20}$$

where i can be ethanol (x_e), dioxane (x_d), or nitromethane (x_n). Other solvents were also used for obtaining an x_i value, such as toluene (x_t) or methyl ethyl ketone (x_m) [2]. It can be assumed that the larger is the x_i value for a specific compound, the higher is the similarity with the comparing solvent. However, the value of x_i also depends on P', and relatively large K''_S do not necessarily lead to large x_i values. For this reason, the values for x_i were used to group the solvents into nine main groups, the solvents in the same group having similar properties. These nine groups are the following: (0) solvents with very low P' values (nonpolar), (1) aliphatic ethers, tetramethylguanidine, hexamethylphosphoric acid triamide, (2) aliphatic alcohols, (3) pyridine derivatives, tetrahydrofuran, amides, glycol ethers, sulfoxides, (4) glycols, benzyl alcohol, acetic acid, formamide, (5) methylene chloride, ethylene chloride, (6) tricresyl phosphate, aliphatic ketones and esters, dioxane, (7) aromatic hydrocarbons, halo-substituted aromatic hydrocarbons, nitro compounds, aromatic ethers, and (8) fluoroalkanols, *m*-cresol, water, chloroform. Values for P' for some solvents classified in these groups are given in Appendix 7c.

The parameters x_e for ethanol, x_d for dioxane, and x_n for nitromethane from Appendix 7c can be used to illustrate the separation of solvents in several groups, since in each group, the x_i values are close to each other. This can be illustrated in a triangular diagram. Since $x_e + x_d + x_n = 1$, their graphic representation can be done in a planar triangular diagram, and a representation in a tridimensional space is not necessary [6]. The polarity parameter P' is correlated, as expected, with Hildebrand solubility parameter δ. The correlation is positive, and for a set of 38 compounds gives $R^2 = 0.7486$. The relatively good correlation between δ and P' indicates that the two parameters provide basically similar information. Better guidance regarding solvent properties is obtained when using x_e, x_d, and x_n values for describing solvent similarities. The separation of solvent in classes based on these values indicates that specific types of interactions of analytes with solvent molecules, which are more prominent for a specific solvent than for another, are important in solvent characterization. These interactions include dispersion, dipole–dipole, hydrogen bonding, charge transfer, and ionic.

The utilization of polarity P' as well as of the parameters x_e, x_d, and x_n for solvent characterization is also basically only qualitative. It allows the placement of the solvent in a specific group of solvents and characterizes its polarity and type of polarity (basic–polar for ethanol, acidic–polar for dioxane, and dipole–dipole for nitromethane). Depending on the type of solvent desired, the selection should be made based on a higher or lower P', x_e, x_d, and x_n parameter. The strength of a solvent should be related to the type of stationary phase and the polarity of the analyte. For RP-HPLC, the solvents with similar properties as the analyte are typically stronger eluents as mobile phase than those that are dissimilar.

Solvatochromic Model and Kamlet–Taft Parameters

Among other procedures for assessing the polarity of a solvent are the spectroscopic measurements. Such measurements can be based, for example, on the chemical shift in a nuclear magnetic resonance (NMR) experiment or a change in the absorption spectrum in infrared (IR) or ultraviolet and visible spectroscopy (UV–Vis) for a compound used as a "molecular probe." For example, solvents with a polar character can produce a bathochromic effect on the UV spectrum generated from a $\pi \rightarrow \pi^*$ transition in a compound. Other such effects under the influence of a solvent are known. The change in the position, intensity, or band width that occurs when a solute is transferred from the gas phase to a solvent is known as the *solvatochromic* effect. One common scale to evaluate solvent polarity known as $E_T(30)$ scale is based on the variations in the maximum wavelength of absorption in visible range of 2,6-diphenyl-4(2,4,6-triphenyl-N-pyridinium)phenolate (compound E_T-30). This test compound has a large solvatochromic effect, changing the absorption from 453 nm for water to 810 nm for diphenyl ether as a solvent (solutions are red in methanol and blue in acetonitrile). The polarity $E_T(30)$ is further calculated from the maximum of the wavelength absorption λ_{max} from the expression [16]:

$$E_T(30) = \frac{28591}{\lambda_{max}(\text{nm})} \tag{13.1.21}$$

From $E_T(30)$ values, a normalized parameter E_T^N can be obtained and is utilized for solvent characterization. This parameter is obtained from the formula:

$$E_T^N = \frac{E_T(30) - 30.7}{32.4} \tag{13.1.22}$$

The normalized scale gives $E_T^N = 1$ for water and $E_T^N = 0$ for tetramethylsilane, which are the two extremes for $E_T(30)$ values. The $E_T(30)$ polarity has been reported for a considerable number of solvents [17,18]. The values for several common solvents are given in Appendix 7d.

The effects involved in producing solvatochromic effects, which are the base of $E_T(30)$ and E_T^N scales describe the polarity and polarizability but also the donor hydrogen bond formation ability of a solvent. Other compounds presenting solvatochromic effects accounting for other interactions are available. A set of parameters known as Kamlet–Taft solvatochromic parameters (π^*, α, and β) can be used for solvent characterization. A π^* scale was developed accounting for polarity and polarizability, which is not affected by hydrogen bonding or ion dipole interactions based on the use of 4-ethyl-nitrobenzene as a molecular probe. The $E_T(30)$ scale and the π^* scale are "single parameter" polarity scales and are based on the measurements of the property change of a single compound. Other such scales are known, some still based on the test of one compound (such as Nile Red), and others being "multiparameter" and based on the measurements on more than one compound as a probe. For example, the solvent hydrogen–bond donor interactions can be described by an α scale developed on compounds such as several common dyes [19]. Hydrogen–bond acceptor interactions can be described by a β scale that was developed based on measurements on compounds such as 4-nitroaniline, N,N-diethyl-4-nitroaniline, 4-nitrophenol, and 4-nitroanisole [20]. Values for the parameters α, β, and π^* are available in the literature [21–23]. Some of these values for specific solvatochromic parameters α, β, and π^* are listed in Appendix 7e.

As expected, solvatochromic parameter $E_T(30)$ is related to π^* and α, as well as to a polarizability correction parameter δ^* specific for different classes of solvents [24]. Solvatochromic parameter π^* has some correlation with the polarity P' obtained from liquid gas partition, but $R^2 = 0.6475$. Also, π^* was found to be correlated with Hildebrand solubility parameter δ^2. However, the correlation remains acceptable only for compounds with low or medium polarity and is not applicable to polar compounds such as water or methanol.

Differently from Hildebrand solubility parameter or octanol/water distribution constant, solvatochromic parameters can be measured directly for solvent mixtures [25]. These studies showed that the variations of E_T^N, α, β, and π^* are not linearly dependent with solvent composition ϕ for mixtures such as methanol/water or acetonitrile/water.

The values for α, β, and π^* were used in attempts to calculate capacity factors in RP-HPLC, although the precision of the results was not satisfactory [26–29]. These parameters remain valuable as a guidance for selecting a solvent that displays stronger polarity and polarizability, stronger hydrogen–bond donor interactions or hydrogen–bond acceptor interactions.

Elutropic Strength

Besides solubility parameter δ, octanol/water log K_{ow}, and polarity P', some other parameters were developed for the characterization of the behavior of certain solvents, in particular related to HPLC applications. One such parameter is the elutropic strength ε^0 [30]. This parameter has been developed in connection with the adsorption-type equilibrium taking place in HPLC. However, attempts were made to apply this parameter to reversed-phase chromatography for which the partition model is more appropriate [31].

13.2 ADDITIONAL PROPERTIES OF LIQUIDS AFFECTING SEPARATION

Depending on the HPLC type, specific physicochemical properties of the solvent used as mobile phase play a role in the separation. Among these properties are the following: (1) solvent viscosity η, (2) dielectric constant ε, (3) superficial tension γ', (4) dipole moment μ, and (5) polarizability α. The importance of these parameters in the selection of a mobile phase may not be critical. However, their role in the separation (partially discussed in Chapters 4 and 7) should be understood for better tuning of mobile phase selection.

Solvent Viscosity

Viscosity of a fluid is a measure of its resistance to gradual deformation. Typically the dynamic viscosity, which is usually of interest, is the measure of fluid resistance to shearing flow and it is measured in poise P. The ratio between dynamic viscosity and density of the fluid is known as kinematic viscosity. Solvent dynamic viscosity η is an important parameter for the mobile phase, affecting the separation in various ways. One effect is related to the backpressure of the column. As shown by expression (6.2.9) (see Section 6.2) the increase in mobile phase viscosity leads to an increase in the backpressure. The increase in the backpressure may not be extremely important for some columns and when using modern UPLC instrumentation, but it may become a limiting factor when the column is sensitive to higher backpressures, when columns with very small particle sizes are used, or when longer columns may need to be used for a better separation.

The other effect of mobile phase viscosity is related to the impact on column plate number N. A more viscous solvent reduces the diffusion coefficient of sample components and slows down the mass transfer process. In van Deemter equation (4.2.49), it is shown that the theoretical plate height depends on the diffusion coefficient of the mobile phase D. A decrease in D leads to an increase in the plate height and therefore a decrease in column efficiency (decrease in N). Diffusion coefficient D of a liquid depends on viscosity following Stokes equation:

$$D = \frac{RT}{6\pi\beta\eta r}\mathcal{N} \quad (13.2.1)$$

In expression (13.2.1), β is a correction constant (close to 1), η is the dynamic viscosity, r is the radius of molecules of the mobile phase, and $\mathcal{N}$ is the Avogadro number. As shown by expression (13.2.1), a lower viscosity of the mobile phase has as effect an increase in the column efficiency (increase in $\mathcal{N}$). Experimental studies [32] show that about a twofold loss in $\mathcal{N}$ takes place for a 2.5-fold increase in viscosity.

Viscosity varies with temperature. Several models were developed for describing the variation of viscosity with the temperature, such as the Arrhenius model where the variation in the viscosity follows an equation similar to that for the rate constant:

$$\eta(T) = \eta_0 \exp(E/RT) \quad (13.2.2)$$

where η_0 and E are constants. As an example, the variation of viscosity of water (in cP) with inverse of the temperature (K) is given in Fig. 13.2.1.

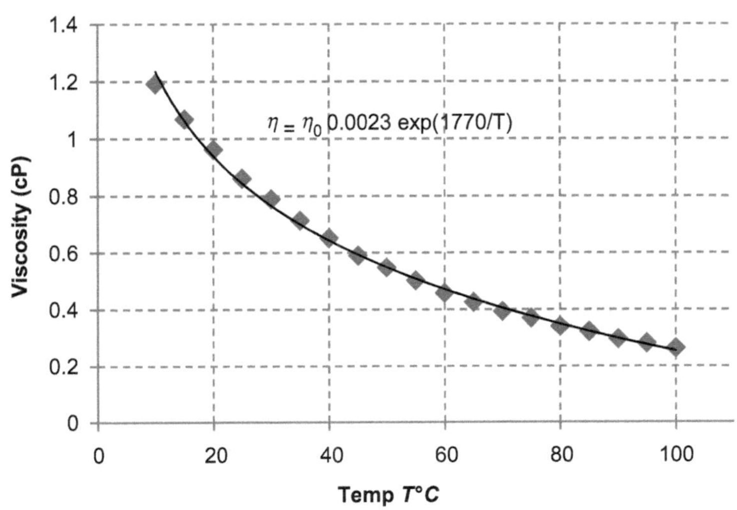

FIGURE 13.2.1 Variation of viscosity of water (in cP) with temperature.

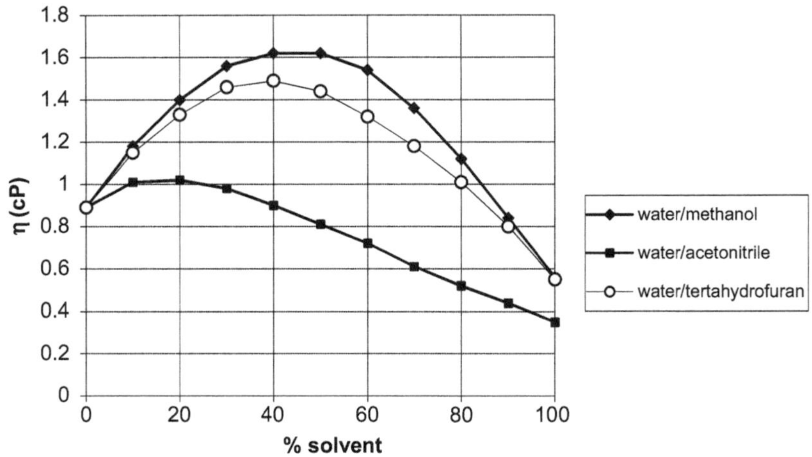

FIGURE 13.2.2 Variation of viscosity (in cP) for mixtures water/methanol, water/tetrahydrofuran, and water/acetonitrile as a function of composition, at 25°C.

Since solvent mixtures are frequently used as a mobile phase, the variation of viscosity with solvent composition is of special interest, in particular when the separation is performed under gradient conditions, and the viscosity changes during the chromatographic run. The HPLC instruments capable of generating high pressures with their pumping systems (e.g., in some UPLC instruments pressures up to 18,000 psi can be generated) can easily generate excessive pressures unless a specific limit is set. High backpressures can damage the chromatographic column, in particular for SEC separations. The variation of viscosity with mobile phase composition is difficult to predict and depends on the interactions at the molecular level between the solvents. As an example, the variation of viscosity as a function of composition for water/methanol, water/tetrahydrofuran, and water/acetonitrile at 25°C is shown in Fig. 13.2.2 [6]. The variation of viscosity with mobile phase composition may be significant, as shown in Fig. 13.2.2. The variation of viscosity due to the presence of an eluted fraction of polymers can be used as a principle for the detection in size-exclusion chromatography [33].

Superficial Tension

The importance of superficial tension in HPLC separations has been discussed in relation to the evaluation of retention coefficient k' in reversed-phase chromatography. Based on expression (7.3.1) generated by the solvophobic theory, when other parameters are kept constant, the following expression can be established between $\log k'$ and γ':

$$\log k' = a + b\gamma' \tag{13.2.3}$$

where a and b are parameters depending on several properties of the separation system. Expression (13.2.3) indicates that higher superficial tension of the mobile phase leads to stronger retention. However, expression (13.2.3) cannot be used as a guide for estimating $\log k'$ since a change in the mobile phase with the purpose of increasing γ' would affect parameters a and b.

Superficial tension decreases linearly with the temperature. This dependence can be obtained from Eötvös rule, which can be written as follows:

$$\gamma'(T) = C_t V^{-2/3}(T_c - T) \tag{13.2.4}$$

where C_t is a constant, V is the molar volume, T_c is the critical temperature for the liquid, and T is the temperature of interest (in Kelvin). From Eq. (13.2.4), the following expression of temperature dependence can be obtained:

$$\gamma'(T_2) = \gamma'(T_1) + C_t'(T_1 - T_2) \tag{13.2.5}$$

Expression (13.2.5) (where C_t' is a temperature change coefficient) indicates that the increase in temperature ($T_2 > T_1$) leads to a decrease in superficial tension.

The decrease of γ' with temperature and its implication in the decrease of retention factor k' in RP-HPLC shows that temperature variation of the mobile phase affects the separation in a more complex way than predicted by formula (4.4.9) with ΔH^0 and ΔS^0 assumed constant. The values of γ' for several common solvents and the coefficient C_t' for the temperature dependence are given in Appendix 7f.

The superficial tension of solvent mixtures is difficult to predict. Several studies are reported in the literature providing results on γ' for different solvent mixtures [34,35]. As an example, the variation of surface tension with the organic phase concentration in water for several common solvents is shown in Fig. 13.2.3 [6].

Changes in dynamic surface tension (DST) due to the presence of the analyte in the mobile phase found application in a detection technique indicated as DSTD (DST detection) [36].

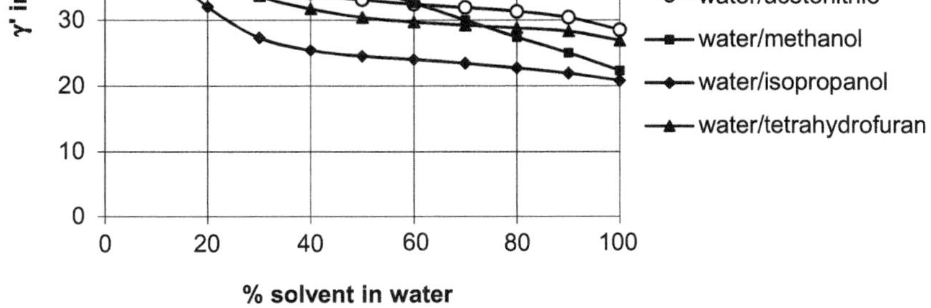

FIGURE 13.2.3 Variation of surface tension γ' (in mN/m = dyne/cm) with the organic solvent concentration in water for several common solvents used in HPLC.

Dielectric Constant, Dipole Moment, and Polarizability

Several solvent properties influence the interactions taking place in solutions and implicitly influence the elution properties of a mobile phase made with these solvents. Among these properties are the dielectric constant ε of the solvent, the dipole moment μ, and the polarizability α. Dielectric constant is the ratio of the permittivity of a substance to the permittivity of vacuum (permittivity is the measure of resistance that is encountered when forming an electric field in a medium). Dielectric constant is dimensionless (vacuum permittivity is $\varepsilon_0 = 8.854 \times 10^{-12}$ C/V m). Dielectric constant of the solvent affects interactions in solution that involve ions and polar molecules, decreasing the intermolecular energy when the dielectric constant increases. Polar compounds are usually more soluble in solvents with higher dielectric constant. Also, it affects solvophobic interactions (as indicated in Section 7.3) and in RP-HPLC solvents with higher ε are usually weaker eluents. Some values of the dielectric constant for several common solvents at 25°C are given in Appendix 7g.

Dipole moment of a molecule has been discussed in Section 5.1, where it was also given its definition (see expression (5.1.1)). The dipole moment of solvent molecules can be used for the characterization of a solvent polarity. The mechanisms of a number of HPLC types of separation including HILIC and to a lower extent RP involve polar interactions. For this reason, the dipole moment μ is a parameter useful in solvent characterization. Dipole moment varies as a function of temperature, and for some compounds it is different for the molecules in gas form and in liquid form. For molecular interactions used as a model in gas phase for the understanding of the separation process, the values of μ in gas phase can be used. However, the values for the dipole moment in the liquid form seem more appropriate for solvent properties characterization. Values for the dipole moment (in debye, D) of several common liquids used in HPLC as mobile phase (or mobile phase additives) are listed in Appendix 7g.

Similar to dipole moment, polarizability in an electric field of a solvent molecule provides some information regarding the intensity of interactions in solution. Polarizability affects the interaction of polar molecules in a similar manner as the dipole moment. The polarizabilities for several common solvents (expressed in $4\pi\varepsilon_0(\text{Å})^3$) are listed in Appendix 7g.

Hydrogen Bonding of Solvent Molecules

Hydrogen bonding is based on electrostatic attraction between a hydrogen atom bound to a highly electronegative atom and another nearby electronegative atom. Hydrogen bonding is known for the hydrogen bound, for example, to oxygen, nitrogen, or fluorine interacting with molecules that contains electronegative atoms (e.g., O, N, etc.). The molecules capable of forming hydrogen bonds contain donor, acceptor, or both types of groups for hydrogen bonding. The energy (enthalpy) of hydrogen bonds varies between 5 and 40 kJ/mol (see Section 4.3), being stronger than van der Waals interactions. Hydrogen bonding may be intermolecular but also intramolecular between different parts of the same molecule. For solvents that have small molecules, hydrogen bonding is typically intermolecular. The enthalpy of hydrogen bonding depends on the nature of the atoms to which the hydrogen is bound and also of the acceptor atom (e.g. O–H···N with about 29 kJ/mol, O–H···O with about 21 kJ/mol, N–H···N with about 13 kJ/mol, N–H···O with about 8 kJ/mol, etc.), but also on the specific molecules that interact, and of temperature. The enthalpy of hydrogen bonding can also be different for the same type of molecules in liquid state or gas state. Hydrogen bonding influences the boiling point and viscosity of a solvent. The hydrogen bonds energy for liquid water (at 0°C) is estimated to about 23 kJ/mol, and it is lower in alcohols and even lower in amines. The enthalpy of hydrogen bonding in solvents plays an important role in their "elution strength," in specific types of chromatography. As already indicated, the water having strong intermolecular interactions is a weak eluent for RP-HPLC and a strong eluent for HILIC. For RP-HPLC, alcohols are stronger eluents than water, and they are weaker eluents in HILIC. This character is partly generated by the strength of the hydrogen bond enthalpy in each solvent [37–39].

Solvent Boiling Point

Solvent boiling point is not a frequent parameter of interest in HPLC. However, HPLC performed with the chromatographic column set at a higher temperature than ambient is not uncommon. The temperature set for the chromatographic process influences the separation, and can be used to the advantage of a more convenient separation [40]. Among the advantages provided by elevated temperature are the increased speed of separation, the decrease of mobile phase viscosity, improved efficiency, changes in selectivity with the temperature [41,42], and the possibility to use only water as a mobile phase [43]. The boiling point of various solvents should be known in case the HPLC is performed at a higher than ambient temperature. Some boiling point values for several common solvents are given in Appendix 7h.

13.3 PROPERTIES OF SOLVENTS OF IMPORTANCE FOR DETECTION

Besides the importance of the mobile phase in the separation process, the selection of mobile phase also depends on the detection system used for the analysis. The mobile phase must differ from eluted molecules by the physicochemical properties planned to be used for detection such that the mobile phase should not interfere with the detection, and in some cases, must help the detection process (such as in MS detection). For each property used for detection, specific requirements are applied to the mobile phase properties. The properties used for detection can be UV-absorption, fluorescence, molecular fragmentation in a mass spectrometer, refractive index, etc. (see Section 4.1). Several physicochemical properties of the mobile phase associated with the detection in HPLC are further discussed.

Refractive Index

Refractive index (RI) detection is commonly used for the analysis of compounds that do not have good absorption bands for UV light. For such compounds, other detection techniques can be used as well, such as MS or evaporative light scattering detection (ELSD). However, RI is a convenient detection procedure since it offers good reproducibility, and does not require expensive equipment. The sensitivity of the RI detection depends on the difference in the refractive index of the mobile phase and that of the analyte (besides other parameters related to the instrument construction). A list of values for the refractive index for a number of common solvents is given in Appendix 7h. The refractive index depends on the wavelength of the incident beam, and the most accurate RI measurements are done with monochromatic light (usually 589 nm, the sodium D line), although white light is still commonly used for the measurements. The refractive index depends on temperature, and typically decreases as temperature increases (for organic solvents this decrease is about 0.0005 for 1°C and for water is about 0.0001). When RI is used as a detector in HPLC, only isocratic separations can be applied. Pure solvents and premixed solvents can be used as the mobile phase. For RI detection, it is not recommended to generate the specific composition for the mobile phase by using two pumps and a mixing device. Small fluctuations in the RI of the mobile phase typically generate large oscillations in the RI detector response. The refractive index of a binary solvent mixture is typically given by the expression:

$$\frac{n_{mix}^2 - 1}{n_{mix}^2 + 2} = \phi \frac{n_1^2 - 1}{n_1^2 + 2} + (1 - \phi) \frac{n_2^2 - 1}{n_2^2 + 2} \qquad (13.3.1)$$

In expression (13.3.1), n_1 and n_2 are the refractive indexes of the individual solvents, n_{mix} is the refractive index of the mixture, and ϕ is the volume fraction of solvent 1.

UV Cut-Off

The measurement of UV absorption of the analytes is probably the most frequent detection technique used in HPLC. This technique is successfully applied for the measurement of analytes containing unsaturated bonds, aromatic groups, and functional groups containing heteroatoms. The influence of the solvent on this type of measurement has two aspects. One is the requirement of the solvent to be "transparent" (to absorb very little light) in the region where the absorption of the analyte is measured. This "transparency" is characterized by the UV cut-off value, defined as the wavelength at which the absorbance A of the solvent versus air, in a 1-cm cell, is equal to unity (see expression (2.2.2)). The UV absorption typically increases at wavelength approaching 200–210 nm, and the cut-off value indicates that at lower wavelength the absorption of light is too strong to allow the utilization of the solvent as a mobile phase. The UV cut-off values for several common solvents are given in Appendix 7h. A more accurate description of solvent absorption is obtained from the UV spectrum of the solvent. A contribution to the cut-off value for a solvent may come not only from the solvent itself, but also from certain impurities or additives that may be present in the solvent (such as butylated hydroxytoluene, phthalates, acetamide, and ethyleneimine). The potential UV absorption of the additives used as buffers should also be considered when choosing the mobile phase appropriate for detection [44].

The second aspect regarding the solvent influence on the UV detection in HPLC is related to the possible modification of the absorption bands of the analyte. This effect is not seen for some analytes, but it is present for other analytes. For example, the solvents with a polar character may produce a bathochromic (higher wavelength) effect on the UV spectrum of the analytes, with the absorption bands corresponding to a $\pi \rightarrow \pi^*$ transition (see Section 13.1). For such compounds, the maximum of the absorption band can increase with 1–20 nm when changing, for example, from hexane to ethanol as solvent. Compounds that form H-bonds with the solvent molecule will also exhibit a bathochromic effect. The increase of the accepting capacity in H-bond of the solvent will increase the maximum wavelength in the absorption spectrum of the analyte. On the other hand, the bands corresponding to $n \rightarrow \pi^*$ transitions may suffer a hypsochromic effect when the polarity of the solvent increases. The $n \rightarrow \pi^*$ transitions can also be influenced by pH changes due to the structural modification of the analyte. For example, a bathochromic shift can be seen for compounds with phenolic hydroxyls when the pH is modified from acid to basic values due to the modification in the compound dissociation status. For aromatic amines the effect is hypsochromic when pH is modified from basic to acidic values (due to protonation). Besides the change of the maximum absorbance wavelength, the change in the value of the absorbance coefficient ε_λ is seen. Such changes may affect the selection of the optimum wavelength for detection or in using UV spectra for identification of separated species [45].

Fluorescence

The fluorescence of a compound is a property frequently affected by the solvent composition. Both emission wavelength and fluorescence intensity for many compounds containing dissociable functional groups are dependent on pH. The fluorescence intensity can be influenced with as much as one order of magnitude and may shift the emission maximum when strong interactions with solvent occur. A shift of the emission band toward higher wavelengths is observed with the increase of the dielectric constant of the solvent. When the solvent absorbs radiation at the emission or absorption wavelengths, a significant decrease of detection sensitivity can be noticed. The presence of impurities in mobile phase, mainly the O_2, can produce a quenching of the fluorescence signal, mainly when the analyte concentration is close to the detection limit. Solvent-mediated fluorescence enhancement and quenching effects can be used in developing HPLC methods based on gradient elution [46].

Solvent Influence in MS Detection

Solvent (mobile phase) selection for MS detection [electrospray ionization (ESI) and atmospheric pressure chemical ionization (APCI)] is very important since the ionization efficiency is strongly influenced by the solvent [47]. Different analytes have different ionization characteristics, and compounds with basic character (even very weak basic character) are typically analyzed in positive mode, while compounds with acidic character (even very weak acidic character) are typically analyzed in negative mode (see Section 2.3). The formation of ions in ESI and in APCI depend on gas-phase basicity (GPB) or gas phase acidity (GPA) of the analyte, but also of the mobile phase molecules. The competing of solvent in the process of ion formation is not desirable, but on the other hand, the formation of analyte ions by proton transfer from the ionized solvent can be an important process.

The MS sensitivity depends on various parameters such as the type of detection (ESI or APCI), the nature of analytes, the nature and proportion of the solvents used in the mobile phase, the additives from the mobile phase (including those influencing the pH), and the instrumental conditions applied to the MS. For a specific analyte, different combinations of these parameters may lead to different results, such that the optimization process of MS sensitivity cannot be obtained by a simple change of one parameter at a time. For this reason, the selection of the mobile phase with the purpose of obtaining good sensitivity in MS detection is a complex problem. A detailed discussion of the subject is beyond the purpose of the present book. For this reason, only general aspects of the problem are further discussed.

Several characteristics of the mobile phase influence ionization of the analytes. Among these are volatility, surface tension, viscosity, conductivity, ionic strength, dielectric constant, electrolyte concentration, pH, and the potential of gas-phase ion–molecule reactions. These properties must be considered in relation to the chemical and physical properties of the analyte, including pK_a, hydrophobicity, surface activity, ion solvation energy, and GPB or GPA, depending on the use of positive or negative ionization mode. Also, the operational parameters of the instrument such as the flow rate and temperature of mobile phase are important. An arbitrary selection of the mobile-phase composition in liquid chromatography (LC)-ESI/MS is not possible since only polar solvents and volatile additives can be used for obtaining good sensitivity. This requirement limits the use of additives present in the mobile phase to a few acids, bases, and salts. These additives may be necessary either for achieving the separation in the HPLC column or for enhancing ionization in the MS source or for both. Among the additives frequently used in LC-ESI/MS are HCOOH, CH_3COOH, CF_3COOH, salts such as $HCOONH_4$, CH_3COONH_4, $(NH_4)_2CO_3$, or basic compounds such as ammonia or volatile amines. It is not possible to use in LC-MS additives/buffers containing H_3PO_4, H_2SO_4 or their salts with ions such as NH_4^+, Na^+, K^+, etc. Such additives produce small nonvolatile particles in the ion source, affecting the MS detection.

Regarding the solvents from the mobile phase, the following are compatible with both APCI and ESI interfaces: water, alcohols (as long as viscosity does not increase too much such that to require unacceptable reduction of the flow rate), acetonitrile, tetrahydrofuran, acetone, dimethylformamide, and to a lesser extent dicholoromethane, and chloroform. For positive ionization mode, the addition to the mobile phase of a volatile acid, usually HCOOH at 0.01–0.2% levels, is common, while for negative ionization mode the addition of a volatile salt such as $HCOONH_4$, CH_3COONH_4, $(NH_4)_2CO_3$ at 40–50 mM level is typically practiced. Addition of a low level of HCOOH can also be used for negative ionization mode. An organic solvent in mixture with water is common as a mobile phase. Generally, the use of only aprotic solvents is not suitable with the ESI interface. Pure water is also a poorer solvent for ESI than water mixed with organic solvents such as methanol and acetonitrile, even if the level of the organic component is as low as 5%. Due to pure water's relative high viscosity, the electrophoretic mobility of ions is lower, leading to inefficient charge separation and difficulties in producing a stable spray. Also, the evaporation of water from the charged droplet is slower than the evaporation of an organic solvent.

Unlike ESI, which requires mobile phases based on polar or medium polar solvents, in APCI both polar and nonpolar solvents can be used. For this reason, APCI can be chosen as an interface in NP-LC or in convergence chromatography (CC) with MS detection, while ESI is not recommended for this type of HPLC. The solvents commonly used in NP-LC/MS-APCI include n-hexane, 2-propanol, methanol, ethanol, iso-hexane, iso-octane, tetrahydrofuran, chloroform, ethoxynonafluorobutane, with additives such as formic acid, acetic acid, trifluoroacetic acid, or ammonia, diethylamine, triethylamine, dimethylethylamine (depending on positive or negative ionization mode). The possible suppression effect of strongly acidic or basic additives depends on the analytes, but in general it must be avoided. In APCI it is recommended, when possible, to replace acetonitrile with methanol in order to enhance detection sensitivity. This can be explained by the stronger basicity character of acetonitrile compared to that of methanol, which competes with target analytes for protonation. Additionally, acetonitrile tends to polymerize in APCI plasma, coating the corona needle with an insulating layer after several hours in operation. Dimethylformamide content in mobile phase is recommended to be lower than 10% when using API electrospray, and a high signal background can be noticed when using this solvent with the APCI interface. Chlorinated hydrocarbons can enhance the ionization yield only for the APCI interface.

The formation of multiple molecular ions, especially due to the formation of sodium ion adducts is commonly observed in electrospray mass spectrometry and may make it difficult to obtain good reproducibility and sensitive quantitation. In negative ionization mode, alkylamine additives could improve detection by suppression of multiple molecular ions through preferential formation of a predominant alkylamine adduct ion [48].

Other Properties Related to Detection

Several detection techniques that depend on an evaporative process [ELSD, charged aerosol detector (CAD), condensation nucleation light scattering detection (CNLSD), etc.] also require a totally volatile mobile phase [49]. In general, the easier it is to evaporate the mobile phases, the higher is the sensitivity for those detectors. However, the detection process in ELSD and mainly in cCAD is more complex than simple evaporation, and the sensitivity of the detection may depend on a number of additional factors [50].

13.4 BUFFERS AND ADDITIVES

Retention in HPLC of organic compounds that have acidic, basic, or amphoteric character is highly dependent on pH. This is mainly caused by the change in the compound structure due to the pH changes (see Section 4.4). It is common that within 2 pH units the structure of a compound is changed by 98%, for example, from a completely neutral state into an ionized form. The change in the molecular structure of a molecule in solution as a result of solution pH has been previously discussed in Section 4.4, and the modification of log D_{ow} as a result of pH modification was discussed in Section 5.1 (see, e.g., Fig. 5.1.1). The change in the retention of a compound at different pH values of the mobile phase can be an undesirable effect, and then effort will be made to maintain a constant pH during the separation. In many cases, however, the pH change of the mobile phase can be used to the advantage of achieving a specific separation.

The value of pH of the mobile phase is frequently controlled with buffers. Buffers are added for maintaining a specific pH of the mobile phase, and several aspects of the use of buffers in HPLC are presented in this section. For the selection of a buffer, the properties that must be considered include the following: (1) pH of utilization, (2) buffer stability, (3) buffer concentration limited to solubility in organic solvents, (4) impact on detection, and (5) effect on column stability in time.

Besides buffers, the pH of the mobile phase can be adjusted by simply adding a base or an acid. The compounds used in such additions can be classified as additives. Besides additives used for pH modification, other compounds can be added to the mobile phase and they do not affect directly the pH. However, they are necessary for other purposes either related to separation or to other aspects of the HPLC analysis. These compounds are also indicated as additives. In the case of ion pair, hydrophobic interaction, or displacement chromatography, the additives are part of the separation process, and such additives are discussed in association with each specific technique. Some aspects related to additives are also discussed in this section.

Buffer pH

A solution of a weak acid HA and its conjugate base A^-, or a solution of a weak base B and its conjugated acid BH^+, has the capability to show resistance to pH changes upon the addition of a strong acid or base. These types of solutions are known as buffers. The weak acids and bases involved in buffer preparation can be monoprotic or polyprotic and can be inorganic or organic. In water solutions, strong acids and strong bases are completely dissociated and the concentration of H^+ (or OH^-) is practically equal to the concentration of the acid (or base). Weak acids and weak bases are only partially dissociated, and for a weak acid, for example, the H^+ concentration is given by expression (4.4.7). Based on this concept, the addition of a small amount of a strong acid to a buffer solution will not change the pH in accordance to the change in strong acid concentration, but in accordance to a weak acid addition. This process can be followed quantitatively considering that for a mixture of a weak acid HA and its salt (e.g., as Na^+ salt), the mass balance for the solution requires that:

$$C_{HA} + C_{NaA} = [HA] + [A^-] \tag{13.4.1}$$

In expression (13.4.1), C_{HA} and C_{NaA} are the analytical molar concentrations and [HA] and $[A^-]$ are the molar concentrations in the solution upon dissociation and after equilibrium is established (the salt NaA is assumed completely dissociated in solution). On the other hand, the electrical neutrality (for a monoprotic acid) of the solution requires that:

$$[Na^+] + [H^+] = [A^-] + [OH^-] \tag{13.4.2}$$

Since the salt is assumed to be completely dissociated, $[Na^+] = C_{NaA}$ and Eq. (13.4.2) gives:

$$[A^-] = C_{NaA} + [H^+] - [OH^-] \tag{13.4.3}$$

From expressions (13.4.1) and (13.4.3) it is easy to obtain the expression for [HA]:

$$[HA] = C_{HA} - [H^+] + [OH^-] \tag{13.4.4}$$

Since the molar concentration of an acid and its conjugate base in practical applications is typically much larger than the difference $[H^+] - [OH^-]$, it is common to use for expressions (13.4.3) and (13.4.4) the simplifications:

$$[A^-] = C_{NaA} \quad \text{and} \quad [HA] = C_{HA} \tag{13.4.5}$$

With these simplifications, the expression for the dissociation constant of an acid can be written in the form:

$$K_a = \frac{[H^+][A^-]}{[HA]} = [H^+]\frac{C_{NaA}}{C_{HA}} \tag{13.4.6}$$

Expression (13.4.6) gives the pH value for a buffer made as a mixture of a weak acid and its salt and can be written in the form:

$$pH = pK_a + \log\frac{C_{NaA}}{C_{HA}} \quad (13.4.7)$$

Expression (13.4.7) is known as Henderson–Hasselbach equation and it gives the pH of a mixture of a weak acid and its salt. For a weak base in the presence of its salt a relation similar to Eq. (13.4.7) can be written as follows:

$$pH = 14 - pK_b - \log\frac{C_{BHX}}{C_B} \quad (13.4.8)$$

Different from the pH of a strong acid that is completely dissociated and has pH = −log C_{HA}, or a strong base that has pH = 14 + log C_B, the buffer solution changes very little upon the addition of a small amount of a strong acid or base. Also, formula (13.4.7) (and Eq. (13.4.8)) indicates that the pH of the buffer does not depend on the concentration of the two buffer components (C_{NaA} and C_{HA}) but only on their ratio. Therefore the buffer pH is not affected (within a certain range) by dilution. Formulas similar to Eqs. (13.4.7) or (13.4.8) can be developed for polyprotic acids and bases.

The variation of the pH of a buffer solution when a strong acid or a strong base is added is described by *buffer capacity*. Several definitions of this property are known. Buffer capacity β can be defined as the number of moles of a strong acid or base that causes 1.00 L of buffer to change the pH by one unit. Another definition that describes buffer capacity even for small changes is given by the formula:

$$\beta = \frac{dn}{dpH} \quad (13.4.9)$$

where n is the number of equivalents (moles for monoprotic bases) of a strong base added to the buffer in the infinitesimal amount dn to change the pH by d pH. It can be assumed that the addition of a base (e.g., NaOH) leads to the increase of C_{NaA} and $dn = dC_{NaA}$. With the notation $C_{buf} = [HA] + [A^-]$ and the value for $[HA] = C_{HA}$ obtained from expression (13.4.6), the value for C_{buf} can be written as follows:

$$C_{buf} = \frac{[H^+][A^-]}{K_a} + [A^-] \quad (13.4.10)$$

Expression (13.4.10) can be rearranged to give the concentration of $[A^-]$ as a function of C_{buf}, $[H^+]$ and K_a. From $dn = dC_{NaA}$ where C_{NaA} is given by Eq. (13.4.3), and by replacing in C_{NaA} the value for $[A^-]$ from Eq. (13.4.10), it can be concluded that:

$$\beta = \frac{dC_{NaA}}{dpH} = \frac{d\left(\frac{K_w}{[H^+]} - [H^+] + \frac{C_{buf}K_a}{K_a + [H^+]}\right)}{d[H^+]}\,\frac{d[H^+]}{dpH} \quad (13.4.11)$$

The calculation of the derivatives leads to the formula:

$$\beta = 2.303\left(\frac{K_w}{[H^+]} + [H^+] + \frac{C_{buf}K_a[H^+]}{(K_a + [H^+])^2}\right) \quad (13.4.12)$$

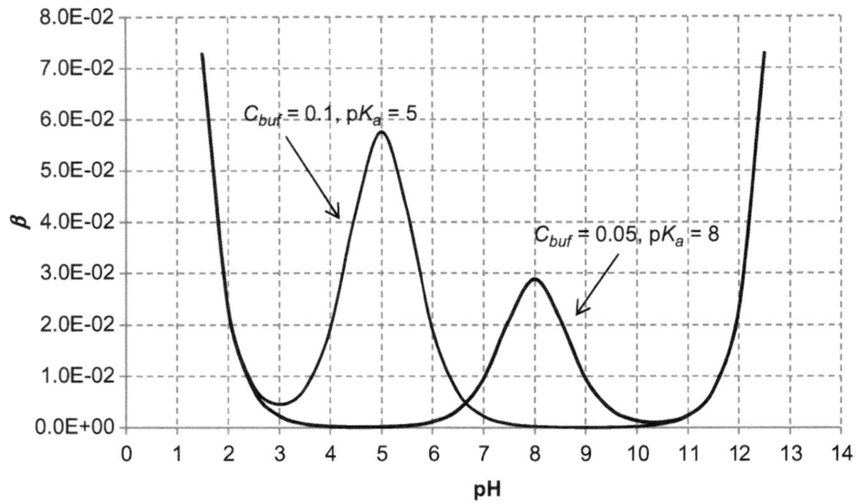

FIGURE 13.4.1 Variation of β with the solution pH and concentration.

Expression (13.4.12) indicates that the buffer capacity β has a relatively complicated dependence on pH and acid dissociation constant K_a, and shows that a higher buffer concentration C_{buf} is associated with a higher buffer capacity. Fig. 13.4.1 shows the variation of buffer capacity β with the pK_a and with the buffer concentration C_{buf}.

From expression (13.4.9) (and considering that $dn = dC_{NaA}$) the following formula can be obtained:

$$dpH = \frac{dC_{NaA}}{\beta} \tag{13.4.13}$$

Formula (13.4.13) indicates that a higher buffer capacity β is associated with a smaller variation in pH for an addition of a strong base NaA. For concentrated solution of an acid or base when a small addition of acid or base is made, the pH does not change and this explains the large β values at the extreme pH values. For a middle pH range, β has a local maximum centered at the pK_a value of the weak acid making the buffer, and the maximum is higher when the concentration of the buffer C_{buf} is higher.

Common Buffers Used in HPLC

The pH of a buffer solution is selected to maintain a desired pH value for the mobile phase. However, the stability of the stationary phase, which is frequently limited to the range 2–8 is also imposing the utilization of buffers in the pH range 2–8, although buffers in the range 1–12.9 are reported in the literature (e.g., HCl + glycine for pH as low as 1.04 and NaOH + glycine for pH as high as 12.97) [51]. The pH range of utilization for a buffer should be around the value of the pK_a of the weak acid or base used in making the buffer. From expression (13.4.7) it can be seen that for equal concentrations of the weak acid and of its salt, the pH of the buffer solution is pH = pK_a. Also, the maximum buffer capacity occurs for pH = pK_a, as indicated in Fig. 13.4.1. For these reasons, the pK_a of the common weak acids and bases used for buffer preparation are of interest. Such values for pK_a are given in Appendix 7i. The same appendix gives a list of some common buffers used in HPLC, and their pH working range.

One common problem related to buffers is their stability in time. Some buffers are stable, but the stability of the pH of other buffer solutions is not always good. This is, for example, the case of buffers that contain NH_4OH as a base. Due to the volatility of ammonia it can be eliminated in time from the solution, and the pH of such buffers may decrease (significantly). Ammonia can also be eliminated from a buffer solution during the solvent sparging (if this is performed). In such situations, the pH of the buffer is not the same at different times during its utilization, which can produce significant variations in the separation. The pH of buffers containing NH_4OH, for example, can vary as much as one pH unit in several days. Fresh buffers must be prepared at set time intervals in order to avoid this problem. Besides the variation in the pH, the stability of buffers in time may also be related to the growth of microorganisms. Some buffers may act as media for microorganism growth, and the same solution of a buffer has only a limited utilization time. This growth can be delayed by the addition of small amounts of NaN_3 and/or by refrigerating the buffer solutions during storage. The growth of microorganisms may slightly modify the buffer pH, but the main problem is related to the clogging of the filters (frits) present along the path of the mobile phase, even when the mobile phase was initially free of particles.

One important aspect when using buffers in HPLC is the buffer concentration. Buffer concentration in HPLC is usually made with the salt concentration of 50 mM or lower, and buffers of 10 mM concentration are common. However, addition of larger concentrations of acids or bases to achieve the desired pH is sometimes utilized. The buffer concentration has an effect on the separation, on column stability, and also on the detector response. In the description of HPLC methods that use buffers, in addition to the buffer pH, the concentration must be indicated.

Buffers in Partially Aqueous Solvent Mixtures

Mobile phases in several types of HPLC usually consist of an aqueous component and an organic phase component. When a specific pH is desired for the mobile phase, it is common practice to make only the aqueous component with a desired pH, while the organic component is typically used without a buffer addition. The addition of a buffer that contains an inorganic salt to the organic component of the mobile phase may encounter solubility problems. For organic additives that are soluble in the organic component of the mobile phase, the addition can be done to both the aqueous phase and organic phase. The change in buffer pH and in buffer concentration in the mobile phase must always be considered for gradient separations. Such changes may occur because the organic content influences the pH, and also because the components of the mobile phase are absent in one of the mobile phase solvents. In some cases, the change in additive concentration during gradient can be detrimental, for example, when MS detection uses an organic acid such as HCOOH or CH_3COOH. In such cases, it is typically recommended to add the acid at the same concentration in both (or all) components of the mobile phase.

When the buffer is added only to the aqueous component of the mobile phase, the final pH of the mobile phase is modified from the nominal value of the pH of the buffer, although in water a dilution of the buffer (C_{buf}) affects only the buffer capacity but not the pH (as indicated by expression (13.4.7)). The cause of this change is that the buffer pH calculated for an aqueous solution is not the same as the pH in a solution that is partially composed of water and an organic miscible solvent (such as methanol, ethanol, or acetonitrile). For water the dissociation constant (ionic product) K_w is given by $K_w = [H^+][OH^-] = 10^{-14}$ at 25°C (more correctly $K_w = a_{H^+}a_{OH^-}$). Neutral is defined as the state at which $[H^+]$ equals $[OH^-]$, which occurs when $[H^+] = 10^{-7}$ equivalent with a pH of 7. For methanol, for example, the autoprotolysis constant is $K_{CH_3OH} = [H^+][CH_3O^-] = 10^{-16.6}$. In methanol, neutral pH should be taken when $[H^+]$ equals $[CH_3O^-]$, which occurs when $[H^+] = 10^{-8.3}$ or a pH of 8.3. In conclusion, methanol–water mixtures have autoprotolysis constants K_{H_2O/CH_3OH} between 10^{-14} (water) and $10^{-16.6}$ (methanol), and the neutral in these mixtures ranges from pH = 7 to pH = 8.3. In aqueous basic solutions, the anion is OH^-, and in basic solutions that contain high concentrations of methanol, the anion will be a mixture of OH^- and CH_3O^- [52].

A common procedure for determining the pH of aqueous/organic solutions is to directly measure it after mixing the aqueous buffer and the organic modifier. For a pH measured in a 100% aqueous solution with the electrode calibrated with aqueous standard buffers, the pH is accurate (and can be indicated as ${}^{w}_{w}\text{pH}$). For the pH measured in an organic/aqueous mixture with the electrode calibrated with aqueous buffers, a value indicated as ${}^{s}_{w}\text{pH}$ is obtained. However, this is not a correct value. A correct value would be obtained only if the pH is measured with an electrode calibrated with buffers prepared in the same solvent as the one used for the mobile phase. In such case, a ${}^{s}_{s}\text{pH}$ value would be obtained. This measurement requires the knowledge of the pH value of the reference buffers prepared at different partial aqueous compositions, which is not usually available [53]. A correction can be obtained for ${}^{s}_{w}\text{pH}$ since between the values ${}^{s}_{s}\text{pH}$ and ${}^{s}_{w}\text{pH}$ there is a difference given by a term δ that is constant for each mobile phase composition. The following relation can be used for ${}^{s}_{s}\text{pH}$ calculation:

$$ {}^{s}_{s}\text{pH} = {}^{s}_{w}\text{pH} - \delta \qquad (13.4.14)$$

The parameter δ for methanol–water-based mobile phases can be estimated from the solvent composition with the following empirical equation [54]:

$$\delta = \frac{0.09\phi_{MeOH} - 0.11\phi^2_{MeOH}}{1 - 3.15\phi_{MeOH} + 3.51\phi^2_{MeOH} - 1.35\phi^3_{MeOH}} \qquad (13.4.15)$$

In expression (13.4.5), ϕ is the volumetric organic phase content expressed by the formula $\phi = \frac{V_{organic}}{V_{organic}+V_{water}}$. The parameter δ for the acetonitrile–water-based mobile phase is the following:

$$\delta = \frac{-0.446\phi^2_{AcCN}}{1 - 1.316\phi_{AcCN} + 0.433\phi^2_{AcCN}} \qquad (13.4.16)$$

The variation of δ for methanol and acetonitrile is shown in Fig. 13.4.2.

The organic content ϕ in a mobile phase also influences the dissociation of acids and bases. For example, methoxide ion is a more potent nucleophile than hydroxide, such that basic compounds in methanol–water mixtures can show different chemical behavior than in water alone, even when the hydrogen ion activity is the same. This fact is relevant to column stability and sample stability in basic methanol–water mixtures. When the content of the

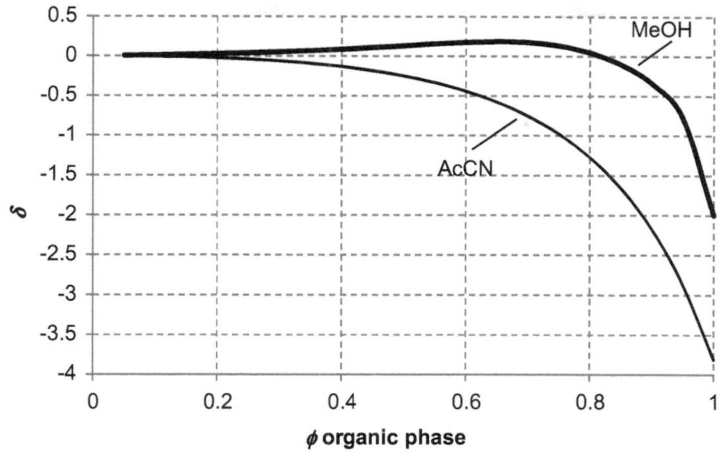

FIGURE 13.4.2 Vartiation of pH correction δ for methanol and acetonitrile at different volume fractions ϕ of the organic solvent.

organic solvent in solution increases, the dielectric constant and the activity coefficients decrease. In the presence of the organic component from mobile phase, the acidic/basic properties of solutes are modified in different proportions, depending on solute structure and the nature of the organic modifier. A linear relationship between $pK_a(w)$ of a solute in pure water and $pK_a(s)$ of the same solute in a solvent s, which may contain water and another solvent, can be written as the following empirical relation:

$$pK_a(s) = \alpha_s pK_a(w) + \beta_s \tag{13.4.17}$$

In Eq. (13.4.17) the intercept β_s is related to the differences in basic character, dielectric constants, and specific solvation interactions (e.g., hydrogen bonding) of the solute and the solvent s and water, respectively. The slope α_s is related to the differences between specific solvation interactions, which depend on the solvent and solute. Thus, for an organic modifier the values of α_s and β_s are calculated for sets of compounds according to the following equations dependent on volume fraction of the solvent (ϕ_s) in the mobile phase composition:

$$\alpha_s = \frac{1 + \alpha_1 \phi_s + \alpha_2 \phi_s^2}{1 + \alpha_3 \phi_s + \alpha_4 \phi_s^2} \tag{13.4.18}$$

and

$$\beta_s = \frac{1 + \beta_1 \phi_s + \beta_2 \phi_s^2}{1 + \beta_3 \phi_s + \beta_4 \phi_s^2} \tag{13.4.19}$$

where α_1, α_2, α_3, α_4, β_1, β_2, β_3, and β_4 are parameters obtained by numerical best-fit techniques for all acids or bases from the same family. For example, the values of the best-fit parameters for several classes of compounds and a mobile phase consisting of water–acetonitrile are given in Table 13.4.1 [55].

Besides dependences as given by formula (13.4.17), different other approximations were proposed to estimate the $pK_a(s)$ values in water/organic solvent mixtures. Simpler formulas

TABLE 13.4.1 The Fitting Parameters for Predicting the Slope (α_s) and the Intercept (β_s) of the Correlation Between $pK_a(w)$ and $pK_a(s)$, as Given in Formula (13.4.17) [55]

Compounds	α_1	α_2	α_3	α_4	β_1	β_2	β_3	β_4
Aliphatic carboxylic acids	9.97	−8.59	8.83	−8.72	−0.68	9.94	8.45	−8.59
Aromatic carboxylic acids	−2.42	3.14	−1.98	2.12	9.97	−9.12	5.96	−6.90
Phenols	10.05	−10.04	7.97	−8.37	−5.33	9.95	0.19	−0.70
Amines	−0.73	−0.27	−0.87	−0.12	−1.82	2.25	−1.75	0.90
Pyridines	−1.67	0.67	−1.66	0.67	−1.78	1.89	−0.58	−0.40

such as linear or quadratic dependencies were suggested for this purpose, with expressions dependent on $pK_a(w)$ and volume fraction of the solvent (ϕ_s) as given below:

$$pK_a(s) = A' + B'\phi_s \tag{13.4.20}$$

$$pK_a(s) = A' + B'\phi_s + C'\phi_s^2 \tag{13.4.21}$$

where $A' \approx pK_a(w)$ and where B' and C' are empirical parameters dependent on a specific compound for which $pK_a(s)$ is needed [56]. Examples of the values of these parameters are given in Table 13.4.2 for methanol/water solutions, as a function of methanol content.

Other such estimations for $pK_a(s)$ based on $pK_a(w)$ and ϕ are reported in the literature [55].

The Influence of Temperature on the pH of Buffers

The pH of a buffer is also influenced by temperature. The first pH variation is related to the ionic product of water, which depends on temperature and $pK_w = 13.99$ only at 25°C. At 0°C $pK_w = 14.95$ and at 75°C $pK_w = 12.70$ (with a nonlinear variation). This variation indicates that the neutral pH is 14 only at 25°C. The dissociation of weak acids and bases also depends

TABLE 13.4.2 Polynomial Equations Describing the Dependences of $pK_a(s)$ of Some Buffers Currently Used in RP-LC as a Function of the MeOH Content in Mobile Phase

pK_a	Equation in solvent/water	R^2
pK_1 (H_3PO_4)	$pK(s) = 2.127 + 2.16 \times 10^{-2}\phi_{MeOH} + 6.81 \times 10^{-5}\phi^2_{MeOH}$	0.9982
pK_2 ($H_2PO_4^-$)	$pK(s) = 7.202 + 1.27 \times 10^{-2}\phi_{MeOH} + 2.14 \times 10^{-4}\phi^2_{MeOH}$	0.9997
pK_1 (citric acid)	$pK(s) = 3.121 + 1.56 \times 10^{-2}\phi_{MeOH} + 6.47 \times 10^{-5}\phi^2_{MeOH}$	0.9992
pK_2 (citric acid)	$pK(s) = 4.756 + 1.62 \times 10^{-2}\phi_{MeOH} + 9.33 \times 10^{-5}\phi^2_{MeOH}$	0.9986
pK_3 (citric acid)	$pK(s) = 6.391 + 2.15 \times 10^{-2}\phi_{MeOH} + 7.65 \times 10^{-5}\phi^2_{MeOH}$	0.9996
pK (acetic acid)	$pK(s) = 4.757 + 1.26 \times 10^{-2}\phi_{MeOH} + 1.09 \times 10^{-4}\phi^2_{MeOH}$	0.9998
pK (NH_4^+)	$pK(s) = 9.238 - 5.87 \times 10^{-3}\phi_{MeOH} - 2.22 \times 10^{-5}\phi^2_{MeOH}$	0.9991

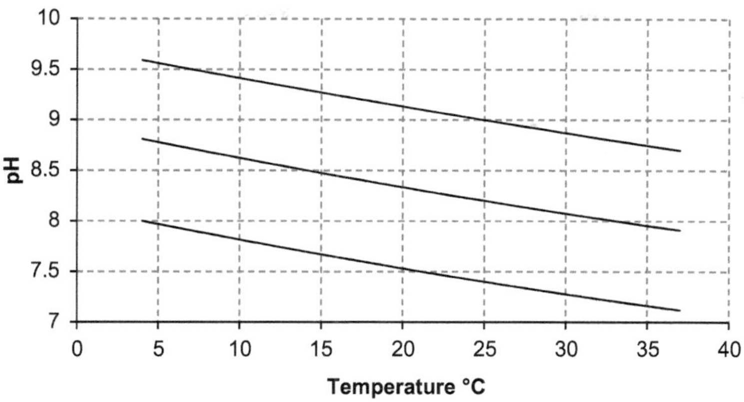

FIGURE 13.4.3 Variation of pH with temperature for three different solutions of tris/HCl buffers with different pH and a concentration of 50 mM.

on temperature, such that pK_a typically listed for 25°C is not the same at a different temperature. As a result, depending on the nature of the buffer, the temperature change may influence the pH differently.

Another aspect is related to the measurement of pH using a glass electrode. As indicated by expression (2.4.3), the electrode potential used for pH measurement is temperature-dependent. The result is a combination of effects, and a detailed description of pH variation with temperature is not usually practical. However, it is common that the pH of buffer solution decreases as the temperature increases. An example of the variation of pH with temperature is given in Fig. 13.4.3 for three different solutions of tris/HCl buffers with different pH and a concentration of 50 mM.

Additional complexity of pH variation with temperature appears when the mobile phase is partially organic. In such cases, only experimental measurements performed considering the influence of temperature and of organic composition on both the solution and the measuring electrode (typically glass electrode) must be considered. This variation with temperature of buffer pH is not usually indicated in HPLC analytical procedures with the goal of measuring specific analytes. In most HPLC separations, the pH for the buffer solutions is indicated for room temperature (25°C). Since the pH varies with the temperature, this effect must be considered when performing a separation at a different temperature than the one where the buffer pH was measured, and any adverse effect of temperature increase must be avoided.

Besides pH the addition of an organic modifier also affects the buffer capacity β. Organic solvents have the effect of decreasing buffer capacity and also of shifting the pH where β has a maximum. For methanol, for example, this shift is toward higher pH values [52].

Solubility of Buffers in Partially Organic Mobile Phases

The choice of a mobile phase composition (nature and concentrations) containing buffers must take into consideration the solubility of the buffer in the presence of the organic modifier. Binary mobile phases, which are the most common in HPLC (see Section 13.5) are frequently made having one aqueous or mainly aqueous component and an organic component. It is common that the buffer is made only in the aqueous or mainly aqueous component of the mobile phase. For the aqueous component, the buffer is made in water. For making the partially aqueous components, there are two alternatives: (1) the buffer is made in water and the organic component is added afterward, (2) the mainly aqueous component is made by mixing water with an organic solvent, and the buffer components are subsequently added. Since the pH of a partially aqueous solution is improperly measured with a common glass electrode, it is preferable to use alternative (1), with a precise pH measurement, although changes in this pH occur when the organic component is added. Nevertheless, both procedures are sometimes described for different HPLC methods. Further pH change may occur during a gradient separation, when more organic component is added to the mobile phase.

Since buffers are made using acids, bases, and salts, and these are sometimes inorganic compounds, their solubility in the organic or partially organic phase can be low. This solubility depends on the nature of the buffer, its concentration, and the nature and percentage of the organic modifier in the mobile phase composition. For inorganic salts, the solubility depends mainly on the nature of the cation, and their solubility trend in partial organic solvents follows approximately the solubility in water: $NH_4^+ > K^+ > Na^+$. The alkylammonium cation has a higher solubility in organic solvents, such as methanol or acetonitrile, due to affinity of the alkyl chain toward these solvents. As an example, several solubility values for potassium phosphate-type buffers in solvents that are commonly used in HPLC are given in Appendix 7j [57]. As a result of lower solubility of many buffers in organic solvents, the buffer concentration should be selected at the lowest acceptable concentration (although this leads to a decrease in buffer capacity). Particular attention must be given to buffer solubility when used in gradient separations and the organic content of the mobile phase is increased during the chromatographic run (e.g., in RP-HPLC). A higher content of organic component is usually associated with a decrease in buffer solubility, and the buffer concentration must be adjusted appropriately to avoid any precipitation of salts in the chromatographic system.

Additives

A variety of compounds can be added to the mobile phase for modifying its properties. The additives have various functions and they can be classified into several groups: (1) additives used for the modification of the pH of mobile phase, (2) additives for the modification of ion strength, (3) ion pairing compounds used in IPC, (4) chiral additives in the mobile phase for chiral separations on nonchiral columns, (5) other compounds that form adducts or complexes with the analytes to make them amenable for detection, (6) compounds added to maintain a specific form of stationary phase, (7) preservatives, (8) additives used for enhancing MS ionization, and (9) other additive types.

1) Certain acids or bases can be added to the mobile phase for pH modification, without participating in a buffer system. Examples of such acids include H_3PO_4, H_2SO_4, HCOOH, CH_3COOH, CHF_2COOH, CF_3COOH, C_2F_5COOH, ammonium hydroxide, pyrrolidine, pyridine, triethylamine, ethanolamine, or quaternary ammonium hydroxides, etc. Such additions are sometimes necessary for keeping the mobile phase very acidic or very basic.

2) Ion strength may be important in some separations, and this can be modified by adding certain salts to the mobile phase. Among these types of additives are certain chaotropes that can disrupt the hydrogen bonding and solvation of ions (see Section 4.4).

3) In ion pairing chromatography, an ion pairing agent (IPA or hetaeron) is added for producing ion pairs with the analyte (or to attach to the stationary phase) (see Section 4.3). A variety of ion pairing agents are reported in the literature (see, e.g., Ref. [58]).

4) Chiral additives can be used in the mobile phase, with separation on nonchiral stationary phases (see Section 4.3). These additives are used to form diastereoisomeric complexes with the chiral analytes. Different such agents are reported in the literature [59].

5) Formation of the ion pairs for IPC analysis and chiral additives are in fact part of a larger group of additives that form complexes or adducts with the analytes making them amenable for detection. Some of these additives are derivatization reagents added in the mobile phase that can contribute to formation of colored or fluorescent compounds (see, e.g., Ref. [60]). Other compounds form adducts that can be detected, for example, by LC-MS/MS [61].

6) Some additives are used for maintaining a specific structure of the stationary phase, such as in ion-mediated chromatography where the stationary phase is a cation exchange material in metal form, and the mobile phase must contain a low level of that metal ion. Also, a low level of a strong acid can be used in the mobile phase, for example, for maintaining an ion exchange column in H^+ form.

7) Preservatives are sometimes added to the mobile phase for avoiding the growth of microorganisms (e.g., in buffered solution of phosphates or acetates) or for avoiding the formation of epoxides (e.g., in mobile phases containing tetrahydrofuran or ethers). As a preservative against microorganism growth, it is common to use a small amount of NaN_3 dissolved in the mobile phase (5–10 mg/L).

8) It is common to use special additives for enhancing MS ionization in LC-MS and LC-MS/MS techniques. The proton transfer process is strongly influenced by the presence of additives, and as a result the detection sensitivity is significantly affected by them. The most common such additives typically added at concentrations around 0.1% are HCOOH when the ionization is performed in positive mode, and $HCOONH_4$ when ionization is performed in negative mode. However, other additives such as CHF_2COOH, CF_3COOH, CCl_3COOH, and CH_3COOH, are used for positive ionization mode, and $HCOONH_4$, CH_3COONH_4, NH_4HCO_3, and even HCOOH are used in negative ionization mode. Some additives such as CF_3COOH are sometimes indicated as suppressing ionization (in positive mode), but this compound can be necessary in the mobile phase for decreasing the pH, and the result in affecting the ionization can be positive or negative, depending on the analyte.

9) Among other additives in the mobile phase are certain organic modifiers that affect the "strength" of the mobile phase (see, e.g., Ref. [62]). For example, small amounts of alcohols with a long aliphatic chain can be added in a mobile phase containing methanol and water, with the result of obtaining a "stronger" mobile phase for RP separations [63]. Such additives can be considered as solvents making the mobile phase, but their low level of addition classifies them more like additives.

Influence of the Buffer and Additives on Column Stability and Properties

The properties of the stationary phases are critical regarding their resilience to extreme pH values of the mobile phase (see Chapter 6). However, the chemical stability of the silica-based stationary phase is also affected by the type and concentration of the used buffers. The effect of different buffers leading to the deterioration of silica backbone may be caused by the combined effects of pH and the capacity of complexation of the buffer. A study performed to evaluate column stability [64], used 6 L of eluent containing 50% CH_3OH and 50% buffer with pH = 10 (v/v) with a column packed with a C18 stationary phase. The effect on column chemical stability was measured as the amount of dissolved silica. A mobile phase based on phosphate buffer dissolved about 110 mg/column, a carbonate-based buffer dissolved about 40 mg/column, and for borate- and glycine-based buffers the dissolution was almost unobservable. This showed that the nature of the buffer composition can influence substantially the longevity of RP columns. Different other studies [65] have shown that column deterioration is best prevented by sodium as the buffering cation. In contrast, potassium is a more aggressive buffering cation compared to sodium and ammonium. These effects seem to be caused by an increase in the pH for the aqueous/organic solution upon the addition of the organic phase, which is different for buffers with different chemical nature, although in water their pH is the same [66]. The longevity of silica-based RP columns depends also on the concentration (ion strength) of the buffer. Even at pH = 7 it was proved that by passing 10 L of mobile phase containing buffer/ACN = 50/50 (v/v), the dissolution of silica support is about 175 mg/column for a buffer concentration of 10 mM/L, 220 mg/column for a concentration of 50 mM/L, and 325 mg/column for 250 mM/L. Therefore, in order to prevent early column failure, low buffer concentrations are recommended [65].

In order to maintain the longevity of silica-based columns apart from the choice of the resistant columns (see Chapter 6), the selection of the nature and concentration of the buffer in the eluent is very important. Typically, borate and organic-based buffers such as glycine, in low concentration combined with the properly selected counterion substantially prevent early column degradation of silica-based RP columns. On the other hand, the use of phosphate and carbonate as buffering anions leads to a faster dissolution of the silica support. Specific columns may display a different type of interaction depending on the mobile phase pH. This is, for example, the case of weak cation exchange columns and weak anion exchange columns. The ionic character of the stationary phase containing, for example, carboxyl groups can be basically eliminated by adjusting the pH of the mobile phase within two pH units of pK_a of the acidic groups. The column will act as a polar-type column and can be used in HILIC-type separations.

Suitability of the Buffers and Additives for Detection in HPLC

In any HPLC method, the buffers and the additives from the mobile phase must be acceptable for the detection procedure used in the method. Even if a separation may be done preferably with the mobile phase containing a specific buffer and/or additives, the mobile phase is rendered useless if the detection cannot be performed properly. Most types of detection can be affected by buffers, including the main detection techniques such as UV, fluorescence, and all techniques based on evaporative processes (MS, ELSD, and cCAD). Even in RI detection, a high concentration of a buffer may reduce the technique sensitivity.

The presence of a buffer in the mobile phase may produce a change in the UV transparency of the mobile phase, modifying the UV cut-off value and also may change the wavelength of absorption for a specific compound. Since many HPLC methods use UV detection, the UV cut-off for specific buffers and at a specific concentration is important. The cut-off UV values for several buffers are given in Appendix 7k.

For fluorescence detection, solution pH is also very important. Both emission wavelength and fluorescence intensity for many compounds are dependent on pH. When different structures are possible for a compound, it is very common that the fluorescence of one species is different from that of the other. For this reason, the mobile phase pH must be carefully controlled in most methods using fluorescence detection. Also, fluorescence may be quenched by specific additives. Oxidants must be avoided from the mobile phase, and specific trace metals may influence fluorescence.

For all detection techniques that use an evaporative step, the use of nonvolatile buffers in the mobile phase is not acceptable even when used at low concentrations. Among these nonvolatile materials are acids such as H_3PO_4, H_2SO_4, H_3BO_3, the salts of these acids, nonvolatile salts of volatile acids such as HCOOK or CH_3COOK, and nonvolatile bases. MS detection, in particular, being a common detection technique and having excellent qualities regarding sensitivity and selectivity, requires volatile buffers. Formation of aerosol particles during the evaporative step in ESI and APCI for MS detection disturbs the formation of ions. In ELSD, the nonvolatile buffers interfere with the quantitation. Chemical properties and concentration of the buffers, as well as pH, have a significant effect on analyte response in ESI. Some buffers acceptable or not for use with evaporative detectors are indicated in Appendix 7k.

Although nonvolatile additives at typical concentrations used for buffers (e.g., 10 mM) are not usable in techniques that use an evaporative step, traces of nonvolatile additives (e.g., in the range of 10–50 μM) can be tolerated in techniques such as mass spectrometry with ESI or APCI ionization (see, e.g., Ref. [61]). Buffers at these concentrations would have an extremely low buffer capacity and therefore are not utilized.

13.5 SELECTION OF MOBILE PHASE IN HPLC

The mobile phase in HPLC can have a constant composition (in isocratic separations) or can have a composition that changes during the chromatographic run (in gradient separations) [67]. The chromatographic column is typically conditioned with the mobile phase before starting a chromatographic run. This assures a proper chromatographic medium that includes a stationary phase in intimate contact with the mobile phase. This section is related to general properties of the mobile phase in different chromatographic types, while specifics about the gradient conditions are presented in Chapter 14.

The main aspects regarding the mobile phase are related to its composition, including the solvents, the buffers, and other additives that made it up [68]. There are also other aspects regarding the mobile phase besides its composition, and these are related to the mobile phase utilization such as the flow rate, the temperature utilized for the separation, and degassing of mobile phase.

Solvent Purity in HPLC

Solvents used in HPLC must be very pure unless a known impurity is present in the solvent and it does not affect in any way the HPLC analysis. Solvent impurities may affect the HPLC analysis in various ways. The impurity may generate the following: (1) interaction with the analytes, (2) problems with the separation, (3) problems with the detection, and (4) deterioration of the HPLC equipment (metal parts, column stationary phase, etc.).

Solvent impurities may come from additives, compounds from solvent decomposition (e.g., peroxide formation in ethers, or hydrolysis of esters), or can be present from the solvent synthesis. As an example, chlorinated compounds may hydrolyze to form HCl, which can produce some decomposition of the analytes. The presence of impurities may also affect the separation by producing changes in the retention time compared to a pure solvent. Impurities frequently affect detection, either by increasing the signal background or by quenching the signal (such as in the case of fluorescence). Equipment deterioration is also possible, for example from the HCl present in some old chlorinated solvents.

Besides the impurities dissolved in solvents, small insoluble particles may be present in some solvents. These impurities must be eliminated by filtration, which is typically performed using 0.45-μm pore filters. The filters must be made from materials perfectly inert to the solvent (e.g., PTFE). However, many prefiltered solvents for HPLC are commercially available.

Mobile Phases Used in RP-HPLC

Mobile phases used in RP-HPLC typically contain water with a certain proportion of one or more organic solvents (besides buffers and additives). For this reason, water can be considered as the most important solvent used in RP-HPLC. As organic solvents, methanol or acetonitrile are usually added to water. Other solvents with solubility in water can also be used as an organic component. Water with no added organic solvent may be used as mobile phase on stationary phases specifically designed to be used in 100% water mobile phase (e.g., Synergy Hydro RP from Phenomenex, Intesil ODS-4 from GL Science, and Acquity UPLC HSS T3 and Atlantis T3 from Waters) for the separation of highly hydrophilic compounds. Many other RP columns may suffer dewetting in mobile phases with a very high content of water, and they must be used only with a certain proportion of organic solvent (e.g., higher than 10%) in the mobile phase. The only exception of water not being used in an RP-type separation is in NARP.

Before each run, the column is conditioned with the same mobile phase used for elution for isocratic separations, or with the initial composition of the mobile phase when gradient elution is used. In gradient elution HPLC, mobile phases with high water content are used at the beginning of gradient programs for achieving retention of the sample at the column head. Following the high water content, the organic component is increased for accelerating the elution from the RP column.

Water has a high polarity ($\delta = 21$, experimental log $K_{ow} = -1.38$, $P' = 10.2$, $\pi^* = 1.09$ as shown in Appendix 7), has a high hydrogen-bond donor capability, and an average hydrogen-bond acceptor capability. With these characteristics, water is a "weak" eluent in RP-HPLC, and the compounds with some hydrophobic character are strongly retained on the stationary phase from mobile phases with high water content.

Besides acting as a polar solvent with "weak" elution character in RP-HPLC, water is the ideal solvent for buffers and ionic additives in the mobile phase. Also, water is a good solvent for polar samples including amino acids, carbohydrates, proteins, and many other compounds. The purity of water for HPLC, in particular when high water content is necessary in the mobile phase and when sensitive detection is utilized (e.g., MS), is very important. Water is also used in superheated water chromatography, at higher temperatures water becoming less polar [69].

Water is also an excellent solvent for buffers and many additives. However, the concentrations of buffer higher than 50 mM are not in general recommended. The pH of buffers in water can be directly measured, for example, with a glass electrode.

For most detection techniques, water is an adequate solvent. It has a very low UV cut-off value (<190 nm), does not interfere with fluorescence, and for evaporative detection techniques can be volatilized. For MS detection, the mobile phase of pure water is not recommended, but a level of as little as 5% organic modifier (e.g., methanol or acetonitrile) in water can be used successfully to achieve good sensitivity (additives such as HCOOH are typically used in the mobile phase for MS detection). Often the best sensitivity in ESI is achieved when the analyte is ionized already in a liquid phase by using an acidic mobile phase for basic analytes, such as amines (pH two units below pK_a of the analyte), and basic conditions for acidic analytes, such as carboxylic acids and phenols (pH two units above pK_a of the analyte) [70].

One group of solvents very common in RP-HPLC is that of alcohols, with methanol the most frequently utilized. Alcohols are typically utilized in mixture with water. The alcohols' polarity is lower than that of water, but they have a relatively high capability to form hydrogen bonds, both as donors and as acceptors, usually with higher hydrogen-bond acceptor capability compared to water. Besides methanol, the alcohols more frequently utilized are isopropanol and ethanol. For the decrease of mobile phase polarity (making the mobile phase "stronger" for RP), it is sufficient in many separations only to increase the content of the organic phase (ϕ), not changing the chemical nature of the alcohol. However, if this does not provide enough "strength" for the mobile phase, alcohols with a higher number of carbon atoms in the molecule can be used (C2, C3, C4) as stronger eluents [63,71,72]. Alcohols, as the length of their hydrocarbon chain increases, also have higher solubility for less polar solvents such as hexane, heptane, or cyclohexane. This allows the use of mixtures of alcohol/hydrocarbon as solvent phase in NARP when very hydrophobic analytes have to be separated. Alcohols in general have a low UV cut-off value. Methanol (and other alcohols) presence in an aqueous mobile phase decreases the solubility of inorganic salts, acids, and bases typically used for adjusting the pH. The amount of buffers and/or additives allowed should be selected such that they are completely soluble in the mobile phase even when the organic content of the mobile phase increases (see Chapter 14).

Acetonitrile is another very common solvent used in RP-HPLC. This solvent has medium polarity, a weak hydrogen-bond acceptor, and very weak donor capability. Compared to methanol, acetonitrile has slightly higher values for P' and π^* but a higher K_{ow}. It typically acts as a stronger eluent compared to methanol. As an example, the correlation between log k' for a set of 30 various compounds on a Lichrosphere 100 RP-18e with 5.0 μm particles (from Merck) in 30% methanol/70% water and log k' for the same compounds in 30% acetonitrile/70% water is shown in Fig. 13.5.1 (log k' data from Ref. [73]). The slope of the dependence shown in Fig. 13.5.1 indicates that 30% acetonitrile in water leads to smaller values for log k' (with a coefficient of about 0.8). However, different compounds have different responses to the two mobile phases and the correlation between log k' in methanol and in acetonitrile is not perfect ($R^2 = 0.8973$).

Although acetonitrile has log $K_{ow} = -0.34$, pure acetonitrile can elute from a hydrophobic chromatographic column even molecules with log K_{ow} as high as 7. Mixtures of water/

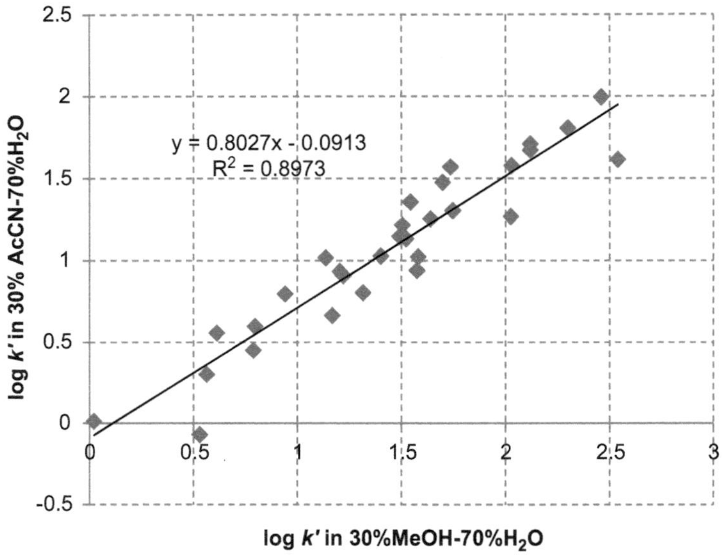

FIGURE 13.5.1 Correlation between log k' for a set of 30 various compounds on a Lichrosphere 100 RP-18e (from Merck) in 30% methanol/70% water and log k' for the same compounds in 30% acetonitrile/70% water.

acetonitrile are common solvents in RP-HPLC, and the eluting strength of this mobile phase increases upon acetonitrile addition. From the log K_{ow} of the analyte an empirical formula can be given, which provides guidance regarding the volume fraction of water in acetonitrile for achieving elution within a 25-min window from a common 150-mm RP column:

$$\phi = \frac{7 - \log K_{ow}}{7.5} \tag{13.5.1}$$

Expression (13.5.1) does not provide any definite value for the retention time (compounds with low log K_{ow} will elute close to the dead volume of the column, while those with large log K_{ow} toward the end of the time window), but it indicates the range of necessary composition for acetonitrile/water mixtures. Acetonitrile presence in an aqueous mobile phase decreases the solubility of inorganic additives.

The UV cut-off of acetonitrile is very low (see Appendix 7h), which is an important characteristic when UV detection is utilized. For MS detection, acetonitrile is adequate in water/acetonitrile mobile phases. The response intensity in MS is usually dependent on the mobile phase composition, and the use of methanol or acetonitrile as an organic modifier in the mobile phase can generate very different results. In a number of cases, acetonitrile favors detection in positive mode by increasing sensitivity, and methanol favors detection in negative mode ionization. However, only direct experiments may verify which is the optimum solvent for MS detection.

Among other solvents used in RP-HPLC and NARP are ethers, ketones, and esters. Ethers are compounds typically used in RP-HPLC for further decreasing the polarity of the mobile phase compared to short-chain aliphatic alcohols or acetonitrile. Tetrahydrofuran, dioxane, tert-butyl methyl ether, and diethyl ether are among the ethers used in RP-HPLC. Ethers also have very low donor hydrogen bonding capability and medium acceptor hydrogen-bonding capability. Except for NARP where ethers may be more frequently used, in conventional RP-HPLC, the ethers, when used, are typically at relatively low proportions. These compounds have a UV cut-off value around 210–215 nm.

Ketones are not very different in their solvent capabilities compared to ethers. However, ketones typically have relatively high UV cut-off values (around 330 nm) and therefore are less amenable for UV detection, which is one of the most common detection techniques. For this reason, the use of ketones as modifiers of the mobile phase in RP-HPLC is less common. However, since some ketones may have strong eluting properties in RP-HPLC, they can be used, for example, for the elution of hydrophobic compounds from graphitic columns. Acetone is considered as a greener alternative to replace acetonitrile in reversed-phase liquid chromatography, but its applicability with UV-based detectors is limited [74].

Esters, such as methyl or ethyl acetate are sometimes used as mobile phase modifiers in RP-HPLC, but due to their low solubility in water a third component (methanol) is necessary for mixture miscibility. However, their low water solubility, potential to hydrolyze in acidic or basic media, and their relatively high UV cut-off value makes these solvents less useful in RP-HPLC [75].

Hydrocarbons as solvents have applications in NARP. Their immiscibility with water makes them useless in conventional RP-HPLC, which usually requires water as a polar component in the mobile phase. However, in NARP, mixtures of alcohols and hydrocarbons are common components of the mobile phase added with the purpose of lowering the polarity of the mobile phase. Hydrocarbons such as hexane, heptane, or cyclohexane are good solvents for highly hydrophobic analytes such as triglycerides or carotenoids [76].

Halogenated hydrocarbons have medium polarity (except for CCl_4) but low hydrogen-bonding capability. They are very good solvents for many compounds, but are not water-soluble, which makes them of little use in RP-HPLC when water is present in the mobile phase. However, they may be used in NARP [77]. Among the solvents used more frequently in this technique are methylene chloride, chloroform, and carbon tetrachloride. Other chlorinated solvents are seldom utilized. However, for some solubility reasons, special chlorinated solvents such as chlorobenzene can be used, for example when the separation must be performed at temperatures higher than 70°C and a less-volatile solvent must be used as an eluent. Some concerns related to the use of chlorinated compounds as mobile phase may come from their potential hydrolysis with the formation of HCl.

Solvents, Ion Pairing Agents, and Additives Used in Ion-Pair HPLC

In ion pairing (IP), the mobile phase contains three types of components with different roles. These components can be classified as follows: (1) the solvents used in the mobile phase, (2) the ion pairing agents (IPAs) or the hetaerons (see Section 4.3), and (3) the buffers and additives used for adjusting pH, ion strength.

The solvents used for mobile phase in IP are very similar to those used in RP-HPLC, with the mobile phase typically consisting of water plus an organic modifier. Methanol and acetonitrile are the most utilized organic component of the mobile phase, although other solvents were reported in the literature. The effect of the organic modifier is similar to that of the organic modifier in RP separations. However, the change of organic content in the mobile phase also causes changes in the mobile phase dielectric constant, which can influence the strength of the ionic interactions between charged analyte and IPA. For this reason, the typical linear dependence of log k' on the content of organic solvent ϕ (as indicated by expression (4.3.13)) is not as common as it is for typical RP-HPLC. In IP-HPLC, the dependence of capacity factor log k' on the content of the organic modifier in mobile phase can significantly deviate from linear.

A large variety of ionic compounds that also have a hydrophobic moiety were used as IPAs [78]. The IPAs are always selected to have the opposite charge to the analyte. The mechanism of IPA retention as discussed in Section 4.3 showed that the hydrophobic chain of the IP agent plays a similar role regarding retention as the hydrophobic moiety of the solutes in RP-HPLC. As the hydrophobic character of the IPA increases, the retention of the analyte (solute) is stronger. However, the analyte structure, in particular on columns where different interactions besides the hydrophobic ones are possible, also plays a significant role in the separation (see expression (4.3.24)).

The most common IPAs for acidic analytes are quaternary amines. The counterion of the quaternary amine can be sulfate, chloride, bromide, iodide, hydroxide, or dihydrogenphosphate. A number of alkyl substituents (hydrophobic chains) at the nitrogen can be chosen for these IPA molecules, and the most common are trimethylalkyl, with the alkyl ranging from C4 to C10. For basic compounds, among the common IP agents are sulfonates $R-SO_2-O^-$ and sulfates $R-O-SO_2-O^-$, with the R group alkyl, aryl, aryl with an alkyl side chain (C4 to C8), and in some cases with other substituents [79]. Also, fluorinated carboxylic acids $CF_3(CF_2)_nCOOH$ with $n = 0$ (trifluoroacetic acid), $n = 1, 2, 3, 4$ are common IPA for basic compounds. The counterions for these reagents are usually either H^+ or Na^+ ions, but other counterions are possible. Besides the typical IPAs, numerous other IPAs were reported in the literature [80]. For the separation of inorganic ions, for example, crystal violet can be used as IPA, with the eluting order obtained on a hydrophobic column being: $Cl^- < NO_2^- < Br^- < NO_3^- < I^- < SO_4^{2-} < S_2O_3^{2-}$, which is the same as that observed in anion exchange LC [81]. Among other IP agents are ionic liquids that are playing either the role of an IPA or that of an additive [82]. Both the cationic and the anionic participants of the ionic liquid (IL) contribute to peak shape and to the improvement of resolution.

Common buffers and additives can be used in IPC, for example for adjusting the mobile phase pH. However, in some instances it is difficult to differentiate the role of additives from that of adjusting the pH or being an IPA. For example, in a method for amino acid analysis on a Dionex Acclaim RSLC PolarAdvantage II column, the mobile phase uses a gradient with Solution A 1% acetonitrile/0.5% heptafluorobutyric acid/0.02% trifluoroacetic acid (TFA) in water, and Solution B 0.1% TFA in acetonitrile [83]. The role of TFA can be that of a hetaeron as well as that of a pH adjusting agent.

Regarding detection in IPC, the UV absorption is widely used. For this reason, the ion-pair agents must have a low UV cut-off wavelength. The UV absorption of sodium alkanesulfonates and quaternary ammonium salts is low for the typical UV range of detection in HPLC. A higher absorption takes place when ion liquids containing pyridinium or imidazolium moieties are used. Fluorescence is also frequently used, and most ion-pairing reagents do not affect it negatively [84].

A number of ion-pair reagents used in IP-LC separations of highly polar analytes are not suited for use in LC-MS. These reagents are nonvolatile and reduce the detection sensitivity.

When MS detection is employed, volatile IP reagents such as aliphatic amines or perfluorinated organic acids at the lowest possible concentration in mobile phase are recommended. TFA, pentafluoropropanoic acid (PFPA), and heptafluorobutanoic acid (HFBA) have commonly been used in the analysis of polar basic compounds. These ion-pairing agents form relatively stable ion-pairs with basic compounds, decreasing the secondary interactions between free silanol groups of the stationary phase and resulting in decreased peak tailing, improved resolution, and better retention. However, acidic ion-pairing agents may suppress ionization in the MS source. Alkylamines (triethylamine, N,N-dimethyl-n-butylamine, N,N-dimethylhexylamine, or tri-n-butylamine) are commonly used as ion-pairing reagents in RP-LC MS detecti/MS detection of acidic compounds, such as nucleoside mono-, di- and triphosphates and sulfonates, sulfates, sulfonated dyes, and halogenated acids [47].

The evaporative light scattering detector is another example of a detector that requires volatile mobile phases, and the choice of IPA and buffers with this detection must avoid the nonvolatile additives such as alkylsulfonates.

Mobile Phases Used in HILIC and NPC

The mobile phase used in HILIC and NPC is characterized by its lower polarity compared to the stationary phase. Typical mobile phases in HILIC are made of an aqueous buffer (10–40% of the mobile phase) with a controlled pH and an organic solvent such as acetonitrile or methanol (60–90%). Other solvents such as ethanol or 2-propanol are sometimes used as organic modifiers. The solvent "strength" in HILIC is reversed from that in RP-HPLC, and has the following order: water > methanol > ethanol > 2-propanol > acetonitrile > acetone > tetrahydrofuran. For this reason, in HILIC separations using gradient conditions, the separation starts with a solvent low in water, and the water content is increased for the elution of the analytes that are more strongly retained [85].

One particular aspect of the mobile phase in HILIC is that compared to the mobile phase in RP, it typically contains a lower level of water (maintaining a lower polarity of the mobile phase compared to the stationary phase). This characteristic may pose problems with buffer solubility. For this reason, buffers with good solubility in mixtures containing a higher content of organic solvents are necessary for HILIC separations, such as ammonium acetate or formate. Buffer concentration is also important in separations by this mechanism: the increase of the buffer concentration leads to a decrease in the retention when the ion exchange mechanism controls the retention, while the opposite effect may occur in the absence of the ion exchange mechanism under HILIC conditions [86].

Among the organic solvents used in HILIC, acetonitrile is the preferred organic solvent, while the other solvents may lead to insufficient sample retention and broad or nonsymmetrical peak shapes. This is, for example, the case of methanol, which is less used as the organic component in HILIC separations. The relatively poor performance of methanol in HILIC may be due to its similarity to water, both methanol and water being protic solvents that are able to form hydrogen bonds. Methanol can compete to the solvation of the surface of silica or of other polar stationary phases used in HILIC and provide strong hydrogen-bonding interactions [87]. Other solvents that can be used in NPC include alkanes (n-pentane, n-hexane, n-heptane, i-octane), cycloalkanes (cyclopentane, cyclohexane), fluoroalkanes, chlorinated alkanes (dichloromethane, chloroform, carbon tetrachloride, propylchloride), ethers (diethyl ether, di-i-propyl ether), esters (methyl acetate, ethyl acetate), other alcohols (ethanol, 1-propanol, 2-propanol), amines (pyridine, propylamine, triethylamine), and carboxylic acids or their derivatives (e.g., dimethylformamide).

The mobile phase in HILIC and NPC seems to play a more important role than simply elution medium. The polar solvent molecules from the mobile phase (e.g., water) can be adsorbed onto the polar sites of the surface of the stationary phase changing its interacting properties with analytes. In HILIC this effect can be diminished when enough water is present in the mobile phase and it is adsorbed on the stationary phase surface, assuring that no significant changes in surface nature take place during the separation. The same role as the water can be played in HILIC by other components of the mobile phase, for example, the acidic buffers.

In case of separations with NP mechanism practiced on a silica column, the control of water content in the mobile phase is essential to maintain constant silica activity. Dry solvents may dissolve some of the water present on the silica surface and modify its structure such that column reproducibility is not very good, exposing to the analytes either immobilized water or silanol groups. To achieve column stability, the silica columns are usually equilibrated with a standardized solvent (e.g., ethyl acetate containing 0.06% water) [88].

Regarding the detection in HILIC and NPC, the solvents utilized in the mobile phase are not different in nature from those used in RP-HPLC. Only solvent proportions are different since the mobile phase must be less polar than the stationary phase. As a result, the comments regarding the influence of solvents on UV detection in RP-HPLC are applicable to HILIC and NPC (solvents must have a cut-off UV value below the wavelength selected for detection). For MS detection, in HILIC and NPC, the polar analytes are typically eluted with higher organic modifier content than in RP-HPLC, which may modify the MS response. For ESI-type ionization, it is difficult to predict if a higher organic content generates better sensitivity or not. In APCI, typically a higher organic solvent content leads to an increase in sensitivity.

Highly volatile NP-LC solvents are also well suited for atmospheric pressure photoionization (APPI). Lower vaporization temperatures can be used with easily vaporizable solvents, and this may be useful when analyzing thermolabile compounds. Many NP-LC solvents possess ionization energies below the 10.6 eV (e.g., 2-propanol 10.17 eV, n-hexane 10.13 eV, i-octane 9.89 eV, tetrahydrofuran 9.40 eV) and can be directly ionized by a krypton discharge lamp without any dopant addition. The use of low proton affinity NP-LC solvents (hexane, chloroform) with toluene as a dopant can enhance the ionization through charge exchange, and thereby they improve the ionization efficiency for nonpolar compounds. NP-LC solvents successfully applied to APPI analysis include ethanol, 2-propanol, hexane, heptane, cyclohexane, i-octane, tetrahydrofuran, ethylacetate, and chloroform [47].

The use of a mobile phase rich in organic volatile solvents, as practiced in HILIC and NPC, may also be favorable for enhancing the sensitivity of detection in techniques such as ELSD and corona charged aerosol detection (cCAD), which involves an evaporative process (see Section 4.1). This has been proved for specific analytes as reported in the literature [48].

Mobile Phases Used in Ion Exchange and Ion-Moderated Chromatography

The main component of the mobile phase in IC is typically water in which a specific buffer, acid, or base is dissolved. The main elution procedure in IC is not based on changes in solvent composition but in the concentration of the competing (driving) ions *Y* that replace reversibly the analyte M retained on the stationary phase. The dissolved additives in the aqueous mobile phase are selected depending on whether anionic or cationic separation is practiced. A special solvent delivery system can be used in ion chromatography. This system is known as an eluent generator (Thermo Scientific/Dionex). In an eluent generator the HPLC pump(s) deliver water while the reagents necessary for the elution are generated electrochemically from special cartridges. For this purpose, DC current is applied to a special cartridge to produce either methanesulfonic acid for cation exchange eluent, or KOH for anion exchange eluents. This type of eluent generator is offered as a whole assembly including other parts such as a degasser, and it is installed before the injector in the LC system.

Similar to other HPLC separations, the column is conditioned before each run and in isocratic separations the same mobile phase is used for the whole run and for conditioning the column. In gradient separations, the conditioning is performed with the initial composition of the mobile phase. For a wide range of pH values for the mobile phase, the retention capacity for strong ion exchange stationary phases remains unchanged. However, the analytes are significantly influenced by the pH values 2 pH units larger than the analyte pK_a. For weak ion exchangers, the mobile phase also influences the stationary phase ionization and therefore its retention capacity. The capacity factor for an analyte M is given in IC by empirical expression (4.3.55), written below for species Y as driving ion (instead of H^+) [89]:

$$\log k'_M = \alpha - \beta \log C(Y) \tag{13.5.2}$$

In expression (13.5.2) $C(Y)$ is the concentration of the driving ions used for elution. The values for the parameters α and β in Eq. (13.5.2) depend on the equilibrium constant $K_{M,Y}^{\text{therm}}$ for the analyte versus driving ions, the charges of the ions involved in the separation, the ion-exchange capacity of the stationary phase, the weight of the stationary phase, and the volume of the mobile phase in the column. The nature of the stationary phase, mobile phase, and of the analytes determines these parameters. The mobile phase properties can be modified using additives, selecting the pH, and even by adding an organic modifier such as acetonitrile or methanol. Also, addition of ligands that can interact with the analytes may be used for facilitating the analyte elution.

Most mobile phases used in cation exchange chromatography are aqueous, frequently containing diluted HCl, HNO_3, H_2SO_4, or CH_3SO_3H (concentration range between 2 and 50 mM/L). The separation of different ions can be achieved based on their different retention constants $K_{M,Y}^{\text{therm}}$ (where Y is the driving cation and M the analyte). For monovalent ions, the retention follows the order $Li^+ < H^+ < Na^+ < NH_4{}^+ < K^+ < Rb^+ < Ag^+$, and the elution can be done using acids (H^+ driving ion) that generate a high enough concentration of H^+ to assure elution (see formula (13.5.2) where $Y = H^+$). However, the retention of divalent ions (or trivalent ions) is much stronger than that of monovalent ions, and the elution with acids requires a higher concentration of H^+ ions. This requirement is difficult to fulfill because in conductometric detections the background signal (conductance) caused by the mobile phase becomes very high. The use of a solution of $AgNO_3$ as mobile phase may elute divalent ions since Ag^+ does have a higher affinity than H^+ for the stationary phase. This procedure requires a suppressor column in Cl^- form for eliminating Ag^+ ions [90]. In the case of strongly retained ions, such as transitional metallic ions, the elution can be achieved using a complexing agent, such as a solution of a weak organic acid (e.g., tartaric, or α-hydroxyisobutyric acid) that forms complexes with the analyte and in conductometric detection does not generate a large signal background (low H^+ concentration). This procedure has been applied even for the separation of ions such as Ca^{2+}, Sr^{2+}, Pb^{2+}, and Ba^{2+} with 2,3-diaminopropionic acid and HCl in the mobile phase [91].

Besides metallic cations, IC has been applied successfully for the analysis of organic cations, such as those from biogenic amines (e.g., putrescine, cadaverine, and histamine). In this case a diluted solution of H_2SO_4 (5.0 mM) can be used as a mobile phase [92]. The extension of ion chromatography to the analysis of various ionic molecules requires in many cases a specific pH of the mobile phase. This pH is specifically chosen to perform the separation and also to provide a mobile phase with electric conductivity that can be easily suppressed. In the case of aromatic amines the addition of an organic modifier like acetonitrile is recommended, under gradient elution, when the hydrophobicities of amines are very different [93]. Specific buffers are recommended to be used in IC depending on the nature of the ionic species, some of these being listed in Appendix 71.

For anion exchange chromatography, water is also typically used as solvent, and ions such as OH^- (KOH) can be electrochemically generated (e.g., in a Dionex EG40 eluent generator). Buffers based on CO_3^{2-}/HCO_3^- are also widely used for separation of inorganic and organic anions. The major advantage of this buffer is related to the suppressor reaction, which leads to H_2CO_3, which is weakly dissociated and consequently has a very low contribution to the background signal. There are alternatives to carbonate/bicarbonate buffer, such as solutions of amino acids. The suppressor reaction carried out at a pH corresponding to the isoelectric point of the amino acid converts it into a zwitterionic form with low contribution to the background signal. Another additive in anion exchange mobile phases is the tetraborate ion $(B_4O_7^{2-})$. The suppression of tetraborates is based on its change to H_3BO_3, which is weakly dissociated and thus does not contribute to the background conductivity. However, $B_4O_7^{2-}$ has a low affinity for the stationary phases, and for this reason it is only used for the elution of F^- and short-chain R-COO^- anions.

The mobile phases with low background conductivity can be used for nonsuppressed anion chromatography. Such phases are usually based on diluted aqueous solutions of organic salts, for example, benzoates, phthalates, or sulfobenzoates. These anions are also characterized by a significant affinity toward the stationary phase, and meanwhile they produce a relatively low conductivity of the mobile phase. The pH of mobile phase must be adjusted to 4–7 in order to favor the dissociation of the weak acid groups of the additive, which in turn influence the retention process of the inorganic anionic species (e.g., F^-, Cl^-, Br^-, I^-, NO_2^-, NO_3^-, PO_4^{3-}, SO_4^{2-}, $S_2O_3^{2-}$, SCN^-). However, the background conductivities of these mobile phases are higher than the conductivity of carbonate/bicarbonate buffer after passing the suppressor column, which is of the 15–20 μS/cm level. Thus, at concentration of 0.5–1 mM/L of the mentioned organic salts, the background conductivity is situated between 60 and 160 μS/cm, which is high and affects the detection performances.

Besides inorganic ions, IC is also used for the separation of numerous other analytes that can be present in ionic form. For these compounds, the use of buffers that assure the formation of ionic form of the analytes is necessary. Similar to the case of cations, the easy elimination of the conductivity created by the mobile phase components using suppressors is one criterion for selecting such buffers. Appendix 71 lists several buffers used in anion exchange LC.

Other mobile phases in anion exchange chromatography have been reported. Some such phases may have a multipart composition that can favor a complex separation process with ion-exchange, ion-exclusion, and ion-pairing principles for the separation.

The mobile phase in IC must be carefully selected in relation to the detection system. The most common detection technique in IC is based on conductivity, and the presence of acids, bases, or salts in the mobile phase may produce a large background signal for this type of detection. For this reason, two types of detection based on conductivity are practiced in IC: (1) with chemical suppression of the background signal, and (2) without chemical suppression of the background signal. Chemical suppression can be achieved using a resin or a semipermeable membrane that eliminates mobile phase ions that cause high conductivity background (see Section 4.1). The mobile phase must be selected in accordance with the type of conductivity detection [94].

When UV detection is used in ion chromatography, solutions of salts of phosphoric, sulfuric or perchloric acid are suitable as mobile phase because these anions do not absorb strongly radiation in this spectral domain. When amperometric detection is chosen, the mobile phase acts as a support electrolyte and the electrolyte concentration must be about 50–100 times higher than the concentration of the anion analytes. In this case, hydroxide, chloride, chlorate, or perchlorate of alkali metals is used as supporting electrolyte for anion elution. Ethylene-diaminotetraacetic acid (EDTA) can be used for the elution of very strongly retained polyvalent anions such as polyphosphates. Besides that, EDTA can form anionic complexes with many metallic ions by the control of pH of mobile phase, a property that can be used in separating metallic cations by anion exchange LC [95].

The choice for mobile phase in ion-moderated chromatography is rather limited. The simplest mobile phase is pure deionized water, which has been proved useful, for example, in the analysis of carbonate ion. For some columns, a specific ion must be present in the mobile phase for maintaining column integrity. Examples are ions such as H^+, Ca^{2+}, Pb^{2+} (e.g., diluted H_2SO_4 is used to generate H^+ ions). For the separation of organic acids, solutions of inorganic acids are usually used as the mobile phase. This depends also on the type of the suppressor column employed in the case of conductivity detection. In case of UV-detection, H_2SO_4 solution is also frequently used. The high retention of some aliphatic and aromatic carboxylic acids can be prevented by the addition in mobile phase of a small content of miscible solvents, such as methanol, ethanol, i-propanol, or acetonitrile [96].

Mobile Phase in Chiral Chromatography

Most chiral separations are performed on chiral stationary phases. However, separations on achiral stationary phases with a chiral agent added to the mobile phase are also practiced (see Section 4.3). The mobile phases used in the two types of chiral separations are different.

Chiral phases can be utilized in normal-phase (NP) mode, reversed-phase (RP) mode, corresponding to HILIC nonchiral chromatography mode, and applicable to ionizable compounds (corresponding to IC nonchiral chromatography) (see Section 10.1). Corresponding to each of those modes of utilization, the mobile phase must have an appropriate composition [97].

Since hydrogen bonding is one common type of interaction utilized for differentiating the enantiomers, the presence of water in the mobile phase in a chiral separation may disturb such interactions. For this reason, the NP-type utilization of chiral phases is the most common. For normal-phase utilization, the mobile phase is made from apolar solvents such as hexane or heptane with a polar component such as methanol, ethanol, 2-propanol, chloroform, methylene chloride, lower ethers, or esthers. In general, better separations are obtained at longer retention times, and therefore, lower levels of polar components in the mobile phase are typically used (in some separation as low as 2–3%). The same effect of longer retention times is obtained using "weak" polar solvents such as chloroform, methylene chloride, tetrahydrofuran, or ethyl acetate, which are less polar than methanol, ethanol, or acetonitrile [98].

The pH of the mobile phase is very important for determining selectivity [99–102]. The control of pH can be achieved using a certain level of additives such as diethylamine (DEA), triethylamine, butylamine, or trifluoroacetic acid. The pH can also be controlled with buffers such as ammonium acetate, citrate buffers, and triethylammonium phosphate [103]. The pH influences molecular structure as well as the formation/elimination of hydrogen bonds and plays an important role in the separation. Column stability is also affected by the mobile phase pH, many chiral phases being less resilient than RP columns to extreme pH values of the mobile phase. The typical pH range of utilization of many chiral phases is between 3 and 7.

Even when the phase is used in RP mode, the content of the polar solvent is typically low and water is frequently absent in the mobile phase. Commonly used solvents are methanol, acetonitrile, tetrahydrofuran, 2-propanol, and ethanol. For cyclodextrin-based stationary phases, dimethylformamide and dimethylsulfoxide are also possible polar components of the mobile phase. The solvent selection may determine which interaction type is more important for the separation, e.g., $\pi-\pi$ stacking, H-bonding, or hydrophobic [104]. Enantioseparations can be strongly influenced by the pH of the mobile phase [105,106].

For HILIC- and IC-type utilization of chiral phases, the mobile phase must contain polar components such as methanol and acetonitrile and only 2–5% water. Additives such as ammonium formate or low level of buffers such as acetic acid/trietanolamine or trifluoroacetic acid/trietanolamine, or specific chaotropic salts can also be added to enhance the mobile phase polarity.

Besides the control of pH, the buffers may have additional roles in chiral separations. For example, the use of triethylamine/acetate buffer is preferred for separations on most cyclodextrin-based stationary phases, leading to better separations and better peak shapes. This can be explained by the special effect of the buffer on various types of interactions determining enantiomer separation. Some buffers such as triethylamine/acetate, formate, or citrate are considered capable of forming inclusion complexes with the cyclodextrins and affect separation not only due to pH control. Buffer concentration is also important for achieving specific separations, and at high enough concentrations (e.g., higher than 1.5%) some buffers affect negatively enantioselectivity [105].

For the chiral separations that use a chiral additive to the mobile phase, there are two possible mechanisms of action as indicated in Section 4.3. One mechanism is the formation of diastereoisomers in solution between the analyte and the chiral additive, and the other is the retention of the chiral additive on the stationary phase, creating in this way a chiral environment for the separation. The solvents making the mobile phase can still be similar to those from chiral NPC or RP-HPLC. Among the solvents utilized in such separations are: acetonitrile, methanol, ethanol, mixture of such solvents with low levels of water, or nonpolar solvents such as hexane, heptane, or cyclohexane.

A variety of additives can be used in the mobile phase for creating a chiral environment for the separation. Among these are cyclodextrins, cationic β-cyclodextrins, bovine serum albumin (BSA), alpha-1-acid glycoprotein (ORM), N-alkyl-L-hydroxyproline plus copper acetate added in the mobile phase, and Cu^{2+}-L-phenylalanine in the mobile phase [107].

In the case of cyclodextrin used as a chiral modifier, solvents with higher hydrophobicity such as ethanol and isopropanol are typically used in the mobile phase. In the case of copper complexes with a chiral amino acid used as additive, the separation takes place by ligand exchange chromatography, and more polar solvents even having some water can be used as the mobile phase.

One type of additive successfully utilized in chiral separation of the enantiomers of compounds that contain ionizable or strongly polar groups is chiral IPAs followed by an ion pairing chromatographic separation. Among common chiral IP agents used for this purpose are, for example, (+)-10-camphor-sulfonic acid or (−)-10-camphor-sulfonic acid for the analysis of cationic compounds, and quinine, quinidine, cinchonidine, and cinchonine as counterion for the separation of acids [108,109]. The mobile phases typically used in such separations should be high in the organic phase component. The ion-pair separation in NP mode can be performed on a silica stationary phase.

Mobile Phase for Size-Exclusion Separations

In most cases in size-exclusion chromatography (in both modes GFC and GPC), the mobile phase serves mainly to dissolve the sample and carry it through the column. Compared to RP-HPLC or HILIC, the mobile phase in SEC has a lower role in modulating the separation (see Section 4.3). However, polymer dissolution can be a challenging task. Polymers may be difficult to solubilize, in particular for producing more concentrated solutions that are usual in SEC injections (e.g., several mg/mL). Also, since the stationary phases in SEC are frequently polymers, the solvents used in SEC separations should not affect the structure of the stationary phase. Column manufacturers typically indicate the range of solvents allowed on a specific column.

Dissolution of polymers is mainly determined by enthalpic interactions, since the gain in mixing entropy for polymers during dissolution is typically small. The intermolecular interactions between the polymer chain segments and solvent molecules have an associated energy, which can be positive or negative. For a "thermodynamic good solvent," interactions between polymer segments and solvent molecules are energetically favorable and will cause polymers to expand their conformation for example of a coil. For a "poor solvent," polymer–polymer self-interactions are preferred, and the polymer coils will contract. The fact that the polymer may contract or expand in a specific solvent plays an important role when attempting to measure the molecular weight Mw of the polymer using SEC. Because K_{SEC}^{pure} depends, as indicated by expression (4.3.61), on the mean radius $\bar{r}$ of the polymer, in order to generate a correct result for the Mw it is necessary to keep intact the polymer shape. In some specific solvents indicated as *theta* solvents, the polymer behaves in solution the same as in the bulk polymer without changing its conformation. In such solvent the measurement of Mw can be considered more accurate [110,111]. Polymer shape in solution also depends on the solution temperature, and this factor also must be considered when measuring the molecular weight.

When the purpose of SEC separations is the measurement of the macromolecular size/weight/hydrodynamic volumes of proteins or protein complexes, or in the case of polymer purification, it is important for the chosen solvent to not affect the macromolecular conformation [112]. Also, the elution process should be controlled only by entropy, and therefore the solvent should assure the minimization of the enthalpic interactions (ionic and hydrophobic) of macromolecular species with stationary phase. The selection can be tedious, and it depends on the investigated sample and SEC column. Adjustments to the ionic strength and pH of the mobile phase are the primary means of reducing electrostatic interactions between the analyte and the SEC column [113]. However, the elimination of electrostatic interactions can be associated with the development of some hydrophobic interactions [114]. Usually the elution mode is isocratic such that possible changes of the stationary phase with the modification of mobile phase compositions can be avoided. In the case of separation of proteins, the selection of the mobile phase takes into consideration possible detrimental effects of changing their biological activity [115].

For gel filtration, water is the most common solvent, although for some separations a limited level of an organic modifier miscible with water can be used. This can be necessary for less hydrophilic polymers that may have some solubility problems in pure water. The organic solvents may be methanol, acetonitrile, isopropanol. For ionic polymers, buffers and additives can be beneficial for the separation. Maintaining a constant ionic strength of the mobile phase is also sometimes necessary for reproducible results of molecular weight measurements. Some typical solvents used in water-soluble polymers are given in Appendix 7m. The salts used for obtaining a constant ionic strength may include Na_2SO_4, $NaNO_3$, CH_3COONa [116]. The buffers that can be used include acetic acid/sodium acetate and other common buffers. The manufacturers of the SEC columns usually indicate restrictions regarding the solvents that can be used for a separation.

Specific hydrophilic polymers may not be soluble in water although they also do not dissolve in organic solvents. This is the case, for example, for cellulose and other polysaccharides, and for specific proteins. In such cases, special solvents are necessary. For example, cellulose can be solubilized in dimethylacetamide + LiCl, certain cationic polymers can be solubilized in formamide + LiCl or in dimethylformamide/triethylamine/pyridine [117].

In the case of proteins, specific restrictions are sometimes imposed to the mobile phase in order to promote solubilization, avoid protein adsorption on the stationary phase, and also to prevent protein denaturation. For example, addition of salts such as NaCl or Na_2HPO_4, of amino acids (glycine, alanine, or arginine), or of organic solvents (methanol, acetonitrile) may preclude protein adsorption, but higher contents than 5−10% of organic solvents or more than 0.1−0.4 mM/mL salts must be avoided in order to maintain protein biological activity. Sometimes, for protein solubilization a detergent such as sodium dodecylsulfate (SDS) is added. This compound leads to the dissociation of protein aggregates, but also increases protein hydrodynamic volume by formation of protein−SDS complexes [118−120].

For GPC, the solubilization of the polymer can be a considerable challenge. The same solvent used for solubilization is typically used for the separation. Among the common solvents in GPC is tetrahydrofuran which dissolves polystyrene, poly-(methylmethacrylate), epoxy resins, polycarbonates, polyvinylchloride, and polystyrene/acrylonitrile. Other solvents used in polymer analysis include toluene, chloroform, and benzene [121]. Some common synthetic polymers such as polyethylene or polypropylene are not easily solubilized and they are soluble only in solvents like 1,2,4-trichlorobenzene, methylcyclohexane, dibutoxymethane (butylal) and only at temperatures above 90−100°C [122,123].

Flow Rate, Temperature, and Degassing of the Mobile Phase

The flow rates for the mobile phase in analytical HPLC can be kept constant during the chromatographic run, or they may be part of a gradient program where the flow rate is increased or decreased. The values for the flow rate are selected depending on several constraints: (1) the type of chromatographic column, (2) the mobile phase properties, and (3) the detector constraints.

1) In theory, the volumetric flow rate U in a chromatographic column should be selected close to the optimum for efficiency as indicated by van Deemter equation (see formula (6.2.5)). By construction, the commercial columns usually provide good efficiency in a range of flow rates that also accommodates other requirements such as acceptable run times (not longer than 15–25 min), backpressures in the range of capabilities for the HPLC or UPLC system. Chromatographic columns of UPLC type (small particles 1.6–2.7 μm) are typically used with flow rates in the range of 0.1 mL/min to 0.6 mL/min. Lower or higher flow rates are sometimes utilized. Common columns utilized in HPLC (e.g., with particles of 3 μm–10 μm) allow larger flow rates in the range of 0.5 mL/min to 2–3 mL/min. Flows lower than 0.1 mL/min in common columns are limited by too much deviation from optimum flow rate indicated by van Deemter and also by the precision of delivering a correct mobile phase composition when the flows are lower than 0.1 mL/min. For larger flow rates, limitations also come from deviation from the optimum flow rate indicated by van Deemter, but mainly from maximum acceptable backpressure of the HPLC system (see Section 6.2). The limits in backpressure and the relation with the flow rate in the column were discussed in Section 6.2.

2) The mobile phase properties also must be considered when selecting a flow rate. For example, mobile phase viscosity is important regarding the maximum flow rate in the chromatographic system because of the limitations in backpressure (see formula (6.2.9)). Other properties such as volatility are also important related to the detector limitations, in particular when evaporative techniques are used in the detector.

3) The detectors may impose limitations on flow rate, especially when evaporative techniques are used. Large flow rate cannot be accommodated by some detectors such as MS with ESI or APCI ionization sources, where the mobile phase must be transferred as dry aerosols. It is common for LC-MS systems, for example, to require flow rates in the range of 0.1–0.6 mL/min, although some systems can work with flow rates around 1 mL/min. Such detectors are well suited for being used with UPLC systems where the lower flow rates are also required by the high backpressure generated by the columns with small particles (see expression (6.2.10)).

The influence of temperature on a separation has been discussed in Section 4.4, and the selection of a specific temperature is usually done for one of the following reasons: (1) it improves separation, (2) it decreases the solvent viscosity and therefore reduces backpressure, and (3) it is necessary for maintaining specific analytes in solution (such as certain polymers). The limitations in the choice of a specific temperature for the mobile phase are a result of the mobile phase boiling point, mobile phase chemical stability, as well as column stability at elevated temperatures. Common HPLC columns are designed to work in the range between 10 and 70°C, and the manufacturers typically specify the temperature limits. A mobile phase can be used at a desired temperature when it is stable and does not boil at the pressure from the chromatographic column.

Most modern HPLC instruments have a "degasser" unit that assures the elimination of dissolved gases from the mobile phase. However, the degassing of initial solvents by sparging an inert gas like helium may be necessary in some instances. This is the case, for example, for low-pressure mixing of solvents when some gas bubbles may be formed in the mobile phase after mixing and before the high-pressure pumps (see Section 4.1). However, initial degassing may eliminate some volatile components from the solvents. For example, when buffers are made using ammonia, sparging is not recommended since it may change the buffer pH.

13.6 SELECTION OF A SOLVENT FOR SAMPLE INJECTION

The introduction of sample in the HPLC is typically performed placing a precisely measured volume of a solution of the sample in the mobile phase generated by the pumps (see Section 4.1). The sample must be completely dissolved, and for this purpose a single solvent or a mixture of miscible solvents can be used. The sample solution must be free of any particles and should have a high stability, not undergoing solvolysis/hydrolysis. For improving sample stability in time, it is common that HPLC autosamplers also have the capability of cooling. This may also help in reducing evaporation when the sample solvent is volatile and the septum of the sample vial has been already punctured (see Section 4.1).

The Role of Sample Solvent in the Chromatographic Process

Besides the role of dissolving and loading the sample, the liquid used for sample dissolution also may have a contribution to the chromatographic process. The injection volume V_{inj} is directly related to the amount of sample delivered to the HPLC system and therefore to the detector response (see Section 4.2). Also, the peak width (characterized for example by peak width W_b) is affected by the injection volume. The theory about peak broadening in HPLC is developed considering that no contribution to W_b is due to injection volume, which is assumed to be extremely small. However, the correct value for W_b should include the width of the injected sample besides the broadening due to various random processes taking place in the chromatographic separation (see expression (4.2.36)). As a result, the larger injection volumes produce a widening of the chromatographic peaks. Larger injection volumes may be necessary to place a larger amount of sample in the chromatographic system in order to improve sensitivity.

The use of a sample solvent (diluent) with the same composition as the mobile phase (initial mobile phase composition in the case of gradient separation) is common practice in HPLC. With this choice, the mobile phase composition is not modified by the injection, and no additional effect regarding the chromatographic process take place. Volumes typically used in HPLC in the range up to 25 mL affect very little the value of W_b or that of the peak shape. For UPLC where the peaks are narrower, the injection volumes should be kept relatively low in order to preserve efficiency, and injections smaller than 3–4 μL do not visibly affect the peak width.

However, the use of mobile phase as sample diluent is not always possible due to the low solubility of analytes or due to sample preparation procedures that deliver the processed sample in a specific solvent. For example, in sample processing such as liquid–liquid extraction (LLE), solid-phase extraction (SPE), QuEChERS ("Quick, Easy, Cheap, Effective, Rugged, and Safe"), stir bar sorptive extraction (SBSE) and others, the analytes can be present in a different solvent than the mobile phase.

When the sample solvent is different from the mobile phase, an initial requirement is that the sample solvent must be miscible with the mobile phase [124]. Also, it is important that the analytes are also soluble in the mobile phase at the concentration reached when placing them in it. When these two requirements are not fulfilled, the separation and quantitation in HPLC can be significantly affected.

The differences between the sample solvent and mobile phase may affect peak retention time and peak shape, and this depends on the nature of the two liquids (chemical and physical properties), the injected volume, and to a smaller extent the flow rate of the mobile phase. Regarding the physical properties, the viscosity of the two solvents plays an important role, and sample solvents with high viscosity seem to affect the most the peak shape. The mutual solubility of the mobile phase and sample solvent also plays a role in affecting the peak shape. Regarding the chemical nature of the solvents, there are three model possibilities regarding the differences between the nature of mobile phase and sample solvent: (1) the sample solvent is an eluent similar to the mobile phase, (2) the sample solvent is a "weaker" eluent, or (3) the sample solvent is a "stronger" eluent.

1) When the sample solvent is a similar eluent to the mobile phase, only small differences regarding the elution process will occur, and these are typically not significant. In UPLC, larger volumes of the injected sample (e.g., 10 μL) can produce a noticeable peak broadening because in UPLC the peaks are narrow (e.g., W_b of 2–3 s). In common HPLC volumes up to 20–25 μL usually do not produce a significant peak broadening, but larger injection volumes may increase W_b.

2) For the sample solvent a "weaker" eluent than the mobile phase, a concentration of the sample at the column head may take place. This effect depends on the injection volume and sample solvent nature. Typically, small injection volumes such as 5–10 μL (for HPLC) do not produce a noticeable concentration effect. However, for larger injection volumes (used, e.g., with diluted samples), and with a very "weak" sample solvent, this effect may reduce the peak broadening caused by the large injection (sample on-column focusing). This property can be used for injecting larger volumes of a diluted sample, when a larger amount of sample is necessary for increasing sensitivity.

3) The use of a "stronger" solvent for sample injection may generate significant problems regarding the peak shape. The sample solvent plug may act as a strong eluent before it is diluted enough by the flowing mobile phase, and may produce a widening of the region where the sample is distributed at the head of the chromatographic column. This process generates wider chromatographic peaks and very frequently peak shape distortion [6,125]. Peak distortion was noticed in particular when the strong eluting solvent has a very different viscosity to the mobile phase.

A sample solvent that is a strong eluent is frequently used because stronger eluents are frequently good solvents for the sample. For example, in RP-HPLC, acetonitrile or methanol are commonly used as sample solvents. In the case of some hydrophobic compounds analyzed by RP-HPLC that are not soluble or are poorly soluble in a partial water solvent, the use of "strong" eluents as sample solvent is necessary. In order to avoid visible peak broadening or peak distortion, such solvents must be injected only at low volume (e.g., 2–5 μL). Solvents with high viscosity that mix more slowly with the mobile phase require even smaller injection volumes.

One possibility to improve the peak shape when this is distorted because of a "strong" sample solvent is the dilution of the sample with mobile phase and injecting a proportionally larger volume [126]. This method is known as precolumn dilution, and a mixer chamber can be utilized, followed by larger volume injection in order to reach low detection limits [127].

One alternative to avoid the use of strong solvents for carrying the sample is to change the solvent and redissolve the sample in a "weaker" solvent. Also, for the reduction of the sample volume when a larger volume is necessary because the sample is too diluted and a specific sensitivity must be attained, the sample solution may need to be concentrated. Evaporation of the diluted sample to concentrate the target analytes can be applied at this point of sample preparation. However, this operation can be time-consuming and can affect precision and analyte recovery (e.g., some analyte evaporation may occur simultaneously with the solvent or the analyte may suffer decomposition when evaporating temperature is too high). In these cases, other alternative paths may be selected for sample preparation, or higher sensitivity for the detector may be evaluated such that the injection volume does not need to be increased [128].

The problem of strong eluting solvents used for sample dissolution is common in RP-HPLC. However, in HILIC the separations can also be influenced by the sample solvent [129]. In HILIC separations, it is common to use pure organic solvents as sample solvent. Organic solvents are weak eluents under HILIC conditions, so that polar analytes are accumulated in a narrow zone on the head of the column (sample on-column focusing). This is an advantage for the analysis of polar drugs in biological samples, for instance, when the proteins are precipitated with acetonitrile or with methanol. The organic supernatant can then be directly injected into the HILIC column, avoiding the step of solvent evaporation and the residue reconstitution. Injection of a large volume of aqueous sample solvent of high elution strength should be avoided in HILIC. The volume of polar solvents for the sample should be limited by the same rules as nonpolar ones in the case of RP-HPLC. The use of large volumes of polar solvents for the sample injection may lead to broad or split peaks. An organic content of more than 50% for the sample solvent is typically recommended in HILIC [130,131].

For other types of chromatographic separations, such as SEC, sample viscosity and injection volume play an even more important role in the separation than for RP or HILIC [132]. In SEC, sample concentration can be high and polymeric content can significantly influence the viscosity of the sample. Also, larger volumes than in other types of chromatography are typically utilized in SEC. For this reason, some specific restrictions are necessary regarding injection in SEC. For example, the relative viscosity of the sample in SEC should not exceed double that of the mobile phase (for a dilute aqueous buffer this corresponds to a concentration of protein of about 70 mg/mL). The volume load in SEC should not exceed 1–5% of the total column volume, although larger volumes are sometimes injected. Only injections that do not exceed 2% of the column volume were usually proven to maintain good resolution [133].

Sample solvent may also be important for chiral separations. Some samples can be dissolved in only a specific solvent, and the solvent may not be adequate for being injected in the column. For example, proteins are usually soluble in water and in DMSO. Since water can disturb the chiral separation, DMSO must be used for protein dissolution when injected in a chiral column. However, the chiral stationary phase must be stable under repeated DMSO injection. Some chiral columns such as Chiralpak IA are recommended as fulfilling such a requirement [134].

Focusing of Sample at the Column Head by Other Procedures

The focusing of the sample at the head of the chromatographic column using a weak sample solvent is an attractive procedure for improving the peak shape, and at the same time of using a larger volume for the sample injection that increases sensitivity. However, weak solvents are not always good solvents for the sample, and in many cases they are not the final solvent delivered by the sample preparation procedure. Another possibility to focus the analyte has been more recently explored [128,135] and consists of using as sample solvent a compound that is very strongly retained at the column head with the analytes less retained than the solvent itself. During the elution process, the sample solvent is retained at the column head longer than the analytes. For achieving focusing by this procedure, the following conditions are necessary: (1) the sample solvent must be hydrophobic (such as i-octane, hexane, heptane, octane, benzene), (2) the mobile phase must have a high content of water in order to avoid the dissolution of the solvent, (3) the analytes must have a lower hydrophobicity than the sample solvent [136,137]. An example of column focusing by this approach is given for the injection of several volumes between 50 and 500 μL of i-octane solutions, each one containing the same amount of 50 μg 4-hydroxy-3-t-butyl anisole. The column utilized was a Zorbax Eclipse XDB-C18 column (150 × 4.6 mm, 5 μm particle size) with the mobile phase containing 40% ACN and 60% water with 0.1% H_3PO_4, at 25°C. The flow rate for the separation was 1.5 mL/min and the detection was performed at 291 nm. After each chromatographic run the column was washed with 100% AcCN and conditioned at the initial mobile phase composition [138]. The results are shown in Fig. 13.6.1 [139]. Larger injection

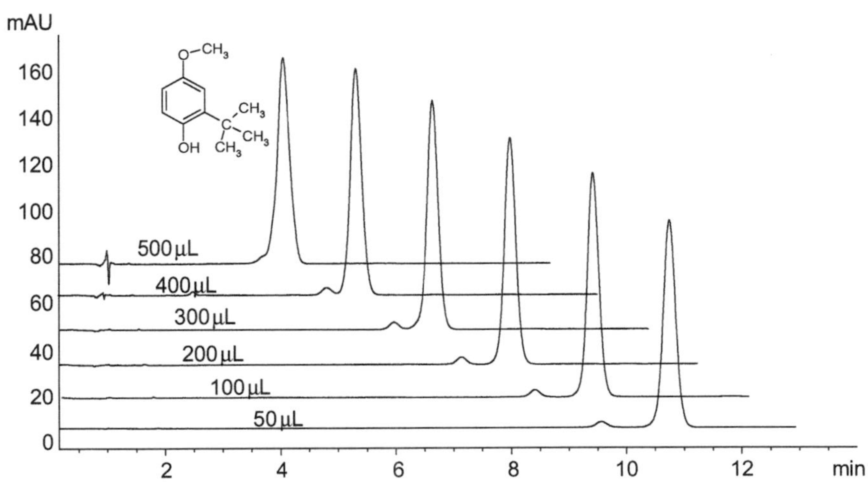

FIGURE 13.6.1 Separation of 50 μg 4-hydroxy-3-t-butyl anisole on a Zorbax Eclipse XDB-C18 column with large injections (50–500 μL) of the sample in i-octane in a partially aqueous mobile phase (traces offset by 15 mAU).

volumes reduce the retention time of the analyte, probably because the portion of the analytical column covered by the sample solvent does not retain the analyte.

Effect of Sample Solvent on Detection

Sample solvent may also have an effect on the response of specific detectors. For example, the signal of RI or of electrochemical detectors can be significantly influenced by the sample solvent. In some cases, the "elution" of sample solvent can generate a large signal (sometimes negative), which can adversely affect the signal measurement for the analytes, in particular when the sample solvent elutes close to the analyte. The problem of detector response to the sample solvent is addressed in various studies (see, e.g., Refs. [89,140]).

13.7 SELECTION OF A SOLVENT FOR THE NEEDLE WASH

In most HPLC (and UPLC) systems, the autosamplers have the capability to wash the needle utilized for taking the sample from the sample vial and placing it in the flow of the mobile phase. The needle wash is necessary for avoiding carryover from sample to sample. The presence of a carryover problem can be seen (e.g., as a ghost peak) in the chromatogram of a blank sample injected after a series of samples that may contain high concentrations of analytes [141]. Carryover depends on the type of analysis, HPLC system, and the type of detection [142]. Most carryover problems come from the injection part of the chromatographic system, and they can have several sources: (1) adsorption of components from sample to the needle, (2) needle seat design, (3) improper or worn out sealing of the needle, (4) worn out rotor/stator of injection valve, (5) tubing fittings contaminated with sample, and (6) contaminated parts of the column (fittings, frits, and connection with detector). The limit for carryover should be less than 0.05% for good quantification results. The needle is washed inside with the mobile phase. However, for external needle wash an auxiliary solvent is used. This solvent selected for the external needle-wash depends on the sample and mobile phase chemistries, making sure that all solutions/buffers are miscible and soluble. The composition of the needle-wash solvent should be the most solubilizing compatible solvent. It is common to select for the needle wash a solvent or a mixture of solvents similar to the mobile phase composition, but with higher content in the stronger eluent. It should be verified experimentally that in using the needle wash, no carryover of the samples take place.

References

[1] J.H. Hildebrand, R.I. Scott, The Solubility of Non-electrolytes, Dover Pub, New York, 1964.
[2] L.R. Snyder, Classification of the solvent properties of common liquids, J. Chromatogr. Sci. 16 (1978) 223–234.
[3] L.R. Snyder, The role of the mobile phase in liquid chromatography, in: J.J. Kirkland (Ed.), Modern Practice of Liquid Chromatography, Wiley-Interscience, New York, 1971.
[4] M.H. Abraham, A. Ibrahim, A.M. Zissimos, Determination of sets of solute descriptors from chromatographic measurements, J. Chromatogr. A 1007 (2004) 29–47.
[5] P.C. Sadek, The HPLC Solvent Guide, second ed., Wiley & Sons, New York, 2002.
[6] S.C. Moldoveanu, V. David, Essentials in Modern HPLC Separations, Elsevier, Amsterdam, 2013.
[7] C. Hansch, A. Leo, D. Hoekman, Exploring QSAR, Hydrophobic, Electronic and Steric Constants, ACS, Washington, 1995.

[8] http://www.epa.gov/oppt/exposure/pubs/episuite.htm.
[9] http://www.chemaxon.com.
[10] V.N. Viswanadhan, A.K. Ghose, G.R. Revankar, R.K. Robins, Atomic physicochemical parameters for three dimensional structure direct quantitative structure-activity relationships. 4. Additional parameters for hydrophobic and dispersive interactions and their application for an automated superposition of certain naturally occurring nucleoside antibiotics, J. Chem. Inf. Comput. Sci. 29 (1989) 162–172.
[11] S.C. Moldoveanu, V. David, Sample Preparation for Chromatography, Elsevier, Amsterdam, 2002.
[12] N. El Tayar, H. van de Waterbeemd, B. Testa, The prediction of substituent interactions in the lipophilicity of disubstituted benzenes using RP-HPLC, Quant. Struct. Act. Relat. 4 (1985) 69–77.
[13] S. Moldoveanu, V. David, Estimation of the phase ratio in reversed-phase high-performance liquid chromatography, J. Chromatogr. A 1381 (2015) 194–201.
[14] E. Caiali, V. David, H.Y. Aboul-Enein, S.C. Moldoveanu, Evaluation of the phase ratio for three C18 high performance liquid chromatographic columns, J. Chromatogr. A 1435 (2016) 85–91.
[15] L. Rohrschneider, Solvent characterization by gas-liquid partition coefficients of selected solutes, Anal. Chem. 45 (1973) 1241–1247.
[16] B.P. Johnson, M.G. Khaledi, J.G. Dorsey, Solvatochromic solvent polarity measurements and retention in reversed-phase liquid chromatography, Anal. Chem. 58 (1986) 2354–2365.
[17] C. Reichardt, S. Asharin-Fard, A. Blum, M. Eschner, A.-M. Mehranpour, P. Milart, T. Niem, G. Schäfer, M. Wilk, Solute/solvent interactions and their empirical determination by means of solvatochromic dyes, Pure Appl. Chem. 65 (1993) 2593–2601.
[18] http://www.uni-marburg.de/fb15/ag-reichardt/et-30-values?language_sync=1.
[19] R.W. Taft, M.J. Kamlet, The solvatochromic comparison method 2. The alpha scale of solvent hydrogen-bond donor (HBD) acidities, J. Amer. Chem. Soc. 98 (1976) 2886–2894.
[20] M.J. Kamlet, R.W. Taft, The solvatochromic comparison method 1. The beta scale of solvent hydrogen-bond acceptor (HBa) basicities, J. Amer. Chem. Soc. 98 (1976) 377–383.
[21] http://www.stenutz.eu/chem/solv26.php.
[22] M.J. Kamlet, J.L.M. Abboud, M.H. Abraham, R.W. Taft, Linear solvation energy relationships. A comprehensive collection of the solvatochromic parameters, π^*, α and β, and some methods for simplifying the generalized solvatochromic equation, J. Org. Chem. 48 (1983) 2877–2887.
[23] D.C. Luehrs, D.S. Chesney, K.A. Godbole, Correlation of the chromatographic eluting strength of solvents with solvent polarity and polarizability parameters, J. Chromatogr. Sci. 29 (1991) 463–466.
[24] M. Vitha, P.W. Carr, The chemical interpretation and practice of linear solvation relationship in chromatography, J. Chromatogr. A 1126 (2006) 143–194.
[25] A. Habibi-Yangjeh, A model for correlation of various solvatochromic parameters with composition of aqueous and organic binary solvent systems, Bull. Korean Chem. Soc. 25 (2004) 1165–1170.
[26] L.R. Snyder, J.J. Kirkland, J.W. Dolan, Introduction to Modern Liquid Chromatography, third ed., Wiley, Hoboken, 2010.
[27] M. Rosés, E. Bosch, Linear solvation energy relationship in reversed-phase liquid chromatography. Prediction of retention from a single solvent and a single solute parameter, Anal. Chim. Acta 274 (1993) 147–162.
[28] J.R. Torres-Lapasio, M. Rosés, E. Bosch, M.C. Garcıa-Alvarez-Coque, Interpretive optimisation strategy applied to the isocratic separation of phenols by reversed-phase liquid chromatography with acetonitrile–water and methanol–water mobile phase, J. Chromatogr. A 886 (2000) 31–46.
[29] P.C. Sadek, P.W. Carr, R.M. Doherty, M.J. Kamlet, R.W. Taft, M.H. Abrams, Study of retention process in reversed-phase high-performance liquid chromatography by the use of the solvatochromic comparison model, Anal. Chem. 57 (1985) 2971–2978.
[30] L.R. Snyder, Principles of Adsorption Chromatography, M. Dekker, New York, 1968.
[31] H.B. Patel, T.M. Jefferies, Eluotropic strength of solvents, Prediction and use in reversed-phase high-performance liquid chromatography, J. Chromatogr. 389 (1987) 21–32.
[32] L.R. Snyder, Column efficiencies in liquid adsorption chromatography: past, present and future, J. Chromatogr. Sci. 7 (1969) 352–360.
[33] A.M. Striegel, Viscometric detection in size-exclusion chromatography: principles and select applications, Chromatographia 79 (15) (2016) 945–960.

[34] W.J. Cheong, P.W. Carr, The surface tension of mixtures of methanol, acetonitrile, tetrahydrofuran, isopropanol, tertiary butanol and dimethyl-sulfoxide with water at 25°C, J. Liq. Chromatogr. 10 (1987) 561–581.
[35] G. Vazquez, E. Alvarez, J.M. Navaza, Surface tension of alcohol water + water from 20 to 50 °C, J. Chem. Eng. Data 40 (1995) 611–614.
[36] E. Bramanti, W.W. Quigley, C. Sortino, F. Beni, M. Onor, G. Raspi, R.E. Synovec, Multidimensional analysis of denatured milk proteins by hydrophobic interaction chromatography coupled to a dynamic surface tension detector, J. Chromatogr. A 1023 (2004) 79–91.
[37] T. Steiner, The hydrogen bond in the solid state, Angew. Chem. Int. Ed. 41 (2002) 48–76.
[38] C.L. Perrin, J.B. Nielson, 'Strong' hydrogen bonds in chemistry and biology, Annu. Rev. Phys. Chem. 48 (1997) 511–544.
[39] E.E. Kwan, An Introduction to Hydrogen Bonding, Evans Group seminar, Sept.11, 2009.
[40] K.B. Sentell, N.I. Ryan, A.N. Henderson, Temperature and solvation effects on homologous series selectivity in reversed phase liquid chromatography, Anal. Chim. Acta 307 (1995) 203–215.
[41] T. Greibrokk, T. Andersen, High-temperature liquid chromatography, J. Chromatogr. A 1000 (2003) 743–755.
[42] G. Vanhoenacker, P. Sandra, High temperature and temperature programmed HPLC: possibilities and limitations, Anal. Bioanal. Chem. 390 (2008) 245–248.
[43] R.M. Smith, Superheated water chromatography – a green technology for the future, J. Chromatogr. A 1184 (2008) 441–455.
[44] T. Taylor, Important aspects in UV detection for HPLC, LCGC North America 33 (2015) 870.
[45] J.O. Ona, M.J. Wornat, The influence of solvents on the ultraviolet-visible absorption spectra of polycyclic hydrocarbons: applications in the identification of fuel product by HPLC/UV/MS, J. Liq. Chromatogr. Rel. Technol. 28 (2008) 15–38.
[46] S. Basavaraj, H.A.E. Benson, D. Havelock Brown, Y. Chen, Application of solvent influenced fluorescence-quenching and enhancement to develop a highly sensitive HPLC methodology for analysis of resveratrol-PEG conjugates, Curr. Pharm. Anal. 9 (2013) 199–207.
[47] R. Kostiainen, T.J. Kauppila, Effect of eluent on the ionization process in liquid chromatography - mass spectrometry, J. Chromatogr. A 1216 (2009) 685–699.
[48] S. Gao, Z.P. Zhang, L.E. Edinboro, L.C. Ngoka, H.T. Karnes, The effect of alkylamine additives on the sensitivity of detection for paclitaxel and docetaxel and analysis in plasma of paclitaxel by liquid chromatography-tandem mass spectrometry, Biomed. Chromatogr. 20 (2006) 683–695.
[49] L.-E. Magnusson, D.S. Risley, J.A. Koropchak, Aerosol-based detectors for liquid chromatography, J. Chromatogr. A 1421 (2015) 68–81.
[50] C.R. Mitchell, Y. Bao, N.J. Benz, S. Zhang, Comparison of the sensitivity of evaporative universal detectors and LC/MS in the HILIC and the reversed-phase HPLC modes, J. Chromatogr. B 877 (2009) 4133–4139.
[51] Y. Lourié, Aide-Mémoire de Chimie Analitique, Editions Mir, Moscou, 1975.
[52] X. Subirats, E. Bosch, M. Roses, Retention of ionisable compounds on HPLC. XVII. Estimation of the pH variation of aqueous buffers with the change of the methanol fraction of the mobile phase, J. Chromatogr. A 1138 (2007) 203–215.
[53] I. Canals, J.A. Portal, E. Bosch, M. Roses, Retention of ionizable compounds on HPLC. 4. Mobile-phase pH measurement in methanol/water, Anal. Chem. 72 (2000) 1802–1809.
[54] I. Canals, F.Z. Oumada, M. Roses, E. Bosch, Retention of ionizable compounds on HPLC. 6. pH measurements with the glass electrode in methanol–water mixtures, J. Chromatogr. A 911 (2001) 191–202.
[55] S. Espinoza, E. Bosch, M. Roses, Retention of ionizable compounds in HPLC. 14. Acid-base pK values in acetonitrile-water mobile phases, J. Chromatogr. A 964 (2002) 55–66.
[56] V. David, F. Albu, A. Medvedovici, Structure – retention correlation of some oxicam drugs in reversed-phase liquid chromatography, J. Liq. Chromatogr. Rel. Technol. 27 (2004) 965–984.
[57] A.P. Schellinger, P.W. Carr, Solubility of buffers in aqueous – organic eluents for reversed-phase liquid chromatography, LCGC North America 22 (2004) 544–548.
[58] T. Cecchi, Ion pairing chromatography, Crit. Rev. Anal. Chem. 38 (2008) 161–213.
[59] L. Yu, S. Wang, S. Zeng, Chiral mobile phase additives in HPLC enantioseparations, Methods Mol. Biol. 970 (2012) 221–231.
[60] G. Lunn, L.C. Hellwig, Handbook of Derivatization Reactions for HPLC, J. Wiley & Sons, New York, 1998.

[61] S. Moldoveanu, W. Scott, J. Zhu, Analysis of small carbohydrates in several bioactive botanicals by gas chromatography with mass spectrometry and liquid chromatography with tandem mass spectrometry, J. Sep. Sci. 38 (2015) 3677–3686.
[62] V. David, T. Galaon, E. Caiali, A. Medvedovici, Competitional hydrophobicity driven separations under RP-LC mechanism: application to sulphonylurea congeners, J. Sep. Sci. 32 (2009) 3099–3106.
[63] P. Nikitas, A. Pappa-Louisi, P. Agrafiotou, S. Fasoula, Simple models for the effect of aliphatic alcohol additives on the retention in reversed-phase liquid chromatography, J. Chromatogr. A 1218 (2011) 3616–3623.
[64] H.A. Claessens, M.A. van Straten, J.J. Kirkland, Effect of buffers on silica-based column stability in reversed-phase high-performance liquid chromatography, J. Chromatogr. A 728 (1996) 259–270.
[65] H.A. Claessens, M.A. van Straten, Review on the chemical and thermal stability of stationary phases for reversed-phase liquid chromatography, J. Chromatogr. A 1060 (2004) 23–41.
[66] G.W. Tindall, R.L. Perry, Explanation for the enhanced dissolution of silica column packing in high pH phosphate and carbonate buffers, J. Chromatogr. A 988 (2003) 309–312.
[67] A.P. Schellinger, P.W. Carr, Isocratic and gradient elution chromatography: a comparison in terms of speed, retention reproducibility and quantitation, J. Chromatogr. A 1109 (2006) 253–266.
[68] P.J. Schoenmakers, H.A.H. Billiet, L. De Galan, Description of solute retention over the full range of mobile phase compositions in reversed-phase liquid chromatography, J. Chromatogr. 282 (1983) 107–121.
[69] S. Heinisch, J.-L. Rocca, Sense and nonsense of high-temperature liquid chromatography, J. Chromatogr. A 1216 (2009) 642–658.
[70] V. Houbart, G. Cobraiville, G. Nys, M.-P. Merville, M. Fillet, Impact of injection solvent composition on protein identification in column-switching chip-liquid chromatography/mass spectrometry, J.Chromatogr. A 1445 (2016) 27–35.
[71] C. Liteanu, S. Gocan, Gradient Liquid Chromatography, Ellis Horwood Limited, Chichester, 1974.
[72] F. Momenbeik, N. Bagheri, Optimization of fat-soluble vitamins separations by reversed-phase liquid chromatography with the use of aliphatic alcohols as mobile phase additives, J. Liq. Chromatogr. Rel. Technol. 38 (2015) 1355–1360.
[73] Á. Sándi, L. Szepesy, Evaluation and modulation of selectivity in reversed-phase high-performance liquid chromatography, J. Chromatogr. A 845 (1999) 113–131.
[74] C.S. Funari, R.L. Carneiro, M.M. Khandagale, A.J. Cavalheiro, E.F. Hilder, Acetone as a greener alternative to acetonitrile in liquid chromatographic fingerprinting, J. Sep. Sci. 38 (2015) 1458–1465.
[75] J.L. Bernal, M.J. Del Nozal, G.A. García Buj, J. Martín Juárez, Use of methyl and ethyl acetate as organic modifiers in reversed-phase high-performance liquid chromatography: application to impurity control in bulk drug steroids, J. Chromatogr. A 607 (1992) 175–181.
[76] P. Damiani, L. Cossignani, M.S. Simonetti, F. Santinelli, Prediction of isocratic nonaqueous reversed-phase high-performance liquid chromatography retention parameters and response factors of triacylglycerols detected by an ultraviolet-diode array – evaporative light-scattering on-line system, J. Chromatogr. Sci. 38 (2000) 195–199.
[77] F.W. Quackenbush, R.L. Smallidge, Nonaqueous reverse phase liquid chromatographic system for separation and quantitation of provitamins A, J. Assoc. Off. Anal. Chem. 69 (1986) 767–772.
[78] M. Radulescu, V. Voicu, A. Medvedovici, V. David, Retention study of some cation-type compounds using bile acid sodium salts as ion pairing agents in liquid chromatography, Biomed. Chromatogr. 25 (2011) 873–878.
[79] G. Szasz, K. Tokacs-Novak, J. Kökösi, HPLC study on ion-pairing ability of deoxycholic acids epimers, J. Liq. Chromatogr. Rel. Technol. 24 (2001) 173–185.
[80] D.C. Woollard, E. Indyk-Harvey, Rapid determination of thiamine, riboflavin, pyridoxine, and niacinamide in infant formulas by liquid chromatography, J. AOAC Int. 85 (2002) 945–951.
[81] D. Tonelli, S. Zappoli, B. Ballarin, Dye-coated stationary phases: a retention model for anions in ion-interaction chromatography, Chromatographia 48 (1998) 190–196.
[82] J.J. Fernandez-Navarro, M.C. García-Álvarez-Coque, M.J. Ruiz-Ángel, The role of the dual nature of ionic liquids in the reversed-phase liquid chromatographic separation of basic drugs, J. Chromatogr. A 1218 (2011) 398–407.
[83] S.C. Moldoveanu, J. Zhu, N. Qian, Free amino acids analysis by liquid chromatography with tandem mass spectrometry in several botanicals with antioxidant character, J. Sep. Sci. 38 (2015) 2208–2222.

[84] D.I. Sora, S. Udrescu, V. David, A. Medvedovici, Validated ion pair liquid chromatography/fluorescence detection method for assessing the variability of the loratadine metabolism occurring in bioequivalence study, Biomed. Chromatogr. 21 (2007) 1023–1029.
[85] P. Hemström, K. Irgum, Hydrophilic interaction chromatography, J. Sep. Sci. 29 (2006) 1784–1821.
[86] B. Buszewski, S. Noga, Hydrophilic interaction liquid chromatography (HILIC) – a powerful separation technique, Anal. Bioanal. Chem. 402 (2012) 231–247.
[87] P. Jandera, Stationary and mobile phases in hydrophilic interaction chromatography: a review, Anal. Chim. Acta 692 (2011) 1–25.
[88] M. Caude, A. Jardy, Normal phase liquid-chromatography, in: E. Katz, R. Eksteen, P. Schoenmakers, N. Miller (Eds.), Handbook of HPLC, Marcel Dekker, New York, 1998, p. 325.
[89] J.S. Fritz, D.T. Gjerde, Ion Chromatography, Wiley-VCH, Weinheim, 2009, p. 356.
[90] H. Small, T.S. Stevens, W.C. Bauman, Novel ion exchange chromatographic method using conductometric detection, Anal. Chem. 47 (1975) 1801–1809.
[91] J. Weiss, T. Weiss, Handbook of Ion Chromatography, Wiley-VCH, Weinheim, 2016, p. 449.
[92] B.S. Liao, J. Sram, T.T. Cain, K.R. Halcrow, Aqueous sulfuric acid as the mobile phase in cation ion chromatography for determination of histamine, putrescine, and cadaverine in fish samples, J. AOAC Int. 94 (2011) 565–571.
[93] Y. Zhu, M. Wang, H. Du, F. Wang, S. Mou, P.R. Haddad, Organic analysis by ion chromatography. 1. Determination of aromatic amines and aromatic diisocyanates by cation-exchange chromatography with amperometric detection, J. Chromatogr. A 956 (2002) 215–220.
[94] J. Weiss, Ion Chromatography, VCH, Weinheim, 1995, p. 66.
[95] P.R. Haddad, P.E. Jackson, Ion Chromatography. Principles and Applications, Elsevier, Amsterdam, 1990.
[96] K. Robards, P.E. Jackson, P.A. Haddad, Principles and Practice of Modern Chromatographic Methods, Elsevier, Amsterdam, 2004, p. 365.
[97] G. Gübitz, M.G. Schmid, Chiral separation by chromatographic and electromigration techniques. A review, Biopharm. Drug Dispos. 22 (2001) 291–336.
[98] K.G. Lynam, R.W. Stringham, Chiral separations on polysaccharide stationary phases using polar organic mobile phases, Chirality 18 (2006) 1–9.
[99] Y.K. Ye, R.W. Stringham, The effect of acidic and basic additives on the enantioseparation of basic drugs using polysaccharide-based chiral stationary phases, Chirality 18 (2006) 519–530.
[100] S.C. Change, G.L. Reid, S. Chen, C.C. Chang, D.W. Armstrong, Evaluation of a new polar-organic high performance liquid chromatographic mobile phase for cyclodextrin bonded chiral stationary phases, Trends Anal. Chem. 12 (1993) 144–153.
[101] Y. Tang, W.L. Zielinski, H.M. Bigott, Separation of nicotine and nornicotine enantiomers via normal phase HPLC on derivatized cellulose chiral stationary phases, Chirality 10 (1998) 364–369.
[102] A.A. Younes, D. Mangelings, Y. Vander Heyden, Chiral separations in normal phase liquid chromatography: enantioselectivity of recently commercialized polysaccharide-based selectors. Part I: enantioselectivity under generic screening conditions, J. Pharm. Biomed. Anal. 55 (2011) 414–423.
[103] T. O'Brien, L. Crocker, R. Thompson, K. Thompson, P.H. Toma, D.A. Conlon, B. Feibush, C. Moeder, G. Bicker, N. Grinberg, Mechanistic aspects of chiral discrimination on modified cellulose, Anal. Chem. 69 (1997) 1999–2007.
[104] I. Ilisz, R. Berkecz, A. Peter, Retention mechanism of high-performance liquid chromatographic enantioseparation on macrocyclic glycopeptide-based chiral stationary phases, J. Chromatogr. A 1216 (2009) 1845–1860.
[105] G. Gubitz, M.S. Schmid, Chiral Separations. Methods and Protocols, Humana Press, New Jersey, 2004.
[106] N. Matthijs, M. Maftouh, Y. Vander Heyden, Screening approach for chiral separation of pharmaceuticals IV. Polar organic solvent chromatography, J. Chromatogr. A 1111 (2006) 48–61.
[107] R. Wernicke, Separation of underivatised amino acid enantiomers by means of a chiral solvent-generated phase, J. Chromatogr. Sci. 23 (1985) 39–48.
[108] A. Karlsson, C. Pettersson, Separation of enantiomeric amines and acids using chiral ion-pair chromatography on porous graphitic carbon, Chirality 4 (1992) 323–332.
[109] C. Pettersson, Chromatographic separation of enantiomers of acids with quinine as chiral counter ion, J. Chromatogr. A 316 (1984) 553–567.

[110] J. Brandrup, E.H. Immergut, E.A. Gruelke, A. Abe, D.R. Bloch (Eds.), Polymer Handbook, Wiley, New York, 1999.
[111] F.W. Billmeyer Jr., Texbook of Polymer Science, Wiley-Interscience, New York, 1962.
[112] E.R. Kunji, M. Harding, P.J. Butler, P. Akamine, Determination of the molecular mass and dimensions of membrane proteins by size exclusion chromatography, Methods 46 (2008) 62–72.
[113] R.D. Ricker, L.A. Sandoval, Fast, reproducible size-exclusion chromatography of biological macromolecules, J. Chromatogr. A 743 (1996) 43–50.
[114] G.B. Irvine, High-performance size-exclusion chromatography of polypeptides on a TSK G2000SW column in acidic mobile phases, J. Chromatogr. A 404 (1987) 215–222.
[115] E.S.P. Bouvier, S.M. Koza, Advances in size-exclusion separations of proteins and polymers by UHPLC, Trends Anal. Chem. 63 (2014) 85–94.
[116] P. Tavlarakis, J.J. Urban, N. Snow, Determination of total polyvinylpyrrolidone(PVP) in ophthalmic solutions by size exclusion chromatography with ultraviolet-visible detection, J. Chromatogr. Sci. 49 (2011) 457–462.
[117] M. Gaborieau, P. Castignolles, Size-exclusion chromatography (SEC) of branched polymers and polysaccharides, Anal. Bioanal. Chem. 399 (2011) 1413–1423.
[118] T. Arakawa, D. Ejima, T. Li, J.S. Philo, The critical role of mobile phase composition in size exclusion chromatography of protein pharmaceuticals, J. Pharm. Sci. 99 (2010) 1674–1692.
[119] M. Potschka, Size-exclusion chromatography of polyelectrolytes: experimental evidence for a general mechanism, J. Chromatogr. 441 (1988) 239–260.
[120] T. Mizutani, A. Mizutani, Prevention of adsorption of protein on controlled-pore glass with amino acid buffer, J. Chromatogr. 111 (1975) 214–216.
[121] Polymer and Hydrocarbon Processing Solutions with HPLC, Agilent Solutions Guide, 1999. Publ. No. 5968–7020E.
[122] B. Rao, S.T. Balke, T.H. Mourey, T.C. Schunk, Methylcyclohexane as a new eluting solvent for the size-exclusion chromatography of polyethylene and polypropylene at 90°C, J. Chromatogr. A 755 (1996) 27–35.
[123] A. Boborodea, A. Luciani, Assessing the suitability of a green solvent for GPC and TREF analyses of polyethylene, LC-GC Europe 27 (2014) 621–623.
[124] B.J. VanMiddlesworth, J.G. Dorsey, Quantifying injection solvent effects in reversed-phase liquid chromatography, J. Chromatogr. A 1236 (2012) 77–89.
[125] S. Keunchkarian, M. Reta, L. Romero, C. Castells, Effect of sample solvent on the chromatographic peak shape of analytes eluted under reversed-phase liquid chromatographic conditions, J. Chromatogr. A 1119 (2006) 20–28.
[126] J. Layne, T. Farcas, I. Rustamov, F. Ahmed, Volume-load capacity in fast-gradient liquid chromatography Effect of sample solvent composition and injection volume on chromatographic performance, J. Chromatogr. A 913 (2001) 233–242.
[127] Q. Zhong, L. Shen, J. Liu, D. Yu, S. Li, J. Yao, S. Zhan, T. Huang, Y. Hashi, S. Kawano, Z. Liu, T. Zhou, Precolumn dilution large volume injection ultra-high performance liquid chromatography-tandem mass spectrometry for the analysis of multi-class pesticides in cabbages, J. Chromatogr. A 1442 (2016) 53–61.
[128] E. Loeser, S. Babiak, P. Drumm, Water-immiscible solvents as diluents in reversed-phase liquid chromatography, J. Chromatogr. A 1216 (2009) 3409–3412.
[129] J. Ruta, S. Rudaz, D.V. McCalley, J.-L. Veuthey, D. Guillarme, A systematic investigation of the effect of sample diluent on peak shape in hydrophilic interaction liquid chromatography, J. Chromatogr. A 1217 (2010) 8230–8240.
[130] T. Ikegami, K. Tomomatsu, H. Takubo, K. Horie, N. Tanaka, Separation efficiencies in hydrophilic interaction chromatography, J. Chromatogr. A 1184 (2008) 474–503.
[131] J.R. Johnson, D. Karlsson, M. Dalene, G. Skarping, Determination of aromatic amines in aqueous extracts of polyurethane foam using hydrophilic interaction liquid chromatography and mass spectrometry, Anal. Chim. Acta 678 (2010) 117–123.
[132] P. Hong, S. Koza, E.S.P. Bouvier, Size-exclusion chromatography for the analysis of protein biotherapeutics and their aggregates, J. Liq. Chromatogr. Rel. Technol. 35 (2012) 2923–2950.
[133] S. Kromidas, HPLC Made to Measure. A Practical Handbook for Optimization, Wiley-VCH, Weinheim, 2006, p. 395.
[134] www.crawfordscientific.com/downloads/pdf_new/Daicel/DAICEL_ChiralPakIA_MD.pdf.

[135] S. Udrescu, A. Medvedovici, V. David, Effect of large volume injection of hydrophobic solvents on the retention of less hydrophobic pharmaceutical solutes in RP-LC, J. Sep. Sci. 31 (2008) 2939–2945.
[136] T. Galaon, E. Bacalum, M. Cheregi, V. David, Retention studies for large volume injection of aromatic solvents on phenyl-silica based stationary phase in RP-LC, J. Chromatogr. Sci. 51 (2013) 166–172.
[137] V. David, T. Galaon, H.Y. Aboul-Enein, Effects of large volume injection of aliphatic alcohols as sample diluents on the retention of low hydrophobic solutes in reversed-phase liquid chromatography, J. Chromatogr. A 1323 (2014) 115–122.
[138] A. Medvedovici, Va David, Vi David, C. Georgita, Retention phenomena induced by large volume injection of organic solvents non-miscible with mobile phase in reversed-phase liquid chromatography, J. Liq. Chromatogr. Rel. Technol. 30 (2007) 199–213.
[139] V. David, M. Cheregi, A. Medvedovici, Alternative sample diluents in bioanalytical LC-MS, Bioanalysis 5 (2013) 3051–3061.
[140] W. Kleiböhmer (Ed.), Environmental Analysis, Elsevier, Amsterdam, 2001, p. 187.
[141] G. Mitulovic, C. Stingl, I. Steinmacher, O. Hudecz, J.R.A. Hutchins, J.-M. Peters, K. Mechtler, Preventing carry-over of peptides and proteins in nano LC-MS separations, Anal. Chem. 81 (2009) 5955–5960.
[142] W. Zeng, D.G. Musson, A.L. Fisher, A.Q. Wang, A new approach for evaluating carryover and its influence on quantitation in high-performance liquid chromatography and tandem mass spectrometry assay, Rapid Commun. Mass Spectrom. 20 (2006) 635–640.

CHAPTER

14

Gradient Elution

14.1 THE USE OF GRADIENT IN HPLC

The separations in high-performance liquid chromatography (HPLC) can be performed without changing the composition of the mobile phase during the chromatographic run, or with changing the mobile phase composition. The first type of separation is indicated as isocratic, and the second as gradient (see also Section 4.2). The composition change may be in the percent of solvents making the mobile phase, in pH, in ionic strength, in the level of specific additives, or a combination of these changes. Gradient separations are common in reversed-phase (RP)-HPLC, ion pair chromatography, hydrophilic interaction chromatography (HILIC), and normal-phase chromatography (NPC). In ion chromatography, the concentration of competing ions is usually modified during gradient separations, and not the solvent composition which is frequently water. In size exclusion, the gradients are not usually applied. When repeated runs are performed using gradient separation, the column must be equilibrated with the initial mobile phase composition before each run.

Purposes for the Use of Gradients

Gradient elution is basically used for four main purposes: (1) reduction of the total run time of separations, (2) modification of retention times in a separation that does not provide a good separation between specific compounds, (3) narrowing of the chromatographic peaks, and (4) cleaning and/or regeneration of the chromatographic column. Other more uncommon utilizations of gradient can be mentioned, such as loading of the column with a specific reagent.

1) In common HPLC separation, the values for the capacity factors k' for different solutes with a given stationary phase and mobile phase should be in the range $1 \leq k' \leq 10$ such that retention times lower than 20–30 min can be obtained with common chromatographic columns (see Section 4.2). Once the column is selected, the k' value in the desired range can be adjusted by modifying the mobile phase composition. However, it is common that in a sample are present compounds that have "low affinity" for the stationary phase and some with "very high affinity" for the same composition of the mobile phase. For example, in RP-HPLC with a mobile phase with a high organic content, some compounds with small hydrophobic moieties are poorly retained and may have $k' < 1$. For these k' values, the retention is

Selection of the HPLC Method in Chemical Analysis
http://dx.doi.org/10.1016/B978-0-12-803684-6.00014-7

very sensitive to small (unintentional) changes in the mobile phase composition, leading to changes in the retention times, and interferences from unretained materials from the sample matrix are possible. For increasing k' values in these cases, a "weak" mobile phase is necessary. However, the use of such a mobile phase for the whole chromatogram may not be acceptable for the elution of more hydrophobic compounds that will have very large k' in a weak mobile phase. With a weak polar mobile phase these compounds will be eluted late in the chromatogram and also will have broad peaks. Although the peaks will be well separated, the separations are unnecessarily large, the peak integration is less accurate, and the run time is too long. The decrease of k' for strongly retained compounds can be achieved using a "strong" eluting mobile phase, but the use of such a mobile phase is inadequate for the compounds with low affinity for the stationary phase. The answer to this problem is the use of a gradient elution. In the gradient elution, a weak mobile phase is used at the beginning of the chromatographic run for assuring the retention of the compounds with low affinity for the stationary phase, and the composition of the mobile phase is changed into a stronger one during the chromatographic run, for a faster elution of the compounds with a high affinity for the stationary phase. The decrease in the retention factor with the mobile phase composition given by expression (4.2.69) can be written for an RP-HPLC separation in the form:

$$\log k_2'(X) = \log k_1'(X) - S(X)(\phi_2 - \phi_1) \tag{14.1.1}$$

where k_1' and k_2' are the retention factors in the two mobile phases, and parameter $S(X)$ is specific for the solute X, solvent mixture, and column, not depending on ϕ. For reversed-phase chromatography, ϕ represents the volumetric fraction of organic component and for $\phi_2 > \phi_1$, the result is $k_2' < k_1'$ (in HILIC, the increase in water content reduces the retention time). Deviations from linearity are sometimes encountered even for RP-HPLC, but more frequently for other separation types. In HILIC, for example, the linearity may follow only a portion of the mobile phase composition, which becomes logarithmic (following expression 4.3.19) for a different concentration range.

One additional advantage of shortening the retention time by using gradient elution is related to the peak shape in the chromatogram. At longer retention times, the peaks are broadened, and shorter times help in obtaining sharper peaks. Gradient elution provides good repeatability of the retention times and also of the peak area, such that no adverse effects appear regarding quantitation with gradient separation as compared to isocratic elution [1]. Contribution to the reduction of the run time can be obtained not only from using a mobile phase with a higher content of organic phase (in RP-HPLC), but also from using sharper gradient changes. The slope of gradient, or its speed, can be optimized in order to assure best results such that the separation is not compromised and the analysis run time is shortened [2].

2) The reduction of k' values for different analytes (X and Y) must be done such that the values for selectivity remain $\alpha(X, Y) > 1.2$ for all the components of interest in the sample, and the value for resolution $R > 1$ (or better $R > 1.5$). However, as indicated by expression (4.2.61), R depends on k' in addition to α and N, and smaller k' values may lead to poor separations. For this reason, when decreasing k' values by using a "stronger" mobile phase, this cannot be done without restrictions, and the gradient must maintain a good separation. The choice of a convenient gradient is relatively simple when an initial isocratic separation (or a gradient with small composition changes) provides a good separation. Starting with this

separation, the slope of gradient is increased such that the change from a "weak eluting" mobile phase to a "strong eluting" mobile phase is done more rapidly. The new separation must remain acceptable regarding selectivity and resolution. The process can also start in reversed order, with a relatively poor separation in a rapid gradient and short retention times, which are changed to slower composition changes that would allow a better separation.

The variation in k' when changing the mobile phase composition depends not only on the mobile phase composition (through ϕ), but also on the nature of the stationary phase, and on the nature of the analytes (through $S(X)$) as shown by formula (14.1.1). When the composition of the mobile phase is changed (within a range) from "weak" to "strong" or when the pH and the ionic strength are changed, different components from the sample may be affected differently, and their retention times, although shortened, may be affected differently. For this reason, the specific values for $\alpha(X, Y)$ can either decrease or increase, although both $k'(X)$ and $k'(Y)$ decrease. The increase of specific $\alpha(X, Y)$ values is not uncommon, and improvements in the separation can sometimes be obtained by increasing the "strength" of the mobile phase, which leads at the same time to shorter retention times. Predictions of such changes are in general difficult to make, and the descriptions of analytical methods from the literature provide details on specific mobile phase compositions and gradients that assure a good separation. This is the main cause for the need of a "trial and error" strategy for optimizing some separations. The selection of a specific stationary phase in obtaining a desired separation is very important in such cases.

3) The use of gradient may allow narrowing of the chromatographic peaks by two mechanisms. The first is the use of a very weak mobile phase that allows the complete retention of the analytes at the head of the chromatographic column. Larger injections would generate wider peaks, but if the analytes are focused at the head of the chromatographic column, they may be eluted as a very narrow zone when a stronger eluent passes through the column. Another effect that produces some narrowing, in particular for broader peaks and in steep gradient, is the faster elution of the front of the peak as compared to the back of the peak. In gradient, the front of a peak is exposed to a slightly stronger eluent compared to the back of the peak, and therefore it elutes faster [3].

4) In most gradients, a purge region is included for cleaning the column from any remaining solutes, not necessarily analytes. This region may have the highest concentration of the "strong" eluent. For example, in RP-HPLC the purge region may consist of a high concentration of organic phase or even of pure organic component. When pure organic component is not necessary for cleaning the column, a lower concentration should be used such that the re-equilibration of the column is done faster. The cleaning of the stationary phase from all injected sample components is part of the good care of the chromatographic column. However, this is not always convenient, and in some applications cleaning is recommended after a number of injections (analyzed samples) and not after each sample. In such cases, a separate gradient run (or isocratic run) with strong (possibly pure) solvents is applied, followed by a longer re-equilibration of the stationary phase to initial condition.

Besides the advantages, some problems may appear in gradient elution. One problem is related to the inability to use refractive index (RI) detection with gradients. Another problem may be the appearance of a drift in the baseline of the chromatogram. This drift depends on the selected solvents and on the detector, and it is not a common problem. Also, the gradient is not necessary when the analytes elute close to each other in isocratic separation.

Practice of Gradient Elution

The gradient mobile phase is usually obtained by mixing in different proportions two solutions (solution A and solution B) with the help of the pumping system of the HPLC instrument (see Section 4.1). Gradients made with three or even four solutions are possible with some HPLC instruments (e.g., with low-pressure mixing), but such gradients are not too frequently necessary. A typical gradient elution starts with a short isocratic range where the weaker mobile phase passes through the chromatographic column. This portion assures that the analytes are retained at the head of the chromatographic column and possibly undergo a focusing process. After this isocratic hold, the gradient starts. The change in the mobile phase composition can be done in a linear mode (see expression 4.1.1), nonlinear mode, or even as step gradient (sudden change in mobile phase composition). Multiple gradient ramps are also possible during one chromatographic run. The gradient slopes are designed to achieve the desired separation in the optimal (or close to optimal) retention times. Following the gradient slope, the gradient program typically includes a section with the "strongest" mobile phase, with the role of eluting all sample components from the stationary phase. After this section, the mobile phase composition is restored to the initial conditions to make the system ready for the next injection. The gradient is controlled in modern HPLC systems by a gradient program (gradient timetable) stored in the computer that operates the HPLC (see Section 4.1). The schematic diagram of a simple gradient program for an RP-HPLC separation is shown in Fig. 14.1.1.

The gradient change in mobile phase composition is performed with a specific slope characterized by the gradient slope $\Delta\phi/t_{grad}$ (see expression 4.2.70) or by the gradient steepness b given by expression (4.2.72). The change $\Delta\phi/t_{grad}$ may indicate for RP-HPLC the change in organic phase content, or for HILIC the change in water content, but also other changes such as pH or concentration of a competing ion can be represented by $\Delta\phi$. The change in concentration of an organic phase may range from 0% to 100%, depending on the requirements of the separation, and t_{grad} can be relatively long (a good part of the chromatographic run) or very steep ($t_{grad} \approx 0$ in step gradient) [4–6]. In practice, mainly when the gradients are very steep (sharp), a gradient distortion can be noticed. This is manifested by large deviations from the sharpness of a required gradient and may be different from instrument to instrument. Low-pressure mixing instruments tend to have more gradient distortions than high-pressure mixing systems. Typically the newer ultra-performance liquid chromatography (UPLC) systems have a better controlled gradient sharpness. Also, the change in the mobile phase composition takes

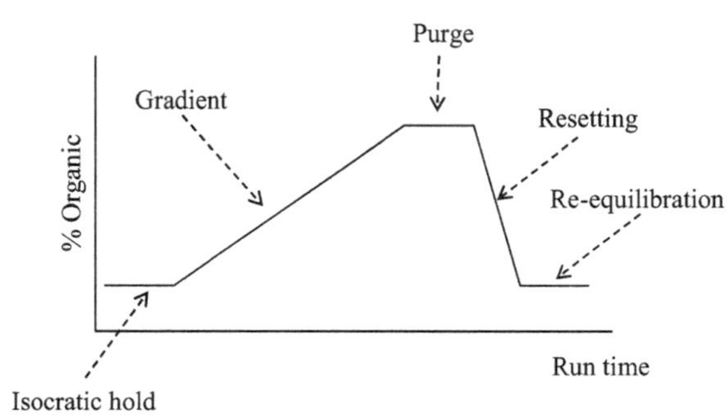

FIGURE 14.1.1 Diagram of a simple gradient variation in an RP-HPLC separation.

place with a certain delay (dwell time t_D) caused by small spaces (dwell volume) existent in the HPLC instruments before the column itself. The dwell volume for common HPLC systems is between 500 μL and 3 mL, and it is smaller for UPLC systems.

When gradients are used, a re-equilibrating time is necessary for restoring the stationary phase to the initial conditions before a new chromatographic run is begun. This re-equilibration time depends on the column dimensions and also on the type of column and mobile phase. It is common to divide the volume necessary for column re-equilibration into two portions, the system washout and the "true" column re-equilibration volumes. One empirical rule to establish these necessary volumes for re-equilibration uses the formula:

$$V_{re-eq} = 3V_{solv} + 5V_0 \quad (14.1.2)$$

In expression (14.1.2), V_{solv} is a volume of solvent necessary for column conditioning and $V_{solv} = 650-3000$ μL depending on the column dimension and properties, and V_0 is the void volume of the column ($V_0 = \varepsilon^*(\pi/4)d^2L$). The re-equilibration time is obtained from $V_{re\text{-}eq}$ based on the flow rate from the formula $t_{re\text{-}eq} = V_{re\text{-}eq}/U$. Shorter times than those recommended by expression (14.1.2) may be sufficient for some analyses.

Gradient of Solvent Composition

The modification of solvent composition of the mobile phase is the most common type of gradient and this is typically done by mixing two solvents. In RP-HPLC the starting composition has a low organic content, which is increased during the chromatographic run. In multistep gradients, it is possible to have regions with a reversed gradient but this is uncommon. In HILIC the starting composition has a high organic content. During the modification of organic phase composition, it should be verified that the solvents are perfectly miscible, and also that their mixture is still a good solvent for the buffers and additives. In some applications it is recommended that solution A contains a small proportion of solution B from the start of the gradient. For example, in a gradient made with two solvents such as water and methanol, solution A contains 5% methanol, while solution B is pure methanol. This assures an easier mixing of the gradient components.

As the organic content increases, the solubility of some additives decreases, in particular that of inorganic salts. Since some buffers and additives are frequently added in only one of the mobile phase components, it should be understood that the concentration of these additives and buffers changes when the solvent composition changes. In some applications, such as in the case of using MS detection with electrospray ionization (ESI) or atmospheric pressure chemical ionization (APCI) detection, special additives such as HCOOH or $HCOONH_4$ are necessary for enhancing analyte ionization. For assuring a constant concentration of such additives, they must be added in equal concentration in both components of the mobile phase.

Gradient of pH

The variation of the pH of a mobile phase is done with a gradient separation and can be achieved by different procedures. For RP-HPLC, for example, one procedure is the modification of the ratio of two solutions, one being an aqueous (or partially aqueous) buffer solution

A, and the other an organic (or partially organic) solution B that does not contain buffers or additives. Another procedure consists of the modification of the ratio of two partially aqueous solutions A and B, each one buffered at a different pH. These solutions with different pH may have the same or different contents of the organic phase. Gradients with different pH and different organic contents can also be achieved by mixing more than two solutions [7]. However, three or four solution gradients, although sometimes utilized, are not common. The ionization status of a compound can be even more important regarding its retention in a polar (HILIC) or in an ion exchange separation than it is in RP-HPLC. When the mobile phase does not contain an organic component (e.g., in ion chromatography), the gradient is achieved by mixing two solutions with different pH (and possibly different inorganic additive contents).

The pH change of the mobile phase is usually performed intentionally, but in some application the change is the result of mixing one solvent that contains the buffer, and another solvent not having an adjusted pH [8]. In order to avoid an undesired pH change it should be assured that the buffer is present in both gradient components. Although the increase in organic constituent still modifies the final pH, the change is expected to be smaller.

Gradient in Flow Rate

Besides composition change, a change in the flow rate of the mobile phase may be practiced in some applications. The need for such a change may appear as a result related, for example, from the excessive increase in backpressure as the mobile phase composition changes. Some solvent mixtures have an increased viscosity as compared to the initial viscosity of the mobile phase. For avoiding excessive backpressure, a slower flow rate can be used for the region with higher viscosity of the mobile phase. Another reason for changing the flow rate during the chromatographic run may be related to the requirements of the detector. For example, the MS detectors when using ESI type ionization must use a specific flow rate for assuring a maximum yield of ions (see Section 4.1). The same flow required for the detector optimum may not be necessary or convenient for the whole chromatogram. It is possible to adjust a specific flow rate only for the time window where the analytes of interest are eluting, while the rest of the separation is performed at a different flow rate.

14.2 PARAMETERS CHARACTERIZING THE GRADIENT SEPARATION

The characterization of gradient separations using the same parameters used for isocratic separation (see Section 4.2) encounters the problem of variability along the chromatographic run of key parameters such as retention factor k', selectivity α, and resolution R. For this reason, the introduction of several *effective parameters* was necessary for the global characterization of the separation in gradient conditions (see Section 4.2). Among effective parameters are the gradient retention factor k^*, effective selectivity α^*, and effective resolution R^*. Such parameters are the equivalent of those used for isocratic separation, but their utility is lower, and they were developed for model systems [9–11].

Retention Factor in Gradient Separations

A common procedure to obtain an expression for retention factor k^* in gradient separations starts with the assumption that the analyte X is present in the chromatographic column in a very narrow zone, moving with the instantaneous linear velocity u_X. The expression for instantaneous linear velocity u_X can be obtained with the formula:

$$u_X = \frac{dx}{dt} \tag{14.2.1}$$

The time t spent by compound X in the column after traveling the distance x is given by formula (4.2.19) (with index $_R$ and X omitted): $t = t_0 + k't_0$ where t_0 is the dead time for the column and k' is the retention factor, which depends on the compound X, the stationary phase, and the mobile phase composition ϕ. During the gradient, the mobile phase composition $\phi(t)$ changes with time. For $k'(\phi)$ taken as the instantaneous capacity factor, expression (14.2.1) is equivalent with the formula:

$$u_X = \frac{u_0}{1 + k'(\phi)} \tag{14.2.2}$$

where u_0 is the linear flow rate of the mobile phase. In relation (14.2.1), the following substitution can be made:

$$t = z + \frac{x}{u_0} \tag{14.2.3}$$

With this substitution, the solvent composition $\phi(t)$ becomes a function of z ($\phi(z)$). The expression for dt from Eq. (14.2.1) with the substitution of u_x given by formula (14.2.2) (and ϕ a function of z) can be written in the form:

$$dt = \frac{1 + k'(\phi(z))}{k'(\phi(z))} dz = \left(1 + \frac{1}{k'(\phi(z))}\right) dz \tag{14.2.4}$$

An assumption can be made that the solute is injected at $x = 0$ and $t = 0$, and expression (14.2.4) can be integrated for the interval from $z = 0$ to $z = t_R(X) - t_0$ (where $t_R(X)$ is the peak retention). The result of the integration is the following:

$$t_0 = \int_0^{t_R(X)-t_0} \frac{1}{k'(\phi(z))} dz \tag{14.2.5}$$

For continuing the integration, the explicit dependence of k' on $\phi(z)$ must be known. Making the same assumptions as in Section 4.2 of a linear gradient starting at $t = 0$ with $\phi = \phi_0 + \frac{b}{t_0} t$ and the gradient steepness $b = \frac{\Delta\phi}{t_{grad}} S(X) t_0$, the dependence of k' on t is given by formula (4.2.73), which can be written in the form:

$$\log k'(X) = \log k'_0(X) - \frac{b}{t_0} t \tag{14.2.6}$$

where $k'_0(X)$ is the capacity factor at initial mobile phase composition ϕ_0. With expression (14.2.6) (and formula 14.2.3 for t), the result of the integration is the following:

$$t_R(X) - t_0 = \frac{t_0}{b} \log(2.303 k'_0(X) b + 1) \tag{14.2.7}$$

In the process of obtaining expression (14.2.11) for t_R, it was assumed that the gradient starts changing at the head of the chromatographic column at $t = 0$, or in other words that the dwell volume $V_D = 0$, and as a result the dwell time $t_D = V_D/U = 0$. However, the dwell volume in common HPLC should not be disregarded. Including the dwell time of the system, the correct expression for $t_R(X)$ will become:

$$t_R(X) = t_0 + \frac{t_0}{b}\log\left(2.303k'_0(X)b + 1\right) + t_D \quad (14.2.8)$$

Expression (14.2.8) is the equivalent for gradient separations of expression (4.2.19) for isocratic separations. For a gradient separation described by the gradient steepness b, a capacity factor $k'_e(X, b)$ can be defined using the expression:

$$k'_e(X, b) = \frac{1}{b}\log\left(2.303k'_0(X)b + 1\right) \quad (14.2.9)$$

The *effective* gradient retention factor k^* usually applied for the characterization of gradient separations (see Section 4.2), being an "average" value, corresponds to a peak that would migrate half way through the chromatographic column in isocratic conditions. For this reason, the value for k^* is usually taken as the double of k'_e, with the following formula:

$$k^*(X) = 2k'_e(X, b) \quad (14.2.10)$$

Other Parameters for the Characterization of Chromatograms in Gradient Separations

The theory of parameters describing the gradient elution has been the subject of numerous developments that attempted either to improve the approximations used for obtaining previous relations or to extend the results for different gradient profiles (e.g., gradient starting at $t \neq 0$ or multiple gradient slopes). Expressions other than Eq. (14.2.10) that approximate a k^* for gradient separations are reported in the literature [12–14].

The parameter $k^*(X)$ can be used for defining other effective parameters used for the characterization of gradient separations. Among these are the effective selectivity α^* and effective resolution R^*. These parameters are defined by the formulas:

$$\alpha^* = \frac{k^*(X)}{k^*(Y)} \quad (14.2.11)$$

$$R^* = \frac{1}{4}\left(\frac{\alpha^* - 1}{\alpha^*}\right)\left(\frac{k^*}{1 + k^*}\right)N^{1/2} \quad (14.2.12)$$

The number of theoretical plates N is also compound-dependent, and relation (14.2.12) would be more correctly written as dependent on N^*. However, $N^* \approx N$ can be taken with good approximation.

The effective retention factor k^* also has been used for the description of the peak width in gradient HPLC. The peaks undergo a "narrowing" effect in gradient separations due to the acceleration of elution as the concentration of stronger eluent increases. The peak width in gradient HPLC is given by the expression:

$$W_b = 4Gt_0(1 + k'_e)N^{-1/2} \tag{14.2.13}$$

where G is a "compressing factor" approximated by the expression:

$$G \approx \frac{1 + k^*}{1 + 2k^*} \tag{14.2.14}$$

Expression (14.2.13) is equivalent with formula (4.2.82) for $G = 1$.

During a gradient separation, one additional parameter that changes is the viscosity of the mobile phase. This viscosity change may lead to undesirable increases in the column backpressure. The column backpressure can be calculated for a column filled with porous particles using relation (6.2.9). Replacing in this formula the linear flow rate u with the volumetric flow rate U given by formula (4.2.1) (where $A_c = \varepsilon^*(\pi/4)d^2$), the expression becomes:

$$\Delta p = \frac{4\eta U \phi_r L}{\pi \varepsilon^* d^2 d_p^2} \approx \frac{2500\,\eta UL}{d^2 d_p^2} \quad (psi) \tag{14.2.15}$$

where common notations were used (η is the solvent viscosity, d is column diameter, d_p is the particle diameter, and ϕ_r is the column flow resistance factor). The maximum pressure during the separation is obtained using maximum η during the gradient (relation 14.2.15 tends to predict lower-pressure values than experimentally obtained). As shown in Section 13.2, the viscosity of solvent mixtures can be significantly higher than that of pure solvents, and an increase in the backpressure during the gradient is very common (see Fig. 13.2.2).

14.3 SELECTION OF GRADIENT IN DIFFERENT CHROMATOGRAPHIC TYPES

Gradient separations are very common in some chromatographic types, and are less useful in some other. Also, different types of gradients are used in different types of chromatography. Some chromatographic techniques such as size exclusion (gel filtration chromatography [GFC] and gel permeation chromatography [GPC]) rarely use gradient separations. Also, gradient separations are not very common in chiral chromatography. Some particularities regarding the use of gradient in specific types of chromatography are briefly discussed in the section.

Gradient in RP-HPLC and Nonaqueous Reversed-Phase Chromatography

In RP-HPLC, gradient separations are very common. Water, methanol, and acetonitrile are the most common solvents used for the mobile phase. Many RP-HPLC methods use as solution A an aqueous solution containing the buffers and desired additives, with a low content of organic solvent (e.g., 5%), and as solution B a pure organic solvent, or an organic solvent containing some additives. The content in the organic solvent may go as high as 100%. Problems such as dewetting of stationary phase in 100% water or buffers/additives precipitating in high organic solvent content must be avoided. At the beginning of a new method development, gradients with a low slope are usually used. This may generate longer run times, and the gradient may be subsequently modified to a steeper slope, which reduces the retention time and the peak separation (resolution). The gradient slope may be increased as long as the separation is still good. Also, multistep gradients can be used for reducing the retention time on a selected portion of the chromatogram. When the first peaks of interest in the chromatogram have relatively long retention times, the starting mobile phase composition can be changed to a stronger eluent. A purging region of gradient is typically useful for "cleaning" the chromatographic column or for the elution of strongly retained analytes. The change in the content of additives during the gradient must always be considered when the additives are not present at the same concentration in both gradient components. Such changes may affect sensitivity.

In some separations, more than two solvent gradients are necessary. This can be the case in some nonaqueous reversed phase separations (NARP). For example, for the separation in the same chromatographic run of compounds with relatively low hydrophobic character (low log K_{ow}) from compounds with very high hydrophobic character, the range of "strength" of two solvents may not be sufficient. In such cases, three solvent gradients can be used, starting for example with the methanol/ethanol gradient, followed by the ethanol/heptane gradient. In such a case, miscibility of the solvents must be assured (methanol and heptane are not miscible).

Gradient in Ion Pair Chromatography

Ion pair separations can be performed either in isocratic or in gradient conditions. For gradient, all the components in the mobile phase can be modified, including organic modifier, IPA concentration, as well as the pH of the mobile phase. When some additive, IPA, and the buffer are present only in one mobile phase component, the change in their concentration must be taken into account as the solvent composition changes.

Gradient in HILIC

Gradient elution is commonly used in HILIC. Gradient elution in HILIC starts from a composition of mobile phase rich in organic constituent (e.g., acetonitrile or sometimes methanol), and the concentration of the polar (aqueous) component is increased in time. The high initial acetonitrile content in the mobile phase will assure sufficient retention for the analytes with low affinity for the stationary phase. Running the gradient toward a high water concentration (e.g., 90%) favors the desorption of strongly retained analytes on the stationary phase. The gradient type depends on the nature of the stationary phase, organic component chosen for mobile phase composition, and on the nature of the analytes in the injected sample. Buffers are used frequently in the mobile phase, and a good practice is to have the same buffer in the two phases that make the gradient such that the pH of the mobile phase is kept constant during the run. The pH of the buffers for HILIC may be selected to enhance the dissociation of sample analytes ($pH > 7$ for acids and $pH < 7$ for bases). Selectivity in HILIC can be well controlled with the pH of the mobile phase, the difference in retention of different analytes in HILIC being more affected by pH as compared to RP-HPLC. The hydrophobic interactions are less sensitive to pH modifications because hydrophobic surfaces are usually not changed with pH changes. On the other hand, the pH of the mobile phase can strongly affect molecular polarity. Since the mobile phase in HILIC (and NPC) plays a role in the nature of stationary phase surface, its content in water (for HILIC) or other polar additives must be carefully considered. The changes in the mobile phase pH or ionic strength should be avoided in cases of gradient separations with two solvents A and B, where A is mainly organic and B is mainly aqueous. For this reason, the same buffer/additive content should be used in both solutions. For the gradient separations with the initial mobile phase with no buffer and with the buffer content increasing during the run, changes in the retention mechanism may occur.

Gradient in Ion Chromatography

The use of gradient elution in ion chromatography (IC) is less common. The conductivity detection, which is usually employed in IC, is sensitive to changes in mobile phase composition, and isocratic separations are more convenient since they produce a constant background. In nonsuppressed IC the use of a so-called isoconductive gradient, in which the conductances of the starting and finishing eluents used to obtain the gradient are equal, partially overcomes this difficulty, but the variation in elutropic strength for isoconductive gradients is quite limited. In suppressed IC, the use of gradient elution is based on the availability of suppressors with sufficient capacity to ensure that the background conductance of the suppressed eluent remains essentially constant over the course of gradient. On the other hand, the eluotropic strength of an eluent with constant composition can be quite limited. For achieving separation, gradients are necessary in many cases, and they can be used conveniently in suppressed IC when good suppressors are available. Such suppressors can provide sufficient capacity to ensure that the background conductance of the suppressed eluent remains essentially constant over the course of the gradient [15]. As indicated in Section 13.5, in IC the structure of the ionic analytes is significantly influenced by pH values with 2 pH units larger than the analyte pK_a. For this reason, the pH must be well controlled in IC.

In gradient IC, similarly to other HPLC types of separation, a gradient capacity factor k'_e can be used to describe the retention. For a linear gradient, the expression for the effective capacity factor can be approximated by an expression of the form:

$$\log k'_e(A) = \alpha' - \beta' \log\left(\frac{\Delta C_X}{\Delta t}\right) \tag{14.3.1}$$

where the intercept α' and slope β' are specific for the given separation system, the analyte A^-, and the competing (driving) ion X^- of the IC separation. The gradient ramp $\Delta C_X/\Delta t$ is expressed in mM/min, and it is assumed to start as the analyte reaches the head of the column.

A special procedure for achieving a gradient in IC is chromatofocusing [16]. In this technique the pH gradient is generated within an ion exchanger column by combining the buffering capacity of the ion exchanger with that of a buffer in the mobile phase, and the result is similar to that obtained when the pH gradient is formed if two buffers at different pH are gradually mixed and sent into the chromatographic column (see, e.g., [17]).

References

[1] A.P. Schellinger, P.W. Carr, Isocratic and gradient elution chromatography: a comparison in terms of speed, retention reproducibility and quantitation, J. Chromatogr. A 1109 (2006) 253–266.
[2] J.W. Dolan, How fast can a gradient be run, LC/GC Eur. 24 (2011) 406–410.
[3] F. Gritti, General theory of peak compression in liquid chromatography, J. Chromatogr. A 1433 (2016) 114–122.
[4] A.P. Schellinger, D.R. Stoll, P.W. Carr, High speed gradient elution reversed-phase liquid chromatography, J. Chromatogr. A 1064 (2005) 143–156.
[5] A.P. Schellinger, D.R. Stoll, P.W. Carr, High-speed gradient elution reversed-phase liquid chromatography of bases in buffered eluents: Part I. Retention repeatability and column re-equilibration, J. Chromatogr. A 1192 (2008) 41–53.
[6] A.P. Schellinger, D.R. Stoll, P.W. Carr, High speed gradient elution reversed phase liquid chromatography of bases in buffered eluents: Part II. Full equilibrium, J. Chromatogr. A 1192 (2008) 54–61.
[7] P. Wiczling, M.J. Markuszewski, M. Kaliszan, R. Kaliszan, pH/organic solvent double-gradient reversed-phase HPLC, Anal. Chem. 77 (2005) 449–458.
[8] R. Kaliszan, P. Wiczling, M.J. Markuszewski, pH gradient high-performance liquid chromatography: theory and applications, J. Chromatogr. A 1060 (2004) 165–175.
[9] W. Hao, X. Zhang, K. Hou, Analytical solutions of the ideal model for gradient liquid chromatography, Anal. Chem. 78 (2006) 7828–7840.
[10] P. Nikitas, A. Pappa-Louisi, Expressions of the fundamental equation of gradient elution and a numerical solution of these equations under any gradient profile, Anal. Chem. 77 (2005) 5670–5677.
[11] M.A. Quarry, R.L. Grob, L.R. Snyder, Prediction of precise isocratic retention data from two or more gradient elution runs. Analysis of some associated errors, Anal. Chem. 58 (1986) 907–917.
[12] C. Liteanu, S. Gocan, Gradient Liquid Chromatography, Ellis Horwood Limited, Chichester, 1974.
[13] L.R. Snyder, J.W. Dolan, High-Performance Gradient Elution. The Practical Application of the Linear-Solvent-Strength Model, Wiley-Interscience, Hoboken, 2007.
[14] J.W. Dolan, L.R. Snyder, Maintaining fixed band spacing when changing column dimensions in gradient elution, J. Chromatogr. A 799 (1998) 21–34.
[15] J.E. Madden, N. Avdalovic, P.R. Haddad, J. Havel, Prediction of retention times for anions in linear gradient elution ion chromatography with hydroxide eluents using artificial neural network, J. Chromatogr. A 910 (2001) 173–179.
[16] Chromatofocusing with Polybuffer and PBE, Amarsham Pharmacia Biotech, Uppsala, 2001.
[17] S.C. Moldoveanu, V. David, Essentials in Modern HPLC Separations, Elsevier, Amsterdam, 2013.

CHAPTER

15

The Practice of HPLC

15.1 THE DEVELOPMENT OF AN HPLC METHOD

Various details related to the development of an high-performance liquid chromatography (HPLC) technique have been discussed so far in this book. The importance of collecting information about the analysis that should be performed was discussed in Chapter 1. The selection of HPLC as a core analytical procedure from various types of available analytical techniques was discussed in Chapters 2 and 3. Basic information about HPLC instrumentation and characterization of the chromatographic process were discussed in Chapter 4. The criteria for selecting a specific HPLC type from several available types, based on analyte and matrix properties, were discussed in Chapter 5. The description of different stationary phases and columns available for HPLC was given in Chapters 6–12. The description of mobile phases and their utilization was given in Chapters 13 and 14 Chapters 13 and 14. From all this material, the task of the analytical chemist is to select the following: (1) a type of HPLC, (2) specific instrumentation, (3) an appropriate analytical column, (4) an appropriate mobile phase, (5) an appropriate detection technique, and (6) an appropriate use of the whole ensemble. An iterative selection process may take place before achieving a successful analytical procedure. For stationary phase and mainly for mobile phase, several cycles of selection are common. This process of selection must be done within a specified time interval, which can be an important factor for the selection, development, and validation of an HPLC method [1]. This process is schematically illustrated in Fig. 15.1.1 (see also Fig. 1.2.4).

Implementation of a Method from the Literature

Scientific literature, including various types of publications (articles in peer-reviewed journals, books, company catalogs and application notes, web information), contains the description of a large number of HPLC analytical methods for almost every possible analyte. For this reason, one convenient way for having a desired HPLC method of analysis is the use of one already described in the literature. The method from literature should describe the analysis of the same analytes and with a matrix close to the one of the samples of interest. Most analytical methods are developed with the capability to perform the analysis on samples where the matrix is not identical from sample to sample but not completely different.

http://dx.doi.org/10.1016/B978-0-12-803684-6.00015-9

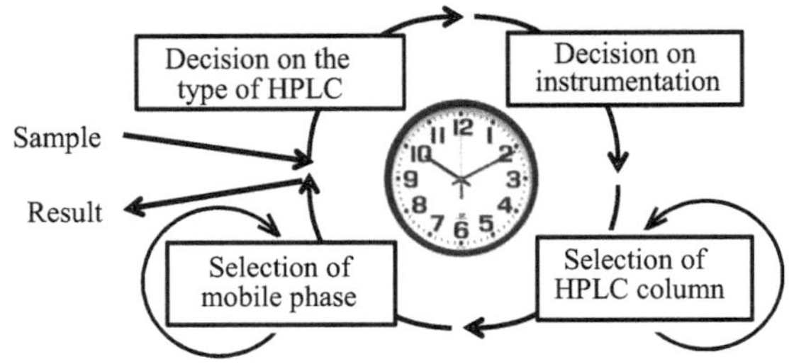

FIGURE 15.1.1 Diagram suggesting the iterative process of establishing an HPLC method. For stationary phase and mainly for mobile phase, several cycles of selection are common.

For this reason, the robustness of the analytical method to matrix variations is an important element to consider when attempting to implement a method from the literature. The implementation of such a method may encounter several problems: (1) differences in the matrix between the method from literature and the necessary method, (2) differences in the level of analytes, (3) differences in the required analysis time, and (4) differences between the recommended instrumentation and the available instrumentation.

1) One potential problem in implementing a method from the literature is related to the differences between the matrix on which the method from the literature has been applied, and the matrix of interest. The matrix differences may be not significant, or they can be critical for the method utilization. Sample preparation (different from the one from the literature) can be used in some cases for simplifying the matrix and possibly for increasing the content of the analyte in a processed sample. Potential matrix interference must be thoroughly verified and eliminated for the implemented method. In the case that the utilization of the method from literature does not work with a particular sample matrix, modifications of the method are first recommended or addition of a sample preparation step, and only if these changes are not sufficient should a radically new method be utilized.

2) Another problem can be related to the content of the analyte in the samples of interest. When the content of the analytes in the samples of interest is too low for the recommended detection, the possibilities of changes to a more sensitive detector should be first evaluated. Also, the addition of a sample concentration step may be utilized.

3) Some methods recommended in the literature may indicate a longer analysis time than acceptable, for example, for the analysis of a large number of samples. In such cases, modifications of the recommended method can be applied, such as using a shorter chromatographic column (if the separation is still acceptable), increase in the flow rate in the HPLC, or attempt to use UPLC instead of HPLC methodology (see Chapter 6).

4) Other problems with the implementation of the method from the literature can be related to the availability of similar instrumentation, or the qualification of the similar HPLC system (which is assumed to be periodically checked). The replacement of a pumping system from high-pressure mixing to low-pressure mixing (see Section 4.1) may generate changes in peak retention times or even in the resolution of certain peaks in the chromatograms. Changes of the chromatographic column may also sometimes be necessary. Columns of the recommended type but with different dimensions can be utilized in some instances, but the change in the retention time of the analytes, loading capacity of the column, and

backpressure changes must be first evaluated. When a replacement with an equivalent but not identical column is intended, the selection of a new column should be made, for example, by using the Product Quality Research Institute (PQRI) approach (see expression 7.3.11) [2], which can indicate which column is expected to provide a similar separation.

The replacement of one type of detection with a different type, although sometimes possible, is not in general straightforward. For example, a change to refractive index (RI) detection requires the use of isocratic separations, which is not necessary for other detection types. The replacement of a more sensitive detector with a less sensitive detector may cause problems with the limit of quantitation (LOQ) value for the analytes. The change from a less sensitive detector to a more sensitive detector also may pose problems. For example, the change from ultraviolet (UV) detection to mass spectrometry (MS) detection (that may provide better sensitivity and selectivity) must assure that the mobile phase utilized with the UV detection is adequate for MS detection (all mobile phase components must be volatile for MS use). Even when changing, for example, from MS detection to UV detection, it must be assured that the UV cut-off of the mobile phase is acceptable.

Improvement of a Method from the Literature

The need for better methods of analysis is common due to new demands in the quality of data and the requirements to analyze a diversity of new types of samples. Also, the HPLC instrumentation, the column quality, and detector performances are continuously improving. On the other hand, a large number of very good methods of analysis by HPLC-type techniques is available in the literature. However, these methods may have not been developed using modern HPLC capabilities, or they may have aspects that can be modified for achieving better selectivity, sensitivity, or shorter run time. The need to apply a method from the literature to a different type of matrix or to add more analytes for quantitation is also common. Many older HPLC methods can be modified and improved to include new requirements and/or new capabilities. The main places where improvement can be made are the following: (1) the injection, (2) the HPLC column with the possibility to move the whole method from HPLC type to UPLC type, (3) the mobile phase, and (4) the detection.

1) The changes regarding the injection may affect the injection volume, the injection solvent, or both. Smaller injection volumes may lead to better separation and when moving from HPLC to UPLC, such changes are usually recommended. A lower amount of analyte in the HPLC system may lead to lower sensitivity of the method, which can be compensated for, if necessary, by improving the detector sensitivity. An increase in sample volume (placing a larger amount of analyte in the system) may be useful in some analyses where better sensitivity is required. Larger volumes of injected sample may reduce resolution, and in some cases a modification of the initial mobile phase composition to a weaker eluent may help to achieve sample focusing at the head of the chromatographic column. Special injection techniques may also allow the injection of larger samples (see Section 13.6).

2) The same types of columns that were previously available with 5 μm diameter particles (or larger) are now available in formats with smaller particles, in core–shell type particles, or as monolithic columns. The new columns offer better efficiency, and for monolithic columns lower backpressure. The new columns having smaller particles may produce higher backpressure and the shift from HPLC to UPLC must be considered if the instrumentation

is available. Such change will offer better separation with more accurate capability for the integration of peaks (see Section 6.3) and much shorter retention times. The change of the column can be done not only regarding the column format, but also regarding the column type. Such change must consider an equivalent column that produces basically the same kind of separation, but with better performance (better resolution, better peak shape, longer column life). Column equivalence can be evaluated using the PQRI approach, or can be recommended in vendor catalogs (see, e.g., Phenomenex, Chromatography Product guide 2016/2017 [3], Waters Chromatography Columns and Supplies Catalog [4], MAC-MOD resources [5], Thermo Scientific HPLC [6], and Restek HPLC Column Selection [7]).

3) Changes in the mobile phase are a useful procedure for improving a reported HPLC method. These changes may involve the following: (1) the modification of the percent composition of the mobile phase using the same solvents as recommended in the initial method, (2) changes of gradient, and (3) changes in the chemical composition including different solvents and additives. The modification of percent composition of mobile phase may improve separation, may provide focusing at the column head of the analytes, and may reduce the run time for the analysis. The improvement of the separation and reduction of the run time can also be achieved by changing the gradient. The changes in the nature of the solvents used for making the mobile phase can bring more significant modifications of the chromatographic results. The effect of such change must be well evaluated before implementation, since it may affect not only the separation but also the detection. For example, the change of acetonitrile with methanol (or the other way around) can affect not only the separation, but also the sensitivity in the ionization of the analytes when MS is used for detection.

4) Changes in the detection type are sometimes necessary for enhancing selectivity or sensitivity. For example, the change from RI to UV detection may improve sensitivity and selectivity. Although some analytes do not have strong chromophores, the detection in UV may still be possible at low wavelength. However, the UV cut-off of the solvent must be verified to be significantly below the setup detection UV wavelength, which is not a requirement for the RI detection. The increased availability of MS detectors and their excellent selectivity and sensitivity provides a good reason to change older methods with different types of detection to MS detection (MS or MS/MS). The change to MS detection may require a number of other changes, in particular to the composition of the mobile phase, which must be completely volatile. With MS detection, common buffers used in HPLC such as KH_2PO_4/K_2HPO_4 cannot be used. Buffers such as HCOOH/$HCOONH_4$ should replace nonvolatile buffers, and this may pose problems regarding the adjustment of the same pH of the mobile phase and the same ionic strength. Also, the use of MS detection may require a reduction in the mobile phase flow rate. This may cause the need to change the larger columns having high k' values for the analytes to smaller columns with lower k' values. The conditions for such transfers are discussed in Section 6.2.

Development of a New HPLC Method

The development of an entirely new HPLC analytical method is sometimes necessary. Besides the requirement to analyze new types of samples, the trend of reducing the sample preparation step requires the development of better HPLC analyses. This book provides detailed information regarding multiple choices available for developing an HPLC method. The development should evaluate the goal of the analysis, the sample characteristics, the type of expected results, the analysis cost, etc., and use the best available instrumentation. The use of the available literature for guidance is the first necessary step for the development of a new method. If possible a set of initial experiments are recommended, to check the chromatographic response to some possible experimental parameters that will affect the results. Individual experience, trial and error experiments, and exchange of information on the web (e.g., using Chromacademy [8]) improve the results in developing a new HPLC method.

Depending on the purpose of HPLC analysis, the method development is performed in different ways. For qualitative analysis, the process of identifying the analytes is less structured. HPLC has some capabilities for the identification of unknowns, but the identification of unknown compounds can be a challenging task. A separation of the analytes is necessary before detection, and this can be done by HPLC either with the intention of identifying the presence of certain expected compounds or just for screening and the detection of unknowns. In the case of screening, the separation should provide differences in the retention times for compounds in a wide range of polarities. When searching for the presence of specific compounds, standard mixtures of these compounds should be used first for assuring that they are well separated. UV detection with the generation of a full spectrum can sometimes be useful for compound characterization, but mass spectrometric detection techniques (MS) and in particular tandem mass spectrometry (MS/MS) may offer more information about the compound nature (see Section 3.6).

HPLC is mainly used for quantitative analysis, and several steps can be described for a common method development: (1) preliminary experiments, (2) establishment of an acceptable separation of the analytes, (3) utilization of the unrefined method on a mixture of pure analytes, (4) first round of the evaluation of separation (good resolution, low peak asymmetry), (5) application of the method on a set of calibration standards, (6) evaluation of linearity and limits of detection, (7) analysis of a set of real samples or on reference materials, (8) evaluation of recovery and practical limit of detection (LOD) and LOQ, (9) assessment of accuracy, robustness, and ruggedness. The process can be iterative as a whole or only for particular steps.

For a new method, the validation process as described in Section 1.3 follows a specific protocol and it should check that the method performs adequately throughout the range of analyte concentrations and test materials to which it is applied [9]. Various levels of validation are necessary, depending on the purpose of analysis and requirements for the quality of results. Variability in the matrix components and verification of robustness of the method to these variations must always be performed for a new method. Because of the interferences that can be generated by the matrix, the quantitation of an analyte, although accurate in samples with one matrix, may generate inaccurate results with a different matrix. This problem may be encountered for example when the MS detection is used to compensate for poor separation in the chromatographic column. The MS detection, being highly selective, does not provide information about the coeluting compounds with the analytes if the detection is not set for that. The good separation in the chromatogram is for this reason always recommended, although the MS detector can differentiate one compound from another without the compounds being chromatographically separated.

Method Optimization

Each newly implemented method, either adopted from the literature or entirely new, is subject to a process of various modifications with the goal of improvement. The extent of this improvement process can range from very limited modifications to a full optimization process. The need for a full optimization should always be critically evaluated, since the process may be time-consuming and expensive and produce only minor improvements. On the other hand, the improvement process should continue until the method becomes entirely adequate for the goal of the analysis.

The improvement process, frequently indicated as optimization (even without reaching an optimized method), consists of a number of trials with changes at various parts of the method or even of the whole method. This process may need to be repeated several times. The optimization of the chromatographic separation is an important part of improving a method, but other parts of the method may be subjected to changes with the goal of improvement. Such parts may include the injection volume, the detector settings, and the length of the chromatographic run.

The improvement process can be done by trial and error, or with multivariate statistical techniques. In a systematic optimization process, it is important to clearly establish the controlled variables (explanatory or independent) and the response variables. The relation between these two sets of variables is indicated in the "response surface" [10]. The statistical techniques utilize a minimum and a maximum value for independent variables, which defines the experimental domain to be investigated during the optimization. Among the designs used to determine response surfaces are the full and fractional factorial designs, central composite design, and Box–Behnken design [11].

The optimization process can be extended to the selection of a calibration type. For establishing a quantitation method, several procedures can be utilized besides simple calibrations with standards. These may include external calibration, internal normalization calibration, matrix-matched calibration, standard addition calibration, and signal-to-ratio calibration [12]. Most common calibration procedures use a simple calibration with a set of standards. There are different recommendations regarding the number of calibration standards and the number of replicates at each calibration level. IUPAC advises for method validation purposes the use of six or more calibration standards that should be run in triplicate in a randomized way [13]. ISO 8466 indicates for an initial assessment of the calibration performance to employ at least five calibration standards, although it recommends 10, and 10 replicates of the lowest and highest standards. Method development performed on matrix reference materials is also preferred, but reference materials are not always available. These materials contain the analytes of interest plus the principal chemical compounds found into the matrix to be matched [14].

15.2 SPECIAL HPLC TECHNIQUES

A large variety of HPLC techniques are reported in the literature. Most methods use a standard setup with a solvent supply system, pumping system, injector, column (in column holder), and detector(s), all controlled by a dedicated computer program. However, more complex

HPLC setups are used for certain analyses. Such systems are sometimes necessary for special HPLC techniques such as multidimensional HPLC, comprehensive HPLC, HPLC methods with column preconcentration, and HPLC methods with postcolumn derivatization. The instrumentation setups for such special techniques were briefly discussed in Section 4.1. Besides instrumentation, special techniques may also require special injection procedures, special selection of chromatographic column, and special requirements for the mobile phase.

Selection of a Column and Mobile Phase in Bidimensional Separations

For bidimensional separations, for example, different selections can be made for the two columns: (1) similar columns but with different efficiency, (2) very different columns (indicated as orthogonal), and (3) one nonchiral column and a chiral column for bidimensional chiral separations. Also, the mobile phase in the two columns for bidimensional separations can be similar or different.

Similar columns with different efficiency can be selected when a complex sample needs to be analyzed regarding only a special group of analytes. The first column in such cases is selected to have a larger sample load capability and average separation power. The second column can be selected with high efficiency and possibly run under a different mobile phase or gradient. Similar columns can be selected also for the application of bidimensional separations in size-exclusion chromatography (SEC). This may be necessary, for example, for a more precise measurement of *Mw* values of certain components of a polymer mixture [15].

For orthogonal separations, the two columns must be very different. The difference in the separation capability of the two columns can be evaluated using procedures specifically developed for this purpose. These procedures are based on the use of test mixtures on the two columns followed by the comparison of separation results [16]. A good correlation between the retention times of the test compounds on the two columns indicates a poor orthogonality. Because the separation selectivity in the RP mode is complementary to that in HILIC or in IC, combinations of the RP with HILIC or RP with IC are attractive for two-dimensional applications. Other criteria for the selection of orthogonal columns are based on comparing column properties such as hydrophobic character and hydrogen-bonding capability.

Chiral separations make use of bidimensional chromatography, in particular for the case of samples with a complex matrix. The chiral separation may require a normal phase (NP) type chromatography, and not all the sample components may be well separated in this way. For such cases, it is common to use a bidimensional system, for example, with an RP-type first column. The separation uses a heart-cut containing the analytes of interest followed by their chiral separation in NP conditions on the second column [17].

Postcolumn Derivatization in HPLC

The derivatization of the analytes for HPLC separation is considered a sample preparation operation, which is discussed in numerous publications (see, e.g., [18,19]). Most derivatizations in HPLC are performed precolumn and HPLC analysis of the derivatized analytes is not different from a typical analysis. In some cases, postcolumn derivatization is applied. Postcolumn derivatization is usually performed "on line" and should be completed in the specific timeframe required by the flow of the mobile phase with the analyte to reach the detector. For such analyses, some requirements are imposed to the HPLC technique. For example, the derivatizing solution and the eluent must be miscible, no precipitation reactions should occur, and the flow rates of different mixing liquids must match. Because no separation is done between the reagents and analytes, the reagent should not interfere with the detector. Various instrumental setups are possible for postcolumn derivatization, such as multiple reagents addition, heating and cooling capability, addition of retention gaps that allow a longer time of contact between the analytes and the reagent, and gas-fragmented flow capability (see, e.g., [20]).

References

[1] J.K. Swadesh (Ed.), HPLC: Practical and Industrial Applications, second ed., CRC Press, Boca Raton, 2001.
[2] http://apps.usp.org/app/USPNF/columnsDB.html.
[3] https://flipflashpages.uniflip.com/2/41719/360817/pub/.
[4] http://www.waters.com/waters/library.htm?lid=10009750&locale=en_US.
[5] http://mac-mod.com/products-ace.php.
[6] https://www.thermofisher.com/us/en/home/industrial/chromatography/liquid-chromatography-lc/hplc-uhplc-columns.html.
[7] www.restek.com/HPLC-Columns.
[8] http://www.chromacademy.com/.
[9] M. Rambla-Alegre, J. Esteve-Romero, S. Carda-Broch, Is it really necessary to validate an analytical method or not? That is the question, J. Chromatogr. A 1232 (2012) 101–109.
[10] G.E.P. Box, N.R. Draper, Response Surfaces, Mixtures, and Ridge Analyses, Wiley, Hoboken, 2007.
[11] S.L.C. Ferreira, R.E. Bruns, E.G. Paranhos da Silva, W.N. Lopes dos Santos, C.M. Quintella, J.M. David, J. Bittencourt de Andrade, M.C. Breitkreitz, I.C. Sales Fontes Jardim, B.B. Neto, Statistical designs and response surface techniques for the optimization of chromatographic systems, J. Chromatogr. A 1158 (2007) 2–14.
[12] L. Cuadros-Rodrıguez, M. Gracia Bagur-Gonzalez, M. Sanchez-Vinas, A. Gonzalez-Casado, A.M. Gomez-Saez, Principles of analytical calibration/quantification for separation sciences, J. Chromatogr. A 1158 (2007) 33–46.
[13] Harmonized guidelines for single-laboratory validation of methods of analysis, IUPAC Technical Report 2002 (prepared by M. Thompson, S.L.R. Ellison and R. Wood), Pure Appl. Chem. 74 (2002) 835–856.
[14] European Co-operation for Accreditation. The Selection and Use of Reference Materials, E.A.-04/14, 2003.
[15] L.K. Kostanski, D.M. Keller, A.E. Hamielec, Size-exclusion chromatography – a review of calibration methodologies, J. Biochem. Biophys. Methods 58 (2004) 159–186.
[16] J. Pellett, P. Lukulay, Y. Mao, W. Bowen, R. Reed, M. Ma, R.C. Munger, J.W. Dolan, L. Wrisley, K. Medwid, N.P. Toltl, C.C. Chan, M. Skibic, K. Biswas, K.A. Wells, L.R. Snyder, Orthogonal separations for reversed-phase liquid chromatography, J. Chromatogr. A 1101 (2006) 122–135.
[17] A. Medvedovici, F. Albu, C. Georgita, D.I. Sora, T. Galaon, S. Udrescu, V. David, Achiral-chiral LC/LC-FLD coupling for determination of carvedilol in plasma samples for bioequivalence purposes, J. Chromatogr. B 850 (2007) 327–335.
[18] G. Lunn, L.C. Hellwig, Handbook of Derivatization Reactions for HPLC, Wiley, Hoboken, 1998.
[19] S.C. Moldoveanu, V. David, Sample Preparation in Chromatography, Elsevier, Amsterdam, 2002.
[20] Y. Qu, R.H. Slocum, J. Fu, W.E. Rasmussen, H.D. Rector, J.B. Miller, J.G. Coldwell, Quantitative amino acid analysis using a Beckman system gold HPLC 126AA analyzer, Clin. Chim. Acta 312 (2001) 153–162.

Appendix 1: USP Classification of HPLC Columns

USP classification of HPLC chromatographic columns (all stationary phases as spherical, unless indicated as (M) for monolith, and (I) for irregular shape) (M. Marques, ed., "USP Chromatographic Columns", U. S. Pharmacopeia, Rockville, 2009–2010.)

Code	Description	Examples
L1	Octadecyl silane (C18) chemically bonded to porous silica or ceramic microparticles, 1.5–10 μm in diameter, or a monolithic rod	Over 250 columns such as: Luna C18(2), Luna C18(2)-HST, Gemini C18, Synergi Hydro-RP, Onyx C18 (M), Aquity UPLC BEH C18, Aquity UPLC Shield RP 18, Atlantis T3, μBondapack C18, Nova-Pak C18, Symmetry C18, ABridge C18, XTerra MS C18
L2	Octadecyl silane (C18) chemically bonded to silica gel of a controlled surface porosity that has been bonded to a solid spherical core, 30–50 μm in diameter	Bondapak Prep C18
L3	Porous silica particles, 1.5–10 μm in diameter, or a monolithic silica rod	Luna Silica(2), Aquity UPLC BEH HILIC, Atlantis HILIC Silica, Onyx Si (M), SunFire Silica, XBridge HILIC, Zorbax SIL
L4	Silica gel of controlled surface porosity bonded to a solid spherical core, 30–50 μm in diameter	Porasil Prep Silica
L5	Alumina of controlled surface porosity bonded to a solid spherical core, 30–50 μm in diameter	—
L6	Strong cation-exchange packing: Sulfonated fluorocarbon polymer coated on a solid spherical core, 30–50 μm in diameter	Adsorbosphere XL SCX, Partisil SCX, Zipax SCX, Zodiac Prep SCX
L7	Octyl silane (C8) chemically bonded to totally porous silica particles, 1.5–10 μm in diameter, or a monolithic silica rod	Luna C8(2), Aquity UPLC BEH C8, Nova-Pak C8, Resolve C8, SunFire C8, Symmetry C8, XBridge C8, XTerra MS C8, XTerra RP8, Onyx C8 (M), Nucleosil C8, Zorbax C8, Zorbax SB-C8, Zorbax C8, Zorbax Eclipse XDB-C8, Hypersil-MOS, LiChrosorb-RP8/RP-SelectB, etc.
L8	An essentially monomolecular layer of aminopropylsilane (NH_2) chemically bonded to totally porous silica gel support, 3–10 μm in diameter	Luna 10 μm NH_2, μBondapak NH_2, Waters Spherisorb NH_2, Zorbax NH_2, etc.
L9	Irregular or spherical, totally porous silica gel having a chemically bonded, strongly acidic cation-exchange coating, 3–10 μm in diameter	Partisil 10 μm SCX (I), Spherisorb SCX, Luna 10 μm SCX, Zorbax SCX

(Continued)

Code	Description	Examples
L10	Nitrile groups (CN) chemically bonded to porous silica particles, 3–10 μm in diameter	Luna CN 100 Å, Capcell CN UG, μBondapak CN, Nova-Pak CN, Resolve CN, Waters Spherisorb CN
L11	Phenyl groups (C6H5) chemically bonded to porous silica particles, 1.5–10 μm in diameter	Synergi Polar-RP, Luna Phenyl-Hexyl, Gemini C6-Phenyl, Prodigy PH-3, Aquity UPLC BEH Phenyl, μBondapak Phenyl, XBridge Phenyl, XTerra Phenyl
L12	Strong anion-exchange packing made by chemically bonding a quaternary amine to a solid silica spherical core, 30–50 μm in diameter	AccelPlus QMA
L13	Trimethylsilane (C1) chemically bonded to porous silica particles, 3–10 μm in diameter	Develosil TMS-UG (C1) 130 Å, TSKgel TMS-250, Waters Spherisorb C1
L14	Silica gel having a chemically bonded, strongly basic quaternary ammonium anion-exchange coating, 5–10 μm in diameter	Partisil 10 μm SAX (I), PartiSphere 5 μm SAX, Waters Spherisorb SAX
L15	Hexylsilane (C6) chemically bonded to totally porous silica particles, 3–10 μm in diameter	PhenoSphere C6, Waters Spherisorb C6
L16	Dimethylsilane (C2) chemically bonded to totally porous silica particles, 5–10 μm in diameter	Maxsil RP2 60 Å (I), Lichrosorb RP2
L17	Strong cation-exchange resin consisting of sulfonated crosslinked styrene-divinylbenzene copolymer in the hydrogen form, 7–11 μm in diameter	Rezex RHM Monosaccharide, IC-Pak Ion Exclusion, IC-Pak Cation, Shodex RSpak DC-613, Rezex ROA
L18	Amino and cyano groups chemically bonded to porous silica particles, 3–10 μm in diameter	Partisil PAC (I)
L19	Strong cation-exchange resin consisting of sulfonated crosslinked styrene-divinylbenzene copolymer in the calcium form, 9 μm in diameter	Rezex RCM, Rezex RCU, Sugar-Pak 1, Shodex SC-1011
L20	Dihydroxypropyl groups chemically bonded to porous silica particles, 3–10 μm in diameter	Luna HILIC Shodex PROTEIN KW-800 series, TSKgel QC-PAK 200 and 300, BioSuite 125, Insulin HMWP (I), Protein-Pak (I)
L21	A rigid, spherical styrene-divinylbenzene copolymer, 3–10 μm in diameter	Polymerx RP-1, Phenogel 100 Å, IC-Pak Ion Exclusion, Shodex SP-0810
L22	A cation exchange resin made of porous polystyrene gel with sulfonic acid groups, about 10 μm in size	Rezex ROA
L23	An anion exchange resin made of porous polymethacrylate or polyacrylate gel with quaternary ammonium groups, about 10 μm in size	Shodex IEC QA-825, TSKgel BioAssist Q, TSKgel SuperQ-5PW, BioSuite Q AXC, BioSuite DEAE, Protein-Pak Q 8HR

Code	Description	Examples
L24	A semirigid hydrophilic gel consisting of vinyl polymers with numerous hydroxyl groups on the matrix surface, 32–63 μm in diameter	YMC-Pac PVA-Sil, Toyopearl HW-type
L25	Packing having the capacity to separate compounds with an MW range from 100 to 5000 Daltons (as determined by polyethylene oxide), applied to neutral, anionic, and cationic water-soluble polymers. A polymethacrylate resin base, crosslinked with poly-hydroxylated ether (surface contained some residual carboxyl functional groups) was found suitable	PolySep-GFC-P2000, Shodex OHpak SB-802.5HQ, Ultrahydrogel DP +120, TSK-gel G1000PW
L26	Butyl silane (C4) chemically bonded to totally porous silica particles, 1.5–10 μm in diameter	Jupiter 300C4, Aquity UPLC BEH300C4, Delta-Pak C4, Symmetry300C4, XBridge BEH300C4
L27	Porous silica particles, 30–50 μm in diameter	Sepra (I), Porasil (I), Nucleodur, YMS-Pack Silica
L28	A multifunctional support, which consists of a high-purity, 100 Å, spherical silica substrate that has been bonded with anionic (amine) functionality in addition to a conventional reversed-phase C8 functionality	Altech mixed mode C8/anion, Generik C8/ Amino, ProTec C8
L29	Gamma alumina, reversed-phase, low carbon percentage by weight, alumina-based polybutadiene spherical particles, 5 μm diameter with a pore diameter of 80 Å	Gamabond ARP-1, Gamabond Alumina Potency, Alusphere RP-Select-B
L30	Ethyl silane (C2) chemically bonded to a totally porous silica particle, 3–10 μm in diameter	Maxsil RP2 60 Å (I), Nucleosuil C2, APEX Prepsil C2
L31	A strong anion-exchange resin-quaternary amine bonded on latex particles attached to a core of 8.5 μm macroporous particles having a pore size of 2000 Å and consisting of ethylvinylbenzene crosslinked with 55% divinyl benzene	Ion Pac AS 10, Ion Pack AS 16
L32	A chiral ligand-exchange packing- L-proline copper complex covalently bonded to irregularly shaped silica particles, 5–10 μm in diameter	CHIRACEL WH, Astec CLD-D (or −L), Nucleosil Chiral-1
L33	Packing having the capacity to separate proteins of 4000 to 400,000 daltons (D). It is spherical, silica-based and processed to provide pH stability	BioSep-SEC-S2000, BioSep-SEC-S3000 BioBasic SEC 120, Nucleosil 125-5 GFC, Shodex KW-404
L34	Strong cation-exchange resin consisting of sulfonated crosslinked styrene-divinylbenzene copolymer in the lead form, about 9 μm in diameter	Aminex Fast Carbohydrate, Rezex RPM Monosaccharide, Shodex Sugar SP0810, Nucleogel Sugar Pb

(Continued)

Code	Description	Examples
L35	A zirconium-stabilized spherical silica packing with a hydrophilic (diol-type) molecular monolayer bonded phase having a pore size of 150 Å	Bio-Sep-SEC-S2000, Zorbax GF-250, Zorbax GF-450
L36	3,5-Dinitrobenzoyl derivative of L-phenylglycine covalently bonded to 5 μm aminopropyl silica	Nucleosil Chiral-3
L37	Polymethacrylate gel packing having the capacity to separate proteins by molecular size over a range of 2000 to 40,000 D	PolySep-GFC-P3000, Shodex OHpak SB-803HQ, Ultrahydrogel 250
L38	Methacrylate-based size-exclusion packing for water-soluble samples	PolySep-GFC-P1000, Shodex OHpak SB-802HQ, Ultrahydrogel
L39	Hydrophilic polyhydroxymethacrylate gel of totally porous spherical resin	PolySep-GFC-P Series, Shodex OHpak SB-800HQ series, Shodex RSpak DM-614
L40	Cellulose tris-3,5-dimethylphenylcarbamate-coated porous silica particles, 5 μm–20 μm in diameter	CHIRACEL OD, Lux cellulose 1, Nucleocel Delta
L41	Immobilized α-acid glycoprotein on spherical silica particles, 5 μm in diameter	Chiral-AGP
L42	Octylsilane and octadecylsilane groups chemically bonded to porous silica particles, 5 μm in diameter	Chromegabond PSC, HiChrom RPB-250A
L43	Pentafluorophenyl groups chemically bonded to silica particles, 5–10 μm in diameter	Curosil-PFP, Ultra-PFP, Pinnacle DB PFP (Restek) Allure PFP Propyl
L44	A multifunctional support, which consists of a high-purity, 60 Å, spherical silica substrate that has been bonded with a cationic exchanger, sulfonic acid functionality in addition to a conventional reversed-phase C8 functionality	Chromegabond RP-SCX, Generik C8/SCX
L45	Beta cyclodextrin bonded to porous silica particles, 5–10 μm in diameter	Astec Cyclobond I, II or II ser., Chiral CD-Ph, Nucleodex Beta-PM, ChiralDex
L46	Polystyrene/divinylbenzene substrate agglomerated with quaternary amine functionalized latex beads, 10 μm in diameter	CarboPac PA1, Transgenomic AN1
L47	High-capacity anion-exchange microporous substrate, fully functionalized with a trimethylamine group, 8 μm in diameter	CarboPac MA1, Hamilton PRP-X100, X110, Hamilton RCX-10
L48	Sulfonated, crosslinked polystyrene with an outer layer of submicron, porous, anion-exchange microbeads, 15 μm in diameter	IonPac AS5, IonPac AS7

Code	Description	Examples
L49	A reversed-phase packing made by coating a thin layer of polybutadiene onto spherical porous zirconia particles, 3–10 μm in diameter	Zirchrom PBD, Discovery ZR-PBD
L50	Multifunction resin with reversed-phase retention and strong anion-exchange functionalities. The resin consists of ethylvinylbenzene, 55% crosslinked with divinylbenzene copolymer, 3–15 μm in diameter, and a surface area of not less than 350 m^2/g, substrate is coated with quaternary ammonium functionalized latex particles consisting of styrene crosslinked with divinylbenzene	OmniPac PAX-500, Proteomix SAX-POR
L51	Amylose tris-3,5-dimethylphenylcarbamate-coated, porous, spherical, silica particles, 5–10 μm in diameter	Chiralpak ADS, Nucleocel Alpha
L52	A strong cation exchange resin made of porous silica with sulfopropyl groups, 5–10 μm in diameter	TSKgel SP-2SW, BioBasic SCX, Supecosil LC-SCX
L53	Weak cation-exchange resin consisting of ethylvinylbenzene, 55% crosslinked with divinylbenzene copolymer, 3–15 μm diameter. Substrate is surface-grafted with carboxylic acid and/or phosphoric acid functionalized monomers. Capacity not less than 500 μEq per column	IonPac CS14
L54	A size exclusion medium made of covalent bonding of dextran to highly crosslinked porous agarose beads, about 13 μm in diameter	Superdex peptide HR 10/30
L55	A strong cation exchange resin made of porous silica coated with polybutadiene-maleic acid copolymer, about 5 μm in diameter	IC-Pak C M/D, Waters Spherisorb SCX, Universal Cation
L56	Isopropyl silane (C3) chemically bonded to totally porous silica particles, 3–10 μm in diameter	Zorbax SB C3
L57	A chiral-recognition protein, ovomucoid, chemically bonded to silica particles, about 5 μm in diameter, with a pore size of 120 Å	Ultron ES-OVM
L58	Strong cation-exchange resin consisting of sulfonated crosslinked styrene-divinylbenzene copolymer in the sodium form, about 7–11 μm diameter	Rezex RNM-Carbohydrate, Aminex HPX-87N, Shodex SUGAR KS-801, -802

(Continued)

Code	Description	Examples
L59	Packing with the capacity to separate proteins by molecular weight over the range of 10–500 kDa. Spherical 10 μm, silica-based, and processed to provide hydrophilic characteristics and pH stability	BioSep-SEC-S3000, Biosuite 125, Nanofilm SEC-150, TSK-GEL G2000SW, G3000SW, etc.
L60	Spherical, porous silica gel, 3–10 μm in diameter, surface has been covalently modified with palmitamidopropyl groups and endcapped	Acclaim polar Advantage, Ascentis RP-Amide, HALO RP-Amide, Prism RP, Supecosil LC-ABZ
L61	Hydroxide-selective, strong anion-exchange resin consisting of a highly crosslinked core of 13 μm microporous particles, pore size less than 10 Å, and consisting of ethylvinylbenzene crosslinked with 55% divinylbenzene with a latex coating composed of 85-nm diameter microbeads bonded with alkanol quaternary ammonium ions (6%)	Ion Pac AS-11, Ion Pac AG-11
L62	C30 silane bonded phase on a fully porous spherical silica, 3–15 μm in diameter	Develosil Combi-RP, Develosil RP-Aqueous, Develosil RP-Aqueous-AR, ProntoSil c30, YMC-Pack Carotenoid, Zodiac 120C30
L63	Glicopeptide teicoplanin linked to spherical silica (chiral phase)	CHIROBIOTIC V, T, TAG, R.
L64	Strongly basic anion exchange resin consisting of 8% crosslinked styrene-divinylbenzene copolymer with a quaternary ammonium group in the chloride form, 45–180 μm in diameter	AG 1-X8
L65	Strongly acidic cation exchange resin, consisting of 8% sulfonated crosslinked styrene-divinylbenzene copolymer with a sulfonic group in hydrogen form, 63–250 μm diameter	AG 50W-X2
L66	Crown ether coated on 5-μm silica gel substrate	CrownPak CR
L67	Porous vinyl alcohol copolymer with C18 alkyl group attached to the hydroxyl group of the polymer, 2–10 μm diameter	Supelcogel ODP-50, apHera C18, Asahipak ODP-40
L68	Spherical porous silica containing a polar group within or intrinsic to the hydrocarbon bonded phase (e.g., alkylamide)	Ultra II IDB
L69	Polymeric (ethylvinylbenzene/divinylbenzene) with strong anion exchange (quaternary amine) on latex beads, about 6.5 μm diameter	CarboPac PA20

Code	Description	Examples
L70	Cellulose tris(phenylcarbamate) coated on 5 μm silica	Chiracel OC-H
L71	A rigid, spherical polymethacrylate, 4–6 μm diameter	RSpak DE-613
L72	(R)-Phenylglycine and 3,5-dinitroaniline urea linkage covalently to silica	Chirex 3012, Sumichiral OA-3300
L73	A rigid, spherical polydivinylbenzene particle 5–10 μm diameter	Jordi-gel DVB
L##	(Dalteparin sodium, anion exchange Dowex 1X8)—strongly basic (type I) anion exchange resin in chloride form	Dowex 1X8
L##	(Dalteparin sodium, cation exchange Dowex 50WX2)—strongly acidic cation exchange resin in H^+ form	Dowex 50WX2
L##	(Glucosamine, Shodex NH_2 P-50) Polyamine chemically bonded to crosslinked polyvinyl alcohol polymer, 5 μm diameter	apHera NH_2 Amino, Shodex NH_2 P50
L##	(Ethylhexyl triazone, FluoFix)—Fluorocarbon chains chemically bonded to 5-μm spherical silica particles	Wakopak FluoFix-II 120E, 120E, 120N

Appendix 2: Hydrophobic Stationary Phases

APPENDIX 2a Common Physical and Chemical Properties of Various Hydrophobic Columns. Retention Factor k' is Given for Toluene in a Mobile Phase Consisting of 80% CH_3OH and 20% Aqueous Buffer 0.25 mM KH_2PO_4 at pH = 6 and at 24°C (in Isocratic Conditions)

No.	Column Name	Particle Diam. (d_p, μm[a])	Pore Size (Å)	Pore Vol. (mL/g)	Surface Area (m^2/g)	C%[b]	Theor. Plate (N/m)[c]	Cover ($\mu M/m^2$)	End-Cap[d]	pH Stabil.	Ret. Factor k' Toluene
1	AccQ (C18)	Sph. 1.7	135	0.70	185	17	–	–	1	–	–
2	ACE C18	5	100	–	300	15.5	111,000	–	1	–	3.56
3	ACE C18-300	5	300	–	100	9	103,500	–	1	–	1.80
4	ACE C18-HL	5	90	–	400	20	102,000	–	2	–	6.36
5	ACE SUPERC18	1.7, 2, 3, 5	90	–	400	14.8	–	–	1	1.5–1.5	–
6	ACE C18-AMIDE	1.7, 2, 3, 5	100	–	300	17.0	–	–	1	2–8	–
7	AQUITY UPLC BEH C18	Sph. 1.7	130	0.70	185	17	–	–	1	–	–
8	AQUITY UPLC CSH C18	Sph. 1.8	100	0.70	230	15	–	–	1	–	–
9	AQUITY UPLC HSS C18	Sph. 1.8	100	0.70	230	15	–	–	1	1–8	–
10	AQUITY UPLC HSS T3	Sph. 1.8	100	0.70	230	11	–	–	1	2–8	–
11	AQUITY UPLC CSH phenyl-Hexyl	Sph. 1.7	130	0.70	185	14	–	–	1	–	–
12	AQUITY UPLC CSH Fluoro Phenyl	Sph. 1.7	130	0.70	185	10	–	–	1	–	–
13	Atlantis T3 (C18)	Sph. 3, 5, 10	100	1.00	330	12	–	1.6	1	–	–
14	Aeris widepore XB-C18	C-S 3.6	–	–	25	–	–	–	3	1.5–9	–
15	Aeris widepore XB-C8	C-S 3.6	–	–	25	–	–	–	4	1.5–9	–
16	Aeris widepore XB-C4	C-S 3.6	–	–	25	–	–	–	2	1.5–9	–
17	Aeris PEPTIDE XB-C18	C-S 1.7, 2.6, 3.6, 5	100	–	200	10	–	–	1	1.5–9	–
18	Aqua C18	Sph. 3, 5	125	1.05	320	15	–	–	2	2.5–7.5	–
19	Aqua C18	Sph. 5	200	1.15	215	11	–	–	4	2.5–7.5	–

(*Continued*)

APPENDIX 2a Common Physical and Chemical Properties of Various Hydrophobic Columns. Retention Factor k' is Given for Toluene in a Mobile Phase Consisting of 80% CH_3OH and 20% Aqueous Buffer 0.25 mM KH_2PO_4 at pH = 6 and at 24°C (in Isocratic Conditions)—cont'd

No.	Column Name	Particle Diam. (d_p, μm[a])	Pore Size (Å)	Pore Vol. (mL/g)	Surface Area (m^2/g)	C%[b]	Theor. Plate (N/m)[c]	Cover (μM/m²)	End-Cap[d]	pH Stabil.	Ret. Factor k' Toluene
20	Boltimate C18	C-S 2.7	90	–	120	9	200,000	–	1	2–8.5	–
21	Boltimate Phenyl-Hexyl	C-S 2.7	90	–	120	7	200,000	–	1	2–8.5	–
22	Boltimate EXT-C18	C-S 2.7	90	–	120	8	200,000	–	1	1.5–12	–
23	Boltimate PFP	C-S 2.7	90	–	120	5	200,000	–	1	1.5–10	–
24	Bondaclone C18	Irreg. 10	148	1.10	300	10 m.	–	1.61	4	2.5–7.5	–
25	μBondapak C18	Irreg. 10	125	–	330	10	36,000	–	3	2.5–7.5	1.68
26	Capcell Pak AG C18	5	120	–	300	15	51,000	–	4	–	3.25
27	Clarity Oligo-RP	Sph. 3, 5, 10	110	–	375	14	–	–	2	1–12	–
28	Clarity Oligo-WAX	Sph. 10	360	–	–	–	–	0.80	2	1–11	–
29	Clarity Oligo-MS	C-S 1.3, 1.7, 2.6, 5	100	–	200	12	–	–	3	1.5–10	–
30	Columbus C8	Sph. 5	110	–	375	13	–	–	1	2.5–7.5	–
31	Columbus C18	Sph. 5	110	–	375	19	–	–	1	2.5–7.5	–
32	CORTEX C18+ (trifunct.)	Sph. 1.6, 2.7	90	0.26	100	5.7	–	–	1	–	–
33	CORTEX C18 (trifunct.)	Sph. 1.6, 2.7	90	0.26	100	6.6	–	–	1	–	–
34	Delta-Pak C4	Sph. 5, 15	100	1.00	300	7.3	–	–	1	–	–
35	Delta-Pak C18	Sph. 5, 15	100	1.00	300	17	–	–	1	–	–
36	Develosil ODS-HG	5	140	–	300	18	85,500	–	3	–	2.97
37	Develosil ODS-MG	5	100	–	450	15	66,000	–	1	–	6.17
38	Develosil ODS-UG	5	140	–	300	18	92,000	–	2	–	3.74
39	Exsil ODS	5	100	–	200	11	93,000	–	2	–	2.97
40	Exsil ODS1	5	100	–	200	11	114,000	–	4	–	1.92
41	Exsil ODSB	5	100	–	200	12	82,000	–	4	–	1.68
42	Gemini C18	Sph. 3, 5, 10	110	1.10	375	14	75,500	–	1	1–12	4.10
43	Gemini C6-Phenyl	Sph. 3, 5	110	–	375	12	–	–	3	1–12	–
44	Gemini NX-C18	Sph. 3, 5, 10	110	–	375	14	–	–	4	1–12	–
45	HyperClone BDS C8	Sph. 3, 5	130	0.60	155	7	–	–	3	2.0–7.5	–
46	HyperClone BDS C18	Sph. 3, 5	130	0.60	155	11	–	–	4	2.0–7.5	–
47	HyperClone MOS (C8)	Sph. 3, 5	120	0.60	155	6.6	–	–	4	2.0–7.5	–

APPENDIX 2a Common Physical and Chemical Properties of Various Hydrophobic Columns. Retention Factor k' is Given for Toluene in a Mobile Phase Consisting of 80% CH_3OH and 20% Aqueous Buffer 0.25 mM KH_2PO_4 at pH = 6 and at 24°C (in Isocratic Conditions)—cont'd

No.	Column Name	Particle Diam. (d_p μm[a])	Pore Size (Å)	Pore Vol. (mL/g)	Surface Area (m^2/g)	C%[b]	Theor. Plate (N/m)[c]	Cover (μM/m^2)	End-Cap[d]	pH Stabil.	Ret. Factor k' Toluene
48	HyperClone ODS (C18)	Sph. 3, 5	120	0.60	155	10	–	–	3	2.0–7.5	–
49	HyperClone CN (CPS)	Sph. 3, 5	120	0.60	155	4	–	–	1	2.0–7.5	–
50	Hichrom RPB	5	110	–	340	14	97,500	–	2	–	3.00
51	Hypersil BDS C18	5	130	–	170	11	76,500	–	2	–	2.66
52	Hypersil GOLD	5	180	–	200	10	91,000	–	4	–	1.99
53	Hypersil HyPurity C18	5	180	–	200	13	73,000	–	1	–	2.34
54	Hypersil ODS	5	120	–	170	10	94,500	–	2	–	2.66
55	IB-Sil C18	Sph. 3, 5	125	0.75	165	11 m.	–	3.27	3	2.5–7.5	–
56	IB-Sil C18	Sph. 5	125	0.75	165	7.5 m.	–	4.29	4	2.5–7.5	–
57	Inertsil ODS	5	100	–	350	14	73,500	–	4	–	3.63
58	Inertsil ODS3	5	100	–	450	15	60,500	–	3	–	8.94
59	Inertsil ODS2	5	150	–	320	18.5	32,000	–	3	–	4.35
60	Jupiter C4	Sph. 5, 10, 15	300	–	170	5	–	6.30	3	1.5–10	–
61	Jupiter C5	Sph. 5, 10	300	–	170	5.5	–	5.30	2	1.5–10	–
62	Jupiter C18	Sph. 5, 10, 15	300	–	170	13.34	–	5.50	2	1.5–10	–
63	Jupiter Proteo	Sph. 4, 10	90	–	475	15	–	–	1	1.5–10	–
64	Kinetex EVO C18	C-S 1.7, 2.6, 5	100	–	200	11	–	–	2	1–12	–
65	Kinetex C18	C-S 1.3, 1.7, 2.6, 5	100	–	200	12	–	–	4	1.5–8.5	–
66	Kinetex XB-C18	C-S 1.7, 2.6, 5	100	–	200	10	–	–	3	1.5–8.5	–
67	Kinetex C8	C-S 1.7, 2.6, 5	100	–	200	8	–	–	3	1.5–8.5	–
68	Kinetex Biphenyl	C-S 1.7, 2.6, 5	100	–	200	11	–	–	2	1.5–8.5	–
69	Kinetex Phenyl-Hexyl	C-S 1.7, 2.6, 5	100	–	200	11	–	–	2	1.5–8.5	–
70	Kinetex F5	C-S 1.7, 2.6	100	–	200	9	–	–	2	1.5–8.5	–
71	Kromasil C18	5	100	–	340	19	99,000	–	4	–	7.32

(Continued)

APPENDIX 2a Common Physical and Chemical Properties of Various Hydrophobic Columns. Retention Factor *k*′ is Given for Toluene in a Mobile Phase Consisting of 80% CH_3OH and 20% Aqueous Buffer 0.25 mM KH_2PO_4 at pH = 6 and at 24°C (in Isocratic Conditions)—cont'd

No.	Column Name	Particle Diam. (d_p μm[a])	Pore Size (Å)	Pore Vol. (mL/g)	Surface Area (m^2/g)	C%[b]	Theor. Plate (N/m)[c]	Cover (μM/m^2)	End-Cap[d]	pH Stabil.	Ret. Factor *k*′ Toluene
72	LiChrosorb RP-18	10	100	–	300	17	74,000	–	1	–	2.99
73	LiChroSph. RP-18	5	100	–	350	21.6	80,000	–	2	–	4.90
74	Luna PFP(2)	Sph. 3, 5	100	1.00	400	11.5	–	2.20	3	1.5–9.0	–
75	Luna Phenyl-Hexyl	Sph. 3, 5, 10, 15	100	1.00	400	17.5	–	4.00	4	1.5–9.0	–
76	Luna C5	Sph. 5, 10	100	1.00	440	12.5	–	7.85	4	1.5–9.0	–
77	Luna C8	Sph. 5, 10	100	1.00	440	14.75	–	5.50	3	1.5–9.0	–
78	Luna C8(2)	Sph. 3, 5, 10, 15	100	1.00	440	13.5	–	5.50	3	1.5–9.0	–
79	Luna C18	Sph. 5, 10	100	1.00	440	19	–	3.00	3	1.5–9.0	–
80	Luna C18(2)-HST	Sph. 2.5	100	1.00	400	17.5	–	3.00	2	1.5–9.0	–
81	Luna C18(2)	Sph. 3, 5, 10, 15	100	1.00	400	17.5	88,000	3.00	2	1.5–9.0	4.85
82	Luna Omega C18	Sph. 1.6	100	–	260	–	350,000	–	1	1.5–8.5	–
83	Luna CN	Sph. 3, 5, 10	100	1.00	400	7	–	3.80	1	1.5–7.0	–
84	Novapak C18	4 (3.9)	60	–	120	7.3	60,000	–	2	–	2.01
85	Nucleosil C18	5	100	–	350	15	101,000	–	4	–	4.14
86	Nucleosil C18AB	5	100	–	350	24	87,000	–	3	–	2.77
87	Nucleoshell RP 18	C-S 2.7	90	–	–	7.5	250,000	–	1	1–11	–
88	Nucleodur C18 Gravity	1.8, 3, 5	110	–	–	18	80,000	–	1	1–11	–
89	Onix C18	Monolith	130	1.00	300	18	–	3.60	3	2.0–7.5	–
90	Partisil ODS	10	85	–	350	5	47,500	–	2	–	1.35
91	Partisil ODS2	10	85	–	350	15	41,000	–	2	–	3.60
92	Partisil ODS3	10	85	–	350	10.5	52,000	–	2	–	2.51
93	PhenoSph.e C1	Sph. 3, 5, 10	80	0.50	220	4 m.	–	1.80	4	2.5–7.5	–
94	PhenoSph.e C6	Sph. 3, 5, 10	80	0.50	220	6 m.	–	2.27	1	2.5–7.5	–
95	PhenoSph.e C8	Sph. 3, 5, 10	80	0.50	220	6 m.	–	3.54	2	2.5–7.5	–
96	PhenoSph.e ODS(1)	Sph. 3, 5, 10	80	0.50	220	7 m.	–	1.74	3	2.5–7.5	–
97	PhenoSph.e ODS(2)	Sph. 3, 5, 10	80	0.50	220	12 m.	–	2.50	4	2.5–7.5	–
98	PhenoSph.e CN	Sph. 3, 5, 10	80	0.50	220	4 m.	–	2.50	4	2.5–7.5	–
99	PhenoSph.e NEXT C8	Sph. 3, 5	120	–	380	10	–	–	3	2.5–7.5	–

APPENDIX 2a Common Physical and Chemical Properties of Various Hydrophobic Columns. Retention Factor k' is Given for Toluene in a Mobile Phase Consisting of 80% CH_3OH and 20% Aqueous Buffer 0.25 mM KH_2PO_4 at pH = 6 and at 24°C (in Isocratic Conditions)—cont'd

No.	Column Name	Particle Diam. (d_p, μm[a])	Pore Size (Å)	Pore Vol. (mL/g)	Surface Area (m^2/g)	C%[b]	Theor. Plate (N/m)[c]	Cover (μM/m^2)	End-Cap[d]	pH Stabil.	Ret. Factor k' Toluene
100	PhenoSph.e NEXT C8	Sph. 3, 5	120	–	380	14	–	–	3	2.5–7.5	–
101	PhenoSph.e NEXT Phenyl	Sph. 5	120	–	380	11	–	–	3	2.5–7.5	–
102	PolymerX RP-1	Sph. 3, 5, 10, 15	100	–	410	0	–	–	2	0–14	–
103	Prodigy ODS(2)	Sph. 5	150	1.10	310	18.4 m.	48,000	3.50	2	2.0–9.0	4.01
104	Prodigy C8	Sph. 5	150	1.10	310	12.5 m.	48,000	5.00	1	2.0–9.0	–
105	Prodigy ODS(3)	Sph. 3, 5, 10	100	1.00	450	15.5 m.	62,000	–	2	2.0–9.0	6.05
106	Prodigy Phenyl (PH-3)	Sph. 5	100	–	450	10.0 poly.	62,000	–	4	2.0–9.0	–
107	Resolve C18	5	90	0.5	200	10.2	45,500	–	4	–	3.16
108	Resolve C8	5	90	0.5	200	5.1	45,500	–	4	–	3.16
109	SiliaChrom AQ C18	3, 5, 10	100	–	380	18	–	–	1	1.5–9	–
110	SiliaChrom AQ C8	3, 5,10	100	–	380	14	–	–	1	1.5–9	–
111	SiliaChrom dt C18	2.5, 3, 5, 10	100	–	410–440	18	–	–	1	1.5–9	–
112	SiliaChrom XT C18	3, 5, 10	150	–	200	15	–	–	1	1.5–12	–
113	SiliaChrom XT Fidelity C18	3, 5, 10	100	–	380	21	–	–	1	1.5–12	–
114	SiliaChrom SB C18	3, 5, 10	150	–	200	12	–	–	1	0.5–7.5	–
115	SiliaChrom SB C18-300	5	300	–	80	5	–	–	1	0.5–7.5	–
116	SiliaChrom SB C8	5	150	–	200	7	–	–	1	1.0–7.5	–
117	SiliaChrom SB C8-300	5	300	–	80	3	–	–	1	1.0–7.5	–
118	SiliaChrom XDB C18	5	150	–	200	15	–	–	1	1.5–9.0	–
119	SiliaChrom XDB C8	3, 5	150	–	200	8	–	–	1	1.5–9.0	–
120	SiliaChrom XDB1 C18	3, 5	100	–	380–400	22	45,000	–	1	1.5–10	2.14
121	SiliaChrom XDB1 C18-300	5, 10	300	–	80	8	–	–	1	1.5–9.0	–
122	SiliaChrom XDB2 C18	3, 5, 10	100	–	380–400	18	28,000	–	1	1.5–9.0	0.61
123	SiliaChrom XDB1 C8	5, 10	100	–	380–400	14	–	–	1	1.5–8.5	–
124	SiliaChrom XDB1 C8-300	5	300	–	80	4	–	–	1	1.5–8.5	–

(Continued)

APPENDIX 2a Common Physical and Chemical Properties of Various Hydrophobic Columns. Retention Factor k' is Given for Toluene in a Mobile Phase Consisting of 80% CH_3OH and 20% Aqueous Buffer 0.25 mM KH_2PO_4 at pH = 6 and at 24°C (in Isocratic Conditions)—cont'd

No.	Column Name	Particle Diam. (d_p μm[a])	Pore Size (Å)	Pore Vol. (mL/g)	Surface Area (m^2/g)	C%[b]	Theor. Plate (N/m)[c]	Cover (μM/m^2)	End-Cap[d]	pH Stabil.	Ret. Factor k' Toluene
125	SiliaChrom XDB1 C4	5	100	–	380–400	7	–	–	1	1.5–8.5	–
126	SiliaChrom XDB1 C4-300	3, 5, 10	300	–	80	3	–	–	1	2.0–8.0	–
127	SiliaChrom XDB1 C1	5	100	–	380–400	3	–	–	1	1.5–8.5	–
128	SiliaChrom XDB1 C1-300	5	300	–	80	1	–	–	1	2.0–8.0	–
129	SiliaChrom XDB1 CN	5, 10	100	–	380–400	5	–	–	1	2.0–8.5	–
130	SiliaChrom XDB1 CN-300	5	300	–	80	3.5	–	–	1	2.0–8.0	–
131	SiliaChrom XDB1 Phenyl	3, 5	100	–	380–400	12	–	–	1	1.5–9.0	–
132	SiliaChrom XDB1 Phenyl-300	3, 5	300	–	80	4.5	–	–	1	2.0–8.0	–
133	SphereClone C6	Sph. 5	80	–	200	6	–	–	2	2.5–7.5	–
134	SphereClone C8	Sph. 3, 5	80	–	200	6	–	–	4	2.5–7.5	–
135	SphereClone ODS (1)	Sph. 3, 5	80	–	200	7	–	–	3	2.5–7.5	–
136	SphereClone ODS (2)	Sph. 3, 5	80	–	200	12	–	–	3	2.5–7.5	–
137	Spherysorb	Sph. 3, 5	80	0.50	220	6.2	–	–	4	–	–
138	SunFire C18	Sph. 2.5, 5	100	0.90	340	16	91,500	–	1	–	6.05
139	SunFire C8	Sph. 2.5, 5	100	0.90	340	12	–	–	1	–	–
140	Synergy Fusion-RP	Sph. 2.5	100	–	400	12	–	–	2	1.5–9.0	–
141	Synergy MAX-RP	Sph. 2.5	100	–	400	17	–	–	2	1.5–9.0	–
142	Synergy Hydro-RP	Sph. 2.5	100	–	400	19	–	–	2	1.5–7.5	–
143	Synergy Polar-RP	Sph. 2.5	100	–	400	11	–	–	4	1.5–7.5	–
144	Synergy Fusion-RP	Sph. 4, 10	80	1.05	475	12	–	–	1	1.5–9.0	–
145	Synergy MAX-RP	Sph. 4, 10	80	1.05	475	17	–	3.21	2	1.5–9.0	–
146	Synergy Hydro-RP	Sph. 4, 10	80	1.05	475	19	–	2.45	3	1.5–7.5	–
147	Synergy Polar-RP	Sph. 4, 10	80	1.05	475	11	–	3.15	4	1.5–7.5	–
148	Symmetry C18	Sph. 3.5, 5	100	0.90	335	19.1	92,000	–	1	–	6.11
149	Symmetry C8	Sph. 3.5, 5	100	0.90	335	11.7	–	–	1	–	–
150	Symmetry300C18	Sph. 3.5, 5	300	0.80	110	8.5	–	–	1	–	–
151	SymmetryShield RP18	Sph. 5	100	0.90	335	17	–	–	1	–	–

APPENDIX 2a Common Physical and Chemical Properties of Various Hydrophobic Columns. Retention Factor *k′* is Given for Toluene in a Mobile Phase Consisting of 80% CH_3OH and 20% Aqueous Buffer 0.25 mM KH_2PO_4 at pH = 6 and at 24°C (in Isocratic Conditions)—cont'd

No.	Column Name	Particle Diam. (d_p, μm[a])	Pore Size (Å)	Pore Vol. (mL/g)	Surface Area (m^2/g)	C%[b]	Theor. Plate (N/m)[c]	Cover (μM/m^2)	End-Cap[d]	pH Stabil.	Ret. Factor *k′* Toluene
152	SymmetryShield RP8	Sph. 3.5, 5	100	0.90	335	15	–	–	1	–	–
153	TSKgel Protein C4-300	3	300	–	–	3	–	–	1	–	–
154	TSKgel ODS-140HTP	2.3	140	–	–	6	–	–	1	–	–
155	TSKgel ODS-100V	3, 5	100	–	–	15 m.	–	–	1	–	–
156	TSKgel ODS-120A	5, 10	150	–	–	22	–	–	4	–	–
157	Ultracarb C8	Sph. 5	60	0.80	550	14 m.	–	2.71	3	2.5–7.5	–
158	Ultracarb ODS (20)	Sph. 3, 5	90	0.75	370	22 m.	–	3.53	3	2.5–7.5	–
159	Ultracarb ODS (30)	Sph. 5	60	0.80	550	31 m.	–	4.06	3	2.5–7.5	–
160	Vydac 218 TP	5	300	–	70	8	63,000	–	2	–	1.49
161	Waters Sph.isorb ODS1	5	80	–	220	6.2	100,500	–	2	–	2.23
162	Waters Sph.isorb ODS2	5	80	–	220	11.5	91,500	–	1	–	3.63
163	Waters Sph.isorb ODSB	5	80	–	220	11.5	92,000	–	2	–	3.03
164	XBridge C18	Sph. 2.5, 3.5, 5	130	0.70	185	18	–	–	1	–	–
165	XBridge C8	Sph. 2.5, 3.5, 5	130	0.70	185	13	–	–	1	–	–
166	XBridge Shield RP18	Sph. 2.5, 3.5, 5	130	0.70	185	17	–	–	1	–	–
167	XBridge BEH C18	Sph. 3.5, 5	130	0.70	185	18	–	–	1	–	–
168	XSelect CSH C18	Sph. 2.5, 3.5, 5	130	0.70	185	15	–	–	1	–	–
169	XSelect CSH Phenyl-Hexyl	Sph. 2.5, 3.5, 5	130	0.70	185	14	–	–	1	–	–
170	XSelect CSH Fluoro-Phenyl	Sph. 2.5, 3.5, 5	130	0.70	185	10	–	–	1	–	–
171	XTerra RP18	Sph. 3.5, 5	125	0.70	175	15	–	–	1	2–12	–
172	XTerra RP8	Sph. 3.5, 5	125	0.70	175	13.5	–	–	1	2–12	–
173	XTerra MS C18	Sph. 2.5, 3.5, 5	125	0.70	175	15.5	–	–	1	2–12	–
174	XTerra MS C8	Sph. 2.5, 3.5, 5	125	0.70	175	12	–	–	1	2–12	–
175	XTerra phenyl	Sph. 3.5, 5	125	0.70	175	12	–	–	1	2–12	–

(Continued)

APPENDIX 2a Common Physical and Chemical Properties of Various Hydrophobic Columns. Retention Factor k' is Given for Toluene in a Mobile Phase Consisting of 80% CH_3OH and 20% Aqueous Buffer 0.25 mM KH_2PO_4 at pH = 6 and at 24°C (in Isocratic Conditions)—cont'd

No.	Column Name	Particle Diam. (d_p μm[a])	Pore Size (Å)	Pore Vol. (mL/g)	Surface Area (m^2/g)	C%[b]	Theor. Plate (N/m)[c]	Cover (μM/m^2)	End-Cap[d]	pH Stabil.	Ret. Factor k' Toluene
176	YMC J'Sph.e ODS H80	4	80	–	510	22	64,500	–	4	–	7.03
177	YMC J'Sph.e ODS M80	4	80	–	510	14	58,000	–	3	–	2.66
178	YMC ODS A	5	120	–	300	17	99,500	–	3	–	3.53
179	YMC ODS AM	5	120	–	300	17	83,500	–	2	–	3.49
180	YMC Pro C18	5	120	–	335	16	105,000	–	4	–	3.53
181	Zorbax Extend C18	5	80	–	180	12.5	80,500	–	3	–	4.57
182	Zorbax ODS	5	70	–	330	20	85,500	–	3	–	5.47
183	Zorbax Rx-C18	5	80	–	180	12	90,500	–	2	–	3.71
184	Zorbax SB-C18	5	80	–	180	10	103,000	–	2	–	3.16
185	Zorbax XDB-C18	5	80	–	180	10	96,000	–	2	–	4.06

[a] *C-S indicates core–shell, Sph. indicates spherical particles, Irreg. indicates irregular particles, no specification for unknown particle shape.*
[b] *m. after the value of C% indicates monomeric type derivatization for the bonded phase.*
[c] *Plate number N/m refers to 5-μm particles if more than one particle dimension is indicated. However, many columns are available in more than one particle size format. For the estimation of N for columns with fully porous particles, the following expression can be used:* $N \approx 1000L/Ct.d_p$ *with* Ct. $\approx$ *2, 2.5, or 3 and* d_p *the diameter of the particle.*
[d] *1 indicates very low silanol activity to 4 high silanol activity.*
– *Indicates no information available.*

APPENDIX 2b List of Various Columns Used in RP-HPLC With the Values for log α for the Separation of Amitryptiline/Acenaphthene in a Mobile Phase 65% CH_3OH/35% Aqueous Buffer 20 mM K_2HPO_4/KH_2PO_4 at pH = 7.0 (v/v), and the Values for log k' for Acenapththene

No.	Column Description	log α	log k'
1	Waters Nova-Pak HP CN	−0.56	1.06
2	Waters Spherisorb S5 CN RP	−0.46	1.17
3	Supelco Discovery Cyano	−0.30	0.37
4	Thermo Hypersil CPS CN	−0.27	0.60
5	Waters Spherisorb S5 Ph	−0.12	1.62
6	Keystone Fluofix 120N	−0.02	1.00
7	Phenomenex Luna CN	−0.02	0.95
8	YMC Pack CN	0.02	0.49
9	Restek Ultra PFP	0.04	0.45
10	Supelco Discovery HS PEG	0.14	−0.07
11	Agilent Zorbax SB-CN	0.16	0.72
12	Waters XSelect HSS Cyano	0.17	0.55
13	Waters Acquity UPLC HSS Cyano	0.17	0.56
14	Thermo Hypersil BDS Phenyl	0.21	0.46
15	Keystone Fluophase RP	0.22	0.84
16	Thermo Hypersil Phenyl	0.23	0.96
17	Varian Pursuit Diphenyl	0.23	0.34
18	GL Science Inertsil 3 CN	0.26	0.61
19	GL Sciences Inertsil CN-3	0.28	0.47
20	Agilent Zorbax SB-Aq	0.38	0.96
21	GL Sciences Inertsil Ph-3	0.46	1.03
22	YMC Pack Ph	0.46	0.59
23	Alltech Platinum EPS C18	0.49	1.20
24	Waters Acquity UPLC CSH Fluoro-Phenyl	0.51	0.47
25	Waters XSelect CSH Fluoro-Phenyl	0.51	0.46
26	Restek Ultra Phenyl	0.51	0.41
27	W. R. Grace Platinum EPS C18	0.52	1.20
28	Agilent Zorbax SB- Phenyl	0.55	1.05
29	Waters Nova-Pak Phenyl	0.59	0.57

(Continued)

APPENDIX 2b List of Various Columns Used in RP-HPLC With the Values for log α for the Separation of Amitryptiline/Acenaphthene in a Mobile Phase 65% CH_3OH/35% Aqueous Buffer 20 mM K_2HPO_4/KH_2PO_4 at pH = 7.0 (v/v), and the Values for log k' for Acenapththene—cont'd

No.	Column Description	log α	log k'
30	Waters Xbridge C8 (Acquity UPLC BEH C8)	0.63	0.19
31	Waters Xterra RP8 (Embedded)	0.65	−0.12
32	Waters Xterra MS C8	0.65	0.16
33	Phenomenex Prodigy C8	0.67	0.19
34	Agilent Zorbax Rx C8	0.69	0.80
35	Agilent Zorbax Eclipse XBD Phenyl	0.69	0.55
36	Supelco Supelcosil LC DB-C8	0.70	0.46
37	ZirChrom PBD	0.71	0.43
38	Waters Xterra Phenyl	0.71	0.22
39	YMC J'Sphere L80	0.73	0.65
40	Waters Bondapack C18	0.75	0.81
41	Mac-Mod HydroBond AQ C8	0.75	0.62
42	Waters μBondapack C18	0.75	0.80
43	Merck Lichrosphere RP Select B	0.76	0.58
44	Restek Allure Ultra IBD	0.76	0.54
45	YMC Basic	0.76	0.26
46	Waters Nova-Pak C8	0.77	0.32
47	Alltech Platinum C18	0.77	0.94
48	Thermo Hypersil BDS C8	0.77	0.24
49	Waters Xbridge Phenyl (AcquityUPLC BEH Phenyl)	0.77	0.28
50	W. R. Grace Altima C8	0.77	0.64
51	W. R. Grace Platinum C18	0.77	0.93
52	Waters XSelect HSS C18 SB	0.78	1.24
53	Alltech Alltima C8	0.78	0.67
54	Agilent Zorbax Stable-Bond C8	0.78	0.74
55	Shiseido Capcell Pack C18	0.78	0.07
56	Supelco Discovery RP Amide C16 (embedded)	0.79	−0.04
57	Waters Acquity UPLC CSH Phenyl-Hexyl	0.81	0.09
58	ACT Ace C8	0.81	0.29

APPENDIX 2b List of Various Columns Used in RP-HPLC With the Values for log α for the Separation of Amitryptiline/Acenaphthene in a Mobile Phase 65% CH_3OH/35% Aqueous Buffer 20 mM K_2HPO_4/KH_2PO_4 at pH = 7.0 (v/v), and the Values for log k' for Acenapththene—cont'd

No.	Column Description	log α	log k'
59	Waters XSelect CSH Phenyl-Hexyl	0.82	0.07
60	Waters Acquity UPLC HSS C18 SB	0.82	1.24
61	Thermo Hypersil GOLD (C12)	0.83	0.20
62	Keystone Spectrum (Embedded)	0.85	0.20
63	Waters XSelect HSS PFP	0.86	0.91
64	Waters Acquity UPLC HSS PFP	0.86	0.91
65	GL Science Inertsil ODS-SP	0.86	0.35
66	Waters Spherisorb ODS-1	0.86	1.63
67	Waters Xterra RP18 (Embedded)	0.88	−0.10
68	Agilent Zorbax Bonus RP (Embedded)	0.88	0.02
69	GL Science Intersil C8	0.89	0.32
70	Phenomenex Gemini NX C18	0.91	0.03
71	Phenomenex Luna C8 (2)	0.91	0.16
72	Waters SymmetryShield RP8 (Embedded)	0.91	−0.02
73	Merck Lichrososorb RP Select B	0.92	0.61
74	Agilent Zorbax Eclipse XDB C8	0.93	0.29
75	Phenomenex Gemini C6-Phenyl	0.94	0.23
76	Azko Nobel Kromasil C8	0.94	0.33
77	Phenomenex Synergy Polar-RP	0.94	0.54
78	Keystone Prism (Embedded)	0.94	0.06
79	Thermo Hypersil Prism	0.96	0.02
80	Waters SunFire C8	0.96	0.20
81	Agilent TC-C18	0.96	0.34
82	Supelco Supelcosil LC-ABZ+ (Embedded)	0.96	−0.08
83	YMC Pro C8	0.97	0.22
84	YMC Carotenoid C30	0.98	0.00
85	Waters Symmetry C8	0.99	0.23
86	Supelco Supelcosil LC-ABZ (Embedded)	0.99	−0.09
87	Thermo Hypersil HyPurity C18	1.00	0.09

(Continued)

APPENDIX 2b List of Various Columns Used in RP-HPLC With the Values for log α for the Separation of Amitryptiline/Acenaphthene in a Mobile Phase 65% CH_3OH/35% Aqueous Buffer 20 mM K_2HPO_4/KH_2PO_4 at pH = 7.0 (v/v), and the Values for log k' for Acenapththene—cont'd

No.	Column Description	log α	log k'
88	Supelco Discovery C18	1.01	0.11
89	Varian Pursuit C18	1.01	0.15
90	Thermo Hypersil ODS	1.02	0.44
91	Phenomenex Kinetex XB-C18	1.03	0.20
92	Supelco Supelcosil LC DB-C18	1.04	0.36
93	Waters XBridge Shield RP18 (Embedded)	1.04	−0.06
94	Phenomenex Luna Phenyl Hexyl	1.05	0.37
95	Phenomenex Kinetex C18	1.06	0.21
96	Machery Nagel Nucleodur Sphinx RP	1.06	0.33
97	Metachem Polaris C18-A	1.06	0.05
98	AMT RP-Amide	1.08	−0.02
99	Thermo Hypersil BDS C18	1.08	0.11
100	Phenomenex Kinetex PFP	1.08	0.69
101	Waters Aquity UPLC CSH C18	1.08	0.11
102	Waters XSelect CSH C18	1.08	0.05
103	Phenomenex Synergy Fusion RP (Embedded)	1.08	0.34
104	Waters XTerra MS C18	1.09	0.18
105	Phenomenex Aqua C18	1.09	0.29
106	Waters Spherisorb ODSB	1.10	0.57
107	Supelco Discovery HS F5	1.10	0.47
108	Supelco Ascentis Expres RP-Amide (Embedded)	1.10	−0.02
109	Waters Xbridge C18 (Acquity UPLC BEH C18) (Embedded)	1.11	0.13
110	YMC Hydrosphere C18	1.12	0.25
111	YMC J'Sphere M80	1.12	0.54
112	TSK-Gel 80 Ts	1.13	0.41
113	Waters Atlantis dC18	1.14	0.28
114	Waters Nova-Pak C18	1.14	0.34
115	Keystone Fluophase PFP	1.14	0.38
116	Machery Nagel Nucleosil C18	1.16	0.66

APPENDIX 2b List of Various Columns Used in RP-HPLC With the Values for log α for the Separation of Amitryptiline/Acenaphthene in a Mobile Phase 65% CH_3OH/35% Aqueous Buffer 20 mM K_2HPO_4/KH_2PO_4 at pH = 7.0 (v/v), and the Values for log k' for Acenapththene—cont'd

No.	Column Description	log α	log k'
117	Phenomenex Gemini C18	1.16	0.26
118	Waters Acquity UPLC HSS T3	1.16	0.31
119	ACT Ace C18	1.17	0.14
120	Thermo Hypersil Elite C18	1.18	0.06
121	Waters Atlantis T3	1.18	0.33
122	Agilent Zorbax Stable-Bond C18	1.18	0.37
123	Waters XSelect HSS T3	1.18	0.27
124	Phenomenex Synergy Max RP	1.18	0.22
125	Merck Purosphere RP18	1.19	0.48
126	Waters SymmetryShield RP18 (Embedded)	1.19	−0.05
127	Agilent Zorbax Rx C18	1.19	0.17
128	YMC ODS AQ	1.20	0.21
129	Phenomenex Prodigy C18	1.20	0.08
130	Agilent Zorbax Eclipse XDB C18	1.21	0.10
131	Shiseido Capcell Pack C18 AQ	1.21	0.33
132	Phenomenex Luna C18 (2)	1.21	0.22
133	Agilent HC-C18	1.21	0.19
134	Nomura Devosil C30 UG 5	1.23	0.14
135	Supelco Ascentis RP-Amide (Embedded)	1.23	0.01
136	Phenomenex Luna C18	1.23	0.17
137	Waters Spherisorb ODS-2	1.24	0.85
138	AMT Halo C18	1.24	0.07
139	YMC Pack Pro C18	1.25	0.19
140	Waters XSelect HSS C18	1.26	0.02
141	Waters Acquity UPLC HSS C18	1.26	0.22
142	Shiseido Capcell Pack SG120C18	1.26	0.23
143	GL Sciences Intersil ODS-2	1.26	0.12
144	Dionex Acclaim PA (Embedded)	1.27	−0.01
145	Supelco Ascentis Express C18	1.27	0.06

(Continued)

APPENDIX 2b List of Various Columns Used in RP-HPLC With the Values for log α for the Separation of Amitryptiline/Acenaphthene in a Mobile Phase 65% CH_3OH/35% Aqueous Buffer 20 mM K_2HPO_4/KH_2PO_4 at pH = 7.0 (v/v), and the Values for log k' for Acenapththene—cont'd

No.	Column Description	log α	log k'
146	Agilent Zorbax Eclipse Plus	1.27	0.19
147	Restek Allure PFP Propyl	1.28	0.58
148	Agilent Zorbax Extend C18	1.28	0.15
149	Waters Resolve C18	1.29	1.32
150	Shiseido Capcell Pak MGIII	1.30	0.19
151	Supelco Discovery HS C18	1.31	0.08
152	Waters SunFire C18	1.32	0.15
153	Shiseido Capcell Pak MGII	1.33	0.21
154	Imtakt Cadenza CD-C18	1.33	0.12
155	Merck Purosphere RP 18e	1.34	0.27
156	Dionex Acclaim C18	1.34	0.17
157	ES Industries EPIC C18	1.34	−0.01
158	Mackery Nagel Nucleodur Gravity C18	1.34	0.15
159	Alltech Alltima C18	1.36	0.54
160	GL Sciences Intersil ODS-3	1.36	0.13
161	W. R. Grace Altima C18	1.39	0.52
162	Azko Nobel Kromasil C18	1.39	0.11
163	Shodex Silica C18P	1.39	−0.11
164	CITI L-Column ODS	1.39	0.16
165	Waters Symmetry C18	1.39	0.09
166	Supelco Ascentis C18	1.41	0.10
167	GL Sciences Intersil ODS 3V	1.42	0.15
168	GL Sciences Intersil ODS-EP (Embedded)	1.43	−0.26
169	YMC J'Sphere H80	1.48	0.43
170	Nomura Develosil ODS SR 5	1.50	0.27
171	YMC Pack Pro C18 RS	1.55	0.10

APPENDIX 2c List of C18 Columns With UPS Parameters[a] (Mobile Phase 80/20 Methanol/Aqueous Buffer 60 mM Phosphate v/v, pH = 7.0) (http://apps.usp.org/app/USPNF/columnsDB.html)

Column	H_y	*CTF*	*CFA*	*TFA*	*BD*	Manuf.[b]
218 TP 300 C18	–	1.4	7	1.3	5	*6*
238 MS 300 C18	0.4	–	0.7	1.6	2.5	*6*
Acclaim 120 C18	2.3	1.2	5.6	1.6	3.2	*4*
Acclaim 120 C8 (for comparison)	1.1	1.5	2.5	1.5	3.4	*4*
Acclaim 300 C18	0.7	1.2	1.7	1.3	3.3	*4*
Acclaim Polar Advantage	1.5	1.5	3.2	1.6	2.6	*4*
Acquity UPLC BEH C18	1.6	1.1	3.8	1.1	3.1	*18*
Acquity UPLC BEH Shield RP18	1.3	1.4	2.4	1.1	3.3	*18*
Alltima C18	3.4	1.1	13.5	2.5	2.8	*5*
Alltima C18 WP	1.4	1.3	4.3	5.3	5.2	*5*
Alltima HP C18	1.4	1.3	3.5	1.2	2.6	*5*
Alltima HP C18 Amide	1.1	1.5	2.2	1.1	2.8	*5*
Alltima HP C18 AQ	3.1	1	26.1	1.5	2.1	*5*
Alltima HP C18 EPS	0.4	1	8.4	1.2	1.3	*5*
Alltima HP C18 High load	3.6	1.2	6.9	1.2	2.4	*5*
Allure	1.4	1.9	12.7	4.4	4.1	*11*
Aquasil C18	1.5	1.9	23	2.8	2.2	*16*
Ascentis C18	2.7	1	6.2	1.3	3.6	*15*
Atlantis dC18	1.7	1.1	5.1	2.4	1.6	*18*
BetaBasic 18	1.6	6.6	4.1	2.7	3.2	*16*
BetaMax Neutral	4.2	1.6	11	3.9	3.6	*16*
BETASIL C18	3.2	1.6	7.6	2	3.6	*16*
Bio Basic 18	0.9	1.3	2.2	2.1	4.2	*2*
Capcell C18 ACR	2.9	1.2	6.3	3.1	3.5	*14*
Capcell C18 AG 120	2.4	1.7	7.8	5.4	2.3	*14*
Capcell C18 AQ	1.8	1.8	5.5	2.1	1.6	*14*
Capcell C18 MG	2.5	1.2	6.2	1.5	2.8	*14*
Capcell C18 MGII	2.5	1.3	6.1	1.1	2.8	*14*
Capcell C18 SG 120	1.9	1.3	4.7	2.3	2.1	*14*
Capcell C18 UG 120	1.9	1.1	4	2	2.4	*14*

(*Continued*)

APPENDIX 2c List of C18 Columns With UPS Parameters[a] (Mobile Phase 80/20 Methanol/Aqueous Buffer 60 mM Phosphate v/v, pH = 7.0) (http://apps.usp.org/app/USPNF/columnsDB.html)—cont'd

Column	H_y	*CTF*	*CFA*	*TFA*	*BD*	Manuf.[b]
Capcell C18 UG 80	3	1.7	6.4	5.3	2.6	*14*
Chromolith Perf RP-18e	0.8	1.5	3.1	2.9	3.3	*7*
Cosmosil 5C18-AR-II	2.4	1.2	14.4	2.3	3.4	*8*
Cosmosil 5C18-MS-II	2	1.1	4.7	1.7	2.8	*8*
Cosmosil 5C18-PAQ	1.2	2.2	2.5	1.5	2	*8*
Delta-Pak C18 100A	2	2	6.3	1.9	3.5	*18*
Denali C18	1.7	1.5	3.9	1.2	3.1	*5*
Discovery C18	1.1	1.1	2.8	1.6	3	*15*
Discovery C18 WP	0.8	1.1	1.8	1.4	3.6	*15*
Discovery HS C18	2.4	1	5.4	1.3	3.8	*15*
Everest C18 300A	0.8	1.5	2.3	3	4.3	*5*
Genesis C18	1.8	1.2	4.7	2	3.8	*5*
Genesis C18 AQ	1.8	1.1	5.4	1.9	4	*5*
Hydrosphere C18	1.5	1.3	4.4	1.9	1.9	*19*
Hypersil BDS 18	1.5	1.5	3.5	2	3.1	*16*
Hypersil ODS	1.5	3.4	4.3	3.6	3.2	*16*
Hypersil ODS-2	1.5	2	5.6	4.1	2.4	*16*
Hypersil PAH	1.2	2.2	12	2.6	4.6	*16*
HyPurity C18	1.3	1.4	3.5	2.1	3.3	*16*
Jsphere H80	3.3	1.3	8.8	2.9	3.2	*19*
Jsphere L80	0.9	1.4	3	2.8	0.9	*19*
Jsphere M80	1.6	1.4	5	2.7	1.4	*19*
Luna 5 μ C18(2)	2.2	1.2	5.3	1.1	3.4	*10*
Matrix C18	2.65	1.557	6.78	1.445	3.144	*9*
MicroBondapak C18	1	6	7.5	4	1.1	*18*
Monitor C18	2.1	1.438	5.71	1.966	3.34	*9*
Nova-Pak C18	1.5	7.5	4.6	3	2.7	*18*
Orosil C18	2.82	1.072	7.54	1.34	2.542	*9*
Pinnacle DB	1.4	1.8	4.3	2.8	3.7	*11*
Pinnacle II	1.4	3.4	5.1	3.4	3.7	*11*

APPENDIX 2c List of C18 Columns With UPS Parameters[a] (Mobile Phase 80/20 Methanol/Aqueous Buffer 60 mM Phosphate v/v, pH = 7.0) (http://apps.usp.org/app/USPNF/columnsDB.html)—cont'd

Column	H_y	*CTF*	*CFA*	*TFA*	*BD*	Manuf.[b]
Platinum EPS C18	0.4	1.4	14.5	3.5	1.2	5
Prestige C18	2.21	1.194	6.61	2.176	3.084	9
Prevail C18	2	1	23	2	2.6	5
Prevail Select C18	2	1.1	4	1.1	3.2	5
ProntoSil 120-5-C18-ace-EPS	2	1.2	3.8	1.6	2.8	3
ProntoSil 120-5-C18-AQ	2	1	6.7	2.6	2.1	3
ProntoSil 120-5-C18-AQ Plus	2.2	1.4	14.2	3.5	3.2	3
ProntoSil 120-5-C18-H	2.3	1	6.1	1.8	2.9	3
ProntoSil 120-5-C18-SH	2.2	1.8	10.2	2.2	3	3
ProntoSil 200-5-C18-ace-EPS	1.1	1.2	2.2	1.2	3.2	3
ProntoSil 200-5-C18-AQ	1	2.2	3.1	2.4	2.1	3
ProntoSil 200-5-C18-H	1.1	1.7	2.9	2.2	2.9	3
ProntoSil 300-5-C18-ace-EPS	0.7	1.3	1.4	1.8	3.2	3
ProntoSil 300-5-C18-H	0.7	1.6	2.2	2.7	2.9	3
ProntoSil 60-5-C18-H	3.9	1.7	12.5	4	2.9	3
Purospher RP-18	1.5	1.2	7.5	1.8	1.6	7
Purospher RP-18 endcapped	2.4	1.1	7.1	1	1.7	7
PurospherSTAR RP-18e 3 μm	2.1	1.2	5.7	1.5	3.3	7
PurospherSTAR RP-18e 5 μm	2.6	1.4	7.8	1.9	3.3	7
Reliachrom C18	1.64	3.047	2.65	2.43	3.751	9
Reliasil C18	2.24	6.906	10.92	7.123	2.637	9
SepaxGP-C18	2.5	1.2	7.7	2.54	3.3	12
SepaxHP-C18	2	1.2	6.8	1.81	3.3	12
Shim-pack VP-ODS	2.2	1.4	5.7	1.5	2.4	13
Shim-pack XR-ODS	2.1	1.2	5.1	1.4	2.6	13
Shim-pack XR-ODS II	3.4	2.5	8.5	4.9	2.4	13
Supelcosil LC18	1.3	2	4.5	13	3.1	15
Superiorex	2.8	1.7	7.5	3	2.8	14
Superspher 100 RP-18 endcapped	2.3	7.9	6	4.1	2.6	7
Symmetry C18	2.2	1.7	5.1	1.7	3.2	18

(*Continued*)

APPENDIX 2c List of C18 Columns With UPS Parameters[a] (Mobile Phase 80/20 Methanol/Aqueous Buffer 60 mM Phosphate v/v, pH = 7.0) (http://apps.usp.org/app/USPNF/columnsDB.html)—cont'd

Column	H_y	*CTF*	*CFA*	*TFA*	*BD*	Manuf.[b]
SymmetryShield RP18	1.6	1.5	3.1	1.2	3.3	*18*
Synergy 4μ Hydro-RP	2.6	1.2	-	3.3	4.0	*1*
TSKgel Octyl-80Ts	1.2	1.7	2.9	2.8	2.8	*17*
TSKgel ODS-100S	2.4	1.2	5.2	2	3	*17*
TSKgel ODS-100V 3 μm	1.9	1	6	1.4	1.8	*17*
TSKgel ODS-100V 5 μm	2.1	1.1	5.9	1.3	1.8	*17*
TSKgel ODS-100Z	3	1.1	7.2	1.3	2.6	*17*
TSKgel ODS-120A	1.6	1.7	16.4	6.2	2.7	*17*
TSKgel ODS-120T	2	2.8	9.8	8.3	2.8	*17*
TSKgel ODS-80TM	2.1	2.9	11.1	4.7	2.1	*17*
TSKgel ODS-80Ts	2.2	1.2	6.4	2.8	2.1	*17*
TSKgel Super-Octyl	0.5	1.5	0.9	1.7	1.9	*17*
TSKgel Super-ODS	1.1	2	3.7	4.4	2.7	*17*
TSKgel Super-Phenyl	0.2	1.9	1.9	3.6	3.9	*17*
Ultra	1.4	1.2	7.7	2.7	3.7	*11*
Viva 300	1.4	1.1	2.7	2.1	3.5	*11*
Xbridge C18	1.6	1.1	3.8	1.1	3.1	*18*
Xbridge Shield RP18	1.3	1.4	2.4	1.1	3.3	*18*
Xterra MS C18	1.5	1.1	3.3	1.3	2.2	*18*
Xterra RP18	1	1.2	1.7	1.1	2.3	*18*
YMC ODS-A	1.9	1.5	5	2.4	2.6	*19*
YMC ODS-AL	1.8	1.5	13.6	2.8	2.6	*19*
YMC ODS-AM	1.9	1.5	5.1	2.4	2.7	*19*
YMC ODS-AQ	1.9	1.5	6	2.9	2.2	*19*
YMC Pro C18	1.9	1.3	5	1.5	2.5	*19*
YMC Pro C18 RS	3.3	1.2	7.5	1.3	3	*19*
Zorbax Eclipse XDB C18	2.4	1	6.1	1.8	4	*1*

APPENDIX 2c List of C18 Columns With UPS Parameters[a] (Mobile Phase 80/20 Methanol/Aqueous Buffer 60 mM Phosphate v/v, pH = 7.0) (http://apps.usp.org/app/USPNF/columnsDB.html)—cont'd

Column	H_y	*CTF*	*CFA*	*TFA*	*BD*	Manuf.[b]
Zorbax Extend C18	2.4	1.1	5.9	3.4	3.8	*1*
Zorbax Rx C18	2.1	1.4	8.2	6.7	3.5	*1*
Zorbax StableBond C18	2	1.1	7.3	2.3	2	*1*

[a]*H_y is the capacity factor k′ for ethylbenzene, **CTF** is the chelating tailing factor for quinizarin, **CFA** is the capacity factor k′ for amitriptyline, **TFA** is the tailing factor for amitriptyline, and **BD** is the bonding density in μmol/m².*
[b]*Manufacturers: **1**. Agilent Technologies, **2**. Bio-Rad, **3**. Bischoff, **4**. Dionex, **5**. Grace/Davison, **6**. Grace/Vydac, **7**. Merck KGaA (EDM Millipore), **8**. Nacalai Tesque, **9**. Orochem Technologies, **10**. Phenomenex, **11**. Restek, **12**. Sepax Technologies, **13**. Shimadzu, **14**. Shiseido, **15**. Supelco, **16**. Thermo Scientific, **17**. Tosoh Bioscience, **18**. Waters, **19**. YMC.*

APPENDIX 2d List of Available Columns With PQRI Parameters (Mobile Phase 50/50 Acetonitrile/Aqueous Buffer 60 mM Phosphate v/v) (http://apps.usp.org/app/USPNF/columnsDB.html)

Column	*H**[a]	*S**[b]	*A**[c]	*B**[d]	*C*(2.8)*[e]	*C*(7.0)*[f]	Silica*[g]	UPS	$\log k'_{EB}$	Manuf.[h]
Acclaim 120 C18	1.032	0.018	−0.143	−0.027	0.086	−0.002	B	L1	1.002	*14*
Acclaim 120 C8	0.857	0.004	−0.274	0.011	0.086	0.016	B	L7	0.780	*14*
Acclaim C300C18	0.957	−0.018	−0.170	0.004	0.261	0.222	B	L1	0.462	*46*
Acclaim Organic Acid	0.833	−0.063	−0.385	−0.001	−0.316	0.349	Other	–	–	*14*
Acclaim Phenyl-1	0.689	−0.15	−0.548	0.068	0.013	0.15	Phenyl	L11	–	*46*
Acclaim Polar Advantage	0.855	−0.068	−0.116	0.023	−0.27	0.357	EP	L60	–	*14*
Acclaim PolarAdvantage II	0.74	0.015	−0.554	0.214	−0.225	0.67	EP	L60	–	*14*
Acclaim300 C18	0.957	−0.018	−0.17	0.019	0.261	0.222	B	L1	–	*14*
Accucore 150-C18	1.027	0.046	−0.037	−0.009	0.2	0.171	B	L1	–	*46*
Accucore 150-C4	0.738	−0.008	−0.309	0.023	0.13	0.224	B	L26	–	*46*
Accucore aQ	1.071	0.05	0.002	−0.059	0.06	0.508	EP	L60	–	*46*
Accucore C18	1.092	0.054	0.055	−0.04	0.072	0.095	B	L1	–	*46*
Accucore C30	0.978	−0.02	−0.143	−0.002	0.321	0.462	B	L62	–	*46*
Accucore C8	0.924	0.015	−0.084	−0.003	0.002	0.186	B	L7	–	*46*
Accucore PFP	0.729	−0.098	−0.249	−0.036	0.873	1.655	F	L43	–	*46*
Accucore Phenyl-Hexyl	0.786	−0.047	−0.223	0.006	0.055	0.574	Phenyl	L11	–	*46*
Accucore Phenyl-X	0.781	−0.103	−0.554	0.044	−0.294	−0.082	Phenyl	L11	–	*46*
Accucore Polar Premium	0.871	0.103	−0.567	0.217	−0.207	0.787	EP	L60	–	*46*
Accucore RP-MS	1	0.024	−0.015	−0.006	0.027	0.057	B	–	–	*46*

(Continued)

APPENDIX 2d List of Available Columns With PQRI Parameters (Mobile Phase 50/50 Acetonitrile/Aqueous Buffer 60 mM Phosphate v/v) (http://apps.usp.org/app/USPNF/columnsDB.html)—cont'd

Column	H^{*a}	S^{*b}	A^{*c}	B^{*d}	C(2.8)*[e]	C(7.0)*[f]	Silica*[g]	UPS	$\log k'_{EB}$	Manuf.[h]
Accucore XL C18	1.089	0.059	0.051	−0.038	0.086	0.384	B	L1	–	***46***
Accucore XL C8	0.926	0.02	−0.161	−0.001	0.02	−0.432	B	L7	–	***46***
ACE 5 AQ	0.804	−0.051	−0.129	0.034	0.009	0.167	EP	–	–	***1***
ACE 5C18	1	0.027	−0.096	−0.007	0.143	0.096	B	L1	0.895	***1***
ACE 5 C18-300	0.983	0.025	0.046	0.012	0.262	0.237	B	L1	–	***1***
ACE 5 C18-AR	0.881	−0.059	−0.208	−0.016	0.079	0.153	Phenyl	L1	–	***1***
ACE 5 C18-HL	1.045	0.052	−0.088	−0.031	−0.038	−0.11	B	L1	–	***1***
ACE 5C18-PFP	0.899	−0.021	−0.246	−0.08	−0.001	−0.995	B	L1	–	***1***
ACE 5 C4	0.674	−0.018	−0.178	0.026	0.09	0.316	B	L26	–	***1***
ACE 5 C4-300	0.71	−0.014	−0.183	0.039	0.166	0.356	B	L26	–	***1***
ACE 5 C8	0.83	−0.004	−0.268	−0.017	−0.334	−0.298	B	L7	0.693	***1***
ACE 5 C8-300	0.786	−0.003	−0.112	0.032	0.145	0.387	B	L7	–	***1***
ACE 5 CN	0.409	−0.107	−0.729	−0.008	−0.086	0.441	CN	L10	−0.019	***1***
ACE 5 CN-300	0.46	−0.074	−0.165	0.03	0.151	0.856	CN	L10	–	***1***
ACE 5 Phenyl	0.647	−0.138	−0.296	0.027	0.132	0.466	Phenyl	L11	0.445	***1***
ACE Phenyl-300	0.599	−0.105	−0.234	0.032	0.164	0.548	Phenyl	L11	–	***1***
Acquity UPLC BEH C8	0.855	−0.008	0.095	0.056	0.22	0.777	B	L7	–	***48***
Acquity UPLC BEH Phenyl	0.764	0.077	−0.051	0.062	0.292	0.586	Phenyl	L11	–	***48***
Acquity UPLC BEH Shield RP-18	0.907	0.016	−0.031	0.133	−0.055	0.416	EP	–	–	***48***
Adsorbosphere C18	0.989	−0.073	0.07	−0.044	1.496	1.683	A	L1	–	***20***
Adsorbosphere UHS C18	1.103	0.004	0.402	−0.046	−0.125	−0.125	B	L1	–	***20***
Advantage 300	0.867	−0.001	0.123	0.02	0.597	1.11	A	L1	–	***6***
Advantage Armor C18 120A	0.962	−0.014	−0.076	−0.004	0.077	0.261	B	L1	–	***6***
Aeris PEPTIDE XB-C18	0.992	−0.021	−0.095	0.016	0.071	0.314	B	L1	–	***37***
Aeris WIDEPORE XB-C18	0.934	−0.032	−0.134	0.068	0.073	0.36	B	L1	–	***37***
Aeris WIDEPORE XB-C4	0.699	−0.021	−0.31	0.017	0.084	0.195	B	L26	–	***37***
Aeris WIDEPORE XB-C8	0.788	−0.038	−0.169	0.073	−0.042	0.518	B	L7	–	***37***
Allsphere ODS1	0.733	−0.16	0.387	0.002	0.846	1.142	A	L1	–	***20***

APPENDIX 2d List of Available Columns With PQRI Parameters (Mobile Phase 50/50 Acetonitrile/Aqueous Buffer 60 mM Phosphate v/v) (http://apps.usp.org/app/USPNF/columnsDB.html)—cont'd

Column	H^{*a}	S^{*b}	A^{*c}	B^{*d}	$C(2.8)^{*e}$	$C(7.0)^{*f}$	Silica*[g]	UPS	$\log k'_{EB}$	Manuf.[h]
Allsphere ODS2	1.004	−0.04	0.243	−0.028	0.96	1.281	A	L1	–	***20***
Alltima AQ	0.882	−0.07	0.301	0.016	0.158	1.157	B	–	–	***20***
Alltima C18	0.993	−0.014	0.035	−0.013	0.092	0.391	B	L1	–	***20***
Alltima C18-LL	0.78	0.085	−0.165	0.041	−0.056	0.367	B	L1	–	***20***
Alltima C18-WP	0.938	−0.062	0.027	0.002	−0.079	−0.081	B	L1	–	***20***
Alltima C8	0.756	0.015	−0.279	0.009	−0.062	0.288	B	L7	–	***20***
Alltima HP C18	0.985	−0.02	−0.04	0.006	0.177	0.199	B	L1	–	***20***
Alltima HP C18 Amide	0.497	−0.026	0.357	0.124	−0.019	0.926	EP	L60	–	***20***
Alltima HP C18 EPS	0.655	−0.104	0.401	0.036	0.459	0.955	EP	L1	–	***20***
Alltima HP C18 High Load	1.08	−0.066	0.066	−0.04	−0.322	−0.244	B	L1	–	***20***
Alltima HP C8	0.834	0.01	−0.116	0.035	0.122	−0.418	B	L7	–	***20***
Alltima HP CN	0.469	−0.012	−0.556	0.009	−0.187	0.44	CN	L10	–	***20***
Allure Aqueous C18	0.968	−0.105	0.395	−0.044	0.006	1.389	B	L1	–	***38***
Allure Basix	0.438	−0.077	−0.423	−0.003	0.016	0.865	CN	Other	–	***38***
Allure Biphenyl	0.716	−0.192	−0.298	0.029	−0.01	0.694	Phenyl	–	–	***38***
Allure C18	1.131	0.052	0.046	−0.049	−0.037	0.02	B	L1	1.195	***38***
Allure Organic Acids	0.91	−0.125	0.53	−0.022	0.191	1.502	B	L1	–	***38***
Allure PFP Propyl	0.833	−0.265	0.051	0.348	1.109	1.659	F	L43	0.833	***38***
Alphabond (C18)	0.845	−0.094	0.061	0.001	0.579	1.76	A	L1	–	***20***
Apex C18	0.985	−0.035	0.013	0.042	1.246	2.311	A	L1	–	***22***
Apex C8	0.869	−0.071	0.235	0.177	1.364	1.373	A	L1	–	***22***
Apex II C18	1.008	−0.074	0.235	0.123	2.039	2.69	A	L1	–	***22***
apHera C18 Polymer	0.838	−0.01	−1.106	0.001	−0.554	7.511	B	L67	–	***45***
Aqua C18	0.966	−0.03	0.033	0.009	0.068	0.276	B	L1	–	***37***
Aquasil C18	0.805	0.114	0.265	0.011	0.23	1.196	EP	–	–	***46***
Armor C18 3μm	0.964	−0.016	−0.079	−0.002	0.122	0.296	B	L1	–	***6***
Ascentis C18	1.077	0.058	0.03	−0.042	−0.088	−0.084	B	L1	–	***45***
Ascentis C8	0.899	0.024	−0.18	−0.002	−0.124	−0.035	B	L7	–	***45***
Ascentis ES Cyano	0.541	−0.101	−0.495	0.013	0.133	−0.449	CN	L10	–	***45***

(*Continued*)

APPENDIX 2d List of Available Columns With PQRI Parameters (Mobile Phase 50/50 Acetonitrile/Aqueous Buffer 60 mM Phosphate v/v) (http://apps.usp.org/app/USPNF/columnsDB.html)—cont'd

Column	H^{*a}	S^{*b}	A^{*c}	B^{*d}	$C(2.8)^{*e}$	$C(7.0)^{*f}$	Silica*[g]	UPS	$\log k'_{EB}$	Manuf.[h]
Ascentis Express 5 C18	1.125	0.053	0.053	−0.05	0.084	0.197	B	L1	–	*45*
Ascentis Express 5 F5	0.711	−0.118	−0.097	−0.067	1.056	1.705	F	L43	–	*45*
Ascentis Express C18	1.136	0.053	0.023	−0.052	0.067	0.107	B	L1	–	*45*
Ascentis Express C8	0.915	0.015	−0.117	−0.005	−0.002	0.172	B	L7	–	*45*
Ascentis Express ES-CN	0.565	−0.098	−0.415	0.014	0.031	1.043	CN	L10	–	*45*
Ascentis Express F5	0.721	−0.124	0.067	−0.074	1.059	1.839	F	L43	–	*45*
Ascentis Express Peptide ES-C18	0.918	−0.042	0.263	0.019	0.192	0.813	B	–	–	*45*
Ascentis Express Phenyl-Hexyl	0.797	−0.088	−0.263	−0.01	0.094	0.483	Phenyl	L11	–	*45*
Ascentis Express RP-Amide	0.848	0.063	−0.416	0.198	−0.471	0.266	EP	L60	–	*45*
Ascentis Phenyl	0.721	−0.108	−0.288	−0.008	0.001	0.579	Phenyl	L11	–	*45*
Ascentis RP-Amide	0.843	0.078	−0.496	0.183	−0.781	0.087	EP	L60	–	*45*
Athena C18	0.997	0.002	−0.03	−0.013	0.094	0.265	B	L1	–	*11*
Athena C18-WP	0.953	−0.03	−0.203	−0.003	−0.052	0.066	B	L1	–	*11*
Athena C8	0.831	−0.021	−0.166	0.009	0.026	0.274	B	L7	–	*11*
Athena Phenyl	0.587	−0.114	−0.396	0.033	0.067	0.345	Phenyl	L11	–	*11*
Atlantis dC18	0.917	−0.031	−0.193	0.001	0.036	0.087	B	L1	0.908	*48*
Atlantis T3	0.941	−0.035	−0.181	0.006	0.029	0.713	B	L1	–	*48*
Balancil C18	0.918	−0.037	−0.17	0.025	0.109	0.734	B	L1	–	*8*
BAS MF-8954	0.979	−0.069	0.181	0.022	1.081	1.397	A	L1	–	*6*
BetaBasic CN	0.426	−0.043	−0.453	0.014	0.014	0.904	CN	L10	–	*46*
BetaBasic Phenyl	0.582	−0.159	−0.411	0.049	0.104	0.758	Phenyl	L11	0.234	*46*
BetaMax Acid	0.635	0.057	−0.597	0.376	−2.064	−0.51	EP	–	–	*46*
BetaMax Base	0.47	−0.06	−0.391	0.01	0.014	1.146	EP	–	–	*46*
Betasil C18	1.056	0.049	−0.1	−0.032	−0.035	−0.044	B	L1	–	*46*
Betasil Phenyl-Hexyl	0.707	−0.053	−0.294	0.028	0.054	0.357	Phenyl	L11	0.637	*46*
BioBasic 4	0.691	−0.009	−0.191	0.032	0.188	0.39	B	L26	–	*46*
BioBasic CN	0.39	−0.033	−0.823	−0.005	−0.07	0.537	CN	L10	–	*46*
BioBasic Phenyl	0.493	−0.233	−0.671	0.217	0.014	0.39	Phenyl	L11	–	*46*
Biobond C18	0.968	−0.021	−0.154	0.004	0.224	0.189	B	L1	–	*13*

APPENDIX 2d List of Available Columns With PQRI Parameters (Mobile Phase 50/50 Acetonitrile/Aqueous Buffer 60 mM Phosphate v/v) (http://apps.usp.org/app/USPNF/columnsDB.html)—cont'd

Column	H^{*a}	S^{*b}	A^{*c}	B^{*d}	C(2.8)*[e]	C(7.0)*[f]	Silica*[g]	UPS	$\log k'_{EB}$	Manuf.[h]
Biobond C4	0.706	−0.001	−0.315	0.024	0.14	0.181	B	L26	–	*13*
Biobond C8	0.823	−0.008	−0.264	0.028	0.172	0.155	B	L7	–	*13*
Bondclone C18	0.824	−0.056	−0.125	0.044	0.078	0.347	B	L1	–	*37*
Brava BDS C18	0.935	−0.033	0.033	0.012	0.281	0.768	B	L1	–	*20*
Cadenza 5CL-C18	1.016	0.004	0.093	−0.046	0.148	0.307	B	L1	–	*28*
Cadenza CD-C18	1.057	0.031	0.083	−0.028	0.113	0.042	B	L1	–	*28*
CAPCELL C18 ACR	1.025	0.045	0.073	−0.015	0.037	0.111	B	L1	–	*42*
CAPCELL C18 AG120	1.03	0.06	0.122	−0.065	0.543	0.628	A	L1	–	*42*
CAPCELL C18 AQ	0.867	−0.046	−0.068	0.014	−0.093	0.402	EP	L1	–	*42*
CAPCELL C18 MG	1.005	0.01	0.042	−0.007	0.079	0.007	B	L1	–	*42*
CAPCELL C18 SG120	0.987	0.031	0.093	−0.023	0.121	0.197	B	L1	–	*42*
CAPCELL C18 UG120	1.007	0.036	0.037	−0.012	0.016	0.001	B	L1	–	*42*
Capcell Pak C18 IF	0.957	0.025	−0.201	−0.001	−0.206	−0.01	B	L1	–	*42*
Capcell Pak C18 MG III	0.957	0.002	−0.116	−0.014	−0.115	0.046	B	L1	–	*42*
Capcell Pak C18 MGII	1.011	0.011	0.047	−0.006	0.007	−0.009	B	L1	–	*42*
CAPCELL PAK C8 DD	0.836	0.02	−0.154	0.015	−0.111	−0.075	B	L7	–	*42*
CAPCELL PAK C8 UG120	0.854	0.037	−0.097	−0.013	−0.046	−0.01	B	L7	–	*42*
Chromegabond WR C18	0.979	0.026	−0.159	−0.003	0.32	0.283	B	L1	0.732	*16*
Chromegabond WR C8	0.855	0.025	−0.279	0.024	0.2	0.144	B	L7	0.554	*16*
Chromolith HighResolut. RP-18e	0.999	0.019	0.017	−0.001	0.206	0.285	B	L1	–	*31*
Chromolith Performance RP-8e	0.833	−0.014	−0.01	0.014	0.071	0.572	B	L1	–	*31*
Chromolith RP18e	1.003	0.029	0.008	−0.014	0.103	0.187	B	L1	0.493	*31*
Clipeus C18	1.002	0.003	−0.043	−0.01	0.079	0.341	B	L1	–	*26*
Clipeus C8	0.822	−0.014	−0.18	0.023	0.095	0.241	B	L7	–	*26*
Clipeus Cyano	0.423	−0.065	−0.424	0.01	0.055	0.786	CN	L10	–	*26*
Clipeus Phenyl	0.586	−0.113	−0.303	0.031	0.073	0.488	Phenyl	L11	–	*26*
Cogent Bidentate C18	0.95	0.059	−0.13	0.004	0.785	2.266	A	L1	–	*32*
Cogent Bidentate C8	0.681	−0.113	0.369	0	0.195	1.351	B	L7	–	*32*
Cogent HPS C18	1.021	−0.011	−0.071	−0.014	0.106	0.089	B	L1	–	*32*

(Continued)

APPENDIX 2d List of Available Columns With PQRI Parameters (Mobile Phase 50/50 Acetonitrile/Aqueous Buffer 60 mM Phosphate v/v) (http://apps.usp.org/app/USPNF/columnsDB.html)—cont'd

Column	H^{*a}	S^{*b}	A^{*c}	B^{*d}	$C(2.8)^{*e}$	$C(7.0)^{*f}$	Silica*[g]	UPS	$\log k'_{EB}$	Manuf.[h]
Cogent hQ C18	0.908	−0.066	0.377	0.005	0.19	2.18	B	L1	–	32
Cogent UDC Cholesterol	0.625	0.227	0.528	0.069	0.745	1.212	Other	–	–	32
Cortecs C18	1.075	0.043	−0.108	−0.037	0.063	0.035	B	L1	–	48
Cortecs C18+	1.037	−0.005	0.175	0.026	−0.033	0.537	B	L1	–	48
Cosmicsil Abra C18	0.81	0.031	−0.482	0.214	−0.178	0.139	C18	L1	–	18
Cosmicsil Adore 100 CN	0.339	−0.067	−0.488	0.008	0.099	0.901	B	L10	–	18
Cosmicsil Adze C18	0.997	−0.019	−0.138	−0.013	−0.082	0.069	B	L1	–	18
Cosmicsil Agate 100C18	1.024	−0.026	−0.07	0.009	0.048	0.263	B	L1	–	18
Cosmicsil Agate RP C18	1.021	0.017	−0.007	−0.007	0.185	0.23	C18	L1	–	18
Cosmicsil Agate RP C8	0.846	−0.008	−0.178	0.022	0.034	0.262	C8	L7	–	18
Cosmicsil APT C18	1.023	0.004	−0.064	−0.007	0.135	0.196	C18	L1	–	18
Cosmicsil AQ C18 120	0.958	−0.039	−0.072	0.013	0.133	0.295	B	L1	–	18
Cosmicsil Aster C18 XD	1.076	0.038	0.042	−0.024	−0.086	0.084	B	L1	–	18
Cosmicsil Aura ODS	0.948	−0.04	−0.185	0.009	−0.047	0.089	B	L1	–	18
Cosmicsil BDS C18	0.854	−0.035	0.195	−0.012	0.476	0.929	B	L1	–	18
Cosmicsil Glory C18	1.001	0.007	−0.17	0.003	0.172	0.222	C18	L1	–	18
COSMOSIL 5-C18-PAQ	0.822	−0.027	−0.342	0.053	−0.353	0.047	EP	–	–	33
Cosmosil 5PYE	0.671	−0.271	−0.283	0.092	0.521	1.318	Other	–	–	33
COSMOSIL C18-AR-II	1.017	0.011	0.126	−0.029	0.116	0.494	B	L1	0.907	33
COSMOSIL C18-MS-II	1.031	0.042	−0.132	−0.014	−0.118	−0.027	B	L1	0.908	33
Cosmosil piNap	0.665	−0.176	−0.258	0.034	0.188	0.789	Phenyl	–	–	33
Curosil-PFP	0.695	−0.079	−0.267	−0.004	0.119	0.379	F	L43	–	37
DeltaPak C18 100A	1.028	0.019	−0.018	−0.011	−0.051	0.024	B	L1	0.956	48
DeltaPak C18 300A	0.955	−0.013	−0.105	0.016	0.235	0.286	B	L1	0.481	48
Denali 120A C18	1.052	0.042	0.125	−0.014	0.143	0.222	B	L1	–	23
Develosil C30-UG-5	0.976	−0.036	−0.196	0.011	0.158	0.176	B	L62	0.982	35
Develosil ODS-HG-5	0.98	0.015	−0.172	−0.008	0.187	0.221	B	L1	0.911	35
Develosil ODS-MG-5	0.963	−0.036	−0.165	−0.003	−0.012	0.051	B	L1	1.051	35
Develosil ODS-UG-5	0.996	0.025	−0.146	−0.004	0.15	0.155	B	L1	0.926	35

APPENDIX 2d List of Available Columns With PQRI Parameters (Mobile Phase 50/50 Acetonitrile/Aqueous Buffer 60 mM Phosphate v/v) (http://apps.usp.org/app/USPNF/columnsDB.html)—cont'd

Column	H^{*a}	S^{*b}	A^{*c}	B^{*d}	$C(2.8)^{*e}$	$C(7.0)^{*f}$	Silica*[g]	UPS	$\log k'_{EB}$	Manuf.[h]
Discovery Amide C16	0.72	0.013	−0.625	0.218	−0.092	−0.025	EP	L60	–	*45*
Discovery BIO Wide pore C18	0.836	0.014	−0.254	0.028	0.121	0.119	B	L1	0.528	*45*
Discovery BIO Wide pore C5	0.654	−0.019	−0.305	0.029	0.091	0.219	B	–	0.059	*45*
Discovery BIO Wide pore C8	0.839	0.018	−0.224	0.034	0.206	0.194	B	L7	0.345	*45*
Discovery C18	0.984	0.027	−0.128	0.004	0.176	0.153	B	L1	0.683	*45*
Discovery C8	0.832	0.011	−0.238	0.029	0.119	0.143	B	L7	0.522	*45*
Discovery CN	0.397	−0.11	−0.615	−0.002	−0.035	0.513	CN	L10	−0.198	*45*
Discovery HS F5	0.673	0.084	−0.284	0.008	0.912	1.429	F	L43	0.603	*45*
Discovery HS PEG	0.318	0.027	−0.713	0.128	−0.531	0.387	EP	–	–	*45*
Durashell C18	1.0424	0.0405	−0.066	−0.038	−0.191	−0.174	B	L1	–	*8*
EC Nucleosil 100-5 Protect 1	0.544	0.048	−0.411	0.309	−3.213	−0.573	EP	–	–	*29*
Econosil C18	0.966	−0.066	0.376	−0.032	1.026	1.339	A	L1	–	*20*
Econosphere C18	0.818	−0.128	0.036	−0.017	1.046	1.522	A	L1	–	*20*
Endeavorsil C18	1.021	0.043	−0.09	−0.02	−0.078	−0.009	B	L1	–	*13*
Epic C18	0.95	−0.027	−0.203	−0.007	−0.131	−0.041	B	L1	–	*16*
Epic C4	0.779	0.019	−0.315	−0.004	−0.2	0.061	B	L26	–	*16*
Epic C8	0.893	0.022	−0.194	−0.001	−0.102	0.038	B	L7	–	*16*
Epic Phenyl Hexyl	0.67	−0.088	−0.379	0.004	−0.148	0.305	Phenyl	L11	–	*16*
EU Reference Column	1.004	0.001	0.264	0.006	0.178	0.449	B	L1	–	*7*
Excel C18-Amide	0.791	0.023	−0.489	0.199	−0.06	0.176	EP	–		*1*
Excel CN-ES	0.811	−0.005	−0.429	−0.01	−0.052	0.153	CN	L10	–	*1*
Excel SuperC18	0.997	−0.003	−0.201	−0.013	0.03	0.009	C18	L1	–	*1*
Exsil C8	0.756	−0.076	−0.044	−0.014	0.472	0.974	A	L7	–	*34*
Exsil ODS	0.992	−0.036	0.292	−0.04	0.836	1.229	A	L1	–	*34*
Flare C18	0.806	−0.073	−0.47	0.841	−1.48	2.41	Other	–	–	*12*
Flare C18+	1.137	−0.308	0.73	0.966	−0.507	1.178	Other	–	–	*12*
Fluophase PFP	0.675	−0.129	−0.311	0.065	0.817	1.375	F	L43	0.653	*46*
Fluophase RP	0.698	0.028	0.103	0.039	1.034	1.417	F	L43	0.532	*46*
Fortis C18	0.96	−0.023	−0.18	−0.009	−0.167	0.111	B	L1	–	*17*

(*Continued*)

APPENDIX 2d List of Available Columns With PQRI Parameters (Mobile Phase 50/50 Acetonitrile/Aqueous Buffer 60 mM Phosphate v/v) (http://apps.usp.org/app/USPNF/columnsDB.html)—cont'd

Column	H^{*a}	S^{*b}	A^{*c}	B^{*d}	$C(2.8)^{*e}$	$C(7.0)^{*f}$	Silica*[g]	UPS	$\log k'_{EB}$	Manuf.[h]
Fortis C8	0.876	0.02	−0.223	0.001	−0.148	−0.041	B	L7	–	*17*
Fortis Cyano	0.383	−0.051	−0.399	0.016	0.011	0.872	CN	L10	–	*17*
Fortis H_2O	1.038	0.032	0.141	−0.025	0.006	0.101	EP	–	–	*17*
Fortis phenyl	0.615	−0.157	−0.402	0.037	0.084	0.772	Phenyl	L11	–	*17*
Fortis UniverSil C18	0.925	−0.027	−0.124	0.048	0.233	0.254	B	L1	–	*17*
Fortis UniverSil HS C18	0.992	−0.012	−0.117	−0.012	−0.065	0.082	B	L1	–	*17*
Gemini C18 110A	0.967	−0.008	0.027	0.013	−0.091	0.195	B	L1	–	*37*
Gemini C6-Phenyl 110A	0.803	0.032	−0.348	−0.005	−0.406	−0.266	Phenyl	L11	–	*37*
Gemini-NX C18	0.969	0.01	−0.204	−0.018	−0.184	−0.135	B	L1	–	*37*
Genesis AQ 120A (C18)	0.96	−0.036	−0.157	0.007	0.06	0.233	B	L1	0.981	*22*
Genesis C18 120A	1.005	0.004	−0.069	−0.007	0.139	0.124	B	L1	0.993	*22*
Genesis C18 300A	0.974	0.005	−0.086	0.013	0.266	0.27	B	L1	0.543	*22*
Genesis C4 300A	0.615	−0.057	−0.397	0.036	0.143	0.249	B	L26	0.059	*22*
Genesis C4 EC 120A	0.646	−0.058	−0.331	0.027	0.063	0.4	B	L26	0.526	*22*
Genesis C8 120A	0.829	−0.016	−0.082	0.018	0.055	0.3	B	L7	0.795	*22*
Genesis CN 120A	0.424	−0.114	−0.681	−0.013	−0.001	0.573	CN	L10	0.134	*22*
Genesis CN 300A	0.397	−0.108	−0.645	−0.009	0.025	0.397	CN	L10	−0.340	*22*
Genesis EC C8 120A	0.863	0.005	−0.174	0.023	0.064	0.141	B	L7	0.837	*22*
Genesis Phenyl	0.609	−0.14	−0.368	0.031	0.133	0.588	Phenyl	L11	0.459	*22*
GraceSmart RP 18	0.832	−0.035	0.07	0.002	0.071	0.74	B	L1	–	*20*
GraceSmart RP 18 5u	0.832	−0.035	0.07	0.002	0.071	0.74	B	L1	–	*20*
GROM Sapphire 110 C18	1.055	−0.002	0.085	0	−0.03	0.115	B	L1	–	*21*
GROM Sapphire 110 C8	0.835	−0.032	−0.103	0.031	−0.093	0.255	B	L7	–	*21*
GROM-SIL 120 Octyl-6 MB	0.872	0.001	−0.007	0.029	−0.017	0.135	B	L7	–	*21*
GROM-SIL 120 ODS-3 CP	1.029	0.019	0.093	−0.005	0.099	0.123	B	L1	–	*21*
Grom-Sil 120 ODS-4 HE	0.97	−0.037	−0.089	0.01	−0.037	0.263	B	L1	–	*21*
GROM-SIL 120 ODS-5 ST	1.035	−0.001	0.134	−0.005	0.135	0.121	B	L1	–	*21*
Haisil 300C18	0.946	−0.003	−0.035	−0.001	0.428	0.683	A	L1	–	*26*
Haisil HL C18	1.045	−0.039	−0.078	−0.029	0.041	−0.057	B	L1	–	*26*

APPENDIX 2d List of Available Columns With PQRI Parameters (Mobile Phase 50/50 Acetonitrile/Aqueous Buffer 60 mM Phosphate v/v) (http://apps.usp.org/app/USPNF/columnsDB.html)—cont'd

Column	H^{*a}	S^{*b}	A^{*c}	B^{*d}	$C(2.8)^{*e}$	$C(7.0)^{*f}$	Silica*[g]	UPS	$\log k'_{EB}$	Manuf.[h]
Halo 5C18	1.148	0.055	0.025	−0.055	0.073	0.1	B	L1	–	2
Halo 5 C8	0.92	0.017	−0.118	−0.002	−0.002	0.205	B	L7	–	2
Halo 5 ES-CN	0.558	−0.102	−0.492	0.012	0.081	0.967	CN	L10	–	2
Halo 5 PFP	0.711	−0.122	−0.07	−0.052	1.159	1.847	F	L43	–	2
Halo 5 Phenyl-Hexyl	0.799	−0.091	−0.295	−0.004	0.115	0.429	Phenyl	L11	–	2
Halo C18	1.107	0.048	0.006	−0.05	0.056	0.04	B	L1	–	2
Halo C8	0.913	0.028	−0.132	−0.008	−0.011	0.188	B	L7	–	2
Halo ES-CN	0.566	−0.11	−0.344	0.021	0.126	1.152	CN	L10	–	2
HALO Peptide ES-C18	0.89	−0.059	0.25	0.024	0.177	0.819	B	L1	–	2
HALO PFP	0.702	−0.117	−0.073	−0.062	1.168	0.972	F	L43	–	2
Halo Phenyl-Hexyl	0.789	−0.094	−0.233	−0.006	0.101	0.456	Phenyl	L11	–	2
Halo RP-Amide	0.859	0.08	−0.384	0.19	−0.417	0.312	EP	L60	–	2
Heavy C18	1.04	0.036	−0.137	−0.035	−0.083	−0.128	B	L1	–	26
Hichrom 300 5 RPB	0.944	0.028	0.044	0.015	0.226	0.216	B	L1	–	25
Hichrom RPB	0.964	0.027	0.106	0.003	0.153	0.143	B	L1	–	25
Hitachi LaChrom C18-PM	1.127	0.069	−0.019	−0.068	−0.267	−0.144	B	L1	–	27
HSS C18	1.022	0.039	−0.136	−0.02	0.059	−0.009	B	L1	–	48
HSS C18 SB	0.73	−0.12	0.4	0.02	0.4	1.41	B	L1	–	48
HSS T3	0.949	−0.021	−0.173	−0.002	0.031	0.18	B	L1	–	48
HxSil C18	0.848	−0.077	0.303	0.017	0.23	1.054	B	L1	–	24
HxSil C8	0.684	−0.075	0.089	0.03	0.066	0.856	B	L7	–	24
HyperClone BDS C18 130A	0.988	0.032	0.019	−0.016	0.194	0.425	B	L1	–	37
HyperClone BDS C8 130A	0.847	0.008	−0.146	0.016	0.231	0.261	B	L7	–	37
HyperClone CN (CPS)	0.408	−0.086	−0.55	−0.003	0.92	1.045	CN	L10	–	37
HyperClone MOS C8 120A	0.847	−0.057	0.043	0.13	1.142	1.12	EP	L60	–	37
HyperClone ODS C18 120A	1.03	−0.02	0.06	0.09	1.03	0.96	A	L1	–	37
HyperClone PAH	0.98	−0.008	0.187	−0.024	1.169	1.116	Other	–	–	37
Hypersil 100 C18	1.048	0.022	0.118	0.031	0.405	0.348	A	L1	–	46
Hypersil BDS C18	0.993	0.016	−0.095	−0.009	0.337	0.281	A	L1	–	46

(Continued)

APPENDIX 2d List of Available Columns With PQRI Parameters (Mobile Phase 50/50 Acetonitrile/Aqueous Buffer 60 mM Phosphate v/v) (http://apps.usp.org/app/USPNF/columnsDB.html)—cont'd

Column	H^{*a}	S^{*b}	A^{*c}	B^{*d}	$C(2.8)^{*e}$	$C(7.0)^{*f}$	Silica*[g]	UPS	$\log k'_{EB}$	Manuf.[h]
Hypersil Beta Basic-18	0.993	0.033	−0.099	0.001	0.163	0.126	B	L1	0.808	***46***
Hypersil Beta Basic-8	0.834	0.016	−0.248	0.029	0.11	0.115	B	L7	0.619	***46***
Hypersil Betamax Neutral (C18)	1.098	0.036	0.067	−0.031	−0.038	0.012	B	L1	1.231	***46***
Hypersil Bio Basic-18	0.974	0.025	−0.1	0.007	0.253	0.217	B	L1	0.512	***46***
Hypersil Bio Basic-8	0.821	0.012	−0.233	0.029	0.231	0.21	B	L7	0.253	***46***
Hypersil Elite	0.958	0.031	0.151	−0.01	0.314	0.739	A	L1	–	***34***
Hypersil GOLD	0.881	0.002	−0.017	0.036	0.162	0.479	B	L1	–	***46***
Hypersil GOLD aQ	0.915	−0.01	−0.065	−0.019	0.371	0.638	A	L1	–	***46***
Hypersil GOLD C4	0.705	−0.003	−0.285	0.026	0.11	0.235	B	L26	–	***46***
Hypersil GOLD C8	0.82	0.01	−0.15	0.03	0.09	0.21	B	L7	–	***46***
Hypersil GOLD CN	0.397	−0.035	−0.886	−0.019	−0.069	0.66	CN	L10	–	***46***
Hypersil GOLD PFP	0.624	−0.081	−0.116	−0.022	0.379	0.991	F	L43	–	***46***
Hypersil GOLD phenyl	0.65	−0.091	−0.362	0.03	0.161	0.416	Phenyl	L11	–	***46***
Hypersil ODS	0.974	−0.026	−0.122	0.02	0.913	0.974	A	L1	–	***46***
Hypersil ODS-2	0.985	0.016	0.139	−0.011	0.254	0.37	B	L1	–	***46***
Hypersil PAH	0.949	−0.057	0.234	−0.017	1.439	1.724	A	L1	–	***46***
Hypersil Prism C18 RP	0.645	0.089	−0.459	0.301	−2.817	−0.716	EP	–	–	***46***
Hypersil Prism C18 RPN	0.678	−0.001	−0.068	0.23	−0.544	0.625	EP	–	–	***46***
Hypurity Advance	0.412	−0.056	−0.095	0.249	−1.332	0.785	EP	L60	–	***46***
Hypurity C18	0.98	0.025	−0.091	0.003	0.192	0.167	B	L1	0.744	***46***
HyPurity C4	0.713	0	−0.291	0.028	0.121	0.252	B	L26	–	***46***
Hypurity C8	0.833	0.011	−0.201	0.035	0.157	0.161	B	L7	0.546	***46***
Hypurity Cyano	0.451	−0.049	−0.492	0.021	−0.016	0.839	CN	L10	–	***46***
Inertsil C8-3	0.83	−0.004	−0.268	−0.017	−0.334	−0.362	B	L7	0.849	***19***
Inertsil C8-4	0.678	−0.039	−0.386	−0.014	−0.094	−0.154	B	L7	–	***19***
Inertsil CN-3	0.369	0.049	−0.808	0.083	−2.607	−1.297	CN	L10	0.050	***19***
Inertsil ODS-2	1.007	0.045	−0.079	−0.014	−0.139	0.446	B	L1	–	***19***
Inertsil ODS-3	0.99	0.022	−0.146	−0.023	−0.474	−0.334	B	L1	1.037	***19***
Inertsil ODS-4	0.911	−0.026	−0.226	−0.03	−0.029	−0.143	B	L1	–	***19***

APPENDIX 2d List of Available Columns With PQRI Parameters (Mobile Phase 50/50 Acetonitrile/Aqueous Buffer 60 mM Phosphate v/v) (http://apps.usp.org/app/USPNF/columnsDB.html)—cont'd

Column	H^{*a}	S^{*b}	A^{*c}	B^{*d}	C(2.8)*[e]	C(7.0)*[f]	Silica*[g]	UPS	$\log k'_{EB}$	Manuf.[h]
Inertsil ODS-EP	0.8	0.06	−1.52	0.05	−0.62	−0.07	EP	L60	–	*19*
Inertsil ODS-P	0.978	−0.028	0.611	−0.039	0.234	1.479	B	L1	1.048	*19*
Inertsil ODS-SP	0.858	−0.027	−0.221	−0.023	−0.048	−0.073	B	L1	–	*19*
Inertsil Ph-3	0.526	−0.179	−0.133	0.04	0.121	0.735	Phenyl	L11	0.409	*19*
Inertsil WP300 C18	0.938	−0.015	−0.117	0.001	0.202	0.163	B	L1	–	*19*
Inertsil WP300 C8	0.793	−0.015	−0.212	0.013	0.122	0.069	B	L7	–	*19*
InertSustain C18	1.01	0.0546	−0.152	−0.087	−0.081	−0.195	B	L1	–	*19*
InertSustain C8	0.868	−0.005	−0.27	−0.021	−0.035	−0.122	B	L7	–	*19*
InertSustain Phenyl	0.569	−0.135	−0.375	0.035	−0.044	0.32	Phenyl	L11	–	*19*
Innoval C18	1.046	0.049	−0.024	−0.034	−0.066	0.027	B	L1	–	*8*
Inspire C18	1.063	−0.052	−0.078	−0.04	−0.081	−0.106	B	L1	–	*13*
Inspire C8	0.889	−0.025	−0.212	−0.004	−0.193	−0.014	B	L7	–	*13*
J'Sphere H80 (C18)	1.132	0.059	−0.023	−0.068	−0.242	−0.161	B	L1	1.124	*51*
J'Sphere L80 (C18)	0.762	−0.036	−0.216	−0.001	−0.4	0.345	B	L1	0.764	*51*
J'Sphere M80 (C18)	0.926	−0.026	−0.123	−0.004	−0.294	0.139	B	L1	0.957	*51*
Jupiter 300 C18	0.945	0.031	−0.225	0.008	0.234	0.218	B	L1	0.467	*37*
Jupiter 300 C4	0.698	0.008	−0.426	0.019	0.152	0.141	B	L26	0.126	*37*
Jupiter 300 C5	0.729	0.021	−0.382	0.016	0.129	0.33	B	–	0.183	*37*
Kinetex Biphenyl 100A	0.697	−0.173	−0.583	0.034	0.122	0.817	Phenyl	L11	–	*37*
Kinetex C18 100A	0.963	0.009	−0.137	−0.011	0.007	0.125	B	L1	–	*37*
Kinetex C8	0.864	0.013	−0.208	−0.009	−0.089	0.002	B	L7	–	*37*
Kinetex EVO C18	1.01	−0.006	−0.174	−0.024	−0.108	−0.01	C18	L1	–	*37*
Kinetex F5	0.725	−0.064	−0.32	−0.046	0.11	2.66	Fluoro	L43	–	*37*
Kinetex PFP 100A	0.688	0.089	−0.273	−0.038	0.943	1.538	F	L43	–	*37*
Kinetex Phenyl-Hexyl	0.795	−0.091	−0.258	0.016	0.062	0.236	Phenyl	L11	–	*37*
Kinetex XB-C18	0.975	−0.013	−0.083	0.023	−0.046	0.305	B	L1	–	*37*
Kromasil 100 5 C18	1.051	0.035	−0.07	−0.022	0.039	−0.057	B	L1	1.098	*5*
Kromasil 100 5 C4	0.733	0.003	−0.335	0.015	0.009	−0.004	B	L26	0.700	*5*
Kromasil 100 5 C8	0.864	0.013	−0.213	0.019	0.054	−0.001	B	L7	0.881	*5*

(Continued)

APPENDIX 2d List of Available Columns With PQRI Parameters (Mobile Phase 50/50 Acetonitrile/Aqueous Buffer 60 mM Phosphate v/v) (http://apps.usp.org/app/USPNF/columnsDB.html)—cont'd

Column	H^{*a}	S^{*b}	A^{*c}	B^{*d}	$C(2.8)^{*e}$	$C(7.0)^{*f}$	Silica*[g]	UPS	$\log k'_{EB}$	Manuf.[h]
Kromasil KR60 5 CN	0.44	−0.135	−0.578	−0.014	0.216	1.036	CN	L10	0.306	5
LaChrom C18	0.993	0.013	−0.151	−0.006	−0.278	−0.12	B	L1	–	27
LaChrom C18-AQ	0.907	−0.023	−0.137	0.011	−0.327	0.192	B	L1	–	27
LaChrom C18-NE	0.962	−0.024	0.36	−0.008	0.138	0.772	B	L1	–	27
LaChrom C8	0.856	0.022	−0.199	0.013	−0.414	−0.022	B	L7	–	27
LaChrom CN	0.425	−0.064	−0.372	0.006	−0.02	0.795	CN	L10	–	27
LaChrom Ph	0.594	−0.105	−0.324	0.024	−0.003	0.45	Phenyl	L11	–	27
Leapsil C18	1.052	0.044	−0.072	−0.035	−0.029	−0.055	B	L1	–	13
LiChrosorb RP-18	0.909	−0.07	0.151	−0.08	0.714	0.976	A	L1	–	31
LiChrospher 100 RP-18	1.006	−0.021	0.183	−0.036	0.646	0.896	A	L1	–	31
LiChrospher 60 RP-Select B	0.747	−0.06	−0.042	0.006	0.108	0.864	B	L1	–	31
Luna C18	1.018	0.025	0.072	0.008	−0.361	−0.036	B	L1	–	37
Luna C18(2)	1.002	0.024	−0.124	−0.007	−0.269	−0.174	B	L1	0.983	37
Luna C5	0.8	0.035	−0.252	0.003	−0.278	0.114	B	–	0.770	37
Luna C8	0.875	0.037	−0.015	0.024	−0.4	0.133	B	L7	–	37
Luna C8(2)	0.889	0.041	−0.222	−0.001	−0.3	−0.17	B	L7	0.859	37
Luna CN	0.452	−0.112	−0.323	−0.024	0.439	1.321	CN	L10	0.104	37
Luna PFP(2)	0.753	−0.076	−0.382	−0.051	0.088	0.548	F	L43	–	37
Luna Phenyl-Hexyl	0.782	−0.118	−0.277	−0.004	0.004	0.387	Phenyl	L11	0.718	37
Matrix C18	0.934	−0.046	0.068	−0.003	0.071	0.765	B	L1	–	36
MicroBondapak C18	0.79	−0.07	−0.03	0.01	0.28	0.85	A	L1	–	48
MicroBondapak Phenyl	0.585	−0.152	−0.247	0.021	0.359	0.976	Phenyl	L11	–	48
Microsorb 100-5 C8	0.875	0.061	0.091	0.192	0.613	0.823	A	L7	–	4
Microsorb 100-5 Phenyl	0.711	−0.14	−0.195	0.163	0.604	0.787	Phenyl	L11	–	4
Microsorb 300-5 C4	0.666	−0.028	−0.315	0.036	0.207	0.419	B	L26	–	4
Microsorb-MV 100 CN	0.357	−0.241	−0.852	−0.029	0.148	0.785	CN	L10	–	4
Monitor C18	0.981	−0.004	−0.131	0.001	0.021	0.063	B	L1	–	36
Nova-Pak C18	1.049	0.004	0.098	−0.027	0.546	0.563	A	L1	–	48
Nova-Pak C8	0.899	−0.028	−0.094	0.006	0.611	0.621	A	L7	–	48

APPENDIX 2d List of Available Columns With PQRI Parameters (Mobile Phase 50/50 Acetonitrile/Aqueous Buffer 60 mM Phosphate v/v) (http://apps.usp.org/app/USPNF/columnsDB.html)—cont'd

Column	H^{*a}	S^{*b}	A^{*c}	B^{*d}	$C(2.8)^{*e}$	$C(7.0)^{*f}$	Silica*[g]	UPS	$\log k'_{EB}$	Manuf.[h]
Nova-Pak CN HP 60A	0.362	−0.165	0.1	0	0.691	1.175	CN	L10	−0.413	*48*
Nova-Pak Phenyl	0.704	−0.159	−0.3	0.015	0.767	0.812	Phenyl	L11	–	*48*
Nucelosil 100-5-C8 HD	0.865	−0.008	−0.174	0.029	0.045	0.188	A	L7	–	*29*
Nucleodur 100-5 C18	0.977	−0.024	0.009	−0.03	0.143	0.59	B	L1	–	*29*
Nucleodur 100-5 C8	0.779	−0.042	−0.152	−0.007	0.048	0.825	B	L1	–	*29*
Nucleodur C18 Gravity	1.056	0.041	−0.097	−0.025	−0.08	0.316	B	L1	–	*29*
Nucleodur C8 Gravity 5 micron	0.868	0.032	−0.24	0	−0.158	0.631	B	L1	–	*29*
Nucleodur HTEC C18	1.045	0.038	−0.022	−0.015	−0.482	0.136	B	L1	–	*29*
Nucleodur Isis	1.023	0.055	−0.078	−0.029	−0.019	0.153	B	L1	–	*29*
Nucleodur PAH C18	1.006	0.008	0.574	−0.022	0.343	1.216	B	L1	–	*29*
Nucleodur PFP	0.712	−0.059	−0.265	−0.036	0.023	0.81	F	L43	–	*29*
Nucleodur POLARTEC C18	0.858	0.168	−0.259	0.351	−3.398	−0.787	B	L1	–	*29*
Nucleodur Pyramid	0.958	0.003	−0.13	−0.016	−0.289	0.21	B	L1	–	*29*
Nucleodur Sphinx RP	0.805	−0.071	−0.274	0	0.022	0.722	Phenyl	L1	–	*29*
Nucleoshell RP 18	1.109	0.036	0.077	−0.04	0.082	0.266	B	L1	–	*29*
Nucleosil 100 5 C18 HD	0.961	−0.021	−0.126	0.009	0.089	0.15	A	L1	–	*29*
Nucleosil 100 5 C18 Nautilus	0.702	0.003	−0.483	0.268	−0.441	0.486	EP	L60	–	*29*
Nucleosil 300 5 C18	0.86	−0.081	−0.008	0.014	0.453	0.984	A	L1	–	*29*
Nucleosil C18	0.906	−0.052	0.012	−0.03	0.321	0.73	A	L1	–	*29*
Nucleosil C8	0.575	−0.134	0.038	0.017	0.282	1.122	A	L7	–	*29*
OmniSpher 5 C18	1.055	0.051	−0.033	−0.029	0.122	0.058	B	L1	1.035	*4*
Onyx Monolithic C18	1.012	0.021	0.227	−0.018	0.12	0.43	B	L1	–	*37*
Onyx Monolithic C8	0.824	0.003	−0.006	0.004	−0.02	0.441	B	L7	–	*37*
Orosil C18	0.981	−0.032	−0.137	0.002	−0.048	0.155	B	L1	–	*36*
Partisil C8	0.749	−0.071	−0.099	0.074	0.035	0.546	B	L7	–	*37*
Partisil ODS(3)	0.81	−0.079	−0.007	0.002	0.317	0.902	A	L1	–	*37*
Peerless Basic C-18	0.988	−0.015	−0.156	−0.015	−0.164	−0.096	B	L1	–	*10*
Peerless C-18	0.999	−0.025	0.033	−0.013	−0.003	0.454	B	L1	–	*10*
Phalanx C18	0.953	−0.03	−0.209	−0.006	−0.075	−0.016	B	L1	–	*26*

(Continued)

APPENDIX 2d List of Available Columns With PQRI Parameters (Mobile Phase 50/50 Acetonitrile/Aqueous Buffer 60 mM Phosphate v/v) (http://apps.usp.org/app/USPNF/columnsDB.html)—cont'd

Column	H^{*a}	S^{*b}	A^{*c}	B^{*d}	C(2.8)*[e]	C(7.0)*[f]	Silica*[g]	UPS	$\log k'_{EB}$	Manuf.[h]
Pinnacle DB Aqueous C18	0.698	−0.148	0.228	0.036	0.379	1.138	A	L1	–	*38*
Pinnacle DB Biphenyl	0.641	−0.175	−0.343	0.051	0.193	0.516	Phenyl	L11	–	*38*
Pinnacle DB C18	1.014	0.025	−0.033	−0.005	0.364	0.28	A	L1	–	*38*
Pinnacle DB C8	0.835	0.002	−0.198	0.037	0.269	0.231	A	L7	–	*38*
Pinnacle DB Cyano	0.376	−0.076	−0.38	0.007	0.221	0.722	CN	L10	–	*38*
Pinnacle DB PFP Propyl	0.648	−0.069	−0.253	0.054	0.598	1.106	F	L43	–	*38*
Pinnacle DB Phenyl	0.577	−0.124	−0.465	0.046	0.169	0.289	Phenyl	L11	–	*38*
Pinnacle II Biphenyl	0.642	−0.195	−0.247	0.054	0.465	0.732	Phenyl	L11	–	*38*
Pinnacle II C18	1.038	0.019	−0.03	0.003	0.445	0.349	A	L1	–	*38*
Pinnacle II C8	0.852	−0.006	−0.117	0.035	0.394	0.435	A	L7	–	*38*
Pinnacle II Cyano	0.37	−0.083	−0.213	−0.01	0.2	0.788	CN	L10	–	*38*
Pinnacle II PAH	0.977	−0.023	0.353	−0.02	0.746	1.188	A	L1	–	*38*
Pinnacle II Phenyl	0.594	−0.134	−0.405	0.061	0.37	0.492	Phenyl	L11	–	*38*
Platinum C18	0.786	−0.076	−0.098	0.005	0.4	0.694	A	L1	–	*20*
Platinum C8	0.584	−0.056	−0.225	0.005	0.251	0.391	B	L7	–	*20*
Platinum EPS C18	0.619	−0.17	0.306	0.026	0.688	1.701	A	L1	0.417	*20*
Platinum EPS C18 300	0.45	−0.058	0.379	0.016	0.247	1.291	EP	–	–	*20*
Platinum EPS C8	0.42	−0.152	0.151	0.026	0.509	1.369	A	L7	0.022	*20*
Platinum EPS C8 300	0.584	−0.113	−0.136	0.089	0.481	0.961	EP	–	–	*20*
Polar C18	1.043	−0.086	0.645	−0.035	1.071	1.557	B	L1	–	*36*
Polaris Amide-C18	0.84	0.116	−0.336	0.345	−1.659	−0.556	EP	L60	–	*4*
Polaris C18-A	0.928	0.007	−0.227	0.061	0.149	0.16	EP	L1	–	*4*
Polaris C18-Ether	0.943	−0.013	−0.122	0.027	0.164	0.553	EP	–	–	*4*
Polaris C8-A	0.601	−0.007	−0.609	0.104	−0.074	0.208	EP	L7	–	*4*
Polaris C8-Ether	0.705	−0.023	−0.312	0.04	0.095	0.269	EP	–	–	*4*
Poroshell 120 Bonus-RP	0.686	−0.03	−0.573	0.18	−0.67	−0.017	EP	L60	–	*3*
Poroshell 120 EC-C18	1.023	0.008	−0.13	−0.004	0.161	0.123	B	L1	–	*3*
Poroshell 120 EC-C8	0.877	0.011	−0.232	0.023	0.127	0.09	B	L7	–	*3*
Poroshell 120 EC-CN	0.421	−0.057	−0.476	0.002	0.045	0.87	CN	L10	–	*3*

APPENDIX 2d List of Available Columns With PQRI Parameters (Mobile Phase 50/50 Acetonitrile/Aqueous Buffer 60 mM Phosphate v/v) (http://apps.usp.org/app/USPNF/columnsDB.html)—cont'd

Column	H^{*a}	S^{*b}	A^{*c}	B^{*d}	$C(2.8)^{*e}$	$C(7.0)^{*f}$	Silica*[g]	UPS	$\log k'_{EB}$	Manuf.[h]
Poroshell 120 Phenyl-Hexyl	0.752	−0.083	−0.394	0.018	0.136	0.14	Phenyl	L11	–	*3*
Poroshell 120 SB-AQ	0.581	−0.12	−0.133	0.051	−0.014	0.67	EP	L60	–	*3*
Poroshell 120 SB-C18	0.956	−0.041	0.168	0.025	0.21	0.763	B	L1	–	*3*
Poroshell 120 SB-C8	0.726	−0.087	0.068	0.044	0.087	0.807	B	L7	–	*3*
Precision C18	1.002	0.003	−0.042	−0.01	0.079	0.341	B	L1	0.976	*30*
Precision C18-PE	0.976	−0.018	−0.085	−0.001	0.005	0.168	EP	L1	–	*30*
Precision C8	0.821	−0.014	−0.18	0.022	0.095	0.241	B	L7	0.692	*30*
Precision C8-PE	0.814	−0.021	−0.159	0.017	0.051	0.279	EP	L7	–	*30*
Precision CN	0.431	−0.114	−0.485	0.019	−0.041	0.606	CN	L10	0.111	*30*
Precision Phenyl	0.595	−0.136	−0.296	0.027	0.099	0.508	Phenyl	L11	0.420	*30*
Prevail Amide	0.862	−0.063	0.251	0.033	0.058	1.209	EP	L60	–	*20*
Prevail C18	0.888	−0.07	0.315	0.022	0.107	1.206	B	L1	0.975	*20*
Prevail C8	0.617	−0.089	0.039	0.041	0.081	1.072	B	L7	0.530	*20*
Prevail Select C18	0.822	0.029	−0.368	0.141	−1.057	0.455	B	L1	–	*20*
Primesep A	0.57	−0.199	0.057	0.032	2.732	2.6	Other	–	–	*44*
Primesep B	0.497	−0.004	0.034	0.584	−1.869	−1.357	Other	–	–	*44*
Primesep C	0.513	−0.146	−0.265	0.124	1.038	1.547	Other	–	–	*44*
Primesil C18 3 micron	1.02	0.025	−0.06	−0.001	0.014	0.12	C18	L1	–	*50*
Primesil C18 5 micron	1.04	−0.002	−0.043	−0.013	0.015	0.15	C18	L1	–	*50*
Primesil C18(2)	1	−0.004	−0.057	−0.006	0.109	0.18	C18	L1	–	*50*
Primesil C8	0.857	−0.014	−0.2	0.018	−0.052	0.17	C8	L7	–	*50*
Primesil ODS-P	0.752	0.09	−0.67	0.042	−0.24	−0.015	EP	L60	–	*50*
Prodigy ODS(2)	0.995	0.03	−0.114	−0.001	−0.091	0.237	B	L1	–	*37*
Prodigy ODS(3)	1.023	0.025	−0.131	−0.012	−0.195	−0.134	B	L1	1.03	*37*
Prodigy Phenyl-3	0.529	−0.195	0.055	0.022	0.23	1.467	Phenyl	L11	0.358	*37*
Promosil C18	1.064	0.036	−0.068	−0.012	−0.002	−0.09	B	L1	–	*8*
ProntoSIL 120 C1	0.413	−0.079	−0.085	0.02	0.042	0.656	B	C13	–	*7*
ProntoSIL 120 C18 ace-EPS	0.772	0.042	−0.59	0.228	−0.304	0.041	EP	–	–	*7*
ProntoSIL 120 C18 AQplus	0.947	−0.017	0.214	0.041	−0.133	0.605	EP	–	–	*7*

(Continued)

APPENDIX 2d List of Available Columns With PQRI Parameters (Mobile Phase 50/50 Acetonitrile/Aqueous Buffer 60 mM Phosphate v/v) (http://apps.usp.org/app/USPNF/columnsDB.html)—cont'd

Column	H^{*a}	S^{*b}	A^{*c}	B^{*d}	$C(2.8)^{*e}$	$C(7.0)^{*f}$	Silica*[g]	UPS	$\log k'_{EB}$	Manuf.[h]
ProntoSIL 120C18 H	1.005	0.008	−0.106	−0.004	0.125	0.156	B	L1	0.873	7
ProntoSIL 120 C18 SH	1.031	0.018	−0.109	−0.024	0.113	0.402	B	L1	0.938	7
ProntoSIL 120 C18-AQ	0.973	−0.007	−0.082	0.004	0.137	0.224	B	L1	0.910	7
ProntoSIL 120 C8 ace-EPS	0.532	−0.007	−0.852	0.213	−0.282	0.094	EP	–	–	7
ProntoSIL 120 C8 SH	0.739	−0.062	−0.081	0.013	0.076	0.526	B	L7	0.687	7
ProntoSIL 120 CN EC	0.427	−0.053	−0.32	0.015	0.019	0.768	CN	L10	–	7
ProntoSIL 120 phenyl	0.568	−0.158	−0.201	0.022	0.176	0.712	Phenyl	L11	0.387	7
ProntoSIL 120-3-C30	0.919	−0.13	0.571	−0.003	0.507	1.788	A	L62	–	7
ProntoSIL 200C18 ace-EPS	0.765	0.021	−0.566	0.214	0.026	0.143	EP	–	–	7
ProntoSIL 200C18 AQ	0.973	−0.011	−0.057	0.006	0.125	0.288	B	L1	–	7
ProntoSIL 200C18 H	0.955	−0.001	−0.121	0.016	0.163	0.218	B	L1	0.679	7
ProntoSIL 200C30	0.909	−0.099	0.347	0.007	0.305	1.171	A	L62	–	7
ProntoSIL 200 C4	0.549	−0.063	−0.221	0.038	0.086	0.511	B	L26	–	7
ProntoSIL 200 C8 SH	0.761	−0.026	−0.195	0.024	0.125	0.238	B	L7	0.439	7
ProntoSIL 300C18 ace-EPS	0.762	0.025	−0.579	0.211	−0.054	0.136	EP	–	–	7
ProntoSIL 300 C30	0.893	−0.107	0.322	0.03	0.401	1.547	A	L62	–	7
ProntoSIL 300 C30 EC	0.925	−0.047	−0.018	0.012	0.303	0.458	B	L62	–	7
ProntoSIL 300 C4	0.471	−0.093	−0.074	0.055	0.115	0.786	B	L26	–	7
ProntoSIL 300 C8 SH	0.739	−0.041	−0.131	0.028	0.156	0.405	B	L7	0.260	7
ProntoSIL 300-5-C18 H	0.956	−0.012	−0.09	0.015	0.238	0.249	B	L1	0.511	7
ProntoSIL 60 C18 H	1.158	0.041	0.066	−0.078	0.102	0.263	B	L1	1.087	7
ProntoSIL 60 C4	0.686	−0.072	0.108	0.001	−0.056	1.201	B	L26	–	7
ProntoSIL 60 C8 SH	0.929	−0.015	0.161	−0.017	−0.313	1.005	B	L7	0.922	7
ProntoSIL 60 Phenyl	0.705	−0.194	−0.003	−0.01	0.411	1.51	Phenyl	L11	0.649	7
ProntoSIL CN	0.37	−0.114	−0.414	−0.028	0.168	0.668	CN	L10	−0.041	7
ProntoSIL HyperSORB 120 ODS	0.951	−0.065	0.039	−0.021	0.795	1.315	A	L1	–	7
ProntoSIL SpheriBOND 80 ODS1	0.7	−0.19	0.367	0.01	1.453	2.4	A	L1	–	7
ProntoSIL SpheriBOND 80 ODS2	1.01	−0.026	0.153	−0.037	0.731	1.008	A	L1	–	7
Prosphere 100 C18	0.883	−0.073	0.305	0.017	0.181	1.517	B	L1	–	*20*

APPENDIX 2d List of Available Columns With PQRI Parameters (Mobile Phase 50/50 Acetonitrile/Aqueous Buffer 60 mM Phosphate v/v) (http://apps.usp.org/app/USPNF/columnsDB.html)—cont'd

Column	H^{*a}	S^{*b}	A^{*c}	B^{*d}	$C(2.8)^{*e}$	$C(7.0)^{*f}$	Silica[*g]	UPS	$\log k'_{EB}$	Manuf.[h]
Prosphere 300 C4	0.689	−0.015	−0.059	0.027	0.312	0.684	B	L26	–	***20***
Prosphere C18 300A	0.903	−0.012	0.176	0.013	0.577	1.266	A	L1	–	***20***
Proto 300 C18	0.962	0.016	−0.132	0.005	0.224	0.147	B	L1	–	***26***
Purospher RP-18	0.841	0.235	0.155	0.3	−0.964	0.901	B	L1	–	***31***
Purospher STAR RP18e	1.003	0.013	−0.071	−0.037	0.018	0.044	B	L1	1.0023	***31***
Pursuit C18	1.001	0.004	−0.166	0.012	0.245	0.226	B	L1	–	***4***
Pursuit DP	0.574	−0.133	−0.538	0.047	0.153	0.261	Phenyl	–	–	***4***
Pursuit PFP	0.658	−0.089	−0.253	−0.021	0.05	0.531	F	L43	–	***4***
Pursuit UPS C18	1.064	0.063	−0.05	−0.008	−0.076	0.207	B	L1	–	***4***
Pursuit XRs C-18	1.046	0.039	−0.121	−0.034	−0.014	−0.102	B	L1	–	***4***
Pursuit XRs C-8	0.882	0.02	−0.226	0.001	−0.152	−0.067	B	L7	–	***4***
Pursuit XRs DP	0.63	−0.136	−0.433	0.033	0.045	0.393	Phenyl		–	***4***
Reliasil C18	0.879	−0.004	0.326	0.048	0.592	0.822	B	L1	–	***36***
Resolve C18	0.968	−0.127	0.335	−0.046	1.921	2.144	A	L1	–	***48***
Restek Ultra C18	1.055	0.030	−0.068	−0.021	0.09	−0.066	B	L1	1.101	***38***
Restek Ultra C8	0.876	0.030	−0.229	0.018	0.043	0.011	B	L7	0.883	***38***
Selectosil C18	0.911	−0.054	0.034	−0.009	0.296	0.743	B	L1	–	***37***
Sepax Bio-C18	0.919	−0.026	−0.133	0.018	0.219	0.228	B	L1	–	***39***
Sepax HP-C18(2)	0.959	−0.024	−0.187	−0.007	−0.134	0.055	B	L1	–	***39***
SepaxBio-C4	0.663	−0.014	−0.291	0.022	0.109	0.228	B	L26	–	***39***
SepaxBio-C8	0.774	−0.025	−0.272	0.025	0.164	0.219	B	L7	–	***39***
SepaxBR-C18	1.019	0.037	−0.078	−0.017	−0.019	0.06	B	L1	–	***39***
SepaxGP-C18	1.014	0.014	−0.112	−0.019	0.103	0.096	B	L1	–	***39***
SepaxGP-C4	0.699	−0.014	−0.261	0.009	−0.026	0.204	B	L26	–	***39***
SepaxGP-C8	0.847	0.009	−0.137	0.015	−0.077	0.265	B	L7	–	***39***
SepaxGP-Phenyl	0.571	−0.118	−0.274	0.029	0.055	0.564	Phenyl	L11	–	***39***
SepaxHP-C18	0.951	−0.026	−0.102	0.001	0.07	0.221	B	L1	–	***39***
SepaxHP-Cyano	0.426	−0.056	−0.471	0.005	−0.001	0.756	CN	L10	–	***39***
Shim- pack XR-ODS II	1.086	0.039	0.076	−0.026	−0.006	0.204	B	L1	–	***41***

(Continued)

APPENDIX 2d List of Available Columns With PQRI Parameters (Mobile Phase 50/50 Acetonitrile/Aqueous Buffer 60 mM Phosphate v/v) (http://apps.usp.org/app/USPNF/columnsDB.html)—cont'd

Column	H^{*a}	S^{*b}	A^{*c}	B^{*d}	C(2.8)*[e]	C(7.0)*[f]	Silica*[g]	UPS	$\log k'_{EB}$	Manuf.[h]
Shim-pack XR-C8	0.8618	0.0101	−0.16	0.017	−0.044	0.2172	B	L7	–	***41***
Shim-pack XR-ODS	1.009	0.032	−0.106	−0.018	−0.061	0.17	B	L1	–	***41***
Shim-pack XR-Phenyl	0.705	−0.0797	−0.264	0.01	−0.021	0.4865	Phenyl	L11	–	***41***
Shodex C18-4D	1.0073	0.027	0.0206	−0.01	−0.11	0.1067	B	L1	–	***43***
Sphereclone ODS(2)	0.975	−0.045	0.278	−0.051	0.866	1.326	A	L1	–	***37***
Spherisorb C8	0.763	−0.091	−0.032	0.053	0.737	1.142	A	L7	–	***48***
Spherisorb ODS-1	0.682	−0.186	0.323	0.018	0.843	1.297	A	L1	–	***48***
Spherisorb ODS-2	0.962	−0.076	0.07	0.034	0.908	1.263	A	L1	–	***48***
Spherisorb S5 ODSB	0.975	0.027	0.24	0.384	−0.642	1.68	B	L1	–	***48***
Spursil C18	0.961	0.009	−0.185	−0.049	0.429	0.765	A	L1	–	***13***
Spursil C18-EP	0.832	0.107	−0.509	0.226	−1.123	0.898	EP	L60	–	***13***
Sunfire C18	1.031	0.034	0.044	−0.014	−0.186	−0.099	B	L1	–	***48***
Sunfire C8	0.856	0.036	−0.122	0.006	−0.278	0.006	B	L7	–	***48***
Sunniest C18	1.021	0.01	−0.169	−0.023	0.089	0.803	B	L1	–	***9***
Sunniest C8	0.845	0.005	−0.304	−0.002	0.036	−0.018	B	L7	–	***9***
Sunniest PFP	0.615	−0.153	−0.049	−0.04	1.133	1.827	Fluoro	L43	–	***9***
Sunniest PhE	0.656	−0.085	−0.469	0.008	0.044	−0.022	Phenyl	L11	–	***9***
Sunniest RP-AQUA	0.958	−0.024	−0.21	−0.008	0.142	0.098	EP	L60	–	***9***
Sunshell C18	1.086	0.028	−0.124	−0.056	−0.043	−0.084	B	L1	–	***9***
Sunshell C8	0.907	0.007	−0.235	−0.028	−0.065	−0.126	B	L7	–	***9***
Sunshell PFP	0.663	−0.176	0.14	−0.076	1.106	2.143	Fluoro	L43	–	***9***
Sunshell Phenyl	0.813	−0.076	−0.323	0.017	−0.061	−0.09	Phenyl	L11	–	***9***
Sunshell RP-AQUA	0.898	−0.04	−0.272	0.013	0.127	0.095	EP	L60	–	***9***
Supelcosil LC-18	1.018	−0.047	0.181	0.162	1.595	1.752	A	L1	–	***45***
Supelcosil LC-18-DB	0.979	−0.026	0.047	0.114	0.481	0.531	A	L1	–	***45***
Supelcosil LC-8	0.834	−0.048	−0.027	0.086	1.117	1.094	A	L7	–	***45***
Supelcosil LC-8-DB	0.819	−0.036	−0.072	0.143	0.446	0.554	A	L7	–	***45***
Supelcosil LC-PAH	0.851	−0.025	0.104	−0.03	0.642	0.83	A	L1	–	***45***
Superspher 100 RP-18e	1.03	0.025	−0.028	−0.011	0.352	0.266	B	L1	–	***31***

APPENDIX 2d List of Available Columns With PQRI Parameters (Mobile Phase 50/50 Acetonitrile/Aqueous Buffer 60 mM Phosphate v/v) (http://apps.usp.org/app/USPNF/columnsDB.html)—cont'd

Column	H^{*a}	S^{*b}	A^{*c}	B^{*d}	$C(2.8)^{*e}$	$C(7.0)^{*f}$	Silica*[g]	UPS	$\log k'_{EB}$	Manuf.[h]
Symmetry 300 C18	0.984	0.031	−0.051	0.003	0.228	0.202	B	L1	0.549	***48***
Symmetry 300 C4	0.659	−0.016	−0.428	0.014	0.101	0.184	B	L26	0.157	***48***
Symmetry C18	1.052	0.063	0.018	−0.021	−0.302	0.123	B	L1	0.993	***48***
Symmetry C4	0.687	−0.012	−0.29	0.022	0.217	0.305	B	L26	–	***48***
Symmetry C8	0.893	0.049	−0.205	0.021	−0.509	0.283	B	L7	0.843	***48***
SymmetryShield C18	0.85	0.027	−0.411	0.093	−0.728	0.136	EP	–	–	***48***
SymmetryShield C8	0.73	−0.006	−0.55	0.103	−0.623	0.138	EP	–	–	***48***
SynChropak RP8	0.639	−0.099	0.109	0.029	0.223	0.94	A	L7	–	***15***
SynChropak RPP	0.746	−0.115	0.23	0.033	0.259	1.286	A	L1	–	***34***
SynChropak RPP 100	0.918	−0.059	−0.072	0.123	0.225	0.317	A	L1	–	***34***
Syncronis aQ	0.984	−0.004	−0.1	−0.038	0.066	1.077	EP	L60	–	***46***
Syncronis C18	1.043	−0.003	−0.086	−0.019	−0.026	0.038	B	L1	–	***46***
Syncronis C8	0.84	−0.024	−0.227	0.01	−0.058	0.116	B	L7	–	***46***
Syncronis Phenyl	0.75	−0.082	−0.32	0.008	−0.031	0.176	Phenyl	L11	–	***46***
Synergi Fusion-RP	0.879	−0.03	−0.014	0.008	−0.238	0.362	EP	–	–	***37***
Synergi Hydro-RP	1.022	−0.006	0.169	−0.042	−0.077	0.26	EP	L1	–	***37***
Synergi Max-RP	0.989	0.028	−0.008	−0.013	−0.133	−0.034	B	L1	0.976	***37***
Synergi Polar-RP	0.654	−0.148	−0.257	−0.007	0.057	0.778	EP	L1	–	***37***
Targa C18	0.977	−0.019	−0.07	0	0.013	0.175	B	L1	–	***26***
Targa C8	0.821	−0.023	−0.221	0.004	−0.027	0.174	B	L7	–	***26***
Thermo CN	0.404	−0.111	−0.709	−0.009	−0.029	0.491	CN	L10	−0.088	***46***
Titan C18	1.011	0.004	−0.084	−0.047	−0.052	0.088	C18	L1	–	***45***
Topsil C18	0.972	−0.011	−0.019	0.009	0.124	0.312	B	L1	–	***49***
TSKgel CN-80Ts	0.373	−0.025	−0.617	0.011	−0.369	0.426	CN	L10	–	***47***
TSKgel Octyl-80Ts	0.814	−0.005	−0.253	0.017	0.089	0.456	B	L7	–	***47***
TSKgel ODS-100S	1.032	0.065	−0.092	−0.034	−0.003	−0.032	B	L1	–	***47***
TSKgel ODS-100V	0.901	−0.043	−0.226	−0.009	−0.06	−0.02	B	L1	–	***47***
TSKgel ODS-100Z	1.032	0.018	−0.135	−0.031	−0.064	−0.161	B	L1	–	***47***
TSKgel ODS-120A	0.896	−0.039	0.28	0.001	0.265	0.963	B	L1	–	***47***

(Continued)

APPENDIX 2d List of Available Columns With PQRI Parameters (Mobile Phase 50/50 Acetonitrile/Aqueous Buffer 60 mM Phosphate v/v) (http://apps.usp.org/app/USPNF/columnsDB.html)—cont'd

Column	H^{*a}	S^{*b}	A^{*c}	B^{*d}	$C(2.8)^{*e}$	$C(7.0)^{*f}$	Silica*[g]	UPS	$\log k'_{EB}$	Manuf.[h]
TSKgel ODS-120T	0.977	0.03	0.114	−0.011	0.195	0.227	B	L1	–	*47*
TSK-gel ODS-140HTP	1.002	0.051	−0.251	−0.032	0.134	0.142	B	L1	–	*47*
TSKgel ODS-80T	0.96	−0.037	0.145	−0.01	0.164	0.589	B	L1	–	*47*
TSKgel ODS-80Ts	0.971	−0.015	−0.132	−0.004	0.01	0.292	B	L1	–	*47*
TSKgel ODS-80Ts QA	0.94	−0.03	−0.118	0.005	0.004	0.361	B	L1	–	*47*
TSKgel OligoDNA RP	0.864	0.028	0.143	0.01	0.315	0.652	A	L1	–	*47*
TSKgel Super-Octyl	0.824	0.034	−0.155	0.01	0.126	0.23	B	L7	–	*47*
TSKgel Super-ODS	0.998	0.03	−0.048	−0.019	0.154	0.237	B	L1	–	*47*
TSKgel Super-Phenyl	0.58	−0.107	−0.146	0.016	0.085	0.672	Phenyl	L11	–	*47*
Ultimate AQ-C18	0.868	−0.033	−0.055	0.029	0.058	0.363	B	L1	–	*49*
Ultimate XB-C18	1.005	0.011	−0.046	0.001	0.118	0.133	B	L1	–	*49*
Ultimate XB-C8	0.835	0.001	−0.074	0.028	0.069	0.281	B	L7	–	*49*
Ultimate XB-Phenyl	0.651	−0.108	−0.25	0.032	0.15	0.476	Phenyl	L11	–	*49*
Ultisil AQ-C18	0.872	−0.037	−0.072	0.031	0.054	0.323	B	L1	–	*49*
Ultisil XB-C18	1.009	0.012	−0.016	0.005	0.118	0.142	B	L1	–	*49*
Ultisil XB-C8	0.841	0.001	−0.108	0.027	0.063	0.244	B	L7	–	*49*
Ultisil XB-Phenyl	0.651	−0.108	−0.273	0.032	0.144	0.466	Phenyl	L11	–	*49*
Ultra Aqueous C18	0.808	−0.128	0.378	0.013	0.229	1.255	B	L1	–	*38*
Ultra Aromax	0.741	−0.179	−0.375	0.021	0.195	0.375	Phenyl	L11	–	*38*
Ultra Biphenyl	0.661	−0.189	−0.283	0.042	0.204	0.721	Phenyl	L11	–	*38*
Ultra C1	0.613	−0.054	−0.408	0.016	−0.032	0.055	B	L13	–	*38*
Ultra C18	1.051	0.033	−0.032	−0.023	0.057	−0.003	B	L1	–	*38*
Ultra C4	0.738	−0.01	−0.276	0.019	0.032	0.318	B	L26	–	*38*
Ultra C8	0.871	0.013	−0.199	0.019	−0.032	−0.078	B	L7	–	*38*
Ultra Cyano	0.409	−0.041	−0.801	−0.011	−0.11	0.628	CN	L10	–	*38*
Ultra IBD	0.672	−0.035	−0.052	0.233	−0.564	0.86	EP	L68	–	*38*
Ultra II Aqueous C18	0.784	−0.154	0.321	0.015	0.468	1.163	B	L1	–	*38*
Ultra II Aromax	0.739	−0.193	−0.344	0.022	0.478	0.64	Phenyl	L11	–	*38*
Ultra II Biphenyl	0.652	−0.198	−0.275	0.047	0.434	0.764	Phenyl	L11	–	*38*

APPENDIX 2d List of Available Columns With PQRI Parameters (Mobile Phase 50/50 Acetonitrile/Aqueous Buffer 60 mM Phosphate v/v) (http://apps.usp.org/app/USPNF/columnsDB.html)—cont'd

Column	H^{*a}	S^{*b}	A^{*c}	B^{*d}	$C(2.8)^{*e}$	$C(7.0)^{*f}$	Silica*[g]	UPS	$\log k'_{EB}$	Manuf.[h]
Ultra II C18	1.041	0.021	−0.037	−0.015	0.264	0.181	B	L1	–	*38*
Ultra II IBD	0.674	−0.043	−0.022	0.225	−0.257	0.864	EP	L68	–	*38*
Ultra II PFP Propyl	0.674	−0.084	−0.303	−0.006	0.769	1.213	F	L43	–	*38*
Ultra PFP	0.446	−0.055	−0.245	0.016	−0.221	0.152	F	L43	0.289	*38*
Ultra PFP Propyl	0.622	−0.102	−0.314	−0.035	0.456	1.047	F	L43	–	*38*
Ultra Phenyl	0.58	−0.127	−0.338	0.039	0.041	0.438	Phenyl	L11	–	*38*
Ultra Quat	0.858	0.016	−0.17	0.017	0.026	0.134	Other	–	–	*38*
Ultracarb ODS (30)	1.114	0.016	0.377	−0.05	−0.311	0.731	B	L62	–	*37*
Ultracore SuperC18	1.093	0.03	−0.111	−0.05	−0.029	0.028	C18	L1	–	*1*
UltraCore SuperPhenylHexyl	0.809	−0.074	−0.347	−0.014	−0.014	−0.007	Phenyl	L11	–	*1*
UltraSep ES AMID H RP18P	0.751	−0.013	−0.101	0.259	−0.527	0.855	EP	–	–	*40*
UltraSep ES PHARM RP18	0.953	−0.061	0.435	−0.057	0.593	1.674	A	L1	–	*40*
Ultrasphere Octyl	0.896	0.016	0.003	0.086	0.157	0.547	B	L7	–	*25*
Ultrasphere ODS	1.085	−0.014	0.173	0.068	0.279	0.382	B	L1	–	*25*
Unisol C18	0.9502	−0.0173	−0.108	−0.009	−0.108	0.0916	B	L1	–	*8*
Unison UK-C18	0.981	−0.019	0.015	−0.011	0.11	0.07	B	L1	–	*28*
Venusil ABS C18	0.8244	−0.0664	0.1664	0.05	0.047	0.6437	B	L1	–	*8*
Venusil ABS C8	0.6771	−0.0777	−0.001	0.044	0.119	0.5695	B	L7	–	*8*
Venusil HLP	0.5467	−0.0145	−0.649	0.254	−0.926	1.6362	EP	L60	–	*8*
Venusil PFP	0.519	−0.158	0.12	0.011	0.35	1.144	F	L43	–	*8*
Venusil XBP Phenyl	0.626	−0.118	−0.234	0.02	0.024	0.776	Phenyl	L11	–	*8*
Venusil XBP Phenyl-Hexyl	0.678	−0.17	0.128	0.034	0.156	1.203	Phenyl	L11	–	*8*
Venusil XBP Polar-Phenyl	0.552	−0.214	0.016	0.052	−0.034	1.066	Phenyl	L11	–	*8*
Venusil XPB C18(2)	1.0097	0.0066	−0.081	−0.022	−0.116	0.0725	B	L1	–	*8*
Venusil XPB C18(L)	0.945	0.025	−0.135	−0.002	0.152	0.318	B	L1	–	*8*
Venusil XPB CN	0.4498	−0.0532	−0.631	−0.013	−0.018	0.7373	CN	L10	–	*8*
Vision C18 B	0.689	−0.111	0.35	0.031	0.39	1.41	A	L1	–	*20*
Vision C18 HL	0.992	0.056	0.057	−0.013	0.133	0.143	B	L1	–	*20*
VisionHT C18	0.786	−0.076	−0.098	0.005	0.4	0.694	B	L1	–	*20*

(Continued)

APPENDIX 2d List of Available Columns With PQRI Parameters (Mobile Phase 50/50 Acetonitrile/Aqueous Buffer 60 mM Phosphate v/v) (http://apps.usp.org/app/USPNF/columnsDB.html)—cont'd

Column	H^{*a}	S^{*b}	A^{*c}	B^{*d}	$C(2.8)^{*e}$	$C(7.0)^{*f}$	Silica*[g]	UPS	$\log k'_{EB}$	Manuf.[h]
VisionHT C18-P	0.636	−0.165	0.287	0.033	0.698	1.618	B	L1	–	*20*
Viva Biphenyl	0.605	−0.183	−0.304	0.047	0.337	0.544	Phenyl	L11	–	*38*
Viva C18	0.98	0.016	−0.076	0.013	0.359	0.305	A	L1	–	*38*
Viva C4	0.708	−0.011	−0.285	0.031	0.223	0.311	A	L26	–	*38*
Viva C8	0.83	0.006	−0.222	0.036	0.262	0.222	A	L7	–	*38*
Viva PFP Propyl	0.626	−0.065	−0.252	0.029	0.283	0.762	F	L43	–	*38*
Vydac 201TP	0.901	−0.022	0.409	−0.004	0.394	1.026	A	L1	–	*23*
Vydac 218MS	0.77	0.182	0.111	−0.373	0.659	1.234	A	L1	–	*23*
Vydac 218MSC18	0.881	−0.013	0.295	0.005	0.171	0.823	B	L1	–	*23*
Vydac 218TP	0.909	0.009	0.345	−0.005	0.279	0.67	B	L1	–	*23*
Vydac C18 Monomeric	0.977	0.044	0.031	−0.002	0.139	0.275	B	L1	–	*23*
Vydac Everest	0.993	0.049	0.121	0.004	0.065	0.341	B	L1	–	*23*
Wakosil 5 C8 RS	0.802	−0.008	−0.272	0.001	−0.117	0.097	B	L7	–	*34*
Wakosil II 5 C18 AR	0.998	0.075	−0.055	−0.034	0.07	0.01	B	L1	–	*34*
Wakosil II 5 C18 HG	1.039	0.036	0.015	−0.023	0.009	0.21	B	L1	–	*34*
Wakosil II 5 C18 RS	0.964	−0.008	−0.16	−0.009	−0.07	0.046	B	L1	–	*34*
XBridge C18	1.007	0.028	−0.097	0.009	0.178	0.138	B	L1	–	*48*
XBridge C8	0.805	0.018	−0.296	0.018	0.129	0.063	B	L7	–	*48*
XBridge Phenyl	0.732	−0.078	−0.358	0.037	0.19	0.206	Phenyl	L11	–	*48*
XBridge Shield RP18	0.835	−0.026	−0.372	0.095	−0.122	−0.051	EP	–	–	*48*
XSelect CSH C18	0.9542	−0.0017	−0.179	0.118	0.082	0.1706	B	L1	–	*48*
XSelect CSH Fluoro-Phenyl	0.4977	−0.1377	−0.052	0.055	−0.321	0.7933	F	L43	–	*48*
XSelect CSH Phenyl-Hexyl	0.7079	−0.0594	−0.435	0.129	−0.068	0.223	Phenyl	L11	–	*48*
Xselect HSS Cyano	0.479	−0.116	−0.224	0.043	0.082	1.18	CN	L10	–	*48*
Xselect HSS PFP	0.592	−0.164	0.146	−0.02	0.74	1.774	F	L43	–	*48*
XTerra C18 RP	0.757	−0.043	−0.483	0.097	−0.17	−0.173	EP	–	–	*48*
XTerra C8 RP	0.657	−0.049	−0.604	0.099	−0.187	−0.198	EP	–	–	*48*
XTerra MS C18	0.984	0.012	−0.143	−0.015	0.133	0.051	B	L1	0.803	*48*
XTerra MS C8	0.803	0.005	−0.293	−0.005	0.058	−0.009	B	L7	0.571	*48*

APPENDIX 2d List of Available Columns With PQRI Parameters (Mobile Phase 50/50 Acetonitrile/Aqueous Buffer 60 mM Phosphate v/v) (http://apps.usp.org/app/USPNF/columnsDB.html)—cont'd

Column	H^{*a}	S^{*b}	A^{*c}	B^{*d}	$C(2.8)^{*e}$	$C(7.0)^{*f}$	Silica*[g]	UPS	$\log k'_{EB}$	Manuf.[h]
XTerra Phenyl	0.683	−0.079	−0.363	−0.003	0.119	0.029	Phenyl	L11	0.409	*48*
Xtimate C18	1	−0.001	−0.083	−0.011	0.155	0.283	B	L1	–	*49*
Xtimate C8	0.855	−0.014	−0.185	0.008	0.013	0.173	B	L7	–	*49*
YMC Basic	0.821	−0.006	−0.235	0.028	0.07	0.093	B	L1	–	*51*
YMC Hydrosphere C18	0.937	−0.022	−0.129	0.006	−0.139	0.157	B	L1	–	*51*
YMC ODS-AQ	0.965	−0.036	−0.135	0.004	−0.068	0.1	B	L1	–	*51*
YMC Pack Pro C18 RS	1.114	0.057	−0.061	−0.056	−0.176	−0.224	B	L1	–	*51*
YMC Pro C18	1.015	0.014	−0.12	−0.007	−0.155	−0.006	B	L1	0.939	*51*
YMC Pro C8	0.89	0.014	−0.215	0.007	−0.323	0.019	B	L7	0.814	*51*
YMC-Triart C18	0.929	−0.02	−0.19	−0.033	−0.023	−0.139	B	L1	–	*51*
ZirChrom-EZ	1.04	0.117	−0.999	−0.001	2.089	2.089	Other	–	–	*52*
ZirChrom-MS	0.948	−0.17	−0.451	0.326	0.483	0.55	Other	–	–	*52*
ZirChrom-PBD	1.284	0.158	−0.384	−0.072	2.188	2.188	Other	L49	–	*52*
ZirChrom-PS	0.589	−0.232	−0.477	0.062	1.75	1.75	Other	–	–	*52*
Zodiac C18	1.0661	0.0519	−0.069	−0.039	−0.184	−0.152	B	L1	–	*53*
Zodiac C18(1)	0.973	0.0079	−0.124	0.019	0.204	0.1561	B	L1	–	*53*
ZodiacSil 120-5 C18 AQ	0.974	−0.007	−0.082	0.003	0.139	0.224	B	L1	–	*53*
ZodiacSil 120-5-C18 ace EPS	0.773	0.042	−0.59	0.228	−0.302	0.041	EP	–	–	*53*
ZodiacSil 120-5-C18 AQ Plus	0.948	−0.017	0.214	0.041	−0.131	0.605	EP	–	–	*53*
ZodiacSil 120-5-C18H	1.006	0.008	−0.106	−0.004	0.127	0.156	B	L1	–	*53*
ZodiacSil 120-5-C18SH	1.032	0.018	−0.109	−0.024	0.115	0.402	B	L1	–	*53*
ZodiacSil 200-5-C18 ace EPS	0.766	0.021	−0.566	0.214	0.028	0.143	EP	–	–	*53*
ZodiacSil 200-5-C18AQ	0.974	−0.011	−0.057	0.006	0.127	0.029	B	L1	–	*53*
ZodiacSil 200-5-C18H	0.956	−0.001	−0.121	0.016	0.165	0.218	B	L1	–	*53*
ZodiacSil 300-5-C18ace EPS	0.763	0.025	−0.579	0.211	−0.052	0.136	EP	–	–	*53*
ZodiacSil 300-5-C18H	0.957	−0.012	−0.09	0.015	0.24	0.249	B	L1	–	*53*
ZodiacSil 60-5-C18H	1.159	0.041	0.066	−0.078	0.104	0.263	B	L1	–	*53*
Zorbax Bonus RP	0.654	0.107	−1.046	0.373	−2.971	−1.103	EP	L60	–	*3*
Zorbax C18	1.089	0.055	0.474	0.06	1.489	1.566	A	L1	–	*3*

(Continued)

APPENDIX 2d List of Available Columns With PQRI Parameters (Mobile Phase 50/50 Acetonitrile/Aqueous Buffer 60 mM Phosphate v/v) (http://apps.usp.org/app/USPNF/columnsDB.html)—cont'd

Column	H^{*a}	S^{*b}	A^{*c}	B^{*d}	$C(2.8)^{*e}$	$C(7.0)^{*f}$	Silica*[g]	UPS	$\log k'_{EB}$	Manuf.[h]
Zorbax C8	0.974	−0.041	0.216	0.176	0.974	1.051	A	L7	–	3
Zorbax Eclipse PAH	1.031	−0.015	0.688	−0.051	0.072	1.409	B	L1	–	3
Zorbax Eclipse Plus C18	1.03	0.007	−0.072	−0.02	−0.004	0.02	B	L1	–	3
Zorbax Eclipse Plus C8	0.889	0.017	−0.172	−0.005	−0.042	0.051	B	L1	–	3
Zorbax Eclipse XDB-C18	1.077	0.024	−0.063	−0.033	0.055	0.088	B	L1	0.958	3
Zorbax Eclipse XDB-C8	0.919	0.025	−0.219	−0.008	0.003	0.012	B	L7	0.823	3
Zorbax Eclipse XDB-CN	0.456	−0.068	−0.312	0.003	0.074	0.994	CN	L10	–	3
Zorbax Extend C18	1.098	0.05	0.012	−0.041	0.03	0.016	B	L1	–	3
Zorbax Rx-18	1.077	0.037	0.309	−0.038	0.096	0.415	B	L1	0.886	3
Zorbax Rx-C8	0.792	−0.076	0.116	0.018	0.012	0.948	B	L7	0.703	3
Zorbax SB-AQ	0.593	−0.12	−0.083	0.038	−0.136	0.736	EP	–	–	3
Zorbax SB-CN	0.502	−0.108	−0.224	0.042	−0.146	1.047	CN	L10	–	3
Zorbax SB-Phenyl	0.623	−0.161	0.065	0.038	0.033	1.089	Phenyl	L11	–	3
Zorbax StableBond 300A C18	0.905	−0.05	0.045	0.043	0.254	0.701	B	L1	0.344	3
Zorbax StableBond 300A C3	0.526	−0.122	−0.195	0.047	0.057	0.357	B	L56	−0.151	3
Zorbax StableBond 300A C8	0.701	−0.085	0.002	0.047	0.146	0.82	B	L7	0.106	3
Zorbax StableBond 80A C18	0.996	−0.032	0.264	−0.001	0.136	1.041	B	L1	0.884	3
Zorbax StableBond 80A C3	0.601	−0.124	−0.081	0.038	−0.084	0.81	B	L56	0.450	3
Zorbax StableBond 80A C8	0.795	−0.079	0.137	0.018	0.014	1.02	B	L7	0.710	3
Zorbax XDB-Phenyl	0.665	−0.127	−0.242	0.019	0.063	0.584	Phenyl	L11	–	3

[a]***H****, *hydrophobic character.*
[b]***S****, *steric hindrance.*
[c]***A****, *hydrogen bonding basic solute.*
[d]***B****, *hydrogen bonding acidic solute.*
[e]**C(2.8)***, *cation exchange at pH = 2.8.*
[f]**C(7.0)***, *cation exchange at pH = 7.0.*
[g]A, Type A *silica;* B, Type B *silica;* EP, *embedded phase;* phenyl, *phenyl phase;* fluoro, *fluoronated phase;* CN, *cyano phase.*
[h]*Manufacturer:* **1**. *ACT (ACE, Advanced Chromatography Technologies Ltd.),* **2**. *Advanced Materials Technologiy,* **3**. *Agilent Technologies,* **4**. *Agilent/Varian (currently Agilent Technologies),* **5**. *Akzo Nobel,* **6**. *Analytical Sales & Services,* **7**. *Bischoff,* **8**. *Bonna-Agela Technologies, Inc.,* **9**. *ChromaNik,* **10**. *Chromatopack,* **11**. *CNW Technologies GmbH,* **12**. *Diamond Analytics,* **13**. *Dikma Technologies,* **14**. *Dionex,* **15**. *Eprogen,* **16**. *ES Industries,* **17**. *Fortis Technologies,* **18**. *Genius Technologies,* **19**. *GL Science,* **20**. *Grace/Alltech,* **21**. *Grace/Chrome,* **22**. *Grace/Jones,* **23**. *Grace/Vydac,* **24**. *Hamilton,* **25**. *Hichrom,* **26**. *Higgins Analytical,* **27**. *Hitachi High-Tech,* **28**. *Imtakt,* **29**. *Macherey Nagel,* **30**. *Mac-Mod Analytical,* **31**. *Merck KGaA (EDM Millipore),* **32**. *MicroSolv,* **33**. *Nacalai Tesque,* **34**. *no longer available,* **35**. *Nomura,* **36**. *Orochem Technologies,* **37**. *Phenomenex,* **38**. *Restek,* **39**. *Sepax Technologies,* **40**. *SEPSERV,* **41**. *Shimadzu,* **42**. *Shiseido,* **43**. *Showa Denko,* **44**. *SIELC,* **45**. *Supelco,* **46**. *Thermo/Hypersil,* **47**. *Tosoh Bioscience,* **48**. *Waters,* **49**. *Welch,* **50**. *Wesley Technologies Inc.,* **51**. *YMC,* **52**. *ZirChrom,* **53**. *Zodiac Life Science.*

APPENDIX 2e Several More Recent Hydrophobic Columns[a]

Column Name	Manufacturer	Phase	Particle Size	Type
AdvancedBio Oligonucleotide[b]	Agilent	C18	2.7	Core–shell
Ascentis Express Biphenyl	Milipore Sigma	Biphenyl	2.7	Core–shell
Cortecs	Waters	C8,phenyl	1.6, 2.7	Core–shell
C8-30HT[b]	ChromaNik Technologies Inc.	C8	3.4	Core–shell
DeltaSil 100C18	Watrex Praha	C18	3, 5	Porous
HALO 2 Peptide ES-C18[b]	Advanced Materials Technol.	C18	2	Core–shell
Kinetex EVO C18	Phenomenex	C18	1.7, 2.6	Core–shell
Luna Omega	Phenomenex	C18	1.6	Porous
MAbPac RP[b]	Thermo Fisher Scientific	Phenyl	4	Polymeric
Poroshell HPH C18 and C8	Agilent Technologies	C18, C8	4	Core–shell
Raptor FluoroPhenyl	Restek Corp.	Pentafluorophenyl	2.7, 5	Core–shell
Roc	Restek Corp.	C18, C8, phenyl-hexyl, cyano, silica	3, 5	Porous
Selectra Aqueous C18	UCT Inc.	Polar-enhanced C18	1.8, 3, 5	Porous
Selectra EtG	UCT Inc.	Proprietary	3	Porous
SiliaChrom Plus PFP	SiliCycle Inc.	Pentafluorophenyl	5, 10	Porous
SMT-C30	Separation Methods Technologies. Inc.	C30	5	Porous
Sun Shell	ChromaNik Technologies Inc.	S18, C30	2.6	Core–shell
Sun Shell HFC18–30, C8-30, C4-30[b]	ChromaNik Technologies Inc.	C18, C8, C8	2.6	Core–shell
SunArmor	ChromaNik Technologies Inc.	C18 and RP-AQUA (C28)	3, 5	Porous

[a]*These columns are available in different formats such as 150 × 4.6 mm, 100 × 2.1 mm.*
[b]*Columns used for large molecule separations.*

Appendix 3: HILIC and NPC Stationary Phases

APPENDIX 3a Common Bare Silica Used for in HILIC and NPC

Brand Name	Manufacturer	Phase Type/Pore Size/Particle Size
Accucore	Thermo Scientific	Silica; 80 Å; 2.6 μm
Atlantis	Waters	Silica; 100 Å; 3.0 μm
Ascentis Express HILIC	Supelco	Core-shell silica; 2.7 μm
Betasil	Thermo Hypersil	Silica; 100 Å; 3.0 μm
Boltimate HILIC	Welch	Silica; 90 Å; 2.7 μm
ChromoLith	Merck	Silica monolith; mesopores (average pore size 13 nm)
COSMOSIL SL-II	Nacalai	Silica (silicagel); 120 Å; 3, 5, 15 μm
Halo; Halo-Penta HILIC	Hichrom	Silica; 90 Å; 2.7, 5.0 μm
Hypersil Gold HILIC	Thermo	Silica; 100 Å; 1.9, 3.0, 5.0 μm
Hypersil Silica	Thermo	Silica; 120 Å; 5 μm
Inspire Silica	Dikma	Silica; 100 Å; 3.0, 5.0, 10 μm
Kromasil	EKA Chemicals	Silica; 60 Å; 5.0, 10 μm
Nucleodur unmodified	Macherey–Nagel	Silica; 110 Å; 1.8, 3.0, 5.0 μm
Nucleosil (SIL) unmodified	Macherey–Nagel	Silica; 50, 100 Å; 5.0 μm
Promosil Silica	Bonna Agela	Silica; 100 Å; 5.0 μm
Silia Chrom XDB Si	Greyhound Chromatography	Silica; 100 Å; 3.0, 5.0, 10 μm
Silia Chrom XDB1 Si	Greyhound Chromatography	Silica; 100 Å; 3.0, 5.0, 10 μm
Silia Chrom XDB1 Si-300	Greyhound Chromatography	Silica; 300 Å; 3.0, 5.0, 10 μm
Spheri-5 Silica	Brownlee (Alltech)	Silica; 80 Å; 5.0 μm
Supelcosil LC-Sil	Supelco	Silica; 120 Å; 5.0 μm
Syncronis Silica	Thermo	Silica; 100 Å; 5.0 μm
Venusil XBP Silica	Bonna Agela	Silica; 100 Å; 3.0, 5.0, 10 μm
Venusil XBP-L Silica	Bonna Agela	Silica; 100 Å; 3.0, 5.0, 10 μm
YMC-Pack SIL	YMC	Silica; 60, 120, 200, 300 Å; 3.0, 5.0, 10 μm
Zorbax Rx-Sil	Agilent	Silica; 80 Å; 1.8, 5 μm

APPENDIX 3b Several Commercially Available Neutral HILIC Columns

Brand Name	Manufacturer	Support/Pore Size/Particle Size	Nature of Phase
Acclaim HILIC 10	Thermo (Dionex)	Silica; 120 Å; 3 μm	Proprietary polar group
Accucore 150 Amide HILIC	Thermo	Silica; 150 Å; 2.6 μm	Amide
Accucore Urea HILIC	Thermo	Silica; 80 Å; 2.6 μm	Urea
Alltima Cyano	Grace Alltech	Silica; 190 Å; 3, 5 μm	3-Cyanopropyl
Cogent Type C Silica	Microsolv	Silica; 100 Å; 4 μm	Silica hydride
COSMOSIL CN-MS	Nacalai	Silica; 120 Å; 5 μm	3-Cyanopropyl
Cyclobond I 2000	ASTEC	Silica; 100 Å; 5, 10 μm	β-Cyclodextrin
Epic Diol HILIC	ES Industries	Silica; 120 Å; 1.8 μm	Diol
Epic HILIC FL	ES Industries	Silica; 120 Å; 1.8 μm	Fluorinated
GlycoSep N	ProZyme	Silica; 10 μm	Amide
Hydrolyzed GMA-co-EDMA	Custom synthesis	Methacrylic copolymer	2,3-Dihydroxypropyl
Inertsil Diol	GL Sciences	Silica; 100 Å; 3, 5 μm	2,3-Dihydroxypropyl
Inspire Diol	Dikma	Silica; 100 Å; 3, 5, 10 μm	Diol
Kinetex HILIC Core-shell diol	Phenomenex	Silica core–shell; 1.7, 2.6, 5 μm	Diol
Lichrospher Diol 100	Merck	Silica; 100 Å; 5 μm	2,3-Dihydroxypropyl
Luna HILIC diol	Phenomenex	Silica; 100 Å; 3, 5, 10 μm	Diol and ether embedded
Nucleodex b-OH a-PM, etc.	Macherey–Nagel	Silica; 100 Å; 5 μm	β-Cyclodextrin
Nucleodur CN/CN-RP	Macherey–Nagel	Silica; 110 Å; 3, 5 μm	3-Cyanopropyl
Nucleosil CN	Macherey–Nagel	Silica; 100, 120 Å; 5, 7, 10 μm	3-Cyanopropyl
Nucleosil OH (diol)	Macherey–Nagel	Silica; 100 Å; 5 μm	2,3-Dihydroxypropyl
Perhydroxyl-CB[6]uril silica	Custom synthesis	Silica	Perhydroxyl-cucurbit[6]uril
PolyGlycoplex	PolyLC	Silica (polysuccinimide); 5, 12 μm	Poly(succinimide)
PolyHydroxyethyl A	PolyLC	Silica (polyaspartic acid); 60–1500 Å; 5, 12 μm	Poly(2-hydroxyethyl-aspartamide)
Promosil CN	Bonna Agela	Silica; 100 Å; 5 μm	3-Cyanopropyl
Silasorb Diol	Chemapol	Silica	2,3-Dihydroxypropyl

(Continued)

APPENDIX 3b Several Commercially Available Neutral HILIC Columns—cont'd

Brand Name	Manufacturer	Support/Pore Size/Particle Size	Nature of Phase
Silia Chrom XDB1 Diol-300	Greyhound Chromatography	Silica; 300 Å; 3, 5 μm	Diol
Silia Chrom HILIC	Greyhound Chromatography	Silica; 100 Å; 3, 5 μm	Urea
Silia Chrom HILIC-300	Greyhound Chromatography	Silica; 300 Å; 3, 5 μm	Urea
Silia Chrom XDB1 Diol	Greyhound Chromatography	Silica; 100 Å; 3, 5 μm	Diol
TSKgel Amide-80	Tosoh Bioscience	Silica; 100 Å; 2, 3, 5, 10 μm	Amide
Unisol Amide	Bonna Agela	Silica; 100 Å; 3, 5 μm	Amide
Venusil XBP CN	Bonna Agela	Silica; 100 Å; 5, 10 μm	3-Cyanopropyl
Venusil XBP Diol	Bonna Agela	Silica; 100 Å; 5 μm	Amide
XBridge amide	Waters	Polyetoxysilane (BEH); 130 Å; 3.5 μm	Amide
Xbridge HILIC	Waters	Polyetoxysilane (BEH); 130 Å; 1.7, 2.5, 3.5, 5, 10 μm	Diol
YMC-Pack CN	YMC	Silica; 120, 300 Å; 3, 5 μm	3-Cyanopropyl
YMC-pack Diol 120 NP	YMC	Silica; 120 Å; 1.9, 3, 5 μm	2,3-Dihydroxypropyl
YMC-Pack PVA-Sil	YMC	Silica support; 60, 120, 200, 300 Å; 3, 5 μm	Polyvinyl alcohol polymerized on silica

APPENDIX 3c Several HILIC Columns With Weak Anion Exchange Properties that are Commercially Available

Brand Name	Manufacturer	Support/Pore Size/Particle Size	Nature of Phase
Amino	Jordi	Silica; 120 Å; 3, 5 μm	Aminopropyl
Amino-bonded Zirconia	Custom synthesis	Zirconia	Mono-, di- and triamine
Asahipak NH_2 P	Shodex	Poly(vinyl alcohol) gel; 100 Å; 4, 5 μm	Amine
Astec apHera	ASTEC	PVA copolymer; 300 Å; 5 μm	Polyamine
COSMOSIL DEAE	Nacalai	Porous polymethacrylate	Diethylaminoethyl (DEAE)
COSMOSIL HILIC	Nacalai USA	Silica; 120 Å; 5 μm	Triazole
Durashell NH2	Bonna Agela	Silica; 100 Å; 5, 10 μm	Amino
EPIC-PI	ES Industries	Silica; 120 Å; 1.8 μm	Aromatic amine
GlycoSep C	ProZyme	Polymeric; 10 μm	DEAE
Hypersil APS2	Thermo Scientific	Silica; 120 Å; 3 μm	3-Aminopropyl
Hypersil GOLD PEI HILIC	Thermo Scientific	Silica; 175 Å; 1.9 μm	Polyethylene amine
Luna Amino	Phenomenex	Silica; 100 Å; 3, 5, 10 μm	3-Aminopropyl
Micropellicular AP Silica	Custom synthesis	Silica	3-Aminopropyl
Nucleodur NH2/NH2-RP	Macherey–Nagel	Silica; 110 Å; 3, 5, 7 μm	Amino
Nucleosil Carbohydrate	Macherey–Nagel	Silica; 10 μm	Amino
Nucleosil N(CH3)2	Macherey–Nagel	Silica; 100 Å; 5 μm	Tertiary amine
Nucleosil NH2	Macherey–Nagel	Silica; 100 Å; 3, 5, 10 μm	Amino
PolyWAX LPTM	PolyLC	Silica; 100, 300, 1000, 1500 Å; 3, 5, 12 μm	Linear polyethyleneimine
Promosil NH2	Bonna Agela	Silica; 100 Å; 5 μm	Amino
Silia Chrom XDB1-Amino	Greyhound Chroma.	Silica; 100 Å; 3, 5 μm	Amino
Silia Chrom XDB1-Amino-300	Greyhound Chroma.	Silica; 300 Å; 3, 5 μm	Amino
Spherisorb NH_2	Waters	Silica; 80 Å; 5, 10 μm	3-Aminopropyl
TSK Gel NH2-100	Tosoh Bioscience	Silica (endcapped); 100 Å; 3 μm	Amino
Venusil XBP NH2	Bonna Agela	Silica; 120 Å; 5 μm	Amino
YMC-Pack PA	YMC	Silica; 120 Å; 5 μm	Amino
YMC-Pack Polyamine II	YMC	Silica; 120 Å; 5 μm	Polyamine (sec. and tert.)
Zorbax NH_2	Agilent	Silica; 70 Å; 3.5, 5, 7 μm	3-Aminopropyl

APPENDIX 3d Several HILIC Columns With Cation Exchange Properties that are Commercially Available

Brand Name	Manufacturer	Support/Pore Size/Particle Size	Nature of Phase	Type
Acrylamide CEC phase	Custom synthesis	4% Crosslinked polyacryl-amide	Dodecyl chains and sulfonic acid	SCX
COSMOSIL CM	Nacalai	Silica; 120 Å; 5 μm	Carboxymethyl	WCX
Excelpak CHA-P44	Yokogawa (Agilent)	Styrene/divinyl-benzene	Sulfonic acid	SCX
PolyCAT A	PolyLC	Silica; 300, 1000, 1500 Å; 3, 5, 12 μm	Poly(aspartic acid)	WCX
PolySulfoethyl A	PolyLC	Silica; 200, 300, 1000 Å; 3, 5, 12 μm	Poly(2-sulfonylethyl aspartamide)	SCX
Sulfonated S-DVB	Custom synthesis	Styrene/divinyl-benzene	Sulfonic acid	SCX
SynChropak CM 300	Eichrom (Eprogen)	Silica; 300 Å; 6 μm	Carboxymethyl	WCX

WCX, weak cation exchange, *SCX*, strong cation exchange.

APPENDIX 3e Several Examples of HILIC Columns with Zwitterionic Phases

Brand Name	Manufacturer	Support/Pore Size/Particle Size	Nature of Phase
Nucleodur HILIC	Macherey–Nagel	Silica; 110 Å; 1.8, 3, 5 μm	Dimethylamino and sulfonic
Obelisc R	Sielc	Silica; 100 Å; 5, 10 μm	Zwitterrionic negative charged groups
Obelisc N	Sielc	Silica; 100 Å; 5, 10 μm	Zwitterrionic positive charged groups
PolyCAT A	PolyLC	Silica; 60, 100, 200, 300, 500, 1000, 1500 Å; 3, 5, 12 μm	Poly(aspartic acid)
Syncronis HILIC	Thermo Fisher	Silica; 100 Å, endcapped; 5 μm	Zwitterrionic
ZIC-HILIC	SeQuant (Merck)	Silica with sulphobetaine; 100, 200 Å; 3.5, 5 μm	Polymeric sulfonylalkylbetaine
ZIC-pHILIC	SeQuant (Merck)	Polymer with sulphobetaine; 5 μm	Polymeric sulfonylalkylbetaine
ZIC-cHILIC	SeQuant (Merck)	Silica with phosphorylcholine; 100 Å; 3 μm	Polymeric phosphorylcholine

APPENDIX 3f Examples of HILIC Columns With Hydride-Based Stationary Phase

Brand Name	Manufacturer	Particle Size (μm)	Surface Area (m^2/g)	Support/Pore Size (Å)	pH Range of Use	Max. Temp (°C)
Cogent Diamond Hydride	HiChrom	4	350	Silica 100	2.0–8.0	80
Diamond Hydride	MicroSolv	4	390	Silica C 100	2.5–7.5	60
Diamond Hydride 2.0	MicroSolv	2.2	340	Silica C 120	2.5–7.5	60
Diamond Hydride	HiChrom	4	350	Silica C 100	2.5–7.0	60

APPENDIX 3g Several More Recent HILIC Columns

Column Name	Manufacturer	Phase	Particle Size	Type
Glycoprotein BEH Amide	Waters	Amide	1.7	Porous
HALO Glycan	Advanced Materials Technol.	Proprietary	2.7	Core–shell
HILICpak VG-50	Shodex	Amino	5	Polymeric
HILICpak VT-50	Shodex	Quaternary ammonium	5	Polymeric
iHILIC-Fusion	HILICON-AB	Hydroxyethyl amide, sulfate, phosphate, quaternary ammonium	1.8, 3.5, 5	Porous
iHILIC-Fusion (+)	HILICON-AB	Hydroxyethyl amide, sulfate, quaternary ammonium	1.8, 3.5, 5	Porous
SunShell HILIC Amide	ChromaNik Technologies Inc.	Amide	2.6	Core–shell

Appendix 4: Ion Exchange and Ion-Moderated Stationary Phases

APPENDIX 4a Some Commercial Silica-Based Cation Exchange Columns

Column	Manufacturer	Type	Dimensions (Length × i.d., mm)	Capacity (mEquiv g^{-1})	Particle Size (μm)	Pore Size (Å)	Type of Phase
BioBasic SCX	Thermo	Strong	Various	0.07	5	300	$-SO_3H$
IC YK-421	Shodex	Weak	125 × 4.6	—	5	20	Coated silica polymer-COOH
LiChrosil IC CA	Merck	Weak	100 × 4.6	—	5	—	Coated PBDMA
Luna SCX	Phenomenex	Strong	150 × 4.6	0.15	5, 10	300	$-C_6H_4-SO_3H$
Nucleosil 5 SA	Machery-Nagel	Strong	125 × 4	0.5	5	100	$-SO_3H$
Nucleosil 5 SA	Machery-Nagel	Strong	Various	1	5, 10	100	$-C_3H_6-C_6H_4\text{-}SO_3H$
Partisil 10 SCX	Whatman	Strong	250 × 4.6	0.5	5, 10	85	$-C_6H_4-SO_3H$
PartiSphere	Whatman	Strong	125 × 4.6 250 × 4.6	—	5	120	—
Phenosphere SCX	Phenomenex	Strong	Various	0.6	5, 10	80	$-C_6H_4-SO_3H$
PolyCAT A	PolyLC	Weak	Various	—	3, 5, 12	300, 1000, 1500	Poly(aspartic acid) bonded to silica
PolySULFOETHYL A	PolyLC	Strong	Various	—	3, 5, 12	300, 1000	Sulfoethyl-aspartamide
Spherisorb SCX	Waters	Strong	125 × 4	—	5	80	—
Supelcosil LC-SCX	Sigma	Strong	250 × 4.6	—	5	120	$-C_3H_6-SO_3H$
SynChropak	Lab Unlimited	Weak	—	—	6	300	Carboxymethyl
TSK Gel IC Cation SW	Toyo Soda	Strong	50 × 4.6	0.5	5	—	$-SO_3H$
TSKgel CM-2SW	Tosoh	Weak	Various	0.3	5	125	Carboxymethyl
TSKgel SP-2SW	Tosoh	Strong	Various	0.3	5	125	$-C_3H_6-SO_3H$
Universal Cation	Alltech	Weak	100 × 4.6	—	7	180	Coated PBDMA
Vydac SC	Separation Group	Strong	250 × 4.6	0.1	30–44	300	$-SO_3H$

PBDMA, poly(butandiolmethacrylate).

APPENDIX 4b Some Cation Exchange Polymeric Columns

Column	Manu-facturer	Type	Dimensions (Length × i.d., mm)	Particle Size/Pore Size	Capacity (mEquiv g^{-1})	Technol.	Type of Phase
CM-825	Shodex	Weak	75 × 8	8 μm/5000 Å	0.4	—	Carboxymethyl
Diaion	Mitsubishi	Strong	Various	—	—	—	PS-DVB–SO_3H
Diaion	Mitsubishi	Weak	Various	—	—	Highly porous	Acrylic acid-methacrylate
ES-502C 7 C	Shodex	Weak	100 × 7.5	9 μm/2000 Å	0.55	—	PVA carboxymethyl
IC YS-50	Shodex	Weak	125 × 4.6	—	—	—	PVA
ICT-521	Shodex	Strong	150 × 4.6	—	—	—	PS-DVB–SO_3H
IonPac CS-10	Dionex	Strong	250 × 4	8.5 μm	0.08	Latex	PS-DVB–SO_3H
IonPac CS-11	Dionex	Strong	250 × 2	8 μm	0.035	Latex	PS-DVB–SO_3H
IonPac CS-12, 14, 16, 17, 18, 19	Thermo/ Dionex	Weak	250 × 4 or 250 × 2	8 μm/60 Å	0.7; 2.8	Latex	PS-DVB–COOH
IonPac CS-12A	Thermo/ Dionex	Medium	Various	5; 8 μm	0.7; 0.94; 2.8	Latex	PS-DVB–COOH + PS-DVB–PO_3H
IonPac CS-15	Thermo/ Dionex	Weak	250 × 4 or 250 × 2	8.5 μm	0.7; 2.8	Latex	PS-DVB–COOH, –PO_3H/crown ether
LCA K02	Sykam	Strong	125 × 4.6	5 μm	0.4	—	PMA
MAbPAC SCX-10	Thermo	Strong	Various	3; 5; 10 μm	—	Latex	PS-DVB–SO_3H
PL-SCX	Agilent	Strong	Various	10; 30 μm/1000, 4000 Å	—		PS-DVB–SO_3H
ProPac SCX	Thermo	Strong	250 × 4	10 μm	—	Latex	PS-DVB–SO_3H
ProPac WCX	Thermo	Weak	250 × 4	10 μm	—	Latex	PS-DVB–COOH
PRP-X100, X200, X400	Hamilton	Strong	250 × 4.6	100 Å	0.035–2.5	—	PS-DVB–SO_3H
Shimpack IC-C1	Shimadzu	Strong	150 × 4	10 μm	—	—	PS-DVB–SO_3H
SP-825	Shodex	Strong	75 × 8	8 μm/5000 Å	0.4	—	–C_3H_6–SO_3H
TSKgel BioAssist S	Tosoh	Strong	Various	7; 13 μm/1300 Å	0.1	—	–C_3H_6–SO_3H
TSKgel CM-STAT	Tosoh	Weak	Various	7; 10 μm	0.1	—	Carboxymethyl
TSKgel OApak-A	Tosoh	Weak	Various	5 μm	1.5	—	Methacrylate
TSKgel SP-5PW	Tosoh	Strong	Various	10, 13, 20 μm/ 1000 Å	>0.1	—	—
TSKgel SP-STAT	Tosoh	Strong	Various	7; 10 μm	0.023	—	–C_3H_6–SO_3H
YS-50	Shodex	Weak	125 × 4.6	5 μm	—	—	PVA–COOH

PS-DVB, polystyrene-divinylbenzene; *PVA*, polyvinylalcohol; *PMA*, polymethylacrylate.

APPENDIX 4c Some Commercial Silica-Based Anion Exchange Columns

Column	Manufacturer	Type	Dimensions (Length × i.d., mm)	Capacity (mEquiv g^{-1})	Particle Size (μm)	Type of Phase
Inertsil AX	GL Sciences	Strong	Various from 33 × 1 to 250 × 4.6	0.43	5	Diethylamino group
Luna NH_2	Phenomenex	Weak	Various	—	3, 5, 10	$-NH_2$
Nucleosil 10 Anion	Macherey–Nagel	Medim	250 × 4.0	0.06	10	Trimethylamine methyldiethylamine
Partisil 10 SAX	Whatman	Strong	250 × 4.6	0.5	5, 10	$-NR_3^+$
Partisphere SAX	Whatman	Strong	150 × 4.6 250 × 4.6	—	5	—
Phenosphere SAX	Phenomenex	Strong	Various	0.4	5, 10	$-NR_3^+$
SynChropak SAX	Lab Unlimited	Weak	100 × 4.6 250 × 4.6	—	6.5	Polyethyleneimine
SynChropak SAX	Lab Unlimited	Strong	100 × 4.6 250 × 4.6	—	6.5	NR_3^+
TSKgel QAE-2SW	Tosoh	Strong	Various	>0.3	5	$-N(CH_3)_3^+$
TSKgel DEAE-2SW	Tosoh	Weak	Various	>0.3	5	Diethylaminoethyl
TSKgel Q-STAT	Tosoh	Strong	Various	0.27	7; 10	$-NR_3^+$
TSKgel DNA-STAT	Tosoh	Strong	Various	0.27	5	Acrylate $-NR_3^+$
TSK Gel IC-SW	Toyo Soda	Strong	250 × 4.6	0.4	5	$-N(C_2H_5)_2CH_3^+$
Vydac 302 IC 4.6	Separation Group	Strong	50 × 4.6	0.1	10	Spherical particles with $-NR_3^+$
Vydac 300 IC 4.6	Separation Group	Strong	250 × 4.6	0.1	15	
Wescan 269-001	Wescan	Strong	250 × 4.6	0.08	13	$-NR_3^+$

APPENDIX 4d Some Anion Exchange Polymeric Columns

Column	Manufacturer	Type	Particle Size (μm)	Dimensions (Length × i.d., mm)	Type of Phase
Allsep	Grace – Alltech	Strong	7 μm	Various	PMA gel- NR_3^+
AN1; AN300	Sarasep	Strong	9 μm	250 × 4.6 100 × 7.5	PS-DVB-amine
BioSuite DEAE	Waters	Strong	2.5 μm; 10 μm	35 × 4.6 75 × 7.5	PMA gel- NR_3^+
BioSuite Q-PEEK	Waters	Strong	10 μm	50 × 4.6	PMA gel- NR_3^+
CarboPac MA1	Thermo/Dionex	Strong	7.5 μm	250 × 4.0	Resin with tertiary amine
CarboPac PA1[a]	Thermo/Dionex	Strong	10 μm	250 × 2.0 250 × 4.0	Latex technology
CarboPac SA10	Thermo/Dionex	Strong	6 μm	250 × 2.0 250 × 4.0	Latex nano beeds
CarboPac PA1, PA10, PA20, PA100, PA200	Thermo/Dionex	Strong	Various	Various	Pellicular, nanoprous beads, etc.
Cosmogel QA	Nacalai Tesque, Inc.	Strong	5 μm	75 × 8	PMA gel- NR_3^+
Cosmogel DEAE					PMA gel- $NH(C_2H_5)_2^+$
Diaion	Mitsubishi Chemical	Strong	10 μm	75 × 7.5	Styrenic/acrylic amine
Discovery BIO PolyMA-WAX	Supelco	Strong	5 μm	50 × 4.6	PMA gel- NR_3^+
IC I-524A	Shodex	Strong	12 μm	100 × 4.6	Polyhydroxy-methacrylate $-NR_3^+$
IC NI-424	Shodex	Strong	5 μm	100 × 4.6	Polyhydroxy-methacrylate $-NR_3^+$
IC-Pak anion	Waters	Strong	10 μm	75 × 4.6	PMA gel- NR_3^+
IC SI-35 4D; IC SI-50 4E; IC SI-52 4E; IC SI-90 4E; IC SI-91 4C	Shodex	Strong	3.5; 5; 9 μm	Various	PVA- NR_3^+
IEC DEAE-825	Shodex	Weak	12	75 × 8	PHMA $-NR_3^+$
Ion Swift Max 100 monolithic	Thermo/Dionex	Medium	—	250 × 1 250 × 0.25	PS-DVB-alkanol $-NR_3^+$
Ion Swift Max 200 monolithic	Thermo/Dionex	Medium	—	250 × 0.25	PS-DVB-alkanol $-NR_3^+$

APPENDIX 4d Some Anion Exchange Polymeric Columns—cont'd

Column	Manufacturer	Type	Particle Size (μm)	Dimensions (Length × i.d., mm)	Type of Phase
IonPac AS10[a]	Thermo/Dionex	Strong	8.5 μm	250 × 2	Latex technology
				250 × 4	
IonPac 15[a], 16[a], 20[a], etc.	Thermo/Dionex	Various	9 μm	250 × 0.4	PS-DVB $-NR_3^+$ and alkanol
				250 × 2	
				250 × 4	
IonPac AS11[a]	Thermo/Dionex	Strong	13 μm	250 × 2	Latex technology
				250 × 4	
IonPac AS11-HC[a]	Thermo/Dionex	Strong	9 μm	250 × 0.2	Latex technology
				250 × 2	
				250 × 4	
IonPac AS12 A[a]	Thermo/Dionex	Strong	9 μm	200 × 2	Latex technology
				200 × 4	
IonPac AS14, 14A, 12A	Thermo/Dionex	Medium	7; 9 μm	Various	PS-DVB-alkanol $-NR_3^+$
IonPac AS16[a]	Thermo/Dionex	Strong	9 μm	250 × 0.2	Latex technology
				250 × 2	
				250 × 4	
IonPac AS17-C[a]	Thermo/Dionex	Strong	10.5 μm	250 × 2	Latex technology
				250 × 4	
IonPac AS18	Thermo/Dionex	Strong	7.5; 13 μm	Various	Polyethylvinylbenzene-DVB $-NR_3^+$
IonPac AS22, AS23,	Thermo/Dionex	Medium	6; 11 μm	250 × 0.4	PS-DVB-alkanol $-NR_3^+$
				250 × 2	
				250 × 4	
IonPac AS4A[a]	Thermo/Dionex			Various	Latex technology
IonPac AS4A-SC[a]	Thermo/Dionex	Strong		Various	Latex technology
IonPac AS5[a]	Thermo/Dionex	Medium	15 μm	250 × 4	PS-DVB-alkanol $-NR_3^+$
IonPac AS7[a]	Thermo/Dionex	Strong	10 μm	250 × 2	PS-DVB $-NR_3^+$
				250 × 4	
IonPac AS9 HC	Thermo/Dionex	Strong	9 μm	250 × 2	PS-DVB $-NR_3^+$
				250 × 4	

(Continued)

APPENDIX 4d Some Anion Exchange Polymeric Columns—cont'd

Column	Manufacturer	Type	Particle Size (μm)	Dimensions (Length × i.d., mm)	Type of Phase
IonPac AS9 SC, AS4A-SC	Thermo/Dionex	Medium	9; 13 μm	250 × 2 250 × 4	PS-DVB $-NR_3^+$ and alkanol
IonPac AS9-HC[a]	Thermo/Dionex	Strong	—	—	Latex technology
IonPac AS9-SC[a]	Thermo/Dionex	Strong	—	—	Latex technology
LCA A01	Sykam	Strong	—	200 × 4.0	PS-DVB-amine
MetroSep anion Dual 2	Metrohm-Peak, Inc.	Strong	8 μm	75 × 4.6	PMA gel- NR_3^+
MetroSep anion Dual 3	Metrohm-Peak, Inc.	Strong	6 μm	100 × 4.0	PMA gel- NR_3^+
MetroSep anion Dual 4	Metrohm-Peak, Inc.		—	100 × 4.6	Monolithic silica gel with 2 μm macropores and 13 nm mesopores
Metrosep A Supp 1	Metrohm-Peak, Inc.	Strong	7 μm	50 × 4.6 250 × 4.6	PS-DVB–$N(CH_3)_3^+$-
Metrosep A Supp 3	Metrohm-Peak, Inc.	Strong	9 μm	250 × 4.6	PS-DVB–$N(CH_3)_3^+$
Metrosep A Supp 5	Metrohm-Peak, Inc.	Strong	5 μm	50 × 4.0 100 × 4.0 150 × 4.0 250 × 4.0	PVA gel- NR_3^+
Metrosep A Supp 7	Metrohm-Peak, Inc.	Strong	5 μm	150 × 4.0 250 × 4.0	PVA gel- NR_3^+
Metrosep A Supp 10 Metrosep A Supp 15 Metrosep A Supp 15	Metrohm-Peak, Inc.	Strong	4.6 μm	50 × 4.0 75 × 4.0 100 × 4.0 250 × 4.0	PS-DVB–$N(CH_3)_3^+$
Nucleogel SAX	Macherey–Nagel	Strong	—	50 × 4.0 75 × 4.0 100 × 4.0 250 × 4.0	PMA gel- NR_3^+
ProPac SAX	Thermo/Dionex	Strong	—	250 × 4	$-NR_3^+$
ProPac WAX	Thermo/Dionex	Strong	—	250 × 4	Tertiary amine
Protein-Pak Q 8HR	Waters	Strong	—	250 × 4	PMA gel- NR_3^+

APPENDIX 4d Some Anion Exchange Polymeric Columns—cont'd

Column	Manufacturer	Type	Particle Size (μm)	Dimensions (Length × i.d., mm)	Type of Phase
PRP X100	Hamilton	Strong	10 μm	125 × 4.0	PS-DVB–$N(CH_3)_3^+$
				250 × 4	
PRP X500	Hamilton	Strong	5 μm	150 × 4.6	Poly(methacrylamido-propyl)–$N(CH_3)_3^+$
RCX-30	Hamilton	Strong	7 μm	150 × 4.6	PS-DVB–$N(CH_3)_3^+$
Shim-Pack WAX-1; WAX-2	Shimadzu	Strong	3; 5 μm	50 × 4	–NR_3^+
Shodex IEC QA-825	Shodex	Strong	12 μm	75 × 8	PMA gel- NR_3^+
Si-90 4E	Shodex	Strong	9 μm	250 × 4	PVA- NR_3^+
Star-Ion A300	Phenomenex	Strong	7 μm	100 × 4.6	PS-DVB–NR_3^+
Star-Ion A300 HC (high capacity)	Phenomenex	Strong	7 μm	100 × 10.0	PS-DVB–NR_3^+
Super-Sep IC Anion	Metrohm-Peak, Inc.	Strong	9 μm	250 × 4.6	PVA gel- NR_3^+
TSKgel BioAssist Q	Tosoh Bioscience	Strong	10; 13 μm	50 × 4.6	PMA gel- NR_3^+
TSKgel DNA-STAT	Tosoh Bioscience	Strong	5 μm	100 × 4.6	PMA gel- NR_3^+
TSKgel IC-Anion-PW	Tosoh Bioscience	Strong	10 μm	50 × 4.6	PMA gel- NR_3^+
TSKgel Q-STAT	Tosoh Bioscience	Strong	7; 10 μm	35 × 3	PMA gel- NR_3^+
TSKgel SuperQ-5PW	Tosoh Bioscience	Strong	10; 13 μm	75 × 7.5	PMA gel- NR_3^+
Zodiac IC anion	Zodiac Life Sciences	Strong	5; 10 μm	—	PMA gel–NR_3^+

[a]*Phases made using latex technology.*

APPENDIX 4e Some Ion-Moderated (Ion Exclusion) Polymeric Columns

Column	Manufacturer	Type	Data About Construction
ChromSpher Lipids	Agilent	Silica with sulfonic groups and Ag^+	Pore size: 120 Å; 5 μm particles; 250 × 4.6 mm
HC-75 Ca^{2+}	Hamilton	PS-DVB sulfonic Ca^{2+}	Pore size: 100 Å; exchange capacity: 5 meq/g
HC-75 H^+	Hamilton	PS-DVB sulfonic H^+	
HC-75 Pb^{2+}	Hamilton	PS-DVB sulfonic Pb^{2+}	
IonPac ICE-AS1	Thermo/Dionex	PS-DVB sulfonic H^+	27 meq/column for 250 × 9 mm; 5.3 meq/column for 250 × 4 mm; 7.5 μm bead diam.
IonPac ICE-AS6	Thermo/Dionex	PS-DVB sulfonic and carboxylic H^+	27 meq/column for 250 × 9 mm; 8 μm bead diam.
IonPac ICE-Borate	Thermo/Dionex	PS-DVB sulfonic H^+	27 meq/column for 250 × 9 mm; 7.5 μm bead diam.
MCI CK/CA Series	Mitsubishi Chem.	PS-DVB sulfonic with various ions Ca^{2+} Ag^+, etc.	5–20 μm particle size; various dimensions
RCM-Monosaccharide (L19 packing)	Phenomenex	8% crosslinked PS-DVB sulfonic Ca^{2+}	300 × 7.8 mm; 8 μm particle size
RHM-Monosaccharide (L17 packing)	Phenomenex	8% crosslinked PS-DVB sulfonic H^+	300 × 7.8 mm; 8 μm particle size
RAM-Carbohydrate	Phenomenex	8% crosslinked PS-DVB sulfonic Ag^+	300 × 7.8 mm; 8 μm particle size
RSO-Oligosaccharide	Phenomenex	4% crosslinked PS-DVB sulfonic Ag^+	200 × 10 mm; 12 μm particle size
RNO-Oligosaccharide	Phenomenex	4% crosslinked PS-DVB sulfonic Na^+	200 × 10 mm; 12 μm particle size
RPM-Monosaccharide (L34 packing)	Phenomenex	8% crosslinked PS-DVB sulfonic Pb^{2+}	300 × 7.8 mm; 8 μm particle size
RNM-Carbohydrate (L54 packing)	Phenomenex	8% crosslinked PS-DVB sulfonic Na^+	300 × 7.8 mm; 8 μm particle size
ROA-Organic Acid (L22 packing)	Phenomenex	8% crosslinked PS-DVB sulfonic H^+	300 × 7.8 mm; 8 μm particle size
RFQ-Fast Acid	Phenomenex	8% crosslinked PS-DVB sulfonic H^+	100 × 7.8 mm; 8 μm particle size
RKP-Potassium	Phenomenex	8% crosslinked PS-DVB sulfonic K^+	300 × 7.8 mm
RCU-USP Sugar Alcohols (L19 packing)	Phenomenex	8% crosslinked PS-DVB sulfonic Ca^{2+}	250 × 4.0 mm; 8 μm particle size
SUGARSH1011, SUGARSC1011, etc.	Shodex	PS-DVB sulfonic H+; Ca^{2+}; Pb^{2+}; Zn^{2+}; Na^+	300 × 8 mm; 50 × 8 mm; 250 × 6 mm; 150 × 6 mm; 6–30 μm particles

Appendix 5: Chiral Stationary Phases

APPENDIX 5a Pirkle-Type Stationary Phases

Chiral Selector	Product Line	Maker
(R)-Phenylglycine, 3,5-dinitrobenzoic acid, amide linkage, propyl handle	Chrex (R)-PGLY and DNB (3001)	Phenomenex
(R)-1-Naphtylglycine, 3,5-dinitrobenzoic acid, amide linkage, propyl handle	Chrex (R)-NGLY and DNB (3005)	Phenomenex
(S)-Valine, 3,5-dinitroaniline, urea linkage, propyl handle	Chrex 3010	Phenomenex
(S)-tert-Leucine, 3,5-dinitroaniline, urea linkage, propyl handle	Chrex 3011	Phenomenex
(R)-Phenylglycine, 3,5-dinitroaniline, urea linkage, propyl handle	Chrex 3012	Phenomenex
(S)-Valine, (R)-1-(α-naphthyl)ethylamine, urea linkage, propyl handle	Chrex 3014	Phenomenex
(S)-Proline, (R)-1-(α-naphthyl)ethylamine, urea linkage, propyl handle	Chrex 3018	Phenomenex
(S)-tert-Leucine, (S)-1-(α-naphthyl)ethylamine, urea linkage, propyl handle	Chrex 3019	Phenomenex
(S)-tert-Leucine, (R)-1-(α-naphthyl)ethylamine, urea linkage, propyl handle	Chrex 3020	Phenomenex
(S)-Indoline-2-carboxylic acid, (R)-1-(α-naphthyl)ethylamine, urea linkage, propyl handle	Chrex 3022	Phenomenex
(R) or (S)-Phenylglycine, 3,5-N-dinitrobenzoyl, amide linkage, propyl handle	ChiralCap (R) or (S)-DNBPG	Astec
(S)-Leucine 3,5-N-dinitrobenzoyl, urea linkage, propyl handle to silica	ChiralCap (S)-DNBLeu	Astec
(R)-Phenylmethylurea (propyl handle to silica)	Spherisorb chiral 1	Waters
N-3,5-Dinitrobenzoyl-3-amino-3-phenyl-2-(1,1-dimethylethyl)-propanoate bonded to silica by undecyl handle	(R,R)-, (S,S)-β-GEM 1	Regis Technologies
Dimethyl-N-3,5-dinitro-benzoyl-amino-2,2-dimethyl-4-pentenyl phosphonate silica	(R)-, (S)-α-Burke 2	Regis Technologies
N-(1-Propylsilica-2-oxo-4-phenylazetidin-3-yl)-3,5-dinitrobenzamide	(3R,4S)-, (3S,4R) Pirkle 1-J	Regis Technologies

(*Continued*)

APPENDIX 5a Pirkle-Type Stationary Phases—cont'd

Chiral Selector	Product Line	Maker
3,5-Dintrobenzoyl derivative of diphenylethylenediamine bonded to silica by tetradecanoic acid handle	(R,R)-, (S,S)-ULMO	Regis Technologies
3,5-Dinitrobenzoyl derivative of 1,2-diaminocyclohexyl bonded to silica by ether containing handle	(R,R)-, (S,S)-DACH-DNB	Regis Technologies
3,5-Dinitro-N-(3-propyl-1,2,3,4-tetrahydrophenanthren-4-yl) benzamide (3-propyl handle to silica)	(R,R)-, (S,S)-Whelk 1	Regis Technologies
Similar to Whelk 1 using trifunctional bonding of propyl handle to silica	(R,R)-, (S,S)-Whelk 2	Regis Technologies
(6-Methoxyquinolin-4-yl)[6-(1H-1,2,3-triazol-4-yl)-1-azabicyclo[2.2.2]octan-2-yl]methyl-N-(3,5-dinitrophenyl) carbamate (propyl handle to silica)	Quinine carbamate	[a]
Qinine group bonded by thioether	Chiralpak QN-AX	[a]
1-N-[1-(Naphthalen-1-yl)ethyl]pyrrolidine-1,2-dicarboxamide (propyl handle to silica)	(S)-proline and (R)-1-(α-naphthyl)ethylamine-urea	[b,c]
Naphthylleucine bonded to silica by tetradecyl handle	(R)-naphthylleucine	[c]

[a]*S.C. Moldoveanu, V. David, Essentials in Modern HPLC Separations, Elsevier, Amsterdam, 2013.*
[b]*K.M. Kacprzak, W. Lindner, Novel Pirkle-type quinine 3,5-dinitrophenylcarbamate chiral stationary phase implementing click chemistry, J. Sep. Sci. 34 (2011) 1–6.*
[c]*M. Lämmerhofer, P. Franco, W. Lindner, Quinine carbamate chiral stationary phases: systematic optimization of steric selector-select and binding increments and enantioselectivity by quantitative structure-enantioselectivity relationship studies, J. Sep. Sci. 29 (2006) 1486–1496.*

APPENDIX 5b SUMICHIRAL[a] Stationary Phases Pirkle-Type

Phase	Phase	Chiral Component	Elution Mode[b]
OA-2000	OA-2000S	(R)-phenylglycine, dinitrobenzene	NP
OA-2500	OA-2500S	(R)-1-naphthylglycine, dinitrobenzene	NP
OA-3100	OA-3100R	(S)-valine, dinitrobenzene	NP,RP
OA-3200	OA-3200R	(S)-tert-leucine, dinitrobenzene	NP,RP
OA-3300	OA-3300S	(R)-phenylglycine, dinitrobenzene	NP-RP
OA-4000	OA-4000S	(S)-valine (S)-1-(α-naphthyl)ethylamine	NP
QA-4100	OA-4100R	(S)-valine (R)-1-(α-naphthyl)ethylamine	NP
OA-4400	OA-4400R	(S)-proline (S)-1-(α-naphthyl)ethylamine	NP
OA-4500	OA-4500R	(S)-proline (R)-1-(α-naphthyl)ethylamine	NP
OA-4600	OA-4600R	(S)-tert-leucine (S)-1-(α-naphthyl)ethylamine	NP
OA-4700	OA-4700R	(S)-tert-leucine (R)-1-(α-naphthyl)ethylamine	NP
OA-4800	–	(S)-indoline-2-carboxylic acid (S)-1-(α-naphthyl)ethylamine	NP
OA-4900	–	(S)-indoline-2-carboxylic acid (R)-1-(α-naphthyl)ethylamine	NP
OA-6000	OA-6000R	(L)-tartaric acid (S)-1-(α-naphthyl)ethylamine	RP
OA-6100	OA-6100R	(L)-tartaric acid, (S)-valine (R)-1-(α-naphthyl)ethylamine	RP

[a]*Manufactured by Sumika Chemical Analysis Services.*
[b]NP, *similar to normal phase;* RP, *similar to reversed phase.*

APPENDIX 5c Chiral Columns Based on Derivatized Cellulose

Name	Derivatization Group	Applications
Chiralcel OA(ester)	$-(CO)-CH_3$	Small aliphatic compounds
Chiralcel CA-1(ester)	$-(CO)-CH_3$	Alcohols
Cellulose triacetate	$-(CO-CH_3$	Various
Cellulose Cel-AC-40XF	$-(CO-CH_3$	Various
Chiralcel OB (ester)	−(CO)−Ph	Small aliphatic and aromatic compounds
Chiralcel OB-H	−(CO)−Ph	High-efficiency separations
Chiralcel OC (carbamate)	−CONH−Ph	Cyclopentenones
Chiralcel OJ (ester)	−(CO)−Ph−4-Me	Aromatic compounds
Chiralcel OD (carbamate)	−CONH−Ph−3,5-di-Me	Alkaloids, tropines, amines, beta blockers
Chiralcel OD-R	−CONH−Ph−3,5-di-Me	Pharmaceutical drugs
Kromasil CelluCoat	−CONH−Ph−3,5-di-Me	Pharmaceutical drugs
Chiral Art cellulose C	−CONH−Ph−3,5-di-Me	Pharmaceutical drugs
Chiral Art cellulose SB	−CONH−Ph−3,5-di-Me	Pharmaceutical drugs
Cellulose DMP (Astec)	−CONH−Ph−3,5-di-Me	Beta blockers
ChromegaChiral CCO F4 T3	−CO−Ph−4-F-3-trifluoromethyl	Small molecules
Lux Cellulose-1	−CONH−Ph-3,5-di-Me	Pharmaceutical drugs
Chiralcel OF	−CONH−Ph-*para*-Cl	Beta lactams, dihydroxypyridines, alkaloids
Chiralcel OG	−CONH−Ph-*para*-Me	Beta lactams, alkaloids
Chiralcel OK (ester)	$-CO-(CH_2{=}CH_2)-Ph$	Aryl methyl esters, aryl methoxy esters
Lux Cellulose-2	−CONH−Ph-3-Cl-4-Me	Pharmaceutical drugs
Sepapak-2	−CONH−Ph-3-Cl-4-Me	Chiral pesticides
Lux Cellulose-3	−(CO)−Ph-4-Me	Chiral pesticides
Lux Cellulose-4	−CONH−Ph-4-Cl-3-Me	Chiral pesticides
Sepapak-4	−CONH−Ph-4-Cl-3-Me	Chiral pesticides

NOTE: Chiralcel (Daicel Chemical Industries).

APPENDIX 5d Chiral Columns Based on Derivatized Amylose

Name	Derivatization Group	Support
Chiralpak AD and AD-3	–CONH–Ph-3,5-di-Me	Silica (coated on) 5 μm and 3 μm
Kromasilu AmyCoat	–CONH–Ph-3,5-di-Me	Silica (coated on) 5 μm
Chiralpak AS	–CONH-C[a]H(Me)Ph	Silica (coated on)
Chiralpak AY-3	–CONH–Ph-5-Cl-2-Me	Silica (coated on) 3 μm
Chiralpak AZ-3, IF, IF-3	–CONH–Ph-3-Cl-4-Me	Silica (coated on) 3 μm
Chiralpak IA, IB	–CONH–Ph-3,5-di-Me	Silica (coated on) 5 μm (also used in SFC)
Chiralpak ID and ID3	–CONH–Ph-3-Cl	Silica (coated on) 5 μm and 3 μm
Chiralpak IE and IE3	–CONH–Ph-3,5-di-Cl	Silica (coated on) 5 μm and 3 μm
Lux Amylose-1	–CONH–Ph-3,5-di-Me	Silica (coated on) 5 μm and 3 μm
Lux Amylose-2	–CONH–Ph-5-Cl-2-Me	Silica (coated on) 5 μm and 3 μm
Chiral Art Amylose C	–CONH–Ph-3,5-di-Me	Silica (coated on) 5, 10 and 20 μm
Chiral Art Amylose SA	–CONH–Ph-3,5-di-Me	Silica (coated on) 10 and 20 μm

[a]*Asymmetric carbon.*

APPENDIX 5e Cyclodextrins on Silica

Product line	Chiral Selector	Information About the Column
Astec CYCLOBOND I 2000	β-Cyclodextrin	Particles diameter d_p 5 or 10 μm; pore size: 100 Å; length × i.d.: various
Astec CYCLOBOND I 2000 AC	β-Cyclodextrin peracetylated	
Astec CYCLOBOND I 2000 SP	β-Cyclodextrin-hydroxypropyl	
Astec CYCLOBOND I 2000 DMP	tris-(3,5-Dimethylphenyl) carbamate-β-cyclodextrin	
Astec CYCLOBOND I 2000 DM	Dimethylated β-cyclodextrin	
Astec CYCLOBOND I 2000 DNP	tris-(3,5-Dinitrophenyl) carbamate β-cyclodextrin	
Astec CYCLOBOND I 2000 RSP	(R,S)-Hydroxypropyl modified beta-cyclodextrin	
Astec CYCLOBOND II	γ-Cyclodextrin	
Astec CYCLOBOND II AC	γ-Cyclodextrin peracetylated	
LARIHC	Derivatized cyclofructan 6, cyclofructan 7	[a,b]
LARIHC CF6 RN	Derivatized cyclofructan 6 (R-naphthylethyl)	[a,b]
LARIHC CF6-P	Derivatized cyclofructan 6	[a,b]
FRULIC-C FRULIC-N	Derivatized cyclofructan 6	d_p: 5 μm; length × i.d.: various
LARIHC CF7-DMP	Derivatized cyclofructan 7 (3,5-dimethylphenyl)	[a,b]
SUMICHRAL OA-7000	Novel spacer to β-cyclodextrin (derivatized)	d_p: 5 μm; 150 × 4.6

[a] *P. Sun, C. Wang, Z.S. Breitbach, Y. Zhang, D.W. Armstrong, Development of new HPLC chiral stationary phases based on native and derivatized cyclofructans, Anal. Chem. 81 (2009) 10215–10226.*
[b] *P. Sun, D.W. Armstrong, Effective enantiomeric separations of racemic primary amines by the isopropyl carbamate-cyclofructan6 chiral stationary phase, J. Chromatogr. A 1217 (2010) 4904–4918.*

APPENDIX 5f Proteins on Silica Used as Chiral Phases

Product Line	Chiral Selector	Manufacturer
Bioptic AV-1	Avidin	GL Science
Chiral BSA	Bovine serum albumin (BSA)	Shandon
CHIRAL-HSA	Human serum albumin (HSA)	Shandon
CHIRALPAK-AGP	α1-Acid glycoprotein	Regis (Sigma)
CHIRALPAK-CBH	Cellobiohydrolase	Regis (Sigma)
CHIRALPAK-HSA	Human serum albumin (HSA)	Regis (Sigma)
Chirobiotic V, V2, T, T2, TAG and R	Glycoproteins	Astec
Resolvosil BSA 7 and BSA-7PX	Bovine serum albumin (BSA)	Machery-Nagel
TSKgel Enantio-OVM	Ovomucoid	Tosoh
Ultron ES-BSA	Bovine serum albumin (BSA)	Shinwa Chem. Ind.
Ultron ES-OVM	Ovoglycoprotein	Shinwa Chem. Ind.

APPENDIX 5g Chiral Synthetic Polymers

Chiral Selector	Product Line
Poly(*trans*-1,2-cyclohexanediyl-bis-acrylamide)	Astec P-CAP
Poly(diphenylethylenediamine-bis-acryloyl)	Astec P-CAP-DP
O,O′-Bis (3,5-dimethylbenzoyl)-N,N′-diallyl-L-tartar diamide	Kromasilu Chiral DMB
O,O′-Bis (4-tert-butylbenzoyl)-N,N′-diallyl-L-tartar diamide	Kromasilu Chiral TBB
Polymethacrylate trimethylphenyl on silica	Chiralpak OT
Polymethacrylate diphenyl(pyridyn-2-yl)methyl on silica	Chiralpak OP

Appendix 6: Size-Exclusion Stationary Phases

APPENDIX 6a Silica-Based SEC Commercial Packings (for Aqueous SEC or GFC)

Phase Name (Supplier)	Pore Size Designation	Range of Use	Particle Size	Chemistry on Silica
BioSuite	125, 250	5000–150,000	10, 12, and 17 μm	—
Ultrahydrogel	120, 250, 500, 1000, 2000 Å	5000–7,000,000	6 μm	—
Shodex Protein KW (Showa Denko)	800 Series	1000–10,000,000	5 μm	Diol; fully porous silica
TSK-gel SW, SW_{XL}, SuperSW (Tosoh)	Various G2000SW, G3000SW, G3000SW_{XL},UP-SW3000	Various 5000–70,000	2 (for UP), 5, 8, and 10 μm	Diol
TSK-gel QC-PAK	TSK 200, TSK 300	5000–50,000	5 μm	Diol
TSK-gel BioAssist DS	125, 250, 300, 450	Various	Various: 2–15 μm	Diol
Zorbax GF Ser. (Agilent/ Crawford Scientific)	GF-250, GF-450	4000–400,000, 10,000–900,000	4 and 6 μm	Zr clad diol
Zorbax PSM (Agilent/ Thomas Scientific)	PMS 60, PMS 60S, PMS 300, PMS 1000, etc.	Various	5 μm	Silanized
LiChrosphere Si	Si60, Si100	Also normal phase	5, 10 μm	Silica
UltraSpherogel (Grace/ Beckman)	2000, 4000, etc.	—	5 μm	Polyether
Bio-Sil, Bio-Select (Bio-Rad)	SEC-125-5, 250-5, 400-5	1000–1,000,000	5, 10 μm	Silica
Protein-Pak (Waters)	300SW	1000–20,000	10 μm	Diol
Bio-Sep (Phenomenex)	SEC-S2000, SEC-S3000, SEC-S4000	500–20,000,000	5 μm	Hydrophilic bonded silica
SynChropak (Lab Unlimited)	GPC Peptide, GPC 100 to GPC 4000, CATSEC 100 to 4000	Various ranges	5, 10 μm	Diol

APPENDIX 6b Several Commercially Available Phases Used in Nonaqueous SEC (GPC)

Phase Name (Supplier)	Pore Size Designation	Range of Mw of Use[a]	Particle Size
Jordi Labs GPC DVB Fluorinated	Various from 100 to 100,000 Å	Various between 100 and 10,000,000	5 μm
PLgel (Polymer Lab. Ltd./ Agilent)	Various from 50 Å to 1,000,000 Å	Various from 100–2000 to 100,000–20,000,000	3, 5, 10, and 20 μm
PLgel multipore bed	Various from MIXED-A to MIXED-E	Various from 200–400,000 to 1000–40,000,000	5, 10, 20 μm
PolySep (Phenomenex)	Various from 1000 to 6000 and linear	Various from 100–2000 to 100,000–20,000,000	5, 10, and 20 μm
Shodex K and KF,KD, KL (Showa Denko)	Various indicated as 801, 802, 807	Various from 70,000 average to 200,000,000 average	Nominally 7 μm
Shodex K multipore bed	Various indicated as 803L, to 807L	Various from 1500 average to 200,000,000 average	Various depending on range (6, 10, and 17 μm)
TSK-Gel H_{XL}, SuperHZ (Toyo Soda)	Various indicated as G1000 to G7000	Various from 1000 average to 400,000,000 average	Various depending on range (5–9 μm)
TSK-GEL H_{HR} and H_{XL} (Aldrich)	G1000H to G7000H, GMH-H, GMH-L, GMH-M	Various from 1500 to 1,000,000	5 and 13 μm
TSK-Gel H_{XL}, SuperHZ multipore bed	GMHXL, GMHXL-HT, GMHXL-L	400,000,000 average	9, 13, and 6 μm
Styragel HR (Waters)	Various indicated as HR 0.5, HR 1 to HR 4	Various from small–1000 to 5000–600,000	5 μm
Styragel HR multipore bed	HR 4E HR 5E	50–10,000 2000–4,000,000	5 μm
Styragel HT, Ultrastyragel (Waters)	Various indicated as HT 3 to HT 6, Ultrastyragel	Various from 500–30,000 to 200,000–10,000,000	10 μm
Styragel HMW (Waters)	HMW 7	500,000–100,000,000	20 μm
Styragel HT, Styragel HMW multipore bed	HT 6E HMW 6E	5000–10,000,000	10, 20 μm
Bio-beades, S-X beads Bio-Rad	—	400–14,000	Soft gel
Hydrocell (Biochrom Labs.)	GPC 3000, 3000HS	20,000–1,000,000	5 μm
Phenogel (Phenomenex)	50 Å, 100 Å, 500, 10^3 Å, 10^4 Å, 10^5 Å, 10^6 Å, and linear	Various	5, 10, and 20 μm

[a]*The range of Mw used is based on polystyrene as analyte.*

APPENDIX 6c Several Commercially Available Phases Used in Aqueous or Mixed Polar Solvents SEC (GFC)

Phase Name (Supplier)	Pore Size Designation	Range of Use (Da)	Particle Size	Chemistry
Jordi BGR	100, 500, 10^3, 10^4, and 10^5 Å	Various polymers between 10^2–10^7	5 μm	PSDVB[a] functionalized
Jordi Hydroxylated DVB	100, 500, 10^3, 10^4 and 10^5 Å	Smaller molecules, not proteins	5 μm	PSDVB functionalized
Jordi polar Pac Wax	100, 500, 10^3, 10^4 and 10^5 Å	Various poly-hydroxy 10^2–10^7	5 μm	PSDVB functionalized
PL aquagel-OH (Polymer Lab. Ltd./Agilent	AOH 30, 40, 50, 60	100,000, 1,000,000, 20,000,000	5, 8 and 15 μm	Polyhydroxyl surface
PL aquagel-OH mixed bed	AOH mixed H, mixed M	20,000,000, up to 600,000 respectively	8 μm	Polymer
PolarGel	M, L	Up to 500,000	5, 10, and 20 μm	—
Shodex OHpak (Showa Denko), Protein KW	Various indicated as KB-802, to KB-806, KB-80M, SB-401, etc.	Various from 4000–20,000,000	Various	HPMMA[b]
TSK-GEL PW, PW_{XL}, PW_{XL}-CP (Toyo Soda)	Various G1000 to G6000, GM, PW_{XL} G5000PW, GMPW, etc.	Various from 1000–8,000,000	Various	HPMMA
TSK-GEL H_{HR} and H_{XL} (Aldrich)	G1000H to G7000H, GMH-H, GMH-L, GMH-M	Various from 1500–1,000,000	7 μm	HPMMA
Toyopearl HW	40S, 40F, 40C, 50S, 50F, 55S, 55F, 65S, 65F, 75F (five pore sizes)	100–50,000,000	20–40 Å, 30–60 Å, 50–100 Å	HPMA[c]
TSK-GEL Alpha and Super AW	Alpha-3000, Alpha-5000, Super AW2500 to Super AW6000	Various up to 10,000,000	Various	HPMMA
TSK-GEL Super-Multipore PW	PW-N, PW-M, PW-H	300–50,000 to 1000–10,000,000	4, 5, and 8 μm	HPMMA
Ultrahydrogel (Waters)	Various 120 Å to 2000 Å indicated as 250, 500, etc., linear	Various from 5000–7,000,000	10 μm	HPMMA
Asahipak (Asahi Chemical/Phenomenex)	Various GS-220, GS-320, GS520, GS-620, GS-710, GFA-30, GFA-7M, GF ser.	Various from 3000–10,000,000	Various	PVA[d] copolymer
Asahipak (Phenomenex)	GF-310 HQ to GF710 HQ, and multimode	Various	5, 6 and 9 μm	PVA copolymer

(*Continued*)

APPENDIX 6c Several Commercially Available Phases Used in Aqueous or Mixed Polar Solvents SEC (GFC)—cont'd

Phase Name (Supplier)	Pore Size Designation	Range of Use (Da)	Particle Size	Chemistry
Suprema (PSS)	30 Å, 100 Å, 300 Å, 1000 Å, Linear S, M, XL	Various from 20,000–10,000,000	Various	HPMA
MCI Gel CQP (Mitsubishi Chemical)	Various indicated as CQO06, CQP06 G, CQP10, CQP30, etc.	Various from 1000–1,000,000	10, 30 μm	HPMA
Polyhydroxyethyl aspartamide; Polyhydroxyethyl A (PolyLC Inc.)	200, 1000 Å	100–1,000,000	3 μm	—
Bio-Prep SE (Bio-Rad)	SE100/17, SE1000/17	5000–1,000,000	17 μm	Agarose
PolySep-GFC-P (Phenomenex)	1000 to 6000 and Linear	Various ranges	—	Highly hydrophilic

[a]*PSDVB = Polystyrenedivinylbenzene.*
[b]*HPPMA = Hydroxylated poly(methylmethacrylate).*
[c]*HPMA = Hydroxylated poly(methacrylate).*
[d]*PVA = Polyvinyl alcohol.*

APPENDIX 6d Several More Recent SEC Columns

Column Name	Manufacturer	Phase	Particle Size	Type
AdvancedBio SEC	Agilent	Proprietary	2.7	Core–shell
Yarra 1.8μm SEC-X150	Phenomenex	Proprietary	1.8	Porous
OHpak LB-803 and LB-806M	Shodex	Proprietary	6, 13 (LB-806M)	Polymeric
TSKGel UP-SW3000	Tosoh Bioscience	Diol	2	Porous

Appendix 7: Properties of Mobile Phase Components

APPENDIX 7a Solubility in Water of Several Solvents Commonly Used as Mobile Phases in HPLC

Solvent	Solubility% in Water	Solvent	Solubility% in Water
Acetic acid	100	Formamide	100
Acetone	100	Heptane	0.0003
Acetonitrile	100	Hexane	0.001
Benzene	0.18	Iso-propanol	100
Butyl acetate	7.81	Methanol	100
Carbon tetrachloride	0.08	Methyl ethyl ketone	24
Chloroform	0.815	Methyl *tert*-butyl ether	4.8
Cyclohexane	0.01	Methylene chloride	1.6
1,2-Dichloroethane	0.81	*n*-Butanol	0.43
Diethyl ether	6.89	*n*-Propanol	100
Di-isopropyl ether	0.87	Pentane	0.004
Dimethylformamide	100	Tetrahydrofuran	100
Dimethylsulfoxide	100	Toluene	0.051
Dioxane	100	Trichloroethylene	0.11
Ethanol	100	Xylene	0.018
Ethyl acetate	8.7		

APPENDIX 7b Molar Volumes (cm^3/mol) and Hildebrand Solubility Parameters δ $(cal/cm^3)^{1/2}$ for Some Common Compounds Used as Solvents (at 25°C). Partial δ Values are Indicated as δ_d (for Dispersion), δ_p (for Polar), δ_a (for Proton Acceptor), δ_h (for Proton Donor)

Compound	*V*	δ	δ_d	δ_p	δ_a	δ_h
Acetic acid	71.3	12.4	7	n.r.[1]	n.r.	n.r.
Acetone	73.8	9.4	6.8	5	2.5	0
Acetonitrile	52.7	11.8	6.5	8	2.5	0
Anisole	108.7	9.7	9.1	2.5	2	0
Benzene	89.2	9.2	9.2	0	0.5	0
Benzonitrile	103.3	10.7	9.2	3.5	1.5	0
Bromobenzene	105.3	9.9	9.6	1.5	0.5	0
1-Butanol	91.5	9.6	n.r.	n.r.	n.r.	n.r.
CCl_3-CF_3	119.3	7.1	6.8	1.5	0.5	0
CCl_4	96.9	8.6	8.6	0	0	0
$CFCl_2$-CF_3	80.02	6.2	5.9	1.5	0	0
CH_2Cl_2	64.4	9.6	6.4	5.5	0.5	0
$CHCl_3$	80.4	9.1	8.1	3	0.5	0
Chlorobenzene	102	9.6	9.2	2	0.5	0
CS_2	60.6	10	10	0	0.5	0
Cyclohexane	108.4	8.2	n.r.	0	0	0
Cyclohexanone	103.8	10.4	6.2	n.r.	n.r.	n.r.
Cyclopentane	93.21	8.1	n.r.	0	0	0
1,2-Dichloroethane	64.4	9.7	8.2	4	0	0
1,3-Dicyanopropane	94.6	13	8	8	3	0
Diethyl ether	104.4	7.4	6.7	2	2	0
Diethyl sulfide	109.98	8.6	8.2	2	0.5	0
Di-isopropyl ether	142	7	6.9	0.5	0.5	0
Dimethylformamide	77.3	11.5	7.9	n.r.	n.r.	n.r.
Dimethylsulfoxide	70.9	12.8	8.4	7.5	5	0
Dioxane	85.3	9.8	7.8	4	3	0
Ethanol	58.6	11.2	6.8	4	5	5
Ethanolamine	60.4	13.5	8.3	Large	Large	Large
Ethyl acetate	98.1	8.6	7	3	2	0
Ethyl bromide	75	8.8	7.8	3	0	0

APPENDIX 7b Molar Volumes (cm^3/mol) and Hildebrand Solubility Parameters δ $(cal/cm^3)^{1/2}$ for Some Common Compounds Used as Solvents (at 25°C). Partial δ Values are Indicated as δ_d (for Dispersion), δ_p (for Polar), δ_a (for Proton Acceptor), δ_h (for Proton Donor)—cont'd

Compound	*V*	δ	δ_d	δ_p	δ_a	δ_h
Ethylene glycol	55.8	14.7	8	Large	Large	Large
Formamide	39.7	17.9	8.3	Large	Large	Large
Isooctane	165.1	7	7	0	0	0
Methanol	40.6	12.9	6.2	5	7.5	7.5
Methyl acetate	79.5	9.2	6.8	4.5	2	0
Methyl benzoate	125.6	9.8	9.2	2.5	1	0
Methyl ethyl ketone	89.9	9.5	6.8	5	2.5	0
Methyl iodide	62.2	9.9	9.3	2	0.5	0
Methylene iodide	52.2	11.9	11.3	1	0.5	0
m-Xylene	123.4	8.8	8.8	0	0.5	0
n-Heptane	146.5	7.4	7.4	0	0	0
n-Hexane	131.1	7.3	7.3	0	0	0
Nitrobenzene	102.6	11.1	9.5	4	0.5	0
Nitromethane	53.7	11	7.3	8	1	0
n-Pentane	115.2	7.1	7.1	0	0	0
Octanol	158	10.3	5.2	n.r.	n.r.	n.r.
Perchloroethylene	102.2	9.3	9.3	0	0.5	0
Perfluoroalkanes	—	6	6	0	1	0
Phenol	87.9	11.4	9.5	n.r.	n.r.	n.r.
1-Propanol	75.1	10.2	7.2	2.5	4	4
2-Propanol	72.74	9.7	n.r.	n.r.	n.r.	n.r.
Propyl amine	81.7	8.7	7.3	4	6.5	0.5
Propyl chloride	88.2	8.3	7.3	3	0	0
Propylene carbonate	85.4	13.3	n.r.	n.r.	n.r.	n.r.
p-Xylene	123.4	8.8	8.8	0	0.5	0
Pyridine	80.6	10.4	9	4	5	0
Tetrahydrofuran	81.1	9.9	7.6	4	3	0
Toluene	106.6	8.9	8.9	0	0.5	0
Triethylamine	139.5	7.5	7.5	0	3.5	0
Water	18.02	21	6.3	Large	Large	Large

[1]*n.r. = Not reported in the referenced literature.*

APPENDIX 7c Values of Polarity (Chromatographic Strength) P' Parameter, and of Separation Parameters x_e, x_d and x_n for Several Common Solvents (*e*, ethanol; *d*, dioxane; *n*, nitromethane). The Values for Log K_{ow} Are Also Listed

	Compound	Group	P'	x_e	x_d	x_n	Log K_{ow}
NONPOLAR, HYDROCARBONS							
1	Carbon disulfide	0	0.3	—	—	—	1.95
2	Carbon tetrachloride	0	1.6	—	—	—	3
3	Cyclohexane	0	0.2	—	—	—	2.67
4	*n*-Decane	0	0.4	—	—	—	4.91
5	*n*-Hexane	0	0.1	—	—	—	3.13
6	Isooctane	0	0.1	—	—	—	3.71
7	Squalane	0	1.2	—	—	—	12.86
ALIPHATIC ETHERS, OTHER COMPOUNDS							
8	Di-butyl ether	1	2.1	0.44	0.18	0.38	2.77
9	Di-ethyl ether	1	2.8	0.53	0.13	0.34	0.84
10	Di-isopropyl ether	1	2.4	0.48	0.14	0.38	1.67
11	Hexamethylphosphoramide	1	7.4	0.47	0.17	0.37	−1.4
12	Tetramethylguanidine	1	6.1	0.48	0.18	0.35	−0.16
13	Triethylamine	1	1.9	0.56	0.12	0.32	1.26
ALIPHATIC ALCOHOLS							
14	*n*-Butanol	2	3.9	0.58	0.18	0.24	0.81
15	tert-Butanol	2	4.1	0.56	0.2	0.24	0.54
16	Ethanol	2	4.3	0.52	0.19	0.29	−0.16
17	Isopentanol	2	3.7	0.56	0.19	0.26	1.09
18	Isopropanol	2	3.9	0.55	0.19	0.27	0.25
19	Methanol	2	5.1	0.48	0.22	0.31	−0.52
20	*n*-Octanol	2	3.4	0.57	0.19	0.25	2.58
21	*n*-Propanol	2	4	0.54	0.19	0.27	0.36
VARIOUS, AMIDES, NITROGENOUS							
22	Diethylene glycol	3	5.2	0.44	0.23	0.33	−1.26
23	2,6-Dimethylpyridine	3	4.5	0.45	0.2	0.36	1.02
24	2-[2-(4-Nonylphenoxy)ethoxy]ethanol	3	—	0.38	0.22	0.4	5.15
25	N,N-Dimethylacetamide	3	6.5	0.41	0.2	0.39	−0.58
26	Dimethylformamide	3	6.4	0.39	0.21	0.4	−0.63

APPENDIX 7c Values of Polarity (Chromatographic Strength) P' Parameter, and of Separation Parameters x_e, x_d and x_n for Several Common Solvents (*e*, ethanol; *d*, dioxane; *n*, nitromethane). The Values for Log K_{ow} Are Also Listed—cont'd

	Compound	Group	P'	x_e	x_d	x_n	Log K_{ow}
27	Dimethylsulfoxide	3	7.2	0.39	0.23	0.39	−1.41
28	Methoxyethanol	3	5.5	0.38	0.24	0.38	−0.57
29	Methyl formamide	3	6	0.41	0.23	0.36	−0.86
30	2-Methylpyridine	3	4.9	0.44	0.21	0.36	0.89
31	*N*-Methyl-2-pyrrolidone	3	6.7	0.4	0.21	0.39	−0.36
32	Pyridine	3	5.3	0.42	0.23	0.36	0.76
33	Quinoline	3	5	0.41	0.23	0.36	2.13
34	Tetrahydrofuran	3	4	0.38	0.2	0.42	0.53
35	Tetramethyl urea	3	6	0.42	0.19	0.39	−0.47
36	Triethyleneglycol	3	5.6	0.42	0.24	0.34	−1.3
VARIOUS, ACETIC ACID, FORMAMIDE							
37	Acetic acid	4	6	0.39	0.31	0.3	−0.22
38	Benzyl alcohol	4	5.7	0.4	0.3	0.3	1.21
39	Ethylene glycol (ethane-1,2-diol)	4	6.9	0.43	0.29	0.28	−1.21
40	Formamide	4	9.6	0.37	0.34	0.3	−1.08
CHLORINATED ALIPHATIC							
41	Ethylene chloride	5	3.5	0.3	0.21	0.49	1.5
42	Methylene chloride	5	3.1	0.29	0.18	0.53	0.84
NITRILES, DIOXANES, KETONES, ETHERS							
43	Acetone	6	5.1	0.35	0.23	0.42	0.11
44	Acetonitrile	6	5.8	0.31	0.27	0.42	−0.17
45	Acetophenone	6	4.8	0.33	0.26	0.41	1.53
46	Aniline	6	6.3	0.32	0.32	0.36	1.14
47	Benzonitrile	6	4.8	0.31	0.27	0.42	1.83
48	Bis-(2-cyanoethyl) ether	6	6.8	0.31	0.29	0.4	−0.33
49	Bis-(2-ethoxyethyl) ether	6	4.6	0.37	0.21	0.43	0.03
50	γ-Butyrolactone	6	6.5	0.34	0.26	0.4	0.15
51	Cyano morpholine	6	5.5	0.35	0.25	0.4	−0.05
52	Cyclohexanone	6	4.7	0.37	0.22	0.42	1.49
53	Dioxane	6	4.8	0.36	0.24	0.4	−0.09

(Continued)

APPENDIX 7c Values of Polarity (Chromatographic Strength) P' Parameter, and of Separation Parameters x_e, x_d and x_n for Several Common Solvents (*e*, ethanol; *d*, dioxane; *n*, nitromethane). The Values for Log K_{ow} Are Also Listed—cont'd

	Compound	Group	P'	x_e	x_d	x_n	Log K_{ow}
54	Ethyl acetate	6	4.4	0.34	0.23	0.43	0.28
55	Formyl morpholine	6	6.4	0.37	0.25	0.39	−0.85
56	Methyl ethyl ketone	6	4.7	0.35	0.22	0.43	0.81
57	Propylene carbonate	6	6.1	0.31	0.27	0.42	0.79
58	Tetrahydrothiophene-1,1-dioxide	6	6.9	0.33	0.28	0.39	−0.59
59	Tricresyl phosphate	6	4.6	0.36	0.23	0.41	6.63
60	Tris-(2-cyanoethoxy)propane	6	6.6	0.32	0.27	0.41	−0.59
AROMATIC HYDROCARBONS, AROMATIC ETHERS, NITRO COMPOUNDS							
61	Benzene	7	2.7	0.23	0.32	0.45	1.97
62	Bromobenzene	7	2.7	0.24	0.33	0.43	2.25
63	Chlorobenzene	7	2.7	0.23	0.33	0.44	2.07
64	Di-benzyl ether	7	4.1	0.3	0.28	0.42	3.57
65	Di-pentyl ether	7	3.4	0.27	0.32	0.41	3.66
66	Ethoxybenzene	7	3.3	0.28	0.28	0.44	2.17
67	Fluorobenzene	7	3.2	0.24	0.32	0.45	2.12
68	Iodobenzene	7	2.8	0.24	0.35	0.41	2.9
69	Methoxybenzene	7	3.8	0.28	0.3	0.43	1.82
70	Nitrobenzene	7	4.4	0.26	0.3	0.44	1.91
71	Nitroethane	7	5.2	0.28	0.29	0.43	0.38
72	Nitromethane	7	6	0.29	0.32	0.4	0.02
73	Toluene	7	2.4	0.25	0.28	0.47	2.49
74	*p*-Xylene	7	2.5	0.27	0.28	0.45	3
VARIOUS, WATER							
75	Chloroform	8	4.1	0.26	0.42	0.33	0.84
76	*m*-Cresol	8	7.4	0.38	0.37	0.25	2.18
77	1H,1H,7H-Dodecafluoroheptanol	8	8.8	0.33	0.4	0.27	3.45
78	Tetrafluoropropanol	8	8.6	0.34	0.36	0.3	0.65
79	Water	8	10.2	0.38	0.38	0.25	−0.65

APPENDIX 7d Values for $E_T(30)$ Polarity for Several Common Solvents

Compound	$E_T(30)$	Compound	$E_T(30)$
Acetic acid	51.7	Formic acid	54.3
Acetonitrile	45.6	Glycerin	57
Benzene	34.3	*n*-Hexane	31
(*tert*-Butyl) methyl ether	34.7	*n*-Heptane	31.1
Carbon dioxide (40°C/150 bar)	28.5	Methoxybenzene	37.1
Carbon tetrachloride	32.4	1-Methylpyrrolidin-2-one	42.2
Chloroform	39.1	Nitrobenzene	41.2
Cyclohexane	30.9	Nitromethane	46.3
Deuterium oxide	62.8	*n*-Octane	31.1
1,2-Dichlorethane	41.3	1,2-Propanediol	54.1
Dibenzylether	36.3	Propionic acid	50.5
Dichlormethane	40.7	Propionitrile	43.6
Diethylether	34.5	Pyridine	40.5
Diglyme	38.6	Styrene	34.8
N,N-Dimethylformamide	43.2	Tetrahydrofuran	37.4
Dimethylsulfoxide	45.1	Tetrahydropyran	36.2
1,4-Dioxane	36	Thiophene	35.4
1,3-Dioxolan	43.1	Toluene	33.9
Diphenylether (30°C)	35.3	Trimethylphosphate	43.6
1,2-Ethanediol	56.3	*p*-Xylene	33.1
Formamide	55.8	Water	63.1

APPENDIX 7e Solvatochromic Parameters α, β, and π* for Several Common Solvents

No.	Compound	α	β	π*	No.	Compound	α	β	π*
1	Acetic acid	1.12	0.45	0.64	33	Formamide	0.71	0.48	0.97
2	Acetonitrile	0.19	0.4	0.75	34	Formic acid	1.23	0.38	0.65
3	Aniline	0.26	0.5	0.73	35	Heptane	0	0	0
4	Benzene	0	0.1	0.59	36	Hexamethylphosphoramide	0	1.05	0.87
5	Benzyl alcohol	0.6	0.52	0.98	37	Hexane	0	0	0
6	Butanoic acid	1.1	0.45	0.56	38	Methanol	0.93	0.66	0.6
7	1-Butanol	0.84	0.88	0.47	39	Methyl acetate	0	0.42	0.6
8	2-Butanol	0.69	0.8	0.4	40	2-Methyl-2-propanol	0.68	1.01	0.41
9	2-Butanone	0.06	0.48	0.67	41	Morpholine	0.29	0.7	0.39
10	Carbon disulfide	0	0.07	0.61	42	Octane	0	0	0.01
11	Carbon tetrachloride	0	0.1	0.28	43	Octanol	0.77	0.81	0.4
12	Chlorobenzene	0	0.07	0.71	44	Pentane	0	0	−0.08
13	1-Chlorobutane	0	0	0.39	45	Pentanoic acid	1.19	0.45	0.54
14	Chloroform	0.44	0	0.58	46	Pentanol	0.84	0.86	0.4
15	Cyclohexane	0	0	0	47	Piperidine	0	1.04	0.3
16	Cyclopentane	0	0	−0.087	48	1,2,3-Propanetriol	1.21	0.51	0.62
17	1,2-Dichlorobenzene	0	0.07	0.67	49	1-Propanol	0.78	0.84	0.52
18	1,1-Dichloroethane	0.1	0.1	0.48	50	2-Propanol	0.76	0.95	0.48
19	1,2-Dichloroethane	0	0.1	0.81	51	Propionitrile	0.10	0.37	0.71
20	Dichloromethane	0.13	0.1	0.82	52	2-Propanone	0.08	0.43	0.71
21	Diethyl ether	0	0.47	0.27	53	Propanoic acid	1.12	0.45	0.58
22	Diethyl sulfide	0	0.37	0.46	54	Pyridine	0	0.64	0.87
23	Diethylamine	0.3	0.7	0.24	55	Pyrrolidine	0.16	0.7	0.39
24	Diisopropyl ether	0	0.49	0.27	56	Sulfolane	0	0.39	0.98
25	*N,N*-Dimethylacetamide	0	0.76	0.88	57	Tetrahydrofuran	0	0.55	0.58
26	*N,N*-Dimethylformamide	0	0.69	0.88	58	Tetramethylsilane	0	0.02	−0.09
27	Dimethyl sulfoxide	0	0.76	1	59	Toluene	0	0.11	0.54
28	Dioxane	0	0.37	0.55	60	Triethylamine	0	0.71	0.14
29	Ethanediol	0.9	0.52	0.92	61	2,2,2-Trifluoroethanol	1.51	0	0.73
30	1,3-Dioxolane	0	0.45	0.69	62	*m*-Xylene	0	0.11	0.47
31	Ethanol	0.86	0.75	0.54	63	*p*-Xylene	0	0.12	0.43
32	Ethyl acetate	0	0.45	0.55	64	Water	1.17	0.47	1.09

APPENDIX 7f Values of Surface Tension γ' in mN/m (dyne/cm) and the Corresponding Temperature Coefficient in mN/(m K) for Several Common Solvents

Compound	γ' 20°C mN/m	Temperature Coefficient mN/(m K)
Acetone (2-propanone)	25.2	−0.112
Benzene	28.88	−0.1291
Carbon disulfide	32.3	−0.1484
Chlorobenzene	33.6	−0.1191
Chloroform	27.5	−0.1295
Cyclohexane	24.95	−0.1211
1,2-Dichloroethane	33.3	−0.1428
1,4-Dioxane	33	−0.1391
Ethanol	22.1	−0.0832
Isopropanol	23	−0.0789
Methanol	22.7	−0.0773
Methyl ethyl ketone (MEK)	24.6	−0.1199
N,N-dimethylformamide (DMF)	37.1	−0.1400
n-Decane	23.83	−0.0920
n-Heptane	20.14	−0.0980
n-Hexane	18.43	−0.1022
Methylene chloride	26.5	−0.1284
Nitrobenzene	43.9	−0.1177
Nitromethane	36.8	−0.1678
1-Octanol	27.6	−0.0795
Propanol (25°C)	23.7	−0.0777
Pyridine	38	−0.1372
Toluene	28.4	−0.1189
Water	72.8	−0.1514

APPENDIX 7g Several Properties of Common Solvents Important for the Separation Process

Compound	Formula	Viscosity η (cP)	Dielectric Constant ε	Dipole μ (D) Liq.	Dipole μ (D) Gas	Polariz. $4\pi\ \varepsilon_0$ (Å)^3
Acetic acid	$C_2H_4O_2$	—	6.2	1.92	1.75	5.33
Acetone	C_3H_6O	0.32	20.7	3.11	2.87	6.41
Acetonitrile	C_2H_3N	0.37	37.5	3.39	3.97	4.27
Benzene	C_6H_6	—	2.3	0	0	8.89
n-Butanol	$C_4H_{10}O$	2.95	17.8	2.96	1.6	9.21
Carbon disulfide	CS_2	—	2.6	0	0	6.79
Carbon tetrachloride	CCl_4	0.97	2.24	0	0	10.5
Chlorobenzene	C_6H_5Cl	—	5.5–6.3	1.39	1.72	11.06
Chloroform	$CHCl_3$	0.57	4.81	1.85	1.03	8.52
Cyclohexane	C_6H_{12}	1	18.5	0.2	0.61	11.07
Cyclohexanone	$C_6H_{10}O$	—	18.2	2.94	—	11.15
Cyclopentane	C_5H_{10}	0.47	2	0	0	9.2
Decane	$C_{10}H_{22}$	0.92	1.99	0	0	20.61
1,2-Dichloroethane	$C_2H_4Cl_2$	–	10.4	2.94	1.84	8.5
Di-ethyl ether	$C4H_{10}O$	0.23	4.34	1.27	1.13	9.33
Di-isopropyl ether	$C_6H_{14}O$	0.37	3.88	1.26	1.13	12.94
N,N-Dimethylformamide	C_3H_7NO	0.92	37.6	3.85	—	7.69
Dimethylsulfoxide	C_2H_6SO	2.24	46.2	3.9	—	7.91
Dioxane	$C_4H_8O_2$	1.54	2.21	0.45	0.43	8.97
Ethanol	C_2H_6O	1.2	24.6	1.66	1.69	5.3
Ethyl acetate	$C_4H_8O_2$	0.45	6.02	2.05	1.78	9.28
Ethylene glycol	$C_2H_6O_2$	19.9	37.7	2.2	2.2	6.18
n-Heptane	C_7H_{16}	0.41	1.89	0	0	14.34
n-Hexane	C_6H_{14}	0.33	1.9	0	0	12.3
Isobutyl alcohol	$C_4H_{10}O$	4.7	16.68	2.96	1.64	9.07
Isooctane	C_8H_{18}	0.53	1.94	0	0	16.18
Isopropanol	C_3H_8O	2.37	19.9	3.09	1.59	7.14
Methanol	CH_4O	0.6	32.7	2.97	1.69	3.38
Methyl acetate	$C_3H_6O_2$	0.37	6.68	1.74	1.68	7.36

APPENDIX 7g Several Properties of Common Solvents Important for the Separation Process—cont'd

Compound	Formula	Viscosity η (cP)	Dielectric Constant ε	Dipole μ (D) Liq.	Dipole μ (D) Gas	Polariz. $4\pi\ \varepsilon_0$ (Å)^3
Methyl ethyl ketone	C_4H_8O	0.43	18.5	3.41	—	8.35
Methyl isobutyl ketone	$C_6H_{12}O$	—	13.11	2.68	—	12.04
Methyl *tert*-butyl ether	$C_5H_{12}O$	—	4	1.25	—	10.88
Methylene chloride	CH_2Cl_2	0.44	9.08	1.9	1.54	4.54
Morpholine	C_4H_9NO	7.42	—	1.75	—	9.47
Nitrobenzene	$C_6H_5NO_2$	—	34.8	3.93	4.28	11.22
Nitromethane	CH_3NO_2	—	35.9	4.39	3.44	4.66
n-Octanol	$C_8H_{18}O$	—	3.4	1.61	—	17.42
n-Pentane	C_5H_{12}	0.23	2.1	0	0	10.27
n-Propanol	C_3H_8O	2.27	20.1	3.09	1.65	13.23
Pyridine	C_5H_5N	—	12.4	2.31	2.15	8.25
Tetrahydrofuran	C_4H_8O	0.55	7.58	1.63	1.63	8.14
Tetrahydrothiophene-1,1-dioxide	$C_4H_8O_2S$	—	43.3	4.69	—	11.46
Toluene	C_7H_8	0.59	2.37	0.38	0.37	10.97
Triethylamine	$C_6H_{15}N$	—	2.4	0.75	0.61	13.47
p-Xylene	C_8H_{10}	—	2.4	0.02	0	13.12
Water	H_2O	1.00	77.46	3.12	1.85	1.51

APPENDIX 7h Physical Properties of Interest for Detection for Several Common Solvents

Solvent	Refractive Index	UV Cut-off (nm)	Boiling Point (°C)	Solvent	Refractive Index	UV Cut-off (nm)	Boiling Point (°C)
Acetone	1.395	330	56.3	*n*-Hexane	1.375	195	68.7
Acetonitrile	1.344	190	81.6	Isobutyl alcohol	1.384	220	98
Benzene	1.501	278	80.1	*i*-Octane	1.404	210	126
n-Butanol	1.347	210	117.7	Methanol	1.329	210	64.7
sec-Butanol	1.397	260	107.7	Methyl acetate	1.362	260	57.5
t-Butyl methyl ether	1.369	210	55.5	Methyl ethyl ketone	1.381	330	80
Carbon tetrachloride	1.466	265	76.5	Methyl *i*-butyl ketone	1.394	330	117.5
Chloroform	1.443	245	61.2	Methylene chloride	1.424	245	39.8
Cyclohexane	1.427	200	80.7	Morpholine	—	285	129
Cyclopentane	1.406	200	49.3	*i*-Pentane	1.371	200	27.7
Decalin	1.476	200	193	*n*-Pentane	1.358	210	36.1
n-Decane	1.412	210	174	*i*-Propanol	1.375	210	82.3
Dimethylformamide	1.427	270	153	*n*-Propanol	1.383	210	97.2
Dimethylsulfoxide	1.476	268	189	*i*-Propyl ether	1.368	220	69
Dioxane	1.422	220	101	Pyridine	1.510	330	115.3
Ethanol	1.361	210	78.3	Tetrahydrofuran	1.408	220	66
Ethyl acetate	1.370	260	77.1	Toluene	1.496	285	110.6
Ethyl ether	1.353	220	34.6	Triethylamine	1.400	—	88.8
Ethylene glycol	1.427	210	197.5	Trimethylpentane	1.389	215	115
n-Heptane	1.385	200	98.4	Water	1.323	<190	100

APPENDIX 7i The pK_a Values of the Common Weak Acids and Bases Used for Buffer Preparation and their pH Working Range[a]

Buffer	pK_a of Acid or Base	pH Working Range
CF_3COOH/CF_3COONH_4	0.3	1.5–2.5
H_3PO_4/KH_2PO_4	pK_1 2.15	1.1–3.1
KH_2PO_4/K_2HPO_4	pK_2 7.20	6.2–8.2
K_2HPO_4/K_3PO_4	pK_3 12.33	11.1–13.6
Citric acid K salt (tribasic)/HCl	pK_1 (citric) 3.13	2.1–4.1
Citric acid K salt (tribasic)/HCl	pK_2 (citric) 4.76	3.7–5.7
Citric acid K salt (tribasic)/HCl	pK_3 (citric) 6.40	4.4–6.4
$HCOOH/HCOONH_4$	pK (formic) 3.75	2.8–4.8
HCOOH/HCOOK	pK (formic) 3.75	2.5–5.0
$HCOONH_4/HCOOH$	pK (ammonia) 9.25	8.2–10.2
CH_3COOH/CH_3COOK	pK (acetic) 4.76	3.8–5.8
CH_3COOH/CH_3COONH_4	pK (acetic) 4.76	3.8–5.8
CH_3COONH_4/CH_3COOH	pK (ammonia) 9.25	8.2–10.2
$CH_3CH_2COOH/CH_3CH_2COOK$	4.86	3.6–6.1
$KHC_8H_4O_4$ (phthalate)/HCl	pK_1 2.89	2.2–3.8
$KHC_8H_4O_4$ (phthalate)/NaOH	pK_2 5.51	4.0–6.2
$H_2CO_3/KHCO_3$	pK_1 6.35	5.1–7.6
$KHCO_3/K_2CO_3$	pK_2 10.33	9.1–11.6
NH_4HCO_3/NH_4OH	pK_2 10.33	9.1–11.6
Tris (tris(hydroxymethyl)aminomethane)/HCl	8.06	7.0–9.0
1,3-Bis-tris propane/HCl (1,3-bis(tris (hydroxymethyl)methylamino)propane/HCl)	6.6, 8.3	6.0–9.5
H_3BO_4/NaOH	pK_1 9.24	9.2–11.0
NH_2CH_2COOH/NaOH	pK_2 9.78	8.5–12.9
Diethylamine/HCl	10.5	9.5–11.5
Triethylamine/HCl	pK_2 10.72	9.5–11.9
Pyrrolidine/HCl	11.27	10.0–12.5
1-Methylpiperidine/HCl	11.3	10.2–12.5

[a]*In some potassium salts the cation can be replaced with sodium.*

APPENDIX 7j Solubility (mM/L) of Potassium Phosphate-Based Buffers[a] in Different Mobile Phase Compositions

Organic Content (%, Vol)	Acetonitrile	Methanol	Tetrahydrofuran
0–40	>50	>50	>50
50	>50	>50	25
60	45	>50	15
70	20	35	10
80	5	15	<5
90	0	5	0

[a]*Solutions containing KH_2PO_4 and K_2HPO_4 are used as buffers in the range of pH = 4.8–8.0 (measured in water).*

APPENDIX 7k UV Cut-off Values of Several Common Buffers (for a Specific Concentration) and Their Suitability for LC/MS Detection

Buffer	UV Cut-off (nm)	Concentration in UV	LC/MS Detection Suitability
Potassium formate/formic acid	210	10 mM	No
Potassium acetate/acetic acid	210	10 mM	No
Ammonium formate	210	50 mM	Yes
Ammonium acetate	210	50 mM	Yes
Trifluoroacetic acid	210	10 mM	Yes
KH_2PO_4/H_3PO_4	<200	0.1%	No
KH_2PO_4/K_2HPO_4	<200	0.1%	No
Ammonium hydroxide/ammonia	200	10 mM	Yes
Diethylamine/HCl	200	10 mM	No
Triethylamine/HCl	200	10 mM	No
Borate ($H_3BO_3/Na_2B_4O_7 \cdot 10H_2O$)	210	10 mM	No
Glycine/HCl	230	10 mM	No
Glycine/HCOOH	215	10 mM	Yes
NH_4HCO_3/NH_4OH	210	10 mM	Yes
Tri-K-Citrate/HCl	230	10 mM	No
1,3- Bis-tris propane/HCl (1,3-bis(tris (hydroxymethyl) methylamino)propane/HCl)	225	10 mM	No
1-Methylpiperidine/HCl	215	10 mM	No
Pyrrolidine/HCl	210	10 mM	No
Pyrrolidine/HCOOH	215	10 mM	Yes

APPENDIX 7l Several Buffers Used in Ion-Exchange Chromatography

Buffer/Counterion	pK_a (at 25°C)	pH Interval
Formic acid/Na^+, or Li^+	3.75	3.5–4.5
Acetic acid/Na^+, or Li^+	4.75	4.8–5.2
Maleic acid/Na^+	2.00	1.5–2.5
Malonic acid/Na^+, or Li^+	2.88	2.35–2.40
Citric acid/Na^+	3.13	2.60–3.60
Lactic acid/Na^+	3.80	3.60–4.30
Succinic acid/Na^+	4.20	4.30–4.80
Piperidine/Cl^-	11.12	10.6–11.6
Piperazine/Cl^-, $HCOO^-$	5.67	5.0–6.0
N-methylpiperazine/Cl^-	4.75	4.5–5.0
Ethanolamine/Cl^-	9.50	9.0–9.5
Diethanolamine/Cl^-	8.88	8.4–8.8
Triethanolamine/Cl^-	7.75	7.3–7.7
Histidine/Cl^-	5.95	5.5–6.0
1,3-Diaminopropane/Cl^-	10.46	9.8–10.3

APPENDIX 7m Common Solvents Used in Gel Filtration

Type of Polymer	Typical Eluents
Nonionic	Water, water with 0.1–0.2 M salt for constant ionic strength, water with buffer at pH = 7
Nonionic with some hydrophobicity	Water with up to 20% organic modifier water, organic modifier, 0.1–0.2 M salt and/or buffer pH = 7
Anionic	Water with 0.1–0.2 M salt for constant ionic strength and with buffer at pH = 7–9
Anionic with some hydrophobicity	Water, organic modifier up to 20%, 0.1–0.2 M salt and/or buffer pH = 7–9
Cationic	Water with 0.3–1.0 M salt for constant ionic strength and with buffer at pH = 2–7
Cationic with some hydrophobicity	Water with organic modifier, 0.3–1.0 M salt for constant ionic strength and with buffer at pH = 2–7

Index

'*Note*: Page numbers followed by "f" indicate figures, "t" indicate tables.'

D

E

N

O

Q

T

U

V

FSC
www.fsc.org
MIX
Papier | Fördert
gute Waldnutzung
FSC® C083411

Kolibri 360 GmbH
Ferdinand-Jühlke-Straße 7
99095 Erfurt, Deutschland
produktsicherheit@kolibri360.de